Studies in Fuzziness and Soft Computing

Volume 390

Series Editor

Janusz Kacprzyk, Systems Research Institute, Polish Academy of Sciences, Warsaw, Poland

The series "Studies in Fuzziness and Soft Computing" contains publications on various topics in the area of soft computing, which include fuzzy sets, rough sets, neural networks, evolutionary computation, probabilistic and evidential reasoning, multi-valued logic, and related fields. The publications within "Studies in Fuzziness and Soft Computing" are primarily monographs and edited volumes. They cover significant recent developments in the field, both of a foundational and applicable character. An important feature of the series is its short publication time and world-wide distribution. This permits a rapid and broad dissemination of research results.

Indexed by ISI, DBLP and Ulrichs, SCOPUS, Zentralblatt Math, GeoRef, Current Mathematical Publications, IngentaConnect, MetaPress and Springerlink. The books of the series are submitted for indexing to Web of Science.

More information about this series at http://www.springer.com/series/2941

Muhammad Akram · Anam Luqman

Fuzzy Hypergraphs and Related Extensions

 Springer

Muhammad Akram
Department of Mathematics
University of the Punjab
Lahore, Pakistan

Anam Luqman
Department of Mathematics
University of the Punjab
Lahore, Pakistan

ISSN 1434-9922   ISSN 1860-0808   (electronic)
Studies in Fuzziness and Soft Computing
ISBN 978-981-15-2402-8   ISBN 978-981-15-2403-5   (eBook)
https://doi.org/10.1007/978-981-15-2403-5

This Springer imprint is published by the registered company Springer Nature Singapore Pte Ltd.
The registered company address is: 152 Beach Road, #21-01/04 Gateway East, Singapore 189721,
Singapore

We dedicate this book to the memory of Prof. Lotfi A. Zadeh!

Foreword

It was stated by G. J. Klir that among the various paradigmatic changes in science and mathematics in the twentieth century, one such change concerned the concept of uncertainty. In science, this change has been manifested by a gradual transition from the traditional view, which states that uncertainty is undesirable in science and should be avoided by all possible means, to an alternative view, which is tolerant of uncertainty and insists that science cannot avoid it. Uncertainty is essential to science and has great utility. An important point in the evolution of the modern concept of uncertainty is the publication of a seminal paper by Lotfi Zadeh.

Fuzzy set theory provides a methodology for carrying out approximate reasoning processes when available information is uncertain, incomplete, imprecise, or vague. This is especially true when observations are expressed in linguistic terms. The success of this methodology has been demonstrated in a variety of fields such as control systems where mathematical models are difficult to specify and in expert systems where rules express knowledge and facts are linguistic in nature. The capability of fuzzy sets to express gradual transitions from membership to non-membership and vice-versa has a broad utility. It provides us not only with meaningful and powerful representations of measurement of uncertainties, but also with a meaningful representation of vague concepts expressed in natural language. Concerning the future of fuzzy logic, it might lie in the new ideas arising from computing with words and perceptions.

Based on Zadeh's fuzzy relations, the first definition of fuzzy graphs was given in 1973 by Arnold Kauffman, the French engineer and professor of applied mechanics and operations research, in the world's first textbook on fuzzy sets and systems. An English version of this book, titled *Introduction to the Theory of Fuzzy Subsets,* was published in 1975. This was the same year Azriel Rosenfeld also provided a concept of a fuzzy graph. He introduced fuzzy analogs of several basic graph-theoretic concepts. Rosenfeld's paper presented the concepts of subgraphs, paths, connectedness, cliques, bridges, trees, and forests and established some of their properties. It appeared in the proceedings of the US-Japan Seminar on Fuzzy Sets and Their Applications. At this seminar, an alternative analysis of fuzzy graphs was also presented by Raymond T. Yeh and S. Y. Bang. Their definition of a fuzzy graph was suitable for cluster analysis and also for database theory. Rosenfeld's paper opened the door for the development of an entire new area in graph theory. Fuzzy graph theory is now a discipline in itself. Hundreds of papers and numerous books have been published in this area. Lee-Kwang and Lee extended crisp hypergraphs by introducing the notion of fuzzy hypergraphs.

This book, *Fuzzy Hypergraphs and Related Extensions*, by Prof. Akram and Dr. Luqman is another strong contribution to fuzzy graph theory. Professor Akram has been a leader in this field for some time. He has published several papers on a wide variety of fuzzy graph-theoretic structures. The results in this book should be very useful in modeling various applications. Crisp hypergraphs have found applications in chemistry, psychology, genetics, human activities, optimization, cellular networks, parallel computing, clustering, information system architecture, social networks, traffic control, engineering, and image processing. The potential for applications of fuzzy hypergraphs can be seen by merely pondering the definition of a fuzzy hypergraph.

It is my hope that researchers will apply the concepts of fuzzy hypergraphs to examine the existential problem of climate change. This may be the most serious problem facing the world at this time. All member states of the United Nations adopted the Agenda 2030 and the Sustainable Development Goals (SDGs). The 17 SDGs describe a universal agenda that applies to and must be implemented by all countries. Among these SDGs is SDG 13 Climate Change. Adverse effects of climate change result in a substantial increase in cruel crimes of human trafficking and modern slavery. These crimes should be brought to a halt. Human trafficking is a prime candidate to be studied using techniques from fuzzy logic. Accurate data concerning trafficking in persons is impossible to obtain. The goal of the trafficker is to be undetected. The size of the problem also makes it very difficult to obtain accurate data. There are also many other reasons for the scarcity of data. I thank Prof. Akram and Dr. Luqman for their timely publication.

Omaha, NE, USA   John N. Mordeson

Preface

Hypergraphs are one of the most successful tools for modeling practical problems in different fields, inclusive of computer science, systems modeling, web information system architecture, service-oriented architecture, and social networks. However, crisp hypergraphs are not sufficient to describe all existing relations between objects. Motivated by this concern, Lee-Kwang and Lee redefined and extended crisp hypergraphs by means of the notion of fuzzy hypergraphs, whose inception had been earlier discussed by Kaufmann. Professors Mordeson and Nair made a real contribution by compiling their comprehensive monograph *Fuzzy Graphs and Fuzzy Hypergraphs*, which motivated us to work in this direction.

Fuzzy set theory was introduced by Lotfi Zadeh in 1965 as a generalization of classical set theory that allows us to represent imprecise and vague phenomena. Since then, fuzzy sets and fuzzy logic have been applied in many real situations that implemented uncertainty. The traditional fuzzy set uses one real value from the unit interval [0,1] in order to represent the grade of membership of objects to a fuzzy set defined on the concerned universe. In some applications such as expert systems, belief systems, and information fusion, not only should we consider the truth-membership supported by evidence but also the falsity-membership against such evidence. Similarly, in most real problems, information consistently comes from more than one agent or from various sources. Due to the limitation of human's knowledge to understand the complex problems, one cannot aspire to apply a single type of uncertain methodology to deal with all such situations. It is therefore necessary to develop generalized mathematical models rather than being satisfied with narrow structures of uncertainty. Researchers have put forward several generalized models of fuzzy sets, including intuitionistic fuzzy sets, bipolar fuzzy sets, m-polar fuzzy sets, Pythagorean fuzzy sets, q-rung orthopair fuzzy sets, and single-valued neutrosophic sets. We have applied these generalized models to hypergraphs.

This monograph deals with fuzzy hypergraphs, their related extensions, and applications. It originates from our papers published in various scientific journals. This book may be useful for researchers in mathematics, computer scientists, and social scientists alike. In Chap. 1, we present fundamental and technical concepts like fuzzy hypergraphs, fuzzy column hypergraphs, fuzzy row hypergraphs, fuzzy competition hypergraphs, fuzzy k-competition hypergraphs and fuzzy neighbor-hood hypergraphs, and $\mathcal{N}$-hypergraphs, complex fuzzy hypergraphs, $\mu e^{i\theta}$-level hypergraphs, and C_f-tempered complex fuzzy hypergraphs. We describe applications of fuzzy competition hypergraphs in decision support systems, including predator–prey relations in ecological niches, social networks, and business marketing.

Chapter 2 defines intuitionistic fuzzy hypergraphs, dual intuitionistic fuzzy hypergraphs, intuitionistic fuzzy line graphs, and 2-section of an intuitionistic fuzzy hypergraph. It also includes applications of intuitionistic fuzzy hypergraphs in planet surface networks, selection of authors of intersecting communities in a social network, and grouping of incompatible chemical substances. We have designed certain algorithms to construct dual intuitionistic fuzzy hypergraphs, intuitionistic fuzzy line graphs, and the selection of objects in decision-making problems. Further, we define complex intuitionistic fuzzy hypergraphs, 2-section, and line graphs of complex intuitionistic fuzzy hypergraphs.

In Chap. 3, we present the notion of $A = [\mu^-, \mu^+]$-tempered interval-valued fuzzy hypergraphs and some of their properties. Moreover, we discuss the notions of vague hypergraphs, dual vague hypergraphs, and A-tempered vague hypergraphs. Finally, we describe interval-valued intuitionistic fuzzy hypergraphs and interval-valued intuitionistic fuzzy transversals of H.

Chapter 4 discusses the concept of bipolar fuzzy directed hypergraph. We describe certain operations on bipolar fuzzy directed hypergraphs, which include addition, multiplication, vertex-wise multiplication, and structural subtraction. We discuss the concept of $B = (m^+, m^-)$-tempered bipolar fuzzy directed hypergraphs and investigate some of their basic properties. We present an algorithm to compute the minimum arc length of a bipolar fuzzy directed hyperpath.

Chapter 5 includes the notions of regular m-polar fuzzy hypergraphs and totally regular m-polar fuzzy hypergraphs. We discuss the applications of m-polar fuzzy hypergraphs in decision-making problems. Furthermore, the notion of m-polar fuzzy directed hypergraph is discussed along with the depiction of certain operations on them. We also describe an application of m-polar fuzzy directed hypergraphs in business strategy.

Chapter 6 presents the concepts including q-rung orthopair fuzzy hypergraphs, (α, β)-level hypergraphs, and transversals and minimal transversals of q-rung orthopair fuzzy hypergraphs. We implement some interesting notions of q-rung orthopair fuzzy hypergraphs into decision-making. We describe additional concepts like q-rung orthopair fuzzy directed hypergraphs, dual directed hypergraphs, line graphs, and coloring of q-rung orthopair fuzzy directed hypergraphs. We also apply other interesting notions of q-rung orthopair fuzzy directed hypergraphs to real-life problems. Further, we study complex q-rung orthopair fuzzy hypergraphs with application.

In Chap. 7, we present q-rung picture fuzzy hypergraphs and illustrate the formation of granular structures using q-rung picture fuzzy hypergraphs and level hypergraphs. Moreover, we define q-rung picture fuzzy equivalence relations and its associated q-rung picture fuzzy hierarchical quotient space structures. We also present an arithmetic example in order to demonstrate the benefits and validity of this model.

In Chap. 8, we illustrate the formation of granular structures using m-polar fuzzy hypergraphs and level hypergraphs. Further, we define m-polar fuzzy hierarchical quotient space structures. The mappings between the m-polar fuzzy hypergraphs depict the relationships among granules occurred in different levels. The consequences reveal that the representation of partition of universal set is more efficient through m-polar fuzzy hypergraphs as compared to crisp hypergraphs. We also present some examples and a real world problem to signify the validity of our proposed model.

Chapter 9 discusses the concepts including single-valued neutrosophic hypergraphs, dual single-valued neutrosophic hypergraphs, and transversal single-valued neutrosophic hypergraphs. Additionally, we discuss the notions of intuitionistic single-valued neutrosophic hypergraphs, and dual intuitionistic single-valued neutrosophic hypergraphs. We describe an application of intuitionistic single-valued neutrosophic hypergraphs in a clustering problem. Then, we present other related concepts like single-valued neutrosophic directed hypergraphs, single-valued neutrosophic line directed graphs, and dual single-valued neutrosophic directed hypergraphs. Finally, in this section, we describe the applications of single-valued neutrosophic directed hypergraphs. Additionally, we define complex neutrosophic hypergraphs and T-related complex neutrosophic hypergraphs with applications.

In Chap. 10, we present bipolar neutrosophic hypergraphs and B-tempered bipolar neutrosophic hypergraphs. We describe the concepts of transversals, minimal transversals, and locally minimal transversals of bipolar neutrosophic hypergraphs. Furthermore, we put forward some applications of bipolar neutrosophic hypergraphs in marketing and biology. We also introduce bipolar neutrosophic directed hypergraphs, regular bipolar neutrosophic directed hypergraphs, homomorphism, and isomorphism on bipolar neutrosophic directed hypergraphs. To conclude, we describe an efficient algorithm to solve decision-making problems.

The authors are grateful to the administration of the University of the Punjab, Lahore, Pakistan, particularly Prof. Dr. Niaz Ahmad Akhtar (Vice Chancellor) and Dr. Muhammad Khalid Khan (Registrar) for their encouraging attitude and for providing the state of art research facilities.

Lahore, Pakistan   Muhammad Akram
                                                                     Anam Luqman

Contents

About the Authors

Muhammad Akram is Professor at the Department of Mathematics, University of the Punjab, Lahore, Pakistan. He previously served as Assistant Professor and Associate Professor at Punjab University College of Information Technology, Lahore, Pakistan. He holds a Ph.D. in Fuzzy Mathematics from Government College University, Lahore, Pakistan. His research interests include numerical algorithms, fuzzy graphs, fuzzy algebras, and fuzzy decision-support systems. He has published eight books and over 320 research articles in peer-reviewed international journals.

Anam Luqman received her M.Phil. and M.Sc. degrees in Mathematics from the University of the Punjab, Lahore, Pakistan, and is currently a Ph.D. scholar at the same university. Her research interests include extensions of fuzzy hypergraphs and their applications. She has published over 15 research articles in various scientific journals.

List of Figures

List of Tables

Chapter 1
Fuzzy Hypergraphs

In this chapter, we present fundamental and technical concepts like fuzzy hypergraphs, fuzzy column hypergraphs, fuzzy row hypergraphs, fuzzy competition hypergraphs, fuzzy k-competition hypergraphs, fuzzy neighborhood hypergraphs, and $\mathcal{N}$-hypergraphs. We describe applications of fuzzy competition hypergraphs in decision support systems, including predator–prey relations in ecological niches, social networks, and business marketing. Further, we introduce complex fuzzy hypergraphs, $\mu e^{i\theta}$−level hypergraphs, covering constructions, 2-sections, and L_2-sections of these hypergraphs. We define certain new products, including Cartesian product, minimal rank preserving, and maximal rank preserving direct products on complex fuzzy hypergraphs. Moreover, we present an application of our proposed model to illustrate the publication data in co-authorship networks. This chapter is basically due to [3, 7–9, 12, 25].

1.1 Introduction

In 1965, Zadeh [30] introduced the strong mathematical notion of fuzzy sets in order to discuss the phenomena of vagueness and uncertainty in various real-life problems. Fuzzy sets are a kind of useful mathematical structure to represent a collection of objects whose boundary is vague. The basic idea of fuzzy set is that the element belongs to a fuzzy set with a certain degree of membership. This branch of mathematics has instilled new life into scientific fields that have been dormant for a long time. Examples of fuzziness are words such as red roses, tall men, and beautiful women. Fuzziness can be found in many areas of daily life including, engineering, meteorology, and manufacturing. Fuzzy sets cannot handle imprecise, inconsistent, and incomplete information of periodic nature. This theory is applicable to different areas of science, but there is one major deficiency in fuzzy sets, that is, a lack of capability to model two-dimensional phenomena. To overcome this difficulty, the concept

© Springer Nature Singapore Pte Ltd. 2020
M. Akram and A. Luqman, *Fuzzy Hypergraphs and Related Extensions*,
Studies in Fuzziness and Soft Computing 390,
https://doi.org/10.1007/978-981-15-2403-5_1

of complex fuzzy sets was introduced by Ramot et al. [21]. A complex fuzzy set C is characterized by a membership function $\mu_C(x)$, whose range is not limited to [0, 1] but extends to the unit circle in the complex plane. Hence, $\mu_C(x)$ is a complex-valued function that assigns a grade of membership of the form $r_C(x)e^{iw_C(x)}$, $i = \sqrt{-1}$ to any element x in the universe of discourse. Thus, the membership function $\mu_C(x)$ of complex fuzzy set consists of two terms, i.e., amplitude term $r_C(x)$ which lies in the unit interval [0, 1] and phase term (periodic term) $w_C(x)$ which lies in the interval $[0, 2\pi]$. This phase term distinguishes a complex fuzzy set model from all other models available in the literature. The potential of a complex fuzzy set for representing two-dimensional phenomena makes it superior to handle ambiguous and intuitive information that are prevalent in time-periodic phenomena. Opposing to a fuzzy characteristic function, the range of a complex fuzzy set's membership degrees is not restricted to [0, 1], but extends to the unit circle in the complex plane.

Graph theory has numerous applications to problems in systems analysis, operations research, economics, and transportation. However, in many cases, some aspects of a graph-theoretic problem may be uncertain. For example, the vehicle travel time or vehicle capacity on a road network may not be known exactly. The ambiguousness in the representation of different objects or in the relationships between them generates the essentiality of fuzzy graphs, which were originally studied and developed by Kaufmann [11] in 1977. The fuzzy relations in fuzzy sets were studied by Rosenfeld [22] and he introduced the structure of fuzzy graphs, obtaining an analysis of various graph theoretical concepts. The concept of fuzzy graphs was generalized to complex fuzzy graphs by Thirunavukarasu et al. [28]. They discussed the energy of complex fuzzy graph and defined its lower and upper bounds. They also illustrated these concepts using numeric examples.

Hypergraphs, a generalization of graphs, have been widely and deeply studied in Berge [5], and quite often have proved to be a successful tool to represent and model concepts and structures in various areas of computer science and discrete mathematics. Hypergraphs are able to describe complex systems as their descriptive power is fairly strong because they are one of the most general graphs and mathematical structures for representing relationships. The hypergraphs are the generalization and extension of concept graphs and finite sets. The mathematical theory of hypergraphs was developed in the past decades; the generalizations of definitions as trees, cycles, and coloring for hypergraphs have been elaborated with the accompanying theorems. Hypergraphs have many applications in different fields including computer science, biological sciences, and natural sciences. The crisp hypergraphs are not sufficient to describe all real-world relations between objects. Hypergraphs do not study the degree of dependence of an object to the other. In 1995, Lee-Kwang and Lee [12] generalized the concept of hypergraphs to fuzzy hypergraphs and extended the notion of fuzzy hypergraphs, whose idea was first discussed by Kaufmann [11]. Later, the idea of fuzzy hypergraphs was studied by Goetschel Jr. in [7–9], and the concepts of Hebbian structures, fuzzy colorings, and fuzzy transversals were introduced.

Definition 1.1 A pair $G^* = (X, E)$ is a *crisp graph*, where $E \subseteq \widetilde{X^2}$ is a collection of 2−element subsets of a non-empty set X.

A *crisp hypergraph* on a non-empty set X is a pair $H^* = (X, E)$ such that

  (i) $X = \{x_1, x_2, \ldots, x_n\}$ is non-empty set of vertices,
 (ii) $E = \{E_1, E_2, \ldots, E_r\}$ is a family of non-empty subsets of X,
(iii) $\cup_k E_k = X, k = 1, 2, \cdots, r$.

A hypergraph is a generalized case of a graph in which a hyperedge may have more than two vertices. A hypergraph is called *simple* if for any $E_i \subseteq E_k \Rightarrow i = k$.

A hypergraph is called *linear* if it is simple, and $|E_i \cap E_k| \leq 1$, for each E_i, $E_k \in E$.

The *line graph* $L(H^*)$ of a hypergraph is a graph in which $E = \{E_1, E_2, \ldots, E_r\}$ is the set of vertices and there is an edge between two vertices E_i and E_k if $E_i \cap E_k \neq \emptyset$.

A hypergraph is a generalization of an ordinary undirected graph, such that an edge need not contain exactly two nodes, but can instead contain an arbitrary nonzero number of vertices. An ordinary undirected graph (without self-loops) is, of course, a hypergraph where every edge has exactly two nodes (vertices). Hypergraphs are often defined by an incidence matrix with columns indexed by the edge set and rows indexed by the vertex set.

The *rank* $r(H^*)$ of a hypergraph is defined as the maximum number of nodes in one edge, $r(H^*) = \max_j |E_j|$, and the *anti rank* $s(H^*)$ is defined likewise, i.e., $s(H^*) = \min_j |E_j|$.

We say that a hypergraph is *uniform* if $r(H^*) = s(H^*)$. A uniform hypergraph of rank k is called k-*uniform hypergraph*. Hence, a simple graph is a $2-$uniform hypergraph, and thus all simple graphs are also hypergraphs.

A hypergraph is *vertex* (resp. *hyperedge*) *symmetric* if for any two vertices (resp. hyperedges) x_i and x_j (resp. E_i and E_j), there is an automorphism of the hypergraph that maps x_i to x_j (resp. E_i to E_j).

Example 1.1 Let $H^* = (X, E)$ be a crisp hypergraph as shown in Fig. 1.1 such that $X = \{x_1, x_2, x_3, x_4, x_5, x_6, x_7\}$, and $E = \{E_1, E_2, E_3, E_4\}$.

The incidence matrix of H^* is given in Table 1.1.

Here, $E_1 = \{x_1, x_2, x_3\}$, $E_2 = \{x_4, x_5, x_6\}$, $E_3 = \{x_1, x_6, x_7\}$, and $E_4 = \{x_3, x_4, x_7\}$ are hyperedges of H^* such that $|E_1| = 3$, $|E_2| = 3$, $|E_3| = 3$ and $|E_4| = 3$. Note that $r(H^*) = 3$, and $s(H^*) = 3$. Hence, H^* is $3-$uniform hypergraph. Also, H^* is simple as there are no repeated hyperedges.

Definition 1.2 The *dual of a hypergraph* $H^* = (X, E)$ with vertex set $X = \{x_1, x_2, \ldots, x_n\}$ and hyperedge set $E = \{e_1, e_2, \ldots, e_m\}$ is a hypergraph $H^{*d} = (X^d, E^d)$ with vertex set $X^d = \{x_1^d, x_2^d, \ldots, x_m^d\}$ and hyperedge set $E^d = \{(e_1)^d, (e_2)^d, \ldots, (e_n)^d\}$ such that x_j^d corresponds to e_j with hyperedges $(e_i)^d = \{x_j^d \mid x_i \in e_j$ and $e_j \in E\}$. In other words, H^{*d} is obtained from H^* by interchanging vertices and hyperedges in H^*. The incidence matrix of H^{*d} is the transpose of the incidence matrix of H^*. Thus, $(H^{*d})^d = H^*$.

Fig. 1.1 Uniform crisp
hypergraph

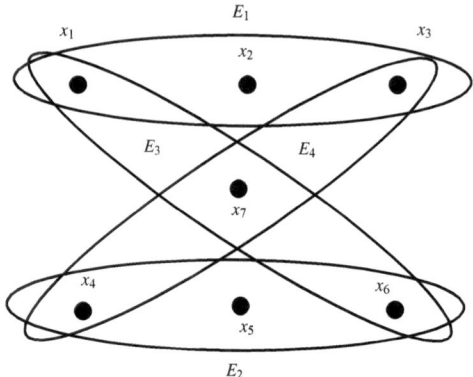

Table 1.1 The incidence matrix of H^*

I_{H^*}	E_1	E_2	E_3	E_4
x_1	1	0	1	0
x_2	1	0	0	0
x_3	1	0	0	1
x_4	0	1	0	1
x_5	0	1	0	0
x_6	0	1	1	0
x_7	0	0	1	1

Table 1.2 Incidence matrix of H^{*d}

$I_{H^{*d}}$	X_1	X_2	X_3	X_4	X_5	X_6	X_7
e_1	1	1	1	0	0	0	0
e_2	0	0	0	1	1	1	0
e_3	1	0	0	0	0	1	1
e_4	0	0	1	1	0	0	1

Example 1.2 Consider a crisp hypergraph $H^* = (X, E)$ as shown in Fig. 1.1. The
dual hypergraph of H^* is given in Fig. 1.2 such that $X^d = \{e_1, e_2, e_3, e_4\}$ are vertices
of H^{*d} corresponding to the edge set of H^* and $E^d = \{X_1, X_2, X_3, X_4, X_5, X_6, X_7\}$
are hyperedges of H^{*d} corresponding to the vertex set of H^*.

Note that $X_1 = \{e_1, e_3\}$, $X_2 = \{e_1\}$, $X_3 = \{e_1, e_4\}$, $X_4 = \{e_2, e_4\}$, $X_5 = \{e_2\}$,
$X_6 = \{e_2, e_3\}$, and $X_7 = \{e_3, e_4\}$ are hyperedges of H^{*d}. The incidence matrix of
H^{*d} is given in Table 1.2, which is the transpose of Table 1.1.

Note that H^{*d} is not a simple and uniform hypergraph.

Proposition 1.1 H^* *is r-uniform if and only if* H^{*d} *is r-regular.*

Fig. 1.2 Dual hypergraph H^{*d}

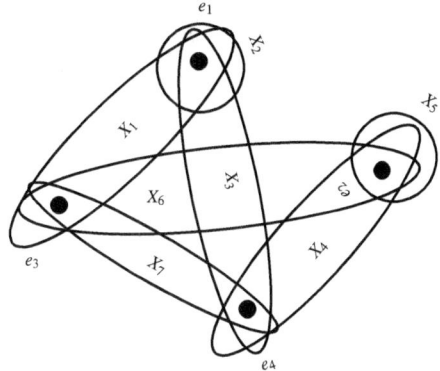

Proposition 1.2 *The dual of a linear hypergraph is also linear.*

Proposition 1.3 *A hypergraph H^* is vertex symmetric if and only if H^{*d} is hyperedge symmetric.*

Proposition 1.4 *The dual of a sub-hypergraph of H^* is a partial hypergraph of the dual hypergraph H^{*d}.*

Definition 1.3 A *fuzzy set* μ on a set X is a map $\mu : X \to [0, 1]$. In clustering , the fuzzy set μ, is called a *fuzzy class*.
We define the *support* of μ by $supp(\mu) = \{x \in X \mid \mu(x) \neq 0\}$, and say μ is nontrivial if $supp(\mu)$ is non-empty.
The *height* of μ is $h(\mu) = \max\{\mu(x) \mid x \in X\}$. We say μ is *normal* if $h(\mu) = 1$.

Definition 1.4 A map $\nu : X \times X \to [0, 1]$ is called a *fuzzy relation* on X if $\nu(x, y) \leq \min(\mu(x), \mu(y))$, for all $x, y \in X$.
A *fuzzy partition* of a set X is a family of nontrivial fuzzy sets $\{\mu_1, \mu_2, \mu_3, \ldots, \mu_m\}$ such that

  (i) $\bigcup_i supp(\mu_i) = X, i = 1, 2, \ldots, m$,
  (ii) $\sum_{i=1}^{m} \mu_i(x) = 1$, for all $x \in X$,
  (iii) $\mu_i \cap \mu_j \neq \emptyset, i \neq j$.

We call a family $\{\mu_1, \mu_2, \mu_3, \ldots, \mu_m\}$ a *fuzzy covering* of X if it verifies only the above conditions (i) and (ii).

Definition 1.5 A *fuzzy graph* on a non-empty universe X is a pair $G = (\mu, \lambda)$, where μ is a fuzzy set on X, and λ is a fuzzy relation on X such that $\lambda(xy) \leq \min\{\mu(x), \mu(y)\}$, for all $x, y \in X$.

Remark 1.1

1. μ is called *fuzzy vertex set* of G, λ is called *fuzzy edge set* of G. If λ is a symmetric fuzzy relation on μ, $G = (\mu, \lambda)$ is called a fuzzy graph on a non-empty universe X.
   A fuzzy relation λ on X is called *symmetric* if $\lambda(x, y) = \lambda(y, x)$, for all $x, y \in X$.

2. If λ is not a symmetric fuzzy relation on μ, $\overrightarrow{G} = (\mu, \overrightarrow{\lambda})$ is called a *fuzzy digraph* on a non-empty universe X.
3. λ is a fuzzy relation on μ, and $\lambda(xy) = 0$ for all $xy \in X \times X - E$, $E \subseteq X \times X$.
4. A fuzzy graph is needed only when vertices and edges are fuzzy. Otherwise, a weighted graph is enough. That is, when there is no precise information about storage/capacity at vertices, or no exact flow through the edges. The flow should not exceed from the capacity of the source. If flow exceeds from the capacity of the source in a network model, is called a *cofuzzy graph* , that is,

$$\lambda(xy) \geq \max\{\mu(x), \mu(y)\}, \quad \forall \, x, y \in X.$$

For further terminologies and studies on fuzzy sets, fuzzy graphs and fuzzy hypergraphs, readers are referred to [1–4, 15–20, 22–27, 29–32].

1.2 Fuzzy Hypergraphs

Definition 1.6 A *fuzzy hypergraph* is defined as an ordered pair $H = (X, \xi)$ such that

(i) $X = \{x_1, x_2, \cdots , x_m\}$ is finite set of vertices,
(ii) $\xi = \{\xi_1, \xi_2, \cdots , \xi_k\}$ is finite family of fuzzy subsets of X,
(iii) $\cup_i supp(\xi_i) = X$, for all $\xi_i \in \xi$.

Note that the hyperedges ξ_i are fuzzy subsets of X. The membership degree of vertex x_i to the hyperedge ξ_j is defined by $\xi_j(x_i)$.

Definition 1.7 Let $H = (X, \xi)$ be a fuzzy hypergraph. The *height*, denoted by $h(H)$, is defined as $h(H) = \max\{h(\mu)|\mu \in \xi\}$.

The *order* of H (number of vertices) is denoted by $|X|$ and the number of edges is denoted by $|\xi|$. The *rank* is the maximal column sum of the incidence matrix and the *anti rank* is the minimal column sum. We say $H = (X, \xi)$ is a *uniform fuzzy hypergraph* if and only if rank(H) = anti rank(H).

Definition 1.8 A fuzzy set $\mu : X \to [0, 1]$ is an *elementary fuzzy set* if μ is single valued on $supp(\mu)$.

An *elementary fuzzy hypergraph* $H = (X, \xi)$ is a fuzzy hypergraph, whose fuzzy hyperedges are all elementary.

Example 1.3 Consider a fuzzy hypergraph $H = (X, \xi)$, where $X = \{x_1, x_2, x_4, x_5\}$ and $\xi = \{\xi_1, \xi_2, \xi_3\}$ such that

$$\xi_1 = \{(x_1, 0.8), (x_2, 0.5)\},$$
$$\xi_2 = \{(x_2, 0.8), (x_3, 1), (x_4, 0.7)\},$$
$$\xi_3 = \{(x_4, 0.8), (x_5, 1)\}.$$

Table 1.3 Incidence matrix of fuzzy hypergraph

$x \in X$	ξ_1	ξ_2	ξ_3
x_1	0.8	0	0
x_2	0.5	0.5	0
x_3	0	1	0
x_4	0	0.8	0.8
x_5	0	0	1

Fig. 1.3 Fuzzy hypergraph

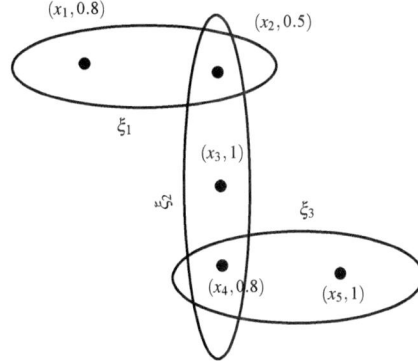

The corresponding incidence matrix is given in Table 1.3.

The fuzzy hypergraph is shown in Fig. 1.3.

Definition 1.9 A fuzzy hypergraph $H = (X, \xi)$ is *simple* if $\xi_i, \xi_j \in \xi$, and $\xi_i \subseteq \xi_j$ imply $\xi_i = \xi_j$.

In particular, a (crisp) hypergraph $H = (X, E)$ is simple if $A, B \in E$, and $A \subseteq B$ imply that $A = B$.

A fuzzy hypergraph $H = (X, \xi)$ is *support simple* if $\xi_i, \xi_j \in \xi$, $supp(\xi_i) = supp(\xi_j)$, and $\xi_i \subseteq \xi_j$ imply $\xi_i = \xi_j$.

A fuzzy hypergraph $H = (X, \xi)$ is *strongly support simple* if $\xi_i, \xi_j \in \xi$, and $supp(\xi_i) = supp(\xi_j)$ imply $\xi_i = \xi_j$.

Remark 1.2 For fuzzy hypergraphs, all three concepts imply no multiple edges. Simple fuzzy hypergraphs are support simple and strongly support simple fuzzy hypergraphs are support simple. Simple and strongly support simple are independent concepts.

Proposition 1.5 *A fuzzy hypergraph* $H = (X, \xi)$ *is a fuzzy graph (with loops) if and only if H is elementary, support simple, and each edge has two (or one) element support.*

Lemma 1.1 *Let* $H = (X, \xi)$ *be an elementary fuzzy hypergraph. Then, H is support simple if and only if H is strongly support simple.*

Proof Suppose that $H = (X, \xi)$ is elementary, support simple, and $supp(\xi_i) = supp(\xi_j)$. We assume that $h(\xi_i) \leq h(\xi_j)$. Since, H is elementary it follows that $\xi_i \subseteq \xi_j$, and since H is support simple, then $\xi_i = \xi_j$. Therefore, H is strongly support simple.

Conversely, by Remark 1.2, it follows that if H is strongly support simple, then H is support simple.

Proposition 1.6 *Let $H = (X, \xi)$ be a simple fuzzy hypergraph of order n. Then, there is no upper bound on $|\xi|$.*

Proof Let $X = \{x, y\}$, and define $\xi_N = \{\mu_i | i = 1, 2, \cdots, N\}$, where

$$\mu_i(x) = \frac{1}{i+1}, \quad \mu_i(y) = 1 - \frac{1}{i+1}.$$

Then, $H_N = (X, \xi_N)$ is a simple fuzzy hypergraph with N edges.

Proposition 1.7 *Let $H = (X, \xi)$ be an elementary, simple fuzzy hypergraph of order n. Then, $|\xi| \leq 2^n - 1$, with equality if and only if $\{supp(\mu) | \mu \in \xi, \mu \neq 0\} = P(X) \setminus \emptyset$.*

Proof Since H is elementary and simple, each nontrivial $A \subseteq X$ can be the support of at most one $\mu \in \xi$. Therefore, $|\xi| \leq 2^n - 1$. To show there exists an elementary, simple H with $|\xi| = 2^n - 1$, let $\xi = \{\mu_A | A \subseteq X\}$ be the set of functions defined by,

$$\mu_A = \begin{cases} \frac{1}{|A|}, & \text{if } x \in A, \\ 0, & \text{otherwise.} \end{cases}$$

Then, each one element set has height 1, each two element set has height 0.5, and so on. H is elementary and simple, and $|\xi| = 2^n - 1$.

Definition 1.10 Let $H = (X, \xi)$ be a fuzzy hypergraph. The *adjacent level* between two vertices is defined as $\gamma(x_i, x_k) = \max_j \min[\xi_j(x_i), \xi_j(x_k)]$, $j = 1, 2, \cdots, k$.

The *adjacent level* between two edges is defined as $\rho(\xi_j, \xi_k) = \max_j \min[\xi_j(x), \xi_k(x)]$, $x \in X$.

Note that in fuzzy hypergraph as given in Fig. 1.3, $\gamma(x_1, x_2) = 0.5$ and $\rho(\xi_1, \xi_2) = 0.5$.

Definition 1.11 Let $H = (X, \xi)$ be a fuzzy hypergraph. The $\alpha-cut\ hypergraph$ of H is defined as an ordered pair $H^\alpha = (X^\alpha, \xi^\alpha)$, where

(i) $X^\alpha = \{x_1, x_2, \cdots, x_m\}$,
(ii) $\xi^\alpha = \cup \xi_i^\alpha$, where $\xi_i^\alpha = \{x | \xi_i(x) \geq \alpha\}$,
(iii) $\xi_{i+1}^\alpha = \{x | \xi_i(x) < \alpha\}$.

Note that $H^\alpha = (X^\alpha, \xi^\alpha)$ is a crisp hypergraph. The edge ξ_{i+1}^α is added to group the elements which are not contained in any edge of H^α.

Fig. 1.4 $H^{0.8}$—level
hypergraph of H

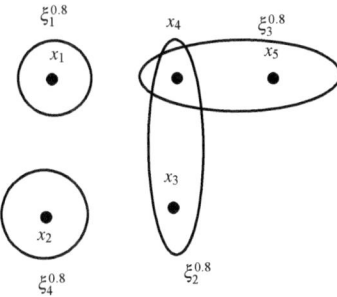

Example 1.4 Consider a fuzzy hypergraph as shown in Fig. 1.3. For $\alpha = 0.8$, $H^{0.8}$—cut hypergraph is shown in Fig. 1.4.

Here,

$$\xi_1^{0.8} = \{x_1\}, \ \xi_2^{0.8} = \{x_3, x_4\}, \ \xi_3^{0.8} = \{x_14, x_5\}, \ \xi_4^{0.8} = \{x_2\}.$$

Note that a new edge $\xi_4^{0.8}$ is added to contain the element x_2.

Definition 1.12 The *strength* of an hyperedge in H is defined as the minimum membership degree of vertices in that hyperedge, i.e., $\beta(\xi_i) = \min\{\xi_i(x_j)|\ \xi_i(x_j) > 0\}$.

Its interpretation is that the edge ξ_i groups elements have participation degree at least $\beta(\xi_i)$ in the hypergraph.

Example 1.5 For example, in fuzzy hypergraph as given in Fig. 1.3, the strength of each hyperedge is given as $\beta(\xi_1) = 0.5$, $\beta(\xi_2) = 0.5$ and $\beta(\xi_3) = 0.8$, respectively.

The edges having high strength are called the *strong hyperedges* because the cohesion in them is strong. In Example 1.5, the hyperedge ξ_3 is stronger than ξ_1 and ξ_2.

Definition 1.13 Let $H = (X, \xi)$ be a fuzzy hypergraph and for each $\alpha \in [0, 1]$, let $H^\alpha = (X^\alpha, \xi^\alpha)$ be the α—level hypergraph of H. The sequence of real numbers $r_1, r_2, \ldots, r_n$, with $1 \geq r_1 > r_2 > \ldots > r_n > 0$, and having the properties,

(i) if $1 \geq s > r_1$, then $\xi_s = \emptyset$,
(ii) if $r_i \geq s > r_{i+1}$, then $\xi_s = \xi_{r_i}$,
(iii) $\xi_{r_i} \subseteq \xi_{r_{i+1}}$,

is called the *fundamental sequence* of H, and is denoted by $f_s(H)$. The corresponding sequence of r_i level hypergraphs $H^{r_1} \subseteq H^{r_2} \subseteq \cdots \subseteq H^{r_n}$ is called the H induced *fundamental sequence* and is denoted by $I(H)$. The r_n level is called the *support level* of H and the hypergraph H^{r_n} is called the *support hypergraph* of H.

Table 1.4 Incidence matrix of fuzzy hypergraph

$x \in X$	ξ_1	ξ_2	ξ_3
x_1	0.8	0.7	0.6
x_2	0.8	0.7	0.6
x_3	0	0.6	0
x_4	0	0	0.6

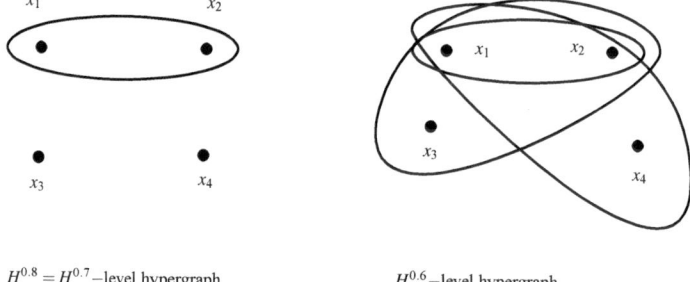

$H^{0.8} = H^{0.7}$ −level hypergraph $H^{0.6}$ −level hypergraph

Fig. 1.5 H-induced fundamental sequence of H

Example 1.6 Let $H = (X, \xi)$ be a fuzzy hypergraph, where $X = \{x_1, x_2, x_3, x_4\}$ and $\xi = \{\xi_1, \xi_2, \xi_3\}$ such that

$$\xi_1 = \{(x_1, 0.8), (x_2, 0.8)\},$$
$$\xi_2 = \{(x_1, 0.7), (x_2, 0.7), (x_3, 0.6)\},$$
$$\xi_3 = \{(x_1, 0.6), (x_2, 0.6), (x_4, 0.6)\}.$$

The hypergraph is represented by the incidence matrix as given in Table 1.4.

By direct calculations, we have $H^{0.8} = (X, \xi^{0.8})$, where $\xi^{0.8} = \{\{x_1, x_2\}\}$, $H^{0.7} = (X, \xi^{0.7})$, where $\xi^{0.7} = \{\{x_1, x_2\}\}$, and $H^{0.6} = (X, \xi^{0.6})$, where $\xi^{0.6} = \{\{x_1, x_2\}, \{x_1, x_2, x_3\}, \{x_1, x_2, x_4\}\}$. We see $H^{0.8} = H^{0.7}$ and so $f_s(H) = \{0.8, 0.6\}$. Note that 0.7 is not an element of $f_s(H)$. The H induced fundamental sequence of H is shown in Fig. 1.5.

Definition 1.14 Let H be a fuzzy hypergraph with $f_s(H) = \{r_1, \cdots, r_n\}$, and let $r_{n+1} = 0$. Then, H is *sectionally elementary* if for each edge $\mu \in \xi$, each $i \in \{1, 2, \cdots, n\}$, and each $c \in (r_{j+1}, r_j]$, we have $\mu^c = \mu^{r_i}$.

Definition 1.15 A fuzzy hypergraph H is *ordered* if the H induced fundamental sequence of hypergraphs is ordered.

The fuzzy hypergraph H is *simply ordered* if the H-induced fundamental sequence of hypergraphs is simply ordered.

Fig. 1.6 Dual fuzzy
hypergraph H^d

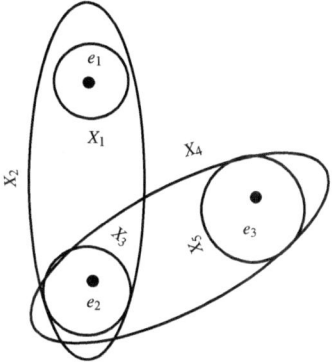

Proposition 1.8 *If $H = (X, \xi)$ is an elementary fuzzy hypergraph, then is ordered.
Also, if $H = (X, \xi)$ is an ordered fuzzy hypergraph with simple support hypergraph,
then H is elementary.*

Definition 1.16 Let $H = (X, \xi)$ be a fuzzy hypergraph. The *dual fuzzy hypergraph*
is defined as $H^d = (\xi, X^d)$, where

(i) $\xi = \{e_1, e_2, \cdots, e_k\}$ is the set of vertices of H^d corresponding to hyperedges
$\{\xi_1, \xi_2, \cdots, \xi_k\}$,

(ii) $X^d = \{X_1, X_2, \cdots, X_m\}$ are the hyperedges corresponding to the vertices of
$\{x_1, x_2, \cdots, x_m\}$, such that $X_i = \{(e_j, \xi_i(e_j)) | \xi_i(e_j) = \xi_j(x_i)\}$.

Example 1.7 Consider a fuzzy hypergraph as shown in Fig. 1.3. Its dual fuzzy hypergraph is given as $H^d = (\xi, X^d)$, where $\xi = \{e_1, e_2, e_3\}$, and

$$X_1 = \{(e_1, 0.8)\},$$
$$X_2 = \{(e_1, 0.5), (e_2, 0.5)\},$$
$$X_3 = \{(e_2, 1)\},$$
$$X_4 = \{(e_2, 0.8), (e_3, 0.8)\},$$
$$X_5 = \{(e_3, 1)\}.$$

The corresponding dual fuzzy hypergraph is shown in Fig. 1.6.

Now, we cut the dual hypergraph at level 0.8 such that

$$X_1^{0.8} = \{e_1\}, \ X_2^{0.8} = \{\emptyset\}, \ X_3^{0.8} = \{e_2\}, \ X_4^{0.8} = \{e_3\}, \ X_5^{0.8} = \{e_2, e_3\}.$$

The corresponding 0.8−cut hypergraph of H^d is given in Fig. 1.7.

Definition 1.17 For a crisp hypergraph $H = (X, E)$, a *transversal* of H is any subset
T of X with the property that for each $A \in E$, $T \cap A \neq \emptyset$. A transversal T of H is
a *minimal transversal* of H, if no proper subset of T is a transversal of H.

Fig. 1.7 0.8−cut
hypergraph of H^d

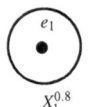
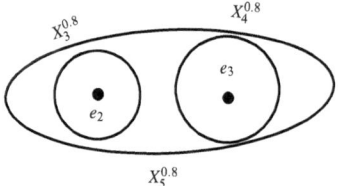

Clearly, a transversal always contains a minimal transversal. The collection of minimal transversals of H can be considered the edge set of a hypergraph where the vertex set is a (perhaps proper) subset of X. Both the set of all minimal transversals of $H = (X, E)$ and the hypergraph defined by this set will be denoted by $T_r(H)$.

Definition 1.18 Let $H = (X, \xi)$ be a fuzzy hypergraph and the height of μ is $h(\mu) = \max\{\mu(x) | x \in X\}$. A *fuzzy transversal* of H is a fuzzy set $\tau \in \mathcal{F}(X)$, where $\mathcal{F}(X)$ is the collection of all fuzzy sets defined on X such that $\tau^{h(\mu)} \cap \mu^{h(\mu)} \neq \emptyset$, for each $\mu \in \xi$. A *minimal fuzzy transversal* of H is a fuzzy transversal τ of H for which $\rho < \tau$ implies ρ is not a fuzzy transversal of H. The set of all minimal fuzzy transversals of H (and the fuzzy hypergraph formed by this set) will be denoted $Tr(H)$.

Proposition 1.9 *Let $H = (X, \xi)$ be a fuzzy hypergraph. Then, the following statements are equivalent,*

(i) *τ is a fuzzy transversal of H,*
(ii) *for each $\mu \in \xi$ and each c with $0 < c \leq h(\mu)$, $\tau^c \cap \mu^c \neq \emptyset$,*
(iii) *for each c with $0 < c \leq r_1$, τ^c is a transversal of H^c.*

Example 1.8 Let $H = (X, \xi)$ be a fuzzy hypergraph, which is defined by the incidence matrix given in Table 1.5.

Note that fuzzy transversals of H are given as

$$\tau_1 = \{(x_1, 0), (x_2, 0.8), (x_3, 0.7), (x_4, 0), (x_5, 0), (x_6, 0.8)\},$$
$$\tau_2 = \{(x_1, 0), (x_2, 0.8), (x_3, 0), (x_4, 0), (x_5, 0.7), (x_6, 0.8)\},$$

which are the minimal transversals of H and H has only these two minimal fuzzy transversals.

Definition 1.19 If τ is a fuzzy set with the property that τ^c is a minimal transversal of H^c for each $c \in (0, 1]$, then τ is called a *locally minimal fuzzy transversal* of H. The set of all locally minimal fuzzy transversals on H is denoted by $Tr^*(H)$.

Table 1.5 Incidence matrix of fuzzy hypergraph

$x \in X$	ξ_1	ξ_2	ξ_3	ξ_4
x_1	0.5	0	0	0
x_2	0.5	0.8	0	0
x_3	0.4	0	0	0.7
x_4	0	0.4	0.7	0
x_5	0	0	0	0.7
x_6	0	0	0.8	0.4

Lemma 1.2 *Let $H = (X, \xi)$ be a fuzzy hypergraph with $f_s(H) = \{r_1, \cdots, r_n\}$. If τ is a fuzzy transversal of H, then $h(\tau) \geq h(\mu)$, for each $\mu \in \xi$. If τ is minimal, then $h(\tau) = \max\{h(\mu)|\mu \in \xi\} = r_1$.*

Construction 1.1 gives an algorithm for finding $Tr(H)$.

Construction 1.1 Let $H = (X, \xi)$ be a fuzzy hypergraph with $I(H) = \{H^{r_1}, H^{r_2}, \cdots, H^{r_n}\}$. We construct a minimal fuzzy transversal τ of H by a recursive process,

1. Find a (crisp) minimal transversal T_1 of H^{r_1}.
2. Find a transversal T_2 of H^{r_2} that is minimal with respect to the property that $T_1 \subseteq T_2$. Equivalently, construct a new hypergraph H_2 with edge set E_{r_2}, augmented by a loop at each $x \in T_1$. Let T_2 be any minimal transversal of H_2.
3. Continue recursively, letting T_j be a transversal of H^{r_j} that is minimal with respect to the property $T_j \subseteq T_{j+1}$.
4. For $1 \leq j \leq n$, let τ_j be the elementary fuzzy set with support T_j and height r_j. Then, $\tau = \max\{\tau_j | 1 \leq j \leq n\}$ is a minimal fuzzy transversal of H.

Lemma 1.3 *For each $\tau \in Tr(H)$, and for each $x \in X$, $\tau(x) \in f_s(H)$. Therefore, the fundamental sequence of $Tr(H)$ is a (possibly proper) subset of $f_s(H)$.*

Proof Let $\tau \in Tr(H)$ and $\tau(x) \in (r_{i+1}, r_i]$. Define ϕ as

$$\phi(y) = \begin{cases} r_1, & \text{if } y = x, \\ \tau(y), & \text{otherwise.} \end{cases}$$

By definition of ϕ, $\phi^{r_i} = \tau^{r_i}$. By Definition 1.13 of $f_s(H)$, $H^c = H^{r_i}$, for each $c \in [r_{i+1}, r_i]$. Therefore, ϕ^{r_i} is a transversal of H for each $c \in [r_{i+1}, r_i]$. Since, τ is a fuzzy transversal, and $\phi^c \neq \tau^c$ for each $c \notin [r_{i+1}, r_i]$, ϕ is a fuzzy transversal as well. Now, $\phi \leq \tau$, and the minimality of τ implies $\phi = \tau$. Hence, $\tau(x) = \phi(x) = r_i$. It follows that for each $\tau \in Tr(H)$, and for each $x \in X$, we have $\tau(x) \in f_s(H)$. Therefore, $f_s(Tr(H)) \subseteq f_s(H)$.

1.3 Fuzzy Competition Hypergraphs

Definition 1.20 A *fuzzy out neighborhood* of a vertex x of a fuzzy digraph $\vec{G} = (\mu, \vec{\lambda})$ is a fuzzy set $\mathcal{N}^+(x) = (X_x^+, \mu_x^+)$, where $X_x^+ = \{y|\vec{\lambda}(xy) > 0\}$, and $\mu_x^+ : X_x^+ \to [0, 1]$ is defined by $\mu_x^+(y) = \vec{\lambda}(xy)$.

Definition 1.21 The *fuzzy in neighborhood* of vertex x of a fuzzy digraph is a fuzzy set $\mathcal{N}^-(x) = (X_x^-, \mu_x^-)$, where $X_x^- = \{y|\vec{\lambda}(yx) > 0\}$, and $\mu_x^- : X_x^- \to [0, 1]$ is defined by $\mu_x^-(y) = \vec{\lambda}(yx)$.

Definition 1.22 Let $\vec{G} = (\mu, \vec{\lambda})$ be a fuzzy digraph. The *underlying fuzzy graph* of $\vec{G}$ is a fuzzy graph $\mathcal{U}(\vec{G}) = (\mu, \lambda)$ such that

$$\lambda(xw) = \begin{cases} \vec{\lambda}(xw), & \text{if } \vec{wx} \notin \vec{E}, \\ \vec{\lambda}(wx), & \text{if } \vec{xw} \notin \vec{E}, \\ \vec{\lambda}(xw) \wedge \vec{\lambda}(wx), & \text{if } \vec{wx}, \vec{xw} \in \vec{E}, \end{cases}$$

where $\vec{E} = supp(\vec{\lambda})$.

Definition 1.23 The *fuzzy open neighborhood* of a vertex y in a fuzzy graph $G = (\mu, \lambda)$ is a fuzzy set $\mathcal{N}(y) = (X_y, \mu_y)$ where $X_y = \{w|\lambda(yw) > 0\}$ and $\mu_y : X_y \to [0, 1]$ a membership function defined by $\mu_y(w) = \lambda(yw)$.

Definition 1.24 The *fuzzy closed neighborhood* $\mathcal{N}[y]$ of a vertex y in a fuzzy graph $G = (\mu, \lambda)$ is defined as $\mathcal{N}[y] = \mathcal{N}(y) \cup \{(y, \mu(y))\}$.

Definition 1.25 Let $A = [x_{ij}]_{n \times n}$ be the adjacency matrix of a fuzzy digraph $\vec{G} = (\mu, \vec{\lambda})$ on a non-empty set X. The *fuzzy row hypergraph* of $\vec{G}$, denoted by $\mathcal{R} \circ \mathcal{H}(\vec{G}) = (\mu, \lambda_r)$, having the same set of vertices as $\vec{G}$ and the set of hyper-edges is defined as

$$\Big\{ \{x_1, x_2, \ldots, x_r\} | A(x_{ij}) > 0, r \geq 2, \text{ for each } 1 \leq i \leq r, x_i \in X, \text{ for some } 1 \leq j \leq n \Big\}.$$

The degree of membership of hyperedges is defined as

$$\lambda_r(\{x_1, x_2, \ldots, x_r\}) = \big[\mu(x_1) \wedge \mu(x_2) \wedge \ldots \wedge \mu(x_s) \big] \times \max_j \{ \vec{\lambda}(x_1 x_j) \wedge \vec{\lambda}(x_2 x_j) \wedge \ldots \wedge \vec{\lambda}(x_r x_j) \}.$$

Definition 1.26 The *fuzzy column hypergraph* of $\vec{G}$, denoted by $\mathcal{C} \circ \mathcal{H}(\vec{G}) = (\mu, \lambda_{cl})$, having the same set of vertices as $\vec{G}$ and the set of hyperedges is defined as

$$\Big\{ \{x_1, x_2, \ldots, x_s\} | A(x_{ji}) > 0, s \geq 2, \text{ for each } 1 \leq i \leq s, x_i \in X, \text{ for some } 1 \leq j \leq n \Big\}.$$

The degree of membership of hyperedges is defined as

$$\lambda_{cl}(\{x_1, x_2, \ldots, x_s\}) = [\mu(x_1) \wedge \mu(x_2) \wedge \ldots \wedge \mu(x_s)] \times \max_j\{\vec{\lambda}(x_jx_1) \wedge \vec{\lambda}(x_jx_2) \wedge \ldots \wedge \vec{\lambda}(x_jx_s)\}.$$

The methods for computing fuzzy row hypergraph and fuzzy column hypergraph are given in Algorithms 1.3.1 and 1.3.2, respectively.

Algorithm 1.3.1 Method for construction of fuzzy row hypergraph

1. Begin
2.   Input the fuzzy set μ on set of vertices $X = \{x_1, x_2, \ldots, x_n\}$.
3.   Input the adjacency matrix $A = [x_{ij}]_{n \times n}$ of fuzzy digraph $\vec{G} = (\mu, \vec{\lambda})$ such that $\vec{\lambda}(x_ix_j) = x_{ij}$ as shown in Table 1.6.
4. **do** j from $1 \to n$
5.   Take a vertex x_j from first jth column.
6.   value1 $= \infty$, value 2 $= \infty$, num $= 0$
7.   **do** i from $1 \to n$
8.   **if** $(x_{ij} > 0)$ **then**
9. x_i belongs to the hyperedge E_j.
10.   num $=$ num $+ 1$
11.   value1 $=$ value1 $\wedge \mu(x_i)$
12.   value2 $=$ value2 $\wedge x_{ij}$
13.   **end if**
14.   **end do**
15.   **if** (num > 1) **then**
16. $\lambda_r(E_j) =$ value1 $\times$ value2, where E_j is a hyperedge.
17.   **end if**
18. **end do**
19. If for some $j, supp(E_j) = supp(E_k), k \in \{j + 1, j + 2, \ldots, n\}$ then, $\lambda_r(E_j) = \max\{\lambda_r(E_j), \lambda_r(E_k), \ldots\}$.

Algorithm 1.3.2 Method for construction of fuzzy column hypergraph

1. Begin
2. Follow steps 2 and 3 of Algorithm 1.3.1.
3. **do** i from $1 \to n$
4.   Take a vertex x_i from first ith row.
5.   value1 $= \infty$, value2 $= \infty$, num $= 0$
6.   **do** j from $1 \to n$
7.   **if** $(x_{ij} > 0)$ **then**
8. x_j belongs to the hyperedge E_i.
9.   num $=$ num $+ 1$
10.   value1 $=$ value1 $\wedge \mu(x_j)$
11.   value2 $=$ value2 $\wedge x_{ij}$
12.   **end if**
13.   **end do**

14.   **if** (num > 1) **then**
15. $\lambda_{cl}(E_i) =$ value1 × value2, where E_i is a hyperedge.
16.   **end if**
17. **end do**
18. If for some $i, supp(E_i) = supp(E_k), k \in \{j+1, j+2, \ldots, n\}$, then $\lambda_{cl}(E_i) = \max\{\lambda_{cl}(E_j), \lambda_{cl}(E_k), \ldots\}$.

Example 1.9 consider the universe $X = \{x_1, x_2, x_3, x_4, x_5, x_6\}$, μ a fuzzy set on X and $\vec{\lambda}$ a fuzzy relation in X as defined in Tables 1.7 and 1.8, respectively. The fuzzy digraph $\vec{G} = (\mu, \vec{\lambda})$ is shown in Fig. 1.8. The adjacency matrix of $\vec{G}$ is given in Table 1.9. Using Algorithm 1.3.1 and Table 1.9, there are three hyperedges $E_2 = \{x_1, x_5, x_6\}$, $E_3 = \{x_2, x_5\}$, and $E_4 = \{x_3, x_5\}$ corresponding to the columns x_2, x_3, and x_4 of adjacency matrix, in fuzzy row hypergraph of $\vec{G}$. The membership degree of the hyperedges is calculated as

$$\lambda_r(E_2) = \left[\mu(x_1) \wedge \mu(x_5) \wedge \mu(x_6)\right] \times \left[x_{12} \wedge x_{52} \wedge x_{62}\right] = 0.3 \times 0.3 = 0.09,$$
$$\lambda_r(E_3) = \left[\mu(x_2) \wedge \mu(x_5)\right] \times \left[x_{23} \wedge x_{53}\right] = 0.4 \times 0.1 = 0.04,$$
$$\lambda_r(E_4) = \left[\mu(x_3) \wedge \mu(x_5)\right] \times \left[x_{34} \wedge x_{54}\right] = 0.4 \times 0.4 = 0.16.$$

The fuzzy row hypergraph is shown in Fig. 1.9. Using Algorithm 1.3.2 and Table 1.9, the hyperedges in fuzzy column hypergraph of $\vec{G}$ are $E_1 = \{x_2, x_6\}$, $E_5 = \{x_2, x_3, x_4\}$, and $E_6 = \{x_2, x_5\}$, corresponding to the rows x_2, x_5, and x_6 of the adjacency matrix. The membership degree of the hyperedges is calculated as

$$\lambda_{cl}(E_5) = \left[\mu(x_2) \wedge \mu(x_3) \wedge \mu(x_4)\right] \times \left[x_{52} \wedge x_{53} \wedge x_{54}\right] = 0.4 \times 0.3 = 0.12,$$
$$\lambda_{cl}(E_1) = \left[\mu(x_2) \wedge \mu(x_6)\right] \times \left[x_{12} \wedge x_{16}\right] = 0.3 \times 0.2 = 0.06,$$
$$\lambda_{cl}(E_6) = \left[\mu(x_2) \wedge \mu(x_5)\right] \times \left[x_{62} \wedge x_{65}\right] = 0.4 \times 0.1 = 0.04.$$

The fuzzy column hypergraph is given in Fig. 1.10.

Table 1.6 Adjacency matrix

A	x_1	x_2	...	x_n
x_1	x_{11}	x_{12}	...	x_{1n}
x_2	x_{21}	x_{22}	...	x_{2n}
$\vdots$	$\vdots$	$\vdots$	...	$\vdots$
x_n	x_{n1}	x_{n2}	...	x_{nn}

Table 1.7 Fuzzy vertex set μ

x	$\mu(x)$	x	$\mu(x)$
x_1	0.5	x_2	0.4
x_3	0.7	x_4	0.6
x_5	0.4	x_6	0.3

Table 1.8 Fuzzy relation $\vec{\lambda}$

x	$\vec{\lambda}(x)$	x	$\vec{\lambda}(x)$
x_1x_2	0.4	x_6x_5	0.1
x_2x_3	0.1	x_1x_6	0.2
x_3x_4	0.6	x_6x_2	0.3
x_5x_4	0.4	x_5x_2	0.4
x_5x_3	0.3		

Fig. 1.8 Fuzzy digraph $\vec{G}$

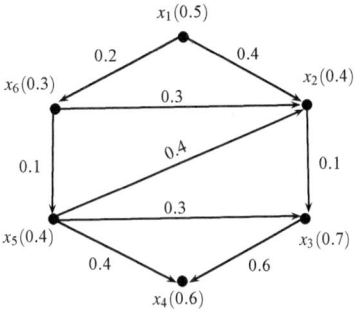

Definition 1.27 A *fuzzy competition graph* of a fuzzy digraph $\vec{G} = (\mu, \vec{v})$ is an undirected fuzzy graph $\mathscr{C}(\vec{G}) = (\mu, v)$, which has the same vertex set as in $\vec{G}$ and there is an edge between two vertices x and y if $\mathscr{N}^+(x) \cap \mathscr{N}^+(y)$ is non-empty. The membership value of the edge xy is defined as

$$v(xy) = \min\{\mu(x), \mu(y)\}h(\mathscr{N}^+(x) \cap \mathscr{N}^+(y)).$$

Table 1.9 Adjacency matrix

A	x_1	x_2	x_3	x_4	x_5	x_6
x_1	0	0.4	0	0	0	0.2
x_2	0	0	0.1	0	0	0
x_3	0	0	0	0.6	0	0
x_4	0	0	0	0	0	0
x_5	0	0.4	0.3	0.4	0	0
x_6	0	0.3	0	0	0.1	0

Fig. 1.9 Fuzzy row
hypergraph $\mathscr{R} \circ \mathscr{H}(\vec{G})$

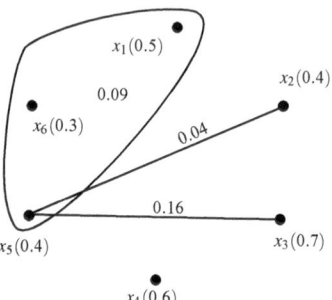

Fig. 1.10 Fuzzy column
hypergraph $\mathscr{C} \circ \mathscr{H}(\vec{G})$

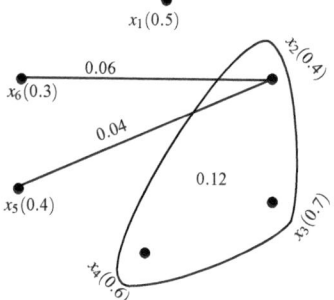

Definition 1.28 Let $\vec{G} = (\mu, \vec{\lambda})$ be a fuzzy digraph on a non-empty set X. The *fuzzy competition hypergraph* $\mathscr{C}\mathscr{H}(\vec{G}) = (\mu, \lambda_c)$ on X having the same vertex set as $\vec{G}$ and there is a hyperedge consisting of vertices $x_1, x_2, \ldots, x_s$ if $\mathscr{N}^+(x_1) \cap \mathscr{N}^+(x_2) \cap \ldots \cap \mathscr{N}^+(x_s) \neq \emptyset$. The degree of membership of hyperedge $E = \{x_1, x_2, \ldots, x_s\}$ is defined as

$$\lambda_c(E) = [\mu(x_1) \wedge \mu(x_2) \wedge \ldots \wedge \mu(x_s)] \times h(\mathscr{N}^+(x_1) \cap \mathscr{N}^+(x_2) \cap \ldots \cap \mathscr{N}^+(x_s)),$$

where $h(\mathscr{N}^+(x_1) \cap \mathscr{N}^+(x_2) \cap \ldots \cap \mathscr{N}^+(x_s))$ denotes the height of fuzzy set $\mathscr{N}^+(x_1) \cap \mathscr{N}^+(x_2) \cap \ldots \cap \mathscr{N}^+(x_s)$.

The method for constructing fuzzy competition hypergraph of a fuzzy digraph is given in Algorithm 1.3.3.

Algorithm 1.3.3 Construction of fuzzy competition hypergraph

1. Begin
2. Input the adjacency matrix $A = [x_{ij}]_{n \times n}$ of a fuzzy digraph $\vec{G}$.
3. Define a relation $f : X \to X$ by $f(x_i) = x_j$, if $x_{ij} > 0$.
4. **do** i from $1 \to n$
5.   **do** j from $1 \to n$
6.   If $x_{ij} > 0$ then (x_j, x_{ij}) belongs to the fuzzy out neighborhood $\mathcal{N}^+(x_i)$.
7.   **end do**
8. **end do**
9. Compute the family of sets $\mathscr{S} = \{E_i = f^{-1}(x_i) : |f^{-1}(x_i)| \geq 2, x_i \in X\}$ where $E_i = \{x_{i_1}, x_{i_2}, \ldots, x_{i_r}\}$ is a hyperedge of $\mathscr{CH}(\vec{G})$.
10. For each hyperedge $E_i \in \mathscr{S}$, calculate the degree of membership of E_i as

$$\lambda_c(E_i) = [\mu(x_{i_1}) \wedge \mu(x_{i_2}) \wedge \ldots \wedge \mu(x_{i_r})] \times h\left(\mathcal{N}^+(x_{i_1}) \cap \mathcal{N}^+(x_{i_2}) \cap \ldots \cap \mathcal{N}^+(x_{i_r})\right).$$

Lemma 1.4 *The fuzzy competition hypergraph of a fuzzy digraph $\vec{G}$ is a fuzzy row hypergraph of $\vec{G}$.*

Proof Let $\vec{G} = (\mu, \vec{\lambda})$ be a fuzzy digraph, then for any hyperedge $E = \{x_1, x_2, \ldots, x_s\}$ of $\mathscr{CH}(\vec{G})$,

$$\lambda_c(E) = [\mu(x_1) \wedge \mu(x_2) \wedge \ldots \wedge \mu(x_s)] \times h(\mathcal{N}^+(x_1) \cap \mathcal{N}^+(x_2) \cap \ldots \cap \mathcal{N}^+(x_s))$$
$$= [\mu(x_1) \wedge \mu(x_2) \wedge \ldots \wedge \mu(x_s)] \times \max_j\{\mathcal{N}^+(x_1) \cap \mathcal{N}^+(x_2) \cap \ldots \cap \mathcal{N}^+(x_s)\}$$
$$= [\mu(x_1) \wedge \mu(x_2) \wedge \ldots \wedge \mu(x_s)] \times \max_j\{\vec{\lambda}(x_1 x_j) \wedge \vec{\lambda}(x_2 x_j) \wedge \ldots \wedge \vec{\lambda}(x_n x_j)\} = \lambda_r(E)$$

It follows that E is a hyperedge of fuzzy row hypergraph.

Example 1.10 Consider the fuzzy digraph given in Fig. 1.8. The fuzzy out neighborhood and fuzzy in neighborhood of all the vertices are given in Table 1.10.

Using Algorithm 1.3.3, the relation $f : X \to X$ of $\vec{G}$ is given in Fig. 1.11. The construction of fuzzy competition hypergraph from $\vec{G}$ is given as follows.

1. Since $f^{-1}(x_2) = E_2 = \{x_1, x_5, x_6\}$, $f^{-1}(x_3) = E_3 = \{x_2, x_5\}$, and $f^{-1}(x_4) = E_4 = \{x_3, x_5\}$, therefore, $\{x_1, x_5, x_6\}$, $\{x_2, x_5\}$, and $\{x_3, x_5\}$ are hyperegdes in $\mathscr{CH}(\vec{G})$.
2. For hyperedge E_2: $\mathcal{N}^+(x_1) \cap \mathcal{N}^+(x_5) \cap \mathcal{N}^+(x_6) = \{(x_2, 0.3)\}$,

$$\lambda_c(E_2) = [\mu(x_1) \wedge \mu(x_5) \wedge \mu(x_6)] \times h\left(\mathcal{N}^+(x_1) \cap \mathcal{N}^+(x_5) \cap \mathcal{N}^+(x_6)\right) = 0.3 \times 0.3 = 0.09.$$

3. Similarly,

$$\lambda_c(E_3) = [\mu(x_2) \wedge \mu(x_5)] \times h\left(\mathcal{N}^+(x_2) \cap \mathcal{N}^+(x_5)\right) = 0.04,$$

$$\lambda_c(E_4) = [\mu(x_3) \wedge \mu(x_5)] \times h\left(\mathcal{N}^+(x_3) \cap \mathcal{N}^+(x_5)\right) = 0.16.$$

Table 1.10 Fuzzy out neighborhood and fuzzy in neighborhood of vertices in $\vec{G}$

$x \in X$	$\mathcal{N}^+(x)$	$\mathcal{N}^-(x)$
x_1	$\{(x_2, 0.4), (x_6, 0.2)\}$	$\emptyset$
x_2	$\{(x_3, 0.1)\}$	$\{(x_1, 0.4), (x_5, 0.4), (x_6, 0.3)\}$
x_3	$\{(x_4, 0.6)\}$	$\{(x_2, 0.1), (x_5, 0.3)\}$
x_4	$\emptyset$	$\{(x_3, 0.6), (x_5, 0.4)\}$
x_5	$\{(x_2, 0.4), (x_3, 0.3), (x_4, 0.4)\}$	$\{(x_6, 0.1)\}$
x_6	$\{(x_2, 0.3), (x_5, 0.1)\}$	$\{(x_1, 0.2)\}$

Fig. 1.11 Representation of fuzzy relation in $\vec{G}$

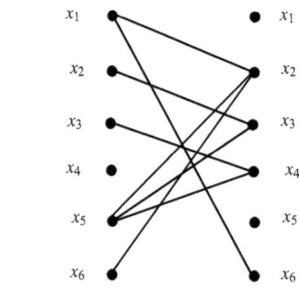

Fig. 1.12 Fuzzy competition hypergraph $\mathscr{C}\mathscr{H}(\vec{G})$

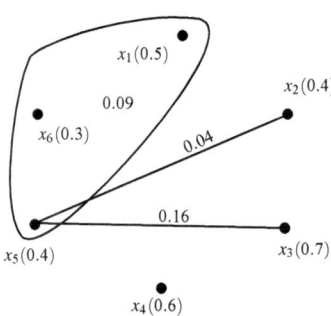

The fuzzy competition hypergraph is given in Fig. 1.12. From Figs. 1.9 and 1.12, it is clear that fuzzy competition hypergraph is a fuzzy row hypergraph.

Definition 1.29 The *fuzzy double competition hypergraph* $\mathscr{D}\mathscr{C}\mathscr{H}(\vec{G}) = (\mu, \lambda_d)$ having same vertex set as $\vec{G}$ and there is hyperedge consisting of vertices x_1, $x_2, \ldots, x_s$ if $\mathcal{N}^+(x_1) \cap \mathcal{N}^+(x_2) \cap \ldots \cap \mathcal{N}^+(x_s) \neq \emptyset$ and $\mathcal{N}^-(x_1) \cap \mathcal{N}^-(x_2) \cap \ldots \cap \mathcal{N}^-(x_s) \neq \emptyset$. The degree of membership of hyperedge $E = \{x_1, x_2, \ldots, x_s\}$ is defined as

$$\lambda_d(E) = [\mu(x_1) \wedge \mu(x_2) \wedge \ldots \wedge \mu(x_s)] \times [h(\mathcal{N}^+(x_1) \cap \mathcal{N}^+(x_2) \cap \ldots \cap \mathcal{N}^+(x_s))$$
$$\wedge h(\mathcal{N}^-(x_1) \cap \mathcal{N}^-(x_2) \cap \ldots \cap \mathcal{N}^-(x_s))].$$

The method for the construction of fuzzy double competition hypergraph is given in Algorithm 1.3.4.

Algorithm 1.3.4 Construction of fuzzy double competition hypergraph

1. Input the adjacency matrix $A = [x_{ij}]_{n \times n}$ of a fuzzy digraph $\vec{G}$.
2. Define a relation $f : X \to X$ by $f(x_i) = x_j$, if $x_{ij} > 0$.
3. Compute the family of sets $\mathscr{S} = \{E_i = f^{-1}(x_i) : |f^{-1}(x_i)| \geq 2, x_i \in X\}$ where $E_i = \{x_{i_1}, x_{i_2}, \ldots, x_{i_r}\}$.
4. If $\mathscr{N}^+(x_{i_1}) \cap \mathscr{N}^+(x_{i_2}) \cap \ldots \cap \mathscr{N}^+(x_{i_r})$ and $\mathscr{N}^-(x_{i_1}) \cap \mathscr{N}^-(x_{i_2}) \cap \ldots \cap \mathscr{N}^-(x_{i_r})$ are non-empty, then $E_i = \{x_{i_1}, x_{i_2}, \ldots, x_{i_r}\}$ is a hyperedge of $\mathscr{D}\mathscr{C}\mathscr{H}(\vec{G})$.
5. For each hyperedge $E_i \in \mathscr{S}$, calculate the degree of membership of hyperedge E_i,
$$\lambda_d(E_i) = [\mu(x_{i_1}) \wedge \mu(x_{i_2}) \wedge \ldots \wedge \mu(x_{i_r})] \times h\left(\mathscr{N}^+(x_{i_1}) \cap \mathscr{N}^+(x_{i_2}) \cap \ldots \cap \mathscr{N}^+(x_{i_r})\right) \wedge h\left(\mathscr{N}^-(x_{i_1}) \cap \mathscr{N}^-(x_{i_2}) \cap \ldots \cap \mathscr{N}^-(x_{i_r})\right).$$

Lemma 1.5 *The fuzzy double competition hypergraph is the intersection of fuzzy row hypergraph and fuzzy column hypergraph.*

Proof Let $\vec{G} = (\mu, \vec{\lambda})$ be a fuzzy digraph, then for any hyperedge $E = \{x_1, x_2, \ldots, x_s\}$ of $\mathscr{C}\mathscr{H}(\vec{G})$,

$$\lambda_d(E) = [\mu(x_1) \wedge \mu(x_2) \wedge \ldots \wedge \mu(x_s)] \times$$
$$[h(\mathscr{N}^+(x_1) \cap \mathscr{N}^+(x_2) \cap \ldots \cap \mathscr{N}^+(x_s)) \wedge h(\mathscr{N}^+(x_1) \cap \mathscr{N}^+(x_2) \cap \ldots \cap \mathscr{N}^+(x_s))].$$
$$= [\mu(x_1) \wedge \mu(x_2) \wedge \ldots \wedge \mu(x_s)] \times$$
$$[\max_j \{\vec{\lambda}(x_1 x_j) \wedge \vec{\lambda}(x_2 x_j) \wedge \ldots \wedge \vec{\lambda}(x_n x_j)\} \wedge \max_k \{\vec{\lambda}(x_k x_1) \wedge \vec{\lambda}(x_k x_2) \wedge \ldots \wedge \vec{\lambda}(x_k x_n)\}].$$
$$= [\{\mu(x_1) \wedge \mu(x_2) \wedge \ldots \wedge \mu(x_s)\} \times \max_j \{\vec{\lambda}(x_1 x_j) \wedge \vec{\lambda}(x_2 x_j) \wedge \ldots \wedge \vec{\lambda}(x_n x_j)\}] \times$$
$$[\{\mu(x_1) \wedge \mu(x_2) \wedge \ldots \wedge \mu(x_s)\} \times \max_k \{\vec{\lambda}(x_k x_1) \wedge \vec{\lambda}(x_k x_2) \wedge \ldots \wedge \vec{\lambda}(x_k x_n)\}]$$
$$= \lambda_r(E) \wedge \lambda_{cl}(E).$$

It follows that fuzzy double competition hypergraph is the intersection of fuzzy row hypergraph and fuzzy column hypergraph.

Example 1.11 Consider the example of fuzzy digraph shown in Fig. 1.8. From Example 1.10, the fuzzy double competition hypergraph of Fig. 1.8 is given in Fig. 1.13. Also Figs. 1.9, 1.10, 1.13 show that the fuzzy double competition hypergraph is the intersection of fuzzy row hypergraph and fuzzy column hypergraph.

Definition 1.30 Let $\vec{G} = (\mu, \vec{\lambda})$ be a fuzzy digraph on a non-empty set X. The *fuzzy niche hypergraph* $\mathscr{N}\mathscr{H}(\vec{G}) = (\mu, \lambda_n)$ has the same vertex set as $\vec{G}$ and there is hyperedge consisting of vertices $x_1, x_2, \ldots, x_s$ if either $\mathscr{N}^+(x_1) \cap \mathscr{N}^+(x_2) \cap \ldots \cap \mathscr{N}^+(x_s) \neq \emptyset$ or $\mathscr{N}^-(x_1) \cap \mathscr{N}^-(x_2) \cap \ldots \cap \mathscr{N}^-(x_s) \neq \emptyset$. The degree of membership of hyperedge $E = \{x_1, x_2, \ldots, x_s\}$ is defined as

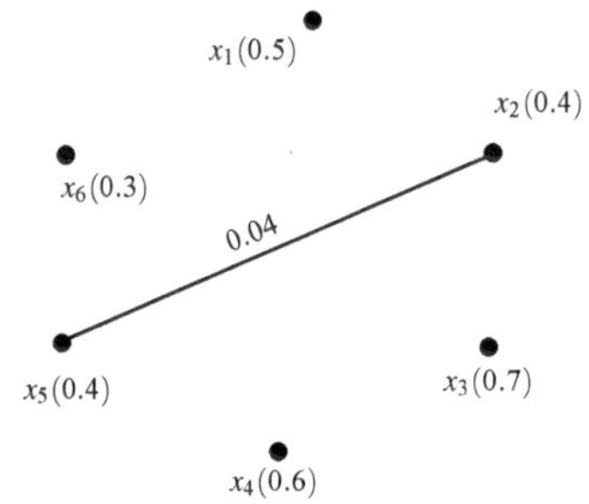

Fig. 1.13 Fuzzy double competition hypergraph $\mathscr{DCH}(\vec{G})$

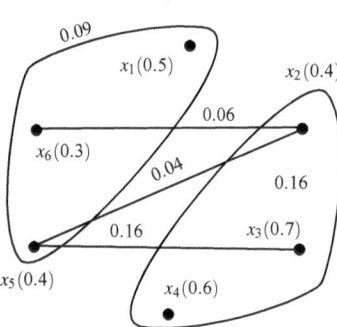

Fig. 1.14 Fuzzy niche hypergraph $\mathscr{NH}(\vec{G})$

$$\lambda_n(E) = [\mu(x_1) \wedge \mu(x_2) \wedge \ldots \wedge \mu(x_s)] \times$$
$$[h\left(\mathscr{N}^+(x_1) \cap \mathscr{N}^+(x_2) \cap \ldots \cap \mathscr{N}^+(x_s)\right) \vee h\left(\mathscr{N}^-(x_1) \cap \mathscr{N}^-(x_2) \cap \ldots \cap \mathscr{N}^-(x_s)\right)].$$

Lemma 1.6 *The fuzzy niche hypergraph is the union of fuzzy row hypergraph and fuzzy column hypergraph.*

Example 1.12 The fuzzy niche hypergraph of Fig. 1.8 is shown in Fig. 1.14, which is the union of Figs. 1.10 and 1.9.

Definition 1.31 Let H be a fuzzy hypergraph and t be the smallest nonnegative number such that $H \cup I_t$ is a fuzzy niche hypergraph of some fuzzy digraph $\vec{G}$, where I_t is a fuzzy set on t isolated vertices X_t, then t is called *fuzzy niche number* of H denoted by $n(H)$.

Lemma 1.7 *Let H be a fuzzy hypergraph on a non-empty set X with $n(H) = t < \infty$ and $H \cup I_t$ is a fuzzy niche hypergraph of an acyclic digraph $\vec{G}$ then for all, $x \in X \cup X_t$,*

$$\mathscr{N}^+(y) \cap I_t \neq \emptyset \Rightarrow \exists\, z \in supp(I_t) \text{ such that } supp(\mathscr{N}^+(y)) = z,$$
$$\mathscr{N}^-(y) \cap I_t \neq \emptyset \Rightarrow \exists\, z \in supp(I_t) \text{ such that } supp(\mathscr{N}^-(y)) = z.$$

Proof On contrary assume that for some $y \in X$ either $supp(\mathscr{N}^+(y)) = \{z\} \cup X'$ or $supp(\mathscr{N}^-(y)) = \{z\} \cup X''$ where $\emptyset \neq X' \subseteq X \cup X_t \setminus \{z\}$. Then, by definition of fuzzy niche hypergraph, z is adjacent to all vertices X' in $H \cup I_t$. A contradiction to the fact that $z \in X_t$.

Lemma 1.8 *Let H be a fuzzy hypergraph with $n(H) = t < \infty$ and $H \cup I_t$ is a fuzzy niche hypergraph of an acyclic fuzzy digraph $\vec{G}$ then for all $z \in X_t$, $\mathcal{N}^+(z) = \emptyset$ and $\mathcal{N}^-(z) = \emptyset$.*

Proof On contrary assume that $X_z^+ = \{y_1, y_2, \ldots, y_s\}$ and $X_z^- = \{y_1', y_2', \ldots, y_r'\}$. Clearly, $\mathcal{N}^+(z) \cap \mathcal{N}^-(z) = \emptyset$ because $\vec{G}$ is acyclic. According to Lemma 1.7, $\mathcal{N}^+(y_i) = \mathcal{N}^+(y_i')$.

Consider another fuzzy digraph $\vec{G}'$ such that $X_{\vec{G}'} = X_{\vec{G}} \setminus \{z\}$ and $E_{\vec{G}'} = (E_{\vec{G}} \setminus \{E_1\}) \cup E_2$ where

$$E_1 = \{\vec{zy_i} : 1 \leq i \leq s\} \cup \{\vec{y_i'z} : 1 \leq i \leq r\},$$
$$E_2 = \{\vec{y_1'y_i} : 1 \leq i \leq s\} \cup \{\vec{y_i'y_1} : 1 \leq i \leq r\}.$$

Clearly, $\mathcal{N}^+(z) = \mathcal{N}^+(y_1)$ and $\mathcal{N}^-(z) = \mathcal{N}^-(y_1')$. Thus $\mathcal{N}\mathcal{H}(\vec{G}') = H \cup I_{t-1}$ which contradicts the fact that $n(H) = t$. Hence, for all $z \in X_t$, $\mathcal{N}^+(z) = \emptyset$ and $\mathcal{N}^-(z) = \emptyset$.

Definition 1.32 Let $H = (\mu, \rho)$ be a fuzzy hypergraph on a non-empty set X. A hyperedge $E_i = \{x_1, x_2, \ldots, x_r\} \subseteq X$ is called *strong* if $\rho(E_i) \geq \frac{1}{2} \bigwedge\limits_{k=1}^{r} \mu_i(x_k)$.

Theorem 1.1 *Let $\vec{G} = (\mu, \vec{\lambda})$ be a fuzzy digraph. If $\mathcal{N}^+(x_1) \cap \mathcal{N}^+(x_2) \cap \ldots \cap \mathcal{N}^+(x_r)$ contains exactly one vertex, then the hyperedge $\{x_1, x_2, \ldots, x_r\}$ of $\mathscr{C}(\vec{G})$ is strong if and only if $|\mathcal{N}^+(x_1) \cap \mathcal{N}^+(x_2) \cap \ldots \cap \mathcal{N}^+(x_r)| > \frac{1}{2}$.*

Proof Assume that $\mathcal{N}^+(x_1) \cap \mathcal{N}^+(x_2) \cap \ldots \cap \mathcal{N}^+(x_r) = \{(u, l)\}$, where l is degree of membership of u. As $|\mathcal{N}^+(x_1) \cap \mathcal{N}^+(x_2) \cap \ldots \cap \mathcal{N}^+(x_r)| = l = h(\mathcal{N}^+(x_1) \cap \mathcal{N}^+(x_2) \cap \ldots \cap \mathcal{N}^+(x_r))$, therefore, $\lambda_c(\{x_1, x_2, \ldots, x_r\}) = (\mu(x_1) \wedge \mu(x_2) \wedge \ldots \wedge \mu(x_r)) \times h(\mathcal{N}^+(x_1) \cap \mathcal{N}^+(x_2) \cap \ldots \cap \mathcal{N}^+(x_r)) = l \times (\mu(x_1) \wedge \mu(x_2) \wedge \ldots \wedge \mu(x_r)\})$. Thus, the hyperedge $\{x_1, x_2, \ldots, x_r\}$ in $\mathscr{C}(\vec{G})$ would be strong if $l > \frac{1}{2}$ by Definition 1.32.

Definition 1.33 Let k be a nonnegative real number, then the *fuzzy k−competition hypergraph* of a fuzzy digraph $\vec{G} = (\mu, \vec{\lambda})$ is fuzzy hypergraph $\mathscr{C}_k(\vec{G}) = (\mu, \lambda_{kc})$ which has the same fuzzy vertex set as in $\vec{G}$ and there is a hyperedge $E = \{x_1, x_2, \ldots, x_r\}$ in $\mathscr{C}_k(\vec{G})$ if $|\mathcal{N}^+(x_1) \cap \mathcal{N}^+(x_2) \cap \ldots \cap \mathcal{N}^+(x_r)| > k$. The membership degree of the hyperedge E is defined as

$$\lambda_{kc}(E) = \frac{l-k}{l}(\mu(x_1) \wedge \mu(x_2) \wedge \ldots \wedge \mu(x_r)) \times h(\mathcal{N}^+(x_1) \cap \mathcal{N}^+(x_2) \cap \ldots \cap \mathcal{N}^+(x_r))$$

where $|\mathcal{N}^+(x_1) \cap \mathcal{N}^+(x_2) \cap \ldots \cap \mathcal{N}^+(x_r)| = l$.

Example 1.13 The fuzzy 0.2−competition hypergraph of Fig. 1.8 is given in Fig. 1.15.

Fig. 1.15 Fuzzy
0.2−competition hypergraph

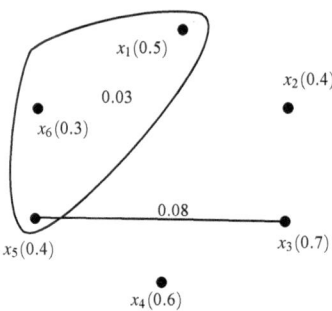

Remark 1.3 For $k = 0$, a fuzzy k−competition hypergraph is simply a fuzzy competition hypergraph.

Theorem 1.2 Let $\vec{G} = (\mu, \vec{\lambda})$ be a fuzzy digraph. If $h(\mathcal{N}^+(x_1) \cap \mathcal{N}^+(x_2) \cap \ldots \cap \mathcal{N}^+(x_r)) = 1$ and $|\mathcal{N}^+(x_1) \cap \mathcal{N}^+(x_2) \cap \ldots \cap \mathcal{N}^+(x_r)| > 2k$ for some $x_1, x_2, \ldots, x_r \in X$, then the hyperedge $\{x_1, x_2, \ldots, x_r\}$ is strong in $\mathscr{C}_k(\vec{G})$.

Proof Let $\mathscr{C}_k(\vec{G}) = (\mu, \lambda_{kc})$ be a fuzzy k−competition hypergraph of fuzzy digraph $\vec{G} = (\mu, \vec{\lambda})$. Suppose for $E = \{x_1, x_2, \ldots, x_r\} \subseteq X$, $|\mathcal{N}^+(x_1) \cap \mathcal{N}^+(x_2) \cap \ldots \cap \mathcal{N}^+(x_r)| = l$. Now,

$$\lambda_{kc}(E) = \frac{l - k}{l}(\mu(x_1) \wedge \mu(x_2) \wedge \ldots \wedge \mu(x_r)) \times h(\mathcal{N}^+(x_1) \cap \mathcal{N}^+(x_2) \cap \ldots \cap \mathcal{N}^+(x_r)),$$

$$\lambda_{kc}(E) = \frac{l - k}{l}(\mu(x_1) \wedge \mu(x_2) \wedge \ldots \wedge \mu(x_r)), \quad \because h(\mathcal{N}^+(x_1) \cap \mathcal{N}^+(x_2) \cap \ldots \cap \mathcal{N}^+(x_r)) = 1,$$

$$\implies \frac{\lambda_{kc}(E)}{\mu(x_1) \wedge \mu(x_2) \wedge \ldots \wedge \mu(x_r)} > \frac{1}{2}, \quad \because l > 2k.$$

Thus, the hyperedge E is strong in $\mathscr{C}_k(\vec{G})$.

1.4 Fuzzy Neighborhood Hypergraphs

Definition 1.34 The *fuzzy open neighborhood hypergraph* of a fuzzy graph $G = (\mu, \lambda)$ is a fuzzy hypergraph $\mathcal{N}(G) = (\mu, \lambda')$ whose fuzzy vertex set is same as G and there is a hyperedge $E = \{x_1, x_2, \ldots, x_r\}$ in $\mathcal{N}(G)$ if $\mathcal{N}(x_1) \cap \mathcal{N}(x_2) \cap \ldots \cap \mathcal{N}(x_r) \neq \emptyset$. The membership function $\lambda' : X \times X \to [0, 1]$ is defined as

$$\lambda'(E) = \big(\mu(x_1) \wedge \mu(x_2) \wedge \ldots \wedge \mu(x_r)\big) \times h\big(\mathcal{N}(x_1) \cap \mathcal{N}(x_2) \cap \ldots \cap \mathcal{N}(x_r)\big).$$

The fuzzy closed neighborhood hypergraph is defined on the same lines in the following definition.

Definition 1.35 The *fuzzy closed neighborhood hypergraph* of $G = (\mu, \lambda)$ is a fuzzy hypergraph $\mathcal{N}[G] = (\mu, \lambda^*)$ whose fuzzy set of vertices is the same as G and there is

Fig. 1.16 Fuzzy graph G

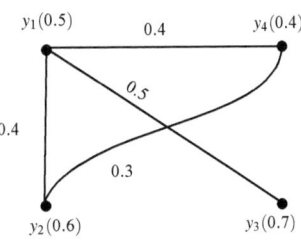

Table 1.11 Fuzzy open neighborhood of vertices

y	$\mathcal{N}(y)$
y_1	$\{(y_2, 0.4), (y_3, 0.5), (y_4, 0.5)\}$
y_2	$\{(y_1, 0.4), (y_4, 0.3)\}$
y_3	$\{(y_1, 0.5)\}$
y_4	$\{(y_1, 0.4), (y_2, 0.3)\}$

Table 1.12 Fuzzy closed neighborhood of vertices

y	$\mathcal{N}[y]$
y_1	$\{(y_1, 0.5), (y_2, 0.4), (y_3, 0.5), (y_4, 0.5)\}$
y_2	$\{(y_2, 0.6), (y_1, 0.4), (y_4, 0.3)\}$
y_3	$\{(y_3, 0.7), (y_1, 0.5)\}$
y_4	$\{(y_4, 0.4), (y_1, 0.4), (y_2, 0.3)\}$

a hyperedge $E = \{x_1, x_2, \ldots, x_r\}$ in $\mathcal{N}[G]$ if $\mathcal{N}[x_1] \cap \mathcal{N}[x_2] \cap X \ldots \cap \mathcal{N}[x_r] \neq \emptyset$. The membership function $\lambda^* : X \times X \to [0, 1]$ is defined as

$$\lambda^*(E) = \big(\mu(x_1) \wedge \mu(x_2) \wedge \ldots \wedge \mu(x_r)\big) \times h\big(\mathcal{N}[x_1] \cap \mathcal{N}[x_2] \cap \ldots \cap \mathcal{N}[x_r]\big).$$

Example 1.14 Consider the fuzzy graph $G = (\mu, \lambda)$ on set $X = \{y_1, y_2, y_3, y_4\}$ as shown in Fig. 1.16. The fuzzy open neighborhoods are given in Table 1.11.

Define a relation $f : X \to X$ by $f(y_i) = y_j$ if $y_j \in supp(\mathcal{N}(y_i))$ as shown in Fig. 1.17. If for $y_i \in X$, $|f^{-1}(y_i)| > 1$, then $f^{-1}(y_i)$ is a hyperedge of $\mathcal{N}[G]$. Since from Fig. 1.17, $f^{-1}(y_1) = \{y_2, y_3, y_4\} = E_1$, $f^{-1}(y_2) = \{y_1, y_4\} = E_2$, and $f^{-1}(y_4) = \{y_1, y_2\}_3$, therefore, E_1, E_2, E_3 are hyperedges of $\mathcal{N}(G)$. The degree of membership of each hyperedge can be computed using Definition 1.34 as follows.

For $f^{-1}(y_1) = E_1 = \{y_2, y_3, y_4\}$, $\lambda'(E_1) = \big(\mu(y_2) \wedge \mu(y_3) \wedge \mu(y_4)\big) \times h\big(\mathcal{N}(y_2)$ $\cap \mathcal{N}(y_3) \cap \mathcal{N}(y_4)\big) = 0.4 \times 0.4 = 0.16$. Similarly, $\lambda'(\{y_1, y_4\}) = 0.4 \times 0.3 = 0.12$ and $\lambda'(\{y_1, y_2\}) = 0.5 \times 0.3 = 0.15$. The fuzzy open neighborhood hypergraph constructed using Definition 1.23 from $\vec{G}$ is given in Fig. 1.17.

The fuzzy closed neighborhoods of all the vertices in G are given in Table 1.12.

Since, $\mathcal{N}[y_1] \cap \mathcal{N}[y_2] \cap \mathcal{N}[y_3] \cap \mathcal{N}[y_4] = \{(y_1, 0.4)\}$, therefore, $E = \{y_1, y_2, y_3, y_4\}$ is a hyperedge of $\mathcal{N}[G]$ and $\lambda^*(E) = 0.4 \times 0.4 = 0.16$. The fuzzy closed neighborhood hypergraph is given in Fig. 1.18.

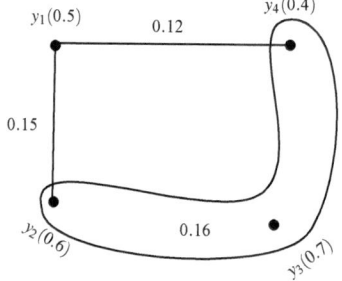 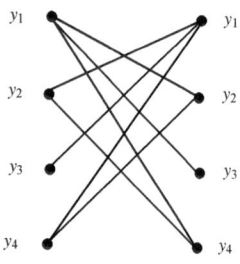

Fig. 1.17 Fuzzy open neighborhood hypergraph of G

Fig. 1.18 Fuzzy closed
neighborhood hypergraph

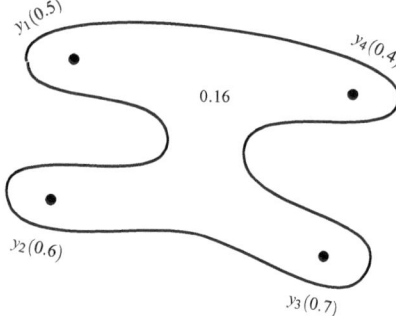

Using different types of fuzzy neighborhood of the vertices, some other types of
fuzzy hypergraphs are defined here.

Definition 1.36 Let k be a nonnegative real number, then the *fuzzy* $(k)-competition$
hypergraph of a fuzzy graph $G = (\mu, \lambda)$ is a fuzzy hypergraph $\mathcal{N}_k(G) = (\mu, \lambda'_{kc})$
having the same fuzzy set of vertices as G and there is a hyperedge $E =$
$\{x_1, x_2, \ldots, x_r\}$ in $\mathcal{N}_k(G)$ if $|\mathcal{N}(x_1) \cap \mathcal{N}(x_2) \cap \ldots \cap \mathcal{N}(x_r)| > k$. The membership value of E is defined as

$$\lambda'_{kc}(E) = \frac{l-k}{l}\big(\mu(x_1) \wedge \mu(x_2) \wedge \ldots \wedge \mu(x_r)\big) \times h\big(\mathcal{N}(x_1) \cap \mathcal{N}(x_2) \cap \ldots \cap \mathcal{N}(x_r)\big),$$

where $|\mathcal{N}(x_1) \cap \mathcal{N}(x_2) \cap \ldots \cap \mathcal{N}(x_r)| = l$.

Definition 1.37 The *fuzzy* $[k]-competition$ *hypergraph* of G is denoted by $\mathcal{N}_k[G] =$
(μ, λ^*_{kc}) and there is a hyperedge E in $\mathcal{N}_k[G]$ if $|\mathcal{N}[x_1] \cap \mathcal{N}[x_2] \cap \ldots \cap \mathcal{N}[x_r]| >$
k. The membership value of E is defined as

$$\lambda^*_{kc}(E) = \frac{p-k}{p}\big(\mu(x_1) \wedge \mu(x_2) \wedge \ldots \wedge \mu(x_r)\big) \times h\big(\mathcal{N}[x_1] \cap \mathcal{N}[x_2] \cap \ldots \cap \mathcal{N}[x_r]\big),$$

where $|\mathcal{N}[x_1] \cap \mathcal{N}[x_2] \cap \ldots \cap \mathcal{N}[x_r]| = p$.

The relations between fuzzy neighborhood hypergraphs and fuzzy competition hypergraphs are given in the following theorems.

Theorem 1.3 *Let $\vec{G} = (\mu, \vec{\lambda})$ be a symmetric fuzzy digraph without any loops, then $\mathscr{C}_k(\vec{G}) = \mathscr{N}_k(\mathscr{U}(\vec{G}))$ where $\mathscr{U}(\vec{G})$ is the underlying fuzzy graph of $\vec{G}$.*

Proof Let $\mathscr{U}(\vec{G}) = (\mu, \lambda)$ corresponds to fuzzy graph $\vec{G} = (\mu, \vec{\lambda})$. Also, let $\mathscr{N}_k(\mathscr{U}(\vec{G})) = (\mu, \lambda'_{kc})$ and $\mathscr{C}_k(\vec{G}) = (\mu, \lambda_{kc})$. Clearly, the fuzzy $k-$competition hypergraph $\mathscr{C}_k(\vec{G})$ and the underlying fuzzy graph have the same fuzzy set of vertices as $\vec{G}$. Hence, $\mathscr{N}_k(\mathscr{U}(\vec{G}))$ has the same vertex set as $\vec{G}$. It remains only to show that $\lambda_{kc}(xw) = \lambda'_{kc}(xw)$ for every $x, w \in X$. So there are two cases.

Case 1: If for each $x_1, x_2, \ldots, x_r \in X$, $\lambda_{kc}(\{x_1, x_2, \ldots, x_r\}) = 0$ in $\mathscr{C}_k(\vec{G})$, then $|\mathscr{N}^+(x_1) \cap \mathscr{N}^+(x_2) \cap \ldots \mathscr{N}^+(x_r)| \leq k$. Since $\vec{G}$ is symmetric, therefore, $|\mathscr{N}(x_1) \cap \mathscr{N}(x_2) \cap \ldots \mathscr{N}(x_r)| \leq k$ in $\mathscr{U}(\vec{G})$. Thus, $\lambda'_{kc}(\{x_1, x_2, \ldots, x_r\}) = 0$ and $\lambda_{kc}(E) = \lambda'_{kc}(E)$ for all $x_1, x_2, \ldots, x_r \in X$.

Case 2: If for some $x_1, x_2, \ldots, x_r \in X$, $\lambda_{kc}(E) > 0$ in $\mathscr{C}_k(\vec{G})$, then $|\mathscr{N}^+(x_1) \cap \mathscr{N}^+(x_2) \cap \ldots \mathscr{N}^+(x_r)| > k$. Thus,

$$\lambda_{kc}(E) = \frac{l-k}{l}[\mu(x_1) \wedge \mu(x_2) \wedge \ldots \wedge \mu(x_r)]h(\mathscr{N}^+(x_1) \cap \mathscr{N}^+(x_2) \cap \ldots \cap \mathscr{N}^+(x_r)),$$

where $l = |\mathscr{N}^+(x_1) \cap \mathscr{N}^+(x_2) \cap \ldots \cap \mathscr{N}^+(x_r)|$. Since, $\vec{G}$ is a symmetric fuzzy digraph, $|\mathscr{N}(x_1) \cap \mathscr{N}(x_2) \cap \ldots \mathscr{N}(x_r)| > k$. Hence, $\lambda_{kc}(E) = \lambda'_{kc}(E)$. Since $x_1, x_2, \ldots, x_r$ were taken to be arbitrary, therefore, the result holds for all hyperedges E of $\mathscr{C}_k(\vec{G})$.

Theorem 1.4 *Let $\vec{G} = (C, \vec{D})$ be a symmetric fuzzy digraph having loops at every vertex, then $\mathscr{C}_k(\vec{G}) = \mathscr{N}_k[\mathscr{U}(\vec{G})]$ where $\mathscr{U}(\vec{G})$ is the underlying fuzzy graph of $\vec{G}$.*

Proof Let $\mathscr{U}(\vec{G}) = (\mu, \lambda)$ be an underlying fuzzy graph corresponding to fuzzy digraph $\vec{G} = (\mu, \vec{\lambda})$. Let $\mathscr{N}_k[\mathscr{U}(\vec{G})] = (\mu, \lambda'_{kc})$ and $\mathscr{C}_k(\vec{G}) = (\mu, \lambda_{kc})$. The fuzzy $k-$competition graph $\mathscr{C}_k(\vec{G})$ as well as the underlying fuzzy graph have the same vertex set as $\vec{G}$. It follows that $\mathscr{N}_k[\mathscr{U}(\vec{G})]$ has the same fuzzy vertex set as $\vec{G}$. It remains only to show that $\lambda_{kc}(\{x_1, x_2, \ldots, x_r\}) = \lambda'_{kc}(\{x_1, x_2, \ldots, x_r\})$ for every $x_1, x_2, \ldots, x_r \in X$. As the fuzzy digraph has a loop at every vertex, therefore, the fuzzy out neighborhood contains the vertex itself. There are two cases.

Case 1: If for all $x_1, x_2, \ldots, x_r \in X$, $\lambda_{kc}(E) = 0$ in $\mathscr{C}_k(\vec{G})$, then $|\mathscr{N}^+(x_1) \cap \mathscr{N}^+(x_2) \cap \ldots \mathscr{N}^+(x_r)| \leq k$. As $\vec{G}$ is symmetric, therefore, $|\mathscr{N}[x_1] \cap \mathscr{N}[x_2] \cap \ldots \mathscr{N}[x_r]| \leq k$ in $\mathscr{U}(\vec{G})$. Hence, $\lambda'_{kc}(E) = 0$ and so $\lambda_{kc}(E) = \lambda'_{kc}(E)$ for all $x_1, x_2, \ldots, x_r \in X$.

Case 2: If for some $x_1, x_2, \ldots, x_r \in X$, $\lambda_{kc}(E) > 0$ in $\mathscr{C}_k(\vec{G})$, then $|\mathscr{N}^+(x_1) \cap \mathscr{N}^+(x_2) \cap \ldots \mathscr{N}^+(x_r)| > k$.

As $\vec{G}$ is symmetric fuzzy digraph and have loops at every vertex, therefore, $|\mathscr{N}[x_1] \cap \mathscr{N}[x_2] \cap \ldots \mathscr{N}[x_r]| > k$. Hence, $\lambda_{kc}(xy) = \lambda'_{kc}(xy)$. As $x_1, x_2, \ldots, x_r$ were taken to be arbitrary, therefore, the result holds for all hyperedges $E = \{x_1, x_2, \ldots, x_r\}$ of $\mathscr{C}_k(\vec{G})$.

1.5 Applications of Fuzzy Competition Hypergraphs

In this section, we present several applications of fuzzy competition hypergraphs in food webs, business marketing, and social network.

1.5.1 Identifying Predator–Prey Relations in Ecosystem

We now present application of fuzzy competition hypergraphs in order to describe the interconnection of food chains between species, flow of energy, and predator–prey relationship in ecosystem. The strength of competition between species represents the competition for food and common preys of species. We will discuss a method to give description of species relationship, danger to the population growth rate of certain species, powerful animals in ecological niches, and lack of food for weak animals.

Competition graphs arose in connection with an application in food webs. However, in some cases, competition hypergraphs provide a detailed description of predator–prey relations than competition graphs. In a competition hypergraph, it is assumed that vertices are defined clearly but in real-world problems, vertices are not defined precisely. As an example, species may be of different types like vegetarian, nonvegetarian, weak, or strong.

Fuzzy food webs can be used to describe the combination of food chains that are interconnected by a fuzzy network of food relationship. There are many interesting variations of the notion of fuzzy competition hypergraph in ecological interpretation. For instance, two species may have a common prey (*fuzzy competition hypergraph*), a common enemy (*fuzzy common enemy hypergraph*), both common prey and common enemy (*fuzzy competition common enemy hypergraph*), either a common prey or a common enemy (*fuzzy niche hypergraph*). We now discuss a type of fuzzy competition hypergraph in which species have common enemies known as *fuzzy common enemy hypergraph*.

Let $\vec{G} = (\mu, \vec{\lambda})$ be a fuzzy food web. The fuzzy common enemy hypergraph $\mathscr{CH}(\vec{G}) = (\mu, \lambda_c)$ has the same vertex set as $\vec{G}$ and there is a hyperedge consisting of vertices $x_1, x_2, \ldots, x_s$ if $\mathscr{N}^+(x_1) \cap \mathscr{N}^+(x_2) \cap \ldots \cap \mathscr{N}^+(x_s) \neq \emptyset$. The degree of membership of hyperedge $E = \{x_1, x_2, \ldots, x_s\}$ is defined as

$$\lambda_c(E) = [\mu(x_1) \wedge \mu(x_2) \wedge \ldots \wedge \mu(x_s)] \times h(\mathscr{N}^+(x_1) \cap \mathscr{N}^+(x_2) \cap \ldots \cap \mathscr{N}^+(x_s)).$$

The strength of common enemies between species can be calculated using Algorithm 1.3.3. Consider the example of a fuzzy food web of 13 species giraffe, lion, vulture, rhinoceros, African skunk, fiscal shrike, grasshopper, baboon, leopard, snake, caracal, mouse, and impala. The degree of membership of each species represents the species' ability to resource defense. The degree of membership of each directed

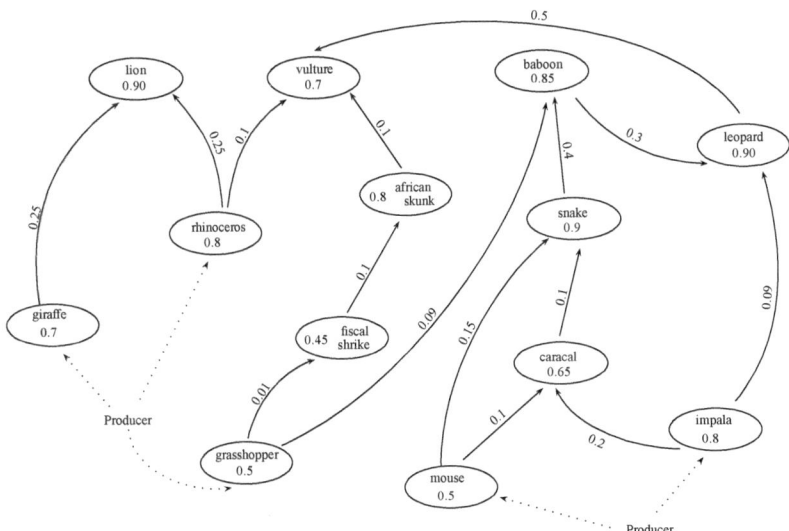

Fig. 1.19 Fuzzy food web

edge represents the strength to which the prey is harmful for predator. The fuzzy food web is shown in Fig. 1.19. The directed edge between giraffe and lion shows that giraffe is eaten by a lion and similarly.

The degree of membership of lion is 0.9, which shows that lion has 90% ability of resource defense, i.e, it can defend itself against other animals as well as can survive many days if the lion doesn't find any food. The directed edge between giraffe and lion has degree of membership 0.25 which represents that giraffe is 25% harmful for lion because a giraffe can kill a lion with its long legs. This is an acyclic fuzzy digraph. The fuzzy out neighborhoods are given in Table 1.13.

The fuzzy common enemy hypergraph is shown in Fig. 1.20. The hyperedges in Fig. 1.20 show that there are common enemies between giraffe and rhinoceros, rhinoceros, African skunk and leopard, grasshopper and snake, mouse and impala, baboon and impala. The membership value of each hyperedge represents the degree of common enemies among the species.

The hyperedge {impala, baboon} has maximum degree of membership which shows that impala and baboon has largest number of common enemies whereas mouse and impala has least number of common enemies.

1.5.2 Identifying Competitors in Business Market

Fuzzy competition hypergraphs are a key approach to study the competition, profit and loss, market power, and rivalry among buyers and sellers using fuzziness in hypergraphical structures. We now discuss a method to study the business competition

Table 1.13 Fuzzy out
neighborhoods of vertices

Species	$\mathscr{N}^+(u) : u$ is a specie
Giraffe	{(lion, 0.25)}
Lion	$\emptyset$
Rhinoceros	{(lion, 0.25), (vulture, 0.1)}
Vulture	$\emptyset$
African skunk	{(vulture, 0.1)}
Fiscal shrike	{(African skunk, 0.1)}
Grasshopper	{(fiscal shrike, 0.01), (baboon, 0.09)}
Baboon	{(leopard, 0.3)}
Leopard	{(vulture, 0.5)}
Snake	{(baboon, 0.4)}
Caracal	{(snake, 0.1)}
Mouse	{(caracal, 0.1), (snake, 0.15)}
Impala	{(caracal, 0.2), (leopard, 0.09)}

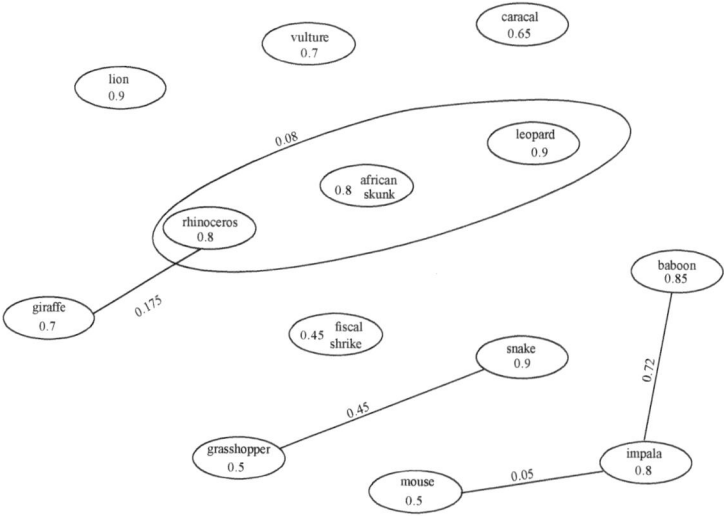

Fig. 1.20 Fuzzy common enemy hypergraph

for power and profit, success and business failure, and demanding products in market.

In business market, there are competitive rivalries among companies which are endeavoring to increase the demand and profit of their product. More than one companies in market sell identical products. Since various companies regularly market identical products, every company wants to attract consumer's attention to its product. There is always a competitive situation in business market. Hypergraph theory is a key approach to study the competitive behavior of buyers and sellers using structures of hypergraphs. In some cases, these structures do not study the level of competition, profit, and loss between the companies. As an example, companies may have different repute in market according to market power and rivalry. These are fuzzy concepts and motivates the necessity of fuzzy competition hypergraphs. The competition among companies can be studied using fuzzy competition hypergraph known as *fuzzy enmity hypergraph*.

We present a method for calculating the strength of competition of companies in the following Algorithm 1.5.1.

Algorithm 1.5.1 Business competition hypergraph

1. Input the adjacency matrix $[x_{ij}]_{n \times n}$ of bipolar fuzzy digraph $\vec{G} = (C, \vec{D})$ of n companies $x_1, x_2, \ldots, x_n$.
2. Construct the table of fuzzy out neighborhoods of all the companies.
3. Construct fuzzy competition hypergraph using Algorithm 1.3.3.
4. **do** i from $1 \rightarrow n$
5.   Calculate the degree of each vertex as $S(x_i) = \sum_{x_i \in E} \lambda_c(E)$ where E is a hyper-edge in fuzzy enmity hypergraph.
6. **end do**
7. $S(x_i)$ denotes the strength of competition of each company x_i, $1 \le i \le n$.

Consider the example of a marketing competition between seven companies DEL, CB, HW, AK, LR, RP, SONY, RA, LR, three retailers, one retailer outlet and one multinational brand as shown in Fig. 1.21. The vertices represent companies, retailers, outlets, and brands. The degree of membership of each vertex represents the strength of rivalry (aggression) of each company in the market. The degree of membership of each directed edge $\vec{xy}$ represents the degree of rejectability of company's x product by company y. The strength of competition of each company can be discussed using fuzzy competition hypergraph known as *fuzzy enmity hypergraph*. The fuzzy out neighborhoods are calculated in Table 1.14.

The fuzzy enmity hypergraph of Fig. 1.21 is shown in Fig. 1.22. The degree of membership of each hyperedge shows the strength of rivalry between the companies.

The strength of rivalry of each company is calculated in Table 1.15 which shows its enmity value within business market. Table 1.15 shows that SONY is the biggest rival company among other companies.

1.5.3 Finding Influential Communities in a Social Network

Fuzzy competition hypergraphs have a wide range of applications in decision-making problems and decision support systems based on social networking. To elaborate on the necessity of the idea discussed in this paper, we apply the notion of fuzzy competition hypergraphs to study the influence, centrality, socialism, and proactiveness of human beings in any social network.

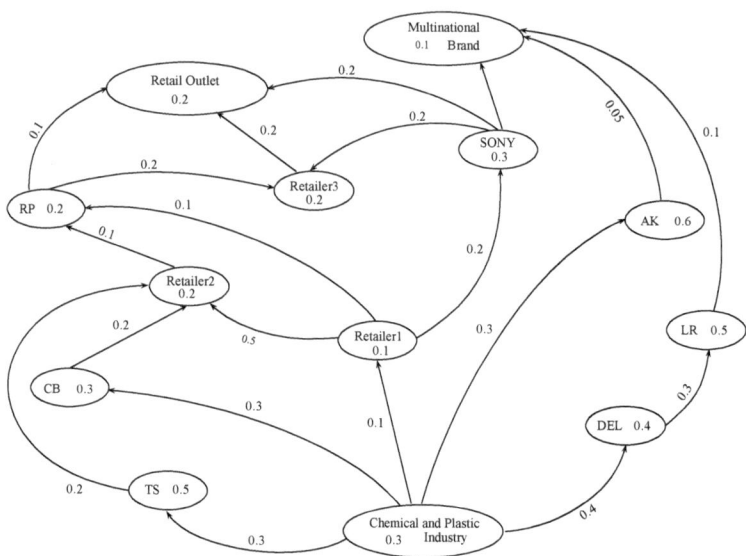

Fig. 1.21 Fuzzy marketing digraph

Table 1.14 Fuzzy out neighborhoods of companies

Company	$\mathscr{N}^+(u) : u$ is a company
Chemical and plastic industries	{(DEL, 0.4), (AK, 0.3), (Retailer1, 0.1), (CB, 0.3), (TS, 0.3)}
DEL	{(LR, 0.3)}
AK	{(Multinational Brand, 0.05)}
LR	{(Multinational Brand, 0.1)}
Retailer1	{(SONY, 0.2), (RP, 0.1), (Retailer2, 0.5)}
CB	{(Retailer2, 0.2)}
TS	{(Retailer2, 0.2)}
Retailer2	{(RP, 0.1)}
SONY	{(Retailer3, 0.2), (R. Outlet, 0.2), (M. Brand, 0.1)}
Retailer3	{(R.Outlet, 0.2)}
RP	{(Retailer3, 0.2), (R. Outlet, 0.1)}
M. Brand	∅
R. Outlet	∅

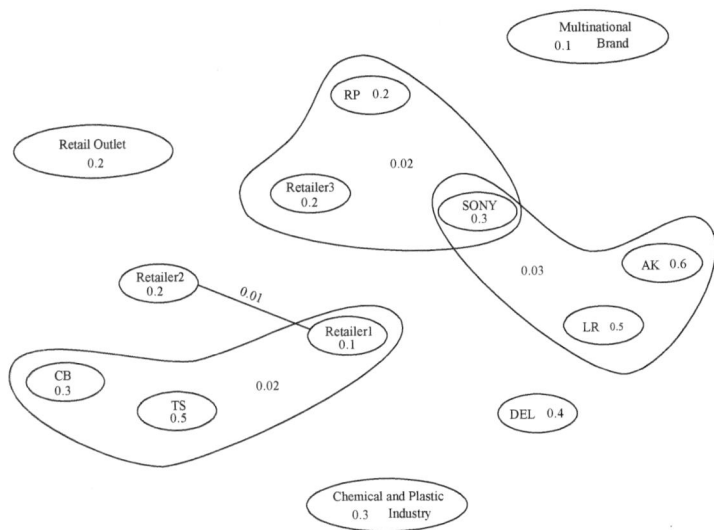

Fig. 1.22 Fuzzy competition hypergraph

Table 1.15 Strength of rivalry between companies

Company	Strength of rivalry
LR	0.03
AK	0.03
SONY	0.05
Retailer3	0.02
RP	0.02
Retailer2	0.01
Retailer1	0.03
CB	0.02
TS	0.02

Social competition is a widespread mechanism to figure out a best-suited group economically, politically, or educationally. Social competition occurs when individual's opinions, decisions, and behaviors are influenced by others. Graph theory is a conceptual framework to study and analyze the units that are intensely or frequently connected in a network. Fuzzy hypergraphs can be used to study the influence and competition between objects more precisely. The social influence and conflict between different communities can be studied using fuzzy competition hypergraph known as *fuzzy influence hypergraph*.

The fuzzy influence hypergraph $G = (\mu, \lambda_c)$ has the same set of vertices as $\vec{G}$ and there is a hyperedge consisting of vertices $x_1, x_2, \ldots, x_r$ if $\mathcal{N}^-(x_1) \cap \mathcal{N}^-(x_2) \cap \ldots \cap \mathcal{N}^-(x_r) \neq \emptyset$. The degree of membership of hyperedge $E = \{x_1, x_2, \ldots, x_r\}$ is defined as

$$\lambda_c(E) = [\mu(x_1) \wedge \mu(x_2) \wedge \ldots \wedge \mu(x_r)] \times h(\mathcal{N}^+(x_1) \cap \mathcal{N}^+(x_2) \cap \ldots \cap \mathcal{N}^+(x_r)).$$

The strength of influence between different objects in a fuzzy influence hypergraph can be calculated by the method presented in Algorithm 1.5.2. The complexity of algorithm is $O(n^2)$.

Algorithm 1.5.2 Fuzzy influence hypergraph

1. Input the adjacency matrix $[x_{ij}]_{n \times n}$ of fuzzy digraph $\vec{G} = (C, \vec{D})$ of n families $x_1, x_2, \ldots, x_n$.
2. Using fuzzy in neighborhoods, construct the fuzzy influence hypergraph following Algorithm 1.3.3.
3. **do** i from $1 \to n$
4.   If x_i belongs to the hyperedge E in fuzzy influence hypergraph, then calculate the degree of each vertex x_i as

$$deg(x_i) = \sum_{x_i \in E} \lambda_c(E) \text{ and } A_i = \sum_{x_i \in E}(|E| - 1).$$

5. **end do**
6. **do** i from $1 \to n$
7.   If $A_i > 1$, then calculate the degree of influence of each vertex x_i a,

$$S(x_i) = \frac{deg(x_i)}{A_i}.$$

8. **end do**

Consider a fuzzy social digraph of Florientine trading families Peruzzi, Lambertes, Bischeri, Strozzi, Guadagni, Tornabuon, Castellan, Ridolfi, Albizzi, Barbadori, Medici, Acciaiuol, Salviati, Ginori, and Pazzi. The vertices in a fuzzy network represent the name of trading families. The degree of membership of each family represents the strength of centrality in that network. The directed edge $\vec{xy}$ indicates that the family x is influenced by y. The degree of membership of each directed edge indicates that to what extent the opinions and suggestions of one family influence the other. The degree of membership of Medici is 0.9 which shows that Medici has 90% central position in trading network. The degree of membership between Redolfi and Medici is 0.6 which indicates that Redolfi follows 60% of the suggestions of Medici. Fuzzy social digraph is shown in Fig. 1.23.

To find the most influential family in this fuzzy network, we construct its fuzzy influence hypergraph. The fuzzy in neighborhoods are given in Table 1.16.

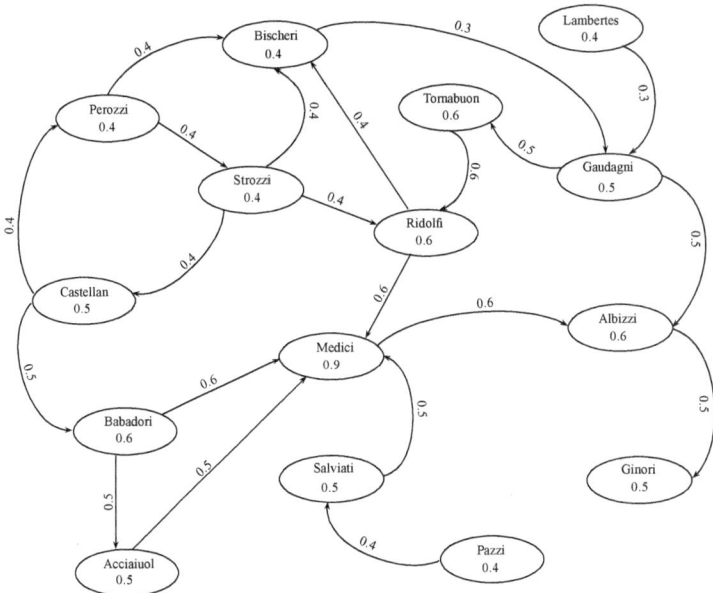

Fig. 1.23 Fuzzy social digraph

Table 1.16 Fuzzy in neighborhoods of all vertices in social network

Family	$\mathcal{N}^-$(family)	Family	$\mathcal{N}^-$(family)
Acciaiuol	{(Babadori, 0.5,)}	Pazzi	∅
Ginori	{(Albizzi, 0.5)}	Salviati	{(Pazzi, 0.4)}
Babadori	{(Castellan, 0.5)}	Castellan	{(Strozzi, 0.4)}
Tornabuon	{(Gaudagni, 0.5)}	Perozzi	{(Castellan, 0.5)}
Lambertes	∅	Strozzi	{(Perozzi, 0.4)}
Medici	{(Babadori, 0.6), (Acciaiuol, 0.5), (Salviati, 0.5), (Ridolfi, 0.6)}		
Bischeri	{(Perozzi, 0.4), (Strozzi, 0.4), (Redolfi, 0.4)}		
Albizzi	{(Medici, 0.6), (Gaudagni, 0.5)}		
Redolfi	{(Strozzi, 0.4), (Tornabuon, 0.6)}		
Gaudgani	{(Bischeri, 0.3), (Lambertes, 0.3)}		

The fuzzy influence hypergraph is shown in Fig. 1.24. The degree of membership of each hyperedge shows the strength of social competition between families to influence the other trading families.

The strength of competition of vertices using Algorithm 1.5.2 is calculated in Table 1.17 where $S(x)$ represents the strength to which each trading family influence the other families. Table 1.17 shows that Acciaiuol and Medici are most influential families in the network.

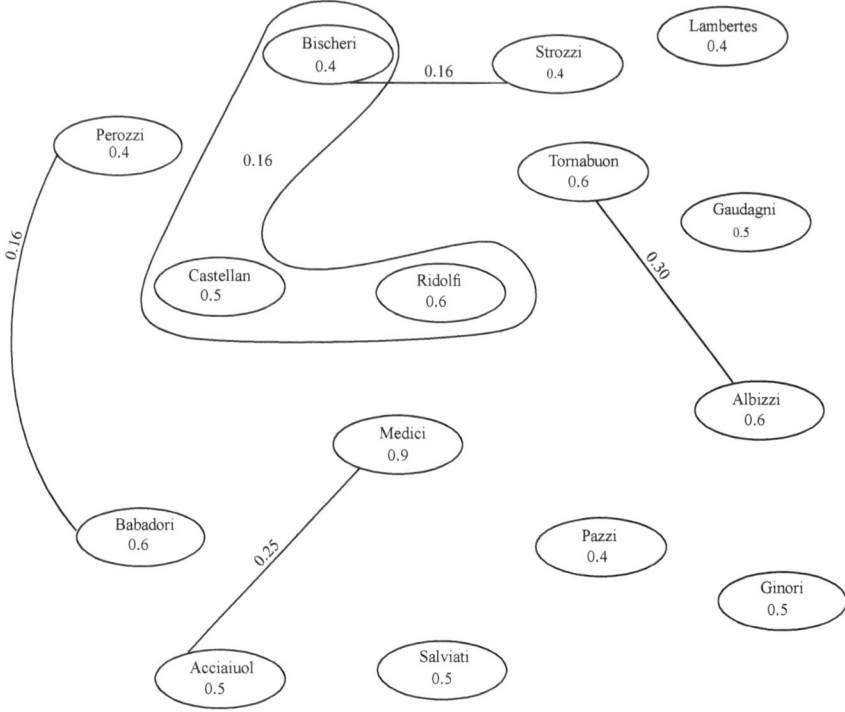

Fig. 1.24 Fuzzy influence hypergraph

Table 1.17 Degree of influence of vertices

x	$deg(x)$	$S(x)$	x	$deg(x)$	$S(x)$
Acciaiuol	0.25	0.25	Medici	0.25	0.25
Babadori	0.16	0.16	Perozzi	0.16	0.16
Castellan	0.16	0.08	Redolfii	0.16	0.08
Strozzi	0.16	0.16	Besceri	0.32	0.12

A View of Fuzzy Competition Hypergraphs in Comparison with Fuzzy Competition Graphs

The concept of fuzzy competition graphs can be utilized successfully in different domains of applications. In the existing methods, we usually consider fuzziness in pair-wise competition and conflicts between objects. But in these representations, we miss some information that whether there is a conflict or a relation among three or more objects. For example, Fig. 1.22 shows strong competition for profit among SONY, LR, and AK. But if we draw the fuzzy competition graph of Fig. 1.21, we cannot discuss the group-wise conflict among companies. Sometimes we are not only interested in pair-wise relations but also in group-wise conflicts, influence, and relations. The novel notion of fuzzy competition hypergraphs is a mathematical tool to overcome this difficulty. We have presented different methods for solving decision-making problems. These methods not only generalize the existing ones but also give better results regarding uncertainty.

1.6 $\mathscr{N}$−Hypergraphs

A (crisp) set A in a universe X can be defined in the form of its characteristic function $\mu_A : X \rightarrow \{0, 1\}$ yielding the value 1 for elements belonging to the set A and the value 0 for elements excluded from the set A. The most of the generalization of the crisp set have been introduced on the unit interval $[0, 1]$ and they are consistent with the asymmetry observation. In other words, the generalization of the crisp set to fuzzy sets relied on spreading positive information that fit the crisp point $\{1\}$ into the interval $[0, 1]$. Because no negative meaning of information is suggested, we now feel a need to deal with negative information. To do so, we also feel a need to supply mathematical tool. To attain such object, Jun et al. [10] have introduced a new function which is called negative-valued function (briefly, $\mathscr{N}$−function) to deal with negative information that fit the crisp point $\{-1\}$ into the interval $[-1, 0]$, and constructed $\mathscr{N}$−structures.

Definition 1.38 Denote by $\mathscr{F}(X, [-1, 0])$ the collection of functions from a non-empty set X to $[-1, 0]$. We say that an element of $\mathscr{F}(X, [-1, 0])$ is a *negative-valued function* from X to $[-1, 0]$ (briefly, $\mathscr{N}$-*function* on X). By an $\mathscr{N}$−*structure*, we mean an ordered pair (X, μ) of X and an $\mathscr{N}$-function μ on X.

The *support* of μ is defined by supp $(\mu) = \{x \in X \mid \mu(x) \neq 0\}$ and say μ is nontrivial if supp(μ) is non-empty. The *height* of μ is $h(\mu) = \min\{\mu(x) \mid x \in X\}$. We say μ is *normal* if $h(\mu) = -1$. By an $\mathscr{N}$−*relation* on X, we mean an $\mathscr{N}$−function ν on $X \times X$ satisfying the following inequality,

$$(\forall x, y \in X)(\nu(x, y) \geq \max\{\mu(x), \mu(y)\}),$$

where $\mu \in \mathscr{F}(X, [-1, 0])$.

Definition 1.39 An $\mathscr{N}$−*graph* with an underlying set X is defined to be a pair $G = (\mu, \nu)$, where μ is an $\mathscr{N}$−function in X and ν is an $\mathscr{N}$−function in $E \subseteq X \times X$ such that
$$\nu(\{x, y\}) \geq \max(\mu(x), \mu(y)),$$

for all $\{x, y\} \in E$. We call μ the $\mathscr{N}$−*vertex function* of X, ν the $\mathscr{N}$−*edge function* of E, respectively. Note that ν is a symmetric $\mathscr{N}$−relation on μ in an undirected graph. But ν may not be a symmetric $\mathscr{N}$−relation on μ in a digraph

Definition 1.40 Let X be a finite set and let E be a finite family of nontrivial $\mathscr{N}$−functions on X such that $V = \bigcup_j \text{supp}(\mu_j)$, $j = 1, 2, \ldots, m$, where μ_j is an

$\mathcal{N}$−function defined on $E_j \in E$. Then, the pair $N = (X, E)$ is a $\mathcal{N}$−*hypergraph* on X, E is the family of $\mathcal{N}$−edges of N and X is the (crisp) vertex set of N. The order of N (number of vertices) is denoted by $|X|$ and the number of edges is denoted by $|E|$.

Let μ be an $\mathcal{N}$−function of X and let E be a collection of $\mathcal{N}$−functions of X such that for each $v \in E$ and $x \in X$, $v(x) \geq \mu(x)$. Then, the pair (μ, v) is an $\mathcal{N}$−hypergraph on the $\mathcal{N}$−function μ. The $\mathcal{N}$−hypergraph (μ, v) is also an $\mathcal{N}$−hypergraph on $X = \text{supp}(\mu)$, the $\mathcal{N}$−function μ defines a condition for membership and nonmembership in the edge set E. This condition can be stated separately, so without loss of generality, we restrict attention to $\mathcal{N}$−hypergraphs on crisp vertex sets.

Throughout this section, H will be a crisp hypergraph, and N an $\mathcal{N}$−hypergraph.

Example 1.15 Consider an $\mathcal{N}$−hypergraph $N = (X, E)$ such that $X = \{a, b, c, d\}$ and $E = \{E_1, E_2, E_3\}$, where

$$E_1 = \{\frac{a}{-0.2}, \frac{b}{-0.4}\}, \ E_2 = \{\frac{b}{-0.4}, \frac{c}{-0.5}\}, \ E_3 = \{\frac{a}{-0.2}, \frac{d}{-0.3}\}.$$

The $\mathcal{N}$−hypergraph is given in Fig. 1.25 and corresponding incidence matrix is given in Table 1.18.

Definition 1.41 An $\mathcal{N}$−function $\mu : X \to [-1, 0]$ is an *elementary $\mathcal{N}$−function* if μ is single valued on $\text{supp}(\mu)$. An *elementary $\mathcal{N}$−hypergraph* $N = (V, E)$ is an $\mathcal{N}$−hypergraph whose edges are elementary.

Proposition 1.10 $\mathcal{N}$−*graphs and* $\mathcal{N}$−*digraphs are special cases of the* $\mathcal{N}$−*hypergraphs.*

Fig. 1.25 $\mathcal{N}$−hypergraph

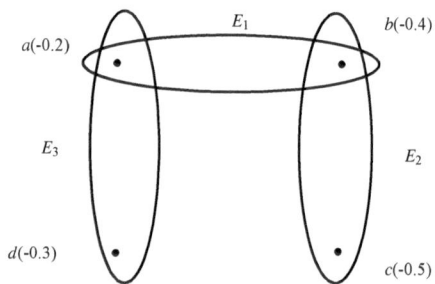

Table 1.18 The corresponding incidence matrix of $\mathcal{N}$−hypergraph

M_N	E_1	E_2	E_3
a	−0.2	0	−0.2
b	−0.4	−0.4	0
c	0	−0.5	0
d	0	0	−0.3

Proof An $\mathcal{N}$ −graph on a set X is a pair $N = (X, E)$, where E is a symmetric $\mathcal{N}$ −function of $X \times X$. That is, $v : X \times X \to [-1, 0]$ and for each x and y in X, we have $v(x, y) = v(y, x)$. An $\mathcal{N}$ −graph on an $\mathcal{N}$ −function $\mu \in V$ is a pair $N = (\mu, v)$ where the symmetric mapping $v : X \times X \to [-1, 0]$ satisfies $v(x, y) \geq \max(\mu(x), \mu(y))$ for all $x, y \in X$. Since, v is well defined, an $\mathcal{N}$ −graph has no multiple edges. An edge is nontrivial if $v(x, y) \neq 0$. A loop at x is represented by $v(x, y) \neq 0$.

Alternately, a nontrivial edge (or loop) represents an elementary $\mathcal{N}$ −function of X with two (or one) element supports. That there are no multiple edges equivalent to the property that each pair of edges have distinct supports. An $\mathcal{N}$ −graph without loops is defined by an antireflexive relation; or equivalently, by not allowing fuzzy subsets with single element support. Therefore, an $\mathcal{N}$ −graph ($\mathcal{N}$ −graph with loops) is an elementary $\mathcal{N}$ −hypergraph for which edges have distinct two vertex (or one element) supports.

Directed $\mathcal{N}$ −graphs ($\mathcal{N}$ −digraphs) on a set X or an $\mathcal{N}$ −function μ of X are similarly defined in terms of a mapping $\lambda : V \times V \to [-1, 0]$ where $\lambda(x, y) \geq \max(\mu(x), \mu(y))$ for all $x, y \in V$. Since λ is well defined, an $\mathcal{N}$ −digraph has at most two edges (which must have opposite orientation) between any two vertices. Therefore, $\mathcal{N}$ −graphs and $\mathcal{N}$ −digraphs are special cases of $\mathcal{N}$ −hypergraphs.

An $\mathcal{N}$ −*multigraph* is a multivalued symmetric mapping $\rho : X \times X \to [-1, 0]$. An $\mathcal{N}$ −multigraph can be considered to be the "disjoint union" or "disjoint sum" of a collection of simple $\mathcal{N}$ −graphs, as is done with crisp multigraphs. The same holds for multidigraphs. Therefore, these structures can be considered as "disjoint unions" or "disjoint sums" of $\mathcal{N}$ −hypergraphs.

Definition 1.42 An $\mathcal{N}$ −hypergraph $N = (X, E)$ is *simple* if $\mu, v \in E$, and $\mu \geq v$ imply that $\mu = \mu, v = v$.

An $\mathcal{N}$ −hypergraph $N = (X, E)$ is *support simple* if $\mu, v \in E$, supp$(\mu) = $ supp(v), and $\mu \geq v$ imply that $\mu = v$.

An $\mathcal{N}$ −hypergraph $N = (X, E)$ is *strongly support simple* if $\mu, v \in E$, and supp$(\mu) = $ supp(v) imply that $\mu = v$.

Remark 1.4 The Definition 1.42 reduces to familiar definitions in the special case where N is a crisp hypergraph. The $\mathcal{N}$ −definition of simple is identical to the crisp definition of simple. A crisp hypergraph is support simple and strongly support simple if and only if it has no multiple edges. For $\mathcal{N}$ −hypergraphs, all three concepts imply no multiple edges. Simple $\mathcal{N}$ −hypergraphs are support simple and strongly support simple $\mathcal{N}$ −hypergraphs are support simple. Simple and strongly support simple are independent concepts.

Definition 1.43 Let $N = (X, E)$ be an $\mathcal{N}$−hypergraph. Suppose that $\alpha \in [-1, 0]$. Let

- $E_{(\alpha)} = \{\mu_\alpha \mid \mu \in E\}, \quad \mu_\alpha = \{x \mid \mu(x) \leq \alpha\}$, and
- $X_\alpha = \bigcup_{\mu \in E} \mu_\alpha.$

If $E_\alpha \neq \emptyset$, then the crisp hypergraph $N_\alpha = (N_\alpha, E_\alpha)$ is the α−*level hypergraph* of N.

Clearly, it is possible that $\mu_\alpha = \nu_\alpha$ for $\mu \neq \nu$, by using distinct markers to identity the various members of E a distinction between μ_α and ν_α to represent multiple edges in N_α. However, we do not take this approach unless otherwise stated, we will always regard N_α as having no repeated edges.

The families of crisp sets (hypergraphs) produced by the α-cuts of an $\mathcal{N}$−hypergraph share an important relationship with each other, as expressed below:

Suppose $\mathbb{X}$ and $\mathbb{Y}$ are two families of sets such that for each set X belonging to $\mathbb{X}$ there is at least one set Y belonging to $\mathbb{Y}$ which contains X. In this case, we say that $\mathbb{Y}$ *absorbs* $\mathbb{X}$ and symbolically write $\mathbb{X} \sqsubseteq \mathbb{Y}$ to express this relationship between $\mathbb{X}$ and $\mathbb{Y}$. Since it is possible for $\mathbb{X} \sqsubseteq \mathbb{Y}$ while $\mathbb{X} \cap \mathbb{Y} = \emptyset$, we have that $\mathbb{X} \subseteq \mathbb{Y} \Rightarrow \mathbb{X} \sqsubseteq \mathbb{Y}$, whereas the converse is generally false. If $\mathbb{X} \sqsubseteq \mathbb{Y}$ and $\mathbb{X} \neq \mathbb{Y}$, then we write $\mathbb{X} \sqsubset \mathbb{Y}$.

Definition 1.44 Let $N = (X, E)$ be an $\mathcal{N}$−hypergraph, and for $0 > s \geq h(N)$. Let N_s be the s−level hypergraph of N. The sequence of real numbers $\{s_1, s_2, \ldots, s_n\}$, $0 > s_1 > s_2 > \cdots > s_n = h(N)$, which satisfies the properties,

- if $s_i > s_{i+1}$, then $E_u = E_{s_i}$, and
- $E_{s_{i+1}} \sqsubset E_{s_i}$,

is called the *fundamental sequence* of N, and is denoted by $F(N)$ and the set of s_i-level hypergraphs $\{N_{s_1}, N_{s_2}, \ldots, N_{s_n}\}$ is called the *set of core hypergraphs* of N or, simply, the *core set* of N, and is denoted by $C(N)$.

Definition 1.45 Suppose $N = (X, E)$ is an $\mathcal{N}$−hypergraph with $F(N) = \{s_1, s_2, \ldots, s_n\}$, and $s_{n+1} = 0$. Then, N is called *sectionally elementary* if for each edge $\mu \in E$, each $i = \{1, 2, \ldots, n\}$, and $s_i \in F(N)$, $\mu_s = \mu_{s_i}$, for all $s \in (s_{i+1}, s_i]$.

Clearly, N is sectionally elementary if and only if $\mu(x) \in F(N)$, for each $\mu \in E$ and each $x \in X$.

Definition 1.46 A sequence of crisp hypergraphs $N_i = (X_i, E_i^*)$, $1 \leq i \leq n$, is said to be *ordered* if $N_1 \subset N_2 \subset \ldots \subset N_n$. The sequence $\{N_i \mid 1 \leq i \leq n\}$ is *simply ordered* if it is ordered and if whenever $E^* \in E_{i+1}^* - E_i^*$, then $E^* \not\subseteq X_i$.

Definition 1.47 An $\mathcal{N}$−hypergraph N is *ordered* if the N-induced fundamental sequence of hypergraphs is ordered. The $\mathcal{N}$−hypergraph N is *simply ordered* if the N-induced fundamental sequence of hypergraphs is simply ordered.

Table 1.19 Incidence matrix of N

N	E_1	E_2	E_3	E_4	E_5
a	-0.7	-0.9	0	0	-0.4
b	-0.7	-0.9	-0.9	-0.7	0
c	0	0	-0.9	-0.7	-0.4
d	0	-0.4	0	-0.4	-0.4

Example 1.16 Consider the $\mathcal{N}$–hypergraph $N = (X, E)$, where $X = \{a, b, c, d\}$, and $E = \{E_1, E_2, E_3, E_4, E_5\}$ which is represented by the following incidence matrix in Table 1.19.

Clearly, $h(N) = -0.9$. Now, $E_{-0.9} = \{\{a, b\}, \{b, c\}\}$, $E_{-0.7} = \{\{a, b\}, \{b, c\}\}$, $E_{-0.4} = \{\{a, b\}, \{a, b, d\}, \{b, c\}, \{b, c, d\}, \{a, c, d\}\}$. Thus, for $-0.4 > s \geq -0.9$, $E_s = \{\{a, b\}, \{b, c\}\}$, and for $0 > s \geq -0.4$, $E_s = \{\{a, b\}, \{a, b, d\}, \{b, c\}, \{b, c, d\}, \{a, c, d\}\}$. We note that $E_{-0.9} \subseteq E_{-0.4}$. The fundamental sequence is $F(N) = \{s_1 = -0.9, s_2 = -0.4\}$, and the set of core hypergraph is $C(N) = \{N_1 = (X_1, E_1) = N_{-0.9}, N_2 = (X_2, E_2) = N_{-0.4}\}$, where $X_1 = \{a, b, c\}$, $E_1 = \{\{a, b\}, \{b, c\}\}$, $X_2 = \{a, b, c, d\}$, $E_2 = \{\{a, b\}, \{a, b, d\}, \{b, c\}, \{b, c, d\}, \{a, c, d\}\}$. N is support simple, but not simple. N is not sectionally elementary since $E_{1(s)} \neq E_{1(-0.9)}$ for $s = -0.7$. Clearly, $\mathcal{N}$–hypergraph N is simply ordered.

Proposition 1.11 *Let $N = (X, E)$ be an elementary $\mathcal{N}$–hypergraph. Then, N is support simple if and only if N is strongly support simple.*

Proof Suppose that N is elementary, support simple, and that $\operatorname{supp}(\mu) = \operatorname{supp}(\nu)$. We assume without loss of generality that $h(\mu) \geq h(\nu)$. Since, N is elementary, it follows that $\mu \geq \nu$ and since N is support simple that $\mu = \nu$. Therefore, N is strongly support simple. The proof of converse part is obvious.

Proposition 1.12 *Let $N = (X, E)$ be a simple $\mathcal{N}$–hypergraph of order n. Then, there is no upper bound on $|E|$.*

Proof Let $X = \{x, y\}$, and define $E_N = \{\mu_i \mid i = 1, 2, \ldots, N\}$, where

$$\mu_i(x) = -1 + \frac{1}{i+1}, \quad \mu_i(y) = -\frac{i}{i+1}.$$

Then, $N_N = (X, E_N)$ is a simple $\mathcal{N}$–hypergraph with N edges. This ends the proof.

Proposition 1.13 *Let $N = (X, E)$ be a support simple $\mathcal{N}$–hypergraph of order n. Then, there is no upper bound on $|E|$.*

Proof The class of support simple $\mathcal{N}$–hypergraphs contains the class of simple $\mathcal{N}$–hypergraphs, thus the result follows from Proposition 1.12.

Proposition 1.14 *Let* $N = (X, E)$ *be a strongly support simple* $\mathcal{N}$*−hypergraph of order n. Then, there is no upper bound on* $|E| \leq 2^n - 1$ *if and only if* $\{supp(\mu) \mid \mu \in E\} = P(V) - \emptyset$.

Proof Each nontrivial $W \subseteq X$ can be the support of at most one $\mu \in E$ and so $|E| \leq 2^n - 1$. The second statement is clear.

Proposition 1.15 *Let* $N = (X, E)$ *be an elementary simple* $\mathcal{N}$*−hypergraph of order n. Then, there is no upper bound on* $|E| \leq 2^n - 1$ *if and only if* $\{supp(\mu) \mid \mu \in E\} = P(V) - \emptyset$.

Proof Since N is elementary and simple, each nontrivial $W \subseteq X$ can be the support of at most one $\mu \in E$. Therefore, $|E| \leq 2^n - 1$. To show there exists an elementary, simple N with $|E| = 2^n - 1$, let $E = \{\mu \mid W \subseteq X\}$ be the set of functions defined by

$$\mu(x) = -1 + \frac{1}{|W|}, \quad \text{if } x \in W, \quad \mu(x) = -1, \text{if } x \notin W.$$

Then, each one element has height $(-1, 0)$, each two elements has height $(-0.5, -0.5)$, and so on. Hence, N is an elementary and simple, and $|E| = 2^n - 1$.

We state the following proposition without proof.

Proposition 1.16 *(a) If* $N = (X, E)$ *is an elementary* $\mathcal{N}$*−hypergraph, then N is ordered.*
(b) If N is an ordered $\mathcal{N}$*−hypergraph with simple support hypergraph, then N is elementary.*

Definition 1.48 *The dual of an* $\mathcal{N}$*−hypergraph* $N = (X, E)$ *is an* $\mathcal{N}$*−hypergraph* $N^D = (E^D, X^D)$ *whose vertex set is the edge set of N and with edges* $X^D : E^D \rightarrow [-1, 0]$ *by* $X^D(\mu^D) = \mu^D(x)$. N^D *is an* $\mathcal{N}$*−hypergraph whose incidence matrix is the transpose of the incidence matrix of N, thus* $N^{DD} = N$.

Example 1.17 Consider an $\mathcal{N}$−hypergraph $N = (X, E)$ such that $X = \{x_1, x_2, x_3, x_4\}$, $E = \{E_1, E_2, E_3, E_4\}$, where $E_1 = \{\frac{x_1}{-0.5}, \frac{x_2}{-0.4}\}$, $E_2 = \{\frac{x_2}{-0.4}, \frac{x_3}{-0.3}\}$, $E_3 = \{\frac{x_3}{-0.3}, \frac{x_4}{-0.5}\}$, $E_4 = \{\frac{x_4}{-0.5}, \frac{x_1}{-0.5}\}$. The corresponding hypergraph is shown in Fig. 1.26.

The corresponding incidence matrix of N is given in Table 1.20.

Consider the dual $\mathcal{N}$−hypergraph $N^D = (E^D, X^D)$ of N such that $E^D = \{e_1, e_2, e_3, e_4\}$, $X^D = \{A, B, C, D\}$, where $A = \{\frac{e_1}{-0.5}, \frac{e_4}{-0.5}\}$, $B = \{\frac{e_1}{-0.4}, \frac{e_2}{-0.4}\}$, $C = \{\frac{e_2}{-0.3}, \frac{e_3}{-0.3}\}$, $D = \{\frac{e_3}{-0.5}, \frac{e_4}{-0.5}\}$. The dual $\mathcal{N}$−hypergraph of N is shown in Fig. 1.27.

The corresponding incidence matrix of N^D is given in Table 1.21.

We see that some edges contain only vertices having high membership degree. We define here the concept of strength of an edge.

Fig. 1.26 $\mathcal{N}$−hypergraph

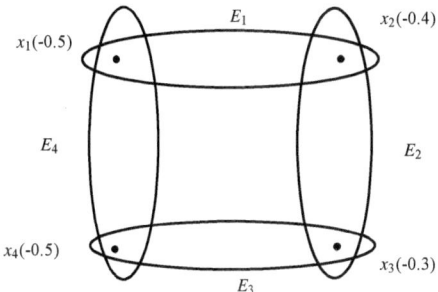

Table 1.20 The corresponding incidence matrix of N

M_N	E_1	E_2	E_3	E_4
x_1	−0.5	0	0	−0.5
x_2	−0.4	−0.4	0	0
x_3	0	−0.3	-0.3	−0
x_4	0	0	−0.5	−0.5

Fig. 1.27 Dual $\mathcal{N}$−hypergraph

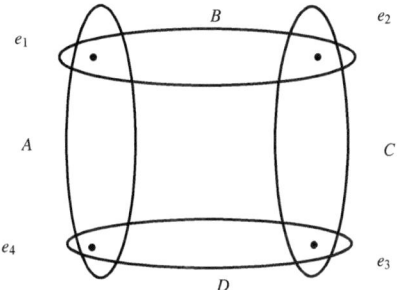

Table 1.21 The incidence matrix of N^D

M_{N^D}	A	B	C	D
e_1	−0.5	−0.4	0	0
e_2	0	−0.4	−0.3	0
e_3	0	0	−0.3	−0.5
e_4	−0.5	0	0	−0.5

Definition 1.49 The *strength* η of an edge E is the maximum membership $\mu(x)$ of vertices in the edge E. That is,

$$\eta(E_j) = \{\max(\mu_j(x) \mid \mu_j(x) < 0)\}.$$

Fig. 1.28 $\mathcal{N}$-hypergraph

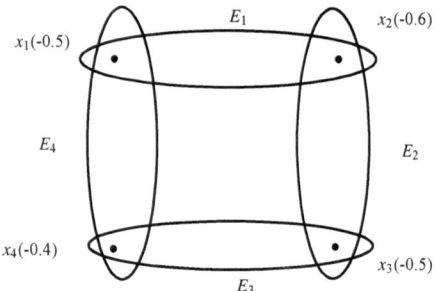

Its interpretation is that the edge E_j groups elements having participation degree at least $\eta(E_j)$ in the hypergraph.

Example 1.18 Consider an $\mathcal{N}$-hypergraph $N = (X, E)$ such that $V = \{a, b, c, d\}$, $E = \{E_1, E_2, E_3, E_4\}$ as shown in Fig. 1.28.

It is easy to see that E_3 is strong than E_1, and E_4 is strong than E_2. We call the edges with high strength the strong edges because the cohesion in them is strong.

1.7 Application of $\mathcal{N}$-Hypergraphs

In this section, we describe an application of $\mathcal{N}$-hypergraphs.

Definition 1.50 An $\mathcal{N}$-*partition* of a set X is a family of nontrivial $\mathcal{N}$-functions $\{\mu_1, \mu_2, \mu_3, \ldots, \mu_m\}$ such that

1. $\bigcup_i \text{supp}(\mu_i) = X$, $\quad i = 1, 2, \ldots, m$,
2. $\sum_{i=1}^m \mu_i(x) = -1$, for all $x \in X$,
3. $supp(\mu_i) \cap supp(\mu_j) = \emptyset$, for $i \neq j$.

We call a family $\{\mu_1, \mu_2, \mu_3, \ldots, \mu_m\}$ a $\mathcal{N}$-*covering* of X if it verifies only the above conditions (1) and (2). An $\mathcal{N}$-partition can be represented by an $\mathcal{N}$-matrix $[a_{ij}]$, where a_{ij} is the negative-value of element x_i in class j. We see that the matrix is the same as the incidence matrix in $\mathcal{N}$-hypergraph. Then, we can represent an $\mathcal{N}$-partition by an $\mathcal{N}$-hypergraph $N = (X, E)$ such that

(1) X: a set of elements x_i, $i = 1, \ldots, n$,
(2) $E = \{E_1, E_2, \ldots, E_m\}$= a set of nontrivial $\mathcal{N}$-classes,
(3) $V = \bigcup_j \text{supp}(E_j)$, $\quad j = 1, 2, \ldots, m$,
(4) $\sum_{i=1}^m \mu_i(x) = -1$, for all $x \in X$.

Note that the last conditions (4) is added to the $\mathcal{N}$-hypergraph for $\mathcal{N}$-partition. If the last condition (4) is eliminated, the $\mathcal{N}$-hypergraph can represent an $\mathcal{N}$-covering. Naturally, we can apply the α-cut to the $\mathcal{N}$-partition.

Table 1.22 $\mathcal{N}$−partition matrix

N	A_t	B_h
x_1	−0.96	−0.04
x_2	−1	0
x_3	−0.61	−0.39
x_4	−0.05	−0.95
x_5	−0.03	−0.97

Table 1.23 Hypergraph $N_{(-0.61)}$

$N_{(-0.61)}$	$A_{t(-0.61)}$	$B_{h(-0.61)}$
x_1	1	0
x_2	1	0
x_3	1	0
x_4	0	1
x_5	0	1

We consider an example of clustering problem. This problem is a typical example of $\mathcal{N}$−partition on the visual image processing. There are five objects and they are classified into two classes: tank and house. To cluster the elements x_1, x_2, x_3, x_4, x_5 into A_t (tank) and B_h (house), an $\mathcal{N}$−partition matrix is given as the form of incidence matrix of $\mathcal{N}$−hypergraph in Table 1.22.

We can apply the α-cut to the hypergraph and obtain a hypergraph N_α which is not $\mathcal{N}$−hypergraph. We denote the edge (class) in α-cut hypergraph N_α as $E_{j(\alpha)}$. This hypergraph N represents generally the covering because of the conditions: (4) $\sum_{i=1}^m \mu_i(x) = -1$ for all $x \in X$ is not always guaranteed. The hypergraph $N_{(-0.61)}$ is shown in Table 1.23.

We obtain dual $\mathcal{N}$−hypergraph $N_{(-0.61)}^D$ of $N_{(-0.61)}$ which is given in Table 1.24.

We consider the strength of edge (class) $E_{j(\alpha)}$, or in the α-cut hypergraph N_α. It is necessary to apply Definition 1.49 to obtain the strength of edge $E_{j(\alpha)}$ in N_α. The possible interpretations of $\eta(E_{j(\alpha)})$ are

- the edge (class) in the hypergraph (partition) N_α, groups elements having at least η membership and nonmembership,
- the strength (cohesion) of edge (class) $E_{j(\alpha)}$ in N_α is η.

Thus, we can use the strength as a measure of the cohesion or strength of a class in a partition. For example, the strengths of classes $A_t(-0.61)$ and $B_h(-0.61)$ at

Table 1.24 Dual $\mathcal{N}$−hypergraph

$N^D(-0.61)$	X_1	X_2	X_3	X_4	X_5
A_t	1	1	1	0	0
B_h	0	0	0	1	1

$s = -0.61$ are $\eta(A_t(-0.61))=(-0.61)$, $\eta(B_h(-0.61)) = (-0.95)$. Thus, we say that the class $\eta(B_h(-0.61))$ is stronger than $\eta(A_t(-0.61))$ because $\eta(B_h(-0.61)) < \eta(A_t(-0.61))$. From the above discussion on the hypergraph $N_{(-0.61)}$ and $N^D_{(-0.61)}$, we can state that

- The $\mathcal{N}$−hypergraph can represent the fuzzy partition visually. The α−cut hypergraph also represents the α-cut partition.
- The dual hypergraph $N^D_{-0.61)}$ can represent elements X_i, which can be grouped into a class $E_{j(\alpha)}$. For example, the edges X_1, X_2, X_3 of the dual hypergraph in Table 9 represent that the elements x_1, x_2, x_3 that can be grouped into A_t at level -0.61.
- In the $\mathcal{N}$−partition, we have $\sum_{i=1}^{m} \mu_i(x) = -1$ for all $x \in X$. If we α−cut at level $\alpha < -0.5$, there is no element which is grouped into two classes simultaneously. That is, if $\alpha < -0.5$, every element is contained in only one class in $N_{-\alpha}$. Therefore, the hypergraph $N_{-\alpha}$, represents a partition (if $s \geq -0.05$, the hypergraph may represent a covering).
- At $\alpha = -0.61$−level, the strength of class $B_h(-0.61)$ is the lowest -0.95, so it is the strongest class. It means that this class can be grouped independently from the other parts. Thus, we can eliminate the class B_h from the others and continue clustering. Therefore, the discrimination of strong classes from others can allow us to decompose a clustering problem into smaller ones. This strategy allows us to work with the reduced data in a clustering problem.

1.8 Complex Fuzzy Hypergraphs

A complex fuzzy set, as an extension of fuzzy set, is used to handle uncertainty and vagueness having the membership degrees which range over complex subset with unit disk instead of real subset with [0, 1]. A complex fuzzy set provides a powerful mathematical framework to describe the membership degrees in the form of a complex number. To illustrate the applicability of this work, we consider a group of three persons working on the same project for a fixed interval of time. To represent the combined work of any two persons, we use a complex fuzzy graph but this graph fails to illustrate the collective efforts of all three persons as an edge can connect only two vertices. To combine these three members through a single edge and to illustrate the level of combined work, we use a fuzzy hypergraph. A fuzzy hypergraph explains the combine efforts of corresponding members by representing the persons as vertices and these vertices are combined by a fuzzy hyperedge to show their collaboration. In this case, we lack the information on the time period which is given to complete that particular task. To handle such type of difficulties occurring in simple graphs and hypergraphs, we propose a novel concept called complex fuzzy hypergraphs.

Definition 1.51 A *complex fuzzy set* on X is defined as $C_f = \{(a, M_{C_f}(a)e^{i\psi_{C_f}(a)}) : a \in X\}$, where $i = \sqrt{-1}$, $M_{C_f} : X \to [0, 1]$, and $\psi_{C_f}(a) \in [0, 2\pi]$. Here, $M_{C_f}(a)$ is named as the amplitude term and $\psi_{C_f}(x)$ is named as the phase term.

Definition 1.52 A *complex fuzzy relation* on $X \times X$ is an object of the form, $R_f = \{(ab, M_{R_f}(ab)e^{i\psi_{R_f}(ab)}) : ab \in X \times X\}$, where $i = \sqrt{-1}$, M_{R_f} is called the amplitude term, and $\psi_{R_f}(ab) \in [0, 2\pi]$ is called the phase term.

Definition 1.53 A *complex fuzzy graph* on a non-empty set X is defined as an ordered pair $G = (A, B)$, where A is a complex fuzzy set on X and B is a complex fuzzy relation on X such that

$$M_B(ab) \leq \min\{M_A(a), M_A(b)\}, \text{ (for amplitude term)}$$
$$\psi_B(ab) \leq \min\{\psi_A(a), \psi_A(b)\}, \text{ (for phase term)}$$

for all $a, b \in X$. Note that A is called the complex fuzzy vertex set and B is called the complex fuzzy edge set of G.

Example 1.19 Let $X = \{a_1, a_2, a_3, a_4\}$ be a non-empty set, we define complex fuzzy set A and complex fuzzy relation B on X such that

$A = \{(a_1, 0.3e^{i\frac{\pi}{2}}), (a_2, 0.4e^{i\frac{\pi}{3}}), (a_3, 0.5e^{i\frac{\pi}{4}}), (a_4, 0.5e^{i\frac{\pi}{3}})\}$,

$B = \{(a_1a_2, 0.3e^{i\frac{\pi}{3}}), (a_1a_3, 0.3e^{i\frac{\pi}{4}}), (a_2a_3, 0.4e^{i\frac{\pi}{4}}), (a_1a_4, 0.3e^{i\frac{\pi}{3}}), (a_2a_4, 0.4e^{i\frac{\pi}{3}}), (a_3a_4, 0.4e^{i\frac{\pi}{3}})\}$.

Then, $G = (A, B)$ is a complex fuzzy graph on X. The corresponding graph is shown in Fig. 1.29.

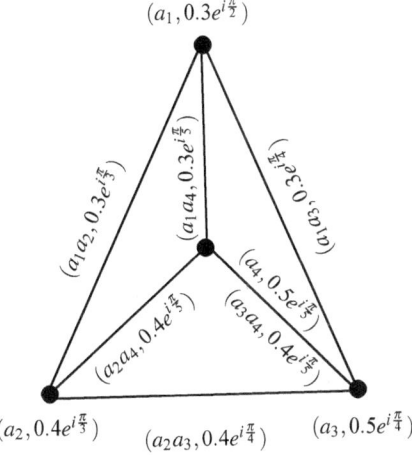

Fig. 1.29 Complex fuzzy graph

Definition 1.54 Let $C_f = \{(a, M_{C_f}(a)e^{i\psi_{C_f}(a)}) : a \in X\}$ be a complex fuzzy set on X. The *support* of C_f, denoted by $supp(C_f)$, is defined as $supp(C_f) = \{c \in X : M_{C_f}(c) \neq 0, \ 0 < \psi_{C_f}(c) < 2\pi\}$.

The *height* of C_f, denoted by $h(C_f)$, is defined as $h(C_f) = \max_{u \in X} M_{C_f}(u)$ $e^{i \max_{u \in X} \psi_{C_f}(u)}$.

If $h(C_f) = 1e^{i2\pi}$, then C_f is called *normal* complex fuzzy set.

Definition 1.55 Let X be a nontrivial set of universe. A *complex fuzzy hypergraph* is defined as an ordered pair $\mathcal{H} = (C, \xi)$, where $C = \{\alpha_1, \alpha_2, \cdots, \alpha_k\}$ is a finite family of complex fuzzy sets on X and ξ is a complex fuzzy relation on complex fuzzy sets α_j's such that

(i)

$$M_\xi(\{r_1, r_2, \cdots, r_l\}) \leq \min\{M_{\alpha_j}(r_1), M_{\alpha_j}(r_2), \cdots, M_{\alpha_j}(r_l)\}, \quad \text{(for amplitude term)}$$
$$\psi_\xi(\{r_1, r_2, \cdots, r_l\}) \leq \min\{\psi_{\alpha_j}(r_1), \psi_{\alpha_j}(r_2), \cdots, \psi_{\alpha_j}(r_l)\}, \quad \text{(for phase term)}$$

for all $r_1, r_2, \cdots, r_l \in X$.

(ii) $\bigcup_j supp(\alpha_j) = X$, for all $\alpha_j \in C$.

Note that $E_k = \{r_1, r_2, \cdots, r_l\}$ is the crisp hyperedge of $\mathcal{H} = (C, \xi)$.

Definition 1.56 A complex fuzzy hypergraph $\mathcal{H} = (C, \xi)$ is *simple* if whenever $\xi_i, \xi_j \in \xi$ and $\xi_i \subseteq \xi_j$, then $\xi_i = \xi_j$.

A complex fuzzy hypergraph $\mathcal{H} = (C, \xi)$ is *support simple* if whenever $\xi_i, \xi_j \in \xi, \xi_i \subseteq \xi_j$, and $supp(\xi_i) = supp(\xi_j)$, then $\xi_i = \xi_j$.

Definition 1.57 Let μ be a complex fuzzy set on X. Then, μ is *elementary* on X if $|\mu(supp(\mu))| = 1$.

An *elementary* complex fuzzy hypergraph is one whose all hyperedges are elementary.

Definition 1.58 Let $\mathcal{H} = (C, \xi)$ be a complex fuzzy hypergraph. Suppose that $\mu \in [0, 1]$ and $\theta \in [0, 2\pi]$. Then, $\mu e^{i\theta}-$*level hypergraph* of $\mathcal{H}$ is defined as an ordered pair $\mathcal{H}_{\mu e^{i\theta}} = (C_{\mu e^{i\theta}}, \xi_{\mu e^{i\theta}})$, where

(i) $\xi_{\mu e^{i\theta}} = \{\rho_{j(\mu e^{i\theta})} : \rho_j \in \xi\}$ and $\rho_{j(\mu e^{i\theta})} = \{d \in X : M_{\rho_j}(d) \geq \mu, \psi_{\rho_j}(d) \geq \theta\}$,

(ii) $C_{\mu e^{i\theta}} = \bigcup_{\rho_j \in \xi} supp\rho_{j(\mu e^{i\theta})}$.

Note that $\mu e^{i\theta}-$level hypergraph of $\mathcal{H}$ is a crisp hypergraph. Here, $M_{\rho_j}(d)$ denotes the membership degree of vertex d to complex fuzzy hyperedge ρ_j.

Remark 1.5 It can be clearly seen that $\rho_{j(\mu e^{i\theta})}$ and $\rho_{k(\mu e^{i\theta})}$ may be the same for $\rho_j \neq \rho_k$. That is, $\mu e^{i\theta}-$level hypergraph of a simple complex fuzzy hypergraph may not be simple in general.

Table 1.25 Complex fuzzy sets on X

$x \in X$	α_1	α_2	α_3
r_1	$0.1e^{i\frac{\pi}{2}}$	$0.1e^{i\frac{\pi}{2}}$	0
r_2	$0.2e^{i\frac{\pi}{2}}$	0	0
r_3	$0.3e^{i\frac{\pi}{3}}$	0	0
r_4	0	$0.4e^{i\frac{\pi}{4}}$	0
r_5	0	0	$0.5e^{i\frac{\pi}{5}}$
r_6	0	0	$0.6e^{i\frac{\pi}{6}}$
r_7	0	$0.7e^{i\frac{\pi}{7}}$	$0.7e^{i\frac{\pi}{7}}$

Fig. 1.30 Complex fuzzy hypergraph

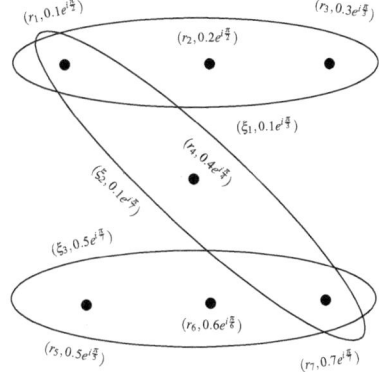

Example 1.20 Let $C = \{\alpha_1, \alpha_2, \alpha_3\}$ be a finite family of complex fuzzy sets on $X = \{r_1, r_2, r_3, r_4, r_5, r_6, r_7\}$ as given in Table 1.25 and ξ be a complex fuzzy relation on α_i, $1 \leq i \leq 3$ such that

$$\xi(\{r_1, r_2, r_3\}) = 0.1e^{i\frac{\pi}{3}},$$
$$\xi(\{r_1, r_4, r_7\}) = 0.1e^{i\frac{\pi}{7}},$$
$$\xi(\{r_5, r_6, r_7\}) = 0.5e^{i\frac{\pi}{7}}.$$

The corresponding complex fuzzy hypergraph $\mathcal{H} = (C, \xi)$ is shown in Fig. 1.30.

Let $\mu = 0.3$ and $\theta = \frac{\pi}{7}$, then $0.3e^{i\frac{\pi}{7}}$ −level hypergraph of $\mathcal{H} = (C, \xi)$ is shown in Fig. 1.31. Note that $\xi_{1(0.3e^{i\frac{\pi}{7}})} = \{r_3\}$, $\xi_{2(0.3e^{i\frac{\pi}{7}})} = \{r_4, r_7\}$, $\xi_{3(0.3e^{i\frac{\pi}{7}})} = \{r_5, r_6, r_7\}$.

Definition 1.59 Let $\mathcal{H} = (C, \xi)$ be a complex fuzzy hypergraph. The *height* of $\mathcal{H}$, denoted by $h(\mathcal{H})$, is defined as $h(\mathcal{H}) = \max \xi_l e^{i \max \psi}$, where $\xi_l = \max M_{\rho_j}(x_k)$ and $\psi = \max \psi_{\rho_j}(x_k)$.

Definition 1.60 Let $\mathcal{H} = (C, \xi)$ be a complex fuzzy hypergraph and for $0 < \mu \leq M(h(\mathcal{H}))$, $0 < \theta \leq \psi(h(\mathcal{H}))$, let $\mathcal{H}_{\mu e^{i\theta}} = (C_{\mu e^{i\theta}}, \xi_{\mu e^{i\theta}})$ be the level hypergraph of $\mathcal{H}$. The sequence of complex numbers $\{\mu_1 e^{i\theta_1}, \mu_2 e^{i\theta_2}, \cdots, \mu_n e^{i\theta_n}\}$ such that

Fig. 1.31 $0.3e^{i\frac{\pi}{7}}$ −level
hypergraph of $\mathcal{H}$

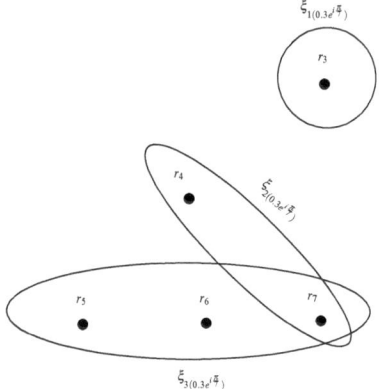

$0 < \mu_1 < \mu_2 < \cdots < \mu_n = M(h(\mathcal{H})), 0 < \theta_1 < \theta_2 < \cdots < \theta_n = \psi(h(\mathcal{H}))$ satisfying the conditions,

(i) if $\mu_{k+1} < v \leq \mu_k, \theta_{k+1} < \phi \leq \theta_k$, then $\xi_{ve^{i\phi}} = \xi_{\mu_k e^{i\theta_k}}$, and
(ii) $\xi_{\mu_k e^{i\theta_k}} \subset \xi_{\mu_{k+1} e^{i\theta_{k+1}}}$,

is called the *fundamental sequence* of $\mathcal{H} = (C, \xi)$, denoted by $\mathscr{F}_s(\mathcal{H})$. The set of $\mu_j e^{i\theta_j}$−level hypergraphs $\{\mathcal{H}_{\mu_1 e^{i\theta_1}}, \mathcal{H}_{\mu_2 e^{i\theta_2}}, \cdots, \mathcal{H}_{\mu_n e^{i\theta_n}}\}$ called the set of core hypergraphs or the *core set* of $\mathcal{H}$, denoted by $Cor(\mathcal{H})$.

Definition 1.61 Let $\mathcal{H} = (C, \xi)$ be a complex fuzzy hypergraph and $\mathscr{F}_s(\mathcal{H}) = \{\mu_1 e^{i\theta_1}, \mu_2 e^{i\theta_2}, \cdots, \mu_n e^{i\theta_n}\}$. Then, $\mathcal{H}$ is *sectionally elementary* if for every $\rho \in \xi$ and $\mu_n e^{i\theta_n} \in \mathscr{F}_s(\mathcal{H})$, we have $\rho_{ve^{i\phi}} = \rho_{\mu_n e^{i\theta_n}}$, for all $v \in (\mu_{n+1}, \mu_n]$ and $\phi \in (\theta_{n+1}, \theta_n]$.

Definition 1.62 Let $\mathcal{H}_1 = (C_1, \xi_1)$ and $\mathcal{H}_2 = (C_2, \xi_2)$ be complex fuzzy hypergraphs on X_1 and X_2, respectively. If $X_1 \subseteq X_2$ and $\xi_1 \subseteq \xi_2$, then $\mathcal{H}_1$ is called *partial hypergraph* of $\mathcal{H}_2$, denoted as $\mathcal{H}_1 \subseteq \mathcal{H}_2$.

Example 1.21 Let $\mathcal{H} = (C, \xi)$ be a complex fuzzy hypergraph on $X = \{u_1, u_2, u_3, u_4\}$ and $C = \{\alpha_1, \alpha_2, \alpha_3, \alpha_4, \alpha_5\}$ be the family of complex fuzzy sets on X as given in Table 1.26.

The corresponding complex fuzzy hypergraph $\mathcal{H} = (C, \xi)$ is shown in Fig. 1.32. Note that $h(\mathcal{H}) = 0.9e^{i\frac{\pi}{4}}$, i.e., $\mu_1 = 0.9, \theta_1 = \frac{\pi}{4}$. Now,

$$\xi_{(0.9e^{i\frac{\pi}{4}})} = \{\{u_1, u_2\}, \{u_2, u_3\}\}, \xi_{(0.7e^{i\frac{\pi}{7}})} = \{\{u_1, u_2\}, \{u_2, u_3\}\},$$
$$\xi_{(0.4e^{i\frac{\pi}{9}})} = \{\{u_1, u_2\}, \{u_1, u_2, u_4\}, \{u_2, u_3\}, \{u_2, u_3, u_4\}, \{u_1, u_3, u_4\}\}.$$

Since, $\xi_{(0.9e^{i\frac{\pi}{4}})} = \xi_{(0.7e^{i\frac{\pi}{7}})}$. Thus, we have for $0.4 < v \leq 0.9, \frac{\pi}{9} < \theta \leq \frac{\pi}{4}, \xi_{ve^{i\theta}} = \{\{u_1, u_2\}, \{u_2, u_3\}\}$, and for $0 < v \leq 0.4, 0 < \theta \leq \frac{\pi}{9}, \xi_{ve^{i\theta}} = \{\{u_1, u_2\}, \{u_1, u_2, u_4\}, \{u_2, u_3\}, \{u_2, u_3, u_4\}, \{u_1, u_3, u_4\}\}$. Furthermore, $\xi_{(0.9e^{i\frac{\pi}{4}})} \subseteq \xi_{(0.4e^{i\frac{\pi}{9}})}$ and $\xi_{(0.9e^{i\frac{\pi}{4}})} \neq \xi_{(0.4e^{i\frac{\pi}{9}})}$. Hence, $\mathscr{F}_s(\mathcal{H}) = \{\mu_1 e^{i\theta_1} = 0.9e^{i\frac{\pi}{4}}, \mu_2 e^{i\theta_2} = 0.4e^{i\frac{\pi}{9}}\}$. The set of core

Table 1.26 Complex fuzzy sets on X

$x \in X$	α_1	α_2	α_3	α_4	α_5
u_1	$0.7e^{i\frac{\pi}{7}}$	$0.9e^{i\frac{\pi}{4}}$	0	0	$0.4e^{i\frac{\pi}{9}}$
u_2	$0.7e^{i\frac{\pi}{7}}$	$0.9e^{i\frac{\pi}{4}}$	$0.9e^{i\frac{\pi}{4}}$	$0.7e^{i\frac{\pi}{7}}$	0
u_3	0	0	$0.9e^{i\frac{\pi}{4}}$	$0.7e^{i\frac{\pi}{7}}$	$0.4e^{i\frac{\pi}{9}}$
u_4	0	$0.4e^{i\frac{\pi}{9}}$	0	$0.4e^{i\frac{\pi}{9}}$	$0.4e^{i\frac{\pi}{9}}$

Fig. 1.32 Complex fuzzy hypergraph

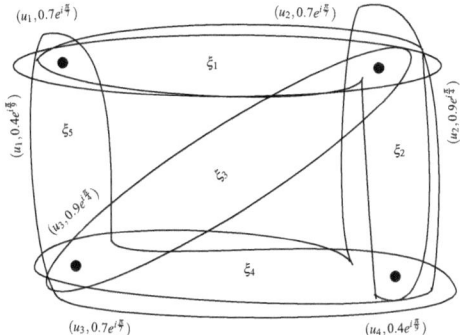

hypergraphs is $Cor(\mathscr{H}) = \{\mathscr{H}_1 = (C_1, \xi_1) = \mathscr{H}_{0.9e^{i\frac{\pi}{4}}}, \mathscr{H}_2 = (C_2, \xi_2) = \mathscr{H}_{0.4e^{i\frac{\pi}{9}}}\}$, where

$$\xi_1 = \{\{u_1, u_2\}, \{u_2, u_3\}\}, \xi_2 = \{\{u_1, u_2\}, \{u_1, u_2, u_4\}, \{u_2, u_3\}, \{u_2, u_3, u_4\}, \{u_1, u_3, u_4\}\}.$$

Note that $\mathscr{H}$ is support simple but not simple. Also $\xi_{(0.9e^{i\frac{\pi}{4}})} \neq \xi_{(0.7e^{i\frac{\pi}{7}})}$, $\mathscr{H}$ is not sectionally elementary.

Definition 1.63 An *ordered sequence* of crisp hypergraphs $H_j = (X_j, E_j), 1 \leq j \leq m$ is defined as the sequence which satisfies $H_1 \subset H_2 \subset \cdots \subset H_m$. The sequence is called *simply ordered* if it is ordered and whenever $E \in E_{k+1} \setminus E_k$, then $E \not\subseteq X_k$.

Definition 1.64 A complex fuzzy hypergraph $\mathscr{H} = (C, \xi)$ is called *ordered* if the corresponding set of core hypergraphs $Cor(\mathscr{H})$ is ordered, i.e., if $Cor(\mathscr{H}) = \{\mathscr{H}_{\mu_1 e^{i\theta_1}}, \mathscr{H}_{\mu_2 e^{i\theta_2}}, \cdots, \mathscr{H}_{\mu_n e^{i\theta_n}}\}$, then $\mathscr{H}_{\mu_1 e^{i\theta_1}} \subset \mathscr{H}_{\mu_2 e^{i\theta_2}} \subset \cdots \subset \mathscr{H}_{\mu_n e^{i\theta_n}}$. The complex fuzzy hypergraph $\mathscr{H} = (C, \xi)$ is said to be *simply ordered* if $Cor(\mathscr{H}) = \{\mathscr{H}_{\mu_1 e^{i\theta_1}}, \mathscr{H}_{\mu_2 e^{i\theta_2}}, \cdots, \mathscr{H}_{\mu_n e^{i\theta_n}}\}$ is simply ordered.

Proposition 1.17 *Let $\mathscr{H} = (C, \xi)$ be an elementary complex fuzzy hypergraph, then $\mathscr{H}$ is ordered. If $\mathscr{H}$ is ordered with $Cor(\mathscr{H}) = \{\mathscr{H}_{\mu_1 e^{i\theta_1}}, \mathscr{H}_{\mu_2 e^{i\theta_2}}, \cdots, \mathscr{H}_{\mu_n e^{i\theta_n}}\}$ and $\mathscr{H}_{\mu_n e^{i\theta_n}}$ is simple, then $\mathscr{H}$ is elementary.*

Definition 1.65 A complex fuzzy hypergraph $\mathscr{H} = (C, \xi)$ is called C_f-*tempered* complex fuzzy hypergraph of $H = (X, E)$, if there exists a crisp hypergraph

Table 1.27 Complex fuzzy sets on X

$x \in X$	α_1	α_2	α_3	α_4
u_1	$0.7e^{i\frac{\pi}{7}}$	0	0	$0.7e^{i\frac{\pi}{7}}$
u_2	$0.7e^{i\frac{\pi}{7}}$	$0.4e^{i\frac{\pi}{9}}$	$0.9e^{i\frac{\pi}{4}}$	0
u_3	0	0	$0.9e^{i\frac{\pi}{4}}$	$0.7e^{i\frac{\pi}{7}}$
u_4	0	$0.4e^{i\frac{\pi}{9}}$	0	0

$H = (X, E)$ and a complex fuzzy set $C_f = \{(a, M_{C_f}(a)e^{i\psi_{C_f}(a)}) : a \in X\}$, where $i = \sqrt{-1}$, $M_{C_f} : X \to [0, 1]$, and $\psi_{C_f}(a) \in [0, 2\pi]$ such that $\xi = \{B'_E | E' \in E\}$, where

$$B'_E(x) = \begin{cases} \min\{M_{C_f}(y)e^{i\,\min\psi_{C_f}(y)}\}, & \text{if } x \in E', \\ 0, & \text{otherwise.} \end{cases}$$

Let us denote by $C_f * H$ the C_f−tempered complex fuzzy hypergraph of H obtained by $H = (X, E)$ and the complex fuzzy set $C_f = \{(a, M_{C_f}(a)e^{i\psi_{C_f}(a)}) : a \in X\}$.

Example 1.22 Let $\mathcal{H} = (C, \xi)$ be a complex fuzzy hypergraph on $X = \{u_1, u_2, u_3, u_4\}$ and $C = \{\alpha_1, \alpha_2, \alpha_3, \alpha_4\}$ be complex fuzzy sets on X as given in Table 1.27. Note that

$$\xi_{(0.9e^{i\frac{\pi}{4}})} = \{\{u_2, u_3\}\}, \xi_{(0.7e^{i\frac{\pi}{7}})} = \{\{u_1, u_2\}, \{u_1, u_3\}, \{u_2, u_3\}\},$$
$$\xi_{(0.4e^{i\frac{\pi}{9}})} = \{\{u_1, u_2\}, \{u_1, u_3\}, \{u_2, u_3\}, \{u_2, u_4\}\}.$$

Let us define a complex fuzzy set $C_f = \{(a, M_{C_f}(a)e^{i\psi_{C_f}(a)}) : a \in X\}$, where $i = \sqrt{-1}$, $M_{C_f} : X \to [0, 1]$, and $\psi_{C_f}(a) \in [0, 2\pi]$ by $C_f(u_1) = 0.7e^{i\frac{\pi}{7}}$, $C_f(u_2) = 0.9e^{i\frac{\pi}{4}}$, $C_f(u_3) = 0.9e^{i\frac{\pi}{4}}$, and $C_f(u_4) = 0.4e^{i\frac{\pi}{9}}$. Note that

$$B'_{\{u_1,u_2\}}(u_1) = \min\{M_{C_f}(u_1), M_{C_f}(u_2)\}e^{i\,\min\{\psi_{C_f}(u_1), \psi_{C_f}(u_2)\}} = 0.7e^{i\frac{\pi}{9}},$$
$$B'_{\{u_1,u_2\}}(u_2) = \min\{M_{C_f}(u_1), M_{C_f}(u_2)\}e^{i\,\min\{\psi_{C_f}(u_1), \psi_{C_f}(u_2)\}} = 0.7e^{i\frac{\pi}{9}},$$
$$B'_{\{u_1,u_2\}}(u_3) = 0, \ B'_{\{u_1,u_2\}}(u_3) = 0.$$

Thus, $B'_{\{u_1,u_2\}} = \rho_1$. Also, $B'_{\{u_2,u_4\}} = \rho_2$, $B'_{\{u_2,u_3\}} = \rho_3$, and $B'_{\{u_1,u_4\}} = \rho_4$. Hence, $\mathcal{H} = (C, \xi)$ is C_f−tempered complex fuzzy hypergraph. The corresponding graph is shown in Fig. 1.33.

Theorem 1.5 *A complex fuzzy hypergraph $\mathcal{H} = (C, \xi)$ is C_f−tempered complex fuzzy hypergraph of some crisp hypergraph $H = (X, E)$ if and only if $\mathcal{H} = (C, \xi)$ is support simple, elementary, and simply ordered.*

Proof Let us suppose that $\mathcal{H} = (C, \xi)$ is C_f−tempered complex fuzzy hypergraph of $H = (X, E)$. Then, $\mathcal{H} = (C, \xi)$ is support simple and elementary. To show

Fig. 1.33 C_f–tempered
complex fuzzy
hypergraph $\mathcal{H}$

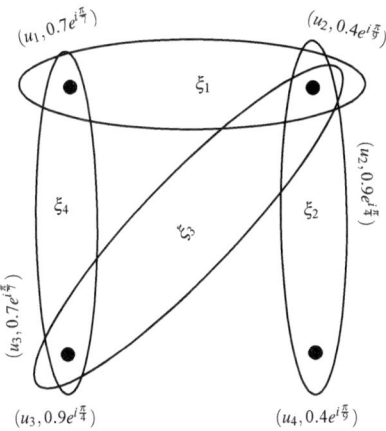

that $H = (X, E)$ is simply ordered, let $Cor(\mathcal{H}) = \{\mathcal{H}_{\mu_1 e^{i\theta_1}} = (C_1, \xi_1), \mathcal{H}_{\mu_2 e^{i\theta_2}} = (C_2, \xi_2), \cdots, \mathcal{H}_{\mu_n e^{i\theta_n}} = (C_n, \xi_n)\}$. Since, $\mathcal{H} = (C, \xi)$ is elementary, Proposition 1.17 follows that $\mathcal{H} = (C, \xi)$ is ordered. We now suppose that $\rho \in \xi_{l+1} \setminus \xi_l$. Then, there exists an element $h \in \rho$ such that $C_f(h) = \mu_{l+1} e^{i\theta_{l+1}} < \mu_l e^{i\theta_l}$, then $h \notin X_l$, $\rho \nsubseteq X_l$, hence $\mathcal{H} = (C, \xi)$ is simply ordered.

Conversely, assume that $\mathcal{H} = (C, \xi)$ is support simple, elementary, and simply ordered. Let $Cor(\mathcal{H}) = \{\mathcal{H}_{\mu_1 e^{i\theta_1}} = (C_1, \xi_1), \mathcal{H}_{\mu_2 e^{i\theta_2}} = (C_2, \xi_2), \cdots, \mathcal{H}_{\mu_n e^{i\theta_n}} = (C_n, \xi_n)\}$. Since, $\mathcal{H} = (C, \xi)$ and $\mathcal{F}_s(\mathcal{H}) = \{\mu_1 e^{i\theta_1}, \mu_2 e^{i\theta_2}, \cdots, \mu_n e^{i\theta_n}\}$ with $0 < \mu_1 < \mu_2 < \cdots < \mu_n = M(h(\mathcal{H})), 0 < \theta_1 < \theta_2 < \cdots < \theta_n = \psi(h(\mathcal{H}))$. Define a complex fuzzy set C_f on X_n by

$$C_f(x) = \begin{cases} \mu_1 e^{i\theta_1}, & \text{if } x \in X_1, \\ \mu_k e^{i\theta_k}, & \text{if } x \in X_k \setminus X_{k-1}, l = 1, 2, \cdots, n. \end{cases}$$

We now prove that $\xi = \{B'_E | E' \in \xi_n\}$, where

$$B'_E(x) = \begin{cases} \min\{M_{C_f}(y)e^{i \min \psi_{C_f}(y)} | y \in E'\}, & \text{if } x \in E', \\ 0, & \text{otherwise.} \end{cases}$$

Let $E' \in \xi_n$, since $\mathcal{H} = (C, \xi)$ is support simple and elementary, then there is a unique CF hyperedge ρ in ξ having support E', i.e., distinct hyperedges in ξ must have distinct supports lying in ξ_n. Thus, it is sufficient to prove that for every $E' \in \xi_n$, $B'_E = \rho$. Since, all hyperedges have distinct supports and are elementary, Definition 1.60 implies that $h(\rho) = \mu_j e^{i\theta_j}$, for some $\mu_j e^{i\theta_j} \in \mathcal{F}_s(\mathcal{H})$. Thus, $E' \subseteq X_j$ and if $j > 1$, then $E' \in \xi_j \setminus \xi_{j-1}$. Since, $E' \subseteq X_j$, definition of C_f implies that for every $e \in E', C_f(e) \geq \mu_j e^{i\theta_j}$. We now claim that $C_f(e) = \mu_j e^{i\theta_j}$, for some $e \in E'$. If not, then definition of C_f implies that $C_f(e) \geq \mu_{j-1} e^{i\theta_{j-1}}$, for all $e \in E'$. This implies that $E' \subseteq X_{j-1}$ and $E' \in \xi_j \setminus \xi_{j-1}$, which is in contradiction to the assumption that $\mathscr{H} = (C, \xi)$ is simply ordered. Hence, for every $E' \in \xi_n$, $B'_E = \rho$.

Proposition 1.18 *Let $\mathscr{H} = (C, \xi)$ be a simply ordered complex fuzzy hypergraph and $\mathscr{F}_s(\mathscr{H}) = \{\mu_1 e^{i\theta_1}, \mu_2 e^{i\theta_2}, \cdots, \mu_n e^{i\theta_n}\}$. If $\mathscr{H}_{\mu_n e^{i\theta_n}}$ is simple, then there exists a partial complex fuzzy hypergraph $\mathscr{H}_1 = (C_1, \xi_1)$ of $\mathscr{H} = (C, \xi)$ such that*

  (i) $\mathscr{H}_1 = (C_1, \xi_1)$ is C_f−tempered complex fuzzy hypergraph of $\mathscr{H}_{\mu_n e^{i\theta_n}}$,
  (ii) $\xi_1 \subseteq \xi$, i.e., for all $\rho \in \xi_1$ there exists $\rho' \in \xi$ such that $\rho \subseteq \rho'$,
  (iii) $\mathscr{F}_s(\mathscr{H}) = \mathscr{F}_s(\mathscr{H}_1)$ and $Cor(\mathscr{H}) = Cor(\mathscr{H}_1)$.

Proof Proposition 1.17 implies that $\mathscr{H} = (C, \xi)$ is an elementary complex fuzzy hypergraph. The partial complex fuzzy hypergraph $\mathscr{H}_1 = (C_2, \xi_1)$ of $\mathscr{H} = (C, \xi)$ is obtained by removing those hyperedges which are properly contained in some other hyperedge of $\mathscr{H}$, where $\xi_1 = \{\rho \in \xi \,|\, \text{if } \rho \subseteq \rho_1 \text{ and } \rho_1 \in \xi, \text{ then } \rho = \rho_1\}$. Since, $\mathscr{H}_{\mu_n e^{i\theta_n}}$ is simple and contains all elementary hyperedges. Hence, (iii) holds. Furthermore, $\mathscr{H}_1 = (C_1, \xi_1)$ is support simple. Thus, $\mathscr{H}_1 = (C_1, \xi_1)$ satisfies all conditions of Theorem 1.5 and other conditions are satisfied.

Definition 1.66 Let $\mathscr{H} = (C, \xi)$ be a complex fuzzy hypergraph. The vertices $a, b \in X$ are said to be *adjacent vertices* if there exists an hyperedge $\rho \in \xi$ such that $a, b \in \rho$. If for $\rho_i, \rho_j \in \xi$, we have $\rho_i \cap \rho_j \neq \emptyset$ then ρ_i, ρ_j are *adjacent hyperedges*.

Definition 1.67 Let $\mathscr{H} = (C, \xi)$ be a complex fuzzy hypergraph. The *rank* of $\mathscr{H}$, denoted by $\triangle(\mathscr{H})$, is defined as $\triangle(\mathscr{H}) = \max\limits_{\rho_k \in \xi} M(\rho_k) e^{i \max \psi(\rho_k)}$, where $\rho_k \in \xi$ such that $|supp(\rho_k)| = \max\limits_{\rho_j \in \xi} |supp(\rho_j)|$.

The *anti rank* of $\mathscr{H}$, denoted by $\nabla(\mathscr{H})$, is defined as $\nabla(\mathscr{H}) = \min\limits_{\rho_k \in \xi} M(\rho_k)$ $e^{i \min \psi(\rho_k)}$, where $\rho_k \in \xi$ such that $|supp(\rho_k)| = \min\limits_{\rho_j \in \xi} |supp(\rho_j)|$. A complex fuzzy hypergraph $\mathscr{H} = (C, \xi)$ is *uniform* if $\triangle(\mathscr{H}) = \nabla(\mathscr{H})$.

1.8.1 Homomorphisms and Covering Constructions of Complex Fuzzy Hypergraphs

Definition 1.68 Let $\mathscr{H}_1 = (C_1, \xi_1)$ and $\mathscr{H}_2 = (C_2, \xi_2)$ be two complex fuzzy hypergraphs, where $C_1 = \{\lambda_{11}, \lambda_{12}, \cdots, \lambda_{1l}\}$ and $C_2 = \{\lambda_{21}, \lambda_{22}, \cdots, \lambda_{2l}\}$. A *homomorphism* of complex fuzzy hypergraphs $\mathscr{H}_1$ and $\mathscr{H}_2$ is a mapping $\kappa : X_1 \to X_2$ which satisfies

  (i) $\min\limits_{k=1}^{l} M_{\lambda_{1k}}(x) \leq \min\limits_{k=1}^{l} M_{\lambda_{2k}}(\kappa(x))$,
  (ii) $M_{\xi_1}(\{r_1, r_2, \cdots, r_j\}) \leq M_{\xi_2}(\{\kappa(r_1), \kappa(r_2), \cdots, \kappa(r_j)\})$, (for amplitude term)
  (iii) $\min\limits_{k=1}^{l} \psi_{\lambda_{1k}}(x) \leq \min\limits_{k=1}^{l} \psi_{\lambda_{2k}}(\kappa(x))$,

(iv) $\psi_{\xi_1}(\{r_1, r_2, \cdots, r_j\}) \leq \psi_{\xi_2}(\{\kappa(r_1), \kappa(r_2), \cdots, \kappa(r_j)\})$, (for phase term)

for all $x \in X$, for all $\{r_1, r_2, \cdots, r_j\} \in E$.

Definition 1.69 A *weak isomorphism* of complex fuzzy hypergraphs $\mathscr{H}_1$ and $\mathscr{H}_2$ is a bijective homomorphism $\kappa : X_1 \to X_2$ which satisfies

(i) $\min\limits_{k=1}^{l} M_{\lambda_{1k}}(x) = \min\limits_{k=1}^{l} M_{\lambda_{2k}}(\kappa(x))$,

(ii) $M_{\xi_1}(\{r_1, r_2, \cdots, r_j\}) \leq M_{\xi_2}(\{\kappa(r_1), \kappa(r_2), \cdots, \kappa(r_j)\})$, (for amplitude term)

(iii) $\min\limits_{k=1}^{l} \psi_{\lambda_{1k}}(x) = \min\limits_{k=1}^{l} \psi_{\lambda_{2k}}(\kappa(x))$,

(iv) $\psi_{\xi_1}(\{r_1, r_2, \cdots, r_j\}) \leq \psi_{\xi_2}(\{\kappa(r_1), \kappa(r_2), \cdots, \kappa(r_j)\})$, (for phase term)

for all $x \in X$, for all $\{r_1, r_2, \cdots, r_j\} \in E$.

Definition 1.70 A *co-weak isomorphism* of complex fuzzy hypergraphs $\mathscr{H}_1$ and $\mathscr{H}_2$ is a bijective homomorphism $\kappa : X_1 \to X_2$ which satisfies

(i) $\min\limits_{k=1}^{l} M_{\lambda_{1k}}(x) \leq \min\limits_{k=1}^{l} M_{\lambda_{2k}}(\kappa(x))$,

(ii) $M_{\xi_1}(\{r_1, r_2, \cdots, r_j\}) = M_{\xi_2}(\{\kappa(r_1), \kappa(r_2), \cdots, \kappa(r_j)\})$, (for amplitude term)

(iii) $\min\limits_{k=1}^{l} \psi_{\lambda_{1k}}(x) \leq \min\limits_{k=1}^{l} \psi_{\lambda_{2k}}(\kappa(x))$,

(iv) $\psi_{\xi_1}(\{r_1, r_2, \cdots, r_j\}) = \psi_{\xi_2}(\{\kappa(r_1), \kappa(r_2), \cdots, \kappa(r_j)\})$, (for phase term)

for all $x \in X$, for all $\{r_1, r_2, \cdots, r_j\} \in E$.

Definition 1.71 An *isomorphism* of complex fuzzy hypergraphs $\mathscr{H}_1$ and $\mathscr{H}_2$ is a mapping $\kappa : X_1 \to X_2$ which satisfies

(i) $\min\limits_{k=1}^{l} M_{\lambda_{1k}}(x) e^{i \min\limits_{k=1}^{l} \psi_{\lambda_{1k}}(x)} = \min\limits_{k=1}^{l} M_{\lambda_{2k}}(\kappa(x)) e^{i \min\limits_{k=1}^{l} \psi_{\lambda_{2k}}(\kappa(x))}$, for all $x \in X$,

(ii) $M_{\xi_1}(\{r_1, r_2, \cdots, r_j\}) e^{i\psi_{\xi_1}(\{r_1, r_2, \cdots, r_j\})} = M_{\xi_2}(\{\kappa(r_1), \kappa(r_2), \cdots, \kappa(r_j)\})$
$e^{i\psi_{\xi_2}(\{\kappa(r_1), \kappa(r_2), \cdots, \kappa(r_j)\})}$, for all $\{r_1, r_2, \cdots, r_j\} \in E$.

Example 1.23 Let $\mathscr{H}_1 = (C_1, \xi_1)$ and $\mathscr{H}_2 = (C_2, \xi_2)$ be two complex fuzzy hypergraphs, where $C_1 = \{\lambda_{11}, \lambda_{12}, \lambda_{13}, \lambda_{14}\}$ and $C_2 = \{\lambda_{21}, \lambda_{22}, \lambda_{23}, \lambda_{24}\}$ are complex fuzzy sets on $X_1 = \{r_1, r_2, r_3, r_4, r_5, r_6\}$ and $X_2 = \{r_1', r_2', r_3', r_4', r_5', r_6'\}$ as given in Tables 1.28, 1.29, respectively.

The complex fuzzy relations ξ_1 and ξ_2 are defined as

$$\xi_{11}(\{r_1, r_3, r_4, r_6\}) = 0.1e^{i\frac{\pi}{5}}, \quad \xi_{12}(\{r_1, r_2, r_3\}) = 0.2e^{i\frac{3\pi}{10}},$$

$$\xi_{13}(\{r_3, r_4\}) = 0.5e^{i\frac{\pi}{5}}, \quad \xi_{14}(\{r_4, r_5, r_6\}) = 0.2e^{i\frac{3\pi}{10}},$$

$$\xi_{21}(\{r_1', r_2', r_3', r_6'\}) = 0.1e^{i\frac{\pi}{5}}, \quad \xi_{22}(\{r_1', r_3', r_4'\}) = 0.2e^{i\frac{3\pi}{10}},$$

$$\xi_{23}(\{r_1', r_2'\}) = 0.5e^{i\frac{\pi}{5}}, \quad \xi_{24}(\{r_2', r_5', r_6'\}) = 0.1e^{i\frac{\pi}{5}}.$$

The complex fuzzy hypergraphs $\mathscr{H}_1 = (C_1, \xi_1)$ and $\mathscr{H}_2 = (C_2, \xi_2)$ are shown in Figs. 1.34 and 1.35, respectively.

Table 1.28 Complex fuzzy sets on X_1

$x \in X$	λ_{11}	λ_{12}	λ_{13}	λ_{14}
r_1	$0.2e^{i\frac{3\pi}{10}}$	$0.2e^{i\frac{3\pi}{10}}$	0	0
r_2	0	$0.5e^{i\frac{2\pi}{5}}$	0	0
r_3	$0.5e^{i\frac{3\pi}{5}}$	$0.5e^{i\frac{3\pi}{5}}$	$0.5e^{i\frac{3\pi}{5}}$	0
r_4	$0.8e^{i\frac{\pi}{5}}$	0	$0.8e^{i\frac{\pi}{5}}$	$0.8e^{i\frac{\pi}{5}}$
r_5	0	0	0	$0.5e^{i\frac{\pi}{2}}$
r_6	$0.1e^{i\frac{\pi}{2}}$	0	0	$0.1e^{i\frac{\pi}{2}}$

Table 1.29 Complex fuzzy sets on X_2

$x \in X$	λ_{21}	λ_{22}	λ_{23}	λ_{24}
r'_1	$0.5e^{i\frac{3\pi}{5}}$	$0.5e^{i\frac{3\pi}{5}}$	$0.5e^{i\frac{3\pi}{5}}$	0
r'_2	$0.8e^{i\frac{\pi}{5}}$	0	$0.8e^{i\frac{\pi}{5}}$	$0.8e^{i\frac{\pi}{5}}$
r'_3	$0.2e^{i\frac{3\pi}{10}}$	$0.2e^{i\frac{3\pi}{10}}$	0	0
r'_4	0	$0.5e^{i\frac{2\pi}{5}}$	0	0
r'_5	0	0	0	$0.5e^{i\frac{\pi}{2}}$
r'_6	$0.1e^{i\frac{\pi}{2}}$	0	0	$0.1e^{i\frac{\pi}{2}}$

Define a mapping $\kappa : X_1 \to X_2$ by $\kappa(r_1) = r'_3, \kappa(r_2) = r'_4, \kappa(r_3) = r'_1, \kappa(r_4) = r'_2,$ $\kappa(r_5) = r'_5,$ and $\kappa(r_6) = r'_6.$ Note that

$$\lambda_{11}(r_1) = 0.2e^{i\frac{3\pi}{10}} = \lambda_{21}(r'_3) = \lambda_{21}(\kappa(r_1)),$$

$$\lambda_{11}(r_2) = 0.5e^{i\frac{2\pi}{5}} = \lambda_{21}(r'_4) = \lambda_{21}(\kappa(r_2)),$$

$$\lambda_{11}(r_3) = 0.5e^{i\frac{3\pi}{5}} = \lambda_{21}(r'_1) = \lambda_{21}(\kappa(r_3)),$$

$$\lambda_{11}(r_4) = 0.8e^{i\frac{\pi}{5}} = \lambda_{21}(r'_2) = \lambda_{21}(\kappa(r_4)),$$

$$\lambda_{11}(r_5) = 0.5e^{i\frac{\pi}{2}} = \lambda_{21}(r'_5) = \lambda_{21}(\kappa(r_5)),$$

$$\lambda_{11}(r_6) = 0.1e^{i\frac{\pi}{2}} = \lambda_{21}(r'_6) = \lambda_{21}(\kappa(r_6)).$$

Similarly, $\lambda_{1j}(x) = \lambda_{2j}(\kappa(x))$ and $\xi_1(\{r_1, r_2, \cdots, r_k\}) = \xi_2(\{\kappa(r_1), \kappa(r_2), \cdots, \kappa(r_k)\}),$ for all $x, r_j \in X$. Hence, $\mathscr{H}_1 = (C_1, \xi_1)$ and $\mathscr{H}_2 = (C_2, \xi_2)$ are isomorphic.

Definition 1.72 Let $\mathscr{H}_1 = (C_1, \xi_1)$ and $\mathscr{H}_2 = (C_2, \xi_2)$ be complex fuzzy hypergraphs on X_1 and X_2, respectively. The complex fuzzy hypergraph $\mathscr{H}_1$ is an $m-fold$ *covering* of $\mathscr{H}_2$ if there exists a surjective homomorphism $\kappa : X_1 \to X_2$ such that

   (i) $|\kappa^{-1}(v)| = m$, for all $v \in X_2$,

   (ii) $|supp(\kappa^{-1}(\rho'_j))| = m$, for all $\rho'_j \in \xi_2$,

   (iii) $supp(\rho_j) \cap supp(\rho_k) = \emptyset$, for all $\rho_j, \rho_k \in \kappa^{-1}(\rho'_l), \rho'_l \in \xi_2$.

Fig. 1.34 Complex fuzzy hypergraph $\mathscr{H}_1 = (C_1, \xi_1)$

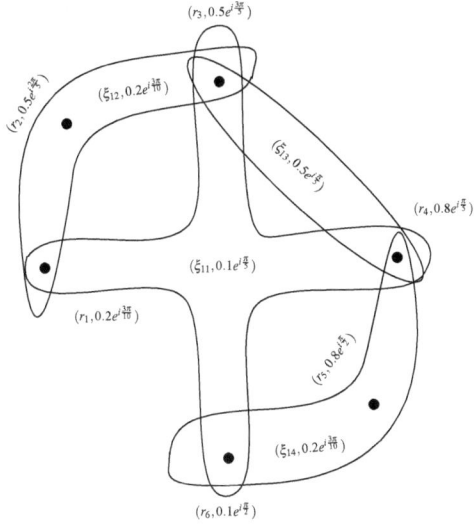

Fig. 1.35 Complex fuzzy hypergraph $\mathscr{H}_2 = (C_2, \xi_2)$

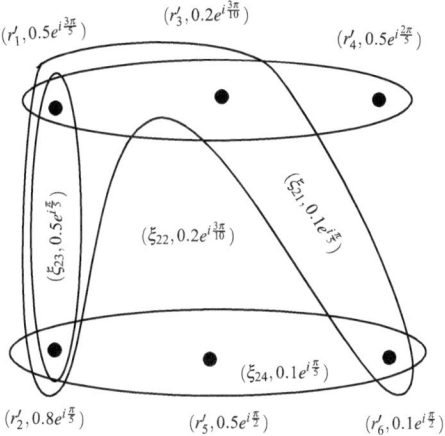

Then, $\mathscr{H}_2 = (C_2, \xi_2)$ is called the *quoitient complex fuzzy hypergraph* of $\mathscr{H}_1 = (C_1, \xi_1)$ and κ is called the *covering projection*. If $m = 2$, $\mathscr{H}_1$ is called the *double covering* of $\mathscr{H}_2$.

Example 1.24 Let $\mathscr{H}_1 = (C_1, \xi_1)$ and $\mathscr{H}_2 = (C_2, \xi_2)$ be complex fuzzy hypergraphs on $X_1 = \{v_1, v_2, v_3, v_4, v_5, v_6, v_7, v_8\}$ and $X_2 = \{v'_1, v'_2, v'_3, v'_4\}$, respectively, where $C_1 = \{\alpha_1, \alpha_2, \alpha_3, \alpha_4\}$ is the family of complex fuzzy sets on X_1 as given in Table 1.30.

The complex fuzzy hypergraph $\mathscr{H}_1 = (C_1, \xi_1)$ is shown in Fig. 1.36.

Let $C_2 = \{\alpha'_1, \alpha'_2, \alpha'_3, \alpha'_4\}$ be the family of complex fuzzy sets on X_2 as given in Table 1.31.

The complex fuzzy hypergraph $\mathscr{H}_2 = (C_2, \xi_2)$ is shown in Fig. 1.37.

Table 1.30 Complex fuzzy sets on X_1

$x \in X_1$	α_1	α_2	α_3	α_4
v_1	$0.2e^{i\frac{3\pi}{10}}$	0	0	0
v_2	$0.5e^{i\frac{2\pi}{5}}$	0	0	0
v_3	0	0	$0.5e^{i\frac{3\pi}{5}}$	0
v_4	0	$0.8e^{i\frac{\pi}{3}}$	0	0
v_5	0	$0.5e^{i\frac{3\pi}{5}}$	0	0
v_6	0	0	$0.1e^{i\frac{\pi}{2}}$	0
v_7	0	0	0	$0.5e^{i\frac{\pi}{3}}$
v_8	0	0	0	$0.2e^{i\frac{\pi}{3}}$

Fig. 1.36 Complex fuzzy hypergraph $\mathcal{H}_1$

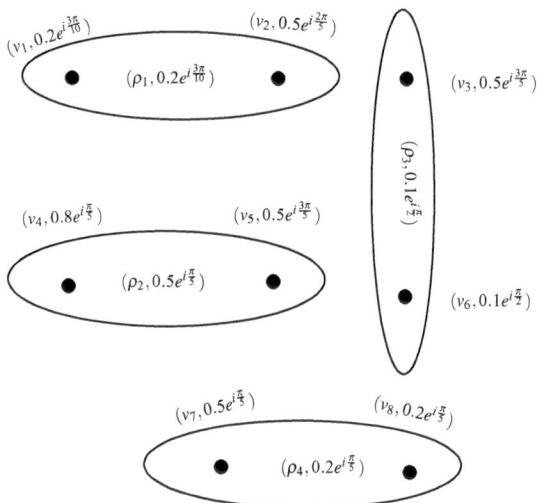

Define a mapping $\kappa : X_1 \to X_2$ as $\kappa(v_1) = v_1'$, $\kappa(v_2) = v_1'$, $\kappa(v_3) = v_2'$, $\kappa(v_4) = v_2'$, $\kappa(v_5) = v_3'$, $\kappa(v_6) = v_3'$, $\kappa(v_7) = v_4'$, and $\kappa(v_8) = v_4'$, which is surjective homomorphism. Note that

$$|\kappa^{-1}(v_1')| = |\{v_1, v_2\}| = 2,$$
$$|\kappa^{-1}(v_2')| = |\{v_3, v_4\}| = 2,$$
$$|\kappa^{-1}(v_3')| = |\{v_5, v_6\}| = 2,$$
$$|\kappa^{-1}(v_4')| = |\{v_7, v_8\}| = 2.$$

Also, $|supp(\kappa^{-1}(\rho_j'))| = 2$, for all $\rho_j' \in \xi_2$, $supp(\rho_j) \cap supp(\rho_k) = \emptyset$, for all $\rho_j, \rho_k \in \kappa^{-1}(\rho_l')$, $\rho_l' \in \xi_2$. Hence, $\mathcal{H}_1$ is the two-fold covering or double covering of $\mathcal{H}_2$.

Table 1.31 Complex fuzzy sets on X_2

$x \in X_1$	α'_1	α'_2	α'_3	α'_4
v'_1	$0.1e^{i\frac{3\pi}{10}}$	0	0	0
v'_2	0	$0.5e^{i\frac{\pi}{5}}$	0	0
v'_3	0	$0.1e^{i\frac{\pi}{2}}$	0	0
v'_4	0	0	0	$0.2e^{i\frac{\pi}{5}}$

Fig. 1.37 Complex fuzzy hypergraph $\mathcal{H}_2$

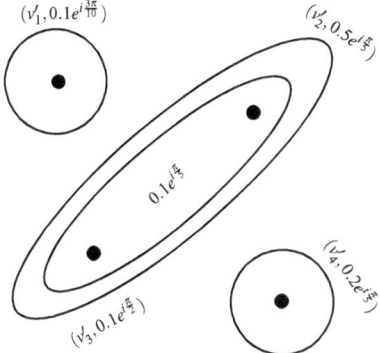

$(v'_1, 0.1e^{i\frac{3\pi}{10}})$ $(v'_2, 0.5e^{i\frac{\pi}{5}})$

$0.1e^{i\frac{\pi}{5}}$

$(v'_4, 0.2e^{i\frac{\pi}{5}})$

$(v'_3, 0.1e^{i\frac{\pi}{2}})$

Definition 1.73 The 2-*section* of a complex fuzzy hypergraph $\mathcal{H} = (C, \xi)$ is a complex fuzzy graph $[\mathcal{H}]_2 = (C, [\xi]_2)$ having same set of vertices as that of $\mathcal{H}$ and $[\xi]_2$ is a complex fuzzy set on $\{\varepsilon = u_j u_k | u_j, u_k \in E_l, l = 1, 2, \cdots \}$, i.e., there is an edge between u_j and u_k in $[\mathcal{H}]_2$ if they are incident to same hyperedge in $\mathcal{H}$ such that

$$[\xi]_2(u_j u_k) = \inf\{\min_l(M_{\alpha_l}(u_j)), \min_l(M_{\alpha_l}(u_k))\}e^{i\inf\{\min_l(\psi_{\alpha_l}(u_j)), \min_l(\psi_{\alpha_l}(u_k))\}}.$$

Definition 1.74 The L_2-*section* of a complex fuzzy hypergraph $\mathcal{H} = (C, \xi)$ is its 2-section complex fuzzy graph $[\mathcal{H}]_2 = (C, [\xi]_2)$ together with a mapping $\delta : \varepsilon \to P(X)$ such that $\delta(\{u_j, u_k\}) = \{supp(\rho_j) | \{u_j, u_k\} \subseteq supp(\rho_j)\}$, for $\rho_j \in \xi$. The L_2-*section* of a complex fuzzy hypergraph is denoted by $\Gamma = (X, [\xi]_2, \delta)$.

Example 1.25 Consider a complex fuzzy hypergraph $\mathcal{H} = (C, \xi)$, where $X = \{r_1, r_2, r_3, r_4\}$, and $C = \{\alpha_1, \alpha_2\}$ be Complex fuzzy sets on X as given in Table 1.32.

The relation ξ on α_j, $j = 1, 2$ is defined as $\xi(\{r_1, r_2, r_4\}) = 0.1e^{i\frac{3\pi}{10}}$ and $\xi(\{r_3, r_4\}) = 0.2e^{i\frac{\pi}{5}}$. The corresponding complex fuzzy hypergraph is shown in Fig. 1.38.

Then, the 2-section $[\mathcal{H}]_2$ of $\mathcal{H}$ is given in Fig. 1.39.

Define a mapping $\delta : \varepsilon \to P(X)$ such that $\delta(r_1 r_2) = E_1(\mathcal{H}), \delta(r_2 r_4) = E_1(\mathcal{H})$, $\delta(r_1 r_4) = E_1(\mathcal{H})$, and $\delta(r_2 r_3) = E_2(\mathcal{H})$. Here, $E_j(\mathcal{H})$ denotes the jth crisp hyperedge of $\mathcal{H}$. Then, the triplet $\Gamma = (X, [\xi]_2, \delta)$ is the L_2-section of $\mathcal{H}$.

Table 1.32 Complex fuzzy sets on X

$r \in X$	α_1	α_2
r_1	$0.1e^{i\frac{3\pi}{10}}$	$0.1e^{i\frac{3\pi}{10}}$
r_2	$0.2e^{i\frac{\pi}{2}}$	0
r_3	0	$0.5e^{i\frac{\pi}{3}}$
r_4	$0.3e^{i\frac{\pi}{3}}$	$0.4e^{i\frac{3\pi}{4}}$

Fig. 1.38 Complex fuzzy hypergraph

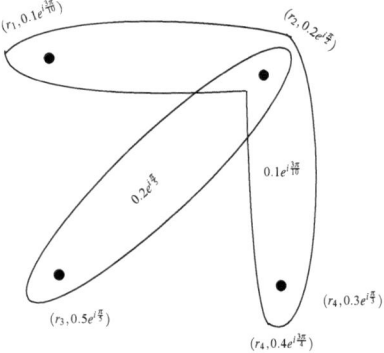

Fig. 1.39 2-section $[\mathscr{H}]_2$ of $\mathscr{H}$

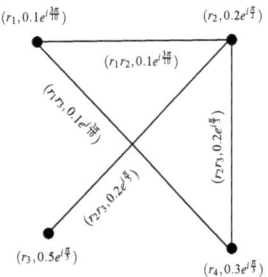

Definition 1.75 A *complex fuzzy hyperpath* H_P of length k in a complex fuzzy hypergraph $\mathscr{H} = (C, \xi)$ is defined as a sequence of distinct nodes and hyperedges, i.e., $s_1, E_1, s_2, E_2, \cdots, s_k, E_k, s_{k+1}$ such that

(i) $M_\xi(E_j) > 0$ and $\psi_\xi(E_j) > 0$, $j = 1, 2, \cdots, k$,
(ii) $s_j, s_{j+1} \in E_j$, $j = 1, 2, \cdots, k$.

If $s_1 = s_{k+1}$, then the complex fuzzy hyperpath is called *complex fuzzy hypercycle*.

A complex fuzzy hypergraph $\mathscr{H} = (C, \xi)$ is *connected* if every pair of distinct vertices is connected through a complex fuzzy hyperpath.

Definition 1.76 The *distance* between two vertices s_i and s_j, denoted by $dis_{\mathscr{H}}(s_i, s_j)$, is defined as $D_{\mathscr{H}}(s_i, s_j) = \sum_j M_\xi(E_j) e^{i \sum_j \psi_\xi(E_j)}$, where E_j are the hyperedges belonging to the shortest hyperpath between s_i and s_j.

Definition 1.77 The *strength of complex fuzzy hyperpath* of length k between u and v is defined as

$$\eta^k(u, v) = \min\{M_\xi(E_1), M_\xi(E_2), \cdots, M_\xi(E_k)\}e^{i \min\{\psi_\xi(E_1), \psi_\xi(E_2), \cdots, \psi_\xi(E_k)\}},$$

$u \in E_1, v \in E_k$, where $E_1, E_2, \cdots, E_k$ are hyperedges. The *strength of connectedness* between u and v is given as

$$\eta^\infty(u, v) = \sup_k\{(\eta^k(u, v)), k = 1, 2, \cdots\}.$$

Theorem 1.6 *A complex fuzzy hypergraph* $\mathcal{H} = (C, \xi)$ *is connected if and only if* $\eta^\infty(u, v) > 0$, *for all* $u, v \in X$.

Definition 1.78 Let $\mathcal{H} = (C, \xi)$ be a complex fuzzy hypergraph, where $C = \{\alpha_1, \alpha_2, \cdots, \alpha_k\}$ is a finite family of complex fuzzy sets on X. Then, $\mathcal{H} = (C, \xi)$ is *strong* if for all $\{r_1, r_2, \cdots, r_l\} \in E$,

$$M_\xi(\{r_1, r_2, \cdots, r_l\})e^{i\psi_\xi(\{r_1, r_2, \cdots, r_l\})} = \min\{M_{\alpha_i}(r_1), M_{\alpha_i}(r_2), \cdots, M_{\alpha_i}(r_l)\}$$
$$\times e^{i \min\{\psi_{\alpha_i}(r_1), \psi_{\alpha_i}(r_2), \cdots, \psi_{\alpha_i}(r_l)\}},$$

Theorem 1.7 *Let* $\mathcal{H}_1 = (C_1, \xi_1)$ *and* $\mathcal{H}_2 = (C_2, \xi_2)$ *be two isomorphic complex fuzzy hypergraphs. Then,* $\mathcal{H}_1$ *is connected if and only if* $\mathcal{H}_2$ *is connected.*

Proof Let $\mathcal{H}_1 = (C_1, \xi_1)$ and $\mathcal{H}_2 = (C_2, \xi_2)$ be two isomorphic complex fuzzy hypergraphs, such that $\xi_1 = \{\mathcal{E}_{11}, \mathcal{E}_{12}, \cdots, \mathcal{E}_{1r}\}$ and $\xi_2 = \{\mathcal{E}_{21}, \mathcal{E}_{22}, \cdots, \mathcal{E}_{2r}\}$ be the sets of hyperedges of $\mathcal{H}_1$ and $\mathcal{H}_2$, respectively. Let $\kappa : \mathcal{H}_1 \to \mathcal{H}_2$ be an isomorphism from $\mathcal{H}_1$ onto $\mathcal{H}_2$. Suppose that $\mathcal{H}_1$ is connected, then we have

$$0 < \eta_1^\infty(u, v) = \sup_r\{\min\{M_{\xi_1}(\mathcal{E}_{11}), M_{\xi_1}(\mathcal{E}_{12}), \cdots, M_{\xi_1}(\mathcal{E}_{1r})\}e^{i \min\{\psi_{\xi_1}(\mathcal{E}_{11}), \psi_{\xi_1}(\mathcal{E}_{12}), \cdots, \psi_{\xi_1}(\mathcal{E}_{1r})\}}\}$$
$$= \sup_r\{\min\{M_{\xi_2}(\kappa(\mathcal{E}_{11})), M_{\xi_2}(\kappa(\mathcal{E}_{12})), \cdots, M_{\xi_2}(\kappa(\mathcal{E}_{1r}))\}$$
$$\times e^{i \min\{\psi_{\xi_2}(\kappa(\mathcal{E}_{11})), \psi_{\xi_2}(\kappa(\mathcal{E}_{12})), \cdots, \psi_{\xi_2}(\kappa(\mathcal{E}_{1r}))\}}\}$$
$$= \eta_2^\infty(\kappa(u), \kappa(v)).$$

Hence, $\mathcal{H}_2 = (C_2, \xi_2)$ is connected. The converse part can be proved on same lines.

Lemma 1.9 *Let* $\mathcal{H} = (C, \xi)$ *be a complex fuzzy hypergraph on X and $s_i, s_j \in X$. Then, the distance between u_i and u_j in $\mathcal{H}$ is same as in* $[\mathcal{H}]_2 = (C, [\xi]_2)$.

Proof Let $\mathcal{H} = (C, \xi)$ be a disconnected complex fuzzy hypergraph. Then, s_i and s_j belong to the different connected components of $\mathcal{H}$ if and only if they are contained in distinct connected components of $[\mathcal{H}]_2$. Hence, $D_{\mathcal{H}}(s_i, s_j) = D_{[\mathcal{H}]_2}(s_i, s_j) = \infty$. Thus, WLOG we assume that $\mathcal{H} = (C, \xi)$ (and therefore $[\mathcal{H}]_2$) is connected. Let $H_P = s_i, E_i, s_{i+1}, E_{i+1}, \cdots, s_{j-1}, E_{j-1}, s_j$ be the shortest CFHP between s_i and s_j. Then, the construction of $[\mathcal{H}]_2$ implies that there exists a walk $H_P^* = s_i, E_i^*, s_{i+1}, E_{i+1}^*, \cdots, s_{j-1}, E_{j-1}^*, s_j$ in $[\mathcal{H}]_2$. Thus, for $D_{\mathcal{H}}(s_i, s_j) = \sum_j M_\xi(E_j)e^{i\sum_j \psi_\xi(E_j)}$ and $D_{[\mathcal{H}]_2}(s_i, s_j) = \sum_j M_\varepsilon(E_j^*)e^{i\sum_j \psi_\varepsilon(E_j^*)}$, we have $M_\xi(E_j) \geq M_\varepsilon(E_j^*)$ and $\psi_\xi(E_j) \geq \psi_\varepsilon(E_j^*)$. Suppose that $M_\xi(E_j) > M_\varepsilon(E_j^*)$ and $\psi_\xi(E_j) > \psi_\varepsilon(E_j^*)$, then there exists a path $Q_P^* = s_i, E_i', s_{i+1}, E_{i+1}', \cdots, s_{j-1}, E_{j-1}', s_j$ in $[\mathcal{H}]_2$, i.e., for all E_i' there is an edge $E_i \in \xi(\mathcal{H})$ such that $E_i' \subseteq E_i$ and we obtain a walk of length $M_\varepsilon(E_j^*)e^{i\psi_\varepsilon(E_j^*)}$ in $\mathcal{H}$, which is a contradiction. Hence, $M_\xi(E_j) = M_\varepsilon(E_j^*)$ and $\psi_\xi(E_j) = \psi_\varepsilon(E_j^*)$.

1.8.2 Products on Complex Fuzzy Hypergraphs

Definition 1.79 Let $\bigotimes_{j=1}^m \mathcal{H}_j = (A, \lambda)$ be an arbitrary product of complex fuzzy hypergraphs $\mathcal{H}_j$, where A is the complex fuzzy set on $\bigotimes_{j=1}^m X_j$. The *projection* $\sigma_j : v \in \bigotimes_{j=1}^m X_j \to V(\mathcal{H}_j)$ is defined as $v = (v_1, v_2, \cdots, v_m) \to v_j$ and v_j is called the jth coordinate of $v \in \bigotimes_{j=1}^m X_j$.

The $\mathcal{H}_k-layer$ hypergraph $\mathcal{H}_k^{w=(w_1, w_2, \cdots, w_m)}$, which is the partial complex fuzzy hypergraph of $\mathcal{H}_j$, determined by $w \in \bigotimes_{j=1}^m X_j$, is defined as

(i) $V(\bigotimes_{j=1}^m \mathcal{H}_j) = V(\mathcal{H}_k^{w=(w_1, w_2, \cdots, w_m)})$, i.e., the vertex set is same.

(ii) $E(\mathcal{H}_k^{w=(w_1, w_2, \cdots, w_m)}) = \{v \in \bigotimes_{j=1}^m X_j | \sigma_j(v) = \sigma_j(w), \ k \neq j\}$ such that

$$M_\lambda(E_l(\mathcal{H}_k^w))e^{i\psi_\lambda(E_l(\mathcal{H}_k^w))} = M_\lambda(E_l(\bigotimes_{j=1}^m \mathcal{H}_j)e^{i\psi_\lambda(E_l(\bigotimes_{j=1}^m \mathcal{H}_j))}.$$

Definition 1.80 Let $\mathcal{H}_1 = (C_1, \xi_1)$ and $\mathcal{H}_2 = (C_2, \xi_2)$ be two complex fuzzy hypergraphs on $X_1 = \{x_1, x_2, \cdots, x_m\}$ and $X_2 = \{y_1, y_2, \cdots, y_n\}$, where $C_1 = \{\alpha_{11}, \alpha_{12}, \cdots, \alpha_{1j}\}$ and $C_2 = \{\alpha_{21}, \alpha_{22}, \cdots, \alpha_{2k}\}$ are complex fuzzy sets on X_1 and X_2, respectively. The *Cartesian product* $\mathcal{H}_1 \square \mathcal{H}_2 = (C_1 \square C_2, \xi_1 \square \xi_2)$ of $\mathcal{H}_1$ and $\mathcal{H}_2$, having the vertex set $X_1 \times X_2$ is defined as

(i) $C_1 \square C_2(x_p, y_q) = \min\{M_{\alpha_{1r}}(x_p), M_{\alpha_{2s}}(y_q)\}e^{i \min\{\psi_{\alpha_{1r}}(x_p), \psi_{\alpha_{2s}}(y_q)\}}$, for all $(x_p, y_q) \in X_1 \times X_2$,

(ii) $\xi_1 \square \xi_2(\{x_p\} \times e_2) = \min\{M_{\alpha_{1r}}(x_p), M_{\xi_2}(e_2)\}e^{i \min\{\psi_{\alpha_{1r}}(x_p), \psi_{\xi_2}(e_2)\}}$, for all $x_p \in X_1, e_2 \in E(\mathcal{H}_2)$,

(iii) $\xi_1 \square \xi_2(e_1 \times \{y_q\}) = \min\{M_{\xi_1}(e_1), M_{\alpha_{2s}}(y_q)\}e^{\min\{\psi_{\xi_1}(e_1), \psi_{\alpha_{2s}}(y_q)\}}$, for all $e_1 \in E(\mathcal{H}_1), y_q \in X_2$,

(a)   **(b)**

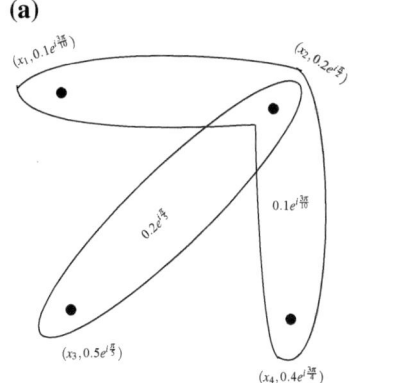   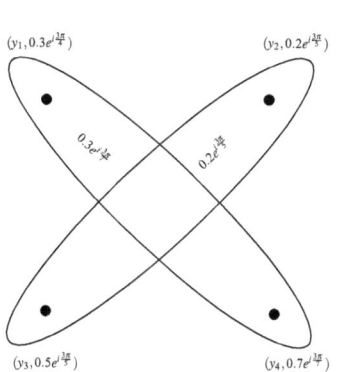

Fig. 1.40 Complex fuzzy hypergraphs $\mathscr{H}_1$ and $\mathscr{H}_2$

where $p = 1, 2, \cdots, m$, $q = 1, 2, \cdots, n$, $r = 1, 2, \cdots, j$, and $s = 1, 2, \cdots, k$.

Remark 1.6 Note that the Cartesian product of two simple complex fuzzy hypergraphs is also simple. Furthermore, $\mathscr{H}_1 \square \mathscr{H}_2$ is connected if and only if $\mathscr{H}_1$ and $\mathscr{H}_2$ both are connected.

Example 1.26 Let Let $\mathscr{H}_1 = (C_1, \xi_1)$ and $\mathscr{H}_2 = (C_2, \xi_2)$ be two complex fuzzy hypergraphs on $X_1 = \{x_1, x_2, x_3, x_4\}$ and $X_2 = \{y_1, y_2, y_3, y_4\}$, respectively. The corresponding complex fuzzy hypergraphs are given in Fig. 1.40a and b, respectively.

Then, the Cartesian product of $\mathscr{H}_1$ and $\mathscr{H}_2$ is given in Fig. 1.41.

Definition 1.81 The *rank* and *anti rank* of $\mathscr{H}_1 \square \mathscr{H}_2$ are defined as

(i) $\triangle(\mathscr{H}_1 \square \mathscr{H}_2) = \max\{\triangle(\mathscr{H}_1), \triangle(\mathscr{H}_2)\}$,
(ii) $\nabla(\mathscr{H}_1 \square \mathscr{H}_2) = \min\{\nabla(\mathscr{H}_1), \nabla(\mathscr{H}_2)\}$, respectively.

Example 1.27 Consider the complex fuzzy hypergraphs as shown in Fig. 1.40a and b, respectively. For $\mathscr{H}_1$, we have $\triangle(\mathscr{H}_1) = \xi(\{x_1, x_2, x_4\}) = 0.1e^{i\frac{3\pi}{10}}$ and $\nabla(\mathscr{H}_1) = \xi(\{x_2, x_3\}) = 0.2e^{i\frac{\pi}{5}}$. For $\mathscr{H}_2$, we have $\triangle(\mathscr{H}_2) = \xi(\{y_1, y_4\}) = 0.3e^{i\frac{3\pi}{5}}$ and $\nabla(\mathscr{H}_1) = \xi(\{y_2, y_3\}) = 0.2e^{i\frac{3\pi}{7}}$. Thus, we have

$$\triangle(\mathscr{H}_1 \square \mathscr{H}_2) = \max\{0.1e^{i\frac{3\pi}{10}}, 0.3e^{i\frac{3\pi}{5}}\} = 0.3e^{i\frac{3\pi}{5}},$$

$$\nabla(\mathscr{H}_1 \square \mathscr{H}_2) = \min\{0.2e^{i\frac{\pi}{5}}, 0.2e^{i\frac{3\pi}{7}}\} = 0.2e^{i\frac{\pi}{5}}.$$

Proposition 1.19 *The 2-section of $\mathscr{H}_1 \square \mathscr{H}_2$ is the Cartesian product of 2-section of $\mathscr{H}_1$ and 2-section of $\mathscr{H}_2$, i.e., $[\mathscr{H}_1 \square \mathscr{H}_2]_2 = [\mathscr{H}_1]_2 \square [\mathscr{H}_2]_2$.*

Definition 1.82 Let $\Gamma_1 = (X_1, [\xi_1]_2, \delta_1)$ and $\Gamma_2 = (X_2, [\xi_2]_2, \delta_2)$ be the L_2-sections of $\mathscr{H}_1$ and $\mathscr{H}_2$, respectively. Then, the Cartesian product of Γ_1 and Γ_2, denoted by

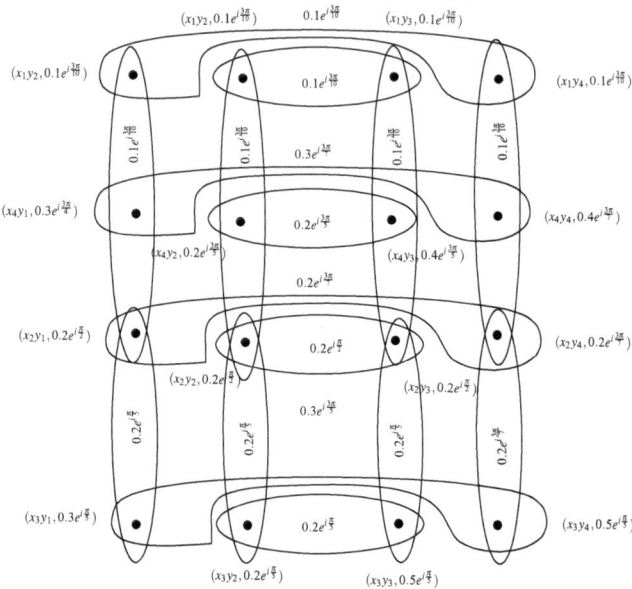

Fig. 1.41 Cartesian product $\mathscr{H}_1\Box\mathscr{H}_2$ of $\mathscr{H}_1$ and $\mathscr{H}_2$

$\Gamma_1\Box\Gamma_2$, is defined as $\Gamma_1\Box\Gamma_2 = (X, [\xi]_2, \Phi)$, which contain a CFG $(C_1, \xi_1)\Box(C_2, \xi_2)$ along with a function $\Phi : [\xi]_2 \to P(X)$ such that

$$\Phi(\{(x_1, y_1), (x_2, y_2)\}) = \begin{cases} \{\{x_1\} \times e_2 | e_2 \in \delta_2(\{y_1, y_2\})\}, & \text{if } x_1 = x_2, \\ \{e_1 \times \{y_1\} | e_1 \in \delta_1(\{x_1, x_2\})\}, & \text{if } y_1 = y_2. \end{cases}$$

Lemma 1.10 *Let $\mathscr{H}_1$ and $\mathscr{H}_2$ be two complex fuzzy hypergraphs. Then, the distance between two arbitrary vertices in $\mathscr{H}_1\Box\mathscr{H}_2$ is given as*

$$D_{\mathscr{H}_1\Box\mathscr{H}_2}(v_i, v_j)(u_k, u_l) = D_{\mathscr{H}_1}(v_i, u_k) + D_{\mathscr{H}_2}(v_j, u_l).$$

Definition 1.83 Let $\mathscr{H}_1 = (C_1, \xi_1)$ and $\mathscr{H}_2 = (C_2, \xi_2)$ be two complex fuzzy hypergraphs on $X_1 = \{x_1, x_2, \cdots, x_m\}$ and $X_2 = \{y_1, y_2, \cdots, y_n\}$, where $C_1 = \{\alpha_{11}, \alpha_{12}, \cdots, \alpha_{1j}\}$ and $C_2 = \{\alpha_{21}, \alpha_{22}, \cdots, \alpha_{2k}\}$ are complex fuzzy sets on X_1 and X_2, respectively. The *minimal crisp rank preserving direct product $\mathscr{H}_1\Diamond\mathscr{H}_2 = (C_1\Diamond C_2, \xi_1\Diamond\xi_2)$* of $\mathscr{H}_1$ and $\mathscr{H}_2$, having the vertex set $X_1 \times X_2$ is defined as

(i) $C_1\Diamond C_2(x_p, y_q) = \min\{M_{\alpha_{1r}}(x_p), M_{\alpha_{2s}}(y_q)\}e^{i\min\{\psi_{\alpha_{1r}}(x_p),\psi_{\alpha_{2s}}(y_q)\}}$, for all $(x_p, y_q) \in X_1 \times X_2$,

(ii) $\xi_1\Diamond\xi_2(\{(x_1, y_1), (x_2, y_2), \cdots, (x_m, y_n)\})=\min\{M_{\xi_1}(\{x_1, x_2, \cdots x_m\}),$
     $\min M_{\alpha_{2s}}(y_q)\}\times$
     $e^{i\min\{\psi_{\xi_1}(\{x_1,x_2,\cdots x_m\}),\min M_{\alpha_{2s}}(y_q)\}}$, for all $\{x_1, x_2, \cdots x_m\} \in E(\mathscr{H}_1)$, $\{y_1, y_2, \cdots y_n\}$
     $\subseteq E(\mathscr{H}_2)$,

(a)   **(b)**

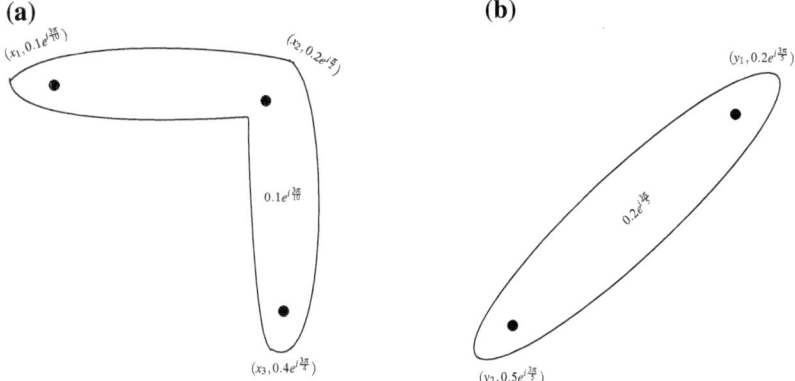

Fig. 1.42 Complex fuzzy hypergraphs $\mathscr{H}_1$ and $\mathscr{H}_2$

Fig. 1.43 Direct product
$\mathscr{H}_1 \Diamond \mathscr{H}_2$ of $\mathscr{H}_1$ and $\mathscr{H}_2$

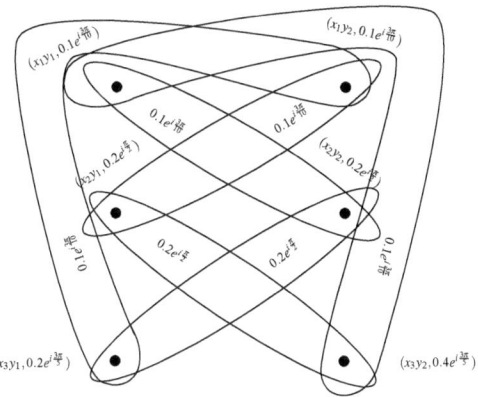

(iii) $\xi_1 \Diamond \xi_2(\{(x_1, y_1), (x_2, y_2), \cdots , (x_m, y_n)\}) = \min\{\min M_{\alpha_{1r}}(x_p), M_{\xi_2}(\{y_1, y_2, \cdots y_n\})\} \times e^{i \min\{\min \psi_{\alpha_{1r}}(x_p), \psi_{\xi_2}(\{y_1, y_2, \cdots y_n\})\}}$, for all $\{x_1, x_2, \cdots x_m\} \subseteq E(\mathscr{H}_1)$, $\{y_1, y_2, \cdots y_n\} \in E(\mathscr{H}_2)$,

where $r = 1, 2, \cdots , j$, and $s = 1, 2, \cdots , k$.

Example 1.28 Let $\mathscr{H}_1 = (C_1, \xi_1)$ and $\mathscr{H}_2 = (C_2, \xi_2)$ be two complex fuzzy hypergraphs on $X_1 = \{x_1, x_2, x_3\}$ and $X_2 = \{y_1, y_2\}$, respectively. The corresponding complex fuzzy hypergraphs are given in Fig. 1.42a and b, respectively.

Then, the direct product (preserving the minimal crisp rank) of $\mathscr{H}_1$ and $\mathscr{H}_2$ is given in Fig. 1.43.

Remark 1.7 For the minimal crisp rank preserving direct product $\mathscr{H}_1 \Diamond \mathscr{H}_2$ holds

(i) max $|supp(\xi_1 \Diamond \xi_2)| = \min\{|supp(\xi_1)|, |supp(\xi_1)|\}$,
(ii) min $|supp(\xi_1 \Diamond \xi_2)| = \min\{|supp(\xi_1)|, |supp(\xi_1)|\}$.

Theorem 1.8 *The (minimal crisp rank preserving) direct product $\mathcal{H}_1 \diamond \mathcal{H}_2$ of simple complex fuzzy hypergraphs is simple.*

Proof Let $\mathcal{H}_1$ and $\mathcal{H}_2$ be two simple complex fuzzy hypergraphs. Then, $\max |supp(\rho_j)| \geq 2$, $\max |supp(\rho_k)| \geq 2$, for $\rho_j \in \xi_1$ and $\rho_k \in \xi_2$, therefore, $\max |supp(\xi_1 \diamond \xi_2)| = \min\{|supp(\xi_1)|, |supp(\xi_1)|\} \geq 2$. This implies that no loops are contained in $E(\mathcal{H}_1 \diamond \mathcal{H}_2)$. Suppose that there exists an edge $\gamma_1 \in \xi_1 \diamond \xi_2$ contained in some other edge $\gamma_2 \in \xi_1 \diamond \xi_2$, i.e., $\gamma_1 \subseteq \gamma_2$. Note that $\sigma_j(\gamma_1) \subseteq \sigma_j(\gamma_2)$, $j = 1, 2$ holds. Suppose that $|supp(\alpha_1)| \leq |supp(\alpha_2)|$, which implies that $\sigma_1(\gamma_1) = \alpha_1$. Then, we have $\alpha_1 = \sigma_1(\gamma_1) \subseteq \sigma_1(\gamma_2) \subseteq \alpha_1^*$. Since, $\mathcal{H}_1$ is simple, $\alpha_1 = \alpha_1^*$ must satisfied and hence $\sigma_1(\gamma_2) = \alpha_1^*$. This implies that $|supp(\gamma_1)| = |supp(\alpha_1)| = |supp(\alpha_1^*)| = |\sigma_1(\gamma_2)| = |\gamma_2|$ and hence $\gamma_1 = \gamma_2$.

Lemma 1.11 *The 2-section of the direct product $\mathcal{H}_1 \diamond \mathcal{H}_2$ is the direct product of 2-section of $\mathcal{H}_1$ and the 2-section of $\mathcal{H}_2$, i.e., $[\mathcal{H}_1 \diamond \mathcal{H}_2]_2 = [\mathcal{H}_1]_2 \diamond [\mathcal{H}_2]_2$.*

Proof Let σ_1 and σ_2 denote the $\sigma_{\mathcal{H}_1}$ and $\sigma_{\mathcal{H}_2}$, respectively. Definitions 1.83 and 1.73 imply that $[\mathcal{H}_1 \diamond \mathcal{H}_2]_2$ and $[\mathcal{H}_1]_2 \diamond [\mathcal{H}_2]_2$ both have same set of vertices. Thus, it is enough to show that the identity mapping given by $V([\mathcal{H}_1 \diamond \mathcal{H}_2]_2) \to V([\mathcal{H}_1]_2 \diamond [\mathcal{H}_2]_2)$ is an isomorphism. Since, we have $\{x_1, x_2\} \in E([\mathcal{H}_1 \diamond \mathcal{H}_2]_2)$ if and only if there exists an edge $e \in E(\mathcal{H}_1]_2 \diamond [\mathcal{H}_2)$ such that $\{x_1, x_2\} \subseteq e$. Also $x_1 \neq x_2$ if and only if there exist $\rho_1 \in \xi_1$ and $\rho_2 \in \xi_2$ such that $\{\sigma_1(x_1), \sigma_1(x_2)\} \subseteq \sigma_1(e) \subseteq \rho_1, \sigma_1(x_1) \neq \sigma_1(x_2)$ and $\{\sigma_2(x_1), \sigma_2(x_2)\} \subseteq \sigma_2(e) \subseteq \rho_2, \sigma_2(x_1) \neq \sigma_2(x_2)$ if and only if $\{\sigma_1(x_1), \sigma_1(x_2)\} \in E(\mathcal{H}_1]_2), \{\sigma_2(x_1), \sigma_2(x_2)\} \in E(\mathcal{H}_2]_2) \Leftrightarrow \{x_1, x_2\} \in E([\mathcal{H}_1]_2 \diamond [\mathcal{H}_2]_2)$. Hence, the result is proved.

Definition 1.84 Let $\mathcal{H}_1 = (C_1, \xi_1)$ and $\mathcal{H}_2 = (C_2, \xi_2)$ be two complex fuzzy hypergraphs on $X_1 = \{x_1, x_2, \cdots, x_m\}$ and $X_2 = \{y_1, y_2, \cdots, y_n\}$, where $C_1 = \{\alpha_{11}, \alpha_{12}, \cdots, \alpha_{1j}\}$ and $C_2 = \{\alpha_{21}, \alpha_{22}, \cdots, \alpha_{2k}\}$ are complex fuzzy sets on X_1 and X_2, respectively. The *maximal crisp rank preserving direct product* $\mathcal{H}_1 \otimes \mathcal{H}_2 = (C_1 \otimes C_2, \xi_1 \otimes \xi_2)$ of $\mathcal{H}_1$ and $\mathcal{H}_2$, having the vertex set $X_1 \times X_2$ is defined as

(i) $C_1 \otimes C_2(x_p, y_q) = \min\{M_{\alpha_{1r}}(x_p), M_{\alpha_{2s}}(y_q)\}e^{i \min\{\psi_{\alpha_{1r}}(x_p), \psi_{\alpha_{2s}}(y_q)\}}$, for all $(x_p, y_q) \in X_1 \times X_2$,

(ii) $\xi_1 \otimes \xi_2(\{(x_1, y_1), (x_2, y_2), \cdots, (x_m, y_n)\}) = \min\{M_{\xi_1}(\{x_1, x_2, \cdots x_m\}), \min M_{\alpha_{2s}}(y_q)\} \times e^{i \min\{\psi_{\xi_1}(\{x_1, x_2, \cdots x_m\}), \min M_{\alpha_{2s}}(y_q)\}}$, for all $\{x_1, x_2, \cdots x_m\} \in E(\mathcal{H}_1)$, there is an edge $e \in E(\mathcal{H}_2)$ such that $\{y_1, y_2, \cdots y_n\}$ is a multiset of elements of e and $e \subseteq \{y_1, y_2, \cdots y_n\}$,

(iii) $\xi_1 \otimes \xi_2(\{(x_1, y_1), (x_2, y_2), \cdots, (x_m, y_n)\}) = \min\{\min M_{\alpha_{1r}}(x_p), M_{\xi_2}(\{y_1, y_2, \cdots y_n\})\} \times e^{i \min\{\min \psi_{\alpha_{1r}}(x_p), \psi_{\xi_2}(\{y_1, y_2, \cdots y_n\})\}}$, for all $\{y_1, y_2, \cdots y_n\} \in E(\mathcal{H}_2)$, there is an edge $f \in E(\mathcal{H}_1)$ such that $\{x_1, x_2, \cdots x_m\}$ is a multiset of elements of f and $f \subseteq \{x_1, x_2, \cdots x_m\}$,

where $p = 1, 2, \cdots, m$, $q = 1, 2, \cdots, n$, $r = 1, 2, \cdots, j$, and $s = 1, 2, \cdots, k$.

Fig. 1.44 Direct product $\mathscr{H}_1 \otimes \mathscr{H}_2$ of $\mathscr{H}_1$ and $\mathscr{H}_2$

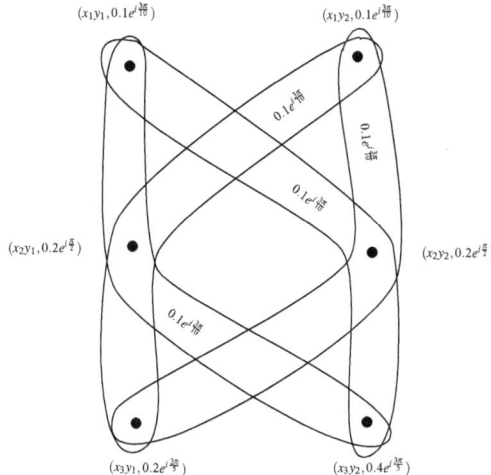

Example 1.29 Let $\mathscr{H}_1 = (C_1, \xi_1)$ and $\mathscr{H}_2 = (C_2, \xi_2)$ be two complex fuzzy hypergraphs on $X_1 = \{x_1, x_2, x_3\}$ and $X_2 = \{y_1, y_2\}$, respectively. The corresponding complex fuzzy hypergraphs are given in Fig. 1.42a and b, respectively. The maximal crisp rank preserving direct product $\mathscr{H}_1 \otimes \mathscr{H}_2$ is given in Fig. 1.44.

Remark 1.8 For the maximal crisp rank preserving direct product $\mathscr{H}_1 \otimes \mathscr{H}_2$ holds,

(i) $\max |supp(\xi_1 \otimes \xi_2)| = \max\{|supp(\xi_1)|, |supp(\xi_1)|\}$,
(ii) $\min |supp(\xi_1 \otimes \xi_2)| = \max\{|supp(\xi_1)|, |supp(\xi_1)|\}$.

Theorem 1.9 *The (maximal crisp rank preserving) direct product $\mathscr{H}_1 \otimes \mathscr{H}_2$ of simple complex fuzzy hypergraphs is simple.*

Proof Let $\mathscr{H}_1$ and $\mathscr{H}_2$ be two simple complex fuzzy hypergraphs. Then, $\max |supp(\rho_j)| \geq 2$, $\max |supp(\rho_k)| \geq 2$, for $\rho_j \in \xi_1$ and $\rho_k \in \xi_2$, therefore, $\max |supp(\xi_1 \lozenge \xi_2)| = \max\{|supp(\xi_1)|, |supp(\xi_1)|\} \geq 2$. This implies that no loops are contained in $E(\mathscr{H}_1 \otimes \mathscr{H}_2)$. Suppose that there exists an edge $\gamma_1 \in \xi_1 \otimes \xi_2$ contained in some other edge $\gamma_2 \in \xi_1 \otimes \xi_2$, i.e., $\gamma_1 \subseteq \gamma_2$. Note that $\sigma_j(\gamma_1) \subseteq \sigma_j(\gamma_2)$, $j = 1, 2$ holds. Suppose that $|supp(\alpha_1)| \leq |supp(\alpha_2)|$, which implies that $\sigma_1(\gamma_1) = \alpha_1$. Then, we have $\alpha_1 = \sigma_1(\gamma_1) \subseteq \sigma_1(\gamma_2) \subseteq \alpha_1^*$. Since, $\mathscr{H}_1$ is simple, $\alpha_1 = \alpha_1^*$ must be satisfied and hence $\sigma_1(\gamma_2) = \alpha_1^*$. This implies that $|supp(\gamma_1)| = |supp(\alpha_1)| = |supp(\alpha_1^*)| = |\sigma_1(\gamma_2)| = |\gamma_2|$ and hence $\gamma_1 = \gamma_2$.

1.9 A Complex Fuzzy Hypergraph Model of Co-authorship Network

Co-authorship is a form of association in which two or more researchers jointly report their research results on some topics. Therefore, *co-authorship networks* can be viewed as social networks encompassing researchers that reflect collaboration among them. Analysis and representation of publications and research articles in the form of a graph network is the most common methodology to illustrate and evaluate the scientific output of a group of researchers. In a co-authorship network, vertices are considered as authors and these authors are connected together if they have published one or more papers together. However, co-authorship networks that are constructed in this way may not fully illustrate the scientific outputs. On the other hand, the networks are constructed by taking vertices as papers and these vertices are connected by hyperedges if they have the same author are more useful in describing the scientific outputs. In this section, we propose a complex fuzzy hypergraph model to represent the publication data by considering the research articles as the vertices of hypergraph. The proposed model includes the information such as number of research articles, number of co-authors as well as collaborations. The authors are represented by hyperedges connecting the papers of corresponding authors. Then, we propose the collaboration measure of authors that illustrate the influence of corresponding author over the publications of their respective co-authors.

Consider a complex fuzzy hypergraph $\mathscr{H} = (C, \xi)$ model of collaborative network in which articles are represented as vertices and these vertices are connected by hyperedges if they have a common author. The corresponding complex fuzzy hypergraph model is shown in Fig. 1.45 in which we have considered nine research articles as the vertices of this hypergraph and six hyperedges representing the authors that have written more than two articles.

Fig. 1.45 Complex fuzzy hypergraph model of co-authorship network

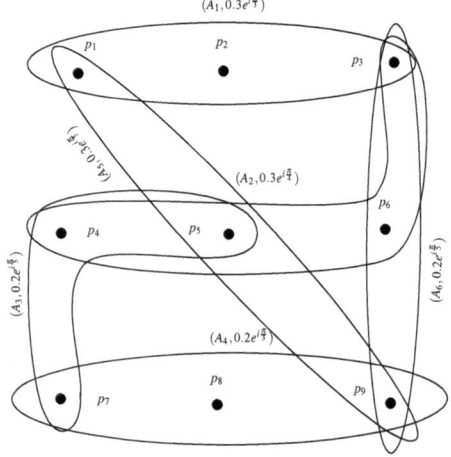

Table 1.33 Complex fuzzy sets on X

$p \in X$	α_1	α_2	α_3
p_1	$0.3e^{i\frac{\pi}{2}}$	$0.4e^{i\frac{2\pi}{3}}$	$0.5e^{i\frac{3\pi}{2}}$
p_2	$0.4e^{i\frac{\pi}{2}}$	$0.5e^{i\frac{2\pi}{4}}$	$0.4e^{i\frac{3\pi}{4}}$
p_3	$0.5e^{i\frac{\pi}{3}}$	$0.4e^{i\frac{2\pi}{5}}$	$0.6e^{i\frac{3\pi}{2}}$
p_4	$0.6e^{i\frac{\pi}{3}}$	$0.3e^{i\frac{2\pi}{6}}$	$0.7e^{i\frac{3\pi}{5}}$
p_5	$0.7e^{i\frac{\pi}{4}}$	$0.5e^{i\frac{2\pi}{7}}$	$0.8e^{i\frac{3\pi}{2}}$
p_6	$0.8e^{i\frac{\pi}{4}}$	$0.3e^{i\frac{2\pi}{8}}$	$0.4e^{i\frac{3\pi}{7}}$
p_7	$0.2e^{i\frac{\pi}{5}}$	$0.7e^{i\frac{2\pi}{9}}$	$0.5e^{i\frac{3\pi}{2}}$
p_8	$0.3e^{i\frac{\pi}{2}}$	$0.5e^{i\frac{2\pi}{6}}$	$0.3e^{i\frac{3\pi}{9}}$
p_9	$0.4e^{i\frac{\pi}{5}}$	$0.4e^{i\frac{2\pi}{5}}$	$0.8e^{i\frac{3\pi}{2}}$

Note that we have six authors $\{A_1, A_2, A_3, A_4, A_5, A_6\}$, represented by the hyperedges of complex fuzzy hypergraph that have published more than two research articles and nine published papers $\{p_1, p_2, p_3, p_4, p_5, p_6, p_7, p_8, p_9\}$. Let $C = \{\alpha_1, \alpha_2, \alpha_3\}$ be the family of complex fuzzy sets on $X = \{p_1, p_2, p_3, p_4, p_5, p_6, p_7, p_8, p_9\}$ as shown in Table 1.33.

The membership degrees $M \in [0, 1]$ of each research article represent contribution of corresponding article to the relative field and the phase angle $\psi \in [0, 2\pi]$ represents the specific period of time in which the corresponding contribution is considered. For example, membership degrees $0.3e^{i\frac{\pi}{2}}, 0.4e^{i\frac{2\pi}{3}}$, and $0.5e^{i\frac{3\pi}{2}}$ illustrate that the contribution of research article p_1 is $0.3, 0.4$, and 0.5 during the corresponding time periods and so on. The complex fuzzy relation ξ representing the authors of corresponding papers is defined as

$$\xi(\{p_1, p_2, p_3\}) = (A_1, 0.3e^{i\frac{\pi}{3}}), \quad \xi(\{p_3, p_4, p_5, p_6\}) = (A_2, 0.3e^{i\frac{\pi}{4}}),$$
$$\xi(\{p_4, p_5, p_7\}) = (A_3, 0.2e^{i\frac{\pi}{5}}), \quad \xi(\{p_7, p_8, p_9\}) = (A_4, 0.2e^{i\frac{\pi}{3}}),$$
$$\xi(\{p_1, p_5, p_9\}) = (A_5, 0.3e^{i\frac{\pi}{5}}), \quad \xi(\{p_3, p_6, p_9\}) = (A_6, 0.2e^{i\frac{\pi}{5}}).$$

A complex fuzzy hypergraph model provides a better illustration to evaluate the collaboration measure of an author by analyzing the condition when we remove the corresponding author from our network. That is removal of an influential author from a co-authorship network will strongly affect its structure. Let $\mathcal{H} = (C, \xi)$ be a complex fuzzy hypergraph model of co-authorship and $A_j \in E$ be the hyperedge representing author j. Then, $n(A_j, A_k) = |supp(A_j) \cap supp(A_k)|$ represents the number of research articles co-authored by A_j and A_k. The magnitude of collaboration measure Π_c of an author relative to the complex fuzzy hypergraph $\mathcal{H} = (C, \xi)$ is defined as

$$\Pi_c(A|\mathscr{H}) = (1 - \frac{\sum\limits_{A_j,A_k \in E \backslash A} \mu_{p_l \in supp(A_j) \cap supp(A_k)}(p_l)}{\sum\limits_{A_j,A_k \in E} \mu_{p_l \in supp(A_j) \cap supp(A_k)}(p_l)}) \times \frac{\sum\limits_{A_j,A_k \in E \backslash A} \theta_{p_l \in supp(A_j) \cap supp(A_k)}(p_l)}{\sum\limits_{A_j,A_k \in E} \theta_{p_l \in supp(A_j) \cap supp(A_k)}(p_l)},$$

where $\mu_{p_l \in n(A_j,A_k)}(p_l) = \min\limits_q M_{\alpha_q}(p_l), \theta_{p_l \in n(A_j,A_k)}(p_l) = \min\limits_q \psi_{\alpha_q}(p_l), l = 1, 2, \cdots, 9.$
The magnitude of collaboration measure illustrates the influence of removal of an author from the network. A higher value of $\Pi_c(A|\mathscr{H})$ indicates that the removal of author A will change the structure of co-authorship network more and may influence possible collaborations among its co-authors. Here, we have considered only absolute value of collaboration measure because we cannot compare the complex values to check the influence of authors in corresponding co-authorship networks. Note that

$$supp(A_1) \cap supp(A_2) = \{p_1\}, \quad supp(A_1) \cap supp(A_3) = \emptyset,$$
$$supp(A_1) \cap supp(A_4) = \emptyset, \quad supp(A_2) \cap supp(A_3) = \{p_4, p_5\},$$
$$supp(A_2) \cap supp(A_4) = \emptyset, \quad supp(A_3) \cap supp(A_4) = \{p_7\},$$
$$supp(A_1) \cap supp(A_5) = \{p_1\}, \quad supp(A_2) \cap supp(A_5) = \{p_5\},$$
$$supp(A_3) \cap supp(A_5) = \{p_5\}, \quad supp(A_4) \cap supp(A_5) = \{p_9\},$$
$$supp(A_6) \cap supp(A_5) = \{p_9\}, \quad supp(A_6) \cap supp(A_1) = \emptyset,$$
$$supp(A_6) \cap supp(A_2) = \{p_6\}, \quad supp(A_6) \cap supp(A_3) = \emptyset,$$
$$supp(A_6) \cap supp(A_4) = \{p_9\}.$$

By routine calculations, we find the magnitudes of collaboration measure of all authors as follows:

$$\Pi_c(A_1|\mathscr{H}) = (1 - \frac{0.3 + 0.5 + 0.2}{0.3 + 0.3 + 0.5 + 0.3 + 0.2 + 0.4})e^{i\frac{\pi}{5}} = 0.5e^{i\frac{\pi}{5}},$$

$$\Pi_c(A_2|\mathscr{H}) = (1 - \frac{0.5 + 0.2 + 0.4}{0.3 + 0.3 + 0.5 + 0.3 + 0.2 + 0.4})e^{i\frac{\pi}{4}} = 0.45e^{i\frac{\pi}{4}},$$

$$\Pi_c(A_3|\mathscr{H}) = (1 - \frac{0.3 + 0.5 + 0.3 + 0.4}{0.3 + 0.3 + 0.5 + 0.3 + 0.2 + 0.4})e^{i\frac{\pi}{3}} = 0.25e^{i\frac{\pi}{3}},$$

$$\Pi_c(A_4|\mathscr{H}) = (1 - \frac{0.3 + 0.3 + 0.5 + 0.3 + 0.4}{0.3 + 0.3 + 0.5 + 0.3 + 0.2 + 0.4})e^{i\frac{\pi}{5}} = 0.1e^{i\frac{\pi}{5}},$$

$$\Pi_c(A_5|\mathscr{H}) = (1 - \frac{0.3 + 0.3 + 0.5 + 0.3 + 0.2 + 0.4}{0.3 + 0.3 + 0.5 + 0.3 + 0.2 + 0.4})e^{i\frac{\pi}{5}} = 0.0e^{i\frac{\pi}{5}},$$

$$\Pi_c(A_6|\mathscr{H}) = (1 - \frac{0.3 + 0.3 + 0.5 + 0.3 + 0.2 + 0.4}{0.3 + 0.3 + 0.5 + 0.3 + 0.2 + 0.4})e^{i\frac{\pi}{5}} = 0.0e^{i\frac{\pi}{5}}.$$

By considering the maximum membership degree and minimum phase angle of collaboration measure, we find out the most influential author in the co-authorship network. The maximum membership value and minimum phase angle $0.5e^{i\frac{\pi}{5}}$ of researcher A_1 indicate that if we remove author A_1 from the structure, it will effect the performance of overall network within a short period of time. Thus, A_1 is the most influential author of under consideration co-authorship network. Similarly, we note that the removal of A_5 and A_6 will affect the structure of this network fewer. The collaboration measure can also be calculated by considering a partial complex fuzzy hypergraph, which illustrates the influence of an author relative to a specific group of research articles. The procedure adopted in our application is described in the following Algorithm 1.9.1.

Algorithm 1.9.1 A complex fuzzy hypergraph model for representing scientific outputs
Input: Complex fuzzy hypergraph model of co-authorship network.
Output: The most influential author of co-authorship network.

1. Input the set of research articles (vertices) $\{p_1, p_2, \cdots, p_r\}$.
2. Combine these research articles (vertices) through an hyperedge if they have a common author.
3. Input the set of authors (hyperedges) $A = \{A_1, A_2, \cdots, A_s\}$.
4. Define complex fuzzy sets $\{\alpha_1, \alpha_2, \cdots, \alpha_t\}$ on the set of vertices $\{p_1, p_2, \cdots, p_r\}$.
5. Calculate the membership degrees of authors using the formula

$$M_A(\{p_1, p_2, \cdots, p_l\})e^{i\psi_A(\{p_1, p_2, \cdots, p_l\})} \leq \min\{M_{\alpha_i}(p_1), M_{\alpha_i}(p_2), \cdots, M_{\alpha_i}(p_l)\}$$

$$\times e^{i \min\{\psi_{\alpha_i}(p_1), \psi_{\alpha_i}(p_2), \cdots, \psi_{\alpha_i}(p_l)\}}.$$

6. Find the research articles which are co-authored by A_j and A_k as $supp(A_j) \cap supp(A_k) = \{p_1, p_2, \cdots, p_m\}$.
7. Compute the collaboration measures of all authors A_j's relative to the complex fuzzy hypergraph $\mathscr{H}$ as follows:

$$\Pi_c(A|\mathscr{H}) = (1 - \frac{\sum\limits_{A_j, A_k \in E \setminus A} \mu_{p_l \in supp(A_j) \cap supp(A_k)}(p_l)}{\sum\limits_{A_j, A_k \in E} \mu_{p_l \in supp(A_j) \cap supp(A_k)}(p_l)}) \times \frac{\sum\limits_{A_j, A_k \in E \setminus A} \theta_{p_l \in supp(A_j) \cap supp(A_k)}(p_l)}{\sum\limits_{A_j, A_k \in E} \theta_{p_l \in supp(A_j) \cap supp(A_k)}(p_l)}.$$

8. Find out the most influential author of co-authorship network having the maximum membership degree and minimum phase angle of collaboration measures, i.e., if

$$\Pi_c(A|\mathscr{H}) = \max\{M_{\Pi_c}(A_j|\mathscr{H})\}e^{i \min \psi_{\Pi_c}(A_j|\mathscr{H})},$$

$j = 1, 2, \cdots, s$, then A is the most influential author and removing the hyperedge of A will change the structure of the network more.

1.9.1   Flow Chart of Proposed Model

The flow chart describing the procedure of our proposed model is given in Fig. 1.46.

Fig. 1.46 Flow chart of our
proposed model

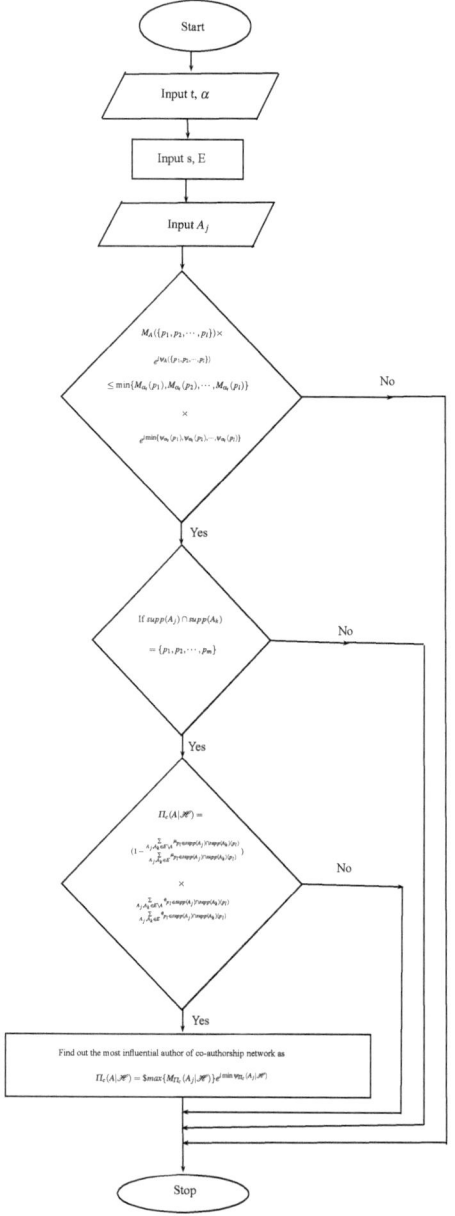

Fig. 1.47 Hypergraph
model of the paper network
representing three authors

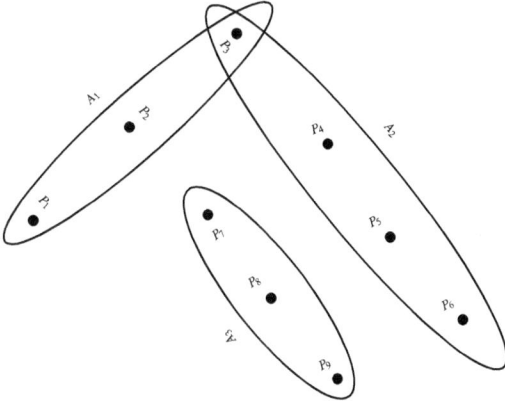

1.9.2 Comparative Analysis

To represent and analyze publication data in the form of a network has become a
common method of illustrating and evaluating the scientific output of a group or of
a scientific field. A hypergraph model illustrates co-authorship data in a simple and
elegant manner. In [13], authors have proposed the use of a crisp hypergraph model to
represent publication data by considering papers as hypergraph nodes. Hyperedges,
connecting the nodes, represent the authors connecting all their papers. In this work,
the number of authors is equal to the number of hyperedges. The degree of its
hyperedge represents the number of papers written by an author. The number of
co-authors is represented by the number of intersecting hyperedges. The strength of
the collaboration is represented by the number of nodes in the intersecting hyperedges. A hypergraph model of co-authorship network represented in [13] is given in
Fig. 1.47.

This model illustrates the number of papers written by authors A_1, A_2, A_3. For
example, hyperedge A_1 shows that author A_1 has written three research papers. Now,
if someone wants to evaluate the worth of these research articles or the collaboration degrees of authors toward their collective papers, then this model fails to fully
illustrate a co-authorship network.

To handle fuzziness in hypergraph models, we use fuzzy hypergraphs. A fuzzy
hypergraphs model of co-authorship network not only describe the number of papers
written by some author but also explain the level of contribution of that author.
This model fails in some circumstances when we have to analyze the co-authorship
networks for a fixed interval of time. For example, a fuzzy hypergraphs model illustrates the worth of research articles and the collaboration degrees of authors toward
their collective papers using membership degrees of vertices and hyperedges, respectively, but it does not tell us about that interval of time for which the corresponding
co-authorship network is being analyzed.

Fig. 1.48 Complex fuzzy
graph model of
co-authorship network

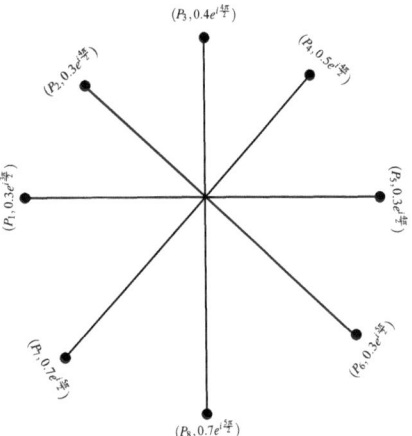

A complex fuzzy graph model is used to represent a simple co-authorship network describing the worth of research articles as well as specific time period. This network is constructed by taking the vertices as articles and two vertices (or articles) are connected through an edge if they have the same author as shown in Fig. 1.48. The membership degrees and phase terms represent the publication data and some specific time interval for which we analyze the co-authorship network, respectively.

The disadvantage of considering a complex fuzzy graph model for co-authorship networks is that it connects only two articles if they have same author. The structure of network model based on a complex fuzzy graph does not reflect the number of papers written by an author and cannot illustrate the information completely if more than two papers are written by the same author as represented by a hyperedge.

To overcome such type of difficulties and deficiencies occurring in crisp hypergraphs, fuzzy hypergraphs, and complex fuzzy graphs models of co-authorship networks, we have constructed a co-authorship network by considering the vertices of a complex fuzzy hypergraph as research articles and these articles are combined through complex fuzzy hyperedges if they have same author. We have generalized the work of [13] using complex fuzzy hypergraph model of co-authorship network. Thus, a complex fuzzy hypergraph is a more generalized framework to deal with vagueness having periodic nature in hypernetworks when the relationships are more generalized rather than the pair-wise interactions.

References

1. Abdul-Jabbar, N., Naoom, J.H., Ouda, E.H.: Fuzzy dual graph. J. Al-Nahrain Univ. **12**(4), 168–171 (2009)
2. Akram, M.: Studies in Fuzziness and Soft Computing. Fuzzy Lie algebras, vol. 9, pp. 1–302. Springer, Berlin (2018)

3. Akram, M., Alshehri, N.O.: Tempered interval-valued fuzzy hypergraphs. Sci. Bull. Ser. A Appl. Math. Phys. **77**(1), 39–48 (2015)
4. Akram, M., Chen, W.J., Davvaz, B.: On $\mathcal{N}$-hypergraphs. J. Intell. Fuzzy Syst. **26**, 2937–2944 (2014)
5. Berge, C.: Graphs and Hypergraphs. North-Holland, Amsterdam (1973)
6. Craine, M.W.L.: Fuzzy hypergraphs and fuzzy intersection graphs. University of Idaho, Ph.D Thesis (1993)
7. Goetschel Jr., R.H.: Introduction to fuzzy hypergraphs and Hebbian structures. Fuzzy Sets Syst. **76**, 113–130 (1995)
8. Goetschel Jr., R.H.: Fuzzy colorings of fuzzy hypergraphs. Fuzzy Sets Syst. **94**, 185–204 (1998)
9. Goetschel Jr., R.H., Craine, W.L., Voxman, W.: Fuzzy transversals of fuzzy hypergraphs. Fuzzy Sets Syst. **84**, 235–254 (1996)
10. Jun, Y.B., Lee, K.J., Song, S.Z.: $\mathcal{N}$-ideals of BCK/BCI-algebras. J. Chungcheong Math. Soc. **22**, 417–437 (2009)
11. Kaufmann, A.: Introduction a la Thiorie des Sous-Ensemble Flous, 1. Masson, Paris (1977)
12. Lee-kwang, H., Lee, K.-M.: Fuzzy hypergraph and fuzzy partition. IEEE Trans. Syst. Man Cybern. **25**(1), 196–201 (1995)
13. Lung, R.I., Gasko, N., Suciu, M.A.: A hypergraph model for representing scientific output. Scientometrics **117**(3), 1361–1379 (2018)
14. Mendel, J.M.: Uncertain Rule-based Fuzzy Logic Systems: Introduction and New Directions. Prentice-Hall, Upper Saddle River, New Jersey (2001)
15. Mordeson, J.N., Chang-Shyh, P.: Operations on fuzzy graphs. Inf. Sci. **79**, 159–170 (1994)
16. Mordeson, J.N.: Fuzzy line graph. Pattern Recognit. Lett. **14**, 381–384 (1993)
17. Mordeson, J.N., Nair, P.S.: Fuzzy Graphs and Fuzzy Hypergraphs, 2nd edn. Physica Verlag, Heidelberg (2001)
18. Mordeson, J.N., Yao, Y.Y.: Fuzzy cycles and fuzzy trees. J. Fuzzy Math. **10**, 189–202 (2002)
19. Radhamani, C., Radhika, C.: Isomorphism on fuzzy hypergraphs. IOSR J. Math. **2**(6), 24–31 (2012)
20. Radhika, C., Radhamani, C., Suresh, S.: Isomorphism properties on strong fuzzy hypergraphs. Int. J. Comput. Appl. **64**(1)(2013)
21. Ramot, D., Milo, R., Friedman, M., Kandel, A.: Complex fuzzy sets. IEEE Trans. Fuzzy Syst. **10**(2), 171–186 (2002)
22. Rosenfeld, A.: Fuzzy graphs. In: Zadeh, L.A., Fu, K.S., Shimura, M. (eds.) Fuzzy Sets and their Applications, pp. 77–95. Academic Press, New York (1975)
23. Samanta, S., Akram, M., Pal, M.: m-Step fuzzy competition graphs. J. Appl. Math. Comput. **47**, 461–472 (2015)
24. Samanta, S., Pal, M.: Fuzzy k-competition graphs and p-competition fuzzy graphs. Fuzzy Inf. Eng. **5**, 191–204 (2013)
25. Sarwar, M., Akram, M., Alshehri, N.O.: A new method to decision-making with fuzzy competition hypergraphs. Symmetry **10**(9), 404 (2018). https://doi.org/10.3390/sym10090404
26. Sonntag, M., Teichert, H.M.: Competition hypergraphs. Discr. Appl. Math. **143**, 324–329 (2004)
27. Sonntag, M., Teichert, H.M.: Competition hypergraphs of digraphs with certain properties II, Hamiltonicity. Discuss. Math. Graph Theory **28**, 23–34 (2008)
28. Thirunavukarasu, P., Suresh, R., Viswanathan, K.K.: Energy of a complex fuzzy graph. Int. J. Math. Sci. Eng. Appl. **10**(1), 243–248 (2016)
29. Yazdanbakhsh, O., Dick, S.: A systematic review of complex fuzzy sets and logic. Fuzzy Sets Syst. **338**, 1–22 (2018)
30. Zadeh, L.A.: Fuzzy sets. Inf. Control **8**(3), 338–353 (1965)
31. Zadeh, L.A.: Similarity relations and fuzzy orderings. Inf. Sci. **3**(2), 177–200 (1971)
32. Zadeh, L.A.: The concept of a linguistic and application to approximate reasoning-I. Inf. Sci. **8**, 199–249 (1975)

Chapter 2
Hypergraphs in Intuitionistic Fuzzy Environment

In this chapter, we define intuitionistic fuzzy hypergraphs, dual intuitionistic fuzzy hypergraphs, intuitionistic fuzzy line graphs, and 2-section of an intuitionistic fuzzy hypergraph. We describe some applications of intuitionistic fuzzy hypergraphs in planet surface networks, intersecting communities in social network, grouping of incompatible chemical substances, and clustering problem. We design certain algorithms to construct dual intuitionistic fuzzy hypergraphs, intuitionistic fuzzy line graphs and the selection of objects in decision-making problems. Further, we present concept of intuitionistic fuzzy directed hypergraphs and complex intuitionistic fuzzy hypergraphs. This chapter is basically due to [2, 3, 6, 12, 14, 15, 17, 18].

2.1 Introduction

Presently, science and technology is featured with complex processes and phenomena for which complete information are not always available. For such cases, mathematical models are developed to handle various types of systems containing elements of uncertainty. A large number of these models is based on an extension of the ordinary set theory, namely, fuzzy sets. The notion of fuzzy sets was introduced by Zadeh [25] as a method of representing uncertainty and vagueness. Since then, the theory of fuzzy sets has become a vigorous area of research in different disciplines, including medical and life sciences, management sciences, social sciences, engineering, statistics, graph theory, artificial intelligence, signal processing, multiagent systems, pattern recognition, robotics, computer networks, expert systems, decision making, and automata theory. In 1983, Atanassov [5] introduced the concept of intuitionistic fuzzy sets as a generalization of fuzzy sets. Atanassov added in the definition of fuzzy set a new component which determines the degree of nonmembership. Fuzzy sets give the degree of membership of an element in a given set (the nonmembership of degree equals one minus the degree of membership), while intuitionistic fuzzy

© Springer Nature Singapore Pte Ltd. 2020
M. Akram and A. Luqman, *Fuzzy Hypergraphs and Related Extensions*,
Studies in Fuzziness and Soft Computing 390,
https://doi.org/10.1007/978-981-15-2403-5_2

sets give both a degree of membership and a degree of nonmembership, which are more or less independent of each other; the only requirement is that the sum of these two degrees is not greater than 1. Intuitionistic fuzzy sets are higher order fuzzy sets. Application of higher order fuzzy sets makes the solution-procedure more complex, but if the complexity in computation-time, computation-volume or memory-space are not the matter of concern then a better result could be achieved. Fuzzy sets and intuitionistic fuzzy sets cannot handle imprecise, inconsistent, and incomplete information of periodic nature. These theories are applicable to different areas of science, but there is one major deficiency in both sets, that is, a lack of capability to model two-dimensional phenomena. To overcome this difficulty, the concept of complex fuzzy sets was introduced by Ramot et al. [20]. A complex fuzzy set C is characterized by a membership function $\mu(x)$, whose range is not limited to [0, 1] but extends to the unit circle in the complex plane. Hence, $\mu(x)$ is a complex-valued function that assigns a grade of membership of the form $r(x)e^{i\alpha(x)}$, $i = \sqrt{-1}$ to any element x in the universe of discourse. Thus, the membership function $\mu(x)$ of complex fuzzy set consists of two terms, i.e., amplitude term $r(x)$ which lies in the unit interval [0, 1] and phase term (periodic term) $w(x)$ which lies in the interval [0, 2π]. This phase term distinguishes a complex fuzzy set model from all other models available in the literature. The potential of a complex fuzzy set for representing two-dimensional phenomena makes it superior to handle ambiguous and intuitive information that are prevalent in time-periodic phenomena. To generalize the concepts of intuitionistic fuzzy sets, complex intuitionistic fuzzy sets were introduced by Alkouri and Salleh [4] by adding nonmembership $v(x) = s(x)e^{i\beta(x)}$ to the complex fuzzy sets subjected to the constraint $r + s \leq 1$.

Graph theory has numerous applications to problems in computer science, electrical engineering, system analysis, operations research, economics, networking routing, and transportation. However, in many cases, some aspects of a graph-theoretic problem may be uncertain. For example, the vehicle travel time or vehicle capacity on a road network may not be known exactly. In such cases, it is natural to deal with the uncertainty using the methods of fuzzy sets and fuzzy logic. Graphs are used to represent the pairwise relationships between objects. However, in many real world phenomena, sometimes relationships are much problematic that they cannot be perceived through simple graphs. By handling such complex relationships by pairwise connections naively, one can face the loss of data which is considered to be worthwhile for learning errands. To overcome these difficulties, we take into account the generalization of simple graphs, named as hypergraphs, to personify the complex relationships. A hypergraph is an extension of a classical graph in this way that a hyperedge can combine two or more than two vertices. Hypergraphs are the generalization of graphs in case of set of multiary relations. It means the expansion of graph models for the modeling complex systems. In case of modeling systems with fuzzy binary and multiary relations between objects, transition to fuzzy hypergraphs, which combine advantages both fuzzy and graph models, is more natural. It allows to realize formal optimization and logical procedures. However, using of the fuzzy graphs and hypergraphs as the models of various systems (social, economic systems, communication networks, and others) leads to difficulties. The graph isomorphic transformations are reduced to redefine the vertices and edges. This redefinition does not change properties the graph determined by an adjacent and an incidence of its vertices and edges. Fuzzy independent sets, domination fuzzy sets, and fuzzy chromatic sets are invariants concerning the isomorphism transformations of the fuzzy graphs and fuzzy hypergraph and allow their structural analysis. Kaufamnn [10] applied the concept of fuzzy sets to hypergraphs. Mordeson and Nair [13] presented fuzzy graphs and fuzzy hypergraphs. Generalization and redefinition of fuzzy hypergraphs were discussed by Lee-Kwang and Lee [11]. The concept of interval-valued fuzzy sets was applied to hypergraphs by Chen [8]. Parvathi et al. [17] established the notion of intuitionistic fuzzy hypergraph, and Myithili and Parvathi [14], Myithili et al. [15] considered intuitionistic fuzzy directed hypergraphs.

Definition 2.1 A mapping $A = (\mu_A, \nu_A) : X \to [0, 1] \times [0, 1]$ is called an *intuitionistic fuzzy set* on X if $\mu_A(x) + \nu_A(x) \leq 1$, for all $x \in X$, where the mappings $\mu_A : X \to [0, 1]$, and $\nu_A : X \to [0, 1]$ denote the *degree of membership* (namely $\mu_A(x)$) and the *degree of nonmembership* (namely $\nu_A(x)$) of each element $x \in X$ to A, respectively.

An intuitionistic fuzzy set A in X can be represented as an object of the form

$$A = (\mu_A, \nu_A) = \{(x, \mu_A(x), \nu_A(x)) \mid x \in X\},$$

where the functions $\mu_A : X \to [0, 1]$ and $\nu_A : X \to [0, 1]$ denote the *degree of membership* (namely $\mu_A(x)$) and the *degree of nonmembership* (namely $\nu_A(x)$) of the element $x \in X$, respectively, and for all $x \in X, 0 \leq \mu_A(x) + \nu_A(x) \leq 1$. Obviously, each fuzzy set maybe written as

$$A = \{(x, \mu_A(x), 1 - \mu_A(x)) \mid x \in X\}.$$

The value

$$\pi_A(x) = 1 - \mu_A(x) - \nu_A(x) \tag{2.1}$$

is called *uncertainty (intuitionistic index)* of the elements $x \in X$ to the intuitionistic fuzzy set A. It represents *hesitancy degree* of x to A.

Clearly, in the case of ordinary fuzzy set, $\pi_A(x) = 0$, for all $x \in X$.

Geometrical Interpretations of an Intuitionistic Fuzzy Set [5]

A geometrical interpretation of an intuitionistic fuzzy set is shown in Fig. 2.1. Atanassov considered a universe X and subset F in the Euclidean plane with the Cartesian coordinates.

This geometrical interpretation can be used as an example when considering a situation at the beginning of negotiations (applications of intuitionistic fuzzy sets for group decision-making, negotiations and other real situations are presented in Fig. 2.2). Each expert i is represented as a point having coordinates $\langle \mu_i, \nu_i, \pi_i \rangle$. Expert

Fig. 2.1 A geometrical
interpretation of an
intuitionistic fuzzy set

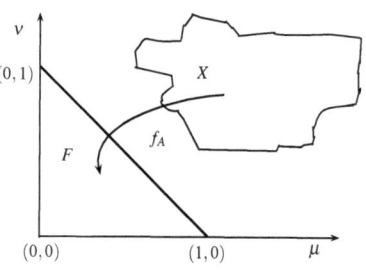

Fig. 2.2 An orthogonal
projection (three dimension)
representation of an
intuitionistic fuzzy set

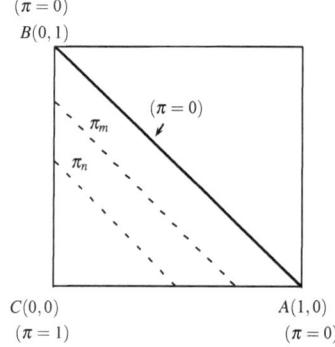

$A : \langle 1, 0, 0 \rangle$—fully accepts a discussed idea. Expert $B : \langle 0, 1, 0 \rangle$—fully rejects it. The experts placed on the segment AB fixed their point of view (their hesitation margins equal zero for segment AB, so each expert is convinced to the extent μ_i, is against to the extent v_i and $\mu_i + v_i = 1$; segment AB represents a fuzzy set). Expert $C : \langle 0, 0, 1 \rangle$ is absolutely hesitant, i.e., undecided he or she is the most open to the influence of the arguments presented. A line parallel to AB describe a set of experts with the same level of hesitancy. For example, in Fig. 2.2, two sets are presented with intuitionistic indices equal to π_m and π_n, where $\pi_n > \pi_m$. In other words, Fig. 2.2 (the triangle ABC) is an orthogonal projection of the real situation (the triangle ABD) presented in Fig. 2.3.

An element of an intuitionistic fuzzy sets has three coordinates $\langle \mu_i, v_i, \pi_i \rangle$, hence the most natural representation of an intuitionistic fuzzy set is to draw a cube (with edge length equal to 1) and because of Eq. (2.1), the triangle ABD (Fig. 2.3) represents an intuitionistic fuzzy set. As before (Fig. 2.2), the triangle ABC is the orthogonal projection of ABD.

Definition 2.2 Let $A = (\mu_A, v_A)$ and $B = (\mu_B, v_B)$ be intuitionistic fuzzy sets on a set X. If $A = (\mu_A, v_A)$ is an intuitionistic fuzzy relation on a set X, then $A = (\mu_A, v_A)$ is called an *intuitionistic fuzzy relation* on $B = (\mu_B, v_B)$ if $\mu_A(x, y) \leq \min(\mu_B(x), \mu_B(y))$ and $v_A(x, y) \leq \max(v_B(x), v_B(y))$, for all $x, y \in X$. An intuitionistic fuzzy relation A on X is called *symmetric* if $\mu_A(x, y) = \mu_A(y, x)$ and $v_A(x, y) = v_A(y, x)$, for all $x, y \in X$.

Fig. 2.3 A three-dimension representation of an intuitionistic fuzzy set

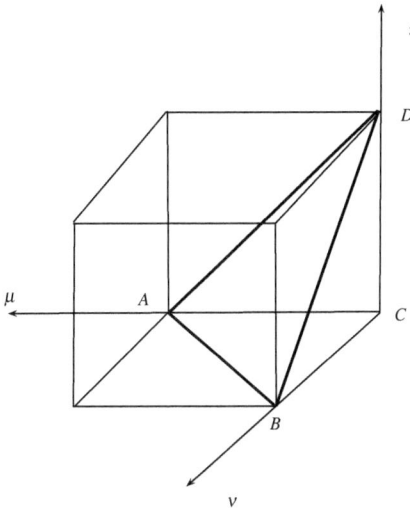

Definition 2.3 The *support* of an intuitionistic fuzzy set $A = (\mu_A, \nu_A)$, denoted by supp(A), is defined by

$$\text{supp}(A) = \{x \mid \mu_A(x) \neq 0 \text{ and } \nu_A(x) \neq 0\}.$$

The support of the intuitionistic fuzzy set is a crisp set.

Definition 2.4 Let $A = (\mu_A, \nu_A)$ be an intuitionistic fuzzy set on X and let $\alpha, \beta \in [0, 1]$ such that $\alpha + \beta \leq 1$. Then, the set $A_{(\alpha,\beta)} = \{x \mid \mu_A(x) \geq \alpha, \ \nu_A(x) \leq \beta\}$ is called an (α, β)-*level subset* of A. $A_{(\alpha,\beta)}$ is a crisp set.

Definition 2.5 The *height* of an intuitionistic fuzzy set $A = (\mu_A, \nu_A)$ is defined as $h(A) = \sup_{x \in X}(A)(x) = (\sup_{x \in X} \mu_A(x), \inf_{x \in X} \nu_A(x))$. We shall say that intuitionistic fuzzy set A is *normal* if there is at least one $x \in X$ such that $\mu_A(x) = 1$.

Definition 2.6 An *intuitionistic fuzzy graph* on X is defined as a pair $\mathscr{G} = (C, D)$, where C is an intuitionistic fuzzy set on X and D is an intuitionistic fuzzy relation in X such that $\lambda_D(yz) \leq \min\{\lambda_C(y), \lambda_C(z)\}$ and $\tau_D(yz) \leq \max\{\tau_C(y), \tau_C(z)\}$, for all $y, z \in X$.

For further terminologies and studies on intuitionistic fuzzy hypergraphs, readers are referred to [1, 7, 9, 16, 19, 21, 22, 24].

2.2 Intuitionistic Fuzzy Hypergraphs

Definition 2.7 An intuitionistic fuzzy hypergraph on a non-empty set X is a pair $\mathscr{H} = (S, R)$ where, $S = \{\eta_1, \eta_2, \ldots, \eta_s\}$ is a family of intuitionistic fuzzy subsets

Table 2.1 Intuitionistic fuzzy subsets on X

$y \in X$	η_1	η_2	η_3	η_4	η_5
a_1	$(0.2, 0.4)$	$(0, 1)$	$(0, 1)$	$(0, 1)$	$(0, 1)$
a_2	$(0.3, 0.5)$	$(0, 1)$	$(0, 1)$	$(0.3, 0.5)$	$(0, 1)$
a_3	$(0, 1)$	$(0.4, 0.6)$	$(0.4, 0.6)$	$(0, 1)$	$(0, 1)$
a_4	$(0.1, 0.3)$	$(0, 1)$	$(0, 1)$	$(0, 1)$	$(0, 1)$
a_5	$(0, 1)$	$(0.3, 0.1)$	$(0, 1)$	$(0, 1)$	$(0, 1)$
a_6	$(0, 1)$	$(0, 1)$	$(0.9, 0.1)$	$(0, 1)$	$(0, 1)$
a_7	$(0, 1)$	$(0, 1)$	$(0.5, 0.4)$	$(0.2, 0.8)$	$(0, 1)$
a_8	$(0, 1)$	$(0, 1)$	$(0, 1)$	$(0, 1)$	$(0.5, 0.3)$

on X and R is an intuitionistic fuzzy relation on intuitionistic fuzzy subsets η_i's such that

1. $\lambda_R(E_i) = \lambda_R(\{y_1, y_2, \ldots, y_r\}) \leq \min\{\lambda_{\eta_i}(y_1), \lambda_{\eta_i}(y_2), \ldots, \lambda_{\eta_i}(y_r)\}$,
2. $\tau_R(E_i) = \tau_R(\{y_1, y_2, \ldots, y_r\}) \leq \max\{\tau_{\eta_i}(y_1), \tau_{\eta_i}(y_2), \ldots, \tau_{\eta_i}(y_r)\}$,
3. $\lambda_R(E_i) + \tau_R(E_i) \leq 1$,   for each $E_i \subset X$,
4. $\bigcup_i supp(\eta_i) = X$, for all $\eta_i \in S$.

Example 2.1 Let $S = \{\eta_1, \eta_2, \eta_3, \eta_4, \eta_5\}$ be a family of intuitionistic fuzzy subsets on $X = \{a_1, a_2, \ldots, a_8\}$ as shown in Table 2.1.

The intuitionistic fuzzy relation R on each $\eta_i, 1 \leq i \leq 5$, is given as $R(\{a_1, a_2, a_4\}) = (0.1, 0.5)$, $R(\{a_3, a_5\}) = (0.3, 0.6)$, $R(\{a_3, a_6, a_7\}) = (0.4, 0.6)$, $R(\{a_2, a_7\}) = (0.2, 0.8)$, and $R(\{a_8\}) = \eta_5(a_8)$. It is clear From Fig. 2.4 that $\mathscr{H}$ is an intuitionistic fuzzy hypergraph.

Example 2.2 Consider another example of an intuitionistic fuzzy hypergraph consisting of nine vertices $X = \{a_1, a_2, \ldots, a_9\}$ and two hyperedges E_1, E_2. The membership values of vertices are given in (Fig. 2.5) and the membership values of hyperedges are $R(\{a_1, a_2, a_3, a_4, a_9\}) = (0.3, 0.6)$ and $R(\{a_5, a_6, a_7, a_8, a_9\}) = (0.2, 0.5)$. The corresponding intuitionistic fuzzy hypergraph in shown in Fig. 2.6.

Definition 2.8 An intuitionistic fuzzy set $C = (\mu_A, \nu_A) : X \rightarrow [0, 1] \times [0, 1]$ is an *elementary intuitionistic fuzzy set* if A is single valued on $supp(A)$. An intuitionistic fuzzy hypergraph $\mathscr{H} = (S, R)$ is *elementary* if each $\eta_i \in A$ and R are elementary otherwise, it is called *nonelementary*.

Proposition 2.1 *Intuitionistic fuzzy graphs are special cases of the intuitionistic fuzzy hypergraphs.*

An *intuitionistic fuzzy multigraph* is a multivalued symmetric mapping $D = (\mu_D, \nu_D) : V \times V \rightarrow [0, 1]$. An intuitionistic fuzzy multigraph can be considered to be the "disjoint union" or "disjoint sum" of a collection of simple intuitionistic fuzzy graphs, as is done with crisp multigraphs. The same holds for multidigraphs. Therefore, these structures can be considered as "disjoint unions" or "disjoint sums" of intuitionistic fuzzy hypergraphs.

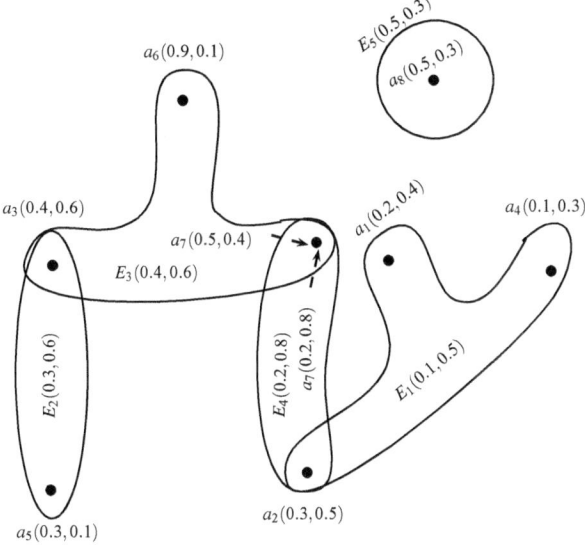

Fig. 2.4 Intuitionistic fuzzy hypergraph

Fig. 2.5 Table of
intuitionistic fuzzy subsets
on X

$y \in X$	η_1	η_2
a_1	$(0.5, 0.4)$	$(0, 1)$
a_2	$(0.6, 0.3)$	$(0, 1)$
a_3	$(0.4, 0.6)$	$(0, 1)$
a_4	$(0.3, 0.5)$	$(0, 1)$
a_5	$(0, 1)$	$(0.6, 0.3)$
a_6	$(0, 1)$	$(0.4, 0.3)$
a_7	$(0, 1)$	$(0.2, 0.4)$
a_8	$(0, 1)$	$(0.4, 0.3)$
a_9	$(0.5, 0.5)$	$(0.5, 0.5)$

Definition 2.9 An intuitionistic fuzzy hypergraph $\mathscr{H} = (S, R)$ is called *simple* if
every $\eta_i, \eta_j \in S$, $\eta_i \subseteq \eta_j$ implies that $\eta_i = \eta_j$.

An intuitionistic fuzzy hypergraph $\mathscr{H} = (S, R)$ is called *support simple* if every
$\eta_i, \eta_j \in S$, $\eta_i \subseteq \eta_j$, and $supp(\eta_i) = supp(\eta_j)$ imply that $\eta_i = \eta_j$.

An intuitionistic fuzzy hypergraph $\mathscr{H} = (S, R)$ is called *support simple* if every
$\eta_i, \eta_j \in S$, $\eta_i \subseteq \eta_j$, and $supp(\eta_i) = supp(\eta_j)$ imply that $\eta_i = \eta_j$.

$\mathscr{H} = (S, R)$ is called *strongly support simple* if every $\eta_i, \eta_j \in S$, $supp(\eta_i) =
supp(\eta_j)$ imply that $\eta_i = \eta_j$.

Remark 2.1 Definition 2.9 reduces to familiar definitions in the special case where
H is a crisp hypergraph. The definition of simple intuitionistic fuzzy hypergraph is
identical to the definition of simple crisp hypergraph. A crisp hypergraph is support
simple and strongly support simple if and only if it has no multiple edges. For
intuitionistic fuzzy hypergraphs all three concepts imply no multiple edges. Any

simple intuitionistic fuzzy hypergraph is support simple and every strongly support simple intuitionistic fuzzy hypergraph is support simple. Simple and strongly support simple are independent concepts in intuitionistic fuzziness.

Definition 2.10 Let $\mathcal{H} = (S, R)$ be an intuitionistic fuzzy hypergraph on X. For $\alpha, \beta \in [0, 1]$, $0 \leq \alpha + \beta \leq 1$, the (α, β)-*level hyperedge* of an intuitionistic fuzzy hyperedge η is defined as

$$\eta_{(\alpha,\beta)} = \{u \in X | \lambda_\eta(u) \geq \alpha, \tau_\eta(u) \leq \beta\}.$$

$\mathcal{H}_{(\alpha,\beta)} = (S_{(\alpha,\beta)}, R_{(\alpha,\beta)})$ is called an (α, β)-*level hypergraph* of $\mathcal{H}$ where, $S_{(\alpha,\beta)}$ is defined as $S_{(\alpha,\beta)} = \cup_{k=1}^{r} \eta_{k(\alpha,\beta)}$.

Definition 2.11 Let $\mathcal{H} = (S, R)$ be an intuitionistic fuzzy hypergraph. The sequence of order pairs $(\alpha_i, \beta_i) \in [0, 1] \times [0, 1]$, $0 \leq \alpha_i + \beta_i \leq 1$, $1 \leq i \leq n$, such that $\alpha_1 > \alpha_2 > \cdots > \alpha_n$, $\beta_1 < \beta_2 < \cdots < \beta_n$ satisfying the properties

1. if $1 \geq \alpha > \alpha_1$ and $0 \leq \beta < \beta_1$ then $R_{(\alpha,\beta)} = \emptyset$,
2. if $\alpha_{i+1} < \alpha \leq \alpha_i$ and $\beta_i \leq \beta < \beta_{i+1}$ then $R_{(\alpha,\beta)} = R_{(\alpha_i,\beta_i)}$,
3. $R_{(\alpha_i,\beta_i)} \sqsubset R_{(\alpha_{i+1},\beta_{i+1})}$,

is called *fundamental sequence* of $\mathcal{H}$, denoted by $f_s(\mathcal{H})$. The corresponding sequence of (α_i, β_i)-level hypergraphs $\mathcal{H}_{(\alpha_1,\beta_1)}, \mathcal{H}_{(\alpha_2,\beta_2)}, \ldots, \mathcal{H}_{(\alpha_n,\beta_n)}$ is called *core set* of $\mathcal{H}$, denoted by $C(\mathcal{H})$. The (α_n, β_n)-level hypergraph, $\mathcal{H}_{(\alpha_n,\beta_n)}$, is called *support level* of $\mathcal{H}$.

Definition 2.12 An intuitionistic fuzzy hypergraph $\mathcal{H} = (S, R)$ is called a *partial intuitionistic fuzzy hypergraph* of $\mathcal{H}' = (S', R')$ if following conditions are satisfied

1. $supp(S) \subseteq supp(S')$ and $supp(R) \subseteq supp(R')$,
2. if $supp(\eta_i) \in supp(S)$ and $supp(\eta_i') \in supp(S')$ such that $supp(\eta_i) = supp(\eta_i')$ then $\eta_i = \eta_i'$.

It is denoted by $\mathcal{H} \subseteq \mathcal{H}'$. An intuitionistic fuzzy hypergraph $\mathcal{H} = (S, R)$ is *ordered* if the core set $C(\mathcal{H}) = \{\mathcal{H}_{(\alpha_1,\beta_1)}, \mathcal{H}_{(\alpha_2,\beta_2)}, \ldots, \mathcal{H}_{(\alpha_n,\beta_n)}\}$ is ordered, that is, $\mathcal{H}_{(\alpha_1,\beta_1)} \subseteq \mathcal{H}_{(\alpha_2,\beta_2)} \subseteq \ldots \subseteq \mathcal{H}_{(\alpha_n,\beta_n)}$. $\mathcal{H}$ is *simply ordered* if $\mathcal{H}$ is ordered and whenever, $R' \subset R_{(\alpha_{i+1},\beta_{i+1})} \setminus R_{(\alpha_i,\beta_i)}$ then $R' \nsubseteq R_{(\alpha_i,\beta_i)}$.

Observation 2.1 *Let $\mathcal{H}$ be an elementary intuitionistic fuzzy hypergraph then $\mathcal{H}$ is ordered. If $\mathcal{H}$ is ordered intuitionistic fuzzy hypergraph and support level $\mathcal{H}_{(\alpha_n,\beta_n)}$ is simple then $\mathcal{H}$ is an elementary intuitionistic fuzzy hypergraph.*

Definition 2.13 Let $\mathcal{H} = (S, R)$ and $\mathcal{H}' = (S', R')$ be any two intuitionistic fuzzy hypergraphs on X and X', respectively, where $S = \{\eta_1, \eta_2, \ldots, \eta_r\}$ and $S' = \{\eta_1', \eta_2', \ldots, \eta_r'\}$. A *homomorphism* between H and H' is a mapping $\psi : X \rightarrow X'$ such that

1. $\wedge_{i=1}^{s}\lambda_{\eta_i}(y) \le \wedge_{i=1}^{s}\lambda_{\eta_i'}(\psi(y))$,
2. $\vee_{i=1}^{s}\tau_{\eta_i}(y) \le \vee_{i=1}^{s}\tau_{\eta_i'}(\psi(y))$, for all $y \in X$,
3. $\lambda_R(\{y_1, y_2, \dots, y_s\}) \le \lambda_{R'}(\{\psi(y_1), \psi(y_2), \dots, \psi(y_s)\})$,
4. $\tau_R(\{y_1, y_2, \dots, y_s\}) \le \tau_{R'}(\{\psi(y_1), \psi(y_2), \dots, \psi(y_s)\})$, for all $y_1, y_2, \dots,$
   $y_s \in X$.

Definition 2.14 A *co-weak isomorphism* of two intuitionistic fuzzy hypergraphs $\mathcal{H}$ and $\mathcal{H}'$ is defined as a bijective homomorphism $\psi : X \to X'$ such that

1. $\lambda_R(\{y_1, y_2, \dots, y_s\}) = \lambda_{R'}(\{\psi(y_1), \psi(y_2), \dots, \psi(y_s)\})$,
2. $\tau_R(\{y_1, y_2, \dots, y_s\}) = \tau_{R'}(\{\psi(y_1), \psi(y_2), \dots, \psi(y_s)\})$, for all $y_1, y_2,$
   $\dots, y_s \in X$.

Definition 2.15 A *weak isomorphism* of two intuitionistic fuzzy hypergraphs $\mathcal{H}$ and $\mathcal{H}'$ is defined as a bijective homomorphism $\psi : X \to X'$ such that

1. $\wedge_{i=1}^{s}\lambda_{\eta_i}(y) = \wedge_{i=1}^{s}\lambda_{\eta_i'}(\psi(y))$,
2. $\vee_{i=1}^{s}\tau_{\eta_i}(y) = \vee_{i=1}^{s}\tau_{\eta_i'}(\psi(y))$, for all $y \in X$.

Definition 2.16 An *isomorphism* of $\mathcal{H}$ and $\mathcal{H}'$ is a mapping $\psi : X \to X'$ such that

1. $\wedge_{i=1}^{s}\lambda_{\eta_i}(y) = \wedge_{i=1}^{s}\lambda_{\eta_i'}(\psi(y))$,
2. $\vee_{i=1}^{s}\tau_{\eta_i}(y) = \vee_{i=1}^{s}\tau_{\eta_i'}(\psi(y))$, for all $y \in X$,
3. $\lambda_R(\{y_1, y_2, \dots, y_s\}) = \lambda_{R'}(\{\psi(y_1), \psi(y_2), \dots, \psi(y_s)\})$,
4. $\tau_R(\{y_1, y_2, \dots, y_s\}) = \tau_{R'}(\{\psi(y_1), \psi(y_2), \dots, \psi(y_s)\})$, for all $y_1, y_2,$
   $\dots, y_s \in X$.

Example 2.3 Assume that $S = \{\eta_1, \eta_2, \eta_3, \eta_4\}$ and $S' = \{\eta_1', \eta_2', \eta_3', \eta_4'\}$ are the families of intuitionistic fuzzy subsets on $X = \{a_1, a_2, \dots, a_6\}$ and $X' = \{a_1', a_2', \dots, a_6'\}$, respectively, as shown in Tables 2.2 and 2.3.

The intuitionistic fuzzy relations R and R' are defined as $R(\{a_1, a_3, a_4, a_6\}) = (0.1, 0.5)$, $R(\{a_1, a_2, a_3\}) = (0.2, 0.5)$, $D(\{a_3, a_4\}) = (0.5, 0.4)$, $R(\{a_4, a_5, a_6\}) = (0.1, 0.8)$ and $R'(\{a_1', a_2', a_3', a_6'\}) = (0.1, 0.5)$, $D'(\{a_1', a_3', a_4'\}) = (0.2, 0.5)$, $D'(\{a_1', a_2'\}) = (0.5, 0.4)$, $R'(\{a_2', a_5', a_6'\}) = (0.1, 0.8)$. The corresponding intuitionistic fuzzy hypergraphs are given in Figs. 2.7 and 2.8.

Table 2.2 Intuitionistic fuzzy subsets on X

$y \in X$	η_1	η_2	η_3	η_4
a_1	(0.2, 0.5)	(0.2, 0.5)	(0, 1)	(0, 1)
a_2	(0, 1)	(0.5, 0.4, 0.4)	(0, 1)	(0, 1)
a_3	(0.5, 0.4)	(0.5, 0.4)	(0.5, 0.4)	(0, 1)
a_4	(0.8, 0.1)	(0, 1)	(0.8, 0.2)	(0.8, 0.2)
a_5	(0, 0, 0, 0)	(0, 1)	(0, 1)	(0.1, 0.8)
a_6	(0.1, 0.2)	(0, 1)	(0, 1)	(0.1, 0.2)

Table 2.3 Intuitionistic fuzzy subsets on X'

$y' \in X'$	η_1'	η_2'	η_3'	η_4'
a_1'	(0.5, 0.4)	(0.5, 0.4)	(0.5, 0.4)	(0, 1)
a_2'	(0.8, 0.2)	(0, 1)	(0.8, 0.2)	(0.8, 0.2)
a_3'	(0.2, 0.5)	(0.2, 0.5)	(0, 1)	(0, 1)
a_4'	(0, 1)	(0.5, 0.4)	(0, 1)	(0, 1)
a_5'	(0, 1)	(0, 0, 1)	(0, 1)	(0.1, 0.8)
a_6'	(0.1, 0.2)	(0, 0, 1)	(0, 1)	(0.1, 0.2)

Fig. 2.6 Intuitionistic fuzzy
hypergraph

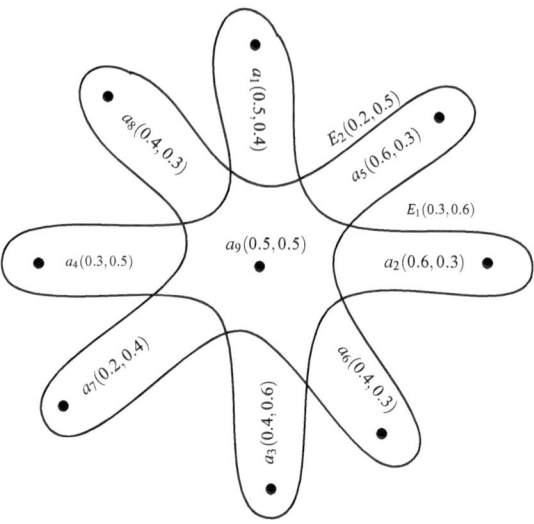

Define a mapping $\psi : X \to X'$ by $\psi(a_1) = a_3', \psi(a_2) = a_4', \psi(a_4) = a_2', \psi(a_3) = a_1', \psi(a_5) = a_5'$, and $\psi(a_6) = a_6'$ then, it can be easily seen that

$$\eta_1(a_1) = (0.2, 0.5) = \eta_1'(a_3') = \eta_1'(\psi(a_1)), \quad \eta_2(a_1) = (0.2, 0.5) = \eta_2'(a_3') = \eta_2'(\psi(a_1)),$$
$$\eta_2(a_2) = (0.5, 0.4) = \eta_2'(a_4') = \eta_1'(\psi(a_2)), \quad \eta_1(a_3) = (0.5, 0.4) = \eta_1'(a_1') = \eta_1'(\psi(a_3)).$$

Similarly, $\eta_i(y) = \eta_i'(\psi(y))$, and $R(\{y_1, y_2, \ldots, y_s\}) = R'(\{\psi(y_1), \psi(y_2), \ldots, \psi(y_s)\})$, for all $y, y_i \in X$. Therefore, $\mathscr{H}$ and $\mathscr{H}'$ are isomorphic.

Definition 2.17 The *order and size* of an intuitionistic fuzzy hypergraph $\mathscr{H} = (S, R)$ can be defined as

$$O(\mathscr{H}) = \sum_{y \in X} (\wedge_j \lambda_{\eta_j}(y), \vee_j \tau_{\eta_j}(y)), \quad S(\mathscr{H}) = \sum_{E_j \subset X} (\lambda_R(E_j), \tau_R(E_j)).$$

Fig. 2.7 Intuitionistic fuzzy hypergraph $\mathscr{H}$

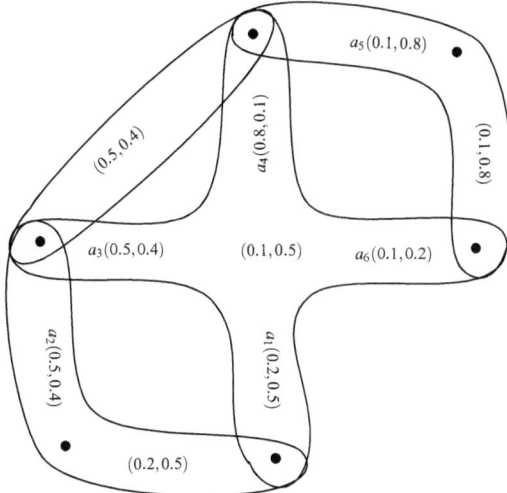

Theorem 2.2 *For any two isomorphic intuitionistic fuzzy hypergraphs, the order and size are same.*

Proof Let $\mathscr{H}_1 = (S_1, R_1)$ and $\mathscr{H}_2 = (S_2, R_2)$ be any two intuitionistic fuzzy hypergraphs where, $S = \{\eta_1, \eta_2, \ldots, \eta_s\}$ and $S' = \{\eta'_1, \eta_2, \ldots, \eta'_s\}$ are the families of intuitionistic fuzzy subsets on X and X', respectively. If $\psi : X \to X'$ is an isomorphism between $\mathscr{H}$ and $\mathscr{H}'$ then

$$O(\mathscr{H}) = \sum_{y \in X}(\wedge_i \lambda_{\eta_i}(y), \vee_i \tau_{\eta_i}(y))$$

$$= \sum_{y \in X}(\wedge_i \lambda_{\eta'_i}(\psi(y)), \vee_i \tau_{\eta'_i}(\psi(y))) = \sum_{y' \in X'}(\wedge_i \lambda_{\eta'_i}(y'), \vee_i \tau_{\eta'_i}(y')) = O(\mathscr{H}').$$

$$S(\mathscr{H}) = \sum_{E_i \subset X} R(E_i) = \sum_{E_i \subset X} R'(\psi(E_i)) = \sum_{E'_i \subset X'} R'(E'_i) = S(\mathscr{H}').$$

It completes the proof.

Remark 2.2 The converse of Theorem 2.2 does not hold, i.e., if the orders and sizes of two intuitionistic fuzzy hypergraphs are same then they may not be isomorphic as given in Example 2.4.

Example 2.4 Consider two intuitionistic fuzzy hypergraphs $\mathscr{H}_1 = (S_1, R_1)$ and $\mathscr{H}_2 = (S_2, R_2)$ given in Figs. 2.9 and 2.10. By Definition 2.17, $O(\mathscr{H}_1) = O(\mathscr{H}_2) = (1.8, 1.3)$ and $S(\mathscr{H}_1) = S(\mathscr{H}_2) = (0.4, 0.9)$. The orders and sizes of intuitionistic fuzzy hypergraphs $\mathscr{H}_1$ and $\mathscr{H}_2$ but $\mathscr{H}_1 \not\approx \mathscr{H}_2$.

Theorem 2.3 *The order of any two weak isomorphic intuitionistic fuzzy hypergraphs is same.*

Fig. 2.8 Intuitionistic fuzzy hypergraph $\mathcal{H}'$

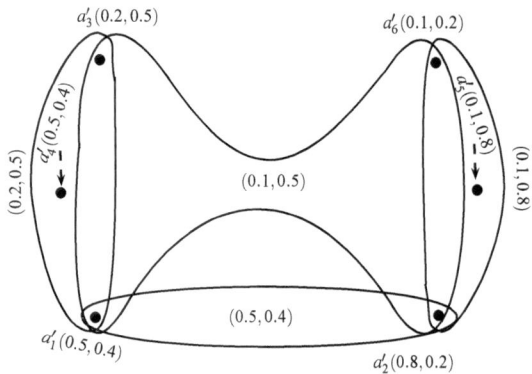

Fig. 2.9 Intuitionistic fuzzy hypergraph $\mathcal{H}_1$

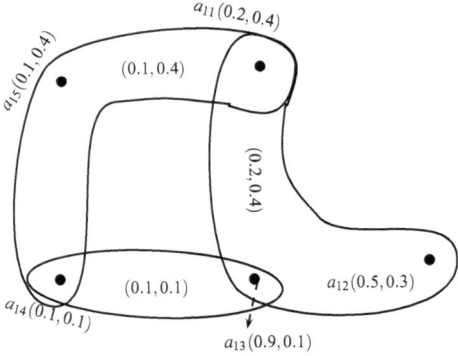

The proof follows from Definition 2.15 and the proof of Theorem 2.2.

Theorem 2.4 *The size of any two co-weak isomorphic intuitionistic fuzzy hypergraphs is same.*

The proof follows from Definition 2.14 and the proof of Theorem 2.2.

Remark 2.3 The intuitionistic fuzzy hypergraphs of same order and size may not be weak isomorphic and co-weak isomorphic, respectively, i.e., the converse of Theorems 2.3 and 2.4 is not true in general as shown in Example 2.5.

Example 2.5 Let $\mathcal{H}_1 = (S_1, R_1)$ and $H_2 = (S_2, R_2)$ be two intuitionistic fuzzy hypergraphs as shown in Figs. 2.11 and 2.12 where $R_1 = \{\eta_{11}, \eta_{12}, \eta_{13}\}$ and $R_2 = \{\eta_{21}, \eta_{22}, \eta_{23}\}$. Clearly, $O(\mathcal{H}_1) = O(\mathcal{H}_2) = (1.6, 1.7)$ and $S(\mathcal{H}_1) = S(\mathcal{H}_2) = (0.5, 1.3)$. Define a mapping $\psi : X_1 \to X_2$ by $\psi(u_1) = u_2$, $\psi(v_1) = v_2$, $\psi(x_1) = x_2$, $\psi(y_1) = y_2$, $\psi(z_1) = z_2$. But $\lambda_{\eta_{12}}(u_1) = 0.5 \nleq 0.2 = \lambda_{\eta_{22}}(u_2)$ so, $\mathcal{H}_1$ and $\mathcal{H}_2$ are not weak isomorphic. Similarly, $\lambda_{R_1}(\{v_1, y_1\}) = 0.2 \nleq \lambda_{R_2}(\{\psi(v_1), \psi(y_1)\}) = 0$. Hence, $\mathcal{H}_1$ and $\mathcal{H}_2$ are not co-weak isomorphic.

Definition 2.18 For any intuitionistic fuzzy hypergraph, the *degree* of a vertex y is defined as, $\deg(y) = \sum_{y \in E_i \subseteq X} R(E_i)$.

Fig. 2.10 Intuitionistic
fuzzy hypergraph $\mathcal{H}_2$

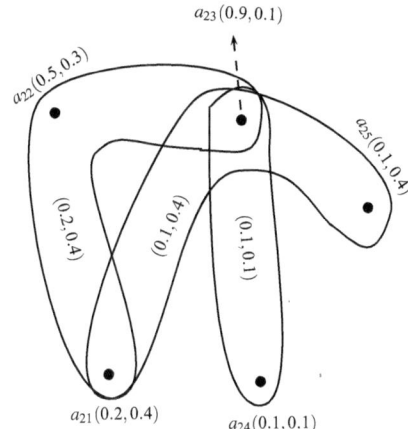

Fig. 2.11 Intuitionistic
fuzzy hypergraph $\mathcal{H}_1$

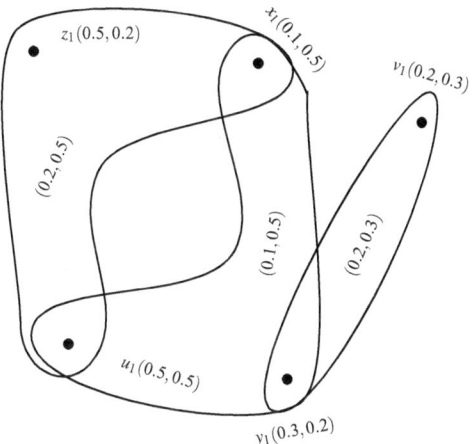

Theorem 2.5 *The degree of vertices of isomorphic intuitionistic fuzzy hypergraphs is preserved.*

Proof Let $\psi : X \to X'$ be an isomorphism of intuitionistic fuzzy hypergraphs $\mathcal{H}$ and $\mathcal{H}'$ where, $X = \{y_1, y_2, \ldots, y_n\}$ and $X' = \{y_1', y_2', \ldots, y_n'\}$. Then Definition 2.18 implies that

$$\deg(y_i) = \sum\nolimits_{y_i \in E_i \subseteq X} R(E_i) = \sum\nolimits_{y_i \in E_i} R'(\phi(E_i)) = \deg(\psi(y_i)).$$

Remark 2.4 If the degrees of vertices of any two intuitionistic fuzzy hypergraphs is preserved then they may not be isomorphic as it is proved in Example 2.6.

Fig. 2.12 Intuitionistic
fuzzy hypergraph $\mathscr{H}_2$

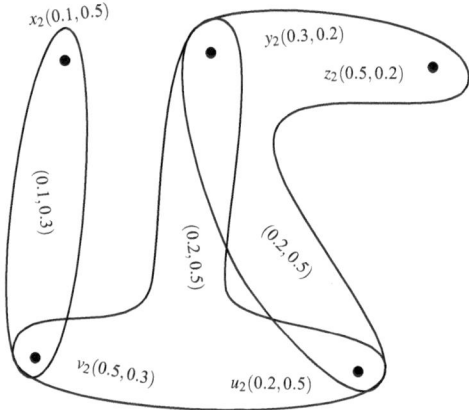

Fig. 2.13 Intuitionistic
fuzzy hypergraph $\mathscr{H}$

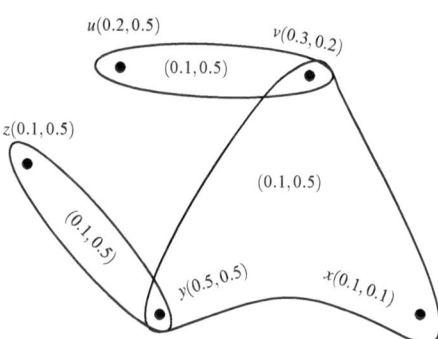

Example 2.6 Consider two intuitionistic fuzzy hypergraphs $\mathscr{H} = (S, R)$ and $\mathscr{H}' = (S', R')$ as given in Figs. 2.13 and 2.14. Define a mapping $\psi : X \to X'$ by $\psi(u) = u'$, $\psi(v) = x'$, $\psi(x) = v'$, $\psi(y) = y'$, $\psi(z) = z'$. Routine calculations show that

$$\deg(u) = (0.2, 0.5) = \deg(\phi(u)), \quad \deg(v) = (0.3, 1.0) = \deg(\phi(v)).$$

The degree of all other vertices is also preserved but $R(\{x, y, v\}) \neq R'(\{\psi(x), \psi(y), \psi(v)\})$. Hence, $\mathscr{H}$ and $\mathscr{H}'$ are not isomorphic to each other.

Theorem 2.6 *The relation of isomorphism between intuitionistic fuzzy hypergraphs is an equivalence relation.*

Proof Assume that $\mathscr{H}_1 = (S_1, R_1)$, $\mathscr{H}_2 = (S_2, R_2)$ and $\mathscr{H}_3 = (S_3, R_3)$ are intuitionistic fuzzy hypergraphs on X_1, X_2 and X_3, respectively, where $S_1 = \{\eta_{11}, \eta_{12}, \ldots, \eta_{1s}\}$, $S_2 = \{\eta_{21}, \eta_{22}, \ldots, \eta_{2s}\}$ and $S_3 = \{\eta_{31}, \eta_{32}, \ldots, \eta_{3s}\}$.

Fig. 2.14 Intuitionistic
fuzzy hypergraph $\mathscr{H}'$

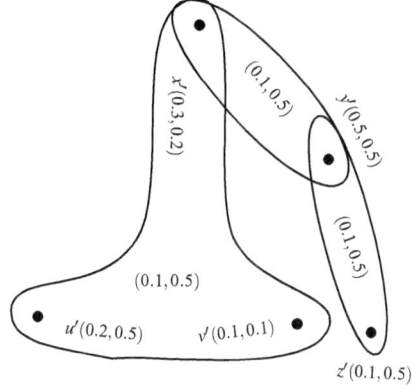

1. Reflexivity: Define an identity mapping $I : X_1 \rightarrow X_1$ by $I(y_1) = y_1$ for all $y_1 \in X_1$. Clearly I is bijective an
   $(\wedge_j \lambda_{\eta_{1j}}(y_1), \vee_j \tau_{\eta_{1j}}(y_1)) = (\wedge_j \lambda_{\eta_{1j}}(I(y_1)), \vee_j \tau_{\eta_{1j}}(I(y_1)))$ and $R_1(E_{1i}) = R_1(I(E_{1i}))$, for all $y_1 \in X_1$, $E_{1i} \subseteq X_1$.
   So, I is an isomorphism of an intuitionistic fuzzy hypergraph to itself.
2. Symmetry: Let $\psi : X_1 \rightarrow X_2$ be an isomorphism defined by $\psi(y_1) = y_2$. Since ψ is bijective therefore, the inverse bijective mapping $\psi^{-1} : X_2 \rightarrow X_1$ exists such that $\psi(y_2) = y_1$ for all $y_2 \in X_2$. Then

$$(\wedge_j \lambda_{\eta_{2j}}(y_2), \vee_j \tau_{\eta_{2j}}(y_2)) = (\wedge_j \lambda_{\eta_{2j}}(\psi(y_1)), \vee_j \tau_{\eta_{2j}}(\psi(y_1))),$$
$$= (\wedge_j \lambda_{\eta_{1j}}(y_1), \vee_j \tau_{\eta_{1j}}(y_1)),$$
$$= (\wedge_j \lambda_{\eta_{1j}}(\psi^{-1}(y_2)), \vee_j \tau_{\eta_{1j}}(\psi^{-1}(y_2))).$$
$$R_2(E_{2j}) = R_2(\psi(E_{1j})) = R_1(E_{1j}) = R_1(\psi^{-1}(E_{2j})), \quad E_{1j} \subseteq X_1, \; E_{2j} \subseteq X_2.$$

   Thus, ψ^{-1} is an isomorphism.
3. Assume that $\psi_1 : X_1 \rightarrow X_2$, $\psi_2 : X_2 \rightarrow X_3$ are isomorphisms of $\mathscr{H}_1$ onto $\mathscr{H}_2$ and $\mathscr{H}_2$ onto $\mathscr{H}_3$, respectively, such that $\psi_1(y_1) = y_2$ and $\psi_2(y_2) = y_3$. By Definition 2.16

$$\wedge_j \lambda_{\eta_{1j}}(y_1) = \wedge_j \lambda_{\eta_{2j}}(y_2) = \wedge_j \lambda_{\eta_{3j}}(\psi(y_2)) = \wedge_j \lambda_{\eta_{3j}}(\psi_2(\psi_1(y_1))) = \wedge_j \lambda_{\eta_{3j}}(\psi_2 \circ \psi_1(y_1)),$$

$$\vee_j \tau_{\eta_{1j}}(y_1) = \vee_j \tau_{\eta_{2j}}(y_2) = \vee_j \tau_{\eta_{3j}}(\psi(y_2)) = \vee_j \tau_{\eta_{3j}}(\psi_2(\psi_1(y_1))) = \vee_j \tau_{\eta_{3j}}(\psi_2 \circ \psi_1(y_1)),$$

$$R_1(E_{1j}) = R_2(E_{2j}) = R_3(\psi_2(E_{2j})) = R_3(\psi_2(\psi_1(E_{1j}))) = R_3(\psi_2 \circ \psi_1(E_{1j})).$$

   where, $E_{1j} \subseteq X_1$, $E_{2j} \subseteq X_2$ and $E_{3j} \subseteq X_3$. Clearly, $\psi_2 \circ \psi_1$ is an isomorphism from $\mathscr{H}_1$ onto $\mathscr{H}_3$. Hence, isomorphism of intuitionistic fuzzy hypergraphs is an equivalent relation.

Theorem 2.7 *The relation of weak isomorphism between intuitionistic fuzzy hypergraphs is a partial order relation.*

Fig. 2.15 Intuitionistic
fuzzy hypergraph $\mathscr{H}$

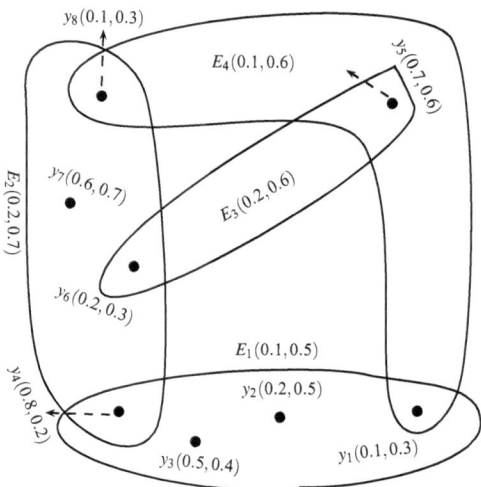

The proof follows from Definition 2.15 and the proof of Theorem 2.6.

Definition 2.19 An *intuitionistic fuzzy hyperpath* $\mathscr{P}$ of length m in an intuitionistic fuzzy hypergraph is defined as a sequence $y_1, E_1, y_2, E_2, \ldots, y_m, E_m, y_{m+1}$ of distinct vertices y_i's and hyperedges E_i's such that

1. $\lambda_R(E_i) > 0,$ for each $1 \le i \le m$,
2. $y_i, y_{i+1} \in E_i,$ for each $1 \le i \le m$.

If $y_m = y_{m+1}$ then, $\mathscr{P}$ is called an *intuitionistic fuzzy hypercycle*.

Example 2.7 Let $\mathscr{H} = (S, R)$ be an intuitionistic fuzzy hypergraph as shown in Fig. 2.15. The sequence $y_5, E_3, y_6, E_2, y_4, E_1, y_1, E_4, y_8$ is an intuitionistic fuzzy hyperpath and y_8, E_2, y_4, E_1, y_8 is an intuitionistic fuzzy hypercycle.

Definition 2.20 An intuitionistic fuzzy hypergraph $\mathscr{H} = (S, R)$ on a non-empty set X is *connected* if every two distinct vertices in $\mathscr{H}$ are joined by an intuitionistic fuzzy hyperpath.

Definition 2.21 Let y and z be two distinct vertices of an intuitionistic fuzzy hypergraph $\mathscr{H}$ which are joined by an intuitionistic fuzzy hyperpath $y = y_1, E_1, y_2, E_2, \ldots, y_p, E_p, y_{p+1} = z$ of length p. The *strength* of intuitionistic fuzzy hyperpath $y - z$ is denoted by $S^P(y, z) = (\lambda_{S^P}(y, z), \tau_{S^P}(y, z))$ and defined as,

$$\lambda_{S^P}(y, z) = \lambda_R(E_1) \wedge \lambda_R(E_2) \wedge \ldots \wedge \lambda_R(E_p),$$
$$\tau_{S^P}(y, z) = \tau_R(E_1) \vee \tau_R(E_2) \vee \ldots \vee \tau_R(E_p), u \in E_1, \ v \in E_p.$$

The *strength of connectedness* between y and z is denoted by $S^\infty(y, z) = (\lambda_{S^\infty}(y, z), \tau_{S^\infty}(y, z))$ and defined as

$$\lambda_{S^\infty}(y, z) = \sup_p \{\lambda_{S^p}(y, z) | p = 1, 2, \ldots\}, \quad \tau_{S^\infty}(y, z) = \inf_p \{\tau_{S^p}(y, z) | p = 1, 2, \ldots\}.$$

Theorem 2.8 *An intuitionistic fuzzy hypergraph $\mathscr{H}$ is connected if and only if $\lambda_{S^\infty}(y, z) > 0$, for all $y, z \in X$.*

Proof Suppose that $\mathscr{H}$ is a connected intuitionistic fuzzy hypergraph then for any two distinct vertices y and z, there exists an intuitionistic fuzzy hyperpath $y - z$ such that

$$\lambda_{S^p}(y, z) > 0 \Rightarrow \sup_p \{\lambda_{S^p}(y, z) | p = 1, 2, \ldots\} > 0 \Rightarrow \lambda_{S^\infty}(y, z) > 0.$$

We can prove the converse part on the same lines as above.

Definition 2.22 A *strong intuitionistic fuzzy hypergraph* on a non-empty set X is an intuitionistic fuzzy hypergraph $\mathscr{H} = (S, R)$ such that for all $E_i = \{y_1, y_2, \ldots, y_r\} \in E$,

$$R(E_i) = (\min_{j=1}^{r}[\wedge_i \lambda_{\eta_i}(y_j)], \max_{j=1}^{r}[\vee_i \tau_{\eta_i}(y_j)]).$$

Definition 2.23 A *complete intuitionistic fuzzy hypergraph* on a non-empty set X is an intuitionistic fuzzy hypergraph $\mathscr{H} = (S, R)$ such that for all $y_1, y_2, \ldots, y_r \in X$,

$$R(E_i) = (\min_{j=1}^{r}[\wedge_i \lambda_{\eta_i}(y_j)], \max_{j=1}^{r}[\vee_i \tau_{\eta_i}(y_j)]).$$

Theorem 2.9 *For any two intuitionistic fuzzy hypergraphs $\mathscr{H}_1$ and $\mathscr{H}_2$, $\mathscr{H}_1$ is connected if and only if $\mathscr{H}_2$ is connected.*

Proof Let $E_1 = \{E_{11}, E_{21}, \ldots, E_{r1}\}$ and $E_2 = \{E_{12}, E_{22}, \ldots, E_{r2}\}$ be the families of hyperedges if intuitionistic fuzzy hypergraphs $\mathscr{H}_1 = (S_1, R_1)$ and $\mathscr{H}_2 = (S_2, R_2)$, respectively. Assume that $\psi : X_1 \to X_2$ is an isomorphism of $\mathscr{H}_1$ onto $\mathscr{H}_2$ and that $\mathscr{H}_1$ is connected then

$$0 < \lambda_{S_1^\infty}(y_1, z_1) = \sup_p \{\wedge_{k=1}^{p} \lambda_{R_1}(E_{k1}), p = 1, 2, \ldots\}$$
$$= \sup_p \{\wedge_{k=1}^{p} \lambda_{R_2}(\psi(E_{k1})), p = 1, 2, \ldots\}$$
$$= \lambda_{S_2^\infty}(\phi(y_1), \phi(z_1)).$$

It follows that $\mathscr{H}_2$ is connected. The converse part can be proved similarly.

Theorem 2.10 *For any two intuitionistic fuzzy hypergraphs $\mathscr{H}_1$ and $\mathscr{H}_2$, $\mathscr{H}_1$ is strong if and only if $\mathscr{H}_2$ is strong.*

Proof Let $\mathscr{H}_1 = (S_1, R_1)$ and $\mathscr{H}_2 = (S_2, R_2)$ be the intuitionistic fuzzy hypergraphs as defined in Theorem 2.9. Assume that $\mathscr{H}_1$ is strong then

$$R_2(E_{i2}) = R_2(\psi(E_{i1})) = R_1(E_{i1}) = (\min_{j=1}^{r}[\wedge_j \lambda_{\eta_{ji}}(y_{j1})], \max_{j=1}^{r}[\vee_j \tau_{\eta_{ji}}(y_{j1})])$$

$$= (\min_{j=1}^{r}[\wedge_j \lambda_{\eta_{j2}}(\psi(y_{j1}))], \max_{j=1}^{r}[\vee_j \tau_{\eta_{j2}}(\psi(y_{j1}))]).$$

$$(2.2)$$

Equation 2.2 clearly shows that $\mathcal{H}_2$ is strong. Similarly, the converse part.

Definition 2.24 An *intuitionistic fuzzy line graph* of an intuitionistic fuzzy hypergraph $\mathcal{H} = (S, R)$ is a pair $L(\mathcal{H}) = (S_l, R_l)$ where, $S_l = R$ and two vertices E_i and E_k in $L(\mathcal{H})$ are connected by an edge if $|supp(\eta_i) \cap supp(\eta_k)| \geq 1$ where, $R(E_i) = \eta_i$ and $R(E_k) = \eta_k$. The membership values of sets of vertices and edges are defined as

1. $S_l(E_i) = R(E_i)$,
2. $R_l(E_i E_k) = (\lambda_R(E_i) \wedge \lambda_R(E_k), \tau_R(E_i) \vee \tau_R(E_k))$.

The method for the construction of an intuitionistic fuzzy line graph from an intuitionistic fuzzy hypergraph is explained in Algorithm 2.2.1.

Algorithm 2.2.1 The construction of an intuitionistic fuzzy line graph

1. Input the number of edges n of an intuitionistic fuzzy hypergraph $\mathcal{H} = (S, R)$.
2. Input the degrees of membership of the hyperedges $E_1, E_2, \ldots, E_s$.
3. Construct an intuitionistic fuzzy graph $L(\mathcal{H}) = (S_l, R_l)$ whose vertices are the s hyperedges $E_1, E_2, \ldots, E_s$ such that $S_l(E_i) = R(E_i)$.
4. If $|supp(\eta_i) \cap supp(\eta_k)| \geq 1$ then, draw an edge between E_i and E_k and $R_l(E_i E_k) = (\lambda_R(E_i) \wedge \lambda_R(E_k), \tau_R(E_i) \vee \tau_R(E_k))$.

Example 2.8 An example of an intuitionistic fuzzy hypergraph is shown in Fig. 2.16. The intuitionistic fuzzy line graph is constructed using Algorithm 2.2.1 and represented with dashed lines.

Definition 2.25 An intuitionistic fuzzy hypergraph is known as *linear intuitionistic fuzzy hypergraph* if

$$supp(\eta_i) \subseteq supp(\eta_k) \Rightarrow i = k \quad \text{and} \quad |supp(\eta_i) \cap supp(\eta_k)| \leq 1.$$

Theorem 2.11 *The intuitionistic fuzzy line graph $L(\mathcal{H})$ of an an intuitionistic fuzzy hypergraph $\mathcal{H}$ is connected if and only if $\mathcal{H}$ is connected.*

Proof Let $\mathcal{H} = (S, R)$ be a connected intuitionistic fuzzy hypergraph and $L(\mathcal{H}) = (S_l, R_l)$. Assume that E_i and E_k are two vertices in $L(\mathcal{H})$ such that $y_i \in E_i$, $y_k \in E_k$ and $y_i \neq y_k$. By Definition 2.20, there exists an intuitionistic fuzzy hyperpath $y_i, E_i, y_{i+1}, E_{i+1}, \ldots, E_k, y_k$ between y_i and y_k. Using Definition 2.21, we have

Fig. 2.16 Intuitionistic fuzzy line graph $L(\mathcal{H})$

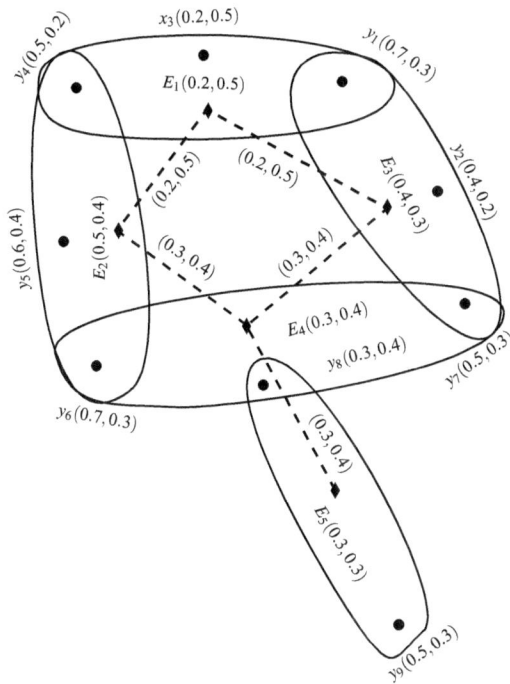

$$\lambda_{S\infty}(E_i, E_k) = \sup\{\lambda_{S^p}(E_i, E_k)|p = 1, 2, \ldots\}$$
$$= \sup\{\lambda_{R_l}(E_i E_{i+1}) \wedge \lambda_{R_l}(E_{i+1} E_{i+2}) \wedge \ldots \wedge \lambda_{R_l}(E_{k-1} E_k)|p = 1, 2, \ldots\}$$
$$= \sup\{\lambda_{R}(E_i) \wedge \lambda_{R}(E_{i+1}) \wedge \ldots \wedge \lambda_{R}(E_j)|p = 1, 2, \ldots\}$$
$$= \sup\{\lambda_{S^p}(y_i, y_k)|p = 1, 2, \ldots\} = \lambda_{S\infty}(y_i, y_k) > 0.$$

Hence, $L(\mathcal{H})$ is connected. Similarly, if $L(\mathcal{H})$ is connected then it can be easily proved that $\mathcal{H}$ is connected.

Definition 2.26 The 2-*section* of an intuitionistic fuzzy hypergraph $\mathcal{H} = (S, R)$ is denoted by $[\mathcal{H}]_2 = (S, U)$ and defined as an intuitionistic fuzzy graph whose set of vertices is same as $\mathcal{H}$ and U is an intuitionistic fuzzy set on $\{y_i y_k | y_i, y_k \in E_p, p = 1, 2, \ldots\}$, i.e., any two vertices of the same hyperedge are joined by an edge and

$$U(y_i y_k) = (\min\{\wedge_p \lambda_{\eta_p}(y_i), \wedge_p \lambda_{\eta_p}(y_k)\}, \max\{\vee_p \tau_{\eta_p}(y_i), \vee_p \tau_{\eta_p}(y_k)\}).$$

Example 2.9 An example of a 2-section of an intuitionistic fuzzy hypergraph is shown in Fig. 2.17. The 2-section of $\mathcal{H}$ is represented with dashed lines.

Definition 2.27 Let $\mathcal{H} = (S, R)$ be an intuitionistic fuzzy hypergraph on X then the *dual* of $\mathcal{H}$ is denoted by $\mathcal{H}^D = (S^D, R^D)$ and it is defined as

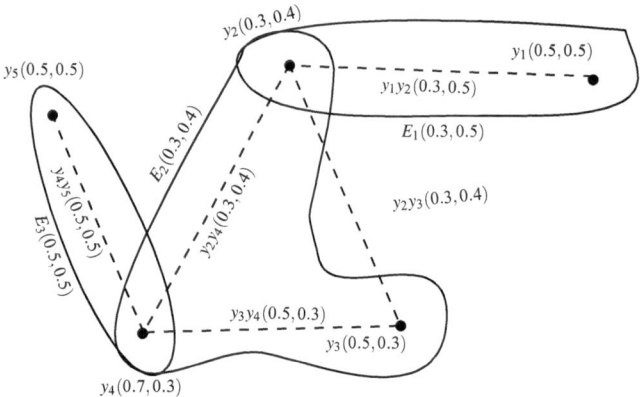

Fig. 2.17 2-section of an intuitionistic fuzzy hypergraph

1. $S^D = R$ is the intuitionistic fuzzy set of vertices of $\mathscr{H}^D$.
2. If $|X| = n$ then, R^D is an intuitionistic fuzzy set on the family of hypeedges $\{X_1, X_2, \ldots, X_n\}$ of $\mathscr{H}^D$ such that $X_k = \{E_i | y_k \in E_i, E_i$ is a hyperedge of $\mathscr{H}\}$. That is, X_k is the collection of those hyperedges which have a common vertex y_k and

$$R^D(X_k) = (\min\{\lambda_R(E_i) | y_k \in E_i\}, \max\{\tau_R(E_i) | y_k \in E_i\}).$$

The method for the construction of dual of an intuitionistic fuzzy hypergraph is presented in Algorithm 2.2.2.

Algorithm 2.2.2 The construction of dual of an intuitionistic fuzzy hypergraph

1. Input $\{y_1, y_2, \ldots, y_n\}$, the set of vertices and $\{E_1, E_2, \ldots, E_r\}$, the set of hyperedges of $\mathscr{H}$.
2. Construct an intuitionistic fuzzy set of vertices of $\mathscr{H}^D$ by defining $S^D = R$.
3. Draw a mapping $g : X \to E$ between sets of vertices and hyperedges. That is, if a vertex y_i belongs to $E_k, E_{k+1}, \ldots, E_r$ then map y_i to $E_k, E_{k+1}, \ldots, E_r$ as drawn in Fig. 2.18.
4. Construct a new family of hyperedges $\{X_1, X_2, \ldots, X_n\}$ of $\mathscr{H}^D$ such that $X_i = \{E_k | g(y_i) = E_k\}$ and $R^D(X_i) = (\min\{\lambda_R(E_k) | g(y_i) = E_k\}, \max\{\tau_R(E_k) | g(y_i) = E_k\})$.

Example 2.10 An intuitionistic fuzzy hypergraph $\mathscr{H}$ on $X = \{y_1, y_2, y_3, y_4, y_5\}$ with a set of hyperedges $E = \{E_1, E_2, E_3, E_4, E_5\}$ is shown in Fig. 2.19. The dual of $\mathscr{H}$ is represented by dashed lines with vertices E_1, E_2, E_3, E_4, E_5 and a family of hyperedges $\{X_1, X_2 = X_3, X_4, X_5\}$.

Theorem 2.12 *For any intuitionistic fuzzy hypergraph $\mathscr{H}$, $[\mathscr{H}^D]_2 = L(\mathscr{H})$.*

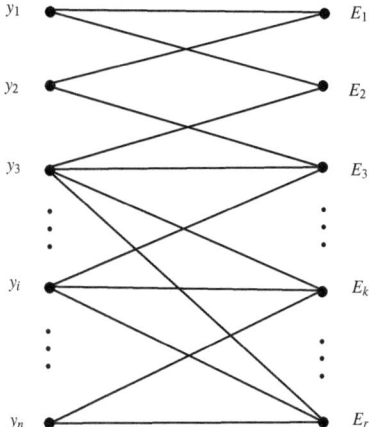

Fig. 2.18 Mapping between sets of vertices and hyperedges

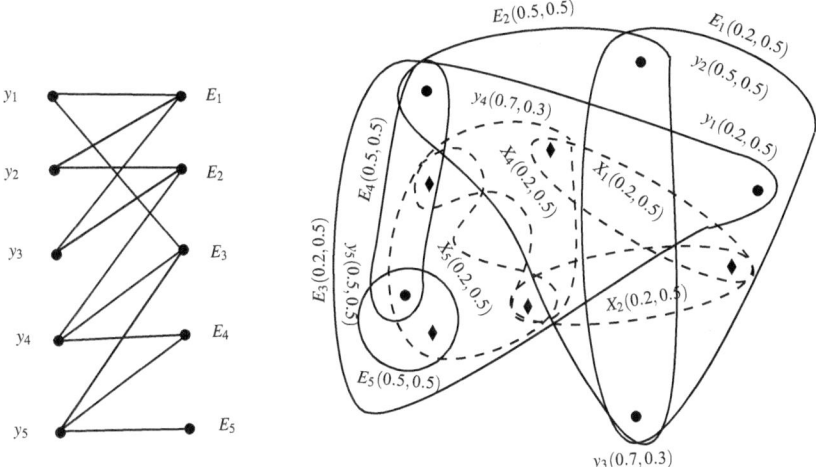

Fig. 2.19 Dual of an intuitionistic fuzzy hypergraph $\mathscr{H}$

Proof Let $\mathscr{H} = (S, R)$ be an intuitionistic fuzzy hypergraph on $X = \{y_1, y_2, \ldots, y_n\}$ with a family of hyperedges $\{E_1, E_2, \ldots, E_s\}$. Assume that $L(\mathscr{H}) = (S_l, R_l)$, $\mathscr{H}^D = (S^D, R^D)$ and $[\mathscr{H}^D]_2 = (S^D, U)$. The 2-section $[\mathscr{H}^D]_2$ has the intuitionistic fuzzy vertex set R which is also an intuitionistic fuzzy vertex set of $L(\mathscr{H})$. Suppose $\{X_1, X_2, \ldots, X_n\}$ is the family of hyperedges of $\mathscr{H}^D$. Clearly $\{E_i E_k | E_i, E_k \in X_i\}$ is the set of edges of $[\mathscr{H}^D]_2$ which the set of edges of $L(\mathscr{H})$. It remains to show that $R_l(E_i E_k) = U(E_i E_k)$.

$$R_l(E_i E_k) = (\lambda_R(E_i) \wedge \lambda_R(E_k), \tau_R(E_i) \vee \tau_R(E_k))$$
$$= (\lambda_{S^D}(E_i) \wedge \lambda_{S^D}(E_k), \tau_{S^D}(E_i) \vee \tau_{S^D}(E_k))$$
$$= U(E_i E_k).$$

Theorem 2.13 *For any two isomorphic intuitionistic fuzzy hypergraphs $\mathscr{H}_1$ and $\mathscr{H}_2$, if $\mathscr{H}_1 \simeq \mathscr{H}_2$ then $\mathscr{H}_1^D \simeq \mathscr{H}_2^D$.*

Proof Let $\mathscr{H}_1 = (S_1, R_1)$ and $\mathscr{H}_2 = (S_2, R_2)$ be two isomorphic intuitionistic fuzzy hypergraphs on $X_1 = \{y_{11}, y_{12}, \ldots, y_{1n}\}$ and $X_2 = \{y_{21}, y_{22}, \ldots, y_{2n}\}$, respectively. Take $\psi : X_1 \to X_2$ as an isomorphism of $\mathscr{H}_1$ onto $\mathscr{H}_2$. Let $\{X_{11}, X_{12}, \ldots, X_{1n}\}$ and $\{X_{21}, X_{22}, \ldots, X_{2n}\}$ be the families of hyperedges of $\mathscr{H}_1^D$ and $\mathscr{H}_2^D$. Also, let E_1 and E_2 be the families of hyperedges of $\mathscr{H}_1$ and $\mathscr{H}_2$ then define a mapping $\phi : E_1 \to E_2$. It is to be shown that ψ is an isomorphism. For each $E_{1k} \in E_1$ and $E_{2k} \in E_2$

$$S_1^D(E_{1k}) = R_1(E_{1k}) = R_2(\psi(E_{1k})) = R_2(E_{2k}) = S_2^D(\phi(E_{1k})).$$

$$R_1^D(X_{1k}) = (\lambda_{R_1}(E_{1k}) \wedge \lambda_{R_1}(E_{1\overline{k+1}}) \wedge \ldots \wedge \lambda_{R_1}(E_{1l}), \tau_{R_1}(E_{1k}) \vee \tau_{R_1}(E_{1\overline{k+1}}) \vee \ldots \vee \tau_{R_1}(E_{1l})),$$
$$= (\lambda_{R_2}(\phi(E_{1k})) \wedge \lambda_{R_2}(\phi(E_{1\overline{k+1}})) \wedge \ldots \wedge \lambda_{R_2}(\phi(E_{1l})),$$
$$\tau_{R_2}(\phi(E_{1k})) \vee \tau_{R_2}(\phi(E_{1\overline{k+1}})) \vee \ldots \vee \tau_{R_2}(\phi(E_{1l}))),$$
$$= R_2^D(X_{2k}) = R_2^D(\psi(X_{1k})).$$

Hence, $\mathscr{H}_1^D \simeq \mathscr{H}_2^D$.

Theorem 2.14 *The dual $\mathscr{H}^D$ of a linear intuitionistic fuzzy hypergraph $\mathscr{H}$ is also linear.*

Proof Let $\mathscr{H} = (S, R)$ and $\mathscr{H}^D = (S^D, R^D)$. On contrary, assume that $\mathscr{H}^D$ is not a linear intuitionistic fuzzy hypergraph then there exist X_i and X_k such that $|supp(\xi_i) \cap supp(\xi_k)| = 2$ where, $R^D(X_i) = \xi_i$ and $R^D(X_k) = \xi_k$. Assume that $supp(\xi_i) \cap supp(\xi_k) = \{E_t, E_s\}$. The definition of duality of $\mathscr{H}^D$ follows that there exist $y_i, y_k \in X$ such that $y_i, y_k \in E_t$ and $y_i, y_k \in E_s$. A contradiction to the given statement that $\mathscr{H}$ is linear. Hence, $\mathscr{H}^D$ is a linear intuitionistic fuzzy hypergraph.

2.3 Applications of Intuitionistic Fuzzy Hypergraphs

Graph theory has proved very useful for solving combinatorial problems of computer science and communication networks. To expand the origin of these applications, graphs were further extended to hypergraphs to model complex systems which arise in operation research, networking, and computer science. In some situations, the given data is fuzzy in nature and contains information about the existence and somewhere non-existence of uncertainty. The intuitionistic fuzzy hypergraphs can be used to formulate these concepts of existence and non-existence of uncertainty in a more generalized form as hypergraphs and intuitionistic fuzzy graphs can do. We now discuss some applications of intuitionistic fuzzy hypergraphs in social networking, chemistry, and planet surface networks.

2.3.1 The Intersecting Communities in Social Network

Nowadays, social networks have become the widely studied areas of research. Social networks are used to represent the lower and higher level interconnections among several communities belonging to social, as well as biological networks. People in society are connected to multiple areas which make them the part of various communities such as companies, universities, colleges, and offices etc. Consider the problem of grouping of authors according to their field of interest. An author can be different from the other regarding his/her critical writing. We present an intuitionistic fuzzy hypergraph $\mathscr{H} = (B, A)$ in which the vertices are authors and membership value of each author represents the degree of good writing and nonmembership value represents that the author's critical writing is not so good. Each hyperedge is the collection of those authors who belong to the same field of interest. The membership value of each hyperedge depicts the common ability of good critical writing of authors and nonmembership value shows the bad writing ability. An example is shown in Fig. 2.20.

The intuitionistic fuzzy hypergraph model can be used for the selection of authors having best writing ability in each field. The method for the selection of authors with best writing is given in Algorithm 2.3.1.

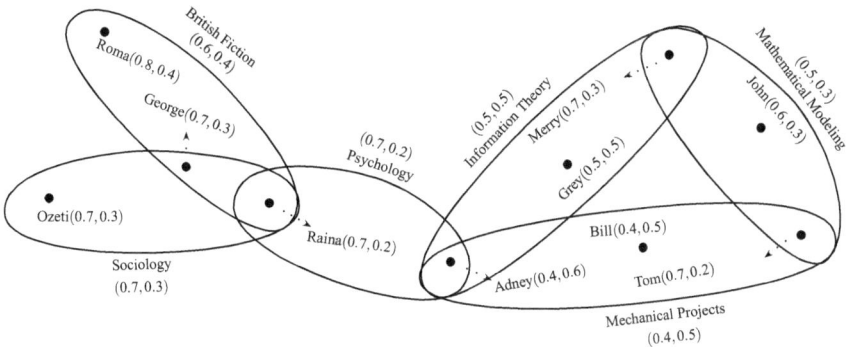

Fig. 2.20 Intuitionistic fuzzy social hypergraph

Algorithm 2.3.1 Selection of authors in an intuitionistic fuzzy social network model

1. Input the set of vertices (authors) $y_1, y_2, \ldots, y_n$.
2. Input the intuitionistic fuzzy set S of vertices such that $S(y_i) = (\lambda_i, \tau_i), 1 \leq i \leq n$.
3. Input the adjacency matrix $\xi = [(\lambda_{ij}, \tau_{ij})]_{n \times n}$ of vertices.
4. **do** i from $1 \to n$
5. $C_i = 0$
6.   **do** j from $1 \to n$
7. $\pi_{ij} = 1 - \lambda_{ij} - \tau_{ij}$
8. $S_{ij} = \sqrt{\lambda_{ij}^2 + \pi_{ij}^2 + (1 - \tau_{ij})^2}$
9. $C_i = C_i + S_{ij}$
10.   **end do**
11. $\pi_i = 1 - \lambda_i - \tau_i$
12. $C_i = C_i + \sqrt{\lambda_i^2 + \pi_i^2 + (1 - \tau_i)^2}$
13. **end do**
14. Using Algorithm 2.2.1, construct the adjacency matrix ξ_l of intuitionistic fuzzy line graph $L(\mathcal{H})$ of intuitionistic fuzzy hypergraph $\mathcal{H}$ whose adjacency matrix is ξ.
15. Compute the score and choice values of all vertices (fields) in $L(\mathcal{H})$ using steps 4–13.
16. Choose a vertex (field) E in $L(\mathcal{H})$ with maximum choice value.
17. Select a vertex of hyperedge E in $\mathcal{H}$ with maximum choice value which is the best option.

The adjacency matrix of Fig. 2.20 is given in Table 2.4. The score values of intuitionistic fuzzy hypergraph are computed using score function $S_{ij} = \sqrt{\lambda_{ij}^2 + \pi_{ij}^2 + (1 - \tau_{ij})^2}$ and the choice values $C_i = \sum_j S_{ij} + \sqrt{\lambda_i^2 + \pi_i^2 + (1 - \tau_i)^2}$ are given in Table 2.5 where, for any two vertices y_i, $y_j \in E$ (E is a hyperedge), $(\lambda_{ij}, \tau_{ij}) = (\lambda_A(E), \tau_A(E))$, $(\lambda_i, \tau_i) = (\lambda_B(y_i), \tau_B y_i)$.

The intuitionistic fuzzy line graph of Fig. 2.20 is shown in Fig. 2.21.

The adjacency matrix of Fig. 2.21 in given in Table 2.6. The score and choice values of Fig. 2.21 are calculated in Table 2.7.

The choice values in Table 2.7 show that the company can gain maximum benefit if it publishes articles and books on Psychology. There are three authors of Psychology, George, Raina, and Adney. The choice values of Table 2.7 show that Adney is the best author of Psychology. The best authors for all the fields are given in Table 2.8 which clearly shows that Raina, Adney, and Merry are the suitable options for all fields.

Table 2.4 Adjacency matrix

ξ	Roma	George	Ozeti	Raina	Adney	Grey	Merry	John	Bill	Tom
Roma	(0, 1)	(0.6, 0.4)	(0, 1)	(0.6, 0.4)	(0, 1)	(0, 1)	(0, 1)	(0, 1)	(0, 1)	(0, 1)
George	(0.6, 0.4)	(0, 1)	(0.7, 0.3)	(0.6, 0.4)	(0, 1)	(0, 1)	(0, 1)	(0, 1)	(0, 1)	(0, 1)
Ozeti	(0, 1)	(0.7, 0.3)	(0, 1)	(0.7, 0.3)	(0, 1)	(0, 1)	(0, 1)	(0, 1)	(0, 1)	(0, 1)
Raina	(0.6, 0.4)	(0.6, 0.4)	(0.7, 0.3)	(0, 1)	(0.7, 0.2)	(0, 1)	(0, 1)	(0, 1)	(0, 1)	(0, 1)
Adney	(0, 1)	(0, 1)	(0, 1)	(0.7, 0.2)	(0, 1)	(0.5, 0.5)	(0.5, 0.5)	(0, 1)	(0.4, 0.5)	(0.4, 0.5)
Grey	(0, 1)	(0, 1)	(0, 1)	(0, 1)	(0.5, 0.5)	(0, 1)	(0.5, 0.5)	(0, 1)	(0, 1)	(0, 1)
Merry	(0, 1)	(0, 1)	(0, 1)	(0, 1)	(0.5, 0.5)	(0.5, 0.5)	(0, 1)	(0.5, 0.3)	(0, 1)	(0.5, 0.3)
John	(0, 1)	(0, 1)	(0, 1)	(0, 1)	(0, 1)	(0, 1)	(0.5, 0.3)	(0, 1)	(0, 1)	(0.5, 0.3)
Bill	(0, 1)	(0, 1)	(0, 1)	(0, 1)	(0.4, 0.5)	(0, 1)	(0, 1)	(0, 1)	(0, 1)	(0.4, 0.5)
Tom	(0, 1)	(0, 1)	(0, 1)	(0, 1)	(0.4, 0.5)	(0, 1)	(0.5, 0.3)	(0.5, 0.3)	(0.4, 0.5)	(0, 1)

Table 2.5 Score and choice values

S_{ij}	Roma	George	Ozeti	Raina	Adney	Grey	Merry	John	Bill	Tom	C_i
Roma	0	0.8485	0	0.8485	0	0	0	0	0	0	2.8284
George	0.8485	0	0.9899	0.8485	0	0	0	0	0	0	3.6768
Ozeti	0	0.9899	0	0.9899	0	0	0	0	0	0	2.9697
Raina	0.8485	0.8485	0.9899	0	1.0677	0	0	0	0	0	4.8223
Adney	0	0	0	1.0677	0	0.7071	0.7071	0	0.6481	0.6481	4.3438
Grey	0	0	0	0	0.7071	0	0.7071	0	0	0	2.1213
Merry	0	0	0	0	0.7071	0.7071	0	0.8832	0	0.8832	4.1705
John	0	0	0	0	0	0	0.8832	0	0	0.8832	2.6938
Bill	0	0	0	0	0.6481	0	0	0	0	0.6481	1.9443
Tom	0	0	0	0	0.6481	0	0.8832	0.8832	0.6481	0	4.1303

2.3.2 Planet Surface Networks

There are various types of satellites in space for network communication and exploration of planets. Hypergraphs are a key tool to model such communication links among surface networks and satellites. There exist disturbance and uncertainty in planet surface communication due to climate change and electrical interference of

Table 2.6 Adjacency matrix of intuitionistic fuzzy line graph

ξ_l	British fiction	Sociology	Psychology	Information theory	Mechanical projects	Mathematical modeling
British Fiction	(0, 1)	(0.6,0.3)	(0.6,0.2)	(0, 1)	(0, 1)	(0, 1)
Sociology	(0.6,0.3)	(0, 1)	(0.7,0.2)	(0, 1)	(0, 1)	(0, 1)
Psychology	(0.6,0.2)	(0.7,0.2)	(0, 1)	(0.5, 0.2)	(0.4, 0.2)	(0, 1)
Information Theory	(0, 1)	(0, 1)	(0.5, 0.2)	(0, 1)	(0.4, 0.5)	(0.5,0.3)
Mechanical Projects	(0, 1)	(0, 1)	(0.4, 0.2)	(0.4, 0.5)	(0, 1)	(0.4,0.3)
Mathematical Modeling	(0, 1)	(0, 1)	(0.4, 0.2)	(0.5, 0.5)	(0.4, 0.5)	(0, 1)

Table 2.7 Score and choice values of intuitionistic fuzzy line graph

S_{ij}	British fiction	Sociology	Psychology	Information theory	Mechanical projects	Mathematical modeling	C_j
British Fiction	0	0.9274	1.0198	0	0	0	2.7957
Sociology	0.9274	0	1.0677	0	0	0	2.9850
Psychology	1.0198	1.0677	0	0.9899	0.9798	0	5.1249
Information Theory	0	0	0.9899	0	0.6481	0.8832	3.2283
Mechanical Projects	0	0	0.9798	0.6481	0	0.8602	3.1362
Mathematical Modeling	0	0	0.9798	0.7071	0.6481	0	3.2010

Table 2.8 Authors with best and critical writing

British Fiction	Sociology	Psychology	Information theory	Mechanical projects	Mathematical modeling
Raina	Raina	Raina	Adney	Adney	Merry

devices. This type of uncertainty in planet surface networks can be modeled using intuitionistic fuzzy hypergraphs as given in Fig. 2.22.

The circular dots denote the wireless devices on Earth, square style vertices show the satellites and diamond style vertices show the Earth gateway links. The membership value of each vertex represents the degree of disturbance in signal communication due to climate change and electrical interference. The nonmembership value shows the falsity of disturbance in signal communication. The membership value of each hyperedge represents the disturbance in corresponding access point. This is an application of intuitionistic fuzzy hypergraphs in planet surface networks and this model can be expanded to large-scale networks.

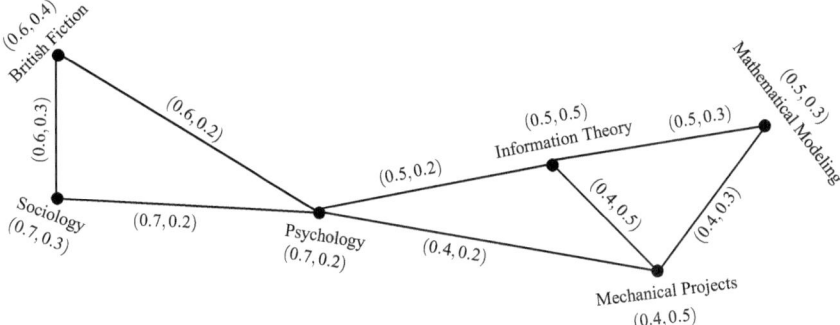

Fig. 2.21 Intuitionistic fuzzy line graph of Fig. 2.20

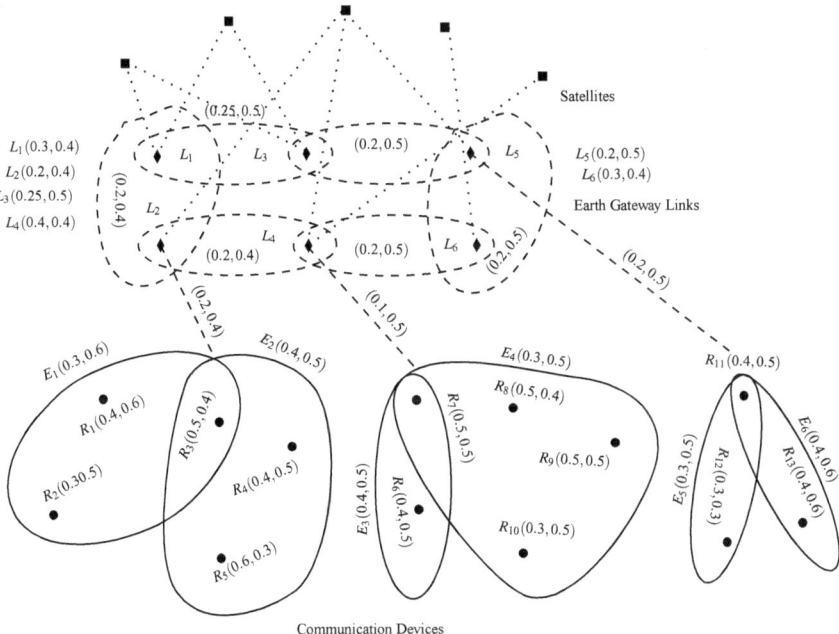

Fig. 2.22 Planet surface communication model

2.3.3 *Grouping of Incompatible Chemical Substances*

In this modern world, chemical engineers are trying day and night to save the economy
by converting raw materials into useful products. Various types of chemicals are used
for this purpose. Chemical industries are producing a variety of chemicals to be used
by other companies to produce different products. But the major problem is to store
the chemicals in order to avoid the accidental mixing to prevent chemical explosions,

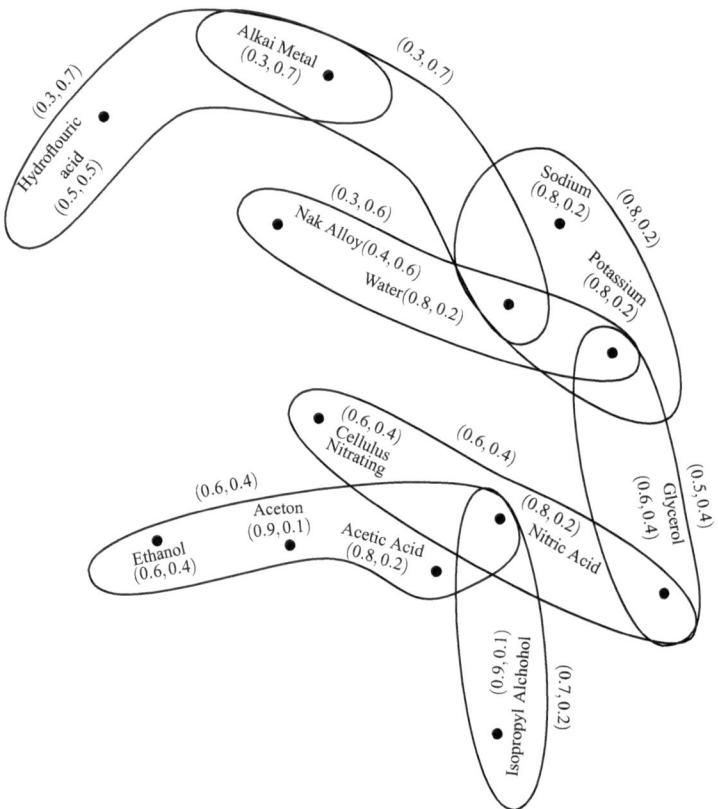

Fig. 2.23 Grouping of incompatible chemicals

oxygen deficiency, and dangerous toxic gases. Intuitionistic fuzzy hypergraphs can be used to model the chemicals in different groups to study the degree of disaster that could happen due to the accidental chemical reactions. An example is shown in Fig. 2.23 in which the vertices represent the chemicals.

Each hyperedge is the collection of those chemicals which can explode when mixed together. The membership value of each chemical show the degree of violent explosion and oxygen deficiency when reacted with various other chemicals. The nonmembership value represents the weakness of disaster. The membership and nonmembership value of each hyperedge represents the strength and weakness of disaster that cause due to chemical reaction. The degree of membership of Sodium is (0.8, 0.2) which shows that sodium is 80% explosive and 20% not explosive when mixed with other chemicals. Intuitionistic fuzzy hypergraphs can also be used for the classification of chemicals which are the most and least destructive in the given group. The method for the computation of such chemicals follows from steps 1–13 of Algorithm 2.3.1. The adjacency matrix of Fig. 2.23 is given in Table 2.9. The score values and choice of chemicals are computed in Table 2.10.

Table 2.9 Adjacency matrix of Fig. 2.23

ξ	Hyd acid	Alkai metal	Sod	Potas	Water	Nak alloy	Glyc	Nitric acid	Cell nitrat	Isop alcho	Acetic acid	Acet	Ethan
Hyd acid	(0, 1)	(0.3, 0.7)	(0, 1)	(0, 1)	(0, 1)	(0, 1)	(0, 1)	(0, 1)	(0, 1)	(0, 1)	(0, 1)	(0, 1)	(0, 1)
Alkai metal	(0.3, 0.7)	(0, 1)	(0, 1)	(0, 1)	(0.3, 0.7)	(0, 1)	(0, 1)	(0, 1)	(0, 1)	(0, 1)	(0, 1)	(0, 1)	(0, 1)
Sod	(0, 1)	(0.3, 0.7)	(0, 1)	(0.8, 0.2)	(0.8, 0.2)	(0, 1)	(0, 1)	(0, 1)	(0, 1)	(0, 1)	(0, 1)	(0, 1)	(0, 1)
Potas	(0, 1)	(0, 1)	(0.8, 0.2)	(0, 1)	(0.3, 0.6)	(0.3, 0.6)	(0.5, 0.4)	(0, 1)	(0, 1)	(0, 1)	(0, 1)	(0, 1)	(0, 1)
Water	(0, 1)	(0.3, 0.7)	(0.8, 0.2)	(0.8, 0.2)	(0, 1)	(0.3, 0.6)	(0, 1)	(0, 1)	(0, 1)	(0, 1)	(0, 1)	(0, 1)	(0, 1)
Nak alloy	(0, 1)	(0, 1)	(0, 1)	(0.3, 0.6)	(0.3, 0.6)	(0, 1)	(0, 1)	(0, 1)	(0, 1)	(0, 1)	(0, 1)	(0, 1)	(0, 1)
Water	(0, 1)	(0.3, 0.7)	(0.8, 0.2)	(0.8, 0.2)	(0, 1)	(0.3, 0.6)	(0, 1)	(0, 1)	(0, 1)	(0, 1)	(0, 1)	(0, 1)	(0, 1)
Glyc	(0, 1)	(0, 1)	(0, 1)	(0.5, 0.4)	(0, 1)	(0, 1)	(0, 1)	(0.6, 0.4)	(0.6, 0.4)	(0, 1)	(0, 1)	(0, 1)	(0, 1)
Nitric acid	(0, 1)	(0, 1)	(0, 1)	(0, 1)	(0, 1)	(0, 1)	(0.6, 0.4)	(0, 1)	(0.6, 0.4)	(0.7, 0.2)	(0.6, 0.4)	(0.6, 0.4)	(0.6, 0.4)
Cell nitrat	(0, 1)	(0, 1)	(0, 1)	(0, 1)	(0, 1)	(0, 1)	(0.6, 0.4)	(0, 1)	(0.6, 0.4)	(0.7, 0.2)	(0.6, 0.4)	(0.6, 0.4)	(0.6, 0.4)
Isop alcho	(0, 1)	(0, 1)	(0, 1)	(0, 1)	(0, 1)	(0, 1)	(0, 1)	(0.7, 0.2)	(0, 1)	(0, 1)	(0, 1)	(0, 1)	(0, 1)
Acetic acid	(0, 1)	(0, 1)	(0, 1)	(0, 1)	(0, 1)	(0, 1)	(0, 1)	(0.6, 0.4)	(0, 1)	(0, 1)	(0, 1)	(0.6, 0.4)	(0.6, 0.4)
Acet	(0, 1)	(0, 1)	(0, 1)	(0, 1)	(0, 1)	(0, 1)	(0, 1)	(0.6, 0.4)	(0, 1)	(0, 1)	(0.6, 0.4)	(0, 1)	(0.6, 0.4)
Ethan	(0, 1)	(0, 1)	(0, 1)	(0, 1)	(0, 1)	(0, 1)	(0, 1)	(0.6, 0.4)	(0, 1)	(0, 1)	(0.6, 0.4)	(0.6, 0.4)	(0, 1)

The score value 0 of some pair of chemicals show that they have no relation in given intuitionistic fuzzy hypergraph. There could be a little hazard due to the mixing of these chemicals. It can be studied on large-scale because it is not in the scope of this article. Table 2.10 shows that Nitric Acid and Cellulus Nitrating are the most explosive chemicals in the given group. These should be stored separately.

2.3.4 Radio Coverage Network

A *hypernetwork* M is a network whose underlying structure is a hypergraph H^*, in which each vertex v_i corresponds to a unique processor p_i of M, and each hyperedge e_j^* corresponds to a connector that connects processors represented by the vertices in e_j^*. A connector is loosely defined as an electronic or a photonic component

Table 2.10 Score and choice values of chemicals

S_{ij}	Hyd acid	Alkai metal	Sod	Potas	Water	Nak alloy	Glyc	Nitric acid	Cell Nitrat	Isop alcho	Acetic acid	Acet	Ethan	C_i
Hyd acid	0	0.4243	0	0	0	0	0	0	0	0	0	0	0	1.1314
Alkai metal	0.4243	0	0	0	0.4243	0	0	0	0	0	0	0	0	1.2729
Sod	0	0.4243	0	1.1314	1.1314	0	0	0	0	0	0	0	0	3.8185
Potas	0	0	1.1314	0	0.5099	0.5099	0.7874	0	0	0	0	0	0	4.07
Water	0	0.4243	1.1314	1.1314	0	0.5099	0	0	0	0	0	0	0	4.3284
Nak alloy	0	0	0	0.5099	0.5099	0	0	0	0	0	0	0	0	1.5855
Glyc	0	0	0	0.7874	0	0	0	0.8485	0.8485	0	0	0	0	3.2718
Nitric acid	0	0	0	0	0	0	0.8485	0	0.8485	1.0677	0.8485	0.8485	0.8485	6.4416
Cell nitrat	0	0	0	0	0	0	0.8485	0.8485	0	1.0677	0.8485	0.8485	0.8485	6.1587
Isop alcho	0	0	0	0	0	0	0	1.0677	0	0	0	0	0	2.4168
Acetic acid	0	0	0	0	0	0	0	0.8485	0	0	0	0.8485	0.8485	3.6769
Acet	0	0	0	0	0	0	0	0.8485	0	0	0.8485	0	0.8485	3.8183
Ethan	0	0	0	0	0	0	0	0.8485	0	0	0.8485	0.8485	0	3.3940

through which messages are transmitted between connected processors, not necessarily simultaneously. We call a connector a *hyperlink*. Unlike a point-to-point network, in which a link is dedicated to a pair of processors, a hyperlink in a hypernetwork is shared by a set of processors. A hyperlink can be implemented by a bus or a crossbar switch. Current optical technologies allow a hyperlink to be implemented by optical waveguides in a foldedbus using time-division multiplexing (TDM). Freespace optical or optoelectronic switching devices such as bulk lens, microlens array, and spatial light modulator (SLM) can also be used to implement hyperlinks. A star coupler, which uses wavelength-division multiplexing (WDM), can be considered either as a generalized bus structure or as a photonic switch, is another implementation of a hyperlink. Similarly, an ATM switch, which uses a variant TDM, is a hyperlink.

In telecommunications, the coverage of a radio station is the geographic area where the station can communicate.

Example 2.11 (**Radio Coverage Network**) Let X be a finite set of radio receivers (vertices); perhaps a set of representative locations at the centroid of a geographic region. For each of m radio transmitters we define the intuitionistic fuzzy set "listening area of station j" where $A_j(x) = (\mu_j(x), v_j(x))$ represents the "quality of reception of station j at location x. The membership and nonmembership values near 1 and 0, respectively, could signify "very clear reception on a very poor radio" while membership and nonmembership values near 0 and 1, respectively, could signify "very poor reception on even a very sensitive radio". Also, for a fixed radio the reception will vary between different stations. The stations can be considered as hyperedges. The membership and nonmembership values of the hyperedge indicate the clear and poor communication between stations. This model uses the full definition of an intuitionistic fuzzy hypergraph. The model could be used to determine station programming or marketing strategies or to establish an emergency broadcast network (is there a minimal subset of stations that reaches every radio with at least strength?). Further variables could relate signal strength to changes in time of day, weather and other conditions.

2.3.5 Clustering Problem

A cluster is two or more interconnected computers that create a solution to provide higher availability, higher scalability or both. The advantage of clustering computers for high availability is seen if one of these computers fails, another computer in the cluster can then assume the workload of the failed computer. The users of the system see no interruption of access. The advantages of clustering computers for scalability include increased application performance and the support of a greater number of users.

Definition 2.28 Let X be a reference set. A family of nontrivial intuitionistic fuzzy sets $\{A_1, A_2, A_3, \ldots, A_m\}$, where $A_i = (\mu_i, v_i)$, is an *intuitionistic fuzzy partition* if

1. $\bigcup_i \text{supp}(A_i) = X, \quad i = 1, 2, \ldots, m,$

Table 2.11 Intuitionistic fuzzy partition matrix

$\mathcal{H}$	A_t	B_h
x_1	(0.96, 0.04)	(0.04, 0.96)
x_2	(1, 0)	(0, 1)
x_3	(0.61, 0.39)	(0.39, 0.61)
x_4	(0.05, 0.95)	(0.95, 0.05)
x_5	(0.03, 0.97)	(0.97, 0.03)

2. $\sum_{i=1}^{m} \mu_i(x) = 1$, for all $x \in X$,
3. There is at most one i such that $v_i(x) = 0$, for all $x \in X$, (there is at most one intuitionistic fuzzy set such that $\mu_i(x) + v_i(x) = 1$, for all x).

Note that, this definition generalizes fuzzy partitions because the definition is equivalent to a fuzzy partition when for all x, $v_i(x) = 0$. We call a family $\{A_1, A_2, A_3, \ldots, A_m\}$ an *intuitionistic fuzzy covering* of X if it satisfies above conditions $1 - 2$.

The concept of intuitionistic fuzzy partition is essential for cluster analysis. An intuitionistic fuzzy partition can be represented by an intuitionistic fuzzy matrix $[a_{ij}]_{5 \times 5}$ where a_{ij} is the membership degree and nonmembership degree of element x_i in class j. We see that this matrix is the same as the incidence matrix in intuitionistic fuzzy hypergraph. Then, we can represent an intuitionistic fuzzy partition by an intuitionistic fuzzy hypergraph $\mathcal{H} = (S, R)$ such that

1. X: A set of elements x_i, $i = 1, 2, \ldots, n$.
2. $S = \{\eta_1, \eta_2, \ldots, \eta_m\}$: A set of nontrivial intuitionistic fuzzy classes.
3. $X = \bigcup_j \text{supp}(\eta_j)$, $j = 1, 2, \ldots, m$.
4. An intuitionistic fuzzy relation R on intuitionistic fuzzy classes η_i's satisfying Definition 2.7.
5. $\sum_{i=1}^{m} \mu_i(x) = 1$, for all $x \in X$.
6. There is at most one i such that $v_i(x) = 0$, for all $x \in X$, that is, there is at most one intuitionistic fuzzy set such that $\mu_i(x) + v_i(x) = 1$ for all $x \in X$.

Note that conditions 5–6 are added to intuitionistic fuzzy hypergraph for intuitionistic fuzzy partition. If these conditions are added, the intuitionistic fuzzy hypergraph can represent an intuitionistic fuzzy covering. Naturally, we can apply the (α, β)-cut to the intuitionistic fuzzy partition.

Example 2.12 We consider the clustering problem which is a typical example of an intuitionistic fuzzy partition on the visual image processing. Let us assume that there are five objects classified into two classes: tank and house. To cluster the elements x_1, x_2, x_3, x_4, x_5 into A_t (tank) and B_h (house), an intuitionistic fuzzy partition matrix is given in Table 2.11 in the form of incidence matrix of an intuitionistic fuzzy hypergraph $\mathcal{H} = (S, R)$ such that $S = A_t, B_h$, $R = \{(x_1 x_3 x_4 x_5, 0.03, 0.97), (x_1 x_3 x_4 x_5, 0.04, 0.96)\}$.

We can apply the (α, β)-cut to intuitionistic fuzzy hypergraph and obtain a crisp hypergraph $\mathcal{H}_{(\alpha, \beta)}$. This hypergraph $\mathcal{H}$ represents, generally, the covering because

Table 2.12 Hypergraph $\mathscr{H}_{(0.61,0.04)}$

$\mathscr{H}_{(0.61,0.04)}$	$A_{t(0.61,0.04)}$	$B_{h(0.61,0.04)}$
x_1	1	0
x_2	1	0
x_3	1	0
x_4	0	1
x_5	0	1

Table 2.13 Dual of the above intuitionistic fuzzy hypergraph

$\mathscr{H}^D_{(0.61,0.04)}$	X_1	X_2	X_3	X_4	X_5
A_t	1	1	1	0	0
B_h	0	0	0	1	1

of condition: 5 $\sum_{i=1}^{m} \mu_i(x) = 1$ for all $x \in X$, and 6 for all $x \in X$, there is at most one i such that $v_i(x) = 0$, is not always guaranteed. The hypergraph $\mathscr{H}_{(0.61,0.04)}$ is shown in Table 2.12.

We obtain the dual of hypergraph $\mathscr{H}_{(0.61,0.04)}$ as $\mathscr{H}^D_{(0.61,0.04)}$ as given in Table 2.13. The strength (cohesion) of an edge (class) $E_j = \eta_{j(\alpha,\beta)} = \{y_1, y_2, \ldots, y_r\}$ in $\mathscr{H}_{(\alpha,\beta)}$ can be used by taking minimum of membership values and maximum of nonmembership values of vertices y_i's in $\mathscr{H}$. Thus, we can use the strength as a measure of the cohesion or strength of a class in a partition. For example, the strengths of classes $A_{t(0.61,0.04)}$ and $B_{h(0.61,0.04)}$ at $s = 0.61, t = 0.04$ are $\beta(A_{t(0.61,0.04)})=(0.61, 0.39)$ and $\beta(B_{h(0.61,0.04)}) = (0.95, 0.05)$, respectively. It can be seen that the class $B_{h(0.61,0.04)}$ is stronger than $A_{t(0.61,0.04)}$ because $\beta(B_{h(0.61,0.04)}) > \beta(A_{t(0.61,0.04)})$. From the above discussion on the hypergraph $\mathscr{H}_{(0.61,0.04)}$ and $\mathscr{H}^D_{(0.61,0.04)}$ we can state that

- The intuitionistic fuzzy hypergraph can represent the fuzzy partition visually. The (α, β)-cut hypergraph also represents the (α, β)-cut partition.
- The dual hypergraph $\mathscr{H}^D_{(0.61,0.04)}$ can represent elements X_i, which can be grouped into a class $\eta_{j(\alpha,\beta)}$. For example, the edges X_1, X_2, X_3 of the dual hypergraph in Table 2.13 represent that the elements x_1, x_2, x_3 that can be grouped into A_t at level $(0.61, 0.04)$.
- In the intuitionistic fuzzy partition, we have $\sum_{i=1}^{m} \mu_i(x) = 1$ for all $x \in X$, and there is at most one i such that $v_i(x) = 0$, for all $x \in X$. If we define (α, β)-cut at level $(\alpha > 0.5$ or $\beta < 0.5)$, there is no element which is grouped into two classes simultaneously. That is, if $\alpha > 0.5$ or $\beta < 0.5$, every element is contained in only one class in $\mathscr{H}_{(\alpha,\beta)}$. Therefore, the hypergraph $\mathscr{H}_{(\alpha,\beta)}$ represents a partition. (If $s \leq 0.05$ or $t \geq 0.05$ the hypergraph may represent a covering).
- If $(\alpha, \beta) = (0.61, 0.04)$ then the strength of class $B_{h(0.61,0.04)}$ is the highest as $(0.95, 0.05)$, so it is the strongest class. It means that this class can be grouped independently from other parts. Thus, we can eliminate the class B_h from other

classes and continue clustering. Therefore, the discrimination of strong classes from others can allow us to decompose a clustering problem into smaller ones. This strategy allows us to work with the reduced data in a clustering problem.

2.4 Intuitionistic Fuzzy Directed Hypergraphs

In this section, certain types of intuitionistic fuzzy directed hypergraphs including core, simple, elementary, sectionally elementary, and (μ, ν)-tempered intuitionistic fuzzy directed hypergraphs are introduced and some of their properties are discussed. The concept of transversals of intuitionistic fuzzy directed hypergraphs has been studied with the notion of fundamental sequence and locally minimal transversals.

Definition 2.29 If $A_1 = (\lambda_{A_1}, \tau_{A_1})$ and $A_2 = (\lambda_{A_2}, \tau_{A_2})$ are two intuitionistic fuzzy sets on a non-empty set X then the *Cartesian product* of A_1 and A_2 is defined as

$$A_1 \times A_2 = \{\langle (x_1, x_2), \lambda_{A_1}(x_1) \wedge \lambda_{A_2}(x_2), \tau_{A_1}(x_1) \vee \tau_{A_2}(x_2)\rangle | x_1, x_2 \in X\}.$$

The Cartesian product of n intuitionistic fuzzy sets $A_1, A_2, \ldots, A_n$ over the non-empty crisp set X can be defined as

$$A_1 \times A_2 \times \ldots \times A_n = \{\langle (x_1, x_2, \ldots, x_n), \wedge_{i=1}^n \lambda_{A_i}(x_i), \vee_{i=1}^n \tau_{A_i}(x_i)\rangle | x_1, x_2, \ldots, x_n \in X\}.$$

Definition 2.30 A *directed hyperarc* on a non-empty set of vertices X is defined as a pair $\vec{E} = (t(\vec{E}), h(\vec{E}))$ where $t(\vec{E})$ and $h(\vec{E})$ are disjoint subsets of X. A vertex x in $\vec{E}$ is said to be a *source vertex* if $x \notin h(\vec{E})$. A vertex d is said to be a *destination vertex* in $\vec{E}$ if $d \notin t(\vec{E})$. An *intuitionistic fuzzy directed hyperedge* or *intuitionistic fuzzy directed hyperarc* is an ordered pair $\vec{\eta} = (t(\vec{\eta}), h(\vec{\eta}))$ of disjoint intuitionistic fuzzy subsets of vertices such that $t(\vec{\eta})$ is the tail of $\vec{\eta}$ while $h(\vec{\eta})$ is its head.

Definition 2.31 An *intuitionistic fuzzy directed hypergraph* on a non-empty set X is a pair $\vec{\mathcal{H}} = (I, R)$, where $I = \{\vec{\zeta}_1, \vec{\zeta}_2, \ldots, \vec{\zeta}_r\}$ is a family of order pairs $\vec{\zeta}_k = (t(\vec{\zeta}_k), h(\vec{\zeta}_k))$, where $t(\vec{\zeta}_k)$ and $h(\vec{\zeta}_k)$ are disjoint intuitionistic fuzzy subsets on X, and R is an intuitionistic fuzzy relation on $\vec{\zeta}_k$'s such that

1. $\lambda_R(\vec{E}_k) = \lambda_R(t(\vec{E}_k), h(\vec{E}_k)) \leq \min\{\wedge_{i=1}^m \lambda_{t(\vec{\zeta}_k)}(x_i), \wedge_{i=1}^n \lambda_{h(\vec{\zeta}_k)}(y_i)\}$,
2. $\tau_R(\vec{E}_k) = \tau_R(t(\vec{E}_k), h(\vec{E}_k)) \leq \max\{\vee_{i=1}^m \tau_{t(\vec{\zeta}_k)}(x_i), \vee_{i=1}^n \tau_{h(\vec{\zeta}_k)}(y_i)\}$,
3. $\lambda_R(\vec{E}_k) + \tau_R(\vec{E}_k) \leq 1$, for each $\vec{E}_k, 1 \leq k \leq r$,
   where $t(\vec{E}_k) = \{x_1, x_2, \ldots, x_m\} \subset X$ and $h(\vec{E}_k) = \{y_1, y_2, \ldots, y_n\} \subset X$.
4. $\bigcup_k supp(t(\vec{\zeta}_k)) \cup \bigcup_k supp(h(\vec{\zeta}_k)) = X$, $k = 1, 2, \ldots r$.

Example 2.13 Let $I = \{\vec{\zeta}_1, \vec{\zeta}_2, \vec{\zeta}_3\}$ be a class of intuitionistic fuzzy directed hyperarcs on $X = \{v_1, v_2, v_3, v_4\}$ as given in Table 2.14 and $\vec{E}_1 = supp(\vec{\zeta}_1) = (\{v_2\}, \{v_4\})$, $\vec{E}_2 = supp(\vec{\zeta}_2) = (\{v_3\}, \{v_4\})$, $\vec{E}_3 = supp(\vec{\zeta}_3) = (\{v_1\}, \{v_2, v_4\})$. R is an intuitionistic fuzzy relation on $\vec{\zeta}_k$'s given as, $R(\vec{E}_1) = (0.5, 0.1)$, $R(\vec{E}_2) = (0.4, 0.3)$ and

Table 2.14 Intuitionistic fuzzy directed hyperarcs on X

$x \in X$	$\vec{\zeta_1}$	$\vec{\zeta_2}$	$\vec{\zeta_3}$
v_1	(0, 1)	(0, 1)	(0.3, 0.4)
v_2	(0.5, 0.1)	(0, 1)	(0.5, 0.1)
v_3	(0, 1)	(0.4, 0.3)	(0, 1)
v_4	(0.5, 0.1)	(0.5, 0.1)	(0.2, 0.5)

Fig. 2.24 Intuitionistic fuzzy directed hypergraph $\vec{\mathscr{H}}$

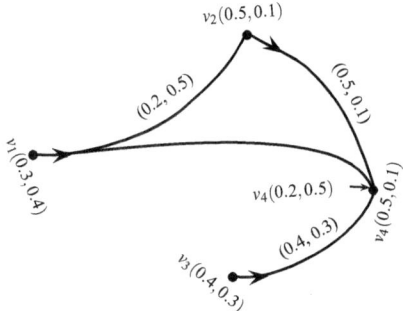

$R(\vec{E_3}) = (0.2, 0.5)$. The corresponding intuitionistic fuzzy directed hypergraph is shown in Fig. 2.24.

Definition 2.32 Let $\vec{\mathscr{H}} = (I, R)$ be an intuitionistic fuzzy directed hypergraph then *height* of an intuitionistic fuzzy directed hyperarc $\vec{\zeta}$ is denoted by $h(\vec{\zeta})$ and defined as

$$h(\vec{\zeta}) = (\lambda_{h(\vec{\zeta})}, \tau_{h(\vec{\zeta})}) = \left(\max\{\vee_{x \in X} \lambda_{t(\vec{\zeta})}(x), \vee_{x \in X} \lambda_{h(\vec{\zeta})}(x)\}, \right.$$
$$\left. \min\{\wedge_{x \in X} \tau_{t(\vec{\zeta})}(x), \wedge_{x \in X} \tau_{h(\vec{\zeta})}(x)\} \right).$$

Definition 2.33 An intuitionistic fuzzy directed hypergraph is called *simple* if for each $\vec{\zeta_i}, \vec{\zeta_j} \in I$, $supp(t(\vec{\zeta_i})) \subseteq supp(t(\vec{\zeta_j}))$, and $supp(h(\vec{\zeta_i})) \subseteq supp(h(\vec{\zeta_j}))$ then $i = j$.

Definition 2.34 An intuitionistic fuzzy directed hypergraph $\vec{\mathscr{H}} = (I, R)$ is *support simple* if whenever $\vec{\zeta_i}, \vec{\zeta_j} \in I$, $t(\vec{\zeta_i}) \subseteq t(\vec{\zeta_j})$, $h(\vec{\zeta_i}) \subseteq h(\vec{\zeta_j})$, and $supp(t(\vec{\zeta_i})) = supp(t(\vec{\zeta_j}))$, $supp(h(\vec{\zeta_i})) = supp(h(\vec{\zeta_j}))$ then $\vec{\zeta_i} = \vec{\zeta_j}$, for all i, j.

Example 2.14 Let $I = \{\vec{\zeta_1}, \vec{\zeta_2}, \vec{\zeta_3}, \vec{\zeta_4}\}$ be a family of intuitionistic fuzzy directed hyperarcs on $X = \{v_1, v_2, v_3, v_4\}$ as shown in Table 2.15. Take $\vec{E_1} = supp(\vec{\zeta_1}) = (\{v_1\}, \{v_2\})$, $\vec{E_2} = supp(\vec{\zeta_2}) = (\{v_1\}, \{v_2, v_4\})$, $\vec{E_3} = supp(\vec{\zeta_3}) = (\{v_2\}, \{v_3\})$ and $\vec{E_4} = supp(\vec{\zeta_4}) = (\{v_2\}, \{v_3, v_4\})$. R is an intuitionistic fuzzy relation on $\vec{\zeta_k}$'s given as, $R(\vec{E_1}) = (0.5, 0.1)$, $R(\vec{E_2}) = (0.4, 0.3)$, $R(\vec{E_3}) = (0.5, 0.2)$, and $R(\vec{E_4}) = (0.4, 0.3)$.

Table 2.15 Intuitionistic fuzzy directed hyperarcs on X

$x \in X$	$\vec{\zeta_1}$	$\vec{\zeta_2}$	$\vec{\zeta_3}$	$\vec{\zeta_4}$
v_1	$(0.7, 0.1)$	$(0.5, 0.2)$	$(0, 1)$	$(0, 1)$
v_2	$(0.7, 0.1)$	$(0.5, 0.2)$	$(0.5, 0.2)$	$(0.5, 0.2)$
v_3	$(0, 1)$	$(0, 1)$	$(0.5, 0.2)$	$(0.5, 0.2)$
v_4	$(0, 1)$	$(0.4, 0.3)$	$(0, 1)$	$(0.4, 0.3)$

Fig. 2.25 Support simple intuitionistic fuzzy directed hypergraph

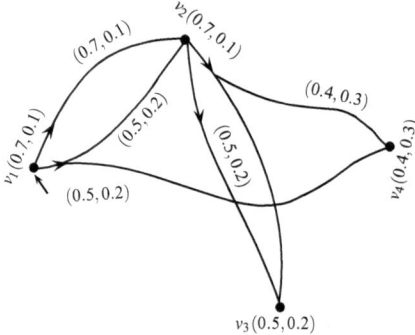

The corresponding support simple intuitionistic fuzzy directed hypergraph is shown in Fig. 2.25.

Definition 2.35 Let $\vec{\mathscr{H}} = (I, R)$ be an intuitionistic fuzzy directed hypergraph on X. For $\alpha, \beta \in [0, 1]$, the (α, β)-*level hyperarc* of an intuitionistic fuzzy directed hyperarc $\vec{\zeta}$ is defined as

$$\vec{\zeta}_{(\alpha,\beta)} = (t(\vec{\zeta}_{(\alpha,\beta)}), h(\vec{\zeta}_{(\alpha,\beta)}))$$
$$= (\{u \in X | \lambda_{t(\vec{\zeta})}(u) \geq \alpha, \tau_{t(\vec{\zeta})}(u) \leq \beta\}, \{v \in X | \lambda_{h(\vec{\zeta})}(v) \geq \alpha, \tau_{h(\vec{\zeta})}(v) \leq \beta\}).$$

$\vec{\mathscr{H}}_{(\alpha,\beta)} = (I_{(\alpha,\beta)}, R_{(\alpha,\beta)})$ is called a (α, β)-*level directed hypergraph* of $\vec{\mathscr{H}}$ where, $I_{(\alpha,\beta)}$ is defined as $I_{(\alpha,\beta)} = \{\cup_{k=1}^r h(\vec{\zeta}_{k(\alpha,\beta)}) \bigcup \cup_{k=1}^r t(\vec{\zeta}_{k(\alpha,\beta)}), 1 \leq k \leq r\}$.

Definition 2.36 Let $\vec{\mathscr{H}} = (I, R)$ be an intuitionistic fuzzy directed hypergraph. The sequence of order pairs $(\alpha_i, \beta_i) \in [0, 1] \times [0, 1], 0 \leq \alpha_i + \beta_i \leq 1, 1 \leq i \leq n$, such that $\alpha_1 > \alpha_2 > \ldots > \alpha_n, \beta_1 < \beta_2 < \cdots < \beta_n$ satisfying the properties

1. if $1 \geq \alpha > \alpha_1$ and $0 \leq \beta < \beta_1$ then $R_{(\alpha,\beta)} = \emptyset$,
2. if $\alpha_{i+1} < \alpha \leq \alpha_i$ and $\beta_i \leq \beta < \beta_{i+1}$ then $R_{(\alpha,\beta)} = R_{(\alpha_i,\beta_i)}$,
3. $R_{(\alpha_i,\beta_i)} \sqsubset R_{(\alpha_{i+1},\beta_{i+1})}$,

is called *fundamental sequence* of $\vec{\mathscr{H}}$, denoted by $f_s(\vec{\mathscr{H}})$. The corresponding sequence of (α_i, β_i)-level directed hypergraphs $\vec{\mathscr{H}}_{(\alpha_1,\beta_1)}, \vec{\mathscr{H}}_{(\alpha_2,\beta_2)}, \ldots, \vec{\mathscr{H}}_{(\alpha_n,\beta_n)}$ is called *core set* of $\vec{\mathscr{H}}$, denoted by $\mathscr{C}(\vec{\mathscr{H}})$. The (α_n, β_n)-level directed hypergraph, $\vec{\mathscr{H}}_{(\alpha_n,\beta_n)}$, is called *support level* of $\vec{\mathscr{H}}$.

Fig. 2.26 Intuitionistic fuzzy directed hypergraph on four vertices

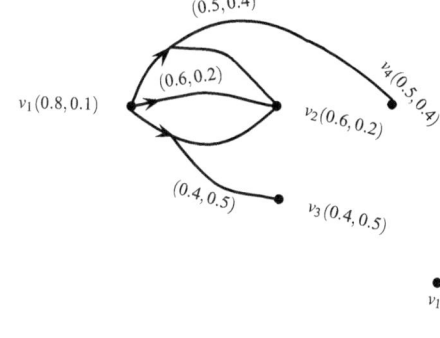

Fig. 2.27 $\vec{\mathscr{H}}_{(0.8,0.1)}$

Fig. 2.28 $\vec{\mathscr{H}}_{(0.6,0.2)}$

Fig. 2.29 $\vec{\mathscr{H}}_{(0.5,0.4)}$

Example 2.15 Let $\vec{\mathscr{H}}$ be an intuitionistic fuzzy directed hypergraph as shown in Fig. 2.26. Take $(\alpha_1, \beta_1) = (0.8, 0.1), (\alpha_2, \beta_2) = (0.6, 0.2), (\alpha_3, \beta_3) = (0.5, 0.4)$ and $(\alpha_4, \beta_4) = (0.4, 0.5)$. Clearly, the set $\{(\alpha_1, \beta_1),(\alpha_2, \beta_2), (\alpha_3, \beta_3), (\alpha_4, \beta_4)\}$ satisfies all conditions of Definition 2.36 and hence is a fundamental sequence of $\vec{\mathscr{H}}$. The corresponding (α_i, β_i)-level directed hypergraphs are shown in Fig. 2.27, 2.28 and 2.29 whereas, $\vec{\mathscr{H}}_{(0.4,0.5)} = supp(\vec{\mathscr{H}}) = (supp(I), supp(R))$.

Definition 2.37 Let $\vec{\mathscr{H}} = (I, R)$ be an intuitionistic fuzzy directed hypergraph on X then $supp(I) = \{(supp(t(\vec{\zeta}_k)), supp(h(\vec{\zeta}_k))) \mid \vec{\zeta}_k \in I\}$. The family of intuitionistic fuzzy directed hyperarcs I is called *elementary* if I is single-valued on $supp(I)$. An intuitionistic fuzzy directed hypergraph $\vec{\mathscr{H}}$ is *elementary* if I and R are elementary, otherwise it is *nonelementary*.

Example 2.16 The elementary and nonelementary intuitionistic fuzzy directed hypergraphs are given in Figs. 2.30 and 2.31, respectively.

Definition 2.38 An intuitionistic fuzzy directed hypergraph $\vec{\mathscr{H}} = (I, R)$ is called a *partial intuitionistic fuzzy directed hypergraph* of $\vec{\mathscr{H}}' = (I', R')$ if following conditions are satisfied

1. $supp(I) \subseteq supp(I')$ and $supp(R) \subseteq supp(R')$,
2. if $supp(\vec{\zeta}_i) \in supp(I)$ and $supp(\vec{\zeta}_i') \in supp(I')$ such that $supp(\vec{\zeta}_i) = supp(\vec{\zeta}_i')$ then $\vec{\zeta}_i = \vec{\zeta}_i'$.

It is denoted by $\vec{\mathscr{H}} \subseteq \vec{\mathscr{H}}'$. An intuitionistic fuzzy directed hypergraph $\vec{\mathscr{H}} = (I, R)$ is *ordered* if the core set $\mathscr{C}(\vec{\mathscr{H}}) = \{\vec{\mathscr{H}}_{(\alpha_1,\beta_1)}, \vec{\mathscr{H}}_{(\alpha_2,\beta_2)}, \ldots \vec{\mathscr{H}}_{(\alpha_n,\beta_n)}\}$ is ordered, that

Fig. 2.30 Elementary
intuitionistic fuzzy directed
hypergraph

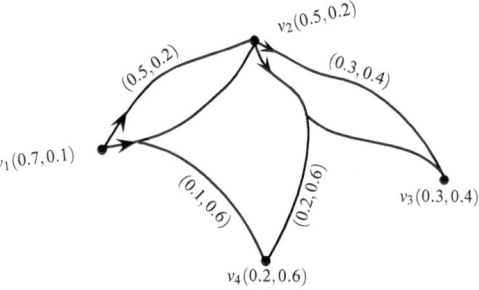

Fig. 2.31 Nonelementary
intuitionistic fuzzy directed
hypergraph

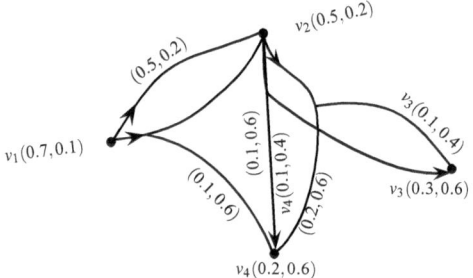

is $\vec{\mathscr{H}}_{(\alpha_1,\beta_1)} \subseteq \vec{\mathscr{H}}_{(\alpha_2,\beta_2)} \subseteq \ldots \subseteq \vec{\mathscr{H}}_{(\alpha_n,\beta_n)}$. $\vec{\mathscr{H}}$ is *simply ordered* if $\vec{\mathscr{H}}$ is ordered and whenever $R' \subset R_{(\alpha_{i+1},\beta_{i+1})} \setminus R_{(\alpha_i,\beta_i)}$ then $R' \not\subseteq R_{(\alpha_i,\beta_i)}$.

Observation 2.15 *Let $\vec{\mathscr{H}}$ be an elementary intuitionistic fuzzy directed hypergraph then $\vec{\mathscr{H}}$ is ordered. If $\vec{\mathscr{H}}$ is ordered intuitionistic fuzzy directed hypergraph and support level $\vec{\mathscr{H}}_{(\alpha_n,\beta_n)}$ is simple then $\vec{\mathscr{H}}$ is an elementary intuitionistic fuzzy directed hypergraph.*

Note 2.1 1. If $\vec{\mathscr{H}} = (I, R)$ is an intuitionistic fuzzy directed hypergraph with $I = \{\vec{\zeta}_1, \vec{\zeta}_2, \ldots, \vec{\zeta}_r\}$ then $I^* = \{\vec{\zeta}_1^{\,*}, \vec{\zeta}_2^{\,*}, \ldots, \vec{\zeta}_r^{\,*}\}$ is the family of crisp directed hyperarcs corresponding to I.
  2. In intuitionistic fuzzy directed hypergraph $\vec{\mathscr{H}} = (I, R)$, if x is a vertex of tail of any intuitionistic fuzzy directed hyperarc $\vec{\zeta}$ then $\vec{\zeta}(x) = (\tau_{t(\vec{\zeta})}, \lambda_{t(\vec{\zeta})})$. If $x \in h(\vec{\zeta})^*$ then $\vec{\zeta}(x) = (\tau_{h(\vec{\zeta})}, \lambda_{t(\vec{\zeta})})$.

Definition 2.39 Let $\vec{\mathscr{H}} = (I, R)$ and $\vec{\mathscr{H}}' = (I', R')$ be any two intuitionistic fuzzy directed hypergraphs on X and X', respectively, where $I = \{\zeta_1, \zeta_2, \ldots, \zeta_r\}$ and $I' = \{\zeta_1', \zeta_2', \ldots, \zeta_r'\}$. A *homomorphism* of intuitionistic fuzzy directed hypergraphs $\vec{\mathscr{H}}$ and $\vec{\mathscr{H}}'$ is a mapping $\phi : X \to X'$ that satisfies

1. $\wedge_{j=1}^r \tau_{\vec{\zeta}_j}(x) \leq \wedge_{j=1}^r \tau_{\vec{\zeta}_j'}(\phi(x))$, $\vee_{j=1}^r \lambda_{\vec{\zeta}_j}(x) \geq \vee_{j=1}^r \lambda_{\vec{\zeta}_j'}(\phi(x))$,   for all $x \in X$.
2. $\tau_R (\{t_1, t_2, \ldots, t_s\}, \{h_1, h_2, \ldots, h_m\}) \leq \tau_{R'} (\{\phi(t_1), \phi(t_2), \ldots, \phi(t_s)\},$
   $\{\phi(h_1), \phi(h_2), \ldots, \phi(h_m)\})$,
   $\lambda_R (\{t_1, t_2, \ldots, t_s\}, \{h_1, h_2, \ldots, h_m\}) \geq \lambda_{R'} (\{\phi(t_1), \phi(t_2), \ldots, \phi(t_s)\},$

$\{\phi(h_1), \phi(h_2), \ldots, \phi(h_m)\}$),

for all $t_1, t_2, \ldots, t_s, h_1, h_2, \ldots, h_m \in X$.

Definition 2.40 A *weak isomorphism* of intuitionistic fuzzy directed hypergraphs $\overrightarrow{\mathscr{H}}$ and $\overrightarrow{\mathscr{H}}'$ is a bijective homomorphism $\phi : X \to X'$ that satisfies

$$\wedge_{j=1}^r \tau_{\vec{\zeta}_j}(x) = \wedge_{j=1}^r \tau_{\vec{\zeta}_j}'(\phi(x)), \quad \vee_{j=1}^r \lambda_{\vec{\zeta}_j}(x) = \vee_{j=1}^r \lambda_{\vec{\zeta}_j}'(\phi(x)), \qquad \text{for all } x \in X.$$

Definition 2.41 A *co-weak isomorphism* of intuitionistic fuzzy directed hypergraphs $\overrightarrow{\mathscr{H}}$ and $\overrightarrow{\mathscr{H}}'$ is a bijective homomorphism $\phi : X \to X'$ that satisfies

$\tau_R (\{t_1, t_2, \ldots, t_s\}, \{h_1, h_2, \ldots, h_m\}) = \tau_{R'} (\{\phi(t_1), \phi(t_2), \ldots, \phi(t_s)\},$
$\{\phi(h_1), \phi(h_2), \ldots, \phi(h_m)\}),$
$\lambda_R (\{t_1, t_2, \ldots, t_s\}, \{h_1, h_2, \ldots, h_m\}) = \lambda_{R'} (\{\phi(t_1), \phi(t_2), \ldots, \phi(t_s)\},$
$\{\phi(h_1), \phi(h_2), \ldots, \phi(h_m)\}),$

for all $t_1, t_2, \ldots, t_s, h_1, h_2, \ldots, h_m \in X$.

Definition 2.42 An *isomorphism* of intuitionistic fuzzy directed hypergraphs $\overrightarrow{\mathscr{H}}$ and $\overrightarrow{\mathscr{H}}'$ is a bijective mapping $\phi : X \to X'$ that satisfies

1. $\wedge_{j=1}^r \tau_{\vec{\zeta}_j}(x) = \wedge_{j=1}^r \tau_{\vec{\zeta}_j}'(\phi(x)), \quad \vee_{j=1}^r \lambda_{\vec{\zeta}_j}(x) = \vee_{j=1}^r \lambda_{\vec{\zeta}_j}'(\phi(x)), \qquad \text{for all } x \in X.$
2. $\tau_R (\{t_1, t_2, \ldots, t_s\}, \{h_1, h_2, \ldots, h_m\}) = \tau_{R'} (\{\phi(t_1), \phi(t_2), \ldots, \phi(t_s)\},$
   $\{\phi(h_1), \phi(h_2), \ldots, \phi(h_m)\}),$
   $\lambda_R (\{t_1, t_2, \ldots, t_s\}, \{h_1, h_2, \ldots, h_m\}) = \lambda_{R'} (\{\phi(t_1), \phi(t_2), \ldots, \phi(t_s)\},$
   $\{\phi(h_1), \phi(h_2), \ldots, \phi(h_m)\}),$
   for all $t_1, t_2, \ldots, t_s, h_1, h_2, \ldots, h_m \in X$.

In this case, $\overrightarrow{\mathscr{H}}$ and $\overrightarrow{\mathscr{H}}'$ are called *isomorphic* to each other.

Definition 2.43 Let $\overrightarrow{\mathscr{H}} = (I, R)$ be an intuitionistic fuzzy directed hypergraph then the *order* $O(\overrightarrow{\mathscr{H}})$ and *size* $S(\overrightarrow{\mathscr{H}})$ of $\overrightarrow{\mathscr{H}}$ are defined as

$$O(\overrightarrow{\mathscr{H}}) = \left(\sum_{x \in X} \wedge_j \tau_{\vec{\zeta}_j}(x), \sum_{x \in X} \vee_j \lambda_{\vec{\zeta}_j}(x) \right), \quad S(\overrightarrow{\mathscr{H}}) = \left(\sum_{\vec{E}_i \in I^*} \tau_R(\vec{E}_i), \sum_{\vec{E}_i \in I^*} \lambda_R(\vec{E}_i) \right).$$

Theorem 2.16 *The order and size of isomorphic intuitionistic fuzzy directed hypergraphs are same.*

Proof Let $\overrightarrow{\mathscr{H}_1} = (I_1, R_1)$ and $\overrightarrow{\mathscr{H}_2} = (I_2, R_2)$ be any two intuitionistic fuzzy directed hypergraphs on X_1 and X_2, respectively, where $I_1 = \{\zeta_{11}, \zeta_{12}, \ldots, \zeta_{1r}\}$ and $I_2 = \{\zeta_{21}, \zeta_{22}, \ldots, \zeta_{2r}\}$ be the classes of intuitionistic fuzzy directed hyperarcs. Let $\phi : X_1 \to X_2$ be an isomorphism from $\overrightarrow{\mathscr{H}_1}$ to $\overrightarrow{\mathscr{H}_2}$ then using Definition 2.42,

$$O(\vec{\mathcal{H}_1}) = \left(\sum_{x_1 \in X_1} \wedge_j \tau^{\rightarrow}_{\zeta_{1j}}(x_1), \sum_{x_1 \in X_1} \vee_j \lambda^{\rightarrow}_{\zeta_{1j}}(x_1) \right)$$

$$= \left(\sum_{x_1 \in X_1} \wedge_j \tau^{\rightarrow}_{\zeta_{2j}}(\phi(x_1)), \sum_{x_1 \in X_1} \vee_j \lambda^{\rightarrow}_{\zeta_{2j}}(\phi(x_1)) \right)$$

$$= \left(\sum_{x_2 \in X_2} \wedge_j \tau^{\rightarrow}_{\zeta_{2j}}(x_2), \sum_{x_2 \in X_2} \vee_j \lambda^{\rightarrow}_{\zeta_{2j}}(x_2) \right) = O(\vec{\mathcal{H}_2}).$$

$$S(\vec{\mathcal{H}_1}) = \left(\sum_{\vec{E}_{1i} \in I_1^*} \tau_{R_1}(\vec{E}_{1i}), \sum_{\vec{E}_{1i} \in I_1^*} \lambda_{R_1}(\vec{E}_{1i}) \right)$$

$$= \left(\sum_{\vec{E}_{1i} \in I_1^*} \tau_{R_2}(\phi(\vec{E}_{1i})), \sum_{\vec{E}_{1i} \in I_1^*} \lambda_{R_2}(\phi(\vec{E}_{1i})) \right)$$

$$= \left(\sum_{\vec{E}_{2i} \in I_2^*} \tau_{R_2}(\vec{E}_{2i}), \sum_{\vec{E}_{2i} \in I_2^*} \lambda_{R_2}(\vec{E}_{2i}) \right) = S(\vec{\mathcal{H}_2}).$$

Remark 2.5 1. The order of weak isomorphic intuitionistic fuzzy directed hypergraphs is same.
2. The size of co-weak isomorphic intuitionistic fuzzy directed hypergraphs is same.

Theorem 2.17 *The relation of isomorphism between intuitionistic fuzzy directed hypergraphs is an equivalence relation.*

Proof Let $\vec{\mathcal{H}_1} = (I_1, R_1)$, $\vec{\mathcal{H}_2} = (I_2, R_2)$ and $\vec{\mathcal{H}_3} = (I_3, R_3)$ be intuitionistic fuzzy directed hypergraphs on X_1, X_2 and X_3, respectively, where, $I_1 = \{\zeta_{11}, \zeta_{12}, \ldots, \zeta_{1r}\}$, $I_2 = \{\zeta_{21}, \zeta_{22}, \ldots, \zeta_{2r}\}$ and $I_3 = \{\zeta_{31}, \zeta_{32}, \ldots, \zeta_{3r}\}$.

1. Reflexive: Define $I : X_1 \rightarrow X_1$ by $I(x_1) = x_1$, for all $x_1 \in X_1$. Then, I is a bijective homomorphism and
   1. $(\wedge_j \tau^{\rightarrow}_{\zeta_{1j}}(x_1), \vee_j \lambda^{\rightarrow}_{\zeta_{1j}}(x_1)) = (\wedge_j \tau^{\rightarrow}_{\zeta_{1j}}(I(x_1)), \vee_j \lambda^{\rightarrow}_{\zeta_{1j}}(I(x_1)))$,
   2. $(\tau_{R_1}(\vec{E}_{1i}), \lambda_{R_1}(\vec{E}_{1i})) = (\tau_{R_1}(I(\vec{E}_{1i})), \lambda_{R_1}(I(\vec{E}_{1i})))$,
   for all $x_1 \in X_1$, $t(\vec{E}_{1i}) \subset X_1$, $h(\vec{E}_{1i}) \subset X_1$.
   I is an isomorphism of an intuitionistic fuzzy directed hypergraph to itself.
2. Symmetric: Let $\phi : X_1 \rightarrow X_2$ be an isomorphism defined by $\phi(x_1) = x_2$. Since, ϕ is a bijective mapping therefore, $\phi^{-1} : X_2 \rightarrow X_1$ exists and $\phi^{-1}(x_2) = x_1$, for all $x_2 \in X_2$. Then

$$(\wedge_j \tau^{\rightarrow}_{\zeta_{2j}}(x_2), \vee_j \lambda^{\rightarrow}_{\zeta_{2j}}(x_2)) = (\wedge_j \tau^{\rightarrow}_{\zeta_{2j}}(\phi(x_1)), \vee_j \lambda^{\rightarrow}_{\zeta_{2j}}(\phi(x_1)))$$

$$= (\wedge_j \tau^{\rightarrow}_{\zeta_{1j}}(x_1), \vee_j \lambda^{\rightarrow}_{\zeta_{1j}}(x_1))$$

$$= (\wedge_j \tau^{\rightarrow}_{\zeta_{1j}}(\phi^{-1}(x_2)), \vee_j \lambda^{\rightarrow}_{\zeta_{1j}}(\phi^{-1}(x_2))).$$

$$R_2(\vec{E}_{2j}) = R_2(\phi(\vec{E}_{1j})) = R_1(\vec{E}_{1j}) = R_1(\phi^{-1}(\vec{E}_{2j})), \ t(\vec{E}_{2j}) \subseteq X_2, \ h(\vec{E}_{2j}) \subseteq X_2.$$

Hence, ϕ^{-1} is an isomorphism.

3. Transitive: Let $\phi : X_1 \to X_2$ and $\psi : X_2 \to X_3$ be the isomorphisms of $\mathcal{H}_1$ onto $\mathcal{H}_2$ and $\mathcal{H}_2$ onto $\mathcal{H}_3$ defined by $\phi(x_1) = x_2$ and $\psi(x_2) = x_3$, respectively. By Definition 2.42

$$(\wedge_j \tau_{\vec{\zeta}_{1j}}(x_1), \vee_j \lambda_{\vec{\zeta}_{1j}}(x_1)) = (\wedge_j \tau_{\vec{\zeta}_{2j}}(x_2), \vee_j \lambda_{\vec{\zeta}_{2j}}(x_2))$$
$$= (\wedge_j \tau_{\vec{\zeta}_{3j}}(\psi(x_2)), \vee_j \lambda_{\vec{\zeta}_{3j}}(\psi(x_2)))$$
$$= (\wedge_j \tau_{\vec{\zeta}_{3j}}(\psi(\phi(x_1))), \vee_j \lambda_{\vec{\zeta}_{3j}}(\psi(\phi(x_1))))$$
$$= (\wedge_j \tau_{\vec{\zeta}_{3j}}(\psi \circ \phi(x_1)), \vee_j \lambda_{\vec{\zeta}_{3j}}(\psi \circ \phi(x_1))).$$

$R_1(\vec{E}_{1j}) = R_2(\vec{E}_{2j}) = R_3(\psi(\vec{E}_{2j})) = R_3(\psi(\phi(\vec{E}_{1j}))) = R_3(\psi \circ \phi(\vec{E}_{1j}))$, where $\vec{E}_{ij} = (t(\vec{E}_{ij}), h(\vec{E}_{ij}), t(\vec{E}_{ij}) \subset X_i, h(\vec{E}_{ij}) \subset X_i$. Clearly, $\psi \circ \phi$ is an isomorphism from $\mathcal{H}_1$ onto $\mathcal{H}_3$. Hence, isomorphism of intuitionistic fuzzy directed hypergraphs is an equivalent relation.

Remark 2.6 The relation of weak isomorphism between intuitionistic fuzzy directed hypergraphs is a partial order relation.

Definition 2.44 Let $\mathcal{H} = (I, R)$ be an intuitionistic fuzzy directed hypergraph. A set of intuitionistic fuzzy directed hyperarcs T with the property that $T_{h(\vec{\zeta}_i)} \cap \zeta_{ih(\vec{\zeta}_i)} \neq \emptyset$, for each $\vec{\zeta}_i \in I$, is called *intuitionistic fuzzy transversal* of $\mathcal{H}$. T is a *minimal intuitionistic fuzzy transversal* of $\mathcal{H}$ if whenever $\rho \subset T$, ρ is not an intuitionistic fuzzy transversal of $\mathcal{H}$. The family of all minimal intuitionistic fuzzy transversals of $\mathcal{H}$ is denoted by $Tr(\mathcal{H})$.

Example 2.17 Consider the intuitionistic fuzzy directed hypergraph as shown in Fig. 2.32 where $I = \{\vec{\zeta}_1, \vec{\zeta}_2, \vec{\zeta}_3\}$ is defined in Table 2.16.

Fig. 2.32 Intuitionistic fuzzy directed hypergraph on four vertices

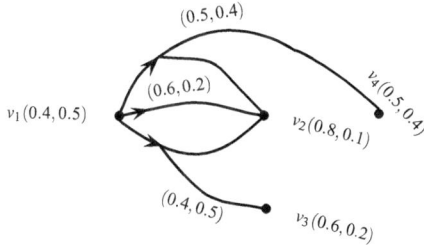

Table 2.16 Intuitionistic fuzzy hyperarcs of $\mathcal{H}$ in Fig. 2.32

Intuitionistic fuzzy hyperarc		$h(\vec{\zeta}_i)$	$\vec{\zeta}_{ih(\vec{\zeta}_i)}$
$\vec{\zeta}_1$	$\{(\{(v_1, 0.4, 0.5)\}, \{(v_2, 0.8, 0.1), (v_4, 0.5, 0.4)\})\}$	$(0.8, 0.1)$	$\{(\{\}, \{v_2\})\}$
$\vec{\zeta}_2$	$\{(\{(v_1, 0.4, 0.5)\}, \{(v_2, 0.8, 0.1)\})\}$	$(0.8, 0.1)$	$\{(\{\}, \{v_2\})\}$
$\vec{\zeta}_3$	$\{(\{(v_1, 0.4, 0.5)\}, \{(v_2, 0.8, 0.1), (v_3, 0.6, 0.2)\})\}$	$(0.8, 0.1)$	$\{(\{\}, \{v_2\})\}$

Clearly, for each $1 \leq i \leq 3$, $\vec{\zeta}_{ih(\vec{\zeta}_i)} \cap T_{ih(\vec{\zeta}_i)} \neq \emptyset$ where, $T = \Big\{ (\{(v_1, 0.8, 0.1)\},$
$\{(v_2, 0.6, 0.2)\}) \Big\}$. Hence, T is an intuitionistic fuzzy transversal of $\vec{\mathcal{H}}$.

We now provide results and discussions of intuitionistic fuzzy transversals.

Lemma 2.1 *Let $\vec{\mathcal{H}} = (I, R)$ be an intuitionistic fuzzy directed hypergraph with fundamental sequence $f_s(\vec{\mathcal{H}}) = \{(\alpha_1, \beta_1), (\alpha_2, \beta_2), \ldots, (\alpha_n, \beta_n)\}$. If T is an intuitionistic fuzzy transversal of $\mathcal{H}$ then $\lambda_{h(T)} \geq \lambda_{h(\vec{\zeta}_i)}$ and $\tau_{h(T)} \leq \tau_{h(\vec{\zeta}_i)}$. If T is a minimal intuitionistic fuzzy transversals of $\vec{\mathcal{H}}$ then $h(T) = (\max\{\lambda_{h(\vec{\zeta}_i)} | \vec{\zeta}_i \in I\}, \min\{\tau_{h(\vec{\zeta}_i)} | \vec{\zeta}_i \in I\}) = (\alpha_1, \beta_1)$.*

Proposition 2.2 *Let $\vec{\mathcal{H}}$ be an intuitionistic fuzzy directed hypergraph then the following statements are equivalent.*
1. T is an intuitionistic fuzzy transversal of $\vec{\mathcal{H}}$.
2. For each $\vec{\zeta}_i \in I$, $(\alpha, \beta) \in [0, 1] \times [0, 1]$, $0 \leq \alpha + \beta \leq 1$ with $\alpha < \lambda_{h(\vec{\zeta}_i)}$ and $\beta > \tau_{h(\vec{\zeta}_i)}$ then $T_{(\alpha, \beta)} \cap \vec{\zeta}_{i(\alpha, \beta)} \neq \emptyset$.
3. $T_{(\alpha, \beta)}$ is a transversal of $\vec{\mathcal{H}}_{(\alpha, \beta)}$.

2.5 Complex Intuitionistic Fuzzy Hypergraphs

To generalize the concepts of intuitionistic fuzzy sets, complex intuitionistic fuzzy sets were introduced by Alkouri and Salleh [4]. Complex intuitionistic fuzzy set is a distinctive intuitionistic fuzzy set in which the membership degrees are determined on the unit disc of the complex plane and can more clearly express the imprecision and ambiguity in the data. Yaqoob et al. [23] defined complex intuitionistic fuzzy graphs and discussed an application of these graphs in cellular networks to test the proposed model.

Definition 2.45 A *complex intuitionistic fuzzy set* I on the universal set X is defined as, $I = \{(u, T_I(u)e^{i\phi_I(u)}, F_I(u)e^{i\psi_I(u)}) | u \in X\}$, where $i = \sqrt{-1}$, $T_I(u), F_I(u) \in [0, 1]$, $\phi_I(u), \psi_I(u) \in [0, 2\pi]$, and for every $u \in X$, $0 \leq T_I(u) + F_I(u) \leq 1$. Here, $T_I(u), F_I(u)$ and $\phi_I(u), \psi_I(u)$ are called the amplitude terms and phase terms for truth membership and falsity membership grades, respectively.

Definition 2.46 A *complex intuitionistic fuzzy graph* on X is an ordered pair $G = (A, B)$, where A is a complex intuitionistic fuzzy set on X and B is complex intuitionistic fuzzy relation on X such that

$T_B(ab) \leq \min\{T_A(a), T_A(b)\}$, $F_B(ab) \leq \max\{F_A(a), F_A(b)\}$, (for amplitude terms)
$\phi_B(ab) \leq \min\{\phi_A(a), \phi_A(b)\}$, $\psi_B(ab) \leq \max\{\psi_A(a), \psi_A(b)\}$, (for phase terms)

$0 \leq T_B(ab) + F_B(ab) \leq 1$, for all $a, b \in X$.

Definition 2.47 Let X be a nontrivial set of universe. A *complex intuitionistic fuzzy hypergraph* is defined as an ordered pair $H = (\mathscr{C}, \mathscr{D})$, where $\mathscr{C} = \{\alpha_1, \alpha_2, \cdots, \alpha_k\}$ is a finite family of complex intuitionistic fuzzy sets on X and $\mathscr{D}$ is a complex intuitionistic fuzzy relation on complex intuitionistic fuzzy sets α_j's such that

(i)

$$T_\mathscr{D}(\{r_1, r_2, \cdots, r_l\}) \le \min\{T_{\alpha_j}(r_1), T_{\alpha_j}(r_2), \cdots, T_{\alpha_j}(r_l)\},$$
$$F_\mathscr{D}(\{r_1, r_2, \cdots, r_l\}) \le \max\{F_{\alpha_j}(r_1), F_{\alpha_j}(r_2), \cdots, F_{\alpha_j}(r_l)\}, \text{ (for amplitude terms)}$$
$$\phi_\mathscr{D}(\{r_1, r_2, \cdots, r_l\}) \le \min\{\phi_{\alpha_j}(r_1), \phi_{\alpha_j}(r_2), \cdots, \phi_{\alpha_j}(r_l)\},$$
$$\psi_\mathscr{D}(\{r_1, r_2, \cdots, r_l\}) \le \max\{\psi_{\alpha_j}(r_1), \psi_{\alpha_j}(r_2), \cdots, \psi_{\alpha_j}(r_l)\}, \text{ (for phase terms)}$$

$0 \le T_\mathscr{D} + F_\mathscr{D} \le 1$, for all $r_1, r_2, \cdots, r_l \in X$.

(ii) $\bigcup_j supp(\alpha_j) = X$, for all $\alpha_j \in \mathscr{C}$.

Note that, $E_k = \{r_1, r_2, \cdots, r_l\}$ is the crisp hyperedge of $H = (\mathscr{C}, \mathscr{D})$.

Example 2.18 Consider a complex intuitionistic fuzzy hypergraph $H = (\mathscr{C}, \mathscr{D})$ on $X = \{v_1, v_2, v_3, v_4\}$. The complex intuitionistic fuzzy relation is defined as $\mathscr{D}(\{v_1, v_2, v_3, v_4\}) = (0.2e^{i(0.4)\pi}, 0.6e^{i(0.3)\pi})$, $\mathscr{D}(\{v_1, v_2\}) = (0.3e^{i(0.6)\pi}, 0.6e^{i(0.3)\pi})$, and $\mathscr{D}(\{v_3, v_4\}) = (0.2e^{i(0.4)\pi}, 0.5e^{i(0.3)\pi})$. The corresponding complex intuitionistic fuzzy hypergraph is shown in Fig. 2.33.

Definition 2.48 A complex intuitionistic fuzzy hypergraph $H = (\mathscr{C}, \mathscr{D})$ is *simple* if whenever $\mathscr{D}_j, \mathscr{D}_k \in \mathscr{D}$ and $\mathscr{D}_j \subseteq \mathscr{D}_k$, then $\mathscr{D}_j = \mathscr{D}_k$.

A complex intuitionistic fuzzy hypergraph $H = (\mathscr{C}, \mathscr{D})$ is *support simple* if whenever $\mathscr{D}_j, \mathscr{D}_k \in \mathscr{D}$, $\mathscr{D}_j \subseteq \mathscr{D}_k$, and $supp(\mathscr{D}_j) = supp(\mathscr{D}_k)$, then $\mathscr{D}_j = \mathscr{D}_k$.

Definition 2.49 Let $H = (\mathscr{C}, \mathscr{D})$ be a complex intuitionistic fuzzy hypergraph. Suppose that $\alpha, \beta \in [0, 1]$ and $\theta, \varphi \in [0, 2\pi]$ such that $0 \le \alpha + \beta \le 1$. The $(\alpha e^{i\theta}, \beta e^{i\varphi})$-*level hypergraph* of H is defined as an ordered pair $H^{(\alpha e^{i\theta}, \beta e^{i\varphi})} = (\mathscr{C}^{(\alpha e^{i\theta}, \beta e^{i\varphi})}, \mathscr{D}^{(\alpha e^{i\theta}, \beta e^{i\varphi})})$, where

(i) $\mathscr{D}^{(\alpha e^{i\theta}, \beta e^{i\varphi})} = \{D_j^{(\alpha e^{i\theta}, \beta e^{i\varphi})} : D_j \in \mathscr{D}\}$ and $D_j^{(\alpha e^{i\theta}, \beta e^{i\varphi})} = \{u \in X : T_{D_j}(u) \ge \alpha, \phi_{D_j}(u) \ge \theta$, and $F_{D_j}(u) \le \beta, \psi_{D_j}(u) \le \varphi\}$,

(ii) $\mathscr{C}^{(\alpha e^{i\theta}, \beta e^{i\varphi})} = \bigcup_{D_j \in \mathscr{D}} D_j^{(\alpha e^{i\theta}, \beta e^{i\varphi})}$.

Note that, $(\alpha e^{i\theta}, \beta e^{i\varphi})$-level hypergraph of H is a crisp hypergraph.

Example 2.19 Consider a complex intuitionistic fuzzy hypergraph $H = (\mathscr{C}, \mathscr{D})$ as shown in Fig. 2.33. Let $\alpha = 0.2$, $\beta = 0.5$, $\theta = 0.5\pi$, and $\varphi = 0.2\pi$. Then, $(\alpha e^{i\theta}, \beta e^{i\varphi})$-level hypergraph of H is shown in Fig. 2.34.

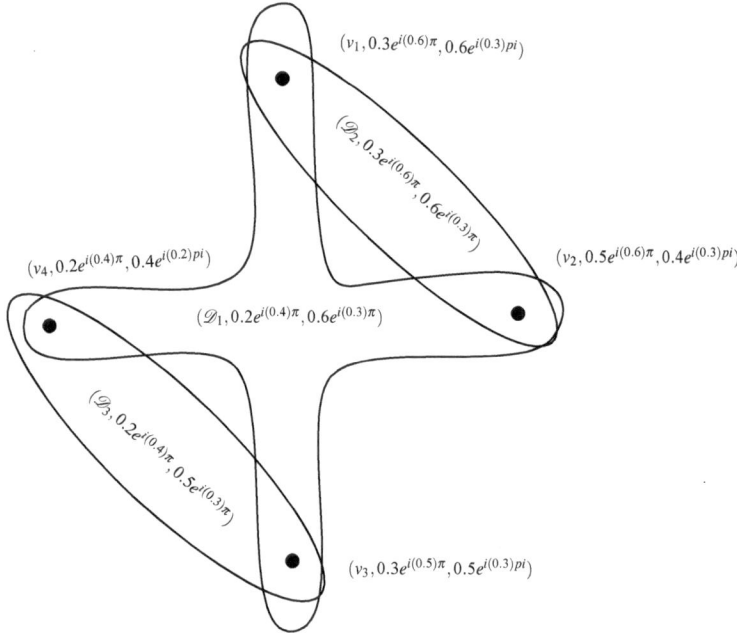

Fig. 2.33 Complex intuitionistic fuzzy hypergraph

Fig. 2.34 $(0.2e^{i(0.5)\pi}, 0.5e^{i(0.2)\pi})$-
level hypergraph of
H

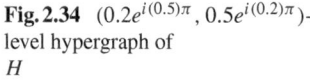

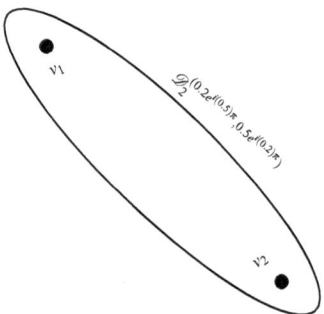

Definition 2.50 Let $H = (\mathscr{C}, \mathscr{D})$ be a complex intuitionistic fuzzy hypergraph. The *complex intuitionistic fuzzy line graph* of H is defined as an ordered pair $l(H) = (\mathscr{C}_l, \mathscr{D}_l)$, where $\mathscr{C}_l = \mathscr{D}$ and there exists an edge between two vertices in $l(H)$ if $|supp(D_j) \cap supp(D_k)| \geq 1$. The membership degrees of $l(H)$ are given as

(i) $\mathscr{C}_l(E_k) = \mathscr{D}(E_k)$,
(ii) $\mathscr{D}_l(E_j E_k) = (\min\{T_{\mathscr{D}}(E_j), T_{\mathscr{D}}(E_k)\}e^{i\min\{\phi_{\mathscr{D}}(E_j), \phi_{\mathscr{D}}(E_k)\}},$
      $\max\{F_{\mathscr{D}}(E_j), F_{\mathscr{D}}(E_k)\}e^{i\max\{\psi_{\mathscr{D}}(E_j), \psi_{\mathscr{D}}(E_k)\}}).$

Definition 2.51 A complex intuitionistic fuzzy hypergraph $H = (\mathscr{C}, \mathscr{D})$ is said to be *linear* if for every $D_j, D_k \in \mathscr{D}$,

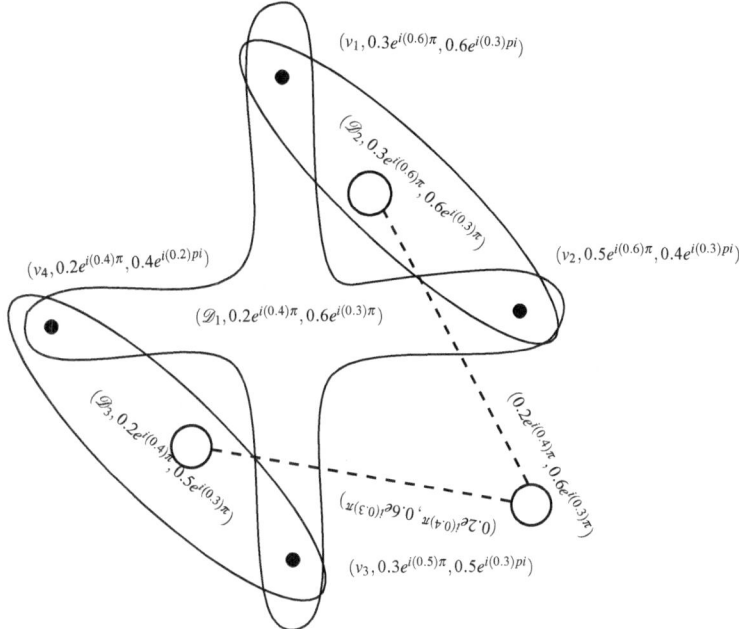

Fig. 2.35 Complex intuitionistic line graph of H

(i) $supp(D_j) \subseteq supp(D_k) \Rightarrow j = k,$
(ii) $|supp(D_j) \cap supp(D_k)| \leq 1.$

Example 2.20 Consider a complex intuitionistic fuzzy hypergraph $H = (\mathscr{C}, \mathscr{D})$ as shown in Fig. 2.33. By direct calculations, we have

$$supp(\mathscr{D}_1) = \{v_1, v_2, v_3, v_4\}, \ supp(\mathscr{D}_2) = \{v_1, v_2\}, \ supp(\mathscr{D}_3) = \{v_3, v_4\}.$$

Note that, $supp(D_j) \subseteq supp(D_k) \Rightarrow j \neq k$ and $|supp(D_j) \cap supp(D_k)| \nleq 1.$ Hence, complex intuitionistic fuzzy hypergraph $H = (\mathscr{C}, \mathscr{D})$ is not linear. The corresponding complex intuitionistic fuzzy hypergraph $H = (\mathscr{C}, \mathscr{D})$ and its line graph is shown in Fig. 2.35.

Theorem 2.18 *A simple strong complex intuitionistic fuzzy graph is the complex intuitionistic line graph of a linear complex intuitionistic fuzzy hypergraph.*

Definition 2.52 The 2-*section* $H_2 = (\mathscr{C}_2, \mathscr{D}_2)$ of a complex intuitionistic fuzzy hypergraph $H = (\mathscr{C}, \mathscr{D})$ is a complex intuitionistic fuzzy graph having same set of vertices as that of H, $\mathscr{D}_2$ is a complex intuitionistic fuzzy set on $\{e = u_j u_k | u_j, u_k \in E_l, \ l = 1, 2, 3, \cdots\}$, and $\mathscr{D}_2(u_j u_k) = (\min\{\min T_{\alpha_l}(u_j), \min T_{\alpha_l}(u_k)\} e^{i \min\{\min \phi_{\alpha_l}(u_j), \min \phi_{\alpha_l}(u_k)\}}, \max\{\max F_{\alpha_l}(u_j), \max F_{\alpha_l}(u_k)\} e^{i \max\{\max \psi_{\alpha_l}(u_j), \max \psi_{\alpha_l}(u_k)\}})$ such that $0 \leq T_{\mathscr{D}_2}(u_j u_k) + F_{\mathscr{D}_2}(u_j u_k) \leq 1.$

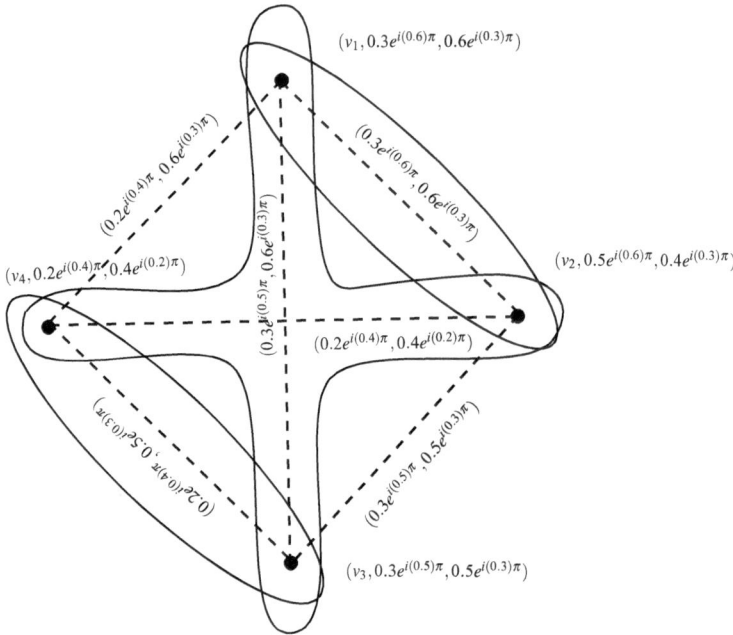

Fig. 2.36 2-section of complex intuitionistic fuzzy hypergraph

Example 2.21 An example of a complex intuitionistic fuzzy hypergraph is given in Fig. 2.36. The 2-section of H is presented with dashed lines.

Definition 2.53 Let $H = (\mathscr{C}, \mathscr{D})$ be a complex intuitionistic fuzzy hypergraph. A *complex intuitionistic fuzzy transversal* τ is a complex intuitionistic fuzzy set of X satisfying the condition $\rho^{h(\rho)} \cap \tau^{h(\rho)} \neq \emptyset$, for all $\rho \in \mathscr{D}$, where $h(\rho)$ is the height of ρ.

A *minimal complex intuitionistic fuzzy transversal* t is the complex intuitionistic fuzzy transversal of H having the property that if $\tau \subset t$, then τ is not a complex intuitionistic fuzzy transversal of H.

References

1. Akram, M., Davvaz, B.: Strong intuitionistic fuzzy graphs. FILOMAT **26**(1), 177–196 (2012)
2. Akram, M., Dudek, W.A.: Intuitionistic fuzzy hypergraphs with applications. Inf. Sci. **218**, 182–193 (2013)
3. Akram, M., Sarwar, M., Borzooei, R.A.: A novel decision-making approach based on hypergraphs in intuitionistic fuzzy environment. J. Intell. Fuzzy Syst. **35**(2), 1905–1922 (2018)
4. Alkouri, A., Salleh, A.: Complex intuitionistic fuzzy sets. AIP Conf. Proc. **14**, 464–470 (2012)
5. Atanassov, K.T.: Intuitionistic fuzzy sets. Fuzzy Sets Syst. **20**(1), 87–96 (1986)

6. Atanassov, K.T.: Intuitionistic fuzzy sets: theory and applications. Studies in Fuzziness and Soft Computing, vol. 35. Physica-Verl, Heidelberg, New York (1999)
7. Berge, C.: Graphs and Hypergraphs. North-Holland, Amsterdam (1973)
8. Chen, S.M., Interval-valued fuzzy hypergraph and fuzzy partition. IEEE Trans. Syst. Man Cybern. (Cybernetics) **27**(4), 725–733 (1997)
9. Gallo, G., Longo, G., Pallottino, S.: Directed hypergraphs and applications. Discret. Appl. Math. **42**, 177–201 (1993)
10. Kaufmann, A.: Introduction a la Thiorie des Sous-Ensemble Flous, vol. 1. Masson, Paris (1977)
11. Lee-Kwang, H., Lee, K.-M.: Fuzzy hypergraph and fuzzy partition. IEEE Trans. Syst. Man Cybern. **25**(1), 196–201 (1995)
12. Luqman, A., Akram, M., Al-Kenani, A.N., Alcantud, J.C.R.: A study on hypergraph representations of complex fuzzy information. Symmetry **11**(11), 1381 (2019)
13. Mordeson, J.N., Nair, P.S.: Fuzzy Graphs and Fuzzy Hypergraphs, 2nd edn. Physica Verlag, Heidelberg (2001)
14. Myithili, K.K., Parvathi, R.: Transversals of intuitionistic fuzzy directed hypergraphs. Notes Intuit. Fuzzy Sets **21**(3), 66–79 (2015)
15. Myithili, K.K., Parvathi, R., Akram, M.: Certain types of intuitionistic fuzzy directed hypergraphs. Int. J. Mach. Learn. Cybern. **7**(2), 287–295 (2016)
16. Parvathi, R., Akram, M., Thilagavathi, S.: Intuitionistic fuzzy shortest hyperpath in a network. Inf. Process. Lett. **113**(17), 599–603 (2013)
17. Parvathi, R., Thilagavathi, S., Karunambigai, M.G.: Intuitionistic fuzzy hypergraphs. Cybern. Inf. Technol. **9**(2), 46–53 (2009)
18. Parvathi, R., Thilagavathi, S.: Intuitionistic fuzzy directed hypergraphs. Adv. Fuzzy Sets Syst. **14**(1), 39–52 (2013)
19. Parvathi, R., Thilagavathi, S., Atanassov, K.T.: Isomorphism on intuitionistic fuzzy directed hypergraphs. Int. J. Sci. Res. Publ. **3**(3) (2013)
20. Ramot, D., Milo, R., Friedman, M., Kandel, A.: Complex fuzzy sets. IEEE Trans. Fuzzy Syst. **10**(2), 171–186 (2002)
21. Ramot, D., Friedman, M., Langholz, G., Kandel, A.: Complex fuzzy logic. IEEE Trans. Fuzzy Syst. **11**(4), 450–461 (2003)
22. Thirunavukarasu, P., Suresh, R., Viswanathan, K.K.: Energy of a complex fuzzy graph. Int. J. Math. Sci. Eng. Appl. **10**(1), 243–248 (2016)
23. Yaqoob, N., Gulistan, M., Kadry, S., Wahab, H.: Complex intuitionistic fuzzy graphs with application in cellular network provider companies. Mathematics **7**(1), 35 (2019)
24. Yazdanbakhsh, O., Dick, S.: A systematic review of complex fuzzy sets and logic. Fuzzy Sets. Syst. **338**, 1–22 (2018)
25. Zadeh, L.A.: Fuzzy sets. Inf. Control **8**(3), 338–353 (1965)

Chapter 3
Hypergraphs for Interval-Valued Structures

In this chapter, we present interval-valued fuzzy hypergraphs, $A = [\mu^-, \mu^+]$– tempered interval-valued fuzzy hypergraphs, and some of their properties. Moreover, we discuss the notions of vague hypergraphs, dual vague hypergraphs, and A-tempered vague hypergraphs. Finally, we describe interval-valued intuitionistic fuzzy hypergraphs and interval-valued intuitionistic fuzzy transversals of $\mathscr{H}$. This chapter is due to [4–6, 8, 11, 22, 27].

3.1 Introduction

Zadeh [27] introduced the notion of interval-valued fuzzy sets as an extension of fuzzy set theory [25] for representing vagueness and uncertainty. Interval-valued fuzzy set theory reflects the uncertainty by the length of the interval membership degree $[\mu_1, \mu_2]$. In intuitionistic fuzzy set theory for every membership degree (μ_1, μ_2), the value $\pi = 1 - \mu_1 - \mu_2$ denotes a measure of non-determinacy (or undecidedness). Interval-valued fuzzy sets provide a more adequate description of vagueness than traditional fuzzy sets. It is, therefore, important to use interval-valued fuzzy sets in applications, such as fuzzy control. One of the computationally most intensive parts of fuzzy control is defuzzification [20]. Since interval-valued fuzzy sets are widely studied and used, we describe briefly the work of Gorzalczany [14, 15] on approximate reasoning, Roy and Biswas [23] on medical diagnosis, Turksen [24] on multivalued logic and Mendel [20] on intelligent control. Atanassov and Gargov [6] introduced the notion of interval-valued intuitionistic fuzzy sets which is a generalization of both intuitionistic fuzzy sets and interval-valued fuzzy sets.

Graph theory has numerous applications to problems in systems analysis, operations research, economics, and transportation. However, in many cases, some aspects of a graph-theoretic problem may be uncertain. For example, the vehicle travel time or vehicle capacity on a road network may not be known exactly. In such cases, it

© Springer Nature Singapore Pte Ltd. 2020
M. Akram and A. Luqman, *Fuzzy Hypergraphs and Related Extensions*,
Studies in Fuzziness and Soft Computing 390,
https://doi.org/10.1007/978-981-15-2403-5_3

is natural to deal with the uncertainty using the methods of fuzzy sets and fuzzy logic. Hypergraph models are more general types of relations than graphs do and can be used to model networks, social networks, biology networks, process scheduling, data structures, computations, and a variety of other systems where complex relationships between the objects in the system play a dominant role. Fuzzy hypergraphs were proposed by Kaufmann [17] and then generalized and redefined by Lee-kwang and Lee [19]. Goetschel Jr. [12] discussed the concept of hypergraphs by initiating a glimpse of what may be done within a fuzzy setting. Also the idea of transversal of a hypergraph has been extended to fuzzy transversal of a fuzzy hypergraph by Goetschel Jr. et al. [13]. Chen [8] presented the notion of the interval-valued fuzzy hypergraph theory which is based on a combination of the interval-valued fuzzy set and hypergraph models. Akram and Dudek [1] presented some properties of intuitionistic fuzzy hypergraphs and provided its application in clustering problem. Naz et al. [22] proposed the concept of the interval-valued intuitionistic fuzzy hypergraphs by combining the interval-valued intuitionistic fuzzy set and hypergraph models.

Definition 3.1 An *interval number* D is an interval $[a^-, a^+]$ with $0 \leq a^- \leq a^+ \leq 1$. The interval $[a, a]$ is identified with the number $a \in [0, 1]$. Let $D[0, 1]$ be the set of all interval numbers. For interval numbers $D_1 = [a_1^-, b_1^+]$ and $D_2 = [a_2^-, b_2^+]$, we define

- $\min\{D_1, D_2\} = \min\{[a_1^-, b_1^+], [a_2^-, b_2^+]\} = [\min\{a_1^-, a_2^-\}, \min\{b_1^+, b_2^+\}]$,
- $\max\{D_1, D_2\} = \max\{[a_1^-, b_1^+], [a_2^-, b_2^+]\} = [\max\{a_1^-, a_2^-\}, \max\{b_1^+, b_2^+\}]$,
- $D_1 + D_2 = [a_1^- + a_2^- - a_1^- \cdot a_2^-, b_1^+ + b_2^+ - b_1^+ \cdot b_2^+]$,
- $D_1 \leq D_2 \iff a_1^- \leq a_2^-$ and $b_1^+ \leq b_2^+$,
- $D_1 = D_2 \iff a_1^- = a_2^-$ and $b_1^+ = b_2^+$,
- $D_1 < D_2 \iff D_1 \leq D_2$ and $D_1 \neq D_2$,
- $kD = k[a_1^-, b_1^+] = [ka_1^-, kb_1^+]$, where $0 \leq k \leq 1$.

Similarly,

$$\sup_{i \in I}\{[a_i^-, b_i^+]\} = [\sup_{i \in I}\{a_i^-\}, \sup_{i \in I}\{b_i^+\}] \quad \text{and} \quad \inf_{i \in I}\{[a_i^-, b_i^+]\} = [\inf_{i \in I}\{a_i^-\}, \inf_{i \in I}\{b_i^+\}].$$

It is known that $(D[0, 1], \leq, \vee, \wedge)$ is a complete lattice with $[0, 0]$ as the least element and $[1, 1]$ as the greatest.

Definition 3.2 The *interval-valued fuzzy set* A in X is defined by, $A = \{(x, [\mu_A^-(x), \mu_A^+(x)]) : x \in X\}$, where $\mu_A^-(x)$ and $\mu_A^+(x)$ are fuzzy subsets of X such that $\mu_A^-(x) \leq \mu_A^+(x)$, for all $x \in X$.

Let X be a non-empty set, then by an *interval-valued fuzzy relation* B on a set X we mean an interval-valued fuzzy set such that

$$\mu_B^-(xy) \leq \min(\mu_A^-(x), \mu_A^-(y)), \mu_B^+(xy) \leq \min(\mu_A^+(x), \mu_A^+(y)),$$

for all $xy \in X \times X$. In the clustering, the interval-valued fuzzy set A, is called an *interval-valued fuzzy class*. We define the *support* of A by supp $(A) = \{x \in X \mid [\mu_A^-(x), \mu_A^+(x)] \neq [0, 0]\}$ and say A is nontrivial if supp(A) is non-empty.

Interval-valued fuzzy relations reflect the idea that membership grades are often not precise and the intervals represent such uncertainty.

Definition 3.3 The *height* of an interval-valued fuzzy set $A = [\mu_A^-(x), \mu_A^+(x)]$ is defined as

$$h(A) = \sup(A)(x) = [\sup_{x \in X} \mu_A^-(x), \sup_{x \in X} \mu_A^+(x)].$$

We shall say that interval-valued fuzzy set A is *normal* if $A = [\mu_A^-(x), \mu_A^+(x)] = [1, 1]$, for all $x \in X$.

Definition 3.4 By an *interval-valued fuzzy graph* on non-empty set X, we mean a pair $G = (A, B)$, where $A = [\mu_A^-, \mu_A^+]$ is an interval-valued fuzzy set on X and $B = [\mu_B^-, \mu_B^+]$ is an interval-valued fuzzy relation on X such that

$$\mu_B^-(xy) \leq \min(\mu_A^-(x), \mu_A^-(y)),$$

$$\mu_B^+(xy) \leq \min(\mu_A^+(x), \mu_A^+(y)),$$

for all $x, y \in X$.

For further terminologies and studies on interval-valued fuzzy hypergraphs, readers are referred to [2, 3, 7, 9, 10, 16, 18, 21, 26, 27].

3.2 Interval-Valued Fuzzy Hypergraphs

Definition 3.5 Let X be a finite set and let $E = \{E_1, E_2, \ldots, E_m\}$ be a finite family of nontrivial interval-valued fuzzy subsets of X such that

$$X = \bigcup_j \text{supp}[\mu_j^-, \mu_j^+], \quad j = 1, 2, \ldots, m,$$

where $A = [\mu_j^-, \mu_j^+]$ is an interval-valued fuzzy set defined on $E_j \in E$. Then, the pair $I = (X, E)$ is an *interval-valued fuzzy hypergraph* on X, E is the family of interval-valued fuzzy edges of I and X is the (crisp) vertex set of I. The order of I (number of vertices) is denoted by $|X|$ and the number of edges is denoted by $|E|$.

Definition 3.6 Let $A = [\mu_A^-, \mu_A^+]$ be an interval-valued fuzzy subset of X and let E be a collection of interval-valued fuzzy subsets of X such that for each $B = [\mu_B^-, \mu_B^+] \in E$ and $x \in X$, $\mu_B^-(x) \leq \mu_A^-(x), \mu_B^+(x) \leq \mu_A^+(x)$. Then the pair (A, B) is an *interval-valued fuzzy hypergraph* on the interval-valued fuzzy set A. The interval-valued fuzzy hypergraph (A, B) is also an interval-valued fuzzy hypergraph on $X = \text{supp}(A)$, the interval-valued fuzzy set A defines a condition for interval-valued in the edge set E. This condition can be stated separately, so without loss of generality we restrict attention to interval-valued fuzzy hypergraphs on crisp vertex sets.

Example 3.1 Consider an interval-valued fuzzy hypergraph $I = (X, E)$ as shown in Fig. 3.1 such that $X = \{a, b, c, d\}$ and $E = \{E_1, E_2, E_3\}$, where

$$E_1 = \left\{ \frac{a}{[0.2, 0.3]}, \frac{b}{[0.4, 0.5]} \right\}, \quad E_2 = \left\{ \frac{b}{[0.4, 0.5]}, \frac{c}{[0.2, 0.5]} \right\}, \quad E_3 = \left\{ \frac{a}{[0.2, 0.3]}, \frac{d}{[0.2, 0.4]} \right\}.$$

The corresponding incidence matrix is given in Table 3.1.

Definition 3.7 An interval-valued fuzzy set $A = [\mu_A^-, \mu_A^+] : X \to D[0, 1]$ is an *elementary interval-valued fuzzy set* if A is single valued on supp(A). An *elementary interval-valued fuzzy hypergraph* $I = (X, E)$ is an interval-valued fuzzy hypergraph whose edges are elementary.

We explore the sense in which an interval-valued fuzzy graph is an interval-valued fuzzy hypergraph.

Proposition 3.1 *Interval-valued fuzzy graphs and interval-valued fuzzy digraphs are special cases of the interval-valued fuzzy hypergraphs.*

An interval-valued fuzzy multigraph is a multivalued symmetric mapping $D = [\mu_D^-, \mu_D^+] : X \times X \to D[0, 1]$. An interval-valued fuzzy multigraph can be considered to be the "disjoint union" or "disjoint sum" of a collection of simple interval-valued fuzzy graphs, as is done with crisp multigraphs. The same holds for multidigraphs. Therefore, these structures can be considered as "disjoint unions" or "disjoint sums" of interval-valued fuzzy hypergraphs.

Fig. 3.1 Interval-valued fuzzy hypergraph

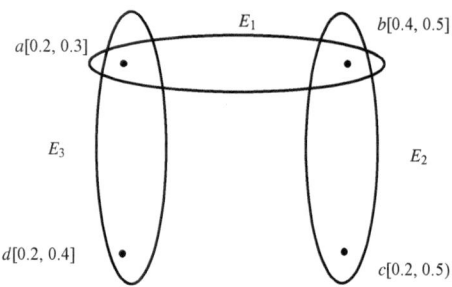

Table 3.1 The corresponding incidence matrix

M_I	E_1	E_2	E_3
a	[0.2, 0.3]	[0, 0]	[0.2, 0.3]
b	[0.4, 0.5]	[0.4, 0.5]	[0, 0]
c	[0, 0]	[0.2, 0.5]	[0, 0]
d	[0, 0]	[0, 0]	[0.2, 0.4]

Definition 3.8 An interval-valued fuzzy hypergraph $I = (X, E)$ is *simple* if $A = [\mu_A^-, \mu_A^+]$, $B = [\mu_B^-, \mu_B^+] \in E$ and $\mu_A^- \leq \mu_B^-, \mu_A^+ \leq \mu_B^+$ imply that $\mu_A^- = \mu_B^-, \mu_A^+ = \mu_B^+$.

An interval-valued fuzzy hypergraph $I = (X, E)$ is *support simple* if $A = [\mu_A^-, \mu_A^+]$, $B = [\mu_B^-, \mu_B^+] \in E$, supp$(A) =$ supp(B), and $\mu_A^- \leq \mu_B^-, \mu_A^+ \leq \mu_B^+$ imply that $\mu_A^- = \mu_B^-, \mu_A^+ = \mu_B^+$.

An interval-valued fuzzy hypergraph $I = (X, E)$ is *strongly support simple* if $A = [\mu_A^-, \mu_A^+]$, $B = [\mu_B^-, \mu_B^+] \in E$ and supp$(A) =$ supp(B) imply that $A = B$.

Remark 3.1 The Definition 3.8 reduces to familiar definitions in the special case where I is a crisp hypergraph. The interval-valued fuzzy definition of simple is identical to the crisp definition of simple. A crisp hypergraph is support simple and strongly support simple if and only if it has no multiple edges. For interval-valued fuzzy hypergraphs all three concepts imply no multiple edges. Simple interval-valued fuzzy hypergraphs are support simple and strongly support simple interval-valued fuzzy hypergraphs are support simple. Simple and strongly support simple are independent concepts.

Definition 3.9 Let $I = (X, E)$ be an interval-valued fuzzy hypergraph. Suppose that $\alpha, \beta \in [0, 1]$. Let

- $E_{[\alpha,\beta]} = \{A_{[\alpha,\beta]} | A \in E\}$, $A_{[\alpha,\beta]} = \{x \mid \mu_A^-(x) \leq \alpha \text{ or } \mu_A^+(x) \leq \beta\}$, and
- $X_{[\alpha,\beta]} = \bigcup_{A \in E} A_{[\alpha,\beta]}$.

If $E_{[\alpha,\beta]} \neq \emptyset$, then the crisp hypergraph $I_{[\alpha,\beta]} = (X_{[\alpha,\beta]}, E_{[\alpha,\beta]})$ is the $[\alpha, \beta]$–*level hypergraph* of I.

Clearly, it is possible that $A_{[\alpha,\beta]} = B_{[\alpha,\beta]}$ for $A \neq B$, by using distinct markers to identity the various members of E a distinction between $A_{[\alpha,\beta]}$ and $B_{[\alpha,\beta]}$ to represent multiple edges in $I_{[\alpha,\beta]}$. However, we do not take this approach unless otherwise stated, we will always regard $I_{[\alpha,\beta]}$ as having no repeated edges.

The families of crisp sets (hypergraphs) produced by the $[\alpha, \beta]$-cuts of an interval-valued fuzzy hypergraph share an important relationship with each other, as expressed below:

Suppose $\mathbb{X}$ and $\mathbb{Y}$ are two families of sets such that for each set X belonging to $\mathbb{X}$ there is at least one set Y belonging to $\mathbb{Y}$ which contains X. In this case, we say that $\mathbb{Y}$ *absorbs* $\mathbb{X}$ and symbolically write $\mathbb{X} \sqsubseteq \mathbb{Y}$ to express this relationship between $\mathbb{X}$ and $\mathbb{Y}$. Since, it is possible for $\mathbb{X} \sqsubseteq \mathbb{Y}$ while $\mathbb{X} \cap \mathbb{Y} = \emptyset$, we have that $\mathbb{X} \subseteq \mathbb{Y} \Rightarrow \mathbb{X} \sqsubseteq \mathbb{Y}$, whereas the converse is generally false. If $\mathbb{X} \sqsubseteq \mathbb{Y}$ and $\mathbb{X} \neq \mathbb{Y}$, then we write $\mathbb{X} \sqsubset \mathbb{Y}$.

Definition 3.10 Let $I = (X, E)$ be an interval-valued fuzzy hypergraph, and for $[0, 0] < [s, t] \leq h(I)$. Let $I_{[s,t]}$ be the $[s, t]$–level hypergraph of I. The sequence of real numbers $\{[s_1, r_1], [s_2, r_2], \ldots, [s_n, r_n]\}$, $[0, 0] < [s_1, r_1] < [s_2, r_2] < \cdots < [s_n, r_n] = h(I)$, which satisfies the properties,

- if $[s_{i+1}, r_{i+1}] < [u, v] \leq [s_i, r_i]$, then $E_{[u,v]} = E_{[s_i, r_i]}$,
- $E_{[s_i, r_i]} \sqsubset E_{[s_{i+1}, r_{i+1}]}$,

is called the *fundamental sequence* of I, and is denoted by $F(I)$ and the set of $[s_i, r_i]$-level hypergraphs $\{I_{[s_1,r_1]}, I_{[s_2,r_2]}, \ldots, I_{[s_n,r_n]}\}$ is called the *set of core hypergraphs* of I or, simply, the *core set* of I, and is denoted by $C(I)$.

Definition 3.11 Suppose $I = (X, E)$ is an interval-valued fuzzy hypergraph with $F(I) = \{[s_1, r_1], [s_2, r_2], \ldots, [s_n, r_n]\}$, and $s_{n+1} = 0$, $r_{n+1} = 0$. Then, I is called *sectionally elementary* if for each edge $A = (\mu_A^-, \mu_A^+) \in E$, each $i = \{1, 2, \ldots, n\}$, and $[s_i, r_i] \in F(I)$, $A_{[s,t]} = A_{[s_i,r_i]}$, for all $[s, t] \in ([s_{i+1}, r_{i+1}], [s_i, r_i]]$.

Clearly I is sectionally elementary if and only if $A(x) = (\mu_A^-(x), \mu_A^+(x)) \in F(I)$ for each $A \in E$ and each $x \in X$.

Definition 3.12 A sequence of crisp hypergraphs $I_i = (X_i, E_i^*)$, $[1, 1] \le i \le [n, n]$, is said to be *ordered* if $I_1 \subset I_2 \subset \cdots \subset I_n$. The sequence $\{I_i \mid [1, 1] \le i \le [n, n]\}$ is *simply ordered* if it is ordered and if whenever $E^* \in E_{i+[1,1]}^* - E_i^*$, then $E^* \not\subseteq X_i$.

Definition 3.13 An interval-valued fuzzy hypergraph I is *ordered* if the I induced fundamental sequence of hypergraphs is ordered. The interval-valued fuzzy hypergraph I is *simply ordered* if the I induced fundamental sequence of hypergraphs is simply ordered.

Example 3.2 Consider the interval-valued fuzzy hypergraph $I = (X, E)$, where $X = \{a, b, c, d\}$ and $E = \{E_1, E_2, E_3, E_4, E_5\}$ which is represented by the following incidence matrix Table 3.2.

Clearly, $h(I) = [0.3, 0.9]$.

Now

$$E_{[0.1,0.9]} = \{\{a, b\}, \{b, c\}\},$$

$$E_{[0.2,0.7)} = \{\{a, b\}, \{b, c\}\},$$

$$E_{[0.3,0.4]} = \{\{a, b\}, \{a, b, d\}, \{b, c\}, \{b, c, d\}, \{a, c, d\}\}.$$

Thus, for $[0.3, 0.4] < [s, t] \le [0.1, 0.9]$, $E_{[s,t]} = \{\{a, b\}, \{b, c\}\}$, and for $[0, 0] < [s, t] \le [0.3, 0.4]$,

$$E_{[s,t]} = \{\{a, b\}, \{a, b, d\}, \{b, c\}, \{b, c, d\}, \{a, c, d\}\}.$$

Table 3.2 Incidence matrix of I

I	E_1	E_2	E_3	E_4	E_5
a	[0.2, 0.7]	[0.0, 0.9]	[0, 0]	[0, 0]	[0.3, 0.4]
b	[0.2, 0.7]	[0.0, 0.9]	[0.0, 0.9]	[0.2, 0.7]	[0, 0]
c	[0, 0]	[0, 0]	[0.0, 0.9]	[0.2, 0.7]	[0.3, 0.4]
d	[0, 0]	[0.3, 0.4]	[0, 0]	[0.3, 0.4]	[0.3, 0.4]

We note that $E_{[0.1,0.9]} \subseteq E_{[0.3,0.4]}$. The fundamental sequence is $F(I) = \{[s_1, r_1] = [0.1, 0.9], [s_2, r_2] = [0.3, 0.4]\}$ and the set of core hypergraph is $C(I) = \{I_1 = (X_1, E_1) = I_{[0.1,0.9]}, I_2 = (X_2, E_2) = I_{[0.3,0.4]}\}$, where

$$X_1 = \{a, b, c\}, \ E_1 = \{\{a, b\}, \{b, c\}, \}$$

$$X_2 = \{a, b, c, d\}, \ E_2 = \{\{a, b\}, \{a, b, d\}, \{b, c\}, \{b, c, d\}, \{a, c, d\}\}.$$

I is support simple, but not simple. I is not sectionally elementary since $E_{1[s,t]} \neq E_{1[0.1,0.9]}$ for $s = 0.2, t = 0.7$. Clearly, interval-valued fuzzy hypergraph I is simply ordered.

Proposition 3.2 *Let $I = (X, E)$ be an elementary interval-valued fuzzy hypergraph. Then, I is support simple if and only if I is strongly support simple.*

Proof Suppose that I is elementary, support simple and that supp(A) = supp(B). We assume without loss of generality that $h(A) \leq h(B)$. Since, I is elementary, it follows that $\mu_A^- \leq \mu_B^-$, $\mu_A^+ \leq \mu_B^+$ and since I is support simple then $\mu_A^- = \mu_B^-$, $\mu_A^+ = \mu_B^+$. Therefore, I is strongly support simple. The proof of converse part is obvious.

The complexity of an interval-valued fuzzy hypergraph depends in part on how many edges it has. The natural question arises: is there an upper bound on the number of edges of an interval-valued fuzzy hypergraph of order n?

Proposition 3.3 *Let $I = (X, E)$ be a simple interval-valued fuzzy hypergraph of order n. Then, there is no upper bound on $|E|$.*

Proof Let $X = \{x, y\}$, and define $E_N = \{A_i = [\mu_{A_i}^-, \mu_{A_i}^+] \mid i = 1, 2, \ldots, N\}$, where

$$\mu_{A_i}^-(x) = \frac{1}{i+1}, \ \mu_{A_i}^+(x) = 1 - \frac{1}{i+1},$$

$$\mu_{A_i}^-(y) = \frac{1}{i+1}, \ \mu_{A_i}^+(y) = \frac{i}{i+1}.$$

Then $I_N = (X, E_N)$ is a simple interval-valued fuzzy hypergraph with N edges. This ends the proof.

Proposition 3.4 *Let $I = (X, E)$ be a support simple interval-valued fuzzy hypergraph of order n. Then, there is no upper bound on $|E|$.*

Proposition 3.5 *Let $I = (X, E)$ be a strongly support simple interval-valued fuzzy hypergraph of order n. Then, there is no upper bound on $|E| \leq 2^n - 1$ if and only if $\{supp(A) \mid A \in E\} = P(X) - \emptyset$.*

Proposition 3.6 *Let $I = (X, E)$ be an elementary simple interval-valued fuzzy hypergraph of order n. Then, there is no upper bound on $|E| \leq 2^n - 1$ if and only if $\{supp(A) \mid A \in E\} = P(X) - \emptyset$.*

Proof Since I is elementary and simple, each nontrivial $W \subseteq X$ can be the support of at most one $A = (\mu_A^-, \mu_A^+) \in E$. Therefore, $|E| \leq 2^n - 1$. To show there exists an elementary, simple I with $|E| = 2^n - 1$, let $E = \{A = (\mu_A^-, \mu_A^+) \mid W \subseteq X\}$ be the set of functions defined by

$$\mu_A^-(x) = \frac{1}{|W|}, \text{ if } x \in W, \quad \mu_A^-(x) = 0, \text{ if } x \notin W,$$

$$\mu_A^+(x) = 1 - \frac{1}{|W|}, \text{ if } x \in W, \quad \mu_A^+(x) = 1, \text{ if } x \notin W.$$

Then, each one element has height $[0, 1]$, each two elements have height $[0.5, 0.5]$ and so on. Hence, I is an elementary and simple, and $|E| = 2^n - 1$.

Proposition 3.7 *(a) If $I = (X, E)$ is an elementary interval-valued fuzzy hypergraph, then I is ordered.*
(b) If I is an ordered interval-valued fuzzy hypergraph with simple support hypergraph, then I is elementary.

Consider the situation where the vertex of a crisp hypergraph is fuzzified. Suppose that each edge is given a uniform degree of membership consistent with the weakest vertex of the edge. Some constructions describe the following subclass of interval-valued fuzzy hypergraphs.

Definition 3.14 An interval-valued fuzzy hypergraph $I = (X, E)$ is called a $A = [\mu_A^-, \mu_A^+]$-*tempered* interval-valued fuzzy hypergraph of $I = (X, E)$ if there is a crisp hypergraph $I^* = (X, E^*)$ and an interval-valued fuzzy set $A = [\mu_A^-, \mu_A^+] :$ $X \to D(0, 1]$ such that $E = \{B_F = [\mu_{B_F}^-, \mu_{B_F}^+] \mid F \in E^*\}$, where

$$\mu_{B_F}^-(x) = \begin{cases} \min(\mu_A^-(y) \mid y \in F), & \text{if } x \in F, \\ 0, & \text{otherwise,} \end{cases}$$

$$\mu_{B_F}^+(x) = \begin{cases} \min(\mu_A^+(y) \mid y \in F), & \text{if } x \in F, \\ 1, & \text{otherwise.} \end{cases}$$

Let $A \otimes I$ denote the A-tempered interval-valued fuzzy hypergraph of I determined by the crisp hypergraph $I^* = (X, E^*)$ and the interval-valued fuzzy set $A : X \to D(0, 1]$.

Example 3.3 Consider the interval-valued fuzzy hypergraph $I = (X, E)$, where $X = \{a, b, c, d\}$ and $E = \{E_1, E_2, E_3, E_4\}$ which is represented by the following incidence matrix given in Table 3.3.

Table 3.3 Incidence matrix of I

I	E_1	E_2	E_3	E_4
a	[0.2, 0.7]	[0.0, 0.0]	[0, 0]	[0.2, 0.7]
b	[0.2, 0.7]	[0.3, 0.4]	[0.0, 0.9]	[0.0, 0.0]
c	[0, 0]	[0, 0]	[0.0, 0.9]	[0.2, 0.7]
d	[0, 0]	[0.3, 0.4]	[0, 0]	[0.0, 0.0]

Then, $E_{[0.0,0.9]} = \{\{b, c\}\}$, $E_{[0.2,0.7)} = \{\{a, b\}, \{a, c\}, \{b, c\}\}$, and $E_{[0.3,0.4]} = \{\{a, b\}, \{a, c\}, \{b, c\}, \{b, d\}\}$. Define $A = [\mu_A^-, \mu_A^+] : X \to D(0, 1]$ by

$$\mu_A^-(a) = 0.2, \ \mu_A^-(b) = \mu_A^-(c) = 0.0, \ \mu_A^-(d) = 0.3,$$

$$\mu_A^+(a) = 0.7, \ \mu_A^+(b) = \mu_A^+(c) = 0.9, \ \mu_A^+(d) = 0.4.$$

Note that

$$\mu_{B_{\{a,b\}}}^-(a) = \min(\mu_A^-(a), \mu_A^-(b)) = 0.0, \ \mu_{B_{\{a,b\}}}^-(b) = \min(\mu_A^-(a), \mu_A^-(b)) = 0.0,$$

$$\mu_{B_{\{a,b\}}}^-(c) = 0.0, \ \mu_{B_{\{a,b\}}}^-(d) = 0.0,$$

$$\mu_{B_{\{a,b\}}}^+(a) = \min(\mu_A^+(a), \mu_A^+(b)) = 0.7, \ \mu_{B_{\{a,b\}}}^+(b) = \min(\mu_A^+(a), \mu_A^+(b)) = 0.7$$

$$\mu_{B_{\{a,b\}}}^+(c) = 1.0, \ \mu_{B_{\{a,b\}}}^+(d) = 1.0.$$

Thus,

$$E_1 = [\mu_{B_{\{a,b\}}}^-, \mu_{B_{\{a,b\}}}^+], \ E_2 = [\mu_{B_{\{b,d\}}}^-, \mu_{B_{\{b,d\}}}^+],$$

$$E_3 = [\mu_{B_{\{b,c\}}}^-, \mu_{B_{\{b,c\}}}^+], \ E_4 = [\mu_{B_{\{a,c\}}}^-, \mu_{B_{\{a,c\}}}^+].$$

Hence, I is A-tempered hypergraph.

Proposition 3.8 *An interval-valued fuzzy hypergraph I is an A-tempered interval-valued fuzzy hypergraph of some crisp hypergraph I^* if and only if I is elementary, support simple, and simply ordered.*

Proof Suppose that $I = (X, E)$ is an A-tempered interval-valued fuzzy hypergraph of some crisp hypergraph I^*. Clearly, I is elementary and support simple. We show that I is simply ordered. Let

$$C(I) = \{(I_1^*)^{r_1} = (X_1, E_1^*), \ (I_2^*)^{r_2} = (X_2, E_2^*), \dots, \ (I_n^*)^{r_n} = (X_n, E_n^*)\}.$$

Since I is elementary, it follows from Proposition 3.7 that I is ordered. To show that I is simply ordered, suppose that there exists $F \in E_{i+1}^* \setminus E_i^*$. Then, there exists $x^* \in F$ such that $\mu_A^-(x^*) = r_{i+1}$, $\mu_A^+(x^*) = \acute{r}_{i+1}$. Since $\mu_A^-(x^*) = r_{i+1} < r_i$ and $\mu_A^+(x^*) = \acute{r}_{i+1} < \acute{r}_i$, it follows that $x^* \notin X_i$ and $F \nsubseteq X_i$, hence I is simply ordered.

Conversely, suppose $I = (X, E)$ is elementary, support simple and simply ordered. Let

$$C(I) = \{(I_1^*)^{r_1} = (X_1, E_1^*),\ (I_2^*)^{r_2} = (X_2, E_2^*), \ldots,\ (I_n^*)^{r_n} = (X_n, E_n^*)\},$$

where $D(I) = \{r_1, r_2, \ldots, r_n\}$ with $0 < r_n < \cdots < r_1$. Since $(I^*)^{r_n} = I_n^* = (X_n, E_n^*)$ and define $A = [\mu_A^-, \mu_A^+] : X_n \to D(0, 1]$ by

$$\mu_A^-(x) = \begin{cases} r_1, & \text{if } x \in X_1, \\ r_i, & \text{if } x \in X_i \setminus X_{i-1}, i = 1, 2, \ldots, n \end{cases} \qquad \mu_A^+(x) = \begin{cases} s_1, & \text{if } x \in X_1, \\ s_i, & \text{if } x \in X_i \setminus X_{i-1}, i = 1, 2, \ldots, n \end{cases}$$

We show that $E = \{B_F = [\mu_{B_F}^-, \mu_{B_F}^+] \mid F \in E^*\}$, where

$$\mu_{B_F}^-(x) = \begin{cases} \min(\mu_A^-(y) \mid y \in F), & \text{if } x \in F, \\ 0, & \text{otherwise}, \end{cases} \qquad \mu_{B_F}^+(x) = \begin{cases} \min(\mu_A^+(y) \mid y \in F), & \text{if } x \in F, \\ 1, & \text{otherwise}. \end{cases}$$

Let $F \in E_n^*$. Since I is elementary and support simple, there is a unique interval-valued fuzzy edge $C_F = [\mu_{C_F}^-, \mu_{C_F}^+]$ in E having support E^*. Indeed, distinct edges in E must have distinct supports that lie in E_n^*. Thus, to show that $E = \{B_F = [\mu_{B_F}^-, \mu_{B_F}^+] \mid F \in E_n^*\}$, it suffices to show that for each $F \in E_n^*$, $\mu_{C_F}^- = \mu_{B_F}^-$ and $\mu_{C_F}^+ = \mu_{B_F}^+$. As all edges are elementary and different edges have different supports, it follows from the definition of fundamental sequence that $h(C_F)$ is equal to some number r_i of $D(I)$. Consequently, $E^* \subseteq X_i$. Moreover, if $i > 1$, then $F \in E^* \setminus E_{i-1}^*$. Since $F \subseteq X_i$, it follows from the definition of $A = [\mu_A^-, \mu_A^+]$ that for each $x \in F$, $\mu_A^-(x) \geq r_i$ and $\mu_A^+(x) \geq s_i$. We claim that $\mu_A^-(x) = r_i$ and $\mu_A^+(x) = s_i$, for some $x \in F$. If not, then by definition of $A = [\mu_A^-, \mu_A^+]$, $\mu_A^-(x) \geq r_i$ and $\mu_A^+(x) \geq s_i$ for all $x \in F$ which implies that $F \subseteq X_{i-1}$ and so $F \in E^* \setminus E_{i-1}^*$ and since I is simply ordered $F \nsubseteq X_{i-1}$, a contradiction. Thus, it follows from the definition of B_F that $B_F = C_F$. This completes the proof. $\qquad \blacksquare$

As a consequence of the above theorem we obtain.

Proposition 3.9 *Suppose that I is a simply ordered interval-valued fuzzy hypergraph and $F(I) = \{r_1, r_2, \ldots, r_n\}$. If I^{r_n} is a simple hypergraph, then there is a partial interval-valued fuzzy hypergraph $\acute{I}$ of I such that the following assertions hold:*

1. *$\acute{I}$ is an A-tempered interval-valued fuzzy hypergraph of I_n.*
2. *$E \sqsubseteq \acute{E}$.*
3. *$F(\acute{I}) = F(I)$ and $C(\acute{I}) = C(I)$.*

3.3 Vague Hypergraphs

Different authors from time to time have made a number of generalizations of Zadeh's [25] fuzzy set theory. The notion of vague set was introduced by Gau and Buehrer [11]. This is because in most cases of judgments, the evaluation is done by human beings and so the certainty is a limitation of knowledge or intellectual functionaries. Naturally, every decision-maker hesitates more or less on every evaluation activity. For example, in order to judge whether a patient has cancer or not, a medical doctor (the decision-maker) will hesitate because of the fact that a fraction of evaluation he thinks in favor of the truthness, another fraction in favor of the falseness and the rest part remains undecided to him. This is the breaking philosophy in the notion of vague set theory introduced by Gau and Buehrer [11].

Definition 3.15 A *vague set* A in the universe of discourse X is a pair (t_A, f_A), where $t_A : X \rightarrow [0, 1]$, $f_A : X \rightarrow [0, 1]$ are true and false memberships, respectively such that $t_A(x) + f_A(x) \leq 1$, for all $x \in X$.

In the above definition, $t_A(x)$ is considered as the lower bound for degree of membership of x in A (based on evidence), and $f_A(x)$ is the lower bound for negation of membership of x in A (based on evidence against). Therefore, the degree of membership of x in the vague set A is characterized by the interval $[t_A(x), 1 - f_A(x)]$. So, a vague set is a special case of interval-valued sets. The interval $[t_A(x), 1 - f_A(x)]$ is called the *vague value* of x in A, and is denoted by $X_A(x)$. We denote zero vague and unit vague value by $\mathbf{0} = [0, 0]$ and $\mathbf{1} = [1, 1]$, respectively. It is worth to mention here that interval-valued fuzzy sets are not vague sets. In interval-valued fuzzy sets, an interval-valued membership value is assigned to each element of the universe considering the "evidence for x" only, without considering "evidence against x". In vague sets both are independently proposed by the decision-maker. This makes a major difference in the judgment about the grade of membership.

Remark 3.2 The intuitionistic fuzzy sets and vague sets look similar, analytically vague sets are more appropriate when representing vague data. The difference between them is discussed below. The membership interval of element x for vague set A is $[t_A(x), 1 - f_A(x)]$. But, the membership value for element x in an intuitionistic fuzzy set B is $< x, \mu_B(x), \nu_B(x) >$. Here the semantics of t_A is the same as with A and μ_B is the same as with B. However, the boundary is able to indicate the possible existence of a data value. This difference gives rise to a simpler but meaningful graphical view of data sets (see Fig. 3.2).

A vague relation is a generalization of a fuzzy relation.

Definition 3.16 Let X and Y be ordinary finite non-empty sets. We call a *vague relation* to be a vague subset of $X \times Y$, that is, an expression R defined by

$$R = \{< (x, y), t_R(x, y), f_R(x, y) > | x \in X, y \in Y\},$$

where $t_R : X \times Y \rightarrow [0, 1]$, $f_R : X \times Y \rightarrow [0, 1]$, which satisfies the condition $0 \leq t_R(x, y) + f_R(x, y) \leq 1$, for all $(x, y) \in X \times Y$. A vague relation R on X is called

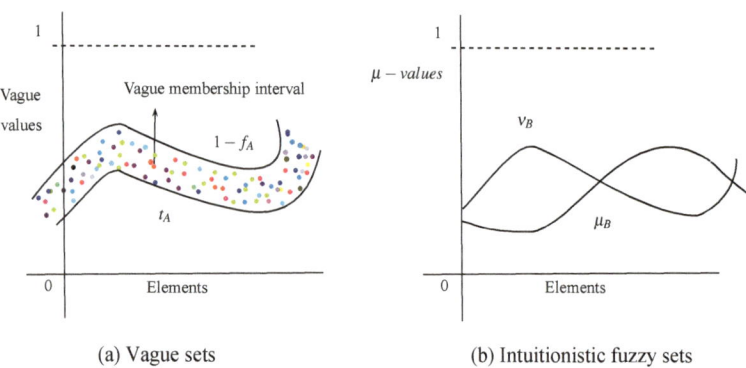

(a) Vague sets   (b) Intuitionistic fuzzy sets

Fig. 3.2 Comparison between vague sets and intuitionistic fuzzy sets

reflexive if $t_R(x, x) = 1$ and $f_R(x, x) = 0$, for all $x \in X$. A vague relation R on X is *symmetric* if $t_R(x, y) = t_R(y, x)$ and $f_R(x, y) = f_R(y, x)$, for all $x, y \in X$.

Definition 3.17 Let $A = (t_A, f_A)$ be a vague set on X and let $\alpha, \beta \in [0, 1]$ be such that $\alpha \leq \beta$. Then, the set $A_{(\alpha, \beta)} = \{x \mid t_A(x) \geq \alpha, \ 1 - f_A(x) \geq \beta\}$ is called a (α, β)-*(weakly) cut set* of A. $A_{(\alpha, \beta)}$ is a crisp set.

Definition 3.18 The *support* of A is defined by $\text{supp}(A) = \{x \in X \mid (t_A(x),$ $f_A(x)) \neq (0, 0)\}$ and we say A is nontrivial if $\text{supp}(A)$ is non-empty. The *height* of a vague set A is defined as $h(A) = \sup_{x \in X} (A)(x)$.

Definition 3.19 A vague relation B on a set X is a vague relation from X to X. If A is a vague set on a set X, then a vague relation B on A is a vague relation which satisfies, $t_B(xy) \leq \min(t_A(x), t_A(y))$ and $f_B(xy) \geq \max(f_A(x), f_A(y))$, for all $xy \in E \subseteq X \times X$.

Definition 3.20 Let X be a non-empty set, members of X are called *nodes*. A *vague graph* $G = (A, B)$ with X as the set of nodes, is a pair of functions A, B, where A is a vague set of X and B is a vague relation on X. We note that vague relation B in vague digraph need not to be symmetric.

We now define vague hypergraph,

Definition 3.21 Let X be a finite set and let $\mathbb{E} = \{\mathbb{E}_1, \mathbb{E}_2, \ldots, \mathbb{E}_m\}$ be a finite family of nontrivial vague subsets of X such that $X = \bigcup_j \text{supp}\mathbb{E}_j, \ j = 1, 2, \ldots, m$. Then, the pair $\mathbb{H} = (X, \mathbb{E})$ is a *vague hypergraph* on X, $\mathbb{E}$ is the family of vague edges of $\mathbb{H}$ and X is the (crisp) vertex set of $\mathbb{H}$.

Definition 3.22 Let $A = (t_A, f_A)$ be a vague subset of X and let $\mathbb{E}$ be a collection of vague subsets of X such that for each $B = (t_B, f_B) \in \mathbb{E}$ and $x \in X, t_A(x) \leq t_B(x)$, $f_B(x) \geq f_A(x)$. Then, the pair (A, B) is a vague hypergraph on the vague set A. The vague hypergraph (A, B) is also a vague hypergraph on $X = \text{supp}(A)$, the vague set A

defines a condition for interval-valued in the edge set $\mathbb{E}$. This condition can be stated separately, so without loss of generality we restrict attention to vague hypergraphs on crisp vertex sets.

Definition 3.23 A vague set A is an elementary vague set if A is single valued on supp(A). An *elementary vague hypergraph* $\mathbb{H} = (X, \mathbb{E})$ is a vague hypergraph whose edges are elementary.

Definition 3.24 A vague hypergraph $\mathbb{H} = (X, \mathbb{E})$ is *simple* if $A = (t_A, f_A)$, $B = (t_B, f_B) \in E$ and $t_A \leq t_B$, $f_A \geq f_B$ imply that $t_A = t_B$, $f_A = f_B$.

A vague hypergraph $\mathbb{H} = (X, \mathbb{E})$ is *support simple* if $A = (t_A, f_A)$, $B = (t_B, f_B) \in E$, supp(A) = supp(B), and $t_A \leq t_B$, $f_A \geq f_B$ imply that $t_A = t_B$, $f_A = f_B$.

A vague hypergraph $\mathbb{H} = (X, \mathbb{E})$ is *strongly support simple* if $A = (t_A, f_A)$, $B = (t_B, f_B) \in E$, and supp(A) = supp(B) imply that $A = B$.

Example 3.4 Consider a vague hypergraph $\mathbb{H} = (X, \mathbb{E})$ as shown in Fig. 3.3 such that $X = \{a, b, c, d\}$ and $\mathbb{E} = \{\mathbb{E}_1, \mathbb{E}_2, \mathbb{E}_3\}$, where

$$\mathbb{E}_1 = \left\{ \frac{a}{(0.2, 0.3)}, \frac{b}{(0.4, 0.5)} \right\}, \quad \mathbb{E}_2 = \left\{ \frac{b}{(0.4, 0.5)}, \frac{c}{(0.2, 0.5)} \right\}, \quad \mathbb{E}_3 = \left\{ \frac{a}{(0.2, 0.3)}, \frac{d}{(0.2, 0.4)} \right\}.$$

The corresponding incidence matrix is given below in Table 3.4.

Fig. 3.3 Vague hypergraph

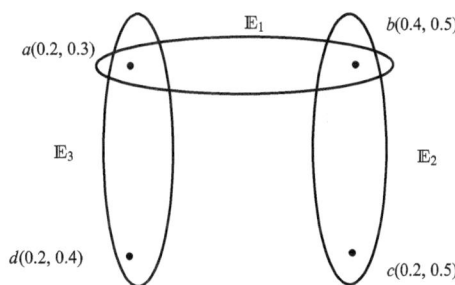

Table 3.4 The incidence matrix of vague hypergraph

$M_{\mathbb{H}}$	$\mathbb{E}_1$	$\mathbb{E}_2$	$\mathbb{E}_3$
a	(0.2, 0.3)	(0, 0)	(0.2, 0.3)
b	(0.4, 0.5)	(0.4, 0.5)	(0, 0)
c	(0, 0)	(0.2, 0.5)	(0,0)
d	(0,0)	(0,0)	(0.2, 0.4)

Definition 3.25 Let $\mathbb{H} = (X, \mathbb{E})$ be a vague hypergraph. Suppose that $\alpha, \beta \in [0, 1]$. Let

- $\mathbb{E}_{(\alpha,\beta)} = \{A_{(\alpha,\beta)} \mid A \in E\}$, $A_{(\alpha,\beta)} = \{x \mid t_A(x) \geq \alpha, \, 1 - f_A(x) \geq \beta\}$, and
- $X_{(\alpha,\beta)} = \bigcup_{A \in \mathbb{E}} A_{(\alpha,\beta)}$.

If $\mathbb{E}_{(\alpha,\beta)} \neq \emptyset$, then the crisp hypergraph $\mathbb{H}_{(\alpha,\beta)} = (X_{(\alpha,\beta)}, \mathbb{E}_{(\alpha,\beta)})$ is the (α, β)–*level hypergraph* of $\mathbb{H}$.

Clearly, it is possible that $A_{(\alpha,\beta)} = B_{(\alpha,\beta)}$ for $A \neq B$, by using distinct markers to identity the various members of $\mathbb{E}$ a distinction between $A_{(\alpha,\beta)}$ and $B_{(\alpha,\beta)}$ to represent multiple edges in $\mathbb{H}_{(\alpha,\beta)}$. However, we do not take this approach unless otherwise stated, we will always regard $\mathbb{H}_{(\alpha,\beta)}$ as having no repeated edges.

The families of crisp sets (hypergraphs) produced by the (α, β)-cuts of a vague hypergraph share an important relationship with each other, as expressed below:

Suppose $\mathbb{X}$ and $\mathbb{Y}$ are two families of sets such that for each set X belonging to $\mathbb{X}$ there is at least one set Y belonging to $\mathbb{Y}$ which contains X. In this case we say that $\mathbb{Y}$ *absorbs* $\mathbb{X}$ and symbolically write $\mathbb{X} \sqsubseteq \mathbb{Y}$ to express this relationship between $\mathbb{X}$ and $\mathbb{Y}$. Since it is possible for $\mathbb{X} \sqsubseteq \mathbb{Y}$ while $\mathbb{X} \cap \mathbb{Y} = \emptyset$, we have that $\mathbb{X} \subseteq \mathbb{Y} \Rightarrow \mathbb{X} \sqsubseteq \mathbb{Y}$, whereas the converse is generally false. If $\mathbb{X} \sqsubseteq \mathbb{Y}$ and $\mathbb{X} \neq \mathbb{Y}$, then we write $\mathbb{X} \sqsubset \mathbb{Y}$.

Definition 3.26 Let $\mathbb{H} = (X, \mathbb{E})$ be a vague hypergraph, and for $(0, 0) < (s, t) \leq h(\mathbb{H})$. Let $\mathbb{H}_{(s,t)}$ be the (s, t)–level hypergraph of $\mathbb{H}$. The sequence of real numbers,

$$\{(s_1, r_1), (s_2, r_2), \ldots, (s_n, r_n)\}, \quad 0 < s_1 < s_2 < \cdots < s_n \text{ and } 0 > r_1 > r_2 > \cdots > r_n,$$

$$\text{where } (s_n, r_n) = h(\mathbb{H}),$$

which satisfies the properties

- if $s_i < u \leq s_{i+1}$ and $r_i > v \geq r_{i+1}$, then $\mathbb{E}_{(u,v)} = \mathbb{E}_{(s_{i+1}, r_{i+1})}$, and
- $\mathbb{E}_{(s_i, r_i)} \sqsubseteq \mathbb{E}_{(s_{i+1}, r_{i+1})}$,

is called the *fundamental sequence* of $\mathbb{H}$, and is denoted by $F(\mathbb{H})$ and the set of (s_i, r_i)-level hypergraphs $\{\mathbb{H}_{((s_1, r_1))}, \mathbb{H}_{(s_2, r_2)}, \ldots, \mathbb{H}_{(s_n, r_n)}\}$ is called the *set of core hypergraphs* of $\mathbb{H}$ or, simply, the *core set* of $\mathbb{H}$, and is denoted by $C(\mathbb{H})$.

Definition 3.27 Suppose $\mathbb{H} = (X, \mathbb{E})$ is vague hypergraph with $F(\mathbb{H}) = \{(s_1, r_1), (s_2, r_2), \ldots, (s_n, r_n)\}$, and $s_{n+1} = 0$, $r_{n+1} = 0$, Then, $\mathbb{H}$ is called *sectionally elementary* if for each edge $A = (t_A, f_A) \in \mathbb{E}$, each $i = \{1, 2, \ldots, n\}$, and $(s_i, r_i) \in F(\mathbb{H})$, $A_{(s,t)} = A_{(s_i, r_i)}$ for all $(s, t) \in ((s_{i+1}, r_{i+1}), (s_i, r_i)]$.

Clearly, $\mathbb{H}$ is sectionally elementary if and only if $A(x) = (t_A(x), f_A(x)) \in F(\mathbb{H})$ for each $A \in \mathbb{E}$ and each $x \in X$.

Definition 3.28 A sequence of crisp hypergraphs $\mathbb{H}_i = (X_i, E_i^*)$, $1 \leq i \leq n$, is said to be *ordered* if $\mathbb{H}_1 \subset \mathbb{H}_2 \subset \ldots \subset \mathbb{H}_n$. The sequence $\{\mathbb{H}_i \mid 1 \leq i \leq n\}$ is *simply ordered* if it is ordered and if whenever $E^* \in E_{i+1}^* - E_i^*$, then $E^* \not\subseteq X_i$.

Definition 3.29 A vague hypergraph $\mathbb{H}$ is *ordered* if the $\mathbb{H}$ induced fundamental sequence of hypergraphs is ordered. The vague hypergraph $\mathbb{H}$ is *simply ordered* if the $\mathbb{H}$ induced fundamental sequence of hypergraphs is simply ordered.

We state the following Propositions without their proofs.

Proposition 3.10 *Let $\mathbb{H} = (X, \mathbb{E})$ be an elementary vague hypergraph. Then, $\mathbb{H}$ is support simple if and only if $\mathbb{H}$ is strongly support simple.*

Proposition 3.11 *Let $\mathbb{H} = (X, \mathbb{E})$ be a simple vague hypergraph of order n. Then, there is no upper bound on $|\mathbb{E}|$.*

Proof Let $X = \{x, y\}$, and define $\mathbb{E}_N = \{A_i = (t_{A_i}, f_{A_i}) \mid i = 1, 2, \ldots, N\}$, where

$$t_{A_i}(x) = \frac{1}{i+1}, \quad f_{A_i}(x) = 1 - \frac{1}{i+1},$$

$$t_{A_i}(y) = \frac{1}{i+1}, \quad f_{A_i}(y) = \frac{i}{i+1}.$$

Then $\mathbb{H}_N = (X, \mathbb{E}_N)$ is a simple vague hypergraph with N edges. This ends the proof.

Proposition 3.12 *Let $\mathbb{H} = (X, \mathbb{E})$ be a support simple vague hypergraph of order n. Then, there is no upper bound on $|\mathbb{E}|$.*

Proposition 3.13 *Let $\mathbb{H} = (X, \mathbb{E})$ be a strongly support simple vague hypergraph of order n. Then, there is no upper bound on $|\mathbb{E}| \leq 2^n - 1$ if and only if $\{supp(A) \mid A \in \mathbb{E}\} = P(X) - \emptyset$.*

Proposition 3.14 *Let $\mathbb{H} = (X, \mathbb{E})$ be an elementary simple vague hypergraph of order n. Then, there is no upper bound on $|\mathbb{E}| \leq 2^n - 1$ if and only if $\{supp(A) \mid A \in \mathbb{E}\} = P(X) - \emptyset$.*

Proof Since $\mathbb{H}$ is elementary and simple, each nontrivial $W \subseteq X$ can be the support of at most one $A = (t_A, f_A) \in \mathbb{E}$. Therefore, $|\mathbb{E}| \leq 2^n - 1$. To show there exists an elementary, simple $\mathbb{H}$ with $|\mathbb{E}| = 2^n - 1$, let $\mathbb{E} = \{A = (t_A, f_A) \mid W \subseteq X\}$ be the set of functions defined by

$$t_A(x) = \frac{1}{|W|}, \text{ if } x \in W, \quad t_A(x) = 0, \text{ if } x \notin W,$$

$$f_A(x) = 1 - \frac{1}{|W|}, \text{ if } x \in W, \quad f_A(x) = 1, \text{ if } x \notin W.$$

Then, each one element has height $(1, 0)$, each two elements have height $(0.5, 0.5)$ and so on. Hence, $\mathbb{H}$ is an elementary and simple, and $|\mathbb{E}| = 2^n - 1$.

Proposition 3.15 (a) If $\mathbb{H} = (X, \mathbb{E})$ is an elementary vague hypergraph, then $\mathbb{H}$ is ordered.

(b) If $\mathbb{H}$ is an ordered vague hypergraph with simple support hypergraph, then $\mathbb{H}$ is elementary.

Definition 3.30 The *dual of a vague hypergraph* $\mathbb{H} = (X, \mathbb{E})$ is a vague hypergraph $\mathbb{H}^D = (\mathbb{E}^D, X^D)$ whose vertex set is the edge set of $\mathbb{H}$ and with edges $X^D : \mathbb{E}^D \to [0, 1] \times [0, 1]$ by $X^D(A^D) = (t_A^D(x), f_A^D(x))$. $\mathbb{H}^D$ is a vague hypergraph whose incidence matrix is the transpose of the incidence matrix of $\mathbb{H}$, thus $\mathbb{H}^{DD} = \mathbb{H}$.

Example 3.5 Consider a vague hypergraph $\mathbb{H} = (X, \mathbb{E})$ as shown in Fig. 3.4 such that $X = \{x_1, x_2, x_3, x_4\}$, $\mathbb{E} = \{\mathbb{E}_1, \mathbb{E}_2, \mathbb{E}_3, \mathbb{E}_4\}$, where $\mathbb{E}_1 = \left\{ \frac{x_1}{(0.5, 0.3)}, \frac{x_2}{(0.4, 0.2)} \right\}$, $\mathbb{E}_2 = \left\{ \frac{x_2}{(0.4, 0.2)}, \frac{x_3}{(0.3, 0.6)} \right\}$, $\mathbb{E}_3 = \left\{ \frac{x_3}{(0.3, 0.6)}, \frac{x_4}{(0.5, 0.1)} \right\}$, $\mathbb{E}_4 = \left\{ \frac{x_4}{(0.5, 0.1)}, \frac{x_1}{(0.5, 0.3)} \right\}$.

The corresponding incidence matrix of $\mathbb{H}$ is given in Table 3.5.

Consider the dual vague hypergraph $\mathbb{H}^D = (\mathbb{E}^D, X^D)$ of $\mathbb{H}$ such that $\mathbb{E}^D = \{e_1, e_2, e_3, e_4\}$, $X^D = \{A, B, C, D\}$, where $A = \{\frac{e_1}{(0.5, 0.3)}, \frac{e_4}{(0.5, 0.3)}\}$, $B = \{\frac{e_1}{(0.4, 0.2)}, \frac{e_2}{(0.4, 0.2)}\}$, $C = \{\frac{e_2}{(0.3, 0.6)}, \frac{e_3}{(0.3, 0.6)}\}$, $D = \{\frac{e_3}{(0.5, 0.1)}, \frac{e_4}{(0.5, 0.1)}\}$. The dual vague hypergraph $\mathbb{H}^D = (\mathbb{E}^D, X^D)$ of $\mathbb{H}$ is shown in Fig. 3.5.

The corresponding incidence matrix of $\mathbb{H}^D$ is given in Table 3.6.

Definition 3.31 A vague hypergraph $\mathbb{H} = (X, \mathbb{E})$ is called $A = (t_A, f_A)$-*tempered vague hypergraph* of $\mathbb{H} = (X, \mathbb{E})$ if there is a crisp hypergraph $H^* = (X, E^*)$ and a vague set $A = (t_A, f_A) : X \to (0, 1]$ such that $\mathbb{E} = \{B_F = (t_{B_F}, f_{B_F}) \mid F \in E^*\}$, where

Fig. 3.4 Vague hypergraph

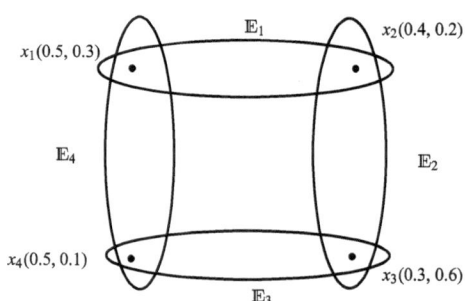

Table 3.5 The corresponding incidence matrix of $\mathbb{H}$

$M_\mathbb{H}$	$\mathbb{E}_1$	$\mathbb{E}_2$	$\mathbb{E}_3$	$\mathbb{E}_4$
x_1	(0.5, 0.3)	(0, 0)	(0, 0)	(0.5, 0.3)
x_2	(0.4, 0.2)	(0.4, 0.2)	(0, 0)	(0, 0)
x_3	(0, 0)	(0.3, 0.6)	(0.3, 0.6)	(0, 0)
x_4	(0, 0)	(0, 0)	(0.5, 0.1)	(0.5, 0.1)

Fig. 3.5 Dual vague
hypergraph

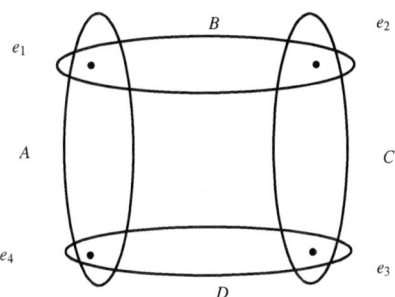

Table 3.6 The incidence matrix of $\mathbb{H}^D$

$M_{\mathbb{H}^D}$	A	B	C	D
e_1	$(0.5, 0.3)$	$(0.4, 0.2)$	$(0, 0)$	$(0, 0)$
e_2	$(0, 0)$	$(0.4, 0.2)$	$(0.3, 0.6)$	$(0, 0)$
e_3	$(0, 0)$	$(0, 0)$	$(0.3, 0.6)$	$(0.5, 0.1)$
e_4	$(0.5, 0.3)$	$(0, 0)$	$(0, 0)$	$(0.5, 0.1)$

$$t_{B_F}(x) = \begin{cases} \min(t_A(y) \mid y \in F), & \text{if } x \in F, \\ 0, & \text{otherwise,} \end{cases} \qquad f_{B_F}(x) = \begin{cases} \max(f_A(y) \mid y \in F), & \text{if } x \in F, \\ 1, & \text{otherwise.} \end{cases}$$

Let $A \otimes \mathbb{H}$ denote the A-tempered vague hypergraph of $\mathbb{H}$ determined by the crisp hypergraph $H^* = (X, E^*)$ and the vague set A.

Example 3.6 Consider the vague hypergraph $\mathbb{H} = (X, \mathbb{E})$, where $X = \{a, b, c, d\}$ and $\mathbb{E} = \{\mathbb{E}_1, \mathbb{E}_2, \mathbb{E}_3, \mathbb{E}_4\}$ which is represented by the following incidence matrix Table 3.7.

We define a vague set $A = (t_A, f_A)$ by

$$t_A(a) = 0.2, \ t_A(b) = t_A(c) = 0.0, \ t_A(d) = 0.3, \ f_A(a) = 0.7, \ f_A(b) = f_A(c) = 0.9, \ f_A(d) = 0.4.$$

Note that

$$t_{B_{\{a,b\}}}(a) = \min(t_A(a), t_A(b)) = 0.0, \ t_{B_{\{a,b\}}}(b) = \min(t_A(a), t_A(b)) = 0.0, \ t_{B_{\{a,b\}}}(c) = 0.0, \ t_{B_{\{a,b\}}}(d) = 0.0,$$

$$f_{B_{\{a,b\}}}(a) = \max(f_A(a), f_A(b)) = 0.9, \ f_{B_{\{a,b\}}}(b) = \max(f_A(a), f_A(b)) = 0.9, \ f_{B_{\{a,b\}}}(c) = 1, \ f_{B_{\{a,b\}}}(d) = 1.$$

Thus,

$$\mathbb{E}_1 = (t_{B_{\{a,b\}}}, f_{B_{\{a,b\}}}), \quad \mathbb{E}_2 = (t_{B_{\{b,d\}}}, f_{B_{\{b,d\}}}), \quad \mathbb{E}_3 = (t_{B_{\{b,c\}}}, f_{B_{\{b,c\}}}), \quad \mathbb{E}_4 = (t_{B_{\{a,c\}}}, f_{B_{\{a,c\}}}).$$

Hence, $\mathbb{H}$ is A-tempered hypergraph.

Theorem 3.1 *A vague hypergraph $\mathbb{H}$ is an A-tempered vague hypergraph of some crisp hypergraph H^* if and only if $\mathbb{H}$ is elementary, support simple, and simply ordered.*

Table 3.7 Incidence matrix of $\mathbb{H}$

$\mathbb{H}$	$\mathbb{E}_1$	$\mathbb{E}_2$	$\mathbb{E}_3$	$\mathbb{E}_4$
a	$(0.2, 0.7)$	$(0, 0)$	$(0, 0)$	$(0.2, 0.7)$
b	$(0.2, 0.7)$	$(0.3, 0.4)$	$(0.0, 0.9)$	$(0, 0)$
c	$(0, 0)$	$(0, 0)$	$(0, 0.9)$	$(0.2, 0.7)$
d	$(0, 0)$	$(0.3, 0.4)$	$(0, 0)$	$(0, 0)$

Proof Suppose that $\mathbb{H} = (X, \mathbb{E})$ is an A-tempered vague hypergraph of some crisp hypergraph H^*. Clearly, $\mathbb{H}$ is elementary and support simple. We show that $\mathbb{H}$ is simply ordered. Let

$$C(\mathbb{H}) = \{(H_1^*)^{r_1} = (X_1, E_1^*), \ (H_2^*)^{r_2} = (X_2, E_2^*), \ldots, \ (H_n^*)^{r_n} = (X_n, E_n^*)\}.$$

Since, $\mathbb{H}$ is elementary, it follows from Proposition 3.15 that $\mathbb{H}$ is ordered. To show that $\mathbb{H}$ is simply ordered, suppose that there exists $F \in E_{i+1}^* \backslash E_i^*$. Then, there exists $x^* \in F$ such that $t_A(x^*) = r_{i+1}$, $f_A(x^*) = \acute{r}_{i+1}$. Since, $t_A(x^*) = r_{i+1} < r_i$ and $f_A(x^*) = \acute{r}_{i+1} < \acute{r}_i$, it follows that $x^* \notin X_I$ and $F \nsubseteq X_i$, hence $\mathbb{H}$ is simply ordered. Conversely, suppose $\mathbb{H} = (X, \mathbb{E})$ is elementary, support simple, and simply ordered. Let

$$C(\mathbb{H}) = \{(H_1^*)^{r_1} = (X_1, E_1^*), \ (H_2^*)^{r_2} = (X_2, E_2^*), \ldots, \ (H_n^*)^{r_n} = (X_n, E_n^*), \}$$

where $D(\mathbb{H}) = \{r_1, r_2, \ldots, r_n\}$ with $0 < r_n < \cdots < r_1$. Since $(H^*)^{r_n} = H_n^* = (X_n, E_n^*)$ and define $A = (t_A, f_A)$ by,

$$t_A(x) = \begin{cases} r_1, & \text{if } x \in X_1, \\ r_i, & \text{if } x \in X_i \backslash X_{i-1}, i = 1, 2, \ldots, n. \end{cases} \qquad f_A(x) = \begin{cases} s_1, & \text{if } x \in X_1, \\ s_i, & \text{if } x \in X_i \backslash X_{i-1}, i = 1, 2, \ldots, n. \end{cases}$$

We show that $\mathbb{E} = \{B_F = (t_{B_F}, f_{B_F}) \mid F \in E^*\}$, where

$$t_{B_F}(x) = \begin{cases} \min(t_A(y) \mid y \in F), & \text{if } x \in F, \\ 0, & \text{otherwise}, \end{cases} \qquad f_{B_F}(x) = \begin{cases} \max(f_A(y) \mid y \in F), & \text{if } x \in F, \\ 1, & \text{otherwise}. \end{cases}$$

Let $F \in E_n^*$. Since, $\mathbb{H}$ is elementary and support simple, there is a unique vague edge $C_F = (t_{C_F}, f_{C_F})$ in $\mathbb{E}$ having support E^*. Indeed, distinct edges in E must have distinct supports that lie in E_n^*. Thus, to show that $E = \{B_F = (t_{B_F}, f_{B_F}) \mid F \in E_n^*\}$, it suffices to show that for each $F \in E_n^*$, $t_{C_F} = t_{B_F}$ and $f_{C_F} = f_{B_F}$. As all edges are elementary and different edges have different supports, it follows from the definition of fundamental sequence that $h(C_F)$ is equal to some number r_i of $D(\mathbb{H})$. Consequently, $E^* \subseteq X_i$. Moreover, if $i > 1$, then $F \in E^* \backslash E_{i-1}^*$. Since $F \subseteq X_i$, it follows from the definition of $A = (t_A, f_A)$ that for each $x \in F$, $t_A(x) \geq r_i$ and $f_A(x) \leq s_i$. We claim that $t_A(x) = r_i$ and $f_A(x) = s_i$, for some $x \in F$. If not, then by definition of $A = (t_A, f_A)$, $t_A(x) \geq r_i$ and $f_A(x) \leq s_i$ for all $x \in F$ which implies that $F \subseteq X_{i-1}$ and so $F \in E^* \backslash E_{i-1}^*$ and since $\mathbb{H}$ is simply ordered $F \subsetneq X_{i-1}$, a contradiction. Thus it follows from the definition of B_F that $B_F = C_F$. This completes the proof.

As a consequence of the above theorem we obtain.

Proposition 3.16 *Suppose that $\mathbb{H}$ is a simply ordered vague hypergraph and $F(\mathbb{H}) = \{r_1, r_2, \ldots, r_n\}$. If $\mathbb{H}^{r_n}$ is a simple hypergraph, then there is a vague subhypergraph $\acute{\mathbb{H}}$ of $\mathbb{H}$ such that the following assertions hold,*

(i) $\acute{\mathbb{H}}$ is an A-tempered vague hypergraph of $\mathbb{H}_n$.
(ii) $\mathbb{E} \subseteq \acute{\mathbb{E}}$.
(iii) $F(\acute{\mathbb{H}}) = F(\mathbb{H})$ and $C(\acute{\mathbb{H}}) = C(\mathbb{H})$.

3.4 Interval-Valued Intuitionistic Fuzzy Hypergraphs

Atanassov and Gargov [6] initiated the concept of interval-valued intuitionistic fuzzy sets as a generalization of intuitionistic fuzzy sets. An interval-valued intuitionistic fuzzy set is characterized by an interval-valued membership degree and an interval-valued nonmembership degree.

Definition 3.32 An *interval-valued intuitionistic fuzzy set* V in X is an object of the form,

$$V = \{\langle x, \mu_V(x), \nu_V(x)\rangle \mid x \in X\},$$

where $\mu_V : X \to \text{Int}([0, 1])$ and $\nu_V : X \to \text{Int}([0, 1])$ such that $\mu_V^+(x) + \nu_V^+(x) \leq 1$ for all $x \in X$.

Definition 3.33 The *support* of an interval-valued intuitionistic fuzzy set $V = \{\langle x, \mu_V(x), \nu_V(x)\rangle \mid x \in X\}$ is defined as, $\text{supp}(V) = \{x \mid \mu_V^-(x) \neq 0, \mu_V^+(x) \neq 0, \nu_V^+(x) \neq 1$ and $\nu_V^+(x) \neq 1\}$.

Definition 3.34 The *height* of an interval-valued intuitionistic fuzzy set $V = \{\langle x, \mu_V(x), \nu_V(x)\rangle \mid x \in X\}$ is defined as, $h(V) = \langle[\sup_{x \in X} \mu_V^-(x), \sup_{x \in X} \mu_V^+(x)], [\inf_{x \in X} \nu_V^-(x), \inf_{x \in X} \nu_V^+(x)]\rangle$.

Definition 3.35 For $\alpha, \beta, \gamma, \delta \in [0, 1]$, the $\langle[\alpha, \beta], [\gamma, \delta]\rangle$-*cut of interval-valued intuitionistic fuzzy set* V is

$$V_{\langle[\alpha,\beta],[\gamma,\delta]\rangle} = \{x \mid \mu_V^-(x) \geq \alpha, \mu_V^+(x) \geq \beta, \nu_V^-(x) \leq \gamma \text{ and } \nu_V^-(x) \leq \delta\}.$$

Definition 3.36 Let $X = \{x_1, x_2, \ldots, x_n\}$ be a finite set of vertices and let $\tau = \{\tau_1, \tau_2, \ldots, \tau_m\}$ be a finite family of nontrivial interval-valued intuitionistic fuzzy sets on X such that

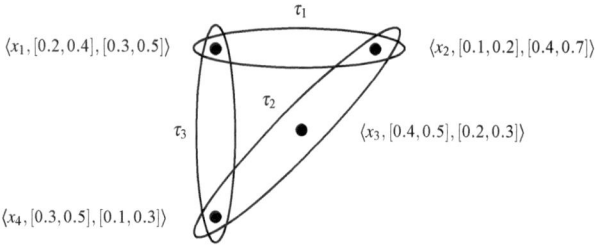

Fig. 3.6 Interval-valued intuitionistic fuzzy hypergraph

$$X = \bigcup_j \text{supp}\langle \mu_j, v_j \rangle, \quad j = 1, 2, \ldots, m,$$

where μ_j, v_j are interval-valued membership and interval-valued nonmembership functions defined on $\tau_j \in \tau$. Then, the pair $\mathscr{H} = (X, \tau)$ denotes an interval-valued intuitionistic fuzzy hypergraph on X, τ is the family of interval-valued intuitionistic fuzzy hyperedges of $\mathscr{H}$.

Example 3.7 Consider an interval-valued intuitionistic fuzzy hypergraph $\mathscr{H} = (X, \tau)$ such that $X = \{x_1, x_2, x_3, x_4\}$ and $\tau = \{\tau_1, \tau_2, \tau_3\}$ as shown in Fig. 3.6, where
$\tau_1 = \{x_1 | \langle [0.2, 0.4], [0.3, 0.5] \rangle, x_2 | \langle [0.1, 0.2], [0.4, 0.7] \rangle\}$,
$\tau_2 = \{x_2 | \langle [0.1, 0.2], [0.4, 0.7] \rangle, x_3 | \langle [0.4, 0.5], [0.2, 0.3] \rangle, x_4 | \langle [0.3, 0.5], [0.1, 0.3] \rangle\}$,
$\tau_3 = \{x_1 | \langle [0.2, 0.4], [0.3, 0.5] \rangle, x_4 | \langle [0.3, 0.5], [0.1, 0.3] \rangle\}$.

The corresponding incidence matrix $M_{\mathscr{H}}$ is as follows:

$$M_{\mathscr{H}} = \begin{array}{c} \\ x_1 \\ x_2 \\ x_3 \\ x_4 \end{array} \begin{pmatrix} \overset{\tau_1}{\langle [0.2, 0.4], [0.3, 0.5] \rangle} & \overset{\tau_2}{\langle [0, 0], [0, 0] \rangle} & \overset{\tau_3}{\langle [0.2, 0.4], [0.3, 0.5] \rangle} \\ \langle [0.1, 0.2], [0.4, 0.7] \rangle & \langle [0.1, 0.2], [0.4, 0.7] \rangle & \langle [0, 0], [0, 0] \rangle \\ \langle [0, 0], [0, 0] \rangle & \langle [0.4, 0.5], [0.2, 0.3] \rangle & \langle [0, 0], [0, 0] \rangle \\ \langle [0, 0], [0, 0] \rangle & \langle [0.3, 0.5], [0.1, 0.3] \rangle & \langle [0.3, 0.5], [0.1, 0.3] \rangle \end{pmatrix}.$$

Definition 3.37 The $\langle [\alpha, \beta], [\gamma, \delta] \rangle$-*cut* of an interval-valued intuitionistic fuzzy hypergraph $\mathscr{H}$, denoted by $H_{\langle [\alpha,\beta],[\gamma,\delta] \rangle}$ and is defined as $H_{\langle [\alpha,\beta],[\gamma,\delta] \rangle} = (X_{\langle [\alpha,\beta],[\gamma,\delta] \rangle}, E_{\langle [\alpha,\beta],[\gamma,\delta] \rangle})$, where

$$X_{\langle [\alpha,\beta],[\gamma,\delta] \rangle} = X,$$
$$E_{j,\langle [\alpha,\beta],[\gamma,\delta] \rangle} = \{x_i \mid \mu_j^-(x_i) \geq \alpha, \mu_j^+(x_i) \geq \beta, v_j^-(x_i) \leq \gamma \text{ and } v_j^-(x_i) \leq \delta, j = 1, 2, \ldots, m\},$$
$$E_{m+1,\langle [\alpha,\beta],[\gamma,\delta] \rangle} = \{x_i \mid \mu_j^-(x_i) < \alpha, \mu_j^+(x_i) < \beta, v_j^-(x_i) > \gamma \text{ and } v_j^-(x_i) > \delta, \forall j\}.$$

The hyperedge $E_{m+1,\langle [\alpha,\beta],[\gamma,\delta] \rangle}$ is added to group the elements which are not contained in any hyperedge $E_{j,\langle [\alpha,\beta],[\gamma,\delta] \rangle}$ of $H_{\langle [\alpha,\beta],[\gamma,\delta] \rangle}$. The hyperedges in the $\langle [\alpha, \beta], [\gamma, \delta] \rangle$-cut hypergraph are now crisp sets.

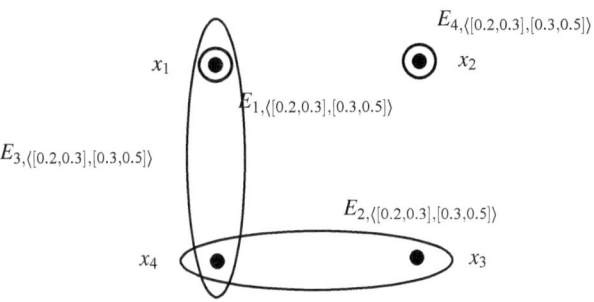

Fig. 3.7 $\langle[0.2, 0.3], [0.3, 0.5]\rangle$-cut hypergraph

Example 3.8 Consider an interval-valued intuitionistic fuzzy hypergraph $\mathscr{H} = (X, \tau)$, where $X = \{x_1, x_2, x_3, x_4\}$ and $\tau = \{\tau_1, \tau_2, \tau_3\}$, given in Example 3.7. Incidence matrix of $H_{\langle[0.2,0.3],[0.3,0.5]\rangle}$

$$M_{H_{\langle[0.2,0.3],[0.3,0.5]\rangle}} = \begin{matrix} & E_{1,\langle[0.2,0.3],[0.3,0.5]\rangle} & E_{2,\langle[0.2,0.3],[0.3,0.5]\rangle} & E_{3,\langle[0.2,0.3],[0.3,0.5]\rangle} & E_{4,\langle[0.2,0.3],[0.3,0.5]\rangle} \\ x_1 & 1 & 0 & 1 & 0 \\ x_2 & 0 & 0 & 0 & 1 \\ x_3 & 0 & 1 & 0 & 0 \\ x_4 & 0 & 1 & 1 & 0 \end{matrix}.$$

The new hyperedge $E_{4,\langle[0.2,0.3],[0.3,0.5]\rangle}$ is added to group the vertex x_2 as shown in Fig. 3.7.

Definition 3.38 The *dual* interval-valued intuitionistic fuzzy hypergraph of an interval-valued intuitionistic fuzzy hypergraph $\mathscr{H} = (X, \tau)$ is defined as $\mathscr{H}^* = (X^*, \tau^*)$, where $X^* = \{e'_1, e'_2, ..., e'_m\}$ is the set of vertices corresponding to $\tau_1, \tau_2, ..., \tau_m$, respectively, and $\{X_1, X_2, ..., X_n\}$ is the set of hyperedges corresponding to $x_1, x_2, ..., x_n$, respectively, where $X_i(e'_j) = \tau_j(x_i), i = 1, 2, ..., n, j = 1, 2, ..., m$.

Example 3.9 Consider the dual interval-valued intuitionistic fuzzy hypergraph $\mathscr{H}^* = (X^*, \tau^*)$ (shown in Fig. 3.8) of an interval-valued intuitionistic fuzzy hypergraph $\mathscr{H} = (X, \tau)$ given in Example 3.7, such that $X^* = \{e'_1, e'_2, e'_3\}$ and $E^* = \{X_1, X_2, X_3, X_4\}$, where
$X_1 = \{e'_1|\langle[0.2, 0.4], [0.3, 0.5]\rangle, e'_3|\langle[0.2, 0.4], [0.3, 0.5]\rangle\}$,
$X_2 = \{e'_1|\langle[0.1, 0.2], [0.4, 0.7]\rangle, e'_2|\langle[0.1, 0.2], [0.4, 0.7]\rangle\}$,
$X_3 = \{e'_2|\langle[0.4, 0.5], [0.2, 0.3]\rangle\}$,
$X_4 = \{e'_2|\langle[0.3, 0.5], [0.1, 0.3]\rangle, e'_3|\langle[0.3, 0.5], [0.1, 0.3]\rangle\}$.
The corresponding incidence matrix $M_{\mathscr{H}^*}$ is as follows:

$$M_{\mathscr{H}^*} = \begin{matrix} & X_1 & X_2 & X_3 & X_4 & X_5 \\ e'_1 & \langle[0.2, 0.4], [0.3, 0.5]\rangle & \langle[0.1, 0.2], [0.4, 0.7]\rangle & \langle[0, 0], [0, 0]\rangle & \langle[0, 0], [0, 0]\rangle \\ e'_2 & \langle[0, 0], [0, 0]\rangle & \langle[0.1, 0.2], [0.4, 0.7]\rangle & \langle[0.4, 0.5], [0.2, 0.3]\rangle & \langle[0.3, 0.5], [0.1, 0.3]\rangle \\ e'_3 & \langle[0.2, 0.4], [0.3, 0.5]\rangle & \langle[0, 0], [0, 0]\rangle & \langle[0, 0], [0, 0]\rangle & \langle[0.3, 0.5], [0.1, 0.3]\rangle \end{matrix}.$$

Fig. 3.8 Dual
interval-valued intuitionistic
fuzzy hypergraph

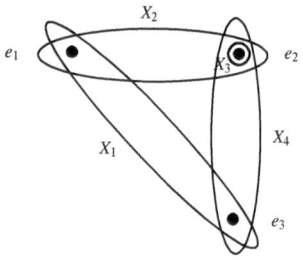

Definition 3.39 The *strength* ρ of a hyperedge τ_j is defined as

$$\rho(\tau_j) = \{\min(\mu_j^-(x) \mid \mu_j^-(x) > 0), \min(\mu_j^+(x) \mid \mu_j^+(x) > 0),$$
$$\max(\nu_j^-(x) \mid \nu_j^-(x) > 0), \max(\nu_j^+(x) \mid \nu_j^+(x) > 0)\}.$$

In other words, the minimum membership values $\mu_j^-(x)$, $\mu_j^+(x)$ of vertices and maximum nonmembership values $\nu_j^-(x)$, $\nu_j^+(x)$ of vertices in the hyperedge τ_j. Its interpretation is that the hyperedge τ_j groups elements having participation degree at least $\rho(\tau_j)$ in the hypergraph. The hyperedges with high strength are called the strong hyperedges because the cohesion in them is strong.

Example 3.10 Consider an interval-valued intuitionistic fuzzy hypergraph $\mathscr{H} = (X, \tau)$, where $X = \{x_1, x_2, x_3, x_4\}$ and $\tau = \{\tau_1, \tau_2, \tau_3, \tau_4\}$ as shown in Fig. 3.9.
   Here, $\rho(\tau_1) = \langle[0.5, 0.7], [0.1, 0.2]\rangle$, $\rho(\tau_2) = \langle[0.2, 0.4], [0.2, 0.3]\rangle$, $\rho(\tau_3) = \langle[0.1, 0.3], [0.5, 0.6]\rangle$ and $\rho(\tau_4) = \langle[0.1, 0.3], [0.5, 0.6]\rangle$, respectively. Therefore, the hyperedge τ_1 is stronger than τ_2, τ_3 and τ_4.

Definition 3.40 An interval-valued intuitionistic fuzzy hypergraph $\mathscr{H}' = (X', \tau')$ is a *partial* interval-valued intuitionistic fuzzy hypergraph of $\mathscr{H} = (X, \tau)$ if $\tau' \subseteq \tau$ and is written as $\mathscr{H}' \subseteq \mathscr{H}$. If $\mathscr{H}' \subseteq \mathscr{H}$ and $\tau' \subset \tau$, we write $\mathscr{H}' \subset \mathscr{H}$.

Definition 3.41 An interval-valued intuitionistic fuzzy hypergraph $\mathscr{H} = (X, \tau)$ is *simple* if τ has no repeated interval-valued intuitionistic fuzzy hyperedges and

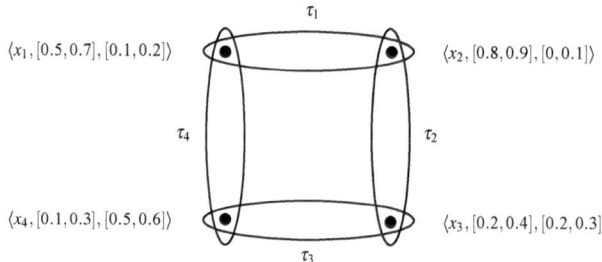

Fig. 3.9 Interval-valued intuitionistic fuzzy hypergraph

whenever $X = \langle \mu_X, \nu_X \rangle, Y = \langle \mu_Y, \nu_Y \rangle \in \tau$ and $\mu_X^-(x) \leq \mu_Y^-(x), \mu_X^+(x) \leq \mu_Y^+(x),$
$\nu_X^-(x) \geq \nu_Y^-(x), \nu_X^+(x) \geq \nu_Y^+(x),$ for all $x \in X,$ then $\mu_X^-(x) = \mu_Y^-(x), \mu_X^+(x) = \mu_Y^+(x), \nu_X^-(x) = \nu_Y^-(x), \nu_X^+(x) = \nu_Y^+(x).$

Definition 3.42 An interval-valued intuitionistic fuzzy hypergraph $\mathscr{H} = (X, \tau)$
is *support simple* if $X = \langle \mu_X, \nu_X \rangle, Y = \langle \mu_Y, \nu_Y \rangle \in \tau, \mu_X^-(x) \leq \mu_Y^-(x), \mu_X^+(x) \leq$
$\mu_Y^+(x), \nu_X^-(x) \geq \nu_Y^-(x), \nu_X^+(x) \geq \nu_Y^+(x),$ for all $x \in X,$ and $supp(X) = supp(Y),$
then $\mu_X^-(x) = \mu_Y^-(x), \mu_X^+(x) = \mu_Y^+(x), \nu_X^-(x) = \nu_Y^-(x), \nu_X^+(x) = \nu_Y^+(x).$ An
interval-valued intuitionistic fuzzy hypergraph $\mathscr{H} = (X, \tau)$ is *strongly support
simple* if $X = \langle \mu_X, \nu_X \rangle, Y = \langle \mu_Y, \nu_Y \rangle \in \tau$ and supp(X)=supp(Y), then $\mu_X^-(x) =$
$\mu_Y^-(x), \mu_X^+(x) = \mu_Y^+(x), \nu_X^-(x) = \nu_Y^-(x), \nu_X^+(x) = \nu_Y^+(x).$

Definition 3.43 An interval-valued intuitionistic fuzzy set $X = \{\langle x, \mu_X(x),$
$\nu_X(x) \rangle \mid x \in X\}$ is an *elementary interval-valued intuitionistic fuzzy set* if X is single
valued on supp(X). An interval-valued intuitionistic fuzzy hypergraph $\mathscr{H} = (X, \tau)$
whose all interval-valued intuitionistic fuzzy hyperedges are elementary is called an
elementary interval-valued intuitionistic fuzzy hypergraph.

Example 3.11 Consider an interval-valued intuitionistic fuzzy hypergraph $\mathscr{H} =$
(X, τ) such that $X = \{x_1, x_2, x_3, x_4\}$ and $\tau = \{\tau_1, \tau_2, \tau_3, \tau_4\},$ represented by the following incidence matrix:

$$M_{\mathscr{H}} = \begin{array}{c} \\ x_1 \\ x_2 \\ x_3 \\ x_4 \end{array} \begin{pmatrix} \overset{\tau_1}{\langle[0.5,0.7],[0.1,0.2]\rangle} & \overset{\tau_2}{\langle[0.8,0.9],[0,0]\rangle} & \overset{\tau_3}{\langle[0,0],[0,0]\rangle} & \overset{\tau_4}{\langle[0,0],[0,0]\rangle} \\ \langle[0.5,0.7],[0.1,0.2]\rangle & \langle[0.8,0.9],[0,0]\rangle & \langle[0.8,0.9],[0,0]\rangle & \overset{\tau_5}{\langle[0.5,0.7],[0.1,0.2]\rangle} \\ \langle[0,0],[0,0]\rangle & \langle[0,0],[0,0]\rangle & \langle[0.8,0.9],[0,0]\rangle & \langle[0.5,0.7],[0.1,0.2]\rangle \\ \langle[0,0],[0,0]\rangle & \langle[0.2,0.4],[0.2,0.3]\rangle & \langle[0,0],[0,0]\rangle & \langle[0.2,0.4],[0.2,0.3]\rangle \end{pmatrix}.$$

Clearly, $\mathscr{H}$ is simple, support simple, and strongly support simple. The partial
interval-valued intuitionistic fuzzy hypergraph $\tau' = \{\tau_1, \tau_3\}$ of $\mathscr{H}$ is elementary.

Theorem 3.2 *Let $\mathscr{H} = (X, \tau)$ be an elementary interval-valued intuitionistic fuzzy
hypergraph. Then, $\mathscr{H}$ is support simple if and only if $\mathscr{H}$ is strongly support simple.*

Proof Suppose that $\mathscr{H}$ is elementary, support simple, and that supp(X) = supp(Y).
Without loss of generality we may assume that $h(X) \leq h(Y).$ Since, $\mathscr{H}$ is elementary, it follows that $\mu_X^-(x) \leq \mu_Y^-(x), \mu_X^+(x) \leq \mu_Y^+(x), \nu_X^-(x) \geq \nu_Y^-(x), \nu_X^+(x) \geq$
$\nu_Y^+(x)$ for all $x \in X,$ and since $\mathscr{H}$ is support simple that $\mu_X^-(x) = \mu_Y^-(x), \mu_X^+(x) =$
$\mu_Y^+(x), \nu_X^-(x) = \nu_Y^-(x), \nu_X^+(x) = \nu_Y^+(x).$ Hence, $\mathscr{H}$ is strongly support simple.

Definition 3.44 Let $\mathscr{H} = (X, \tau)$ be an interval-valued intuitionistic fuzzy hypergraph. Let $\alpha, \beta, \gamma, \delta \in [0, 1]$ and

$$E_{\langle[\alpha,\beta],[\gamma,\delta]\rangle} = \{X_{\langle[\alpha,\beta],[\gamma,\delta]\rangle} \neq \emptyset \mid X \in \tau\}, \quad X_{\langle[\alpha,\beta],[\gamma,\delta]\rangle} = \bigcup_{X \in \tau} X_{\langle[\alpha,\beta],[\gamma,\delta]\rangle}.$$

If $E_{\langle[\alpha,\beta],[\gamma,\delta]\rangle} \neq \emptyset,$ then the crisp hypergraph $H_{\langle[\alpha,\beta],[\gamma,\delta]\rangle} = (X_{\langle[\alpha,\beta],[\gamma,\delta]\rangle},$
$E_{\langle[\alpha,\beta],[\gamma,\delta]\rangle})$ is the $\langle[\alpha, \beta], [\gamma, \delta]\rangle$-*level hypergraph* of $\mathscr{H}.$

The families of crisp sets (hypergraphs) produced by the $\langle[\alpha, \beta], [\gamma, \delta]\rangle$-cuts of an interval-valued intuitionistic fuzzy hypergraph share an important relationship with each other, as expressed below:

Suppose $\mathbb{A}$ and $\mathbb{B}$ are two families of sets such that for each set $A \in \mathbb{A}$ there is at least one set $B \in \mathbb{B}$ which contains A. In this case we say that $\mathbb{B}$ absorbs $\mathbb{A}$ and symbolically write $\mathbb{A} \sqsubseteq \mathbb{B}$. Since it is possible for $\mathbb{A} \sqsubseteq \mathbb{B}$ while $\mathbb{A} \cap \mathbb{B} = \emptyset$, we have that $\mathbb{A} \subseteq \mathbb{B}$ implies $\mathbb{A} \sqsubseteq \mathbb{B}$, whereas the converse is generally false. If $\mathbb{A} \sqsubseteq \mathbb{B}$ and $\mathbb{A} \neq \mathbb{B}$, then we write $\mathbb{A} \sqsubset \mathbb{B}$.

Definition 3.45 Let $\mathscr{H} = (X, \tau)$ be an interval-valued intuitionistic fuzzy hypergraph, and for $\langle[0, 0], [0, 0]\rangle < \langle[\alpha, \beta], [\gamma, \delta]\rangle \leq h(\mathscr{H})$, let $H_{\langle[\alpha,\beta],[\gamma,\delta]\rangle} = (X_{\langle[\alpha,\beta],[\gamma,\delta]\rangle}, E_{\langle[\alpha,\beta],[\gamma,\delta]\rangle})$ be the $\langle[\alpha, \beta], [\gamma, \delta]\rangle$-level hypergraph of $\mathscr{H}$. The sequence of real numbers $\{\langle[r_i, s_i], [t_i, q_i]\rangle \mid 1 \leq i \leq n\}, 0 < r_n < \ldots < r_1, 0 < s_n < \ldots < s_1, 1 > t_n > \ldots > t_1$ and $1 > q_n > \ldots > q_1$, where $h(\mathscr{H}) = \langle[r_1, s_1], [t_1, q_1]\rangle$, which satisfies the properties

(i) if $r_{i+1} < u \leq r_i, \ s_{i+1} < v \leq s_i, \ t_{i+1} > l \geq t_i,$ and $q_{i+1} > m \geq q_i,$ then $E_{\langle[u,v],[l,m]\rangle} = E_{\langle[r_i,s_i],[t_i,q_i]\rangle}, \quad i = 1, 2, \ldots, n,$

(ii) $E_{\langle[r_i,s_i],[t_i,q_i]\rangle} \sqsubset E_{\langle[r_{i+1},s_{i+1}],[t_{i+1},q_{i+1}]\rangle}, \quad i = 1, 2, \ldots, n-1,$

is called the *fundamental sequence* of $\mathscr{H}$, denoted by $F(\mathscr{H})$. The set of $\langle[r_i, s_i], [t_i, q_i]\rangle$-level hypergraphs $\{H_{\langle[r_i,s_i],[t_i,q_i]\rangle} \mid 1 \leq i \leq n\}$ is the *set of core hypergraphs* of $\mathscr{H}$ or, the *core set* of $\mathscr{H}$, denoted by $C(\mathscr{H})$.

Definition 3.46 Let $\mathscr{H} = (X, \tau)$ be an interval-valued intuitionistic fuzzy hypergraph and $F(\mathscr{H}) = \{\langle[r_i, s_i], [t_i, q_i]\rangle \mid 1 \leq i \leq n\}$. Then, $\mathscr{H}$ is called *sectionally elementary* if for each X, where X is an interval-valued intuitionistic fuzzy set defined on $\tau_j \in \tau$ and each $\langle[r_i, s_i], [t_i, q_i]\rangle \in F(\mathscr{H})$, $X_{\langle[\alpha,\beta],[\gamma,\delta]\rangle} = X_{\langle[r_i,s_i],[t_i,q_i]\rangle}$ for all $\langle[\alpha, \beta], [\gamma, \delta]\rangle \in (\langle[r_{i+1}, s_{i+1}], [t_{i+1}, q_{i+1}]\rangle, \langle[r_i, s_i], [t_i, q_i]\rangle]$. (Take $r_{n+1} = 0, s_{n+1} = 0, t_{n+1} = 0, q_{n+1} = 0$.)

Definition 3.47 An interval-valued intuitionistic fuzzy hypergraph $\mathscr{H}$ is *ordered* if $C(\mathscr{H}) = \{H_{\langle[r_i,s_i],[t_i,q_i]\rangle} \mid 1 \leq i \leq n\}$ is ordered, and is *simply ordered* if $C(\mathscr{H})$ is simply ordered.

Example 3.12 Consider an interval-valued intuitionistic fuzzy hypergraph $\mathscr{H} = (X, \tau)$, represented by incidence matrix, as in Example 3.11. Clearly, $h(\mathscr{H}) = \langle[0.8, 0.9], [0, 0]\rangle$. Now

$$E_{\langle[0.8,0.9],[0,0]\rangle} = E_{\langle[0.5,0.7],[0.1,0.2]\rangle} = \{\{x_1, x_2\}, \{x_2, x_3\}\},$$
$$E_{\langle[0.2,0.4],[0.2,0.3]\rangle} = \{\{x_1, x_2\}, \{x_1, x_2, x_4\}, \{x_2, x_3\}, \{x_2, x_3, x_4\}\}.$$

Thus, for $0.2 < \alpha \leq 0.8, 0.4 < \beta \leq 0.9, 0.2 > \gamma \geq 0, 0.3 > \delta \geq 0,$

$$E_{\langle[\alpha,\beta],[\gamma,\delta]\rangle} = \{\{x_1, x_2\}, \{x_2, x_3\}, \}$$

and for $0 < \alpha \leq 0.2, 0 < \beta \leq 0.4, 1 > \gamma \geq 0.2, 1 > \delta \geq 0.3$,

$$E_{\langle[\alpha,\beta],[\gamma,\delta]\rangle} = \{\{x_1, x_2\}, \{x_1, x_2, x_4\}, \{x_2, x_3\}, \{x_2, x_3, x_4\}\}.$$

It is easy to see that, $E_{\langle[0.8,0.9],[0,0]\rangle} \sqsubseteq E_{\langle[0.2,0.4],[0.2,0.3]\rangle}$. Therefore, the fundamental sequence is $F(\mathcal{H}) = \{\langle[r_1, s_1], [t_1, q_1]\rangle = \langle[0.8, 0.9], [0, 0]\rangle, \langle[r_2, s_2], [t_2, q_2]\rangle = \langle[0.2, 0.4], [0.2, 0.3]\rangle\}$ and the set of core hypergraphs is $C(\mathcal{H}) = \{H_{\langle[0.8,0.9],[0,0]\rangle}, H_{\langle[0.2,0.4],[0.2,0.3]\rangle}\}$. $\mathcal{H}$ is not sectionally elementary, as $\tau_{1,\langle[\alpha,\beta],[\gamma,\delta]\rangle} \neq \tau_{1,\langle[0.8,0.9],[0,0]\rangle}$ for $\langle[\alpha, \beta], [\gamma, \delta]\rangle = \langle[0.5, 0.7], [0.1, 0.2]\rangle$. Clearly, $\mathcal{H}$ is simply ordered.

Proposition 3.17 *(i) An elementary interval-valued intuitionistic fuzzy hypergraph $\mathcal{H}(X, \tau)$ is ordered.*

(ii) An ordered interval-valued intuitionistic fuzzy hypergraph $\mathcal{H}(X, \tau)$ with $C(\mathcal{H}) = \{H_{\langle[r_i,s_i],[t_i,q_i]\rangle} \mid 1 \leq i \leq n\}$ and simple $H_{\langle[r_n,s_n],[t_n,q_n]\rangle}$, is elementary.

The complexity of an interval-valued intuitionistic fuzzy hypergraph depends in part on how many hyperedges it has. The natural question arises: is there an upper bound on the number of hyperedges of an interval-valued intuitionistic fuzzy hypergraph of order n?

Proposition 3.18 *Let $\mathcal{H} = (X, \tau)$ be a simple interval-valued intuitionistic fuzzy hypergraph of order n. Then, there is no upper bound on $|\tau|$.*

Proof Let $X = \{x, y\}$, and define $\tau_N = \{X_i = \langle[\mu_{X_i}^-, \mu_{X_i}^+][\nu_{X_i}^-, \nu_{X_i}^+]\rangle \mid i = 1, 2, \ldots, N\}$, where

$$\mu_{X_i}^-(x) = 1/1+i, \quad \mu_{X_i}^+(x) = 1/1+i, \quad \nu_{X_i}^-(x) = 1/1+i, \quad \nu_{X_i}^+(x) = 1/1+i,$$
$$\mu_{X_i}^-(y) = i/1+i, \quad \mu_{X_i}^+(y) = i/1+i, \quad \nu_{X_i}^-(y) = i/1+i, \quad \nu_{X_i}^+(y) = i/1+i.$$

Then, $\mathcal{H}_N = (X, \tau_N)$ is a simple interval-valued intuitionistic fuzzy hypergraph with N hyperedges.

Proposition 3.19 *Let $\mathcal{H} = (X, \tau)$ be a support simple interval-valued intuitionistic fuzzy hypergraph of order n. Then, there is no upper bound on $|\tau|$.*

Proof The proof follows at once from Proposition 3.18, as the class of support simple interval-valued intuitionistic fuzzy hypergraphs contains the class of simple interval-valued intuitionistic fuzzy hypergraphs.

Proposition 3.20 *Let $\mathcal{H} = (X, \tau)$ be a strongly support simple interval-valued intuitionistic fuzzy hypergraph of order n. Then, $|\tau| \leq 2^n - 1$, with equality if and only if $\{\text{supp}(X) \mid X \in \tau\} = P(X) - \emptyset$.*

Proof Each nontrivial $U \subseteq X$ can be the support of at most one $X \in \tau$, therefore $|\tau| \leq 2^n - 1$. The second statement is obvious.

Consider the situation where the node set of a (crisp) hypergraph is fuzzified. Suppose that each hyperedge is given a uniform degree of interval-valued membership and interval-valued nonmembership consistent with the weakest node of the hyperedge. Such constructions describe the following subclass of interval-valued intuitionistic fuzzy hypergraphs.

Definition 3.48 An interval-valued intuitionistic fuzzy hypergraph $\mathcal{H} = (X, \tau)$ is said to be a $V = \langle[\mu_V^-, \mu_V^+], [\nu_V^-, \nu_V^+]\rangle$-*tempered interval-valued intuitionistic fuzzy hypergraph* of H^*, if there is a crisp hypergraph $H^* = (X, E^*)$ and an interval-valued intuitionistic fuzzy set $X = \langle[\mu_X^-, \mu_X^+], [\nu_X^-, \nu_X^+]\rangle : X \to \mathrm{Int}((0, 1])$ such that $\tau = \{Y_e = \langle[(\mu_Y^-)_e, (\mu_Y^+)_e], [(\nu_Y^-)_e, (\nu_Y^+)_e]\rangle \mid e \in E\}$, where

$$(\mu_Y^-)_e(x) = \begin{cases} \min(\mu_X^-(y) \mid y \in e), & \text{if } x \in e, \\ 0, & \text{otherwise,} \end{cases}$$

$$(\mu_Y^+)_e(x) = \begin{cases} \min(\mu_X^+(y) \mid y \in e), & \text{if } x \in e, \\ 0, & \text{otherwise,} \end{cases}$$

$$(\nu_Y^-)_e(x) = \begin{cases} \max(\nu_X^-(y) \mid y \in e), & \text{if } x \in e, \\ 0, & \text{otherwise,} \end{cases}$$

$$(\nu_Y^+)_e(x) = \begin{cases} \max(\nu_X^+(y) \mid y \in e), & \text{if } x \in e, \\ 0, & \text{otherwise,} \end{cases}$$

The V-tempered interval-valued intuitionistic fuzzy hypergraph of H^* will be denoted by $V \otimes H^*$.

Example 3.13 Consider an interval-valued intuitionistic fuzzy hypergraph $\mathcal{H} = (X, \tau)$ such that $X = \{x_1, x_2, x_3, x_4\}$ and $\tau = \{\tau_1, \tau_2, \tau_3, \tau_4\}$, represented by the following incidence matrix:

	τ_1	τ_2	τ_3	τ_4
x_1	$\langle[0.5, 0.7], [0.1, 0.3]\rangle$	$\langle[0, 0], [0, 0]\rangle$	$\langle[0, 0], [0, 0]\rangle$	$\langle[0.5, 0.7], [0.1, 0.3]\rangle$
x_2	$\langle[0.5, 0.7], [0.1, 0.3]\rangle$	$\langle[0.2, 0.4], [0.2, 0.3]\rangle$	$\langle[0.6, 0.8], [0.1, 0.2]\rangle$	$\langle[0, 0], [0, 0]\rangle$
x_3	$\langle[0, 0], [0, 0]\rangle$	$\langle[0, 0], [0, 0]\rangle$	$\langle[0.6, 0.8], [0.1, 0.2]\rangle$	$\langle[0.5, 0.7], [0.1, 0.3]\rangle$
x_4	$\langle[0, 0], [0, 0]\rangle$	$\langle[0.2, 0.4], [0.2, 0.3]\rangle$	$\langle[0, 0], [0, 0]\rangle$	$\langle[0, 0], [0, 0]\rangle$

Define $V = \langle[\mu_V^-, \mu_V^+], [\nu_V^-, \nu_V^+]\rangle : X \to \mathrm{Int}((0, 1])$ by,

$$\mu_V^-(x_1) = 0.5, \quad \mu_X^-(x_2) = \mu_V^-(x_3) = 0.6, \quad \mu_V^-(x_4) = 0.2,$$
$$\mu_V^+(x_1) = 0.7, \quad \mu_X^+(x_2) = \mu_V^+(x_3) = 0.8, \quad \mu_V^+(x_4) = 0.4,$$
$$\nu_V^-(x_1) = 0.1, \quad \nu_X^-(x_2) = \nu_V^-(x_3) = 0.1, \quad \nu_V^-(x_4) = 0.2,$$
$$\nu_V^+(x_1) = 0.3, \quad \nu_X^+(x_2) = \nu_V^+(x_3) = 0.2, \quad \nu_V^+(x_4) = 0.3.$$

Now

$$(\mu_Y^-)_{\{x_1,x_2\}}(x_1) = (\mu_Y^-)_{\{x_1,x_2\}}(x_2) = \min(\mu_V^-(x_1), \mu_V^-(x_2)) = 0.5,$$
$$(\mu_Y^-)_{\{x_1,x_2\}}(x_3) = (\mu_Y^-)_{\{x_1,x_2\}}(x_4) = 0,$$
$$(\mu_Y^+)_{\{x_1,x_2\}}(x_1) = (\mu_Y^+)_{\{x_1,x_2\}}(x_2) = \min(\mu_V^+(x_1), \mu_V^+(x_2)) = 0.7,$$
$$(\mu_Y^+)_{\{x_1,x_2\}}(x_3) = (\mu_Y^+)_{\{x_1,x_2\}}(x_4) = 0,$$
$$(\nu_Y^-)_{\{x_1,x_2\}}(x_1) = (\nu_Y^-)_{\{x_1,x_2\}}(x_2) = \max(\nu_V^-(x_1), \nu_V^-(x_2)) = 0.1,$$
$$(\nu_Y^-)_{\{x_1,x_2\}}(x_3) = (\nu_Y^-)_{\{x_1,x_2\}}(x_4) = 0,$$
$$(\nu_Y^+)_{\{x_1,x_2\}}(x_1) = (\nu_Y^+)_{\{x_1,x_2\}}(x_2) = \max(\nu_V^+(x_1), \nu_V^+(x_2)) = 0.3,$$
$$(\nu_Y^+)_{\{x_1,x_2\}}(x_3) = (\nu_Y^+)_{\{x_1,x_2\}}(x_4) = 0.$$

Therefore, $\tau_1 = \langle [(\mu_Y^-)_{\{x_1,x_2\}}, (\mu_Y^+)_{\{x_1,x_2\}}], [(\nu_Y^-)_{\{x_1,x_2\}}.(\nu_Y^+)_{\{x_1,x_2\}}] \rangle$.

Also, it is easy to see that
$$\tau_2 = \langle [(\mu_Y^-)_{\{x_2,x_4\}}, (\mu_Y^+)_{\{x_2,x_4\}}], [(\nu_Y^-)_{\{x_2,x_4\}}, (\nu_Y^+)_{\{x_2,x_4\}}] \rangle,$$
$$\tau_3 = \langle [(\mu_Y^-)_{\{x_2,x_3\}}, (\mu_Y^+)_{\{x_2,x_3\}}], [(\nu_Y^-)_{\{x_2,x_3\}}, (\nu_Y^+)_{\{x_2,x_3\}}] \rangle,$$
$$\tau_4 = \langle [(\mu_Y^-)_{\{x_1,x_3\}}, (\mu_Y^+)_{\{x_1,x_3\}}], [(\nu_Y^-)_{\{x_1,x_3\}}, (\nu_Y^+)_{\{x_1,x_3\}}] \rangle.$$

Thus, $\mathscr{H}$ is $X = \langle [\mu_V^-, \mu_V^+], [\nu_V^-, \nu_V^+] \rangle$-tempered interval-valued intuitionistic fuzzy hypergraph.

Theorem 3.3 *An interval-valued intuitionistic fuzzy hypergraph $\mathscr{H}$ is a V-tempered interval-valued intuitionistic fuzzy hypergraph of some crisp hypergraph H^* if and only if $\mathscr{H}$ is elementary, support simple, and simply ordered.*

Proof Suppose that $\mathscr{H} = (X, \tau)$ is a V-tempered interval-valued intuitionistic fuzzy hypergraph of $H^* = (X, E^*)$. Clearly, $\mathscr{H}$ is elementary, support simple and ordered (being elementary). To show that $\mathscr{H}$ is simply ordered, let $C(\mathscr{H}) = \{H_{\langle [r_i,s_i],[t_i,q_i] \rangle}$ $(X_i, E_i) \mid 1 \leq i \leq n\}$. Suppose there exists $e \in E_{i+1} \backslash E_i$, then there exists $z \in e$ such that $\mu_X^-(z) = r_{i+1}, \mu_X^+(z) = s_{i+1}, \nu_X^-(z) = t_{i+1}$ and $\nu_X^+(z) = q_{i+1}$. Since $\mu_X^-(z) = r_{i+1} < r_i, \mu_X^+(z) = s_{i+1} < s_i, \nu_X^-(z) = t_{i+1} > t_i$ and $\nu_X^+(z) = q_{i+1} > q_i$, it follows that $z \notin X_i$ and $e \nsubseteq X_i$, hence $\mathscr{H}$ is simply ordered.

Conversely, suppose that $\mathscr{H} = (X, \tau)$ is elementary, support simple, and simply ordered. Define $V = \langle [\mu_V^-, \mu_V^+], [\nu_V^-, \nu_V^+] \rangle : X_n \to \text{Int}((0, 1])$ by

$$\mu_Y^-(x) = \begin{cases} r_1, & \text{if } x \in X_1, \\ r_i, & \text{if } x \in X_i \backslash X_{i-1}, \ i = 2, 3, \ldots, n, \end{cases}$$

$$\mu_Y^+(x) = \begin{cases} s_1, & \text{if } x \in X_1, \\ s_i, & \text{if } x \in X_i \backslash X_{i-1}, \ i = 2, 3, \ldots, n, \end{cases}$$

$$\nu_Y^-(x) = \begin{cases} t_1, & \text{if } x \in X_1, \\ t_i, & \text{if } x \in X_i \backslash X_{i-1}, \ i = 2, 3, \ldots, n, \end{cases}$$

$$v_Y^+(x) = \begin{cases} q_1, & \text{if } x \in X_1, \\ q_i, & \text{if } x \in X_i \backslash X_{i-1}, \ i = 2, 3, \ldots, n. \end{cases}$$

We show that $\tau = \{Y_e = \langle [(\mu_Y^-)_e, (\mu_Y^+)_e], [(\nu_Y^-)_e, (\nu_Y^+)_e] \rangle \mid e \in E_n\}$. Since, $\mathscr{H}$ is elementary and support simple, there is a unique interval-valued intuitionistic fuzzy hyperedge Z_e in τ having support e. Since distinct hyperedges in τ must have distinct supports that lie in E_n. Thus, to show that $\tau = \{Y_e = \langle [(\mu_Y^-)_e, (\mu_Y^+)_e], [(\nu_Y^-)_e, (\nu_Y^+)_e] \rangle \mid e \in E_n\}$, it suffices to show that $Y_e = Z_e$, for each $e \in E_n$.

Since, all hyperedges are elementary and different hyperedges have different supports, it follows from Definition 3.45 that $h(Z_e) = \langle [r_i, s_i], [t_i, q_i] \rangle \in F(\mathscr{H})$. Consequently, $e \subseteq X_i$. Moreover, $e \in E_i \backslash E_{i-1}, i = 2, 3, \ldots, n$. As $e \subseteq X_i$, it follows from the definition of $V = \langle [\mu_V^-, \mu_V^+], [\nu_V^-, \nu_V^+] \rangle$ that $\mu_V^-(x) \geq r_i$, $\mu_V^+(x) \geq s_i$, $\nu_V^-(x) \leq t_i$ and $\nu_V^+(x) \leq q_i$ for each $x \in e$. We claim that $\mu_V^-(x) = r_i$, $\mu_V^+(x) = s_i$, $\nu_V^-(x) = t_i$ and $\nu_V^+(x) = q_i$, for some $x \in e$. For if not, then, by definition of V, $\mu_V^-(x) \geq r_{i-1}$, $\mu_V^+(x) \geq s_{i-1}$, $\nu_V^-(x) \leq t_{i-1}$ and $\nu_V^+(x) \leq q_{i-1}$ for all $x \in e$ which implies that $e \subseteq X_{i-1}$ and so $e \in E_i \backslash E_{i-1}$ and since $\mathscr{H}$ is simply ordered $e \not\subseteq X_{i-1}$, a contradiction. Hence, $Y_e = Z_e$, by definition of Y_e.

Corollary 3.1 *Suppose that $\mathscr{H} = (X, \tau)$ is a simply ordered interval-valued intuitionistic fuzzy hypergraph with $F(\mathscr{H}) = \{ \langle [r_i, s_i], [t_i, q_i] \rangle \mid 1 \leq i \leq n \}$. If $H_{\langle [r_n, s_n], [t_n, q_n] \rangle}$ is a simple hypergraph, then there is a partial interval-valued intuitionistic fuzzy hypergraph $\mathscr{H}' = (X, \tau')$ of $\mathscr{H}$ such that the following assertions hold.*

(i) $\mathscr{H}' = (X, \tau')$ is a V-tempered interval-valued intuitionistic fuzzy hypergraph of H_n.

(ii) $\tau \sqsubseteq \tau'$.

(iii) $F(\mathscr{H}') = F(\mathscr{H})$ and $C(\mathscr{H}') = C(\mathscr{H})$.

Definition 3.49 Let $\mathscr{H} = (X, \tau)$ be an interval-valued intuitionistic fuzzy hypergraph. An *interval-valued intuitionistic fuzzy transversal* $\mathscr{T}$ of $\mathscr{H} = (X, \tau)$ is an interval-valued intuitionistic fuzzy set defined on X such that $\mathscr{T}_{h(\tau_j)} \cap (\tau_j)_{h(\tau_j)} \neq \emptyset$, for each $\tau_j \in \tau$, $j = 1, 2, \ldots, m$.

Definition 3.50 A *minimal interval-valued intuitionistic fuzzy transversal* $\mathscr{T}$ of $\mathscr{H}$ is a transversal of $\mathscr{H}$ such that if $\mathscr{T}' \subset \mathscr{T}$, then $\mathscr{T}'$ is not an interval-valued intuitionistic fuzzy transversal of $\mathscr{H}$. The class of all minimal interval-valued intuitionistic fuzzy transversals of $\mathscr{H}$ will be denoted by $Tr(\mathscr{H})$.

Example 3.14 Consider an interval-valued intuitionistic fuzzy hypergraph $\mathscr{H} = (X, \tau)$ such that $X = \{x_1, x_2, x_3, x_4, x_5\}$ and $\tau = \{\tau_1, \tau_2, \tau_3, \tau_4\}$, represented by the following incidence matrix:

$$
M_{\mathscr{H}} = \begin{array}{c} \\ x_1 \\ x_2 \\ x_3 \\ x_4 \\ x_5 \end{array}
\begin{pmatrix}
\tau_1 & \tau_2 & \tau_3 & \tau_4 \\
\langle[0.3,0.5],[0.2,0.4]\rangle & \langle[0,0],[0,0]\rangle & \langle[0,0],[0,0]\rangle & \langle[0,0],[0,0]\rangle \\
\langle[0.3,0.5],[0.2,0.4]\rangle & \langle[0.6,0.8],[0.1,0.2]\rangle & \langle[0,0],[0,0]\rangle & \langle[0,0],[0,0]\rangle \\
\langle[0.2,0.4],[0.3,0.5]\rangle & \langle[0,0],[0,0]\rangle & \langle[0.4,0.6],[0.2,0.3]\rangle & \langle[0.5,0.7],[0.1,0.2]\rangle \\
\langle[0,0],[0,0]\rangle & \langle[0.2,0.4],[0.3,0.5]\rangle & \langle[0,0],[0,0]\rangle & \langle[0,0],[0,0]\rangle \\
\langle[0,0],[0,0]\rangle & \langle[0,0],[0,0]\rangle & \langle[0,0],[0,0]\rangle & \langle[0.5,0.7],[0.1,0.2]\rangle
\end{pmatrix}.
$$

$$
Tr(\mathscr{H}) = \begin{array}{c} \\ x_1 \\ x_2 \\ x_3 \\ x_4 \\ x_5 \end{array}
\begin{pmatrix}
\mathscr{T} \\
\langle[0,0],[0,0]\rangle \\
\langle[0.6,0.8],[0.1,0.2]\rangle \\
\langle[0.5,0.7],[0.1,0.2]\rangle \\
\langle[0,0],[0,0]\rangle \\
\langle[0,0],[0,0]\rangle
\end{pmatrix}.
$$

Theorem 3.4 *If $\mathscr{T}$ is an interval-valued intuitionistic fuzzy transversal of an interval-valued intuitionistic fuzzy hypergraph $\mathscr{H} = (X, \tau)$, then $h(\mathscr{T}) \geq h(\tau_j)$ for each $\tau_j \in \tau$. Moreover, if $\mathscr{T}$ is a minimal interval-valued intuitionistic fuzzy transversal of $\mathscr{H}$, then $h(\mathscr{T}) = h(\mathscr{H})$.*

Proof The proof follows at once from above definitions.

Theorem 3.5 *Let $\mathscr{H} = (X, \tau)$ be an interval-valued intuitionistic fuzzy hypergraph. Then the following statements are equivalent:*

(i) $\mathscr{T}$ is an interval-valued intuitionistic fuzzy transversal of $\mathscr{H}$,

(ii) for each $\tau_j \in \tau$ and each $\langle[\alpha, \beta], [\gamma, \delta]\rangle$, $\langle[0, 0], [0, 0]\rangle < \langle[\alpha, \beta], [\gamma, \delta]\rangle \leq h(\tau_j)$, $\mathscr{T}_{\langle[\alpha,\beta],[\gamma,\delta]\rangle} \cap (\tau_j)_{\langle[\alpha,\beta],[\gamma,\delta]\rangle} \neq \emptyset$,

(iii) for each $\langle[\alpha, \beta], [\gamma, \delta]\rangle$, $\langle[0, 0], [0, 0]\rangle < \langle[\alpha, \beta], [\gamma, \delta]\rangle \leq h(\mathscr{H})$, $\mathscr{T}_{\langle[\alpha,\beta],[\gamma,\delta]\rangle}$ is a transversal of $H_{\langle[\alpha,\beta],[\gamma,\delta]\rangle}$.

If $\mathscr{T}$ is a minimal interval-valued intuitionistic fuzzy transversal of $\mathscr{H}$, then $\mathscr{T}_{\langle[\alpha,\beta],[\gamma,\delta]\rangle}$ need not be a minimal transversal of $H_{\langle[\alpha,\beta],[\gamma,\delta]\rangle}$ for each $\langle[\alpha, \beta], [\gamma, \delta]\rangle$, $\langle[0, 0], [0, 0]\rangle < \langle[\alpha, \beta], [\gamma, \delta]\rangle \leq h(H)$. However, interval-valued intuitionistic fuzzy transversals satisfying this condition are of interest.

Definition 3.51 An interval-valued intuitionistic fuzzy set $\mathscr{T}$ with the property that $\mathscr{T}_{\langle[\alpha,\beta],[\gamma,\delta]\rangle}$ is a minimal transversal of $H_{\langle[\alpha,\beta],[\gamma,\delta]\rangle}$, for each $\langle[\alpha, \beta], [\gamma, \delta]\rangle$, $\langle[0, 0], [0, 0]\rangle < \langle[\alpha, \beta], [\gamma, \delta]\rangle \leq h(H)$ is called a *locally minimal interval-valued intuitionistic fuzzy transversal* of $\mathscr{H}$. The class of all locally minimal interval-valued intuitionistic fuzzy transversals of $\mathscr{H}$ will be denoted by $Tr^*(\mathscr{H})$. That is

$$Tr^*(\mathscr{H}) = \{\mathscr{T} \mid h(\mathscr{T}) = h(\mathscr{H}) \,\&\, \mathscr{T}_{\langle[\alpha,\beta],[\gamma,\delta]\rangle} \in Tr(H_{\langle[\alpha,\beta],[\gamma,\delta]\rangle})\}.$$

Remark 3.3 For any interval-valued intuitionistic fuzzy hypergraph $\mathscr{H}$, $Tr^*(\mathscr{H}) \subseteq Tr(\mathscr{H})$.

Theorem 3.6 *Suppose $\mathscr{H} = (X, \tau)$ is an ordered interval-valued intuitionistic fuzzy hypergraph with $F(\mathscr{H}) = \{\langle[r_i, s_i], [t_i, q_i]\rangle \mid 1 \leq i \leq n\}$ and $C(\mathscr{H}) = \{\mathscr{H}_{\langle[r_i,s_i],[t_i,q_i]\rangle} \mid 1 \leq i \leq n\}$. Then, $Tr^*(\mathscr{H}) \neq \emptyset$.*

References

1. Akram, M., Dudek, W.A.: Intuitionistic fuzzy hypergraphs with applications. Inf. Sci. **218**, 182–193 (2013)
2. Akram, M., Dudek, W.A.: Interval-valued fuzzy graphs. Comput. Math. Appl. **61**, 289–299 (2011)
3. Akram, M., Feng, F., Sarwar, S., Jun, Y.B.: Certain types of vague graphs. UPB Scientific Bulletin, Series A–Applied Mathematics and Physics, vol. 3, pp. 1–15 (2013)
4. Akram, M., Alshehri, N.O.: Tempered interval-valued fuzzy hypergraphs. Scientific Bulletin Series A–Applied Mathematics and Physics, vol. 77(1), pp. 39–48 (2015)
5. Akram, M., Gani, N., Saeid, A.B.: Vague hypergraphs. J. Intell. Fuzzy Syst. **26**, 647–653 (2014)
6. Atanassov, K.T., Gargov, G.: Interval-valued intuitionistic fuzzy sets. Fuzzy Sets Syst. **31**(3), 343–349 (1989)
7. Berge, C.: Graphs and Hypergraphs. North-Holland, Amsterdam (1973)
8. Chen, S.M.: Interval-valued fuzzy hypergraph and fuzzy partition. IEEE Trans. Syst. Man Cybern. (Cybernetics) **27**(4), 725–733 (1997)
9. Deschrijver, G., Cornelis, C.: Representability in interval-valued fuzzy set theory. Int. J. Uncertain. Fuzziness Knowl.-Based Syst. **15**, 345–361 (2007)
10. Deschrijver, G., Kerre, E.E.: On the relationships between some extensions of fuzzy set theory. Fuzzy Sets Syst. **133**, 227–235 (2003)
11. Gau, W.L., Buehrer, D.J.: Vague sets. IEEE Trans. Syst. Man Cybern. **23**, 610–614 (1993)
12. Goetschel Jr., R.H.: Introduction to fuzzy hypergraphs and Hebbian structures. Fuzzy Sets Syst. **76**, 113–130 (1995)
13. Goetschel Jr., R.H., Craine, W.L., Voxman, W.: Fuzzy transversals of fuzzy hypergraphs. Fuzzy sets Syst. **84**, 235–254 (1996)
14. Gorzalczany, M.B.: A method of inference in approximate reasoning based on interval-valued fuzzy sets. Fuzzy Sets Syst. **21**, 1–17 (1987)
15. Gorzalczany, M.B.: An Interval-valued fuzzy inference method some basic properties. Fuzzy Sets Syst. **31**, 243–251 (1989)
16. Hongmei, J., Lianhua, W.: Interval-valued fuzzy subsemigroups and subgroups associated by interval-valued fuzzy graphs. In: 2009 WRI Global Congress on Intelligent Systems, pp. 484–487 (2009)
17. Kaufmann, A.: Introduction a la Thiorie des Sous-Ensemble Flous, vol. 1. Masson, Paris (1977)
18. Lee, K.M.: Comparison of interval-valued fuzzy sets, intuitionistic fuzzy sets, and bipolar-valued fuzzy sets. J. Fuzzy Log. Intell. Sys. **14**, 125–129 (2004)
19. Lee-kwang, H., Lee, K.-M.: Fuzzy hypergraph and fuzzy partition. IEEE Trans. Syst. Man Cybern. **25**(1), 196–201 (1995)
20. Mendel, J.M.: Uncertain Rule-Based Fuzzy Logic Systems: Introduction and New Directions. Prentice-Hall, Upper Saddle River, NJ (2001)
21. Mordeson, J.N., Nair, P.S.: Fuzzy Graphs and Fuzzy Hypergraphs, 2nd edn. Physica Verlag, Heidelberg (2001)
22. Naz, S., Malik, M.A., Rashmanlou, H.: Hypergraphs and transversals of hypergraphs in interval-valued intuitionistic fuzzy setting. J. Mult.-Valued Log. Soft Comput. **30**, 399–417 (2018)
23. Roy, M.K., Biswas, R.: l-v fuzzy relations and Sanchez's approach for medical diagnosis. Fuzzy Sets Syst. **47**, 35–38 (1992)
24. Turksen, I.B.: Interval valued fuzzy sets based on normal forms. Fuzzy Sets Syst. **20**, 191–210 (1986)
25. Zadeh, L.A.: Fuzzy sets. Inf. Control **8**(3), 338–353 (1965)
26. Zadeh, L.A.: Similarity relations and fuzzy orderings. Inf. Sci. **3**(2), 177–200 (1971)
27. Zadeh, L.A.: The concept of a linguistic and application to approximate reasoning-I. Inf. Sci. **8**, 199–249 (1975)

Chapter 4
Bipolar Fuzzy (Directed) Hypergraphs

In this chapter, we present the concept of bipolar fuzzy hypergraphs and directed hypergraphs. We describe certain operations on bipolar fuzzy directed hypergraphs, which include addition, multiplication, vertex-wise multiplication, and structural subtraction. We discuss the concept of $B = (m^+, m^-)$−tempered bipolar fuzzy directed hypergraphs and investigate some of their basic properties. We present an algorithm to compute the minimum arc length of a bipolar fuzzy directed hyperpath. This chapter is due to [1, 3, 4, 10, 18].

4.1 Introduction

A wide variety of human decision-making is based on double-sided or bipolar judgmental thinking on a positive side and a negative side. For instances, cooperation and competition, friendship and hostility, common interests and conflict interests, effect and side effect, likelihood and unlikelihood, feedforward and feedback. In Chinese medicine, *Yin* and *Yang* are the two sides. *Yin* is the negative side of a system and *Yang* is the positive side of a system. The notion of bipolar fuzzy sets (YinYang bipolar fuzzy sets) was introduced by Zhang [18, 19] in the space $\{\forall\, (x, y) \mid (x, y) \in [-1, 0] \times [0, 1]\}$. Although bipolar fuzzy sets and intuitionistic fuzzy sets look similar to each other, they are essentially different sets [9, 10].

Hypergraphs have many applications in various fields, including biological sciences, computer science, and natural sciences. To study the degree of dependence of an object to the other, Kaufmann [8] applied the concept of fuzzy sets to hypergraphs. Mordeson and Nair [12] presented fuzzy graphs and fuzzy hypergraphs. Generalization and redefinition of fuzzy hypergraphs were discussed by Lee-Kwang and Lee [11]. The concept of interval-valued fuzzy sets was applied to hypergraphs by Chen [6]. Parvathi et al. [13] established the notion of intuitionistic fuzzy hypergraphs.

© Springer Nature Singapore Pte Ltd. 2020
M. Akram and A. Luqman, *Fuzzy Hypergraphs and Related Extensions*,
Studies in Fuzziness and Soft Computing 390,
https://doi.org/10.1007/978-981-15-2403-5_4

Definition 4.1 Let X be a non-empty set. A *bipolar fuzzy set B* in X is an object having the form, $B = (\mu_B^+, \mu_B^-) = \{(x, \mu_B^+(x), \mu_B^-(x)) \mid x \in X\}$, where $\mu_B^+ : X \to [0, 1]$ and $\mu_B^- : X \to [-1, 0]$ are mappings.

Positive membership degree $\mu_B^+(x)$ denotes the satisfaction degree of an element x to the property corresponding to a bipolar fuzzy set B and negative membership degree $\mu_B^-(x)$ denotes the satisfaction degree of x to some implicit counter-property corresponding to B. If $\mu_B^+(x) \neq 0$ and $\mu_B^-(x) = 0$, it is the state when x has only positive satisfaction for B. If $\mu_B^+(x) = 0$ and $\mu_B^-(x) \neq 0$, it is the state when x does not satisfy the property of B but somewhat satisfies the counter property of B. It is possible for an element x to be such that $\mu_B^+(x) \neq 0$ and $\mu_B^-(x) \neq 0$ when the membership function of the property coincides with its counter property over $x \in X$.

Example 4.1 Suppose that there is a fuzzy set "young" defined on the age domain [0, 100] like Fig. 4.1. In that fuzzy set, consider two ages 50 and 95 with membership degree 0. Although both of them do not satisfy the property "young", we may say that age 95 is more apart from the property rather than age 50. Only with the membership degrees ranged on the interval [0, 1], it is difficult to express this kind of meaning.

We define a bipolar fuzzy set as in Fig. 4.2 for the same fuzzy set "young" of Fig. 4.1. The negative membership degrees indicate the satisfaction range of elements to an implicit counter-property (e.g., old against the property young). This kind of bipolar fuzzy set representation enables the elements with 0 degree of membership in traditional fuzzy sets, to be expressed into the elements with zero degree of membership (when irrelevant elements) and negative degree of membership (when contrary elements). The age elements 50 and 95, with membership degree 0 in the fuzzy set of Fig. 4.1, have 0 and a negative membership degree in the bipolar fuzzy set of Fig. 4.2, respectively. Now it is manifested that 50 is an irrelevant age to the property young and 95 is more apart from the property young than 50 (i.e., 95 is a contrary age to the property young).

Example 4.2 Let $X = \{P_1, P_2, P_3, P_4, P_5, P_6\}$ be a set of products manufactured in a company. The products can be categorized according to their profit and loss. The

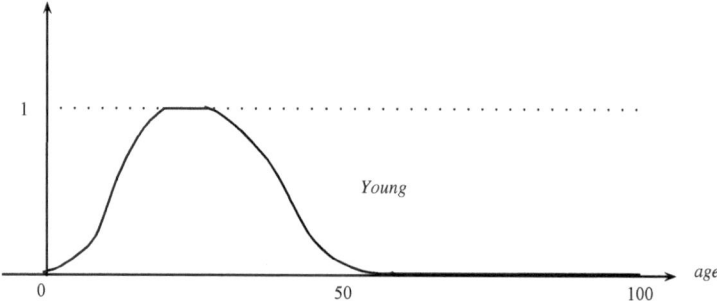

Fig. 4.1 A fuzzy set "young"

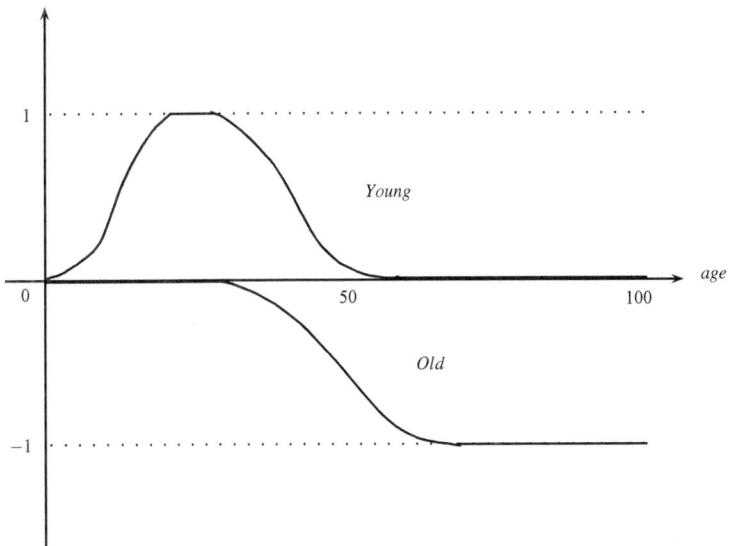

Fig. 4.2 A bipolar fuzzy set "young"

Table 4.1 Profit and loss of products

Product	Profit	Loss
P_1	0.6	0.4
P_2	0.8	0.5
P_3	0.9	0.1
P_4	0.7	0.2
P_5	0.5	0.6
P_6	0.6	0.4

profit and loss of every product varies from time to time. The possibilities of profit and loss of all the products are given in Table 4.1.

Table 4.1 shows that the product P_1 has 60% profit and 40% loss on the average. Profit is the positive and loss is the negative behavior of the product, that is, two-sided behavior. It can be written in the form of a bipolar fuzzy set as, $A = \{(P_1, 0.6, -0.4),$ $(P_2, 0.8, -0.5), (P_3, 0.9, -0.1), (P_4, 0.7, -0.2), (P_5, 0.5, -0.6), (P_6, 0.6, -0.4)\}$.

Example 4.3 Consider a fuzzy set

frog's prey $= \{(\text{mosquito}, 1.0), (\text{dragon fly}, 0.4), (\text{turtle}, 0.0), (\text{snake}, 0.0)\}$.

In this fuzzy set, both turtle and snake have the membership degree 0. It is known that frog and turtle are indifferent from each other concerning the prey-hunting relationship, but snake is a predator of frog. Turtle is an irrelevant animal and snake is related to frog by a counter implicit property but they both seem irrelevant in fuzzy set. As we can see from this example, it is difficult to express the difference of the irrelevant elements in fuzzy sets.

The same fuzzy set "frog's prey" can be redefined in the form of a bipolar fuzzy set, as follows:

frog's prey $= \{$(mosquito, 1, 0), (dragon fly, 0.4, 0), (turtle, 0, 0), (snake, 0, -1)$\}$.

We can see that membership degree 0 and nonmembership degree 0 of turtle means that frog never hunts turtle and turtle never hunts frog. While membership degree 0 and nonmembership degree -1 of snake means that frog never hunts snake but snake always hunts frog. Here the counter implicit property is "predator of frog", which created the difference between fuzzy set and bipolar fuzzy set of frog's prey.

Definition 4.2 For every two bipolar fuzzy sets $A = (\mu_A^+, \mu_A^-)$ and $B = (\mu_B^+, \mu_B^-)$ in X, we define the *intersection* and *union* of A and B as follows:

- $(A \cap B)(x) = (\min\{\mu_A^+(x), \mu_B^+(x)\}, \max\{\mu_A^-(x), \mu_B^-(x)\})$,
- $(A \cup B)(x) = (\max\{\mu_A^+(x), \mu_B^+(x)\}, \min\{\mu_A^-(x), \mu_B^-(x)\})$.

Definition 4.3 Let X be a non-empty set. Then, we call a mapping $A = (\mu_A^+, \mu_A^-) :$ $X \times X \to [-1, 0] \times [0, 1]$ a *bipolar fuzzy relation* on X such that $\mu_A^+(x, y) \in [0, 1]$ and $\mu_A^-(x, y) \in [-1, 0]$.

Definition 4.4 Let $A = (\mu_A^+, \mu_A^-)$ and $B = (\mu_B^+, \mu_B^-)$ be bipolar fuzzy sets on a set X. If $A = (\mu_A^+, \mu_A^-)$ is a bipolar fuzzy relation on a set X, then $A = (\mu_A^+, \mu_A^-)$ is called a *bipolar fuzzy relation* on $B = (\mu_B^+, \mu_B^-)$ if $\mu_A^+(x, y) \leq \min(\mu_B^+(x), \mu_B^+(y))$ and $\mu_A^-(x, y) \geq \max(\mu_B^-(x), \mu_B^-(y))$, for all $x, y \in X$.

A bipolar fuzzy relation A on X is called *symmetric* if $\mu_A^+(x, y) = \mu_A^+(y, x)$ and $\mu_A^-(x, y) = \mu_A^-(y, x)$, for all $x, y \in X$.

Definition 4.5 The *support* of a bipolar fuzzy set $A = (\mu_A^+, \mu_A^-)$, denoted by $supp(A)$, is defined as

$$supp(A) = supp^+(A) \cup supp^-(A), \quad supp^+(A) = \{x \mid \mu_A^+(x) > 0\}, \quad supp^-(A)$$
$$= \{x \mid \mu_A^-(x) < 0\}.$$

We call $supp^+(A)$ as *positive support* and $supp^-(A)$ as *negative support*.

Definition 4.6 Let $A = (\mu_A^+, \mu_A^-)$ be a bipolar fuzzy set on X. Let $\alpha \in [0, 1]$ and $\beta \in [-1, 0]$, then $(\alpha, \beta)-cut$ $A_{(\alpha,\beta)}$ of A can be defined as, $A_{(\alpha,\beta)} = \{x \mid \mu_\alpha^+(x) \geq \alpha, \ \mu_\alpha^-(x) \leq \beta\}$.

Definition 4.7 The *height* of a bipolar fuzzy set $A = (\mu_A^+, \mu_A^-)$ is defined as $h(A) = \max\{\mu_A^+(x) | x \in X\}$.

The *depth* of a bipolar fuzzy set $A = (\mu_A^+, \mu_A^-)$ is defined as $d(A) = \min\{\mu_A^-(x) | x \in X\}$.

We shall say that bipolar fuzzy set A is *normal*, if $h(A) = 1$ and $d(A) = -1$.

Definition 4.8 A *bipolar fuzzy graph* on X is a pair $G = (\mathscr{C}, \mathscr{D})$, where $\mathscr{C} = (\mu_{\mathscr{C}}^+, \mu_{\mathscr{C}}^-)$ is a bipolar fuzzy set on X and $\mathscr{D} = (\mu_{\mathscr{D}}^+, \mu_{\mathscr{D}}^-)$ is a bipolar fuzzy relation on X such that

$$\mu_{\mathscr{D}}^+(xy) \leq \min\{\mu_{\mathscr{C}}^+(x), \mu_{\mathscr{C}}^+(y)\}, \ \mu_{\mathscr{D}}^-(xy) \geq \max\{\mu_{\mathscr{C}}^-(x), \mu_{\mathscr{C}}^-(y)\}, \text{ for all } x, \ y \in X.$$

Note that, $\mathscr{D}$ is a bipolar fuzzy relation on $\mathscr{C}$, and $\mu_{\mathscr{D}}^+(xy) > 0$, $\mu_{\mathscr{D}}^-(xy) < 0$ for $xy \in X \times X$, $\mu_{\mathscr{D}}^+(xy) = \mu_{\mathscr{D}}^-(xy) = 0$, for $xy \in X \times X - E$.

Definition 4.9 A *bipolar fuzzy digraph* on X is a pair $\vec{G} = (\mathscr{C}, \vec{\mathscr{D}})$, where $\mathscr{C} = (\mu_{\mathscr{C}}^+, \mu_{\mathscr{C}}^-)$ is a bipolar fuzzy set on X and $\vec{\mathscr{D}} = (\mu_{\vec{\mathscr{D}}}^+, \mu_{\vec{\mathscr{D}}}^-)$ is a bipolar fuzzy relation on X such that

$$\mu_{\vec{\mathscr{D}}}^+(xy) \leq \min\{\mu_{\mathscr{C}}^+(x), \mu_{\mathscr{C}}^+(y)\}, \ \mu_{\vec{\mathscr{D}}}^-(xy) \geq \max\{\mu_{\mathscr{C}}^-(x), \mu_{\mathscr{C}}^-(y)\}, \text{ for all } x, y \in X.$$

For further terminologies and studies on bipolar fuzzy hypergraphs, readers are referred to [2, 5, 7, 14–17].

4.2 Bipolar Fuzzy Hypergraphs

Definition 4.10 A *bipolar fuzzy hypergraph* on a non-empty set X is a pair $H = (C, D)$, where $C = \{\xi_1, \xi_2, \ldots, \xi_r\}$ is family of bipolar fuzzy subsets on X and $D = (\mu_D^+, \mu_D^-)$ is a bipolar fuzzy relation on the bipolar fuzzy subsets $\xi_i = (\mu_{\xi_i}^+, \mu_{\xi_i}^-)$ such that

(i)
$$\mu_D^+(E_i) = \mu_D^+(\{x_1, x_2, \cdots, x_s\}) \leq \min\{\mu_{\xi_i}^+(x_1), \mu_{\xi_i}^+(x_2), \ldots, \mu_{\xi_i}^+(x_s)\},$$

$$\mu_D^-(E_i) = \mu_D^-(\{x_1, x_2, \cdots, x_s\}) \leq \max\{\mu_{\xi_i}^-(x_1), \mu_{\xi_i}^-(x_2), \ldots, \mu_{\xi_i}^-(x_s)\},$$

for all $x_1, x_2, \ldots, x_s \in X$.

(ii) $\bigcup_i supp(\xi_i) = X$, for all $\xi_i \in C$.

Example 4.4 Let $C = \{\xi_1, \xi_2, \xi_3, \xi_4\}$ be a family of bipolar fuzzy subsets on $X = \{a, b, c, d, e, f\}$ as given in Table 4.2. The bipolar fuzzy relation D on the bipolar fuzzy subsets ξ_i's is defined as $D(\{a, c, d, f\}) = (0.1, -0.2)$, $D(\{a, b, c\}) = (0.2, -0.4)$, $D(\{d, e, f\}) = (0.1, -0.2)$. Routine calculations show that H is a bipolar fuzzy hypergraph as shown in Fig. 4.3.

Definition 4.11 A bipolar fuzzy set $C = (\mu_C^+, \mu_C^-) : X \to [-1, 0] \times [0, 1]$ is an *elementary bipolar fuzzy set* if C is single valued on $supp(C)$. A bipolar fuzzy hypergraph $H = (C, D)$ is *elementary* if each $\xi_i \in C$ and D are elementary otherwise, it is called *non-elementary*.

Table 4.2 Family of bipolar fuzzy subsets on X

$x \in X$	ξ_1	ξ_2	ξ_3	ξ_4
a	$(0.2, -0.5)$	$(0.2, -0.6)$	$(0, 0)$	$(0, 0)$
b	$(0, 0)$	$(0.5, -0.7)$	$(0, 0)$	$(0, 0)$
c	$(0.5, -0.4)$	$(0.5 - 0.4)$	$(0.5, -0.4)$	$(0, 0)$
d	$(0.8, -0.6)$	$(0, 0)$	$(0.8, -0.6)$	$(0.8, -0.6)$
e	$(0, 0)$	$(0, 0)$	$(0, 0)$	$(0.5, -0.8)$
f	$(0.1, -0.2)$	$(0, 0)$	$(0, 0)$	$(0.1, -0.2)$

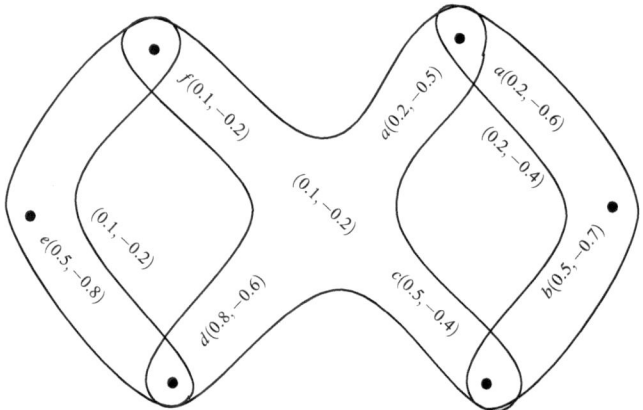

Fig. 4.3 Bipolar fuzzy hypergraph H

We now explore the concept in which a bipolar fuzzy graph is a bipolar fuzzy hypergraph.

Proposition 4.1 *Bipolar fuzzy graphs are special cases of the bipolar fuzzy hypergraphs.*

A *bipolar fuzzy multigraph* is a multivalued symmetric mapping $D = (\mu_D^+, \mu_D^-)$: $X \times X \to [0, 1] \times [-1, 0]$. A bipolar fuzzy multigraph can be considered to be the "disjoint union" or "disjoint sum" of a collection of simple bipolar fuzzy graphs, as is done with crisp multigraphs. The same holds for multidigraphs. Therefore, these structures can be considered as "disjoint unions" or "disjoint sums" of bipolar fuzzy hypergraphs.

Definition 4.12 A bipolar fuzzy hypergraph $H = (C, D)$ is called *simple* if every $\xi_i, \xi_j \in C, \xi_i \subseteq \xi_j$ implies that $\xi_i = \xi_j$.

A bipolar fuzzy hypergraph $H = (C, D)$ is called *support simple* if every $\xi_i, \xi_j \in C, \xi_i \subseteq \xi_j$, and $supp(\xi_i) = supp(\xi_j)$ implies that $\xi_i = \xi_j$.

A bipolar fuzzy hypergraph $H = (C, D)$ is called *strongly support simple* if every $\xi_i, \xi_j \in C, supp(\xi_i) = supp(\xi_j)$ implies that $\xi_i = \xi_j$.

Remark 4.1 Definition 4.12 reduces to familiar definitions in the special case where H is a crisp hypergraph. The definition of simple bipolar fuzzy hypergraph is identical to the definition of simple crisp hypergraph. A crisp hypergraph is support simple and strongly support simple if and only if it has no multiple edges. For bipolar fuzzy hypergraphs all three concepts imply no multiple edges. Any simple bipolar fuzzy hypergraph is support simple and every strongly support simple bipolar fuzzy hypergraph is support simple. Simple and strongly support simple are independent concepts in bipolar fuzziness.

Definition 4.13 Let $H = (C, D)$ be a bipolar fuzzy hypergraph and $(\alpha, \beta) \in [-1, 0] \times [0, 1]$. Define an (α, β)-cut level set of a bipolar fuzzy set ξ_i as, $\xi_{i(\alpha,\beta)} = \{x \mid \mu_{\xi_i}^+(x) \geq \alpha \text{ and } \mu_{\xi_i}^-(x) \leq \beta\}$.

$H_{(\alpha,\beta)} = (C_{(\alpha,\beta)}, D_{(\alpha,\beta)})$ is called an $(\alpha, \beta)-level\ hypergraph$ of H, where $C_{(\alpha,\beta)} = \cup_{i=1}^{r} \xi_{i(\alpha,\beta)}$.

Clearly, it is possible that $A_{(\alpha,\beta)} = B_{(\alpha,\beta)}$ for $A \neq B$, by using distinct markers to identity the various members of E a distinction between $A_{(\alpha,\beta)}$ and $B_{(\alpha,\beta)}$ to represent multiple edges in $H_{(\alpha,\beta)}$. However, we do not take this approach unless otherwise stated, we will always regard $H_{(\alpha,\beta)}$ as having no repeated edges.

The families of crisp sets (hypergraphs) produced by the (α, β)-cuts of a bipolar fuzzy hypergraph share an important relationship with each other, as expressed below:

Suppose $\mathbb{X}$ and $\mathbb{Y}$ are two families of sets such that for each set X belonging to $\mathbb{X}$ there is at least one set X belonging to $\mathbb{Y}$ which contains X. In this case, we say that $\mathbb{X}$ *absorbs* $\mathbb{Y}$ and symbolically write $\mathbb{X} \sqsubseteq \mathbb{Y}$ to express this relationship between $\mathbb{X}$ and $\mathbb{Y}$. Since, it is possible for $\mathbb{X} \sqsubseteq \mathbb{Y}$ while $\mathbb{X} \cap \mathbb{Y} = \emptyset$, we have that $\mathbb{X} \subseteq \mathbb{Y} \Rightarrow \mathbb{X} \sqsubseteq \mathbb{Y}$, whereas the converse is generally false. If $\mathbb{X} \sqsubseteq \mathbb{Y}$ and $\mathbb{X} \neq \mathbb{Y}$, then we write $\mathbb{X} \sqsubset \mathbb{Y}$.

Definition 4.14 Let $H = (C, D)$ be a bipolar fuzzy hypergraph and for each $(\alpha, \beta) \in [0, 1] \times [-1, 0]$, $H_{(\alpha,\beta)} = (C_{(\alpha,\beta)}, D_{(\alpha,\beta)})$ be an $(\alpha, \beta)-level$ hypergraph of H. The sequence of ordered pairs $(r_1^+, r_1^-), (r_2^+, r_2^-), ..., (r_n^+, r_n^-)$ with $1 \geq r_1^+ > r_2^+ > ... > r_n^+ > 0$ and $-1 \leq r_1^- < r_2^- < ... < r_n^- < 0$ satisfying the properties,

(i) If $1 \geq u^+ \geq r_1^+$ and $-1 \leq u^- \leq r_1^-$ then $D_{(u^+, u^-)} = \emptyset$,
(ii) If $r_i^+ \geq u^+ \geq r_{i+1}^+$ and $r_i^- \leq u^- \leq r_{i+1}^-$ then $D_{(u^+, u^-)} = D_{(r_i^+, r_i^-)}$,
(iii) $D_{(r_i^+, r_i^-)} \sqsubset D_{(r_{i+1}^+, r_{i+1}^-)}$,

is called the *fundamental sequence* of bipolar fuzzy hypergraph H, denoted by $f_s(H)$. The corresponding sequence of $(r_i^+, r_i^-)-$level hypergraphs $H_{(r_1^+, r_1^-)}$, $H_{(r_2^+, r_2^-)}$, ..., $H_{(r_n^+, r_n^-)}$ is called the *core set* of H, denoted by $C(H)$. The $(r_n^+, r_n^-)-$level hypergraph, $H_{(r_n^+, r_n^-)}$, is called the *support level* of H.

Definition 4.15 Let $H = (C, D)$ be a bipolar fuzzy hypergraph with fundamental sequence $f_s(H) = \{(r_1^+, r_1^-), (r_2^+, r_2^-), ..., (r_n^+, r_n^-)\}$ and $(r_{n+1}^+, r_{n+1}^-) = (0, 0)$, then H is called *sectionally elementary* if for each $\xi \in C$, $(r_i^+, r_i^-) \in f_s(H)$, $1 \leq i \leq n$ $\xi_{(s^+, s^-)} = \xi_{(r_i^+, r_i^-)}$, for all $(s^+, s^-) \in (r_i^+, r_{i-1}^+] \times (r_{i-1}^-, r_i^-]$.

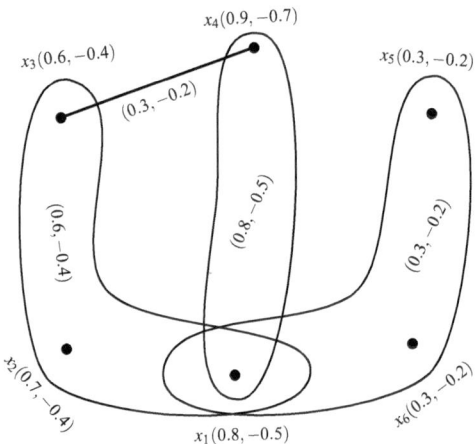

Fig. 4.4 Bipolar fuzzy hypergraph

•

x_4

Fig. 4.5 $H_{(r_1^+,r_1^-)}$

Example 4.5 Let $H = (C, D)$ be a bipolar fuzzy hypergraph as shown in Fig. 4.4. Take $(r_1^+, r_1^-) = (0.9, -0.7)$, $(r_2^+, r_2^-) = (0.8, -0.5)$, $(r_3^+, r_3^-) = (0.6, -0.4)$, and $(r_4^+, r_4^-) = (0.3, -0.2)$. Clearly, the sequence $\{(r_1^+, r_1^-), (r_2^+, r_2^-), (r_3^+, r_3^-), (r_4^+, r_4^-)\}$ satisfies all the conditions of Definition 4.14 and hence it is a fundamental sequence of H. The corresponding sequence of (r_i^+, r_i^-)−level hypergraphs are shown in Figs. 4.5, 4.6, 4.7, 4.8.

Definition 4.16 A bipolar fuzzy hypergraph $H = (C, D)$ is called *ordered* if the core set $C(H) = \{H_{(r_1^+,r_1^-)}, H_{(r_2^+,r_2^-)}, \ldots, H_{(r_n^+,r_n^-)}\}$ is ordered, i.e., $H_{(r_1^+,r_1^-)} \sqsubseteq H_{(r_2^+,r_2^-)} \sqsubseteq \ldots \sqsubseteq H_{(r_n^+,r_n^-)}$.
H is called *simply ordered* if H is ordered and whenever $D' \sqsubset D_{(r_{i+1}^+,r_{i+1}^-)} \setminus D_{(r_i^+,r_i^-)}$ then, $D' \not\subseteq C_{(r_i^+,r_i^-)}$.

Example 4.6 Consider a bipolar fuzzy hypergraph $H = (C, D)$ on the set $X = \{a, b, c, d\}$ and $C = \{\xi_1, \xi_2, \xi_3, \xi_4, \xi_5\}$ be the family of bipolar fuzzy sets as given in Table 4.3. The bipolar fuzzy relation D is given as $D(\{a, b\}) = (0.7, -0.2)$, $D(\{a, b, d\}) = (0.4, -0.2)$, $D(\{b, c\}) = (0.9, -0.2)$, and $D(\{a, c, d\}) = (0.4, -0.3)$.
    Note that,

$$D_{(0.9,-0.1)} = \{\{b, c\}\}, \quad D_{(0.7,-0.2)} = \{\{a, b\}, \{b, c\}\},$$

$$D_{(0.4,-0.2)} = \{\{a, b\}, \{a, b, d\}, \{b, c\}, \{b, c, d\}, \{a, c, d\}\}.$$

x_4

Fig. 4.6 $H_{(r_2^+, r_2^-)}$

x_1

Fig. 4.7 $H_{(r_3^+, r_3^-)}$

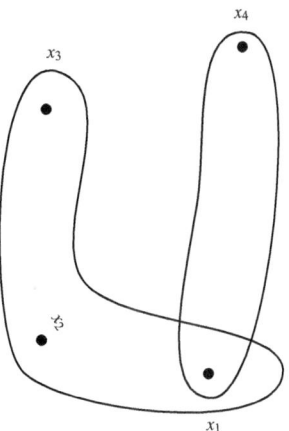

Thus, for $0.4 < s^+ \leq 0.9$ and $-0.1 > s^- \geq -0.3$, $D_{(s^+, s^-)} = \{\{a, b\}, \{b, c\}\}$, and for $0 < s^+ \leq 0.4$ and $-1 < s^- \geq -0.3$, $D_{(s^+, s^-)} = \{\{a, b\}, \{a, b, d\}, \{b, c\}, \{b, c, d\}, \{a, c, d\}\}$.

Note that, $D_{(0.9, -0.1)} \subseteq D_{(0.4, -0.3)}$.

The fundamental sequence is $f_s(H)=\{(s_1, r_1) = (0.9, -0.1), (s_2, r_2) = (0.4, -0.2)\}$ and the set of core hypergraphs is $C(H) = \{H_{(0.9, -0.1)}, H_{(0.4, -0.2)}\}$.

H is support simple, but not simple. H is not sectionally elementary. Clearly, bipolar fuzzy hypergraph H is simply ordered.

Proposition 4.2 *Let $H = (C, D)$ be an elementary bipolar fuzzy hypergraph then H is support simple if and only if H is strongly support simple.*

Fig. 4.8 $H_{(r_4^+, r_4^-)}$

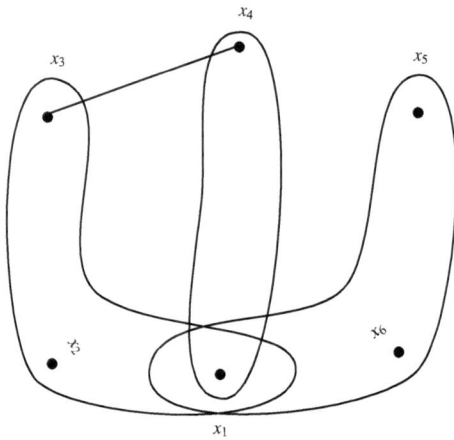

Table 4.3 Bipolar fuzzy subsets on X

C	ξ_1	ξ_2	ξ_3	ξ_4	ξ_5
a	$(0.7, -0.2)$	$(0.9, -0.2)$	$(0, 0)$	$(0, 0)$	$(0.4, -0.3)$
b	$(0.7, -0.2)$	$(0.9, -0.2)$	$(0.9, -0.2)$	$(0.7, -0.2)$	$(0, 0)$
c	$(0, 0)$	$(0, 0)$	$(0.9, -0.2)$	$(0.7, -0.2)$	$(0.4, -0.3)$
d	$(0, 0)$	$(0.4, -0.3)$	$(0, 0)$	$(0.4, -0.3)$	$(0.4, -0.3)$

Proof Suppose that H is elementary support simple bipolar fuzzy graph and that for each $\xi_i, \xi_j \in C$, $\xi_i \subseteq \xi_j$, $supp(\xi_i) = supp(\xi_j)$. Since, H is elementary, therefore, $supp(\xi_i) = supp(\xi_j)$ implies that $\xi_i \not\subset \xi_j$ and H is support simple implies that $\xi_i = \xi_j$. Hence, H is strongly support simple. The proof of converse part is obvious.

The complexity of a bipolar fuzzy hypergraph depends on how many hyperedges it has. The natural question arises: is there an upper bound on the number of edges of a bipolar fuzzy hypergraph of order n?

Proposition 4.3 *Let $H = (C, D)$ be a simple bipolar fuzzy hypergraph of order n. Then, there is no upper bound on $|D|$.*

The Proposition 4.3 is explained in Example 4.7.

Example 4.7 Let $H = (C, D)$ be a bipolar fuzzy hypergraph on $X = \{x, y\}$ such that $C = \{\xi_i = (\mu_{\xi_i}^+, \mu_{\xi_i}^-) \mid i = 1, 2, \ldots, n\}$, where

$$\mu_{\xi_i}^+(x) = \frac{1}{i+1}, \quad \mu_{\xi_i}^-(x) = -1 + \frac{1}{i+1},$$

$$\mu_{\xi_i}^+(y) = \frac{1}{i+1}, \quad \mu_{\xi_i}^-(y) = -\frac{i}{i+1}.$$

Then, $H = (C, D)$ is a simple bipolar fuzzy hypergraph with n edges. Clearly there is no upper bound for $|D|$.

Proposition 4.4 *Let $H = (C, D)$ be a support simple bipolar fuzzy hypergraph of order with n hyperedges. Then there is no upper bound on $|D|$.*

Proof The class of support simple bipolar fuzzy hypergraphs contains the class of simple bipolar fuzzy hypergraphs, thus the result follows from Proposition 4.3.

Proposition 4.5 *Let H be an elementary bipolar fuzzy hypergraph then H is ordered. If H is an ordered bipolar fuzzy hypergraph and support level $H_{(r_n^+, r_n^-)}$ is simple then H is elementary.*

Definition 4.17 A *bipolar fuzzy hyperpath* of length s in a bipolar fuzzy hypergraph can be defined as an alternative sequence $x_1, E_1, x_2, E_2, \ldots, x_s, E_s, x_{s+1}$ of distinct vertices and hyperedges such that

(i) $\mu^+(E_i) > 0$ or $\mu^-(E_i) < 0, i = 1, 2, \ldots, s$,
(ii) $x_i, x_{i+1} \in E_i, i = 1, 2, \ldots, s$.

Definition 4.18 A bipolar fuzzy hypergraph H is said to *connected* if there exists a bipolar fuzzy hyperpath between every pair of distinct vertices.

Definition 4.19 Let u and v be any two distinct vertices of a bipolar fuzzy hypergraph H which are connected by a bipolar fuzzy hyperpath of length l. The *strength of a bipolar fuzzy hyperpath $u - v$* is defined as

$$S^l(u, v) = (S^{l+}, S^{l-}) = (\min\{\mu_D^+(E_1), \mu_D^+(E_2), \ldots, \mu_D^+(E_k)\}, \max\{\mu_D^-(E_1),$$
$$\mu_D^-(E_2), \ldots, \mu_D^-(E_k)\}), u \in E_1, \ v \in E_k,$$

where $E_1, E_2, \ldots, E_k$ are the hyperedges.
The *strength of connectedness* between u and v is defined as

$$S^\infty(u, v) = \{(\max_l S^{l+}(u, v), \min_l S^{l-}(u, v)) | l = 1, 2, \ldots\}.$$

4.3   Bipolar Fuzzy Directed Hypergraphs

Definition 4.20 A *directed hyperedge* (or *hyperarc*) is defined as an ordered pair $Y = (u, v)$, where u and v are disjoint subsets of X. u is taken as the *tail* of Y and v is called its *head*. $t(Y)$ and $h(Y)$ are used to denote the *tail* and *head* of directed hyperarc, respectively.

Definition 4.21 A *bipolar fuzzy directed hypergraph* is a pair $\vec{D} = (T, U)$, where T is a finite family of bipolar fuzzy sets on X and U is a set of bipolar fuzzy directed hyperarcs (hyperedges).

A *bipolar fuzzy directed hyperarc (hyperedge)* $a \in U$ is a pair $(t(a), h(a))$, where $t(a)$ and $h(a)$ are two distinct bipolar fuzzy subsets on X such that $t(a) \subset U, t(a) \neq \phi$ is its *tail* and $h(a) \in U - t(a)$ is called its *head*. A *source vertex* s is defined as a

vertex in $\vec{D}$ if $h(a) \neq s$, for each $a \in U$. A *destination vertex d* is defined as a vertex
if $t(a) \neq d$, for every $a \in U$.

Definition 4.22 A *backward bipolar fuzzy directed hyperarc* or $b-$arc is defined as
a bipolar fuzzy hyperarc $U = (t(U), h(U))$, with $|supp(h(U))| = 1$.

A *forward bipolar fuzzy hyperarc* or $f-$arc is a bipolar fuzzy hyperarc $U = (t(U), h(U))$, with $|supp(t(U))| = 1$.

A bipolar fuzzy directed hypergraph is called a *backward bipolar fuzzy directed
hypergraph*, if its all hyperarcs are $b-$arcs. A bipolar fuzzy directed hypergraph is
said to be *forward bipolar fuzzy directed hypergraph* if its all hyperarcs are $f-$arcs.
A *bf$-$graph (or bf$-$bipolar fuzzy directed hypergraph)* is a bipolar fuzzy directed
hypergraph, whose hyperarcs are either $b-$arcs or $f-$arcs.

Definition 4.23 A *directed hyperpath* between nodes s and d in a bipolar fuzzy
directed hypergraph $\vec{D}$ is an alternating sequence of distinct vertices and bipolar
fuzzy directed hyperedges $s = t_0, e_1, t_1, e_2, ..., e_k = d$, such that $t_{i-1}, t_i \in e_i$, for all
$i = 1, 2, 3, ..., k$.

Example 4.8 A bipolar fuzzy directed hypergraph and a hyperpath between two
nodes s and d is shown in Fig. 4.9.

The path is drawn in thick line.

Definition 4.24 The *incidence matrix representation* of a crisp directed hypergraph
$N = (X, A)$ is given as a matrix $[b_{ij}]$ of order $n \times m$, defined as follow:

$$b_{ij} = \begin{cases} -1, & \text{if } e_i \in t(A_j), \\ 1, & \text{if } e_i \in h(A_j), \\ 0, & \text{otherwise.} \end{cases}$$

Definition 4.25 The *incidence matrix* of a bipolar fuzzy directed hypergraph $\vec{D} = (T, U)$ is characterized by an $n \times m$ matrix $[a_{ij}]$ as follows:

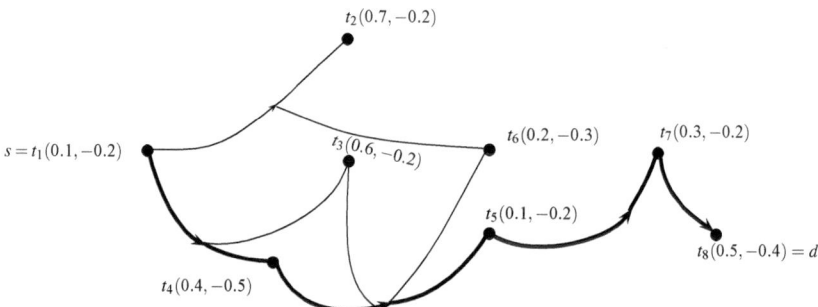

Fig. 4.9 A directed hyperpath in a bipolar fuzzy directed hypergraph

$$a_{ij} = \begin{cases} (m_j^+(n_i), m_j^-(n_i)), & \text{if } n_i \in U_j, \\ 0, & \text{otherwise.} \end{cases}$$

Here, $0 = (0, 0)$.

Definition 4.26 Let $\vec{D} = (T, U)$ be a bipolar fuzzy directed hypergraph. The *height* $h(\vec{D})$ of $\vec{D}$ is defined as

$$h(\vec{D}) = \{\max(U_i), \min(U_j) : U_i, U_j \in U\},$$

where $U_i = \max(m_{ij}^+)$ and $U_j = \min(m_{ij}^-)$, m_{ij}^+ is taken as the positive membership value and m_{ij}^- indicates the negative membership value of vertex i to hyperedge j.

Definition 4.27 A bipolar fuzzy directed hypergraph $\vec{D} = (T, U)$ is *simple* if there are no repeated bipolar fuzzy hyperedges in U and if $U_k, U_j \in U$ and $U_k \subseteq U_j$ then $U_k = U_j$, for each k and j.

A bipolar fuzzy directed hypergraph $\vec{D} = (T, U)$ is called *support simple* if whenever $U_i, U_j \in U$, $U_i \subseteq U_j$ and $supp(U_i) = supp(U_j)$, then $U_i = U_j$, for all i and j.

Then, the hyperedges U_i and U_j are called *supporting edges*.

Definition 4.28 A bipolar fuzzy directed hypergraph is called *elementary* if $m_{ij}^+ :\to [0, 1]$ and $m_{ij}^- :\to [-1, 0]$ are constant functions.

If $|supp(m_{ij}^+, m_{ij}^-)| = 1$, then it is characterized as a *spike*. That is, a bipolar fuzzy subset with singleton support.

Theorem 4.1 *The bipolar fuzzy directed hyperedges of a bipolar fuzzy directed hypergraph are elementary.*

Example 4.9 Consider a bipolar fuzzy directed hypergraph $\vec{D} = (T, U)$, where T is the family of bipolar fuzzy subsets on X and U is the set of bipolar fuzzy relations on T. The corresponding incidence matrix is given in Table 4.4.

The corresponding elementary bipolar fuzzy directed hypergraph is shown in Fig. 4.10.

Definition 4.29 Let $\vec{D} = (T, U)$ be a bipolar fuzzy directed hypergraph. Suppose that $\mu \in [0, 1]$ and $v \in [-1, 0]$. The (μ, v)-level is defined as,

Table 4.4 Elementary bipolar fuzzy directed hypergraph

I	U_1	U_2	U_3	U_4
t_1	$(0.2, -0.3)$	$(0.5, -0.2)$	0	$(0.3, -0.4)$
t_2	0	$(0.5, -0.2)$	$(0.5, -0.2)$	0
t_3	$(0.2, -0.3)$	0	$(0.5, -0.2)$	$(0.3, -0.4)$
t_4	0	0	$(0.5, -0.2)$	$(0.3, -0.4)$

Fig. 4.10 Elementary
bipolar fuzzy directed
hypergraph

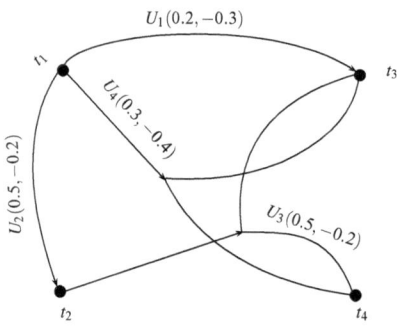

$U_{(\mu,\nu)} = \{v \in X | T^+(v) \geq \mu \text{ and } T^-(v) \leq \nu\}$. The crisp directed hypergraph $\vec{D}_{(\mu,\nu)} = (T_{(\mu,\nu)}, U_{(\mu,\nu)})$ such that

- $U_{(\mu,\nu)} = \{v \in X | U_i^+(v) \geq \mu \text{ and } U_i^-(v) \leq \nu\}$,
- $T_{(\mu,\nu)} = \bigcup U_{i(\mu,\nu)}$, for all $U_i \in U$,

is called the $(\mu, \nu)-level$ hypergraph of $\vec{D}$.

Definition 4.30 Let $\vec{D} = (T, U)$ be a bipolar fuzzy directed hypergraph and $\vec{D}_{(\mu_i,\nu_i)} = (T_{(\mu_i,\nu_i)}, U_{(\mu_i,\nu_i)})$ be the $(\mu_i, \nu_i)-level$ directed hypergraphs of $\vec{D}$. The sequence $\{(\mu_1, \nu_1), (\mu_2, \nu_2), ..., (\mu_n, \nu_n)\}$ of real numbers, where $0 < \mu_1 < \mu_2 < ... < \mu_n$ and $0 > \nu_1 > \nu_2 > ... > \nu_n$, $(\mu_n, \nu_n) = h(\vec{D})$ such that the following properties:

(i) if $(\mu_{i-1}, \nu_{i-1}) < (\alpha, \beta) \leq (\mu_i, \nu_i)$, then $U_{(\alpha,\beta)} = U_{(\mu_i,\nu_i)}$,
(ii) $U_{(\mu_i,\nu_i)} \sqsubset U_{(\mu_{i+1},\nu_{i+1})}$,

are satisfied, is called the *fundamental sequence of* $\vec{D}$. The sequence is denoted by $FS(\vec{D})$. The $(\mu_i, \nu_i)-level$ hypergraphs $\{\vec{D}_{(\mu_1,\nu_1)}, \vec{D}_{(\mu_2,\nu_2)}, ..., \vec{D}_{(\mu_n,\nu_n)}\}$ are called the *core hypergraphs* of $\vec{D}$. It is also called *core set* of $\vec{D}$ and is denoted by $c(\vec{D})$.

Definition 4.31 Let $\vec{D} = (T, U)$ be a bipolar fuzzy directed hypergraph and $FS(\vec{D}) = \{(\mu_1, \nu_1), (\mu_2, \nu_2), ..., (\mu_n, \nu_n)\}$. If for each $E = (m^+, m^-) \in U$ and each $(\mu_i, \nu_i) \in FS(\vec{D})$, $E_{(\mu,\nu)} = U_{(\mu_i,\nu_i)}$, for all $(\mu, \nu) \in ((\mu_{i-1,\nu_{i-1}}), (\mu_i, \nu_i)]$, then $\vec{D}$ is *sectionally elementary*.

Definition 4.32 Let $\vec{D} = (T, U)$ be a bipolar fuzzy directed hypergraph and $c(\vec{D}) = \{\vec{D}_{(\mu_1,\nu_1)}, \vec{D}_{(\mu_2,\nu_2)}, ..., \vec{D}_{(\mu_n,\nu_n)}\}$. $\vec{D}$ is said to be *ordered* if $c(\vec{D})$ is *ordered*. That is, $\vec{D}_{(\mu_1,\nu_1)} \subset \vec{D}_{(\mu_2,\nu_2)} \subset ... \subset \vec{D}_{(\mu_n,\nu_n)}$.

The bipolar fuzzy directed hypergraph is called *simply ordered* if the sequence $\{\vec{D}_{(\mu_1,\nu_1)}, \vec{D}_{(\mu_2,\nu_2)}, ..., \vec{D}_{(\mu_n,\nu_n)}\}$ is *simply ordered*.

Example 4.10 Consider a bipolar fuzzy directed hypergraph $\vec{D} = (T, U)$, where $X = \{t_1, t_2, t_3, t_4, t_5\}$ and T is the family of bipolar fuzzy subsets on X. The corresponding incidence matrix is given in Table 4.5.

Table 4.5 Bipolar fuzzy directed hypergraph

I	U_1	U_2	U_3
t_1	$(0.8, -0.2)$	$(0.6, -0.1)$	$(0.4, -0.3)$
t_2	0	0	$(0.4, -0.3)$
t_3	$(0.8, -0.1)$	0	0
t_4	$(0.8, -0.2)$	$(0.6, -0.1)$	$(0.4, -0.3)$
t_5	0	$(0.5, -0.1)$	$(0.4, -0.3)$

The corresponding hypergraph is shown in Fig. 4.11.

By computing the (μ_i, ν_i)−level bipolar fuzzy directed hypergraphs of $\vec{D}$, we have $U_{(0.8,-0.2)} = \{t_1, t_4\}$, $U_{(0.6,-0.1)} = \{t_1, t_4\}$ and $U_{(0.4,-0.3)} = \{\{t_1, t_2\}, \{t_4, t_5\}\}$. Note that, $\vec{D}_{(0.8,-0.2)} = \vec{D}_{(0.6,-0.1)}$ and $\vec{D}_{(0.8,-0.2)} \subseteq \vec{D}_{(0.4,-0.3)}$. The fundamental sequence is $FS(\vec{D}) = \{(0.8, -0.2), (0.4, -0.3)\}$. The $(0.6, -0.1)$−level is not in $FS(\vec{D})$. Also $\vec{D}_{(0.8,-0.2)} \neq \vec{D}_{(0.4,-0.3)}$.

$\vec{D}$ is not *sectionally elementary* since $U_{2(\mu,\nu)} \neq U_{2(0.8,-0.2)}$ for $\mu = 0.6, \nu = -0.1$. The bipolar fuzzy directed hypergraph is ordered and the set of core hypergraphs is $c(\vec{D}) = \{\vec{D}_1 = \vec{D}_{(0.8,-0.2)}, \vec{D}_2 = \vec{D}_{(0.4,-0.3)}\}$. Induced fundamental sequence of $\vec{D}$ is given in Fig. 4.12.

Theorem 4.2 *(i) If $\vec{D} = (T, U)$ is an elementary bipolar fuzzy directed hypergraph, then $\vec{D}$ is ordered.*

(ii) If $\vec{D}$ is an ordered bipolar fuzzy directed hypergraph with $c(\vec{D}) = \{\vec{D}_{(\mu_1,\nu_1)}, \vec{D}_{(\mu_2,\nu_2)},...,\vec{D}_{(\mu_n,\nu_n)}\}$ and if $\vec{D}_{(\mu_n,\nu_n)}$ is simple, then $\vec{D}$ is elementary.

We now define the index matrix representation and certain operations on bipolar fuzzy directed hypergraphs.

Fig. 4.11 Bipolar fuzzy directed hypergraph

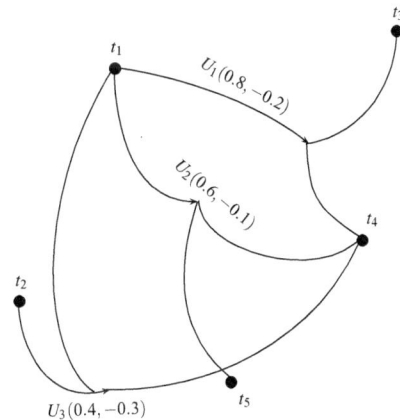

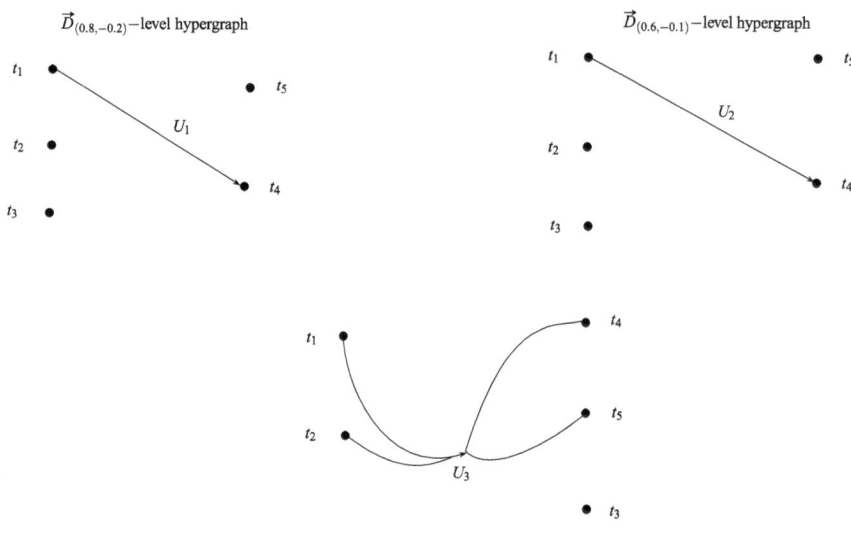

$\vec{D}_{(0.4,-0.3)}$−level hypergraph

Fig. 4.12 $\vec{D}$ induced fundamental sequence

Table 4.6 Index matrix of $\vec{D}$

I	t_1	t_2	$\cdots$	t_k
t_1	(m_{11}^+, m_{11}^-)	(m_{12}^+, m_{12}^-)	$\cdots$	(m_{1k}^+, m_{1k}^-)
t_2	(m_{21}^+, m_{21}^-)	(m_{22}^+, m_{22}^-)	$\cdots$	(m_{2k}^+, m_{2k}^-)
$\vdots$	$\vdots$	$\vdots$	$\vdots$	$\vdots$
t_k	(m_{k1}^+, m_{k1}^-)	(m_{k2}^+, m_{k2}^-)	$\cdots$	(m_{kk}^+, m_{kk}^-)

Definition 4.33 Let $\vec{D} = (T, U)$ be a bipolar fuzzy directed hypergraph. Then the *index matrix* of $\vec{D}$ is of the form $[T, U \subset T \times T]$, as given in Table 4.6, where $X = \{t_1, t_2, t_3, \cdots, t_k\}$, T is the family of bipolar fuzzy sets on X and $U = (m_{ij}^+, m_{ij}^-)$ is the set of bipolar fuzzy relations on t_i such that

Here, $m_{ij}^+ \in [0, 1]$ and $m_{ij}^- \in [-1, 0]$, $i, j = 1, 2, 3, ..., k$. The edge between two vertices v_i and v_j is indexed by (m_{ij}^+, m_{ij}^-), whose values can be found out by using the Cartesian products defined below.

Definition 4.34 Let X be a fixed set of points. The *Cartesian product* of two bipolar fuzzy sets B_1 and B_2 over X is defined as

(i) $B_1 \times_1 B_2 = \{((v_1, v_2), \min(m^+(v_1), m^+(v_2)), \max(m^-(v_1), m^-(v_2)))|v_1 \in B_1, v_2 \in B_2\}$.

(ii) $B_1 \times_2 B_2 = \{((v_1, v_2), \max(m^+(v_1), m^+(v_2)), \min(m^-(v_1), m^-(v_2)))|v_1 \in B_1, v_2 \in B_2\}$.

Note that, the Cartesian product $B_1 \times_i B_2$ is a bipolar fuzzy set, where $i = 1, 2$.

We now define some operations on bipolar fuzzy directed hypergraphs.

Definition 4.35 The *addition* of bipolar fuzzy directed hypergraphs $\vec{D}_1 = (T_1, U_1, (m_i^+, m_i^-), (m_{ij}^+, m_{ij}^-))$ and $\vec{D}_2 = (T_2, U_2, (m_p^+, m_p^-), (m_{pq}^+, m_{pq}^-))$, which is denoted by $\vec{D} = \vec{D}_1 \boxplus \vec{D}_2$, is defined as $\vec{D}_1 \boxplus \vec{D}_2 = [T_1 \cup T_2, (m_r^+, m_r^-), (m_{rs}^+, m_{rs}^-)]$, where

$$
(m_r^+, m_r^-) = \begin{cases}
(m_i^+, m_i^-), & \text{if } v_r \in T_1 - T_2, \\
(m_p^+, m_p^-), & \text{if } v_r \in T_2 - T_1, \\
(\max(m_i^+, m_p^+), \min(m_i^-, m_p^-)), & \text{if } v_r \in T_1 \cap T_2, \\
0, & \text{otherwise.}
\end{cases} \tag{4.1}
$$

$$
(m_{rs}^+, m_{rs}^-) = \begin{cases}
(m_{ij}^+, m_{ij}^-), & \text{if } v_r = v_i \in T_1 \text{ and } v_s = v_j \in T_1 - T_2, \\
& \text{or } v_r = v_i \in T_1 - T_2 \text{ and } v_s = v_j \in T_1, \\
(m_{pq}^+, m_{pq}^-), & \text{if } v_r = v_p \in T_2 \text{ and } v_s = v_q \in T_2 - T_1, \\
& \text{or } v_r = v_p \in T_2 - T_1 \text{ and } v_s = v_q \in T_2, \\
(\max(m_{ij}^+, m_{pq}^+), \min(m_{ij}^-, m_{pq}^-)), & \text{if } v_r = v_i = v_p \in T_1 \cap T_2 \text{ and} \\
& v_s = v_j = v_q \in T_2 \cap T_1, \\
0, & \text{otherwise.}
\end{cases} \tag{4.2}
$$

Example 4.11 Consider the bipolar fuzzy directed hypergraphs $\vec{D}_1 = (T_1, U_1)$ and $\vec{D}_2 = (T_2, U_2)$, where $X_1 = \{t_1, t_2, t_3, \ldots, t_8\}$, $U_1 = \{(\{t_1\}, t_3), (\{t_1, t_2\}, t_4),$ $(\{t_3\}, t_5), (\{t_1, t_5\}, t_6), (\{t_5, t_7\}, t_8)\}$ and $X_2 = \{t_1, t_2, t_3, t_4, t_5\}$, $U_2 = \{(\{t_1, t_3\}, t_2),$ $(\{t_2\}, t_2), (\{t_4\}, t_2), (\{t_3, t_5\}, t_4), (\{t_3\}, t_5)\}$ as shown in Figs. 4.13 and 4.14, respectively.

The index matrix of $\vec{D}_1$ is given in Table 4.7.

The index matrix of $\vec{D}_2$ is in Table 4.8.

The index matrix of $\vec{D}_1 \boxplus \vec{D}_2$ is $[T_1 \cup T_2, (m_r^+, m_r^-), (m_{rs}^+, m_{rs}^-)]$, where $X_1 \cup X_2 = \{t_1, t_2, t_3, t_4, t_5, t_6, t_7, t_8\}$. The membership values (m_r^+, m_r^-) are calculated using Eq. 4.1 and (m_{rs}^+, m_{rs}^-) are calculated using Eq. 4.2. The corresponding matrix is given in Table 4.9.

The graph of $\vec{D}_1 \boxplus \vec{D}_2$ is shown in Fig. 4.15.

Definition 4.36 The *vertexwise multiplication* of two bipolar fuzzy directed hypergraphs $\vec{D}_1$ and $\vec{D}_2$, denoted by $\vec{D}_1 \otimes \vec{D}_2$, is defined as, $[T_1 \cap T_2, (m_r^+, m_r^-), (m_{rs}^+, m_{rs}^-)]$, where

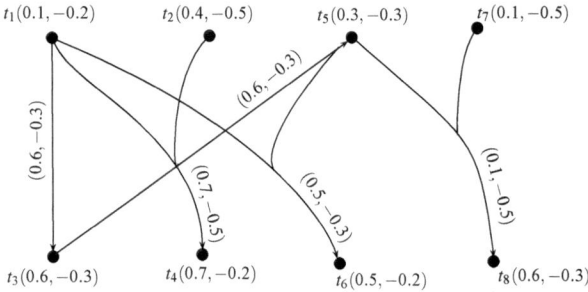

Fig. 4.13 Bipolar fuzzy directed hypergraph $\vec{D}_1$

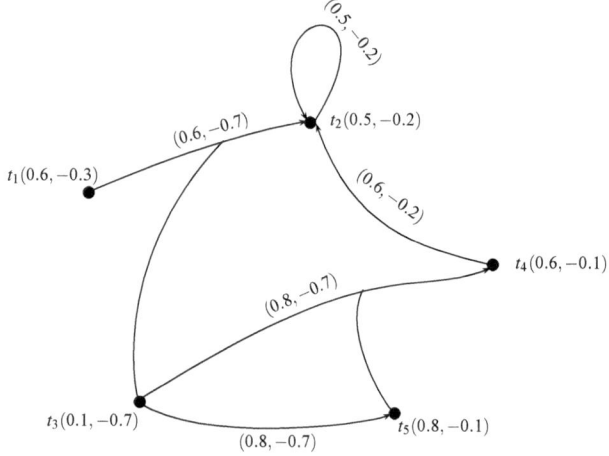

Fig. 4.14 Bipolar fuzzy directed hypergraph $\vec{D}_2$

Table 4.7 Index matrix of $\vec{D}_1$

$I_{\vec{D}_1}$	t_1	t_2	t_3	t_4	t_5	t_6	t_7	t_8
t_1	**0**	**0**	**0**	**0**	**0**	**0**	**0**	**0**
t_2	**0**	**0**	**0**	**0**	**0**	**0**	**0**	**0**
t_3	(0.6, −0.3)	**0**	**0**	**0**	**0**	**0**	**0**	**0**
t_4	(0.7, −0.5)	(0.7, −0.5)	**0**	**0**	**0**	**0**	**0**	**0**
t_5	**0**	**0**	(0.6, −0.3)	**0**	**0**	**0**	**0**	**0**
t_6	(0.5, −0.3)	**0**	**0**	**0**	(0.5, −0.3)	**0**	**0**	**0**
t_7	**0**	**0**	**0**	**0**	**0**	**0**	**0**	**0**
t_8	**0**	**0**	**0**	**0**	(0.1, −0.5)	**0**	(0.1, −0.5)	**0**

Table 4.8 The index matrix of $\vec{D}_2$

$I_{\vec{D}_2}$	t_1	t_2	t_3	t_4	t_5
t_1	0	0	0	0	0
t_2	(0.6, −0.7)	(0.5, −0.2)	(0.6, −0.7)	(0.6, −0.2)	0
t_3	0	0	0	0	0
t_4	0	0	(0.8, −0.7)	0	(0.8, −0.7)
t_5	0	0	(0.8, −0.7)	0	0

Table 4.9 The index matrix of $\vec{D}_1 \boxplus \vec{D}_2$

$I_{\vec{D}_1 \boxplus \vec{D}_2}$	t_1	t_2	t_3	t_4	t_5	t_6	t_7	t_8
t_1	0	0	0	0	0	0	0	0
t_2	(0.6, −0.7)	(0.5, −0.2)	(0.6, −0.7)	(0.6, −0.2)	0	0	0	0
t_3	(0.6, −0.3)	0	0	0	0	0	0	0
t_4	(0.7, −0.5)	(0.7, −0.5)	(0.8, −0.7)	0	(0.8, −0.7)	0	0	0
t_5	0	0	(0.8, −0.7)	0	0	0	0	0
t_6	(0.5, −0.3)	0	0	0	(0.5, −0.3)	0	0	0
t_7	0	0	0	0	0	0	0	0
t_8	0	0	0	0	(0.1, −0.5)	0	(0.1, −0.5)	0

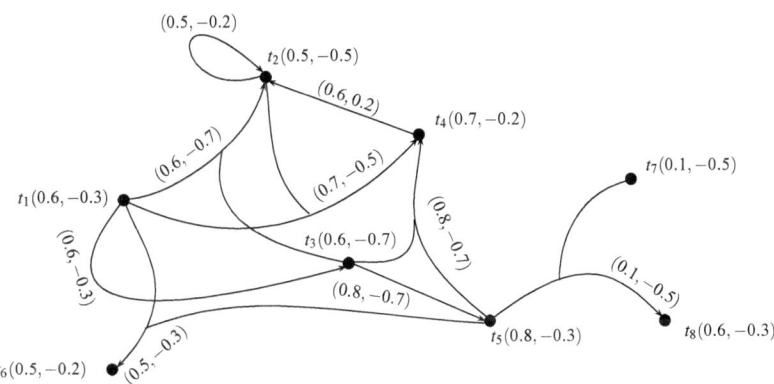

Fig. 4.15 $\vec{D}_1 \boxplus \vec{D}_2$

$$(m_r^+, m_r^-) = (\min(m_i^+, m_p^+), \max(m_i^-, m_p^-)), \text{ if } v_r \in T_1 \cap T_2, \qquad (4.3)$$

$$(m_{rs}^+, m_{rs}^-) = (\min(m_{ij}^+, m_{pq}^+), \max(m_{ij}^-, m_{pq}^-)), \text{ if } v_r = v_i = v_p \in T_1 \cap T_2 \text{ and } v_s = v_j = v_q \in T_1 \cap T_2. \qquad (4.4)$$

Example 4.12 Consider bipolar fuzzy directed hypergraphs $\vec{D}_1$ and $\vec{D}_2$ as shown in Figs. 4.13 and 4.14, respectively. The *index matrix of* $\vec{D}_1 \otimes \vec{D}_2$ is $[T_1 \cap T_2, (m_r^+, m_r^-), (m_{rs}^+, m_{rs}^-)]$, where $X_1 \cap X_2 = \{t_1, t_2, t_3, t_4, t_5\}$. The membership values (m_r^+, m_r^-)

Table 4.10 Index matrix of $\vec{D}_1 \otimes \vec{D}_2$

$I_{\vec{D}_1 \otimes \vec{D}_2}$	t_1	t_2	t_3	t_4	t_5
t_1	0	0	0	0	0
t_2	0	0	0	0	0
t_3	0	0	0	0	0
t_4	0	0	0	0	0
t_5	0	0	$(0.6, -0.3)$	0	0

Fig. 4.16 $\vec{D}_1 \otimes \vec{D}_2$

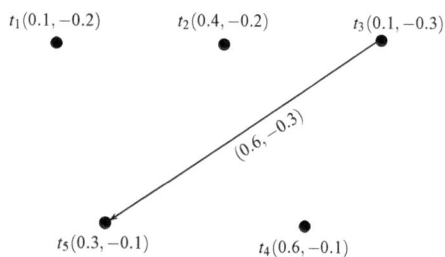

are calculated using Eq. 4.3 and (m_{rs}^+, m_{rs}^-) are calculated using Eq. 4.4. Index matrix of $\vec{D}_1 \otimes \vec{D}_2$ is given in Table 4.10.

The graph of $\vec{D}_1 \otimes \vec{D}_2$ is given in Fig. 4.16.

Definition 4.37 The *multiplication* of two bipolar fuzzy directed hypergraphs $\vec{D}_1$ and $\vec{D}_2$, denoted by $\vec{D}_1 \odot \vec{D}_2$, is defined as $[T_1 \cup (T_2 - T_1), T_2 \cup (T_1 - T_2), (m_r^+, m_r^-), (m_{rs}^+, m_{rs}^-)]$, where

$$(m_r^+, m_r^-) = \begin{cases} (m_i^+, m_i^-), & \text{if } v_r \in T_1, \\ (m_p^+, m_p^-), & \text{if } v_r \in T_2, \\ (\min(m_i^+, m_p^+), \max(m_i^-, m_p^-)), & \text{if } v_r \in T_1 \cap T_2. \end{cases} \quad (4.5)$$

$$(m_{rs}^+, m_{rs}^-) = \begin{cases} (m_{ij}^+, m_{ij}^-), & \text{if } v_r = v_i \in T_1 \text{ and } v_s = v_j \in T_1 - T_2, \\ (m_{pq}^+, m_{pq}^-), & \text{if } v_r = v_p \in T_2 \text{ and } v_s = v_q \in T_2 - T_1, \\ (\max(\min(m_{ij}^+, m_{pq}^+)), \min(\max(m_{ij}^-, m_{pq}^-))), & \text{if } v_r = v_i \in T_1 \cap T_2 \text{ and } v_s = v_q \in T_1 \cap T_2, \\ 0, & \text{otherwise.} \end{cases} \quad (4.6)$$

Remark 4.2 The positive membership and negative membership values of the loops (m_r^+, m_r^-) in the resultant graph (if present) can be calculated as $m_r^+ \leq m_i^+$ or $m_r^+ \leq m_p^+$ and $m_r^- \geq m_i^-$ or $m_r^- \geq m_p^-$.

Example 4.13 The index matrix of graph $\vec{D}_1 \odot \vec{D}_2$ is $[T_1 \cup (T_2 - T_1), T_2 \cup (T_1 - T_2), (m_r^+, m_r^-), (m_{rs}^+, m_{rs}^-)]$, where $X_2 \cup (X_1 - X_2) = \{t_1, t_2, t_3, ..., t_8\}$. The

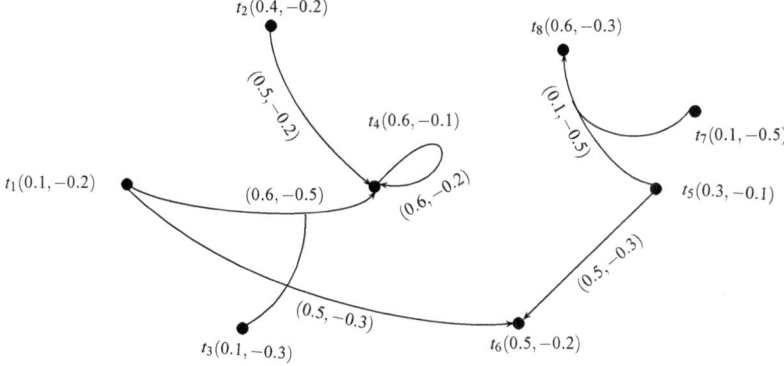

Fig. 4.17 $\vec{D}_1 \odot \vec{D}_2$

Table 4.11 Index matrix of graph $\vec{D}_1 \odot \vec{D}_2$

$I_{\vec{D}_1 \odot \vec{D}_2}$	t_1	t_2	t_3	t_4	t_5	t_6	t_7	t_8
t_1	0	0	0	0	0	0	0	0
t_2	0	0	0	0	0	0	0	0
t_3	0	0	0	0	0	0	0	0
t_4	(0.6, −0.5)	(0.5, −0.2)	(0.6, −0.5)	(0.6, −0.2)	0	0	0	0
t_5	0	0	0	0	0	0	0	0
t_6	(0.5, −0.3)	0	0	0	(0.5, −0.3)	0	0	0
t_7	0	0	0	0	0	0	0	0
t_8	0	0	0	0	(0.1, −0.5)	0	(0.1, −0.5)	0

membership values (m_r^+, m_r^-) are calculated using Eq. 4.5 and (m_{rs}^+, m_{rs}^-) are calculated using Eq. 4.6 as given in Table 4.11.

The graph of $\vec{D}_1 \odot \vec{D}_2$ is shown in Fig. 4.17.

Definition 4.38 The *structural subtraction* of $\vec{D}_1$ and $\vec{D}_2$, denoted by $\vec{D}_1 \boxminus \vec{D}_2$, is defined as $[T_1 - T_2, (m_r^+, m_r^-), (m_{rs}^+, m_{rs}^-)]$, where '−' is the set theoretic difference operation and

$$(m_r^+, m_r^-) = \begin{cases} (m_i^+, m_i^-), & \text{if } v_r \in T_1, \\ (m_p^+, m_p^-), & \text{if } v_r \in T_2, \\ 0, & \text{otherwise.} \end{cases} \tag{4.7}$$

$$(m_{rs}^+, m_{rs}^-) = (m_{ij}^+, m_{ij}^-), \text{ if } v_r = v_s \in T_1 - T_2 \text{ and } v_s = v_j \in T_1 - T_2. \tag{4.8}$$

If $T_1 - T_2 = \emptyset$, then graph of $\vec{D}_1 \boxminus \vec{D}_2$ is also empty.

Example 4.14 Consider bipolar fuzzy directed hypergraphs $\vec{D}_1$ and $\vec{D}_2$ as shown in Figs. 4.13 and 4.14. The *index matrix of* $\vec{D}_1 \boxminus \vec{D}_2$ is $[T_1 - T_2, (m_r^+, m_r^-), (m_{rs}^+, m_{rs}^-)]$,

Fig. 4.18 $\vec{D}_1 \boxminus \vec{D}_2$

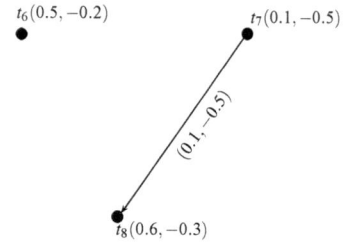

Table 4.12 Index matrix of $\vec{D}_1 \boxminus \vec{D}_2$

$I_{\vec{D}_1 \boxminus \vec{D}_2}$	t_6	t_7	t_8
t_6	**0**	**0**	**0**
t_7	**0**	**0**	**0**
t_8	**0**	$(0.1, -0.5)$	**0**

where $X_1 - X_2 = \{t_6, t_7, t_8\}$. The membership values (m_r^+, m_r^-) of $T_1 - T_2$ are calculated using Eq. 4.7 and (m_{rs}^+, m_{rs}^-) are calculated using Eq. 4.8. The corresponding matrix is given in Table 4.12.

The following Fig. 4.18 shows their structural subtraction.

Definition 4.39 A bipolar fuzzy directed hypergraph $\vec{D} = (T, U)$ is called $B = (m^+, m^-)-$ *tempered bipolar fuzzy directed hypergraph* of $\vec{D} = (T, U)$, if there exists a crisp hypergraph $\vec{D}^* = (X, U^*)$ and a bipolar fuzzy set $B = (m^+, m^-) : X \longrightarrow [0, 1] \times [-1, 0]$ such that $U = \{F_Y = (m^+, m^-)|Y \in U^*\}$, where

$$m^+(x) = \begin{cases} \min(m^+(t)|t \in Y), & \text{if } x \in Y, \\ 0, & \text{otherwise.} \end{cases} \quad m^-(x) = \begin{cases} \max(m^-(t)|t \in Y), & \text{if } x \in Y, \\ 0, & \text{otherwise.} \end{cases}$$

Let $B \otimes \vec{D}$ denotes the $B-$tempered hypergraph of $\vec{D}$, which is formed by the crisp hypergraph $\vec{D}^* = (X, U^*)$ and the bipolar fuzzy set $B : X \longrightarrow [0, 1] \times [-1, 0]$.

Example 4.15 Consider the bipolar fuzzy directed hypergraph $\vec{D} = (T, U)$, where $X = \{t_1, t_2, t_3, t_4\}$ and $U = \{U_1, U_2, U_3\}$, the corresponding incidence matrix is given in Table 4.13.

The corresponding hypergraph is shown in Fig. 4.19.

Then, $U_{(0.5, -0.2)} = \{\{t_1, \{t_2, t_3\}\}, \{t_2, t_3\}\}$, $U_{(0.4, -0.3)} = \{\{t_1, t_4\}\}$.
Define $B = (m^+, m^-) : X \longrightarrow [0, 1] \times [-1, 0]$ by $m^+(t_1) = 0.6$, $m^+(t_2) = 0.5$, $m^+(t_3) = 0.5$, $m^+(t_4) = 0.4$, $m^-(t_1) = -0.3$, $m^-(t_2) = -0.2$, $m^-(t_3) = -0.4$, $m^-(t_4) = -0.3$.

Table 4.13 Incidence matrix of $\vec{D}$

$I_{\vec{D}}$	U_1	U_2	U_3
t_1	(0.5, −0.2)	(0.4, −0.3)	0
t_2	(0.5, −0.2)	0	(0.5, −0.2)
t_3	(0.5, −0.2)	0	(0.5, −0.2)
t_4	0	(0.4, −0.3)	0

Fig. 4.19 B−tempered bipolar fuzzy directed hypergraph

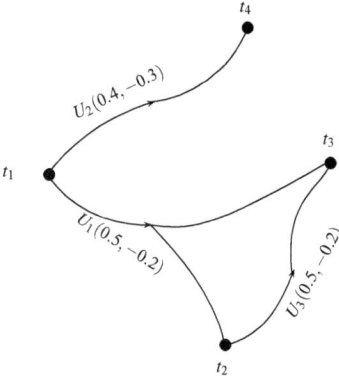

Note that

$$m^+_{F\{t_1,t_2,t_3\}} = \min(m^+(t_1), m^+(t_2), m^+(t_3)) = 0.5,$$
$$m^-_{F\{t_1,t_2,t_3\}} = \max(m^-(t_1), m^-(t_2), m^-(t_3)) = -0.2,$$
$$m^+_{F\{t_1,t_4\}} = \min(m^+(t_1), m^+(t_4)) = 0.4,$$
$$m^-_{F\{t_1,t_4\}} = \max(m^+(t_1), m^+(t_4)) = -0.3,$$
$$m^+_{F\{t_2,t_3\}} = \min(m^+(t_2), m^+(t_3)) = 0.5,$$
$$m^-_{F\{t_2,t_3\}} = \max(m^+(t_2), m^+(t_3)) = -0.2.$$

Thus, we have $U_1 = (m^+_{F\{t_1,t_2,t_3\}}, m^-_{F\{t_1,t_2,t_3\}})$, $U_2 = (m^+_{F\{t_1,t_4\}}, m^-_{F\{t_1,t_4\}})$, $U_3 = (m^+_{F\{t_2,t_3\}}, m^-_{F\{t_2,t_3\}})$.
Hence, $\vec{D}$ is B−tempered bipolar fuzzy directed hypergraph.

Theorem 4.3 *A bipolar fuzzy directed hypergraph $\vec{D} = (T, U)$ is $B = (m^+, m^-)$−tempered bipolar fuzzy directed hypergraph determined by some crisp directed hypergraph $\vec{D}^*$ if and only if $\vec{D}$ is elementary, simply ordered, and support simple.*

Proof Suppose that $\vec{D} = (T, U)$ is a B−tempered bipolar fuzzy directed hypergraph, which is formed by some crisp hypergraph $\vec{D}^*$. Since $\vec{D}$ is B-tempered, then the positive membership values and negative membership values of bipolar fuzzy directed hyperedges are same. Hence, $\vec{D}$ is elementary. If support of two bipolar fuzzy directed hyperedges of the B−tempered bipolar fuzzy directed hypergraph is same then the bipolar fuzzy hyperedges are equal. Hence, $\vec{D}$ is support simple. Let $c(\vec{D}) = \{\vec{D}_{(\mu_1, \nu_1)}, \vec{D}_{(\mu_2, \nu_2)}, \ldots, \vec{D}_{(\mu_n, \nu_n)}\}$. Since, $\vec{D}$ is elementary, it will be ordered.

Claim: $\vec{D}$ is simply ordered.
Let $U \in \vec{D}_{\mu_{i+1}, \nu_{i+1}} - \vec{D}_{\mu_i, \nu_i}$ then there exists $v_i \in U$ such that $m_{ij}^{+}(v_i) = \mu_{i+1}$ and $m_{ij}^{-}(v_i) = \nu_{i+1}$. Since, $\mu_{i+1} < \mu_i$ and $\nu_{i+1} < \nu_i$, it follows that $v_i \notin \vec{D}_{\mu_i, \nu_i}$ and $U \nsubseteq \vec{D}_{\mu_i, \nu_i}$. Hence, $\vec{D}$ is simply ordered.

Conversely, suppose $\vec{D} = (T, U)$ is elementary, simply ordered, and support simple. As we know that $\vec{D}_{\mu_i, \nu_i} = \vec{D}_i = (T_i, U_i)$ and $m_{ij}^{+} : T \longrightarrow [0, 1]$ and $m_{ij}^{-} : T \longrightarrow [-1, 0]$ are defined by,

$$m_{ij}^{+} = \begin{cases} \mu_1, & \text{if } v_i \in T_1, \\ \mu_i, & \text{if } v_i \in T_i - T_{i-1}, i = 1, 2, 3, \cdots, m. \end{cases}$$

$$m_{ij}^{-} = \begin{cases} \nu_1, & \text{if } v_i \in T_1, \\ \nu_i, & \text{if } v_i \in T_i - T_{i-1}, i = 1, 2, 3, \cdots, m. \end{cases}$$

To prove $U = \{m_{ij}^{+}(v_i), m_{ij}^{-}(v_i) | v_i \in U_i\}$, where

$$m_{ij}^{+}(v_i) = \begin{cases} \min m_i^{+}(y) | y \in U, & \text{if } v_i \in U_i, \\ 0, & \text{otherwise.} \end{cases} \qquad m_{ij}^{-}(v_i) = \begin{cases} \max m_i^{-}(y) | y \in U, & \text{if } v_i \in U_i, \\ 0, & \text{otherwise.} \end{cases}$$

Let $U' \in U_i$.
There is a unique bipolar fuzzy hyperedge (a_{ij}, b_{ij}) in U having support U' because $\vec{D}$ is elementary and support simple. Clearly, different edges in U having distinct supports lie in U_i. We have to prove that for each $U' \in U_i$, $m_{ij}^{+}(v_i) = a_{ij}$, $m_{ij}^{-}(v_i) = b_{ij}$. Since, distinct edges have different supports and all edges are elementary, then the definition of the fundamental sequence implies that $h(a_{ij}, b_{ij})$ is same as an arbitrary element of (μ_i, ν_i) of $FS(\vec{D})$. Therefore, $U' \subseteq T_i$. Further, if $i > 1$, then $U' \in U_i - U_{i-1}$. Since $U \subseteq U_i$, the definition of B-tempered indicates that for each $v_i \in U_i$, $m_{ij}^{+}(v_i) \geq \mu_i$ and $m_{ij}^{-}(v_i) \leq \nu_i$.

To prove $m_{ij}^{+}(v_i) = \mu_i$ and $m_{ij}^{-}(v_i) = \nu_i$ for some $X_i \in U_i$. It follows from the definition of B−tempered $m_{ij}^{+}(v_i) \geq \mu_{i-1}$ and $m_{ij}^{-}(v_i) \leq \nu_{i-1}$ for all $v_i \in U_i \Longrightarrow U \subseteq U_{i-1}$ and so $U \in U_i - U_{i-1}$. Since, $\vec{D}$ is simply ordered, therefore $U \nsubseteq U_{i-1}$, which is a contradiction to the definition of B−tempered bipolar fuzzy directed hypergraphs. Thus, from the definition of $m_{ij}^{+}(v_i)$ and $m_{ij}^{-}(v_i)$, we have $m_{ij}^{+} = a_{ij}$, $m_{ij}^{-} = b_{ij}$.

Theorem 4.4 *Let $\vec{D} = (T, U)$ be a simply ordered bipolar fuzzy directed hypergraph and $FS(\vec{D}) = \{\mu_n, \mu_{n-1}, \mu_{n-2}, \cdots, \mu_1 \nu_1, \nu_2, \cdots, \nu_n\}$. If $\vec{D}_{(\mu_n, \nu_n)}$ is simple hypergraph. Then, there exists a partial bipolar fuzzy directed hypergraph $D' = (T, U')$ of $\vec{D}$ such that the conditions given below are satisfied*

(i) D' is a B-tempered bipolar fuzzy directed hypergraph of $\vec{D}$.

(ii) $U \sqsubseteq U'$, that is, for all $(m^+, m^-) \in U$ there exist $(m^+, m^-) \in U'$ such that $(m^+(U) \subseteq m^+(U'))$ and $(m^-(U) \subseteq m^-(U'))$.

(iii) $FS(\vec{D}) = FS(D')$ and $c(\vec{D}) = c(D')$.

Proof By above Theorem 4.3, we have $\vec{D}$ is an elementary bipolar fuzzy directed hypergraph. By the removal of all those edges of $\vec{D}$, which lie in another edge of $\vec{D}$ properly, we attain the partial bipolar fuzzy directed hypergraph $D' = (T, U')$, where $U' = \{U_i \in U | U_i \subseteq U_j$ and $U_j \in U$, then $U_i = U_j\}$. Since, $\vec{D}_{\mu_n, \nu_n}$ is simple and all its edges are elementary, no edges can properly contained in other edges of $\vec{D}$ if they have different support. Hence, (iii) holds. We know that D' is support simple. Thus, all above conditions are satisfied by D'. From the definition of U', D' is elementary and support simple. Thus, D' is B-tempered.

4.4 Algorithm for Computing Minimum Arc Length and Shortest Hyperpath

This section investigates the definition of triangular bipolar fuzzy number. The score and ranking of a bipolar fuzzy numbers are also defined. A triangular bipolar fuzzy number is used to represent the arc length in a hypernetwork. Let L_j denotes arc length of the j-th hyperpath.

Definition 4.40 Let X be a finite set, which is non-empty and $B = (m^+, m^-)$ be a bipolar fuzzy set. Then, the pair $(m_i^+ a), m^-(a))$ is called a *bipolar fuzzy number*, denoted by $((l, m, n), (c, d, e))$, where $(l, m, n) \in F(I^+)$, $(c, d, e) \in F(I^-)$, $I^+ = [0, 1]$, $I^- = [-1, 0]$.

Definition 4.41 A *triangular bipolar fuzzy number* B is denoted by $B = \{(m^+(a), m^-(a)) | a \in X\}$, where $m^+(a)$ and $m^-(a)$ are bipolar fuzzy numbers. So, a triangular bipolar fuzzy number is given by $\tilde{B} = ((l, m, n), (c, d, e))$. The *diagrammatic representation of bipolar fuzzy number* $((l, m, n), (c, d, e))$ is shown in Fig. 4.20.

Definition 4.42 Let $\tilde{B} = ((l, m, n), (c, d, e))$ be a triangular bipolar fuzzy number, then the *score* of $\tilde{B}$ is a bipolar fuzzy set whose positive membership value is $S^+(\tilde{B}) = \frac{l+2m+n}{4}$ and negative membership value is $S^-(\tilde{B}) = \frac{c+2d+e}{4}$.

Definition 4.43 The *accuracy* of a triangular bipolar fuzzy number is defined as $Acc(\tilde{B}) = \frac{1}{2}(S^+(\tilde{B}) + S^-(\tilde{B}))$.

The minimum arc length of bipolar fuzzy directed hypernetwork is calculated by following the procedure given in Algorithm 4.4.1.

Algorithm 4.4.1

Finding the minimum arc length of bipolar fuzzy directed hypernetwork

---

Input. Enter the number of hyperpaths and their membership values, which are taken as triangular bipolar fuzzy number.

Output. Minimum arc length of bipolar fuzzy directed hypernetwork.

1. Calculate the lengths of all possible hyperpaths L_j for $j = 1, 2, 3, ..., k$, where
$$L_j = ((\bar{l}_j, \bar{m}_j, \bar{n}_j)(\bar{c}_j, \bar{d}_j, \bar{e}_j)).$$

2. Initialize $L_{\min} = ((l, m, n)(c, d, e)) = L_1 = ((\bar{l}_1, \bar{m}_1, \bar{n}_1)(\bar{c}_1, \bar{d}_1, \bar{e}_1))$.

3. Set $j = 2$.

4. The positive membership values (l, m, n) are computed as,
$$l = \min(l, \bar{l}_j),$$
$$m = \begin{cases} m, & \text{if } m \le \bar{l}_j, \\ \frac{m\bar{m}_j - l\bar{l}_j}{(m + \bar{m}_j) - (l + \bar{l}_j)}, & \text{if } m > \bar{l}_j, \end{cases}$$
$$n = \min(n, \bar{m}_j).$$

5. The negative membership values (c, d, e) are computed as,
$$c = \min(c, \bar{c}_j),$$
$$d = \begin{cases} d, & \text{if } d \le \bar{c}_j, \\ \frac{d\bar{d}_j - c\bar{c}_j}{(d + \bar{d}_j) - (c + \bar{c}_j)}, & \text{if } d > \bar{c}_j, \end{cases}$$
$$e = \min(e, \bar{d}_j).$$

6. Set $L_{\min} = ((l, m, n)(c, d, e))$, as calculated in Step 4.

7. $j = j + 1$.

8. If $j = k$, then stop the procedure.

9. If $j < k + 1$, then go to Step 3.

Example 4.16 Consider a hypernetwork with a triangular bipolar fuzzy arc lengths shown in Fig. 4.21.

1. From source vertex 1 to destination vertex 8, there are four possible paths ($k = 4$), given as follows:
   Path(1) : $1 \to 2 \to 6 \to 8$, $L_1 = ((10, 14, 18)(-13, -20, -25))$,
   Path(2) : $1 \to 3 \to 6 \to 8$, $L_2 = ((12, 16, 20)(-18, -22, -23))$,
   Path(3) : $1 \to 5 \to 8$, $L_3 = ((13, 17, 21)(-19, -20, -21))$,
   Path(4) : $1 \to 4 \to 7 \to 8$, $L_4 = ((6, 11, 16)(-12, -16, -23))$.
2. Initialize $L_{\min} = ((l, m, n)(c, d, e)) = L_1 = ((\bar{l}_1, \bar{m}_1, \bar{n}_1)(\bar{c}_1, \bar{d}_1, \bar{e}_1))$
   $= ((10, 14, 18)(-13, -20, -25))$.

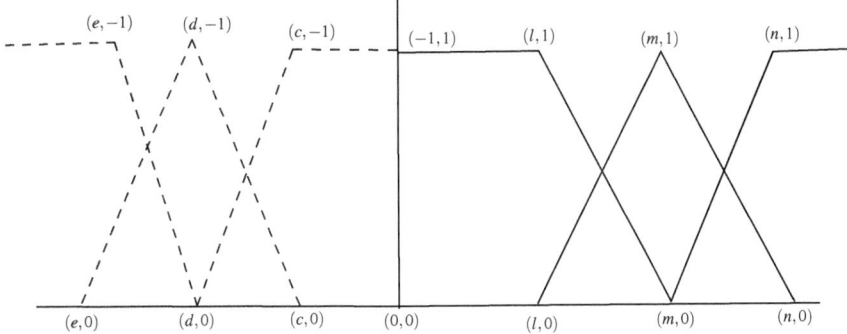

Fig. 4.20 Triangular bipolar fuzzy number

3. Initialize $j = 2$.
4. Let $L_{\min} = (10, 14, 18)(-13, -20, -25)$ and $L_2 = ((\bar{l}_2, \bar{m}_2, \bar{n}_2)(\bar{c}_2, \bar{d}_2, \bar{e}_2)) = (12, 16, 20)(-18, -22, -23)$. Compute the positive membership values (l, m, n) as

$$l = \min(l, \bar{l}_2) = \min(10, 12) = 10,$$
$$m = \begin{cases} \dfrac{(14 \times 16) - (10 \times 12)}{(14 + 16) - (10 + 12)} = 13, & \text{since } m > \bar{l}_2, \end{cases}$$
$$n = \min(n, \bar{m}_2) = \min(18, 16) = 16.$$

The negative membership values (c, d, e) as,

$$c = \min(c, \bar{c}_2) = \min(-13, -18) = -18,$$
$$d = \bar{d}_2 = -22 \text{ since } d < \bar{d}_2,$$
$$e = \min(e, \bar{d}_2) = \min(-25, -22) = -25.$$

5. Set $L_{\min} = ((10, 13, 16)(-18, -22, -25))$.
6. $j = j + 1 = 3$.
7. If $j < k + 1$, go to Step 4.
8. Let $L_{\min} = (10, 13, 16)(-18, -22, -25)$ and $L_3 = ((\bar{l}_3, \bar{m}_3, \bar{n}_3)(\bar{c}_3, \bar{d}_3, \bar{e}_3)) = (13, 17, 21)(-19, -20, -21)$. Calculate the positive membership values as,

$$l = \min(l, \bar{l}_3) = \min(10, 13) = 10,$$
$$m = 13 \text{ since } m = \bar{l}_3,$$
$$n = \min(n, \bar{m}_3) = \min(16, 17) = 16.$$

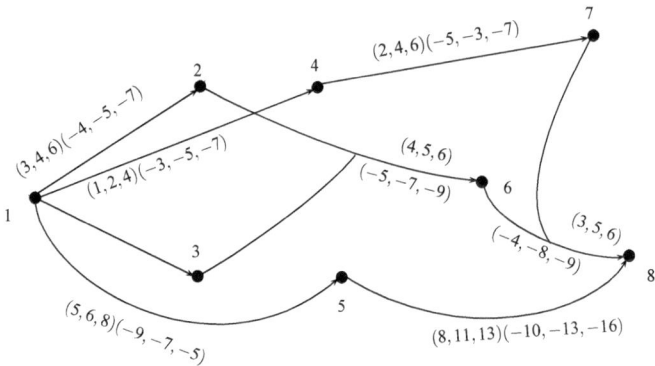

Fig. 4.21 Bipolar fuzzy hypernetwork

The negative membership values (c, d, e) as,

$$c = \min(c, \bar{c}_3) = \min(-18, -19) = -19,$$
$$d = -22 \text{ since } d < \bar{c}_3,$$
$$e = \min(e, \bar{d}_3) = \min(-25, -20) = -25.$$

9. Set $L_{\min} = ((10, 13, 16)(-19, -22, -25))$. Repeat the procedure until $j = 4$. Finally, we get the minimum arc length of bipolar fuzzy hypernetwork as, $L_{\min} = ((6, 10.38, 11)(-19, -22, -25))$.

We now write steps of score-based method to determine a bipolar fuzzy shortest hyperpath.
1. All possible hyperpaths are considered from source point to destination.
2. Compute the scores of the hyperpaths.
3. Find the accuracy of all paths.
4. The shortest hyperpath is obtained with the lowest accuracy.

Example 4.17 Consider the bipolar fuzzy hypernetwork as shown in Fig. 4.21. The bipolar fuzzy shortest hyperpath in this hypernetwork is recognized using score-based method. The scores of hyperpaths can be calculated as

$$S(P_1) = (\tfrac{l+2m+n}{4}, \tfrac{c+2d+e}{4}) = (\tfrac{10+2(14)+18}{4}, \tfrac{-13+2(-20)-25}{4}) = (14, -19.5).$$
$$S(P_2) = (\tfrac{12+2(16)+20}{4}, \tfrac{-18+2(-22)-23}{4}) = (16, -21.25).$$

Similarly, $S(P_3) = (17, -20)$, and $S(P_4) = (11, -16.75)$.
Accuracy of hyperpaths can be computed as
$Acc(P_1) = \tfrac{1}{2}(14 + (-19.5)) = -2.75,$ $Acc(P_2) = \tfrac{1}{2}(16 + (-21.25)) = -2.625,$ $Acc(P_3) = \tfrac{1}{2}(17 + (-20))$, and $Acc(P_4) = \tfrac{1}{2}(11 + (-16.75)) = -2.875$
From the Table 4.14, the path $P_4 : 1 \rightarrow 4 \rightarrow 7 \rightarrow 8$ with minimum accuracy is identified as bipolar fuzzy shortest hyperpath.

Table 4.14 Accuracy of hyperpaths

Path	Score	Accuracy	Rank
P_1	$(14, -19.5)$	-2.75	2
P_2	$(16, -21.25)$	-2.625	3
P_3	$(17, -20)$	-1.50	4
P_4	$(11, -16.75)$	-2.875	1

4.5 Application

A *hypernetwork* M is a network whose underlying structure is a hypergraph H^*, in which each vertex v_i corresponds to a unique processor P_i of M, and each hyper-edge e_j^* corresponds to a connector that connects processors represented by the vertices in e_j^*. A connector is loosely defined as an electronic or a photonic component through which messages are transmitted between connected processors, not necessarily simultaneously. We call a connector a hyperlink. Unlike a point-to-point network, in which a link is dedicated to a pair of processors, a hyperlink in a hyper-network is shared by a set of processors. A hyperlink can be implemented by a bus or a crossbar switch. Current optical technologies allow a hyperlink to be implemented by optical wave guides in a folded bus using time-division multiplexing (TDM). Free space optical or optoelectronic switching devices such as bulk lens, microlens array, and spatial light modulator (SLM) can also be used to implement hyperlinks. A star coupler, which uses wavelength-division multiplexing (WDM), can be considered either as a generalized bus structure or as a photonic switch, is another implemen-tation of a hyperlink. Similarly, an ATM switch, which uses a variant TDM, is a hyperlink.

Definition 4.44 Let X be a reference set. Then, a family of nontrivial bipolar fuzzy sets $\{A_1, A_2, A_3, \ldots, A_m\}$, where $A_i = (\mu_i^+, \mu_i^-)$ is a *bipolar fuzzy partition* if

1. $\bigcup_i supp(A_i) = X, \quad i = 1, 2, \ldots, m,$
2. $\sum_{i=1}^{m} \mu_i^+(x) = 1$, for all $x \in X,$
3. $\sum_{i=1}^{m} \mu_i^-(x) = -1$, for all $x \in X.$

Note that this definition generalizes fuzzy partitions because the definition is equiva-lent to a fuzzy partition when for all x, $v_i(x) = 0$. We call a family $\{A_1, A_2, A_3, \ldots, A_m\}$ a *bipolar fuzzy covering* of X if it satisfies above conditions (1)–(3).

A bipolar fuzzy partition can be represented by a bipolar fuzzy matrix $[a_{ij}]$, where a_{ij} is the positive membership degree and negative membership degree of element x_i in class j. We see that the matrix is the same as the incidence matrix in bipolar fuzzy hypergraph. Then, we can represent a bipolar fuzzy partition by a bipolar fuzzy hypergraph $H = (C, D)$ such that

(i) X is a non-empty set of points,
(ii) $C = \{C_1, C_2, \ldots, C_m\}$ is a set of nontrivial bipolar fuzzy classes,

Table 4.15 Bipolar fuzzy partition matrix

H	A_t	B_h
x_1	$(0.96, -0.04)$	$(0.04, -0.96)$
x_2	$(0.95, -0.5)$	$(0.5, -0.95)$
x_3	$(0.61, -0.39)$	$(0.39, -0.61)$
x_4	$(0.05, -0.95)$	$(0.95, -0.05)$
x_5	$(0.03, -0.97)$	$(0.97, -0.03)$

Table 4.16 Hypergraph $H_{(0.61,-0.03)}$

$H_{(0.61,-0.03)}$	$A_{t(0.61,-0.03)}$	$B_{h(0.61,-0.03)}$
x_1	1	0
x_2	1	0
x_3	1	0
x_4	0	1
x_5	0	1

(iii) $X = \bigcup_j supp(C_j)$, $j = 1, 2, \ldots, m$,
(iv) $\sum_{i=1}^{m} \mu_i^+(x) = 1$, for all $x \in C_i$,
(v) $\sum_{i=1}^{m} \mu_i^-(x) = -1$, for all $x \in C_i$.

Note that, conditions (iv)–(v) are added to the bipolar fuzzy hypergraph for bipolar fuzzy partition. If these conditions are added, the bipolar fuzzy hypergraph can represent a bipolar fuzzy covering. Naturally, we can apply the (α, β)-cut to the bipolar fuzzy partition.

Clustering Problem: We consider the clustering problem, which is a typical example of a bipolar fuzzy partition on the visual image processing. There are five objects and they are classified into two classes: tank and house. To cluster the elements x_1, x_2, x_3, x_4, x_5 into A_t (tank) and B_h (house), a bipolar fuzzy partition matrix is given as the form of incidence matrix of bipolar fuzzy hypergraph as given in Table 4.15.

We can apply the (α, β)-cut to the hypergraph and obtain a hypergraph $H_{(\alpha,\beta)}$ which is not bipolar fuzzy hypergraph. We denote the edge (class) in (α, β)-cut hypergraph $H_{(\alpha,\beta)}$ as $X_{j(\alpha,\beta)}$. This hypergraph H represents generally the covering because the conditions: (iv) $\sum_{i=1}^{m} \mu_i^+(x) = 1$ for all $x \in X$, $(v) \sum_{i=1}^{m} \mu_i^-(x) = -1$ for all $x \in X$, is not always guaranteed. The hypergraph $H_{(0.61,-0.03)}$ is shown in Table 4.16.

We obtain dual bipolar fuzzy hypergraph $H^D_{(0.61,-0.03)}$ of $H_{(0.61,-0.03)}$ which is given in Table 4.17.

We consider the strength of edge (class) $X_{j(\alpha,\beta)}$, or in the (α, β)-cut hypergraph $H_{(\alpha,\beta)}$. It is necessary to apply Definition to obtain the strength of edge $X_{j(\alpha,\beta)}$ in $H_{(\alpha,\beta)}$. The possible interpretations of $\eta(X_{j(\alpha,\beta)})$ are

Table 4.17 Dual bipolar fuzzy hypergraph

$H^D(0.61, -0.03)$	X_1	X_2	X_3	X_4	X_5
A_t	1	1	1	0	0
B_h	0	0	0	1	1

- the edge (class) in the hypergraph (partition) $H_{(\alpha,\beta)}$, groups elements having at least η positive membership and negative membership,
- the strength (cohesion) of edge (class) $X_{j(\alpha,\beta)}$ in $H_{(\alpha,\beta)}$ is η.

Thus, we can use the strength as a measure of the cohesion or strength of a class in a partition. For example, the strengths of classes $A_t(0.61, -0.03)$ and $B_h(0.61, -0.03)$ at $s = 0.61$, $t = -0.03$ are $\eta(A_t(0.61, -0.03)) = (0.96, -0.04)$, $\eta(B_h(0.61, -0.03)) = (0.97, -0.03)$. Thus, we say that the class $\eta(B_h(0.61, -0.03))$ is stronger than $\eta(A_t(0.61, -0.03))$ because $\eta(B_h(0.61, -0.03)) > \eta(A_t(0.61, -0.03))$.

From the above discussion on the hypergraph $H_{(0.61, -0.03)}$ and $H^D_{(0.61, -0.03)}$, we can state that

- The bipolar fuzzy hypergraph can represent the fuzzy partition visually. The (α, β)-cut hypergraph also represents the (α, β)-cut partition.
- The dual hypergraph $H^D_{(0.61, -0.03)}$ can represent elements X_i, which can be grouped into a class $X_{j(\alpha,\beta)}$. For example, the edges X_1, X_2, X_3 of the dual hypergraph in Table 4.17 represents that the elements x_1, x_2, x_3 can be grouped into A_t at level $(0.61, -0.03)$.
- At $(\alpha, \beta) = (0.61, -0.03)$ level, the strength of class $B_h(0.61, -0.03)$ is the highest $(0.95, -0.05)$, so it is the strongest class. It means that this class can be grouped independently from the other parts. Thus, we can eliminate the class B_h from the others and continue clustering. Therefore, the discrimination of strong classes from the others can allow us to decompose a clustering problem into smaller ones.

This strategy allows us to work with the reduced data in a clustering problem.

References

1. Akram, M.: Bipolar fuzzy graphs. Inf. Sci. **181**, 5548–5564 (2011)
2. Akram, M.: Bipolar fuzzy graphs with applications. Knowl. Based Syst. **39**, 1–8 (2013)
3. Akram, M., Dudek, W.A., Sarwar, S.: Properties of bipolar fuzzy hypergraphs. Ital. J. Pure Appl. Math. **31**, 141–160 (2013)
4. Akram, A., Luqman, A.: Certain concepts of bipolar fuzzy directed hypergraphs. Mathematics **5**(1), 17 (2017)
5. Berge, C.: Graphs and Hypergraphs. North-Holland, Amsterdam (1973)
6. Chen, S.M.: Interval-valued fuzzy hypergraph and fuzzy partition. IEEE Trans. Syst. Man Cybern. (Cybernetics) **27**(4), 725–733 (1997)
7. Gallo, G., Longo, G., Pallottino, S.: Directed hypergraphs and applications. Discret. Appl. Math. **42**, 177–201 (1993)

8. Kaufmann, A.: Introduction a la Thiorie des Sous-Ensemble Flous, 1. Masson, Paris (1977)
9. Lee, K.M.: Bipolar-valued fuzzy sets and their basic operations. In: Proceedings of the International Conference, Bangkok, Thailand, pp. 307–317 (2000)
10. Lee, K.-M.: Comparison of interval-valued fuzzy sets, intuitionistic fuzzy sets and bipolar-valued fuzzy sets. J. Fuzzy Logic Intell. Syst. **14**(2), 125–129 (2004)
11. Lee-kwang, H., Lee, K.-M.: Fuzzy hypergraph and fuzzy partition. IEEE Trans. Syst. Man Cybern. **25**(1), 196–201 (1995)
12. Mordeson, J.N., Nair, P.S.: Fuzzy Graphs and Fuzzy Hypergraphs, 2nd edn. Physica Verlag, Heidelberg (2001)
13. Parvathi, R., Thilagavathi, S., Karunambigai, M.G.: Intuitionistic fuzzy hypergraphs. Cybern. Inf. Technol. **9**(2), 46–53 (2009)
14. Rosenfeld, A.: Fuzzy graphs. In: Zadeh, L.A., Fu, K.S., Shimura, M. (eds.) Fuzzy Sets and their Applications, pp. 77–95. Academic Press, New York (1975)
15. Samanta, S., Pal, M.: Bipolar fuzzy hypergraphs. Int. J. Fuzzy Logic Syst. **2**(1), 17–28 (2012)
16. Zadeh, L.A.: Fuzzy sets. Inf. Control **8**(3), 338–353 (1965)
17. Zadeh, L.A.: Similarity relations and fuzzy orderings. Inf. Sci. **3**(2), 177–200 (1971)
18. Zhang, W.R.: Bipolar fuzzy sets and relations: a computational framework forcognitive modeling and multiagent decision analysis. In: Proceedings of the IEEE Conference, pp. 305–309 (1994)
19. Zhang, W.R.: YinYang bipolar fuzzy sets. In: Fuzzy Systems Proceedings, IEEE World Congress on Computational Intelligence, pp. 835–840 (1998)

Chapter 5
Extended Bipolar Fuzzy (Directed) Hypergraphs to m-Polar Information

An m-polar fuzzy set is a useful tool to solve real-world problems that involve multi-agents, multi-attributes, multi-objects, multi-indexes, and multipolar information. In this chapter, we present the notions of regular m-polar fuzzy hypergraphs and totally regular m-polar fuzzy hypergraphs. We discuss applications of m-polar fuzzy hypergraphs in decision-making problems. Furthermore, we discuss the notion of m-polar fuzzy directed hypergraphs and depict certain operations on them. We also describe an application of m-polar fuzzy directed hypergraphs in business strategy. This chapter is based on [7–9, 12].

5.1 Introduction

Fuzzy set theory deals with real-life data incorporating vagueness. Zhang [20] extended the theory of fuzzy sets to bipolar fuzzy sets, which register the bipolar behavior of objects. Nowadays, analysts believe that the world is moving toward multipolarity. Therefore, it comes as no surprise that multipolarity in data and information plays a vital role in various fields of science and technology. In neurobiology, multipolar neurons in brain gather a great deal of information from other neurons. In information technology, multipolar technology can be exploited to operate large-scale systems. Based on this motivation, Chen et al. [12] introduced the concept of m-polar fuzzy set as a generalization of a bipolar fuzzy set and shown that 2-polar and bipolar fuzzy sets are cryptomorphic mathematical notions. The framework of this theory is that "multipolar information" (not like the bipolar information which gives two-valued logic) arises because information for a natural world is frequently from n factors ($n \geq 2$). For example, "Pakistan is a good country". The truth value of this statement may not be a real number in [0, 1]. Being good country may have several properties: good in agriculture, good in political awareness, good in regaining macroeconomic stability, etc. The each component may be a real number in [0, 1].

© Springer Nature Singapore Pte Ltd. 2020
M. Akram and A. Luqman, *Fuzzy Hypergraphs and Related Extensions*,
Studies in Fuzziness and Soft Computing 390,
https://doi.org/10.1007/978-981-15-2403-5_5

If n is the number of such components under consideration, then the truth value of fuzzy statement is a n-tuple of real numbers in $[0, 1]$, that is, an element of $[0, 1]^n$.

Hypergraphs have many applications in various fields, including biological sciences, computer science, and natural sciences. To study the degree of dependence of an object to the other, Kaufamnn [14] applied the concept of fuzzy sets to hypergraphs. Mordeson and Nair [16] presented fuzzy graphs and fuzzy hypergraphs. Generalization and redefinition of fuzzy hypergraphs were discussed by Lee-Kwang and Lee [15]. The concept of interval-valued fuzzy sets was applied to hypergraphs by Chen [11]. Parvathi et al. [17] established the notion of intuitionistic fuzzy hypergraphs.

Definition 5.1 An *m-polar fuzzy set* C on a non-empty set X is a mapping C : $X \to [0, 1]^m$. The membership value of every element $x \in X$ is denoted by $C(x) = (P_1 \circ C(x), P_2 \circ C(x), \ldots, P_m \circ C(x))$, where $P_i \circ C : [0, 1]^m \to [0, 1]$ is defined as the i−th projection mapping.

Note that, $[0, 1]^m$ (mth-power of $[0, 1]$) is considered as a partially ordered set with the point-wise order $\leq$, where m is an arbitrary ordinal number (we make an appointment that $m = \{n|n < m\}$ when $m > 0$), $\leq$ is defined by $x \leq y \Leftrightarrow P_i(x) \leq P_i(y)$ for each $i \in m$ ($x, y \in [0, 1]^m$), and $P_i : [0, 1]^m \to [0, 1]$ is the i−th projection mapping ($i \in m$). $\mathbf{1} = (1, 1, \ldots, 1)$ is the greatest value and $\mathbf{0} = (0, 0, \ldots, 0)$ is the smallest value in $[0, 1]^m$. $m\mathscr{F}(X)$ is the power set of all m-polar fuzzy subsets on X.

1. When $m = 2$, $[0, 1]^2$ is the ordinary closed unit square in $\mathbb{R}^2$, the Euclidean plane. The righter (resp., the upper), the point in this square, the larger it is. Let $x = (0, 0) = \mathbf{0}$ (the smallest element of $[0, 1]^2$), $a = (0.35, 0.85)$, $b = (0.85, 0.35)$, and $y = (1, 1) = \mathbf{1}$ (the largest element of $[0, 1]^2$). Then $x \leq c \leq y$, $\forall$ $c \in [0, 1]^2$, (especially, $x \leq a \leq y$ and $x \leq b \leq y$ hold). It is easy to note that $a \nleq b \nleq a$ because $P_0(a) = 0.35 < 0.85 = P_0(b)$ and $P_1(a) = 0.85 > 0.35 = P_1(b)$ hold. The "order relation $\leq$" on $[0, 1]^2$ can be described in at least two ways. It can be seen in Fig. 5.1.
2. When $m = 4$, the order relation can be seen in Fig. 5.2.

Example 5.1 Suppose that a democratic country wants to elect its leader. Let $C = \{$Irtiza, Moeed, Ramish, Ahad$\}$ be the set of four candidates and $X = \{a, b, c, \ldots, s, t\}$ be the set of voters. We assume that the voting is weighted. A voter in $\{a, b, c\}$ can send a value in $[0, 1]$ to each candidate but a voter in $X - \{a, b, c\}$ can

Fig. 5.1 Order relation when $m = 2$

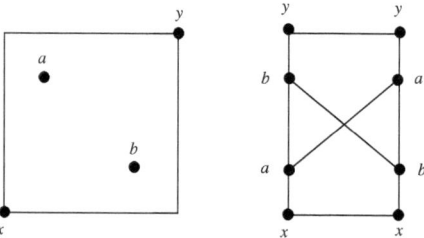

Fig. 5.2 Order relation
when $m = 4$

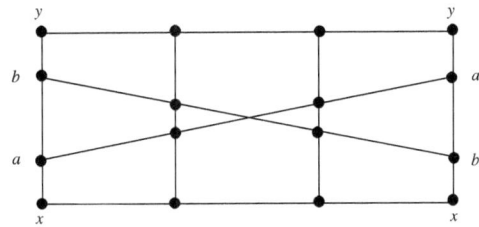

only send a value in $[0.2, 0.7]$ to each candidate. Let $A(a) = (0.8, 0.6, 0.5, 0.1)$ (which shows that the preference degrees of a corresponding to Irtiza, Moeed, Ramish, and Ahad are 0.8, 0.6, 0.5, and 0.1, respectively.), $A(b) = (0.9, 0.7, 0.5, 0.8)$, $A(c) = (0.9, 0.9, 0.8, 0.4)$, ..., $A(s) = (0.6, 0.7, 0.5, 0.3)$, and $A(t) = (0.5, 0.7, 0.2, 0.5)$. Thus, we obtain a 4-polar fuzzy set $A : X \rightarrow [0, 1]^4$ which can also be written as

$$A = \{(a, (0.8, 0.6, 0.5, 0.1)), (b, (0.9, 0.7, 0.5, 0.8)), (c, (0.9, 0.9, 0.8, 0.4)), \dots,$$
$$(s, (0.6, 0.7, 0.5, 0.3)), (t, (0.5, 0.7, 0.2, 0.5))\}.$$

Definition 5.2 Let C and D be two m-polar fuzzy sets on X. Then, the operations $C \cup D, C \cap D, C \subseteq D$, and $C = D$ are defined as

1. $P_i \circ (C \cup D)(x) = \sup\{P_i \circ C(x), P_i \circ D(x)\} = P_i \circ C(x) \vee P_i \circ D(x)$,
2. $P_i \circ (C \cap D)(x) = \inf\{P_i \circ C(x), P_i \circ D(x)\} = P_i \circ C(x) \wedge P_i \circ D(x)$,
3. $C \subseteq D$ if and only if $P_i \circ C(x) \leq P_i \circ D(x)$,
4. $C = D$ if and only if $P_i \circ C(x) = P_i \circ D(x)$,

for all $x \in X$, for each $1 \leq i \leq m$.

Definition 5.3 Let C be an m-polar fuzzy set on a non-empty crisp set X. An *m-polar fuzzy relation* on C is a mapping $(P_1 \circ D, P_2 \circ D, \dots, P_m \circ D) = D : C \rightarrow C$ such that
$$D(xy) \leq \inf\{C(x), C(y)\}, \quad \text{for all } x, y \in X$$

that is, for each $1 \leq i \leq m$,

$$P_i \circ D(xy) \leq \inf\{P_i \circ C(x), P_i \circ C(y)\}, \quad \text{for all } x, y \in X$$

where $P_i \circ C(x)$ denotes the i-th degree of membership of the vertex x and $P_i \circ D(xy)$ denotes the i-th degree of membership of the edge xy. D is also an m-polar fuzzy relation in X defined by the mapping $D : X \times X \rightarrow [0, 1]^m$.

Definition 5.4 An *m-polar fuzzy graph* on a non-empty set X is a pair $G = (C, D)$, where $C : X \rightarrow [0, 1]^m$ is an m-polar fuzzy set on the set of vertices X and $D : X \times X \rightarrow [0, 1]^m$ is an m-polar fuzzy relation in X such that

$$D(xy) \leq \inf\{C(x), C(y)\}, \qquad \text{for all } x, y \in X.$$

Note that, $D(xy) = \mathbf{0}$, for all $xy \in X \times X - E$, where $\mathbf{0} = (0, 0, \ldots, 0)$ and $E \subseteq X \times X$ is the set of edges. C is called an *m-polar fuzzy vertex set* of G and D is an *m-polar fuzzy edge set* of G. An m-polar fuzzy relation D on X is symmetric if $P_i \circ D(xy) = P_i \circ D(yx)$, for all $x, y \in X$.

For further terminologies and studies on m-polar fuzzy hypergraphs, readers are referred to [1–6, 10, 13, 18, 19].

5.2 *m*-Polar Fuzzy Hypergraphs

Definition 5.5 An m-polar fuzzy hypergraph on a non-empty set X is a pair $H = (A, B)$, where $A = \{\zeta_1, \zeta_2, \ldots, \zeta_r\}$ is a family of m-polar fuzzy subsets on X and B is an m-polar fuzzy relation on the m-polar fuzzy subsets ζ_i's such that

1. $B(E_i) = B(\{x_1, x_2, \ldots, x_s\}) \leq \inf\{\zeta_i(x_1), \zeta_i(x_2), \ldots, \zeta_i(x_s)\}$,   for all $x_1, x_2, \ldots, x_s \in X$.
2. $\bigcup_k supp(\zeta_k) = X$, for all $\xi_k \in A$.

Example 5.2 Let $A = \{\zeta_1, \zeta_2, \zeta_3, \zeta_4, \zeta_5\}$ be a family of 4-polar fuzzy subsets on $X = \{a, b, c, d, e, f, g\}$ given in Table 5.1. Let B be a 4-polar fuzzy relation on ζ_j's, $1 \leq j \leq 5$, given as, $B(\{a, c, e\}) = (0.2, 0.4, 0.1, 0.3)$, $B(\{b, d, f\}) = (0.2, 0.1, 0.1, 0.1)$, $B(\{a, b\}) = (0.3, 0.1, 0.1, 0.6)$, $B(\{e, f\}) = (0.2, 0.4, 0.3, 0.2)$, $B(\{b, e, g\}) = (0.2, 0.1, 0.2, 0.4)$. Thus, the 4-polar fuzzy hypergraph is shown in Fig. 5.3.

Example 5.3 Consider a 5-polar fuzzy hypergraph with vertex set $X = \{a, b, c, d, e, f, g\}$ whose degrees of membership are given in Table 5.2 and three hyperedges $\{a, b, c\}$, $\{b, d, e\}$, $\{b, f, g\}$ such that $B(\{a, b, c\}) = (0.2, 0.1, 0.3, 0.1, 0.2)$,

Table 5.1 4-polar fuzzy subsets on $X = \{a, b, c, d, e, f, g\}$

$x \in X$	ζ_1	ζ_2	ζ_3	ζ_4	ζ_5
a	(0.3, 0.4, 0.5, 0.6)	(0, 0, 0, 0)	(0.3, 0.4, 0.5, 0.6)	(0, 0, 0, 0)	(0, 0, 0, 0)
b	(0, 0, 0, 0)	(0.4, 0.1, 0.1, 0.6)	(0.4, 0.1, 0.1, 0.6)	(0, 0, 0, 0)	(0.4, 0.1, 0.1, 0.6)
c	(0.3, 0.5, 0.1, 0.3)	(0, 0, 0, 0)	(0, 0, 0, 0)	(0, 0, 0, 0)	(0, 0, 0, 0)
d	(0, 0, 0, 0)	(0.4, 0.2, 0.5, 0.1)	(0, 0, 0, 0)	(0, 0, 0, 0)	(0, 0, 0, 0)
e	(0.2, 0.4, 0.6, 0.8)	(0, 0, 0, 0)	(0, 0, 0, 0)	(0.2, 0.4, 0.6, 0.8)	(0.2, 0.4, 0.6, 0.8)
f	(0, 0, 0, 0)	(0.2, 0.5, 0.3, 0.2)	(0, 0, 0, 0)	(0.2, 0.5, 0.3, 0.2)	(0, 0, 0, 0)
g	(0, 0, 0, 0)	(0, 0, 0, 0)	(0, 0, 0, 0)	(0, 0, 0, 0)	(0.3, 0.5, 0.1, 0.4)

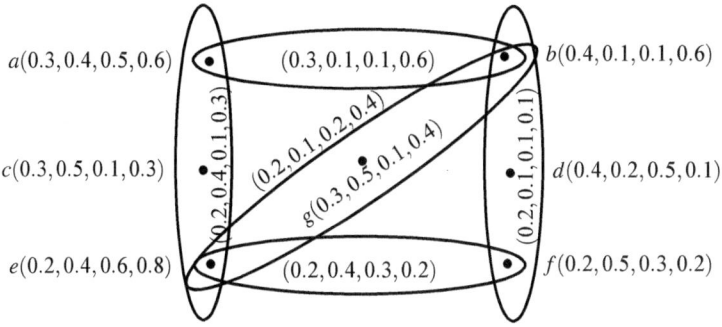

Fig. 5.3 4-polar fuzzy hypergraph

Table 5.2 5-polar fuzzy subsets on X

$x \in X$	ζ_1	ζ_2	ζ_3
a	(0.2, 0.1, 0.3, 0.1, 0.3)	(0, 0, 0, 0, 0)	(0, 0, 0, 0, 0)
b	(0.2, 0.3, 0.5, 0.6, 0.2)	(0.2, 0.3, 0.5, 0.6, 0.2)	(0.2, 0.3, 0.5, 0.6, 0.2)
c	(0.3, 0.2, 0.4, 0.5, 0.2)	(0, 0, 0, 0, 0)	(0, 0, 0, 0, 0)
d	(0, 0, 0, 0, 0)	(0.6, 0.2, 0.2, 0.3, 0.3)	(0, 0, 0, 0, 0)
e	(0, 0, 0, 0, 0)	(0.4, 0.5, 0.6, 0.7, 0.3)	(0, 0, 0, 0, 0)
f	(0, 0, 0, 0, 0)	(0, 0, 0, 0, 0)	(0.1, 0.2, 0.3, 0.4, 0.4)
g	(0, 0, 0, 0, 0)	(0, 0, 0, 0, 0)	(0.2, 0.4, 0.6, 0.8, 0.4)

$B(\{b, d, e\}) = (0.1, 0.2, 0.3, 0.4, 0.2), \qquad B(\{b, f, g\}) = (0.2, 0.2, 0.3, 0.3, 0.2).$
Hence, the 5-polar fuzzy hypergraph is shown in Fig. 5.4.

Example 5.4 Let $A = \{\zeta_1, \zeta_2, \zeta_3, \zeta_4, \zeta_5\}$ be a family of 4-polar fuzzy subsets on $X = \{a, b, c, d, e, f, g\}$ as given in Table 5.3. Let B be a 4-polar fuzzy relation on $\zeta_i's$, $1 \leq i \leq 5$, which is given as follows.

$$B(\{a, c, e\}) = (0.2, 0.4, 0.1, 0.3), \qquad B(\{b, d, f\}) = (0.2, 0.1, 0.1, 0.1),$$
$$B(\{a, b\}) = (0.3, 0.1, 0.1, 0.6), \qquad B(\{e, f\}) = (0.2, 0.4, 0.3, 0.2),$$
$$B(\{b, e, g\}) = (0.2, 0.1, 0.2, 0.4).$$

By routine computations, it is easy to see that $H = (A, B)$ is a 4-polar fuzzy hypergraph as shown in Fig. 5.5.

Definition 5.6 An *m*-polar fuzzy hypergraph $H = (A, B)$ is called *m-polar fuzzy r-uniform hypergraph* if $|supp(B_i)| = r$ for each $\zeta_i \in B$, $1 \leq i \leq r$.

Example 5.5 Consider $H = (A, B)$ is a 3-polar fuzzy hypergraph as shown in Fig. 5.6, where $A = \{(v_1, 0.1, 0.3, 0.2), (v_2, 0.1, 0.1, 0.3), (v_3, 0.2, 0.1, 0.1),$

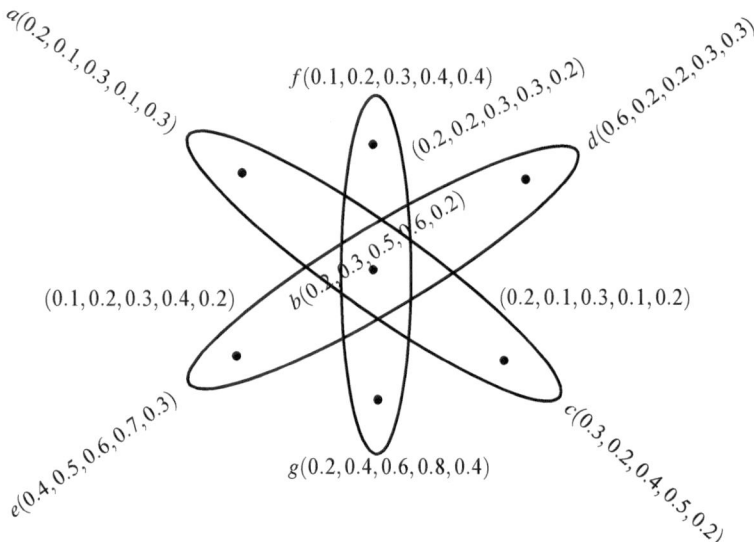

Fig. 5.4 5-polar fuzzy hypergraph

Table 5.3 4-polar fuzzy subsets on $X = \{a, b, c, d, e, f, g\}$

$x \in X$	ζ_1	ζ_2	ζ_3	ζ_4	ζ_5
a	(0.4, 0.5, 0.6, 0.7)	(0, 0, 0, 0)	(0.4, 0.5, 0.6, 0.7)	(0, 0, 0, 0)	(0, 0, 0, 0)
b	(0, 0, 0, 0)	(0.3, 0.2, 0.2, 0.7)	(0.3, 0.2, 0.2, 0.7)	(0, 0, 0, 0)	(0.3, 0.2, 0.2, 0.7)
c	(0.4, 0.6, 0.1, 0.4)	(0, 0, 0, 0)	(0, 0, 0, 0)	(0, 0, 0, 0)	(0, 0, 0, 0)
d	(0, 0, 0, 0)	(0.5, 0.3, 0.6, 0.1)	(0, 0, 0, 0)	(0, 0, 0, 0)	(0, 0, 0, 0)
e	(0.2, 0.4, 0.6, 0.8)	(0, 0, 0, 0)	(0, 0, 0, 0)	(0.2, 0.4, 0.6, 0.8)	(0.2, 0.4, 0.6, 0.8)
g	(0, 0, 0, 0)	(0, 0, 0, 0)	(0, 0, 0, 0)	(0, 0, 0, 0)	(0.4, 0.6, 0.5, 0.5)

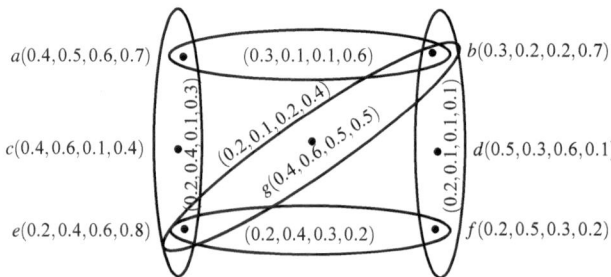

Fig. 5.5 4-polar fuzzy hypergraph

$(v_4, 0.1, 0.1, 0.2)\}$ is a 3-polar fuzzy set of vertices on $X = \{v_1, v_2, v_3, v_4\}$ and the B is defined as $B(\{v_1, v_2\}) = (0.1, 0.1, 0.2)$, $B(\{v_2, v_3\}) = (0.1, 0.1, 0.1)$, $B(\{v_3, v_4\}) = (0.1, 0.1, 0.2)$. Clearly, $|supp(\zeta_i)| = 2$, for each $i = 1, 2, 3$. Thus, $H = (A, B)$ is a 3-polar fuzzy 2-uniform hypergraph, as shown in Fig. 5.6.

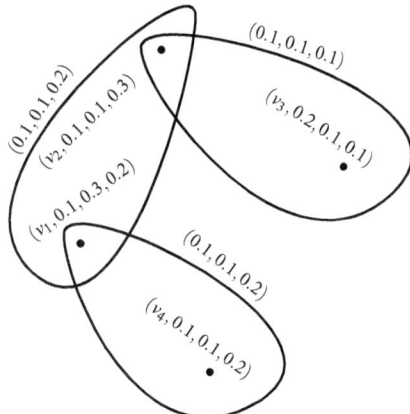

Fig. 5.6 3-polar fuzzy 2-uniform hypergraph

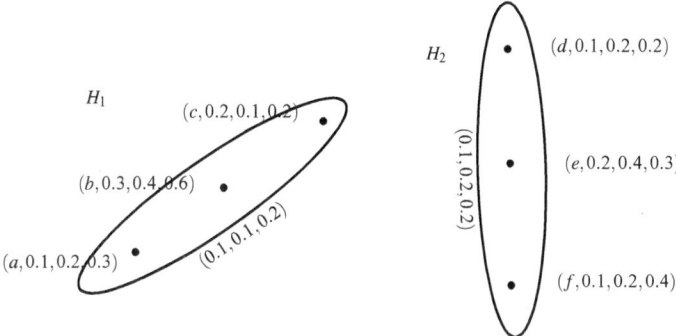

Fig. 5.7 3-polar fuzzy hypergraphs H_1 and H_2

Definition 5.7 Let $H_1 = (A_1, B_1)$ and $H_2 = (A_2, B_2)$ be two *m*-polar fuzzy hypergraphs on X_1 and X_2, respectively. The *Cartesian product* of H_1 and H_2 is an ordered pair $H = H_1 \square H_2 = (A_1 \square A_2, B_1 \square B_2)$ such that

1. $P_i \circ (A_1 \square A_2)(v_1, v_2) = \inf\{P_i \circ A_1(v_1), P_i \circ A_2(v_2)\}$, $\forall$ $(v_1, v_2) \in X_1 \times X_2$,
2. $P_i \circ (B_1 \square B_2)(\{v_1\} \times e_2) = \inf\{P_i \circ A_1(v_1), P_i \circ B_2(e_2)\}$, $\forall$ $v_1 \in X_1$, $\forall$ $e_2 \in E_2$,
3. $P_i \circ (B_1 \square B_2)(e_1 \times \{v_2\}) = \inf\{P_i \circ B_1(e_1), P_i \circ A_2(v_2)\}$, $\forall$ $v_2 \in X_2$, $\forall$ $e_1 \in E_1$.

Example 5.6 Let $H_1 = (A_1, B_1)$ and $H_2 = (A_2, B_2)$ be two 3-polar fuzzy hypergraphs on $X_1 = \{a, b, c\}$ and $X_2 = \{d, e, f\}$, respectively, as shown in Fig. 5.7.
The Cartesian product $H_1 \square H_2$ is shown in Fig. 5.8.

Theorem 5.1 *If H_1 and H_2 are the m-polar fuzzy hypergraphs then $H_1 \square H_2$ is as m-polar fuzzy hypergraph.*

Fig. 5.8 Cartesian product $H_1 \square H_2$

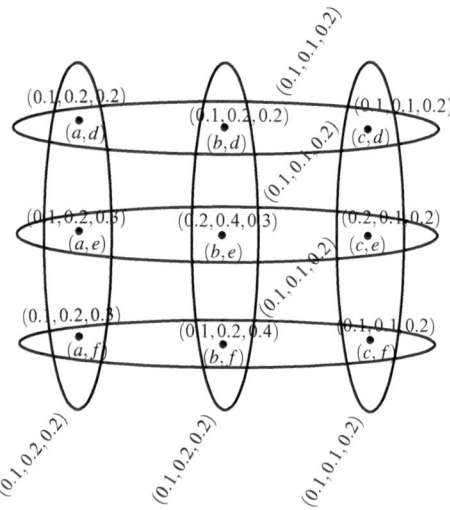

Proof Case (i): Let $v_1 \in X_1$, $e_2 = \{v_{21}, v_{22}, \ldots, v_{2q}\} \subseteq X_2$ then for each $1 \leq i \leq m$,

$$P_i \circ (B_1 \square B_2)(\{v_1\} \times e_2)$$
$$= \inf\{P_i \circ A_1(v_1), P_i \circ B_2(e_2)\}$$
$$\leq \inf\{P_i \circ A_1(v_1), \inf_{v_2 \in e_2} P_i \circ A_2(v_2)\}$$
$$= \inf\{P_i \circ A_1(v_1), \inf\{P_i \circ A_2(v_{21}), P_i \circ A_2(v_{22}), \ldots, P_i \circ A_2(v_{2q})\}\}$$
$$= \inf\{\inf\{P_i \circ A_1(v_1), P_i \circ A_2(v_{21})\}, \inf\{P_i \circ A_1(v_1), P_i \circ A_2(v_{22})\},$$
$$\ldots, \inf\{P_i \circ A_1(v_1), P_i \circ A_2(v_{2q})\}\}$$
$$= \inf\{P_i \circ (A_1 \square A_2)(v_1, v_{21}), P_i \circ (A_1 \square A_2)(v_1, v_{22}), \ldots, P_i \circ (A_1 \square A_2)(v_1, v_{2q})\}$$
$$= \inf_{v_1 \in e_1, v_2 \in e_2} P_i \circ (A_1 \square A_2)(v_1, v_2).$$

Case (ii): Let $v_2 \in X_2$, $e_1 = \{v_{11}, v_{12}, \ldots, v_{1p}\} \subseteq X_1$ then for each $1 \leq i \leq m$,

$$P_i \circ (B_1 \square B_2)(e_1 \times \{v_1\})$$
$$= \inf\{P_i \circ B_1(e_1), P_i \circ A_2(v_2)\}$$
$$\leq \inf\{\inf_{v_1 \in e_1} P_i \circ A_1(v_1), P_i \circ A_2(v_2)\}$$
$$= \inf\{\inf\{P_i \circ A_1(v_{11}), P_i \circ A_1(v_{12}), \ldots, P_i \circ A_1(v_{1p})\}, P_i \circ A_2(v_2)\}$$
$$= \inf\{\inf\{P_i \circ A_1(v_{11}), P_i \circ A_2(v_2)\}, \inf\{P_i \circ A_1(v_{12}), P_i \circ A_2(v_2)\},$$
$$\ldots, \inf\{P_i \circ A_1(v_{1p}), P_i \circ A_2(v_2)\}\}$$
$$= \inf\{P_i \circ (A_1 \square A_2)(v_{11}, v_2), P_i \circ (A_1 \square A_2)(v_{12}, v_2), \ldots, P_i \circ (A_1 \square A_2)(v_{1p}, v_2)\}$$
$$= \inf_{v_1 \in e_1, v_2 \in e_2} P_i \circ (A_1 \square A_2)(v_1, v_2).$$

Definition 5.8 Let $H_1 = (A_1, B_1)$ and $H_2 = (A_2, B_2)$ be two *m*-polar fuzzy hypergraphs on X_1 and X_2, respectively. Then, the *direct product* of H_1 and H_2 is an ordered pair $H = H_1 \times H_2 = (A_1 \times A_2, B_1 \times B_2)$ such that

1. $P_i \circ (A_1 \times A_2)(v_1, v_2) = \inf\{P_i \circ A_1(v_1), P_i \circ A_2(v_2)\}$, $\quad \forall \ (v_1, v_2) \in X_1 \times X_2$,
2. $P_i \circ (B_1 \times B_2)(e_1 \times e_2) = \inf\{P_i \circ B_1(e_1), P_i \circ B_2(e_2)\}$, $\quad \forall \ e_1 \in E_1, e_2 \in E_2$.

Definition 5.9 Let $H_1 = (A_1, B_1)$ and $H_2 = (A_2, B_2)$ be two *m*-polar fuzzy hypergraphs on X_1 and X_2, respectively, then the *strong product* of H_1 and H_2 is an ordered pair $H = H_1 \boxtimes H_2 = (A_1 \boxtimes A, B_1 \boxtimes B_2)$ such that

1. $P_i \circ (A_1 \boxtimes A_2)(v_1, v_2) = \inf\{P_i \circ A_1(v_1), P_i \circ A_2(v_2)\}$, $\quad \forall \ (v_1, v_2) \in X_1 \times X_2$,
2. $P_i \circ (B_1 \boxtimes B_2)(\{v_1\} \times e_2) = \inf\{P_i \circ A_1(v_1), P_i \circ B_2(e_2)\}$, $\quad \forall \ v_1 \in X_1, \forall \ e_2 \in E_2$,
3. $P_i \circ (B_1 \boxtimes B_2)(e_1 \times \{v_2\}) = \inf\{P_i \circ B_1(e_1), P_i \circ A_2(v_2)\}$, $\quad \forall \ v_2 \in X_2, \forall \ e_1 \in E_1$,
4. $P_i \circ (B_1 \boxtimes B_2)(e_1 \times e_2) = \inf\{P_i \circ B_1(e_1), P_i \circ B_2(e_2)\}$, $\quad \forall \ e_1 \in E_1, e_2 \in E_2$.

Theorem 5.2 *If H_1 and H_2 are two m-polar fuzzy r-uniform hypergraphs, then $H_1 \boxtimes H_2$ is a m-polar fuzzy hypergraph.*

Proof Case (i): Let $v_1 \in X_1, e_2 = \{v_{21}, v_{22}, \ldots, v_{2q}\} \subseteq X_2$ then for each $1 \leq i \leq m$,

$$P_i \circ (B_1 \boxtimes B_2)(\{v_1\} \times e_2)$$
$$= \inf\{P_i \circ A_1(v_1), P_i \circ B_2(e_2)\}$$
$$\leq \inf\{P_i \circ A_1(v_1), \inf_{v_2 \in e_2} P_i \circ A_2(v_2)\}$$
$$= \inf\{P_i \circ A_1(v_1), \inf\{P_i \circ A_2(v_{21}), P_i \circ A_2(v_{22}), \ldots, P_i \circ A_2(v_{2q})\}\}$$
$$= \inf\{\inf\{P_i \circ A_1(v_1), P_i \circ A_2(v_{21})\}, \inf\{P_i \circ A_1(v_1), P_i \circ A_2(v_{22})\},$$
$$\ldots, \inf\{P_i \circ A_1(v_1), P_i \circ A_2(v_{2q})\}\}$$
$$= \inf\{P_i \circ (A_1 \boxtimes A_2)(v_1, v_{21}), P_i \circ (A_1 \boxtimes A_2)(v_1, v_{22}),$$
$$\ldots, P_i \circ (A_1 \boxtimes A_2)(v_1, v_{2q})\}$$
$$= \inf_{v_1 \in e_1, v_2 \in e_2} P_i \circ (A_1 \boxtimes A_2)(v_1, v_2).$$

Case (ii): Let $v_2 \in X_2, e_1 = \{v_{11}, v_{12}, \ldots, v_{1p}\} \subseteq X_1$ then for each $1 \leq i \leq m$,

$$P_i \circ (B_1 \boxtimes B_2)(e_1 \times \{v_1\})$$
$$= \inf\{P_i \circ B_1(e_1), P_i \circ A_2(v_2)\}$$
$$\leq \inf\{\inf_{v_1 \in e_1} P_i \circ A_1(v_1), P_i \circ A_2(v_2)\}$$
$$= \inf\{\inf\{P_i \circ A_1(v_{11}), P_i \circ A_1(v_{12}), \ldots, P_i \circ A_1(v_{1p})\}, P_i \circ A_2(v_2)\}$$
$$= \inf\{\inf\{P_i \circ A_1(v_{11}), P_i \circ A_2(v_2)\}, \inf\{P_i \circ A_1(v_{12}), P_i \circ A_2(v_2)\},$$
$$\ldots, \inf\{P_i \circ A_1(v_{1p}), P_i \circ A_2(v_2)\}\}$$

$$= \inf\{P_i \circ (A_1 \boxtimes A_2)(v_{11}, v_2), P_i \circ (A_1 \boxtimes A_2)(v_{12}, v_2),$$
$$\ldots, P_i \circ (A_1 \boxtimes A_2)(v_{1p}, v_2)\}$$
$$= \inf_{v_1 \in e_1, v_2 \in e_2} P_i \circ (A_1 \boxtimes A_2)(v_1, v_2).$$

Case (iii): Let $e_1 = \{v_{11}, v_{12}, \ldots, v_{1p}\} \subseteq X_1$ and $e_2 = \{v_{21}, v_{22}, \ldots, v_{2q}\} \subseteq X_2$ then for each $1 \leq i \leq m$,

$$P_i \circ (B_1 \boxtimes B_2)(e_1 \times e_2)$$
$$= \inf\{P_i \circ B_1(e_1), P_i \circ B_2(e_2)\}$$
$$\leq \inf\{\inf_{v_1 \in e_1} P_i \circ A_1(v_1), \inf_{v_2 \in e_2} P_i \circ A_2(v_2)\}$$
$$= \inf\{\inf\{P_i \circ A_1(v_{11}), P_i \circ A_1(v_{12}), \ldots, P_i \circ A_1(v_{1p})\}$$
$$, \inf\{P_i \circ A_2(v_{21}), P_i \circ A_2(v_{22}), \ldots, P_i \circ A_2(v_{2q})\}\}$$
$$= \inf\{\inf\{P_i \circ A_1(v_{11}), P_i \circ A_2(v_{21})\}, \inf\{P_i \circ A_1(v_{12}), P_i \circ A_2(v_{22})\}$$
$$, \ldots, \inf\{P_i \circ A_1(v_{1p}), P_i \circ A_2(v_{2q})\}\}$$
$$= \inf\{P_i \circ (A_1 \boxtimes A_2)(v_{11}, v_{21}), P_i \circ (A_1 \boxtimes A_2)(v_{12}, v_{22}), \ldots, P_i \circ (A_1 \boxtimes A_2)(v_{1p}, v_{2q})\}$$
$$= \inf_{v_1 \in e_1, v_2 \in e_2} P_i \circ (A_1 \boxtimes A_2)(v_1, v_2).$$

Definition 5.10 Let $H_1 = (A_1, B_1)$ and $H_2 = (A_2, B_2)$ be two m-polar fuzzy hypergraphs on X_1 and X_2, respectively, then composition of H_1 and H_2 is an ordered pair $H = H_1 \diamond H_2 = (A_1 \diamond A_2, B_1 \diamond B_2)$ such that,

1. $P_i \circ (A_1 \diamond A_2)(v_1, v_2) = \inf\{P_i \circ A_1(v_1), P_i \circ A_2(v_2)\}, \ \forall \ (v_1, v_2) \in X_1 \times X_2$,
2. $P_i \circ (B_1 \diamond B_2)(\{v_1\} \times e_2) = \inf\{P_i \circ A_1(v_1), P_i \circ B_2(e_2)\}, \ \forall \ v_1 \in X_1,$
   $\forall \ e_2 \in E_2$,
3. $P_i \circ (B_1 \diamond B_2)(e_1 \times \{v_2\} = \inf\{P_i \circ B_1(e_1), P_i \circ A_2(v_2)\}, \ \forall \ v_2 \in X_2,$
   $\forall \ e_1 \in E_1$,
4. $P_i \circ (B_1 \diamond B_2)((v_{11}, v_{21})(v_{12}, v_{22}) \cdots (v_{1p}, v_{2q})) = \inf\{P_i \circ B_1(e_1),$
   $P_i \circ A_2(v_{21}), P_i \circ A_2(v_{22}), \ldots,$
   $P_i \circ A_2(v_{2q})\}, \ \forall \ e_1 \in E_1, v_{21}, v_{22}, \ldots, v_{2q} \in X_2$.

Theorem 5.3 *If H_1 and H_2 are two m-polar fuzzy hypergraphs, then $H_1 \diamond H_2$ is a m-polar fuzzy hypergraph.*

Proof Case(i): Let $v_1 \in X_1, e_2 \subseteq X_2$ then for each $1 \leq i \leq m$,

$$P_i \circ (B_1 \diamond B_2)(\{v_1\} \times e_2)$$
$$= \inf\{P_i \circ A_1(v_1), P_i \circ B_2(e_2)\}$$
$$\leq \inf\{P_i \circ A_1(v_1), \inf_{v_2 \in e_2} P_i \circ A_2(v_2)\}$$
$$= \inf\{P_i \circ A_1(v_1), \inf\{P_i \circ A_2(v_{21}), P_i \circ A_2(v_{22}), \ldots, P_i \circ A_2(v_{2q})\}\}$$
$$= \inf\{\inf\{P_i \circ A_1(v_1), P_i \circ A_2(v_{21})\}, \inf\{P_i \circ A_1(v_1), P_i \circ A_2(v_{22})\}$$
$$, \ldots, \inf\{P_i \circ A_1(v_1), P_i \circ A_2(v_{2q})\}\}$$
$$= \inf\{P_i \circ (A_1 \diamond A_2)(v_1, v_{21}), P_i \circ (A_1 \diamond A_2)(v_1, v_{22})$$

$$, \ldots, P_i \circ (A_1 \diamond A_2)(v_1, v_{2q})\}$$
$$= \inf_{v_1 \in e_1, v_2 \in e_2} P_i \circ (A_1 \diamond A_2)(v_1, v_2).$$

Case(ii): Let $v_2 \in X_2, e_1 \subseteq X_1$ then for each $1 \le i \le m$,

$$P_i \circ (B_1 \diamond B_2)(e_1 \times \{v_1\})$$
$$= \inf\{P_i \circ B_1(e_1), P_i \circ A_2(v_2)\}$$
$$\le \inf\{\inf_{v_1 \in e_1} P_i \circ A_1(v_1), P_i \circ A_2(v_2)\}$$
$$= \inf\{\inf\{P_i \circ A_1(v_{11}), P_i \circ A_1(v_{12}), \ldots, P_i \circ A_1(v_{1p})\}, P_i \circ A_2(v_2)\}$$
$$= \inf\{\inf\{P_i \circ A_1(v_{11}), P_i \circ A_2(v_2)\}, \inf\{P_i \circ A_1(v_{12}), P_i \circ A_2(v_2)\}$$
$$, \ldots, \inf\{P_i \circ A_1(v_{1p}), P_i \circ A_2(v_2)\}\}$$
$$= \inf\{P_i \circ (A_1 \diamond A_2)(v_{11}, v_2), P_i \circ (A_1 \diamond A_2)(v_{12}, v_2)$$
$$, \ldots, P_i \circ (A_1 \diamond A_2)(v_{1p}, v_2)\}$$
$$= \inf_{v_1 \in e_1, v_2 \in e_2} P_i \circ (A_1 \diamond A_2)(v_1, v_2).$$

Case(iii): Let $e_1 = \{v_{11}, v_{12}, \ldots, v_{1p}\} \subseteq X_1, v_{21}, v_{22}, \ldots, v_{2q} \in X_2$ then for each $1 \le i \le m$,

$$P_i \circ (B_1 \diamond B_2)((v_{11}, v_{21})(v_{12}, v_{22}) \cdots (v_{1p}, v_{2q}))$$
$$= \inf\{P_i \circ B_1(e_1), P_i \circ A_2(v_{21}), P_i \circ A_2(v_{22}), \ldots, P_i \circ A_2(v_{2q})\}$$
$$\le \inf\{\inf_{v_1 \in e_1} P_i \circ A_1(v_1), P_i \circ A_2(v_{21}), P_i \circ A_2(v_{22}), \ldots, P_i \circ A_2(v_{2q})\}$$
$$= \inf\{\inf\{P_i \circ A_1(v_{11}), P_i \circ A_1(v_{12}), \ldots, P_i \circ A_1(v_{1p})\}$$
$$, P_i \circ A_2(v_{21}), P_i \circ A_2(v_{22}), \ldots, P_i \circ A_2(v_{2q})\}$$
$$= \inf\{\inf\{P_i \circ A_1(v_{11}), P_i \circ A_2(v_{21})\}, \inf\{P_i \circ A_1(v_{12}), P_i \circ A_2(v_{22})\}$$
$$, \ldots, \inf\{P_i \circ A_1(v_{1p}), P_i \circ A_2(v_{2q})\}\}$$
$$= \inf\{P_i \circ (A_1 \diamond A_2)(v_{11}, v_{21}), P_i \circ (A_1 \diamond A_2)(v_{12}, v_{22})$$
$$, \ldots, P_i \circ (A_1 \diamond A_2)(v_{1p}, v_{2q})\}$$
$$= \inf_{v_1 \in e_1, v_2 \in e_2} P_i \circ (A_1 \diamond A_2)(v_1, v_2).$$

Definition 5.11 Let $H_1 = (A_1, B_1)$ and $H_2 = (A_2, B_2)$ be two *m*-polar fuzzy hypergraphs on X_1 and X_2, respectively, then the *union* of H_1 and H_2 is an ordered pair $H = H_1 \cup H_2 = (A_1 \cup A_2, B_1 \cup B_2)$ such that

1. $P_i \circ (A_1 \cup A_2)(v) = \begin{cases} P_i \circ A_1(v), & \text{if } v \in X_1 - X_2, \\ P_i \circ A_2(v), & \text{if } v \in X_2 - X_1, \\ \sup\{P_i \circ A_1(v), P_i \circ A_2(v)\}, & \text{if } v \in X_1 \cap X_2. \end{cases}$

2. $P_i \circ (B_1 \cup B_2)(e) = \begin{cases} P_i \circ B_1(e), & \text{if } e \in E_1 - E_2, \\ P_i \circ B_2(e), & \text{if } e \in E_2 - E_1, \\ \sup\{P_i \circ B_1(e), P_i \circ B_2(e)\}, & \text{if } e \in E_1 \cap E_2. \end{cases}$

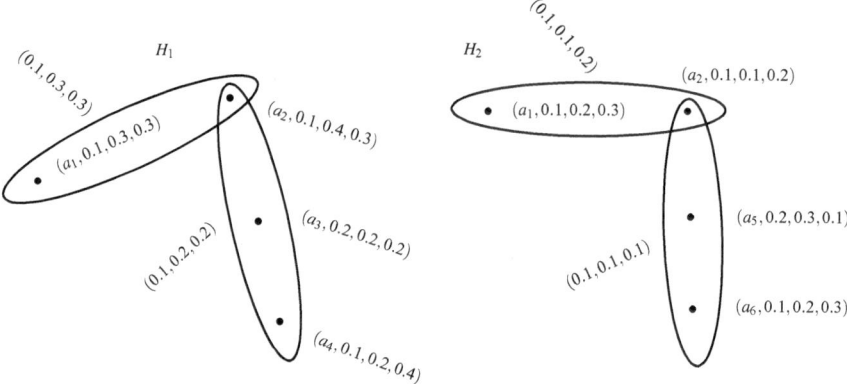

Fig. 5.9 3-polar fuzzy hypergraphs H_1 and H_2

Fig. 5.10 $H_1 \cup H_2$

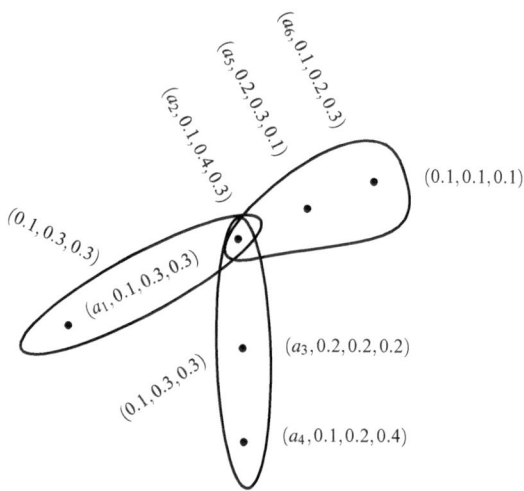

where $E_1 = supp(B_1)$ and $E_2 = supp(B_2)$.

Example 5.7 Consider 3-polar fuzzy hypergraphs $H_1 = (A_1, B_1)$ and $H_2 = (A_2, B_2)$ as shown in Fig. 5.9.

The union of H_1 and H_2 is given in Fig. 5.10.

Theorem 5.4 *The union $H_1 \cup H_2 = (A_1 \cup A_2, B_1 \cup B_2)$ of two m-polar fuzzy hypergraphs $H_1 = (A_1, B_1)$ and $H_2 = (A_2, B_2)$ is an m-polar fuzzy hypergraph.*

Proof Let $H_1 = (A_1, B_1)$ and $H_2 = (A_2, B_2)$ be two m-polar fuzzy hypergraphs on X_1 and X_2, respectively, such that $E_1 = supp(B_1)$ and $E_2 = supp(B_2)$. It is to be shown that that $H_1 \cup H_2 = (A_1 \cup A_2, B_1 \cup B_2)$ is an m-polar fuzzy hypergraph.

Since, all conditions for $A_1 \cup A_2$ are satisfied automatically, therefore, it is enough to show that $B_1 \cup B_2$ is an *m*-polar fuzzy relation on $A_1 \cup A_2$.

Case(i): If $e \in E_1 - E_2$ then for each $1 \leq i \leq m$,

$$
\begin{aligned}
P_i \circ (B_1 \cup B_2)(e) &= P_i \circ B_1(e_1) \\
&\leq \inf_{v_1 \in e_1} P_i o A_1(v_1) \\
&= \inf\{P_i \circ A_1(v_{11}), P_i \circ A_1(v_{12}), \ldots, P_i \circ A_1(v_{1p})\} \\
&= \inf\{P_i \circ (A_1 \cup A_2)(v_{11}), P_i \circ (A_1 \cup A_2)(v_{12}), \ldots, P_i \circ (A_1 \cup A_2)(v_{1p})\}.
\end{aligned}
$$

Case(ii): If $e \in E_2 - E_1$ then for each $1 \leq i \leq m$,

$$
\begin{aligned}
P_i \circ (B_1 \cup B_2)(e) &= P_i \circ B_2(e_2) \\
&\leq \inf_{v_2 \in e_2} P_i o A_2(v_2) \\
&= \inf\{P_i \circ A_2(v_{21}), P_i \circ A_2(v_{22}), \ldots, P_i \circ A_2(v_{2q})\} \\
&= \inf\{P_i \circ (A_1 \cup A_2)(v_{21}), P_i \circ (A_1 \cup A_2)(v_{22}), \ldots, P_i \circ (A_1 \cup A_2)(v_{2q})\}.
\end{aligned}
$$

Case(iii): If $e \in E_1 \cap E_2$ or $v_{j1}, v_{j2}, \ldots, v_{jp} \in X_1 \cap X_2$ then for each $1 \leq i \leq m$,

$$
\begin{aligned}
P_i \circ (B_1 \cup B_2)(e) &= \sup\{P_i \circ B_1(e), P_i \circ B_2(e)\} \\
&\leq \sup\{\inf\{P_i \circ A_1(v_{j1}), P_i \circ A_1(v_{j2}), \ldots, P_i \circ A_1(v_{jp})\} \\
&\quad , \inf\{P_i \circ A_2(v_{j1}), P_i \circ A_2(v_{j2}), \ldots, P_i \circ A_2(v_{jp})\}\} \\
&= \inf\{\sup\{P_i \circ A_1(v_{j1}), P_i \circ A_2(v_{j1})\}, \sup\{P_i \circ A_1(v_{j2}), P_i \circ A_2(v_{j2})\} \\
&\quad , \ldots, \sup\{P_i \circ A_1(v_{jp}), P_i \circ A_2(v_{jp})\} \\
&= \inf\{P_i \circ (A_1 \cup A_2)(v_{11}), P_i \circ (A_1 \cup A_2)(v_{12}), \ldots, P_i \circ (A_1 \cup A_2)(v_{1p})\}.
\end{aligned}
$$

Definition 5.12 Let $H_1 = (A_1, B_1)$ and $H_2 = (A_2, B_2)$ be two *m*-polar fuzzy hypergraphs on X_1 and X_2, respectively, then the *join* $H = H_1 + H_2$ of two *m*-polar fuzzy hypergraphs H_1 and H_2 is defined as follows:

1. $P_i \circ (A_1 + A_2)(v) = P_i \circ (A_1 \cup A_2)(v)$, if $v \in X_1 \cup X_2$,
2. $P_i \circ (B_1 + B_2)(e) = P_i \circ (B_1 \cup B_2)(e)$, if $e \in E_1 \cup E_2$,
3. $P_i \circ (B_1 + B_2)(e) = \inf\{P_i \circ A_1(v_1), P_i \circ A_2(v_2)\}$, if $e \in E'$,

where E' is the set of all the edges joining the vertices of X_1 and X_2 and $X_1 \cap X_2 = \emptyset$.

Example 5.8 Consider $H_1 = (A_1, B_1)$ and $H_2 = (A_2, B_2)$ be two 3-polar fuzzy hypergraphs as shown in Fig. 5.11 then their join is given in Fig. 5.12.

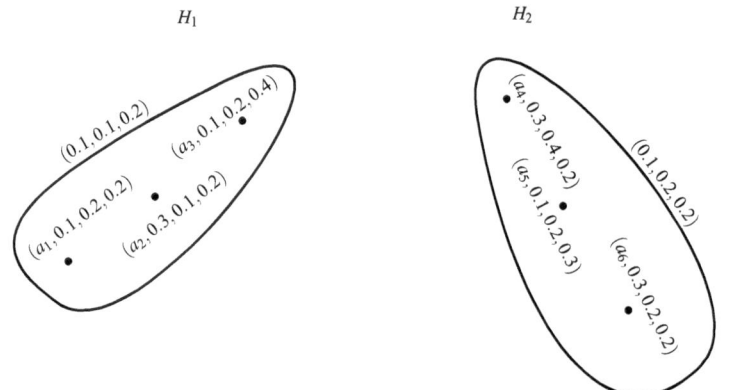

Fig. 5.11 3-polar fuzzy hypergraphs H_1 and H_2

Fig. 5.12 $H_1 + H_2$

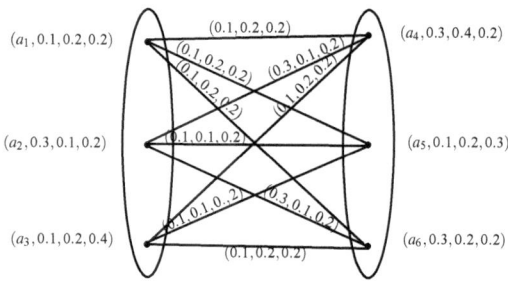

Definition 5.13 Let $H_1 = (A_1, B_1)$ and $H_2 = (A_2, B_2)$ be two m-polar fuzzy hypergraphs on X_1 and X_2, respectively, then the *lexicographic product* of H_1 and H_2 is defined by the ordered pair $H = H_1 \bullet H_2 = (A_1 \bullet A_2, B_1 \bullet B_2)$ such that

1. $P_i \circ (A_1 \bullet A_2)(v_1, v_2) = \inf\{P_i \circ A_1(v_1), P_i \circ A_2(v_2)\}, \ \forall \ (v_1, v_2) \in X_1 \times X_2$,
2. $P_i \circ (B_1 \bullet B_2)(\{v_1\} \times e_2) = \inf\{P_i \circ A_1(v_1), P_i \circ B_2(e_2)\}, \ \forall \ v_1 \in X_1, \ \forall \ e_2 \in E_2$,
3. $P_i \circ (B_1 \bullet B_2)(e_1 \times e_2) = \inf\{P_i \circ B_1(e_1), P_i \circ B_2(v_2)\}, \ \forall \ e_1 \in E_1, \ \forall \ e_2 \in E_2$.

Theorem 5.5 *If H_1 and H_2 are m-polar fuzzy hypergraphs then $H_1 \bullet H_2$ is an m-polar fuzzy hypergraph.*

Proof Case(i): Let $v_1 \in X_1, e_2 = \{v_{21}, v_{22}, \ldots, v_{2q}\} \subseteq X_2$ then for each $1 \leq i \leq m$,

$P_i \circ (B_1 \bullet B_2)(\{v_1\} \times e_2)$

$\quad = \inf\{P_i \circ A_1(v_1), P_i \circ B_2(e_2)\}$

$\quad \leq \inf\{P_i \circ A_1(v_1), \inf_{v_2 \in e_2} P_i \circ A_2(v_2)\}$

$\quad = \inf\{P_i \circ A_1(v_1), \inf\{P_i \circ A_2(v_{21}), P_i \circ A_2(v_{22}), \ldots, P_i \circ A_2(v_{2q})\}\}$

$\quad = \inf\{\inf\{P_i \circ A_1(v_1), P_i \circ A_2(v_{21})\}, \inf\{P_i \circ A_1(v_1), P_i \circ A_2(v_{22})\}$

$\quad\quad, \ldots, \inf\{P_i \circ A_1(v_1), P_i \circ A_2(v_{2q})\}\}$

$\quad = \inf\{P_i \circ (A_1 \bullet A_2)(v_1, v_{21}), P_i \circ (A_1 \bullet A_2)(v_1, v_{22}), \ldots, P_i \circ (A_1 \bullet A_2)(v_1, v_{2q})\}$

$\quad = \inf_{v_1 \in e_1, v_2 \in e_2} P_i \circ (A_1 \bullet A_2)(v_1, v_2).$

Case(ii): Let $e_1 = \{v_{11}, v_{12}, \ldots, v_{1p}\} \subseteq X_1$, $e_2 = \{v_{21}, v_{22}, \ldots, v_{2q}\} \subseteq X_2$ then for each $1 \leq i \leq m$,

$P_i \circ (B_1 \bullet B_2)(e_1 \times e_2)$

$\quad = \inf\{P_i \circ B_1(e_1), P_i \circ B_2(e_2)\}$

$\quad \leq \inf\{\inf_{v_1 \in e_1} P_i \circ A_1(v_1), \inf_{v_2 \in e_2} P_i \circ A_2(v_2)\}$

$\quad = \inf\{\inf\{P_i \circ A_1(v_{11}), P_i \circ A_1(v_{12}), \ldots, P_i \circ A_1(v_{1p})\}$

$\quad\quad, \inf\{P_i \circ A_2(v_{21}), P_i \circ A_2(v_{22}), \ldots, P_i \circ A_2(v_{2q})\}\}$

$\quad = \inf\{\inf\{P_i \circ A_1(v_{11}), P_i \circ A_2(v_{21})\}, \inf\{P_i \circ A_1(v_{12}), P_i \circ A_2(v_{22})\}$

$\quad\quad, \ldots, \inf\{P_i \circ A_1(v_{1p}), P_i \circ A_2(v_{2q})\}\}$

$\quad = \inf\{P_i \circ (A_1 \bullet A_2)(v_{11}, v_{21}), P_i \circ (A_1 \bullet A_2)(v_{12}, v_{22})$

$\quad\quad, \ldots, P_i \circ (A_1 \bullet A_2)(v_{1p}, v_{2q})\}$

$\quad = \inf_{v_1 \in e_1, v_2 \in e_2} P_i \circ (A_1 \bullet A_2)(v_1, v_2).$

Definition 5.14 Let $H = (A, B)$ be an *m*-polar fuzzy hypergraph on a non-empty set X. The dual *m*-polar fuzzy hypergraph of H, denoted by $H^D = (A^*, B^*)$, is defined as

1. $A^* = B$ is the *m*-polar fuzzy set of vertices of H^D.
2. If $|X| = n$ then, B^* is an *m*-polar fuzzy set on the family of hyperedges $\{X_1, X_2, \ldots, X_n\}$ such that, $X_i = \{E_j \mid x_j \in E_j, E_j$ is a hyperedge of $H\}$, i.e., X_i is the *m*-polar fuzzy set of those hyperedges which share the common vertex x_i and $B^*(X_i) = \inf\{E_j \mid x_j \in E_j\}$.

Example 5.9 Consider the example of a 3-polar fuzzy hypergraph $H = (A, B)$ given in Fig. 5.13, where $X = \{x_1, x_2, x_3, x_4, x_5, x_6\}$ and $E = \{E_1, E_2, E_3, E_4\}$. The dual 3-polar fuzzy hypergraph is shown in Fig. 5.14 with dashed lines with vertex set $E = \{E_1, E_2, E_3, E_4\}$ and set of hyperedges $\{X_1, X_2, X_3, X_4, X_5, X_6\}$ such that $X_1 = X_3$.

Definition 5.15 The *open neighborhood* of a vertex x in an *m*-polar fuzzy hypergraph is the set of adjacent vertices of x excluding that vertex and it is denoted by $N(x)$.

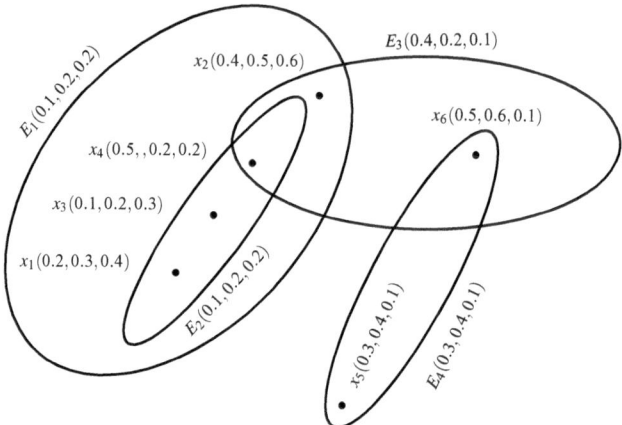

Fig. 5.13 3-polar fuzzy hypergraph

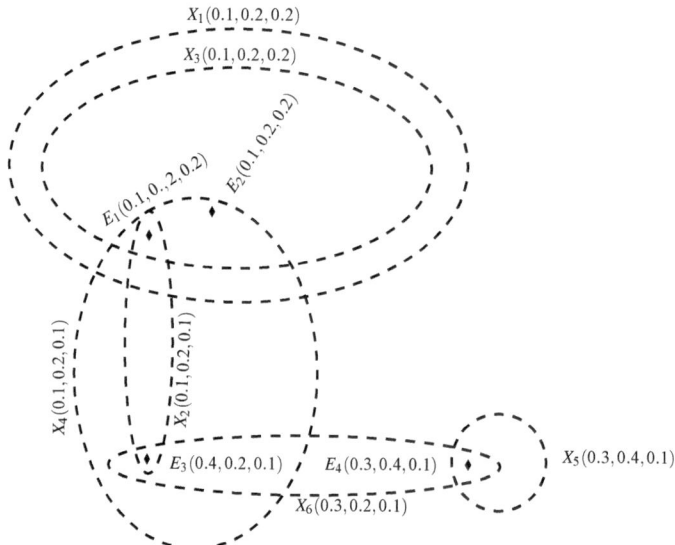

Fig. 5.14 Dual 3-polar fuzzy hypergraph

Example 5.10 Consider the 3-polar fuzzy hypergraph $H = (A, B)$, where $A = \{\zeta_1, \zeta_2, \zeta_3, \zeta_4\}$ is a family of 3-polar fuzzy subsets on $X = \{a, b, c, d, e\}$ and B is a 3polar fuzzy relation on the 3-polar fuzzy subsets ζ_i's such that $\zeta_1 = \{(a, 0.3, 0.4, 0.5), (b, 0.2, 0.4, 0.6)\}, \zeta_2 = \{(c, 0.2, 0.1, 0.4), (d, 0.5, 0.1, 0.1), (e, 0.2, 0.3, 0.1)\}$, $\zeta_3 = \{(b, 0.1, 0.2, 0.4), (c, 0.4, 0.5, 0.6)\},$ $\zeta_4 = \{(a, 0.1, 0.3, 0.2), (d, 0.3, 0.4, 0.4)\}$. In this example, the open neighborhood of the vertex a is $\{b, d\}$ as shown in Fig. 5.15.

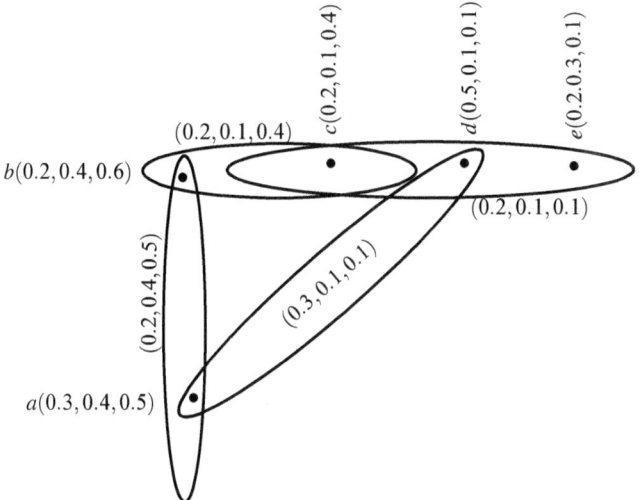

Fig. 5.15 3-polar fuzzy hypergraph

Definition 5.16 The *closed neighborhood* of a vertex x in an m-polar fuzzy hypergraph is the set of adjacent vertices of x including x and it is denoted by $N[x]$.

Example 5.11 Consider a 3-polar fuzzy hypergraph $H = (A, B)$ as shown in Fig. 5.15. In this example, closed neighborhood of the vertex a is $\{a, b, d\}$.

Definition 5.17 The *open neighborhood degree* of a vertex x in H is denoted by $deg(x)$ and defined as an m-tuple $deg(x) = (deg^{(1)}(x), deg^{(2)}(x), deg^{(3)}(x), \ldots, deg^{(m)}(x))$, such that

$$deg^{(1)}(x) = \Sigma_{x \in N(x)} P_1 \circ \zeta_j(x),$$

$$deg^{(2)}(x) = \Sigma_{x \in N(x)} P_2 \circ \zeta_j(x),$$

$$deg^{(3)}(x) = \Sigma_{x \in N(x)} P_3 \circ \zeta_j(x),$$

$$\vdots$$

$$deg^{(m)}(x) = \Sigma_{x \in N(x)} P_m \circ \zeta_j(x).$$

Definition 5.18 Let $H = (A, B)$ be an m-polar fuzzy hypergraph on a non-empty set X. If all vertices in A have the same open neighborhood degree n, then H is called *n-regular m*-polar fuzzy hypergraph.

Definition 5.19 The *closed neighborhood degree* of a vertex x in H is denoted by $deg[x]$ and defined as an m-tuple such that $deg[x] = (deg^{(1)}[x], deg^{(2)}[x],$

Fig. 5.16 Regular and totally regular 4-polar fuzzy hypergraph

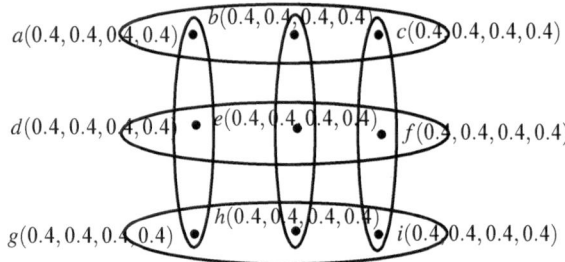

$deg^{(3)}[x], \ldots, deg^{(m)}[x])$, where

$$deg^{(1)}[x] = deg^{(1)}(x) + \wedge_j P_1 \circ \zeta_j(x),$$

$$deg^{(2)}[x] = deg^{(2)}(x) + \wedge_j P_2 \circ \zeta_j(x),$$

$$deg^{(3)}[x] = deg^{(3)}(x) + \wedge_j P_3 \circ \zeta_j(x),$$

$$\vdots$$

$$deg^{(m)}[x] = d_G^{(m)}(x) + \wedge_j P_m \circ \zeta_j(x).$$

Example 5.12 Consider the example of a 3-polar fuzzy hypergraph $H = (A, B)$, where $A = \{\zeta_1, \zeta_2, \zeta_3, \zeta_4\}$ is a family of 3-polar fuzzy subsets on $X = \{a, b, c, d, e\}$ and B is a 3-polar fuzzy relation on the 3-polar fuzzy subsets ζ_j, where $\zeta_1 = \{(a, 0.3, 0.4, 0.5), (b, 0.2, 0.4, 0.6)\}$, $\zeta_2 = \{(c, 0.2, 0.1, 0.4), (d, 0.5, 0.1, 0.1),$ $(e, 0.2, 0.3, 0.1)\}, \zeta_3 = \{(b, 0.1, 0.2, 0.4), (c, 0.4, 0.5, 0.6)\}, \zeta_4 = \{(a, 0.1, 0.3, 0.2),$ $(d, 0.3, 0.4, 0.4)\}$. Then, $deg(a) = (0.5, 0.8, 1)$ and $deg[a] = (0.6, 1.1, 1.2)$.

Definition 5.20 Let $H = (A, B)$ be an m-polar fuzzy hypergraph on X. If all vertices in A have the same closed neighborhood degree m, then H is called m-*totally regular* m-polar fuzzy hypergraph.

Example 5.13 Consider the 3-polar fuzzy hypergraph $H = (A, B)$, where $A = \{\zeta_1, \zeta_2, \zeta_3\}$ is a family of 3-polar fuzzy subsets on $X = \{a, b, c, d, e\}$ and B is a 3-polar fuzzy relation on the 3-polar fuzzy subsets ζ_j such
$\zeta_1 = \{(a, 0.5, 0.4, 0.1), (b, 0.3, 0.4, 0.1), (c, 0.4, 0.4, 0.3)\}$,
$\zeta_2 = \{(a, 0.3, 0.1, 0.1), (d, 0.2, 0.3, 0.2), (e, 0.4, 0.6, 0.1)\}$,
$\zeta_3 = \{(b, 0.3, 0.4, 0.3), (d, 0.4, 0.3, 0.4), (e, 0.4, 0.3, 0.1)\}$.
By routine calculations, it easy to see that the H is neither regular nor totally regular 3-polar fuzzy graph.

Example 5.14 The 4-polar fuzzy hypergraph shown in Fig. 5.16 is both regular and totally regular.

Remark 5.1 (a) For an *m*-polar fuzzy hypergraph $H = (A, B)$ to be both regular and totally regular, the number of vertices in each hyperedge E_j must be same. Suppose that $|E_j| = k$ for every j, then H is said to be k-uniform.

(b) Each vertex lies in exactly same number of hyperedges.

Definition 5.21 Let $H = (A, B)$ be a regular *m*-polar fuzzy hypergraph. The *order* of a regular *m*-polar fuzzy hypergraph H is an *m*-tuple of the form,

$$O(H) = (\Sigma_{x \in X} \wedge P_1 \circ \zeta_j(x), \Sigma_{x \in X} \wedge P_2 \circ \zeta_j(x), \ldots, \Sigma_{x \in X} \wedge P_m \circ \zeta_j(x)).$$

The *size* of a regular *m*-polar fuzzy hypergraph is $S(H) = \sum_{E_j \subseteq X} B(E_j)$.

Example 5.15 Consider the 4-polar fuzzy hypergraph $H = (A, B)$ on $X = \{a, b, c, d, e, f, g, h, i\}$ and $A = \{\zeta_1, \zeta_2, \zeta_3, \zeta_4, \zeta_5, \zeta_6\}$, where
$\zeta_1 = \{(a, 0.4, 0.4, 0.4, 0.4), (b, 0.4, 0.4, 0.4, 0.4), (c, 0.4, 0.4, 0.4, 0.4)\}$,
$\zeta_2 = \{(d, 0.4, 0.4, 0.4, 0.4), (e, 0.4, 0.4, 0.4, 0.4), (f, 0.4, 0.4, 0.4, 0.4)\}$,
$\zeta_3 = \{(g, 0.4, 0.4, 0.4, 0.4), (h, 0.4, 0.4, 0.4, 0.4), (i, 0.4, 0.4, 0.4, 0.4)\}$,
$\zeta_4 = \{(a, 0.4, 0.4, 0.4, 0.4), (d, 0.4, 0.4, 0.4, 0.4), (g, 0.4, 0.4, 0.4, 0.4)\}$,
$\zeta_5 = \{(b, 0.4, 0.4, 0.4, 0.4), (e, 0.4, 0.4, 0.4, 0.4), (h, 0.4, 0.4, 0.4, 0.4)\}$,
$\zeta_6 = \{(c, 0.4, 0.4, 0.4, 0.4), (f, 0.4, 0.4, 0.4, 0.4), (i, 0.4, 0.4, 0.4, 0.4)\}$.
Clearly, $O(H) = (3.6, 3.6, 3.6, 3.6)$ and $S(H) = (7.2, 7.2, 7.2, 7.2)$.

Theorem 5.6 *Let* $H = (A, B)$ *be an m-polar fuzzy hypergraph on* X. *Then,* $A : X \longrightarrow [0, 1]^m$ *is a constant function if and only if the following statements are equivalent,*

(a) H is a regular m-polar fuzzy hypergraph,

(b) H is a totally regular m-polar fuzzy hypergraph.

Proof Suppose that $A : X \longrightarrow [0, 1]^m$, where $A = \{\zeta_1, \zeta_2, \ldots, \zeta_r\}$ is a constant function. That is, $P_i \circ \zeta_j(x) = c_i$, for all $x \in \zeta_j$, $1 \leq i \leq m$, $1 \leq j \leq r$.

(a) $\Rightarrow$ (b) Suppose that H is *n*-regular *m*-polar fuzzy hypergraph. Then $deg^{(i)}(x) = n_i$, for all $x \in X$, $1 \leq i \leq m$. By using Definition 5.19, $deg^{(i)}[x] = n_i + k_i$, for all $x \in X$, $1 \leq i \leq m$. Hence, H is a totally regular *m*-polar fuzzy hypergraph.

(b) $\Rightarrow$ (a) Suppose that H is a *k*-totally regular *m*-polar fuzzy hypergraph. Then, $deg^{(i)}[x] = k_i$, for all $x \in X$, $1 \leq i \leq m$.
$\Rightarrow deg^{(i)}(x) + \wedge_j P_i \circ \zeta_j(x) = k_i$ for all $x \in \zeta_j$,
$\Rightarrow deg^{(i)}(x) + c_i = k_i$, for all $x \in \zeta_j$,
$\Rightarrow deg^{(i)}(x) = k_i - c_i$, for all $x \in \zeta_j$. Thus, H is a regular *m*-polar fuzzy hypergraph. Hence, (a) and (b) are equivalent.

Conversely, suppose that (a) and (b) are equivalent, i.e., H is regular if and only if H is a totally regular. On contrary suppose that A is not constant, that is, $P_i \circ \zeta_j(x) \neq P_i \circ \zeta_j(y)$ for some x and y in A. Let $H = (A, B)$ be a *n*-regular *m*-polar fuzzy hypergraph then, $deg^{(i)}(x) = n_i$ for all $x \in \zeta_j(x)$. Consider,

$$deg^{(i)}[x] = deg^{(i)}(x) + \wedge_j P_i \circ \zeta_j(x) = n_i + \wedge_j P_i \circ \zeta_j(x),$$
$$deg^{(i)}[y] = deg^{(i)}(y) + \wedge_j P_i \circ \zeta_j(y) = n_i + \wedge_j P_i \circ \zeta_j(y).$$

Since, $P_i \circ \zeta_j(x)$ and $P_i \circ \zeta_j(y)$ are not equal for some x and y in X, hence $deg[x]$ and $deg[y]$ are not equals, thus H is not a totally regular m-poalr fuzzy hypergraph, which is a contradiction to our assumption. Next, let H be a totally regular m-polar fuzzy hypergraph, then $deg[x] = deg[y]$, that is,

$$deg^{(i)}(x) + \wedge_j P_i \circ \zeta_j(x) = deg^{(i)}(y) + \wedge_j P_i \circ \zeta_j(y),$$
$$deg^{(i)}(x) - deg^{(i)}(y) = \wedge_j P_i \circ \zeta_j(y) - \wedge_j P_i \circ \zeta_j(x).$$

It follows that $deg(x)$ and $deg(y)$ are not equal, so H is not a regular m-polar fuzzy hypergraph, which is again a contradiction to our assumption. Hence, A must be constant and it completes the proof.

Theorem 5.7 *If an m-polar fuzzy hypergraph is both regular and totally regular then $A : X \longrightarrow [0, 1]^m$ is constant function.*

Proof Let H be a regular and totally regular m-polar fuzzy hypergraph then,

$$deg^{(i)}(x) = n_i \text{ for all } x \in X, 1 \leq i \leq m.$$
$$deg^{(i)}[x] = k_i \text{ for all } x \in \zeta_j(x),$$
$$\Leftrightarrow deg^{(i)}(x) + \wedge_j P_i \circ \zeta_j(x) = k_i, \text{ for all } x \in \zeta_j(x),$$
$$\Leftrightarrow n_1 + \wedge_j P_i \circ \zeta_j(x) = k_i, \text{ for all } x \in \zeta_j(x),$$
$$\Leftrightarrow \wedge_j P_i \circ \zeta_j(x) = k_i - n_i, \text{ for all } x \in \zeta_j(x),$$
$$\Leftrightarrow P_i \circ \zeta_j(x) = k_i - n_i, \text{ for all } x \in X, 1 \leq i \leq m.$$

Hence, $A : X \longrightarrow [0, 1]^m$ is a constant function.

Remark 5.2 The converse of Theorem 5.7 may not be true, in general as it can be seen in the following example.

Consider a 3-polar fuzzy hypergraph $H = (A, B)$ on $X = \{a, b, c, d, e\}$,
$\zeta_1 = \{(a, 0.2, 0, 2, 0.2), (b, 0.2, 0.2, 0.2), (c, 0.2, 0.2, 0.2)\}$,
$\zeta_2 = \{(a, 0.2, 0, 2, 0.2), (d, 0.2, 0.2, 0.2)\}$,
$\zeta_3 = \{(b, 0.2, 0.2, 0.2), (e, 0.2, 0.2, 0.2)\}$,
$\zeta_4 = \{(c, 0.2, 0.2, 0.2), (e, 0.2, 0.2, 0.2)\}$. Then, $A : X \longrightarrow [0, 1]^m$, where $A = \{\zeta_1, \zeta_2, ..., \zeta_r\}$ is a constant function. But $deg(a) = (0.6, 0.6, 0.6) \neq (0.4, 0.4, 0.4) = deg(e)$. Also $(deg[a] = (0.8, 0.8, 0.8) \neq (0.6, 0.6, 0.6) = deg[e])$. So H is neither regular nor totally regular m-polar fuzzy hypergraph.

Definition 5.22 An m-polar fuzzy hypergraph $H = (A, B)$ is called *complete* if for every $x \in X$, $N(x) = \{xy| y \in X - x\}$, that is, $N(x)$ contains all the remaining vertices of X except x.

Example 5.16 Consider a 3-polar fuzzy hypergraph $H = (A, B)$ on $X = \{a, b, c, d\}$ as shown in Fig. 5.17 then $N(a) = \{b, c, d\}, N(b) = \{a, c, d\}$, and $N(c) = \{a, b, d\}$.

Fig. 5.17 Complete 3-polar
fuzzy hypergraph

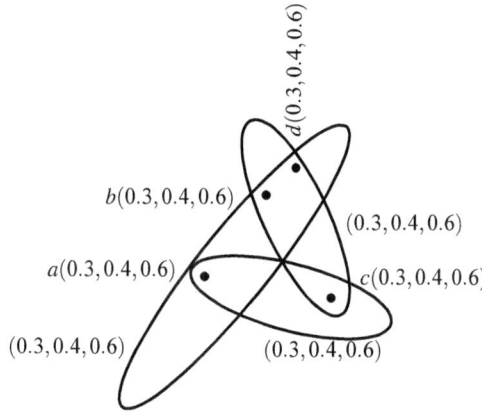

Remark 5.3 For a complete *m*-polar fuzzy hypergraph, the cardinality of $N(x)$ is same for every vertex.

Theorem 5.8 *Every complete m-polar fuzzy hypergraph is a totally regular m-polar fuzzy hypergraph.*

Proof Since given *m*-polar fuzzy hypergraph H is complete, each vertex lies in exactly same number of hyperedges and each vertex have same closed neighborhood degree *m*. That is, $deg[x_1] = deg[x_2]$ for all $x_1, x_2 \in X$. Hence, H is *m*-totally regular.

5.3 Applications of *m*-Polar Fuzzy Hypergraphs

Analysis of human nature and their culture has been tangled with assessment of social networks from many years. Such networks are refined by designating one or more relations on the set of individuals and the relations can be taken from efficacious relationships, facets of some management and from a large range of others means. For super-dyadic relationships between the nodes, network models represented by simple graph are not sufficient. Natural presence of hyperedges can be found in co-citation, e-mail networks, co-authorship, web log networks, and social networks, etc. Representation of these models as hypergraphs maintain the dyadic relationships.

5.3.1 Super-Dyadic Managements in Marketing Channels

In marketing channels, dyadic correspondence organization has been a basic implementation. Marketing researchers and managers are realized that their common engagement in marketing channels is a central key for successful marketing and to yield benefits for company. m-polar fuzzy hypergraphs consist of marketing managers as vertices and hyperedges show their dyadic communication involving their parallel thoughts, objectives, plans, and proposals. The more powerful close relation in the researchers is more beneficial for the marketing strategies and the production of an organization. A 3-polar fuzzy network model showing the dyadic communications among the marketing managers of an organization is given in Fig. 5.18. The membership degrees of each person symbolize the percentage of its dyadic behavior toward the other persons of the same dyad group. Adjacent level between any pair of vertices illustrates that how much their dyadic relationship is proficient. The adjacent levels are given in Table 5.4. It can be seen that the most capable dyadic pair is (Kashif, Kaamil). 3-polar fuzzy hyperedges are taken as the different digital marketing strategies adopted by the different dyadic groups of the same organization. The vital goal of this model is to figure out the most potent dyad of digital marketing techniques. The six different groups are made by the marketing managers and the digital marketing strategies adopted by these six groups are represented by hyperedges, i.e., the 3-polar fuzzy hyperedges $\{T_1, T_2, T_3, T_4, T_5, T_6\}$ show the following strategies {Product pricing, Product planning, Environment analysis and marketing research, Brand name, Build the relationships, Promotions}, respectively. The exclusive effects

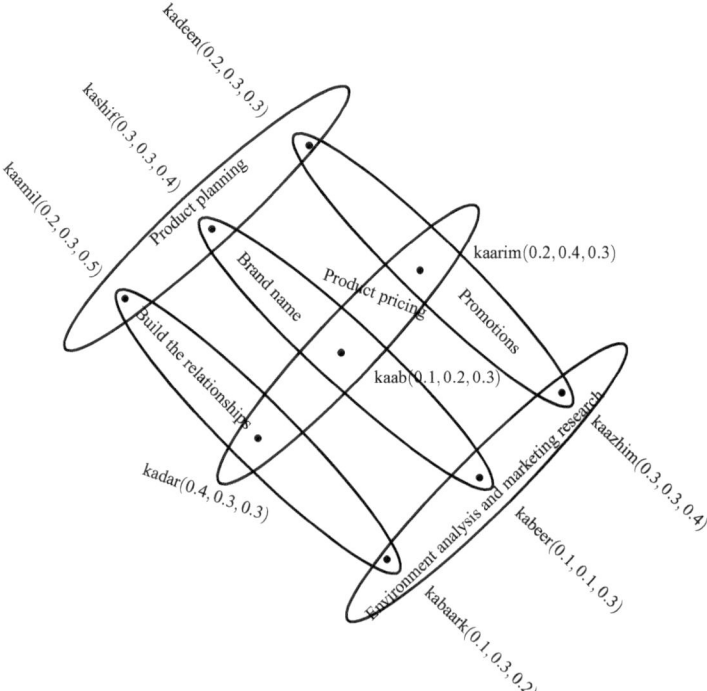

Fig. 5.18 Super-dyadic managements in marketing channels

Table 5.4 Adjacent levels of 3-polar fuzzy hypergraph

Dyad pairs	Adjacent level	Dyad pairs	Adjacent level
γ(Kadeen, Kashif)	(0.2, 0.3, 0.3)	γ(Kaarim, Kaazhim)	(0.2, 0.3, 0.3)
γ(Kadeen, Kaamil)	(0.2, 0.3, 0.3)	γ(Kaarim, Kaab)	(0.1, 0.2, 0.3)
γ(Kadeen, Kaarim)	(0.2, 0.3, 0.3)	γ(Kaarim, Kadar)	(0.2, 0.3, 0.3)
γ(Kadeen, Kaazhim)	(0.2, 0.3, 0.3)	γ(Kaab, Kadar)	(0.1, 0.2, 0.3)
γ(Kashif, Kaamil)	(0.2, 0.3, 0.4)	γ(Kaab, Kabeer)	(0.1, 0.1, 0.3)
γ(Kashif, Kaab)	(0.1, 0.2, 0.3)	γ(Kadar, Kabaark)	(0.1, 0.3, 0.2)
γ(Kashif, Kabeer)	(0.1, 0.1, 0.3)	γ(Kaazhim, Kabeer)	(0.1, 0.1, 0.3)
γ(Kaamil, Kadar))	(0.2, 0.2, 0.3)	γ(Kaazhim, Kabaark)	(0.1, 0.3, 0.2)
γ(Kaamil, Kabaark)	(0.1, 0.3, 0.2)	γ(Kabeer, Kabaark)	(0.1, 0.1, 0.2)

Table 5.5 Effects of marketing strategies

Marketing strategy	Profitable growth	Instruction manual for company success	Create longevity of the business
Product pricing	0.1	0.2	0.3
Product planning	0.2	0.3	0.3
Environment analysis and marketing research	0.1	0.2	0.2
Brand name	0.1	0.3	0.3
Build the relationships	0.1	0.3	0.2
Promotions	0.2	0.3	0.3

of membership degrees of each marketing strategy toward the achievements of an organization are given in Table 5.5. Effective dyads of market strategies enhance the performance of an organization and discover the better techniques to be adopted. The adjacency of all dyadic communication managements is given in Table 5.6. The most dominant and capable marketing strategies adopted mutually are Product planning and Promotions. Thus to increase the efficiency of an organization, dyadic managements should make the powerful planning for products and use the promotions skill to attract customers to purchase their products. The membership degrees of this dyad is (0.2, 0.3, 0.3) which shows that the amalgamated effect of this dyad will increase the profitable growth of an organization up to 20%, instruction manual for company success up to 30%, create longevity of the business up to 30% . Thus, to promote the performance of an organization, super dyad marketing communications are more energetic. The method of finding out the most effective dyads is explained in Algorithm 5.3.1.

Table 5.6 Adjacency of all dyadic communication managements

Dyadic strategies	Effects
σ (Product pricing, Product planning)	(0.1, 0.2, 0.3)
σ (Product pricing, Environment analysis and marketing research)	(0.1, 0.2, 0.2)
σ (Product pricing, Brand name)	(0.1, 0.2, 0.3)
σ (Product pricing, Build the relationships)	(0.1, 0.2, 0.2)
σ (Product pricing, Promotions)	(0.1, 0.2, 0.3)
σ (Product planning, Environment analysis and marketing research)	(0.1, 0.2, 0.2)
σ (Product planning, Brand name)	(0.1, 0.3, 0.3)
σ (Product planning, Build the relationships)	(0.1, 0.3, 0.2)
σ (Product planning, Promotions)	(0.2, 0.3, 0.3)
σ (Environment analysis and marketing research, Brand name)	(0.1, 0.2, 0.2)
σ (Environment analysis and marketing research, Build the relationships)	(0.1, 0.2, 0.2)
σ (Environment analysis and marketing research, Promotions)	(0.1, 0.2, 0.2)
σ (Brand name, Build the relationships)	(0.1, 0.3, 0.2)
σ (Brand name, Promotions)	(0.1, 0.3, 0.3)
σ (Build the relationships, Promotions)	(0.1, 0.3, 0.2)

Algorithm 5.3.1 Finding the most effective dyads

1. Input the membership values $A(x_i)$ of all nodes (marketing managers) $x_1, x_2, ..., x_n$.
2. Input the membership values $B(T_i)$ of all hyperedges $T_1, T_2, ..., T_r$.
3. Find the adjacent level between nodes x_i and x_j as,
4. **do** i from $1 \rightarrow n - 1$
5.   **do** j from $i + 1 \rightarrow n$
6.   **do** k from $1 \rightarrow r$
7.   **if** $x_i, x_j \in E_k$ **then**
8. $\gamma(x_i, x_j) = \sup_k \inf\{A(x_i), A(x_j)\}$.
9.   **end if**
10.   **end do**
11.   **end do**
12. **end do**
13. Find the best capable dyadic pair as $\sup_{i,j} \gamma(x_i, x_j)$.
14. **do** i from $1 \rightarrow r - 1$
15.   **do** j from $i + 1 \rightarrow r$
16.   **do** k from $1 \rightarrow r$
17.   **if** $x_k \in T_i \cap T_j$ **then**
18. $\sigma(T_i, T_j) = \sup_k \inf\{B(T_i), B(T_j)\}$.
19.   **end if**
20.   **end do**
21.   **end do**
22. **end do**
23. Find the best effective super dyad management as $\sup_{i,j} \sigma(T_i, T_j)$.

Description of Algorithm 5.3.1: Lines 1 and 2 passes the input of *m*-polar fuzzy set A on n vertices $x_1, x_2, \ldots, x_n$ and *m*-polar fuzzy relation B on r edges $T_1, T_2, \ldots, T_r$. Lines 3–12 calculate the adjacent level between each pair of nodes. Line 14 calculates the best capable dyadic pair. The loop initializes by taking the value $i = 1$ of do loop which is always true, i.e., the loop runs for the first iteration. For any ith iteration of do loop on line 3, the do loop on line 4 runs $n - i$ times and, the do loop on line 5 runs r times. If there exists a hyperedge E_k containing x_i and x_j then, line 7 is executed otherwise the if conditional terminates. For every ith iteration of the loop on line 3, this process continues n times and then increments i for the next iteration maintaining the loop throughout the algorithm. For $i = n - 1$, the loop calculates the adjacent level for every pair of distinct vertices and terminates successfully at line 12. Similarly, the loops on lines 13, 14 , and 15 maintain and terminate successfully.

5.3.2 *m-Polar Fuzzy Hypergraphs in Work Allotment Problem*

In customer care centers, availability of employees plays a vital to solve people's problems. Such a department should ensure that the system has been managed carefully to overcome practical difficulties. A lot of customers visit such centers to find a solution of their problems. In this part, focus is given to alteration of duties for the employees taking leave. The problem is that employees are taking leave without proper intimation and alteration. We now show the importance of *m*-polar fuzzy hypergraphs for the allocation of duties to avoid any difficulties.

Consider the example of a customer care center consisting of 30 employees. Assuming that six workers are necessary to be available at their duties. We present the employees as vertices and degree of membership of each employee represents the workload, percentage of available time and number of workers who are also aware of the employee's work type. The range of values for present time and the workers knowing the type of work is given in Tables 5.7 and 5.8, respectively. The degree of membership of each edge represents the common work load, percentage of available time and number of workers who are also aware of the employee's work type. This phenomenon can be represented by a 3-polar fuzzy graph as shown in Fig. 5.19. Using Algorithm 5.3.2, the strength of allocation and alteration of duties among employees is given in Table 5.9. Column 3 in Table 5.9 shows the percentage of alteration of duties. For example, in case of leave, duties of a_1 can be given to a_3 and similarly for other employees. The method for the calculation of alteration of duties is given in Algorithm 5.3.2.

Table 5.7 Range of membership values of table time

Time	Membership value
5 h	0.40
6 h	0.50
8 h	0.70
10 h	0.90

Table 5.8 Workers knowing the work type

Workers	Membership value
3	0.40
4	0.60
5	0.80
6	0.90

Fig. 5.19 3-polar fuzzy hypergraph

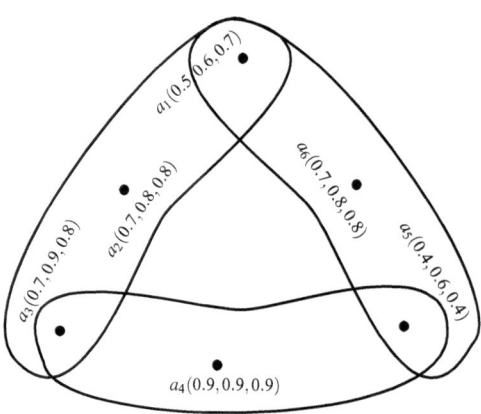

Table 5.9 Alteration of duties

Workers	$A(a_i, a_j)$	$S(a_i, a_j)$
a_1, a_2	(0.7, 0.8, 0.8)	0.77
a_1, a_3	(0.7, 0.9, 0.8)	0.80
a_2, a_3	(0.5, 0.7, 0.7)	0.63
a_3, a_4	(0.7, 0.6, 0.8)	0.70
a_3, a_5	(0.7, 0.9, 0.8)	0.80
a_4, a_5	(0.9, 0.9, 0.9)	0.90
a_5, a_6	(0.7, 0.8, 0.8)	0.77
a_5, a_1	(0.5, 0.6, 0.7)	0.60
a_1, a_6	(0.6, 0.8, 0.5)	0.63

Algorithm 5.3.2 Calculation of alteration of duties

1. Input the n number of employees $a_1, a_2, \ldots, a_n$.
2. Input the number of edges $E_1, E_2, \ldots, E_r$.
3. Input the incident matrix B_{ij} where, $1 \leq i \leq n, 1 \leq j \leq r$.
4. Input the membership values of edges $\xi_1, \xi_2, \ldots, \xi_r$
5. **do** i from $1 \to n$
6.   **do** j from $1 \to n$
7.   **do** k from $1 \to r$
8.   **if** $a_i, a_j \in E_k$ **then**
9.   **do** t from $1 \to m$
10. $P_t \circ A(a_i, a_j) = |P_t \circ B_{ik} - P_t \circ B_{jk}| + P_t \circ \xi_k$
11.   **end do**
12.   **end if**
13.   **end do**
14.   **end do**
15. **end do**
16. **do** i from $1 \to n$
17.   **do** j from $1 \to n$
18.   **if** $A(a_i, a_j) > 0$ **then**
19. $S(a_i, a_j) = \dfrac{P_1 \circ A(a_i, a_j) + P_2 \circ A(a_i, a_j) + \ldots + P_m \circ A(a_i, a_j)}{m}$
20.   **end if**
21.   **end do**
22. **end do**

Description of Algorithm 5.3.2: Lines 1, 2, 3 and 4 passes the input of membership values of vertices, hyperedges and an m-polar fuzzy adjacency matrix B_{ij}. The nested loops on lines 5 to 15 calculate the rth, $1 \leq r \leq m$, strength of allocation and alteration of duties between each pair of employees. The nested loops on lines 16 to 22 calculate the strength of allocation and alteration of duties between each pair of employees. The net time complexity of the algorithm is $O(n^2 rm)$.

5.3.3 Availability of Books in Library

A library in college is a collection of sources of information and similar resources, made accessible to student community for reference and examination preparation. A student preparing for some examination will use the knowledge sources such as
1. Prescribed textbooks (A)
2. Reference books in syllabus (B)
3. Other books from library (C)
4. Knowledgeable study materials (D)
5. E-gadgets and internet (E)

Fig. 5.20 3-polar fuzzy hypergraph

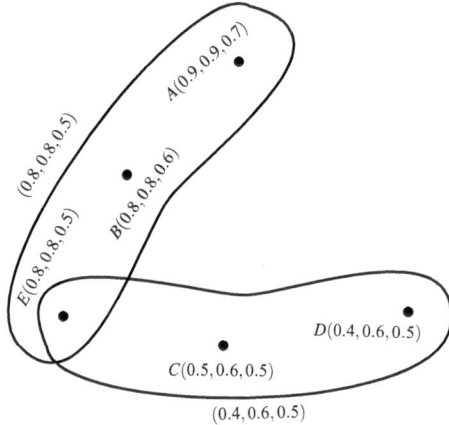

Table 5.10 Library sources

Sources s_i	$T(s_i)$	$S(a_i, a_j)$
A	(1.7, 1.7, 1.4)	1.60
B	(1.6, 1.6, 1.1)	1.43
E	(1.6, 1.6, 1.0)	1.40
C	(0.9, 1.2, 1.0)	1.03
D	(0.8, 1.2, 1.0)	1.0

The important thing is to consider the maximum availability of the sources which students mostly use. This phenomenon can be discussed using m-polar fuzzy hypergraphs. We now calculate the importance of each source in student community.

Consider the example of five library resources $\{A, B, C, D, E\}$ in a college. We represent these sources as vertices in a 3-polar fuzzy hypergraph. The degree of membership of each vertex represents the percentage of students using a particular source for exam preparation, percentage of faculty of members using the sources and number of sources available. The degree of membership of each edge represents the common percentage. The 3-polar fuzzy hypergraph is shown in Fig. 5.20. Using Algorithm 5.3.3, the strength of each library source is given in Table 5.10.

Column 3 in Table 5.10 shows that sources A and B are mostly used by students and faculty. Therefore, these should be available in maximum number. There is also a need to confirm the availability of source E to students and faculty. The method for the calculation of percentage importance of the sources is given in Algorithm 5.3.3 whose net time complexity is O(nrm).

Algorithm 5.3.3 *Calculation of percentage importance of the sources*

1. Input the n number of sources $s_1, s_2, \ldots, s_n$.
2. Input the number of edges $E_1, E_2, \ldots, E_r$.
3. Input the incident matrix B_{ij}, where $1 \leq i \leq n, 1 \leq j \leq r$.
4. Input the membership values of edges $\xi_1, \xi_2, \ldots, \xi_r$
5. **do** i from $1 \to n$
6. $A(s_i) = 1$
7. $C(s_i) = 1$
8.   **do** k from $1 \to r$
9.   **if** $s_i \in E_k$ **then**
10. $A(s_i) = \sup\{A(s_i), \xi_k\}$
11. $C(s_i) = \inf\{C(s_i), B_{ik}\}$
12.   **end if**
13.   **end do**
14. $T(s_i) = C(s_i) + A(s_i)$
15. **end do**
16. **do** i from $1 \to n$
17.   **if** $T(s_i) > 0$ **then**
18. $S(s_i) = \dfrac{P_1 \circ T(s_i) + P_2 \circ T(s_i) + \ldots + P_m \circ T(s_i)}{m}$
19.   **end if**
20. **end do**

Description of Algorithm 5.3.3: Lines 1, 2, 3, and 4 passes the input of membership values of vertices, hyperedges and an m-polar fuzzy adjacency matrix B_{ij}. The nested loops on lines 5 to 15 calculate the degree of usage and availability of library sources. The nested loops on lines 16–20 calculate the strength of each library source.

5.3.4 Selection of a Pair of Good Team for Competition

Competition grants the inspiration to achieve a goal; to demonstrate determination, creativity, and perseverance to overcome challenges; and to understand that hard work and commitment leads to a greater chance of success. It is inarguably accepted that a bit of healthy competition in any field is known to enhance motivation and generate increased effort from those competing. The sporting field is no exception to this rule. While there will always be varying levels of sporting talent and interest across any group of people, the benefits that competitive sport provides are still accessible to all. There is a role for both competitive and noncompetitive sporting pursuits. To get success in any competition, a strong team can be held largely accountable for the success.

The purpose of this application is to select a pair of good player team for competition with other country. For example, we have three teams of players (three

Table 5.11 3-polar subsets of teams

Players	Self confidence	Strong sense of motivation	Adaptability
Adnan	0.5	0.6	0.5
Usman	0.6	0.4	0.8
Awais	0.5	0.8	0.9
Hamza	0.7	0.7	0.6
Waseem	0.3	0.7	0.4
Usama	0.4	0.2	0.3
Iqbal	0.5	0.5	0.5
Noman	0.3	0.6	0.6
Arshad	0.4	0.3	0.7
Saeed	0.4	0.2	0.9
Nawab	0.7	0.5	0.6
Haris	0.6	0.6	0.5

3-polar fuzzy hypergraphs) and we have to select only one pair of team for competition with other country. Then to select it, we use union operation of m-polar fuzzy hypergraphs. Hypergraph is used because there is a link in one team more than two players and m-polar represents different qualities of players and teams. Consider three teams, team 1 consists of players $\{Adnan, Usman, Hamza, Awais\}$. Team 2 consists of players $\{Waseem, Usama, Iqbal, Noman\}$. Team 3 consists of players $\{Arshad, Saeed, Nawab, Haris\}$. The 3-polar fuzzy set of players represent the three different qualities of each player, i.e., self confidence, strong sense of motivation, adaptability. 3-polar fuzzy hyperedges represent the three characteristics of a good team. First membership degree of 3-polar fuzzy hyperedges represents the focus of team on goals, second represents the communication with each other, third represents how much team is organized. We want to select a pair of good team which qualify these three properties with maximum membership degrees values (Tables 5.11 and 5.12).

Let $A = \{(Adnan, 0.5, 0.6, 0.5), (Usman, 0.6, 0.4, 0.8), (Awais, 0.5, 0.8, 0.9),$
$(Hamza, 0.7, 0.7, 0.6),$

$(Waseem, 0.3, 0.7, 0.4),$ $(Usama, 0.4, 0.2, 0.3),$ $(Iqbal, 0.5, 0.5, 0.5),$
$(Noman, 0.3, 0.6, 0.6),$

$(Arshad, 0.4, 0.3, 0.7), (Saeed, 0.4, 0.2, 0.9), (Nawab, 0.7, 0.5, 0.6), (Haris,$
$0.6, 0.6, 0.5)\}$ be a 3-polar fuzzy set of players and $B = \{(Team\ 1, 0.5, 0.4, 0.5),$
$(Team\ 2, 0.3, 0.2, 0.3), (Team\ 3, 0.4, 0.2, 0.6)\}$ is a set of 3-polar fuzzy hyperedges.

We select that pair of team whose union is strong, i.e., we select that union whose edges have maximum membership degrees. It represents the focus of teams on goals, second represents the communication with each other of both teams, and

Table 5.12 3-polar fuzzy qualities of teams

Teams	Focus on goals	Communication skills	Organization
1	0.5	0.4	0.5
2	0.3	0.2	0.3
3	0.4	0.2	0.6

Fig. 5.21 3-polar fuzzy hypergraph H_1

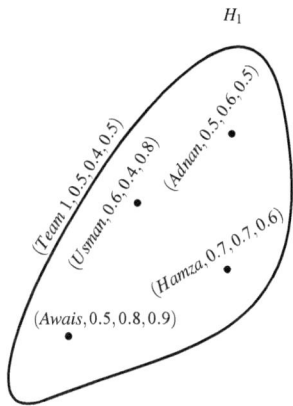

Fig. 5.22 3-polar fuzzy hypergraph H_2

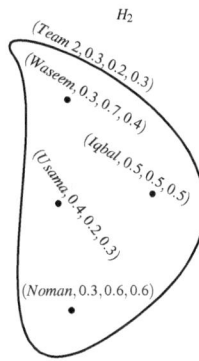

Fig. 5.23 3-polar fuzzy hypergraph H_3

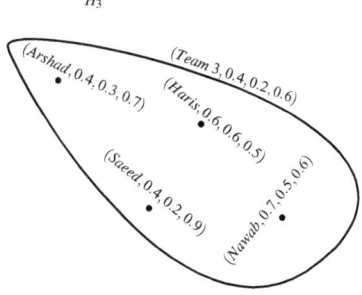

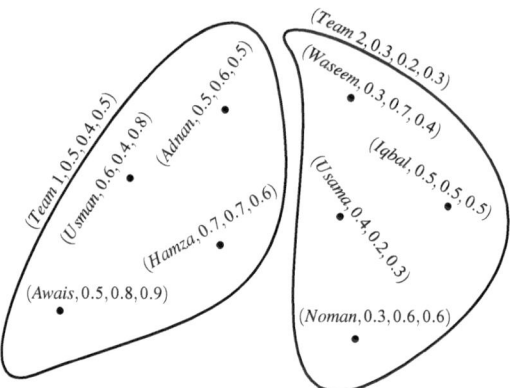

Fig. 5.24 $H_1 \cup H_2$

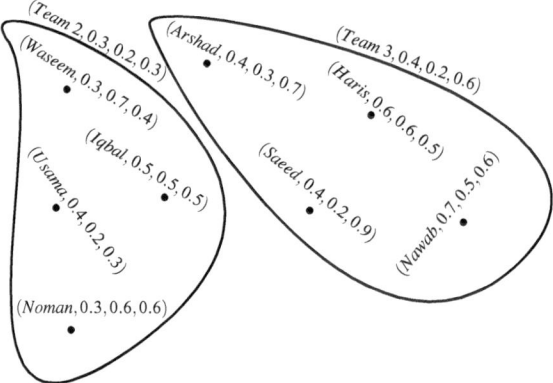

Fig. 5.25 $H_2 \cup H_3$

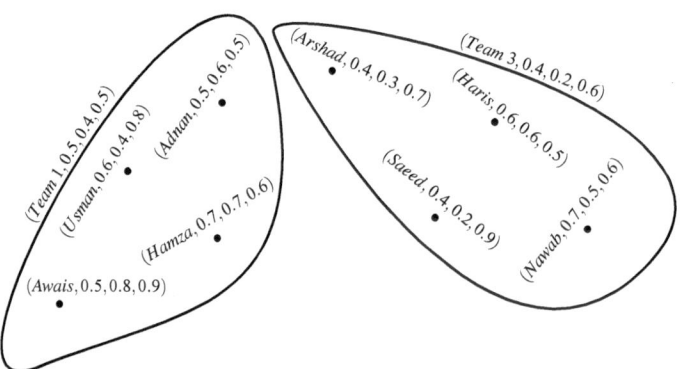

Fig. 5.26 $H_1 \cup H_3$

third represents how much team is organized. So, we select the pair of team 1 and team 3 (Figs. 5.21, 5.22, 5.23, 5.24, 5.25 and 5.26).

We present our proposed method in Algorithm 5.3.4.

Algorithm 5.3.4 *Selection of team for competition*

Step 1: **Input**

The set of players.

Assign the membership values to each player.

Select the players of each team.

Step 2: Compute the membership values of each team(edges) by using the relation

$B(E_i) = B(\{x_1, x_2, \ldots, x_r\}) \leq \inf\{\zeta_i(x_1), \zeta_i(x_2), \ldots, \zeta_i(x_s)\}$, for all x_1, $x_2, \ldots, x_s \in X$.

Step 3: Compute union of teams.

Compute their union by using the relation

(i) $P_i \circ (A_1 \cup A_2)(v) = \begin{cases} P_i \circ A_1(v) \ if \ v \in X_1 - X_2, \\ P_i \circ A_2(v) \ if \ v \in X_2 - X_1, \\ \sup\{P_i \circ A_1(v), P_i \circ A_2(v)\} \ if \ v \in X_1 \cap X_2. \end{cases}$

(ii) $P_i \circ (B_1 \cup B_2)(e) = \begin{cases} P_i \circ B_1(e) \ if \ e \in E_1 - E_2, \\ P_i \circ B_2(e) \ if \ e \in E_2 - E_1, \\ \sup\{P_i \circ B_1(e), P_i \circ B_2(e)\} \ if \ e \in E_1 \cap E_2. \end{cases}$

Step 4: **Output**

Select that pair of team for competition for which edges of union have maximum membership degree.

5.4 m-Polar Fuzzy Directed Hypergraphs

Definition 5.23 A directed hypergraph is a hypergraph with directed hyperedges. A directed hyperedge or hyperarc is an ordered pair $E = (X, Y)$ of (possibly empty) disjoint subsets of vertices. X is the tail of E, while Y is its head. A sequence of crisp hypergraphs $H_i = (V_i, E_i)$, $1 \leq i \leq n$, is said to ordered if $H_1 \subset H_2 \subset \ldots, H_n$. The sequence $\{H_i \mid 1 \leq i \leq n\}$ is said to be simply ordered if it is ordered, and if whenever $E \subset E_{i+1} \backslash E_i$, then $E \not\subset V_i$.

We now define an m-polar fuzzy directed hypergraph.

Definition 5.24 An m-polar fuzzy directed hypergraph with underlying set X is an ordered pair $H = (\sigma, \varepsilon)$, where σ is non-empty set of vertices and ε is a family of m-polar fuzzy (m-polar fuzzy) directed hyperarcs (or hyperedges). An m-polar fuzzy directed hyperarc (or hyperedge) $e_i \in \varepsilon$ is an ordered pair $(t(e_i), h(e_i))$, such that, $t(e_i) \neq \emptyset$, is called its tail and $h(e_i) \neq t(e_i)$ is its head, such that $P_k \circ \varepsilon_i(\{v_1, v_2, \ldots, v_s\}) \leq \inf\{P_k \circ \sigma_i(v_1), P_k \circ \sigma_i(v_2), \ldots, P_k \circ \sigma_i(v_s)\}$, for all $v_1, v_2, \ldots, v_s \in V$, $1 \leq k \leq m$.

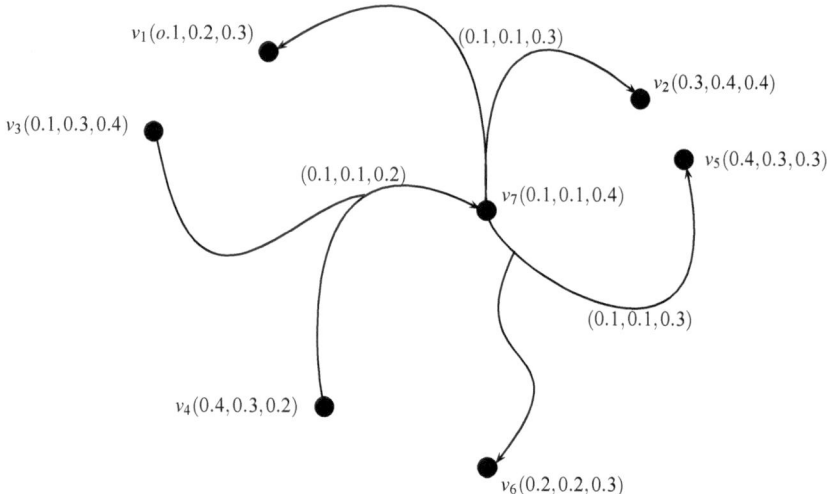

Fig. 5.27 3-polar fuzzy directed hypergraph

Definition 5.25 Let $H = (\sigma, \varepsilon)$ be an m-polar fuzzy directed hypergraph. The order of H, denoted by $O(H)$, is defined as $O(H) = \sum_{x \in V} \wedge \sigma_i(x)$. The size of H, denoted by $S(H)$, is defined by $S(H) = \sum_{e_k \subset V} \varepsilon(e_k)$.

In an m-polar fuzzy directed hypergraph, the vertices v_i and v_j are adjacent vertices if they both belong to the same m-polar fuzzy directed hyperedge. Two m-polar fuzzy directed hyperedges e_i and e_j are called adjacent if they have non-empty intersection. That is, $supp(e_i) \cap supp(e_j) \neq \emptyset, i \neq j$.

Definition 5.26 An m-polar fuzzy directed hypergraph $H = (\sigma, \varepsilon)$ is *simple* if it contains no repeated directed hyperedges, i.e., if $e_j, e_k \in \varepsilon$ and $e_j \subseteq e_k$ then $e_j = e_k$. An m-polar fuzzy directed hypergraph $H = (\sigma, \varepsilon)$ is called *support simple* if $e_j, e_k \in \varepsilon$ and $supp(e_j) = supp(e_k)$ and $e_j \subseteq e_k$, then $e_j = e_k$. An m-polar fuzzy directed hypergraph, $H = (\sigma, \varepsilon)$ is called *strongly support simple* if $e_j, e_k \in \varepsilon$ and $supp(e_j) = supp(e_k)$, then $e_j = e_k$.

Example 5.17 Consider a 3-polar fuzzy directed hypergraph $H = (\sigma, \varepsilon)$, such that $\sigma = \{\sigma_1, \sigma_2, \sigma_3, \sigma_4, \sigma_5\}$ is the family of 3-polar fuzzy subsets on $X = \{v_1, v_2, v_3, v_4, v_5, v_6\}$, as shown in Fig. 5.27, such that
$\sigma_1 = \{(v_1, 0.1, 0.2, 0.3), (v_2, 0.3., 0.4, 0.4), (v_3, 0.1, 0.3, 0.4)\}$,
$\sigma_2 = \{(v_5, 0.4, 0.3, 0.3), (v_6, 0.2, 0.2, 0.3), (v_7, 0.1, 0.1, 0.4)\}$,
$\sigma_3 = \{(v_3, 0.1, 0.3, 0.4), (v_4, 0.4, 0.3, 0.2), (v_7, 0.1, 0.1, 0.4)\}$.
3-polar fuzzy relation ε is defined as, $\varepsilon(v_1, v_2, v_7) = (0.1, 0.1, 0.3)$, $\varepsilon(v_5, v_6, v_7) = (0.1, 0.1, 0.3)$, $\varepsilon(v_3, v_4, v_7) = (0.1, 0.1, 0.2)$.

Clearly, H is simple, strongly support simple, and support simple, that is, it contains no repeated directed hyperedges and if whenever $e_j, e_k \in \varepsilon$ and $supp(e_j) = supp(e_k)$, then $e_j = e_k$. Further, $O(H) = (1.6, 1.8, 2.3)$ and $S(H) = (0.3, 0.3, 0.8)$.

Fig. 5.28 Regular 3-polar fuzzy hypergraph

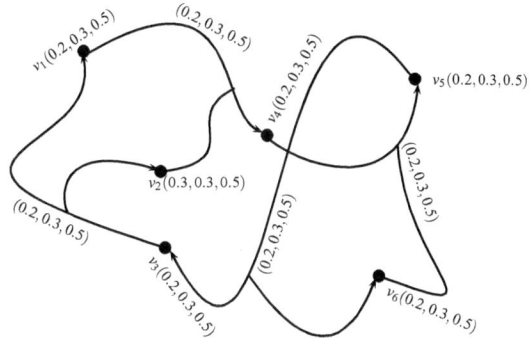

Definition 5.27 Let $\varepsilon = (\varepsilon^-, \varepsilon^+)$ be a directed *m*-polar fuzzy hyperedge in an *m*-polar fuzzy directed hypergraph. Then, the vertex set ε^- is called the *m*-polar fuzzy in-set and the vertex set ε^+ is called the *m*-polar fuzzy out-set of the directed hyperedge ε. It is not necessary that the sets ε^-, ε^+ will be disjoint. The hyperedge ε is called the join of the vertices of ε^- and ε^+.

Definition 5.28 The in-degree $D_H^-(v)$ of a vertex v in an *m*-polar fuzzy directed hypergraph is defined as the sum of membership degrees of all those directed hyperedges such that v is contained in their out-set, that is,

$$D_H^-(v) = \sum_{v \in h(e_i)} \varepsilon(e_i), \ 1 \le k \le m.$$

The out-degree $D_H^+(v)$ of a vertex v in an *m*-polar fuzzy directed hypergraph is defined as the sum of membership degrees of all those directed hyperedges such that v is contained in their in-set, that is,

$$D_H^+(v) = \sum_{v \in t(e_i)} \varepsilon(e_i), \ 1 \le k \le m.$$

Definition 5.29 An *m*-polar fuzzy directed hypergraph $H = (\sigma, \varepsilon)$ is said to be *k*-regular if in-degrees and out-degrees of all vertices in H are same.

Example 5.18 Consider a 3-polar fuzzy directed hypergraph $H = (\sigma, \varepsilon)$ as shown in Fig. 5.28, where $\sigma = \{\sigma_1, \sigma_2, \sigma_3, \sigma_4\}$ is the family of 3-polar fuzzy subsets on $V = \{v_1, v_2, v_3, v_4, v_5, v_6\}$ and

$\sigma_1 = \{(v_1, 0.2, 0.3, 0.5), (v_2, 0.2, 0.3, 0.5), (v_4, 0.2, 0.3, 0.5)\}$,
$\sigma_2 = \{(v_4, 0.2, 0.3, 0.5), (v_5, 0.2, 0.3, 0.5), (v_6, 0.2, 0.3, 0.5)\}$,
$\sigma_3 = \{(v_3, 0.2, 0.3, 0.5), (v_5, 0.2, 0.3, 0.5), (v_6, 0.2, 0.3, 0.5)\}$,
$\sigma_4 = \{(v_1, 0.2, 0.3, 0.5), (v_2, 0.2, 0.3, 0.5), (v_3, 0.2, 0.3, 0.5)\}$. By routine calculations, we see that the 3-polar fuzzy directed hypergraph is regular.

Fig. 5.29 Directed hyperpath (denoted by a thick line)

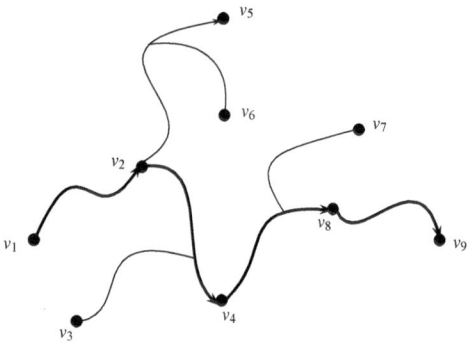

Note that, $D_H^-(v_1) = (0.2, 0.3, 0.5) = D_H^+(v_1)$ and $D_H^-(v_2) = (0.2, 0.3, 0.5) = D_H^+(v_2)$. Similarly, $D_H^-(v_3) = D_H^+(v_3)$, $D_H^-(v_4) = D_H^+(v_4)$, $D_H^-(v_5) = D_H^+(v_5)$. Hence, H is regular 3-polar fuzzy directed hypergraph.

Definition 5.30 An m-polar fuzzy directed hyperpath of length k in an m-polar fuzzy directed hypergraph is defined as a sequence $v_1, e_1, v_2, e_2, \ldots, e_k, v_{k+1}$ of distinct vertices and directed hyperedges such that

1. $\varepsilon(e_i) > 0, i = 1, 2, \ldots, k$,
2. $v_i, v_{i+1} \in e_i$.

The consecutive pairs (v_i, v_{i+1}) are called the directed arcs of the directed hyperpath. The path is shown by a thick line in Fig. 5.29.

Definition 5.31 The incidence matrix of an m-polar fuzzy directed hypergraph $H = (\sigma, \varepsilon)$ is characterized by an $n \times m$ matrix $[a_{ij}]$ as follows:

$$a_{ij} = \begin{cases} P_k o \varepsilon_j(v_i), & \text{if } v_i \in \varepsilon_j, \\ 0, & \text{otherwise.} \end{cases}$$

Definition 5.32 An m-polar fuzzy directed hypergraph is called elementary if $P_k o \varepsilon_{ij} : V \longrightarrow [0, 1]^m$ are constant functions, $P_k o \varepsilon_{ij}$ is taken as the membership degree of vertex i to hyperedge j.

Proposition 5.1 *In an m-polar fuzzy directed hypergraph, when m-polar fuzzy vertices have constant membership degrees, then m-polar fuzzy directed hyperedges are elementary.*

Example 5.19 Consider a 3-polar fuzzy directed hypergraph $H = (\sigma, \varepsilon)$, where $\sigma = \{\sigma_1, \sigma_2, \sigma_3\}$ be the family of 3-polar fuzzy subsets on $V = \{v_1, v_2, v_3, v_4, v_5\}$. The corresponding incidence matrix is given in Table 5.13.

The corresponding elementary 3-polar fuzzy directed hypergraph is shown in Fig. 5.30.

Table 5.13 Elementary 3-polar fuzzy directed hypergraph

I	ε_1	ε_2	ε_3
v_1	(0.1, 0.2, 0.3)	0	(0.1, 0.2, 0.3)
v_2	(0.1, 0.2, 0.3)	0	(0.1, 0.2, 0.3)
v_3	(0.1, 0.2, 0.3)	(0.1, 0.2, 0.3)	0
v_4	0	(0.1, 0.2, 0.3)	(0.1, 0.2, 0.3)
v_5	0	(0.1, 0.2, 0.3)	0

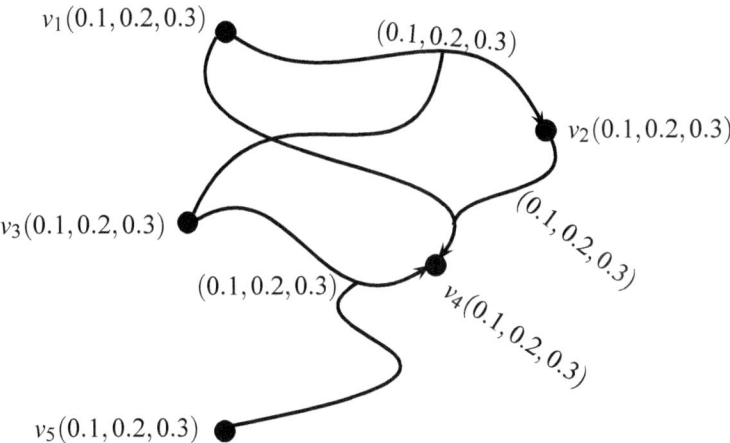

Fig. 5.30 Elementary 3-polar fuzzy directed hypergraph

Definition 5.33 Let $H = (\sigma, \varepsilon)$ be an m-polar fuzzy directed hypergraph. Suppose $\mu = (\mu_1, \mu_2, ..., \mu_m) \in [0, 1]^m$. The μ-level is defined as $\varepsilon_\mu = \{v \in \sigma \mid P_k o\sigma(v) \geq \mu_k\}$. The crisp directed hypergraph $H_\mu = (\sigma_\mu, \varepsilon_\mu)$, such that

- $\varepsilon_\mu = \{v \in \sigma \mid P_k o\sigma(v) \geq \mu_k\}$, $1 \leq k \leq m$.
- $\sigma_\mu = \bigcup \varepsilon_\mu$,

is called the μ-level directed hypergraph of H.

Definition 5.34 Let $H = (\sigma, \varepsilon)$ be an m-polar fuzzy directed hypergraph and $H_{\mu_i} = (\sigma_{\mu_i}, \varepsilon_{\mu_i})$ be the μ_i-level directed hypergraphs of H. The sequence $\{\mu_1, \mu_2, \mu_3, ..., \mu_n\}$ of m-tuples, where $\mu_1 > \mu_2 > ...\mu_n > 0$ and $\mu_n = h(H)$(height of m-polar fuzzy directed hypergraph), such that the following properties,

1. if $\mu_{i+1} < \alpha \leq \mu_i$, then $\varepsilon_\alpha = \varepsilon_{\mu_i}$,
2. $\varepsilon_{\mu_i} \sqsubset \varepsilon_{\mu_{i+1}}$,

are satisfied, is called a *fundamental sequence* of H. The sequence is denoted by $FS(H)$. The μ_i-level hypergraphs $\{H_{\mu_1}, H_{\mu_2}, ..., H_{\mu_n}\}$ are called the core hypergraphs of H. This is also called core set of H and is denoted by $c(H)$.

Table 5.14 3-polar fuzzy directed hypergraph

I	ε_1	ε_2	ε_3
v_1	(0.8, 0.6, 0.1)	0	0
v_2	(0.8, 0.6, 0.5)	(0.6, 0.4, 0.3)	(0.5, 0.3, 0.2)
v_3	(0.8, 0.6, 0.5)	(0.6, 0.4, 0.3)	(0.5, 0.3, 0.2)
v_4	0	(0.6, 0.4, 0.1)	0
v_5	0	0	(0.5, 0.3, 0.2)
v_6	0	0	(0.5, 0.3, 0.2)

Fig. 5.31 3-polar fuzzy directed hypergraph

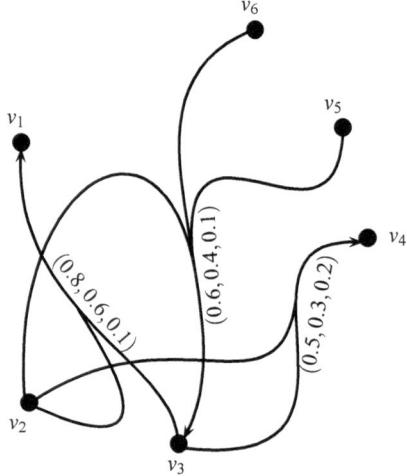

Definition 5.35 Let $H = (\sigma, \varepsilon)$ be an m-polar fuzzy directed hypergraph and $FS(H) = \{\mu_1, \mu_2, \mu_3, ..., \mu_n\}$. If for each $e \in \varepsilon$ and each $\mu_i \in FS(H)$, $e_\mu = \varepsilon_{\mu_i}$, for all $\mu \in (\mu_{i+1}, \mu_i]$, then H is called *sectionally elementary*.

Definition 5.36 Let $H = (\sigma, \varepsilon)$ be an m-polar fuzzy directed hypergraph and $c(H) = \{H_{\mu_1}, H_{\mu_2}, ..., H_{\mu_n}\}$. H is said to be *ordered* if $c(H)$ is ordered. That is, $H_{\mu_1} \subset H_{\mu_2} \subset ... \subset H_{\mu_n}$. The m-polar fuzzy directed hypergraph is called simply ordered if the sequence $\{H_{\mu_1}, H_{\mu_2}, ..., H_{\mu_n}\}$ is simply ordered.

Example 5.20 Consider a 3-polar fuzzy directed hypergraph $H = (\sigma, \varepsilon)$ as shown in Fig. 5.31 and given by incidence matrix in Table 5.14.

By computing the μ_i-level 3-polar fuzzy directed hypergraphs of H, we have $\varepsilon_{(0.8,0.6,0.5)} = \{v_2, v_3\}$, $\varepsilon_{(0.6,0.4,0.3)} = \{v_2, v_3\}$ and $\varepsilon_{(0.5,0.3,0.2)} = \{v_2, v_3, v_5, v_6\}$. Note that, $H_{(0.8,0.6,0.5)} = H_{(0.6,0.4,0.3)}$ and $H_{(0.8,0.6,0.5)} \subseteq H_{(0.5,0.3,0.2)}$. The fundamental sequence is $FS(H) = \{(0.8, 0.6, 0.5), (0.5, 0.3, 0.2)\}$. Furthermore, $H_{(0.8,0.6,0.5)} \neq H_{(0.6,0.4,0.3)}$. H is not sectionally elementary since $\varepsilon_{2(\mu)} \neq \varepsilon_{2(0.8,0.6,0.5)}$ for $\mu = (0.6, 0.4, 0.3)$. The 3-polar fuzzy directed hypergraph is ordered, and the set of core

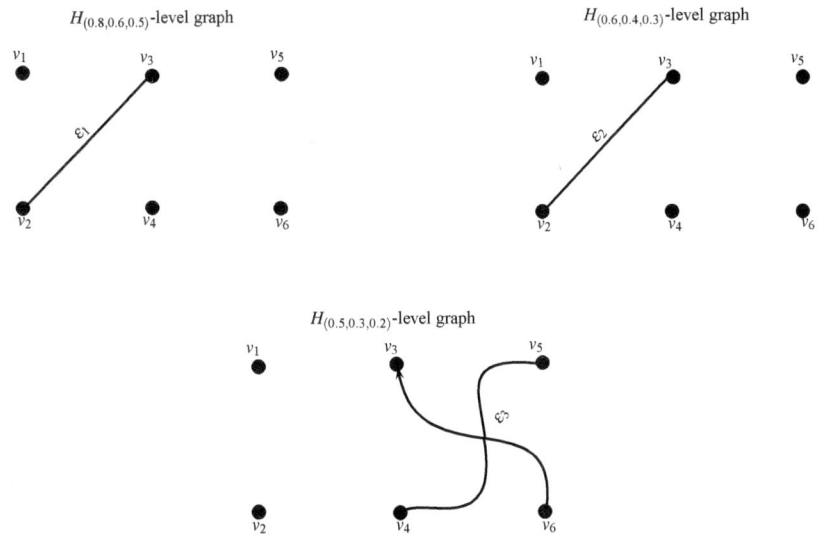

Fig. 5.32 H induced fundamental sequence

Table 5.15 Index matrix of an *m*-polar fuzzy hypergraph

I	t_1	t_2	$\cdots$	t_n
t_1	$\varepsilon(t_1 t_1)$	$\varepsilon(t_1 t_2)$	$\cdots$	$\varepsilon(t_1 t_n)$
t_2	$\varepsilon(t_2 t_1)$	$\varepsilon(t_2 t_2)$	$\cdots$	$\varepsilon(t_2 t_n)$
.	.	.	.	.
.	.	.	.	.
.	.	.	.	.
t_n	$\varepsilon(t_n t_1)$	$\varepsilon(t_n t_2)$	$\cdots$	$\varepsilon(t_n t_n)$

hypergraphs is $c(H) = \{H_1 = H_{(0.8,0.6,0.5)}, H_2 = H_{(0.5,0.3,0.2)}\}$. The induced fundamental sequence of H is given in Fig. 5.32 (Table 5.15).

Proposition 5.2 Let $H = (\sigma, \varepsilon)$ be an m-polar fuzzy directed hypergraph, the following conditions hold

(a) If $H = (\sigma, \varepsilon)$ is an elementary m-polar fuzzy directed hypergraph, then H is ordered.
(b) If H is an ordered m-polar fuzzy directed hypergraph with $c(H) = \{H_{\mu_1}, H_{\mu_2}, ..., H_{\mu_n}\}$ and if H_{μ_n} is simple, then H is elementary.

Definition 5.37 Let $H = (\sigma, \varepsilon)$ be an *m*-polar fuzzy directed hypergraph. The index matrix of H is defined by

Now we present certain operations on *m*-polar fuzzy directed hypergraphs.

Fig. 5.33 3-polar fuzzy
directed hypergraph H_1

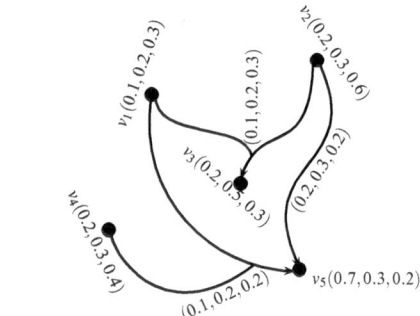

Fig. 5.34 3-polar fuzzy
directed hypergraph H_2

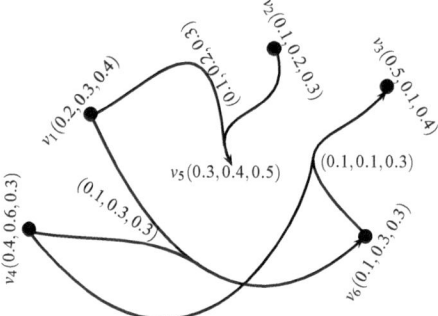

Definition 5.38 Let $H_1 = (\sigma_1, \varepsilon_1)$ and $H_2 = (\sigma_2, \varepsilon_2)$ be two m-polar fuzzy directed
hypergraphs. The *addition* of two m-polar fuzzy directed hypergraphs over a fixed
set X is denoted by $H_1 \boxplus H_2 = (\sigma_1 \cup \sigma_2, \varepsilon_1 \cup \varepsilon_2)$ and defined as

$$
P_k o (\sigma_1 \cup \sigma_2)(v_r) = \begin{cases}
P_k o \sigma_1(v_r), & \text{if } v_r \in \sigma_1 \setminus \sigma_2, \\
P_k o \sigma_2(v_r), & \text{if } v_r \in \sigma_2 \setminus \sigma_1, \\
\sup\{P_k o \sigma_1(v_r), P_k o \sigma_2(v_r)\}, & \text{if } v_r \in \sigma_1 \cap \sigma_2, \\
0, & \text{otherwise.}
\end{cases}
\tag{5.1}
$$

$$
P_k o (\varepsilon_1 \cup \varepsilon_2)(e_{rs}) = \begin{cases}
P_k o \varepsilon_1(e_{ij}), & \text{if } v_r = v_i \in \sigma_1 \text{ and } v_s = v_j \in \sigma_1 \setminus \sigma_2, \\
P_k o \varepsilon_2(e_{pq}), & \text{if } v_r = v_p \in \sigma_2 \text{ and } v_s = v_q \in \sigma_2 \setminus \sigma_1, \\
\sup\{P_k o \varepsilon_1(e_{ij}), P_k o \varepsilon_2(e_{pq})\}, & \text{if } v_r = v_i = v_p \in \sigma_1 \cap \sigma_2, v_s = v_j = v_q \in \sigma_1 \cap \sigma_2, \\
0, & \text{otherwise.}
\end{cases}
\tag{5.2}
$$

Example 5.21 Let $H_1 = (\sigma_1, \varepsilon_1)$ and $H_2 = (\sigma_2, \varepsilon_2)$ be two 3-polar fuzzy directed
hypergraphs, where $\sigma_1 = \{v_1, v_2, ..., v_5\}$, $\varepsilon_1 = \{(\{v_1, v_2\}, v_3), (\{v_1, v_4\}, v_5), \{\{v_2\},$
$v_5\}$ and $\sigma_2 = \{v_1, v_2, ..., v_6\}$, $\varepsilon_2 = \{((\{v_1, v_2\}, v_5), (\{v_4, v_6\}, v_3), \{\{v_1, v_4\}, v_6\}$ as
shown in Figs. 5.33 and 5.34, respectively.

The index matrix of H_1 is given in Table 5.16, where $\sigma_1 = \{v_1, v_2, ..., v_5\}$.
The index matrix of H_2 is given in Table 5.17, where $\sigma_2 = \{v_1, v_2, ..., v_6\}$.

Table 5.16 Index matrix of H_1

I	v_1	v_2	v_3	v_4	v_5
v_1	0	0	0	0	0
v_2	0	0	0	0	0
v_3	$(0.1, 0.2, 0.3)$	$(0.1, 0.2, 0.3)$	0	0	0
v_4	0	0	0	0	0
v_5	$(0.1, 0.2, 0.2)$	$(0.2, 0.3, 0.2)$	0	$(0.1, 0.2, 0.2)$	0

Table 5.17 Index matrix of H_2

I	v_1	v_2	v_3	v_4	v_5	v_6
v_1	0	0	0	0	0	0
v_2	0	0	0	0	0	0
v_3	$(0.1, 0.2, 0.3)$	$(0.1, 0.2, 0.3)$	0	$(0.1, 0.1, 0.3)$	0	$(0.1, 0.1, 0.3)$
v_4	0	0	0	0	0	0
v_5	$(0.1, 0.2, 0.2)$	0	0	$(0.1, 0.2, 0.2)$	0	0
v_6	$(0.1, 0.3, 0.3)$	0	0	$(0.1, 0.3, 0.3)$	0	0

Table 5.18 Index matrix of $H_1 \boxplus H_2$

$H_1 \boxplus H_2$	v_1	v_2	v_3	v_4	v_5	v_6
v_1	0	0	0	0	0	0
v_2	0	0	0	0	0	0
v_3	$(0.1, 0.2, 0.3)$	$(0.1, 0.2, 0.3)$	0	$(0.1, 0.1, 0.3)$	0	$(0.1, 0.1, 0.3)$
v_4	0	0	0	0	0	0
v_5	$(0.1, 0.2, 0.2)$	0	0	$(0.1, 0.2, 0.2)$	0	0
v_6	$(0.1, 0.3, 0.3)$	0	0	$(0.1, 0.3, 0.3)$	0	0

The index matrix of $H_1 \boxplus H_2$ is given in Table 5.18, where $\sigma_1 \cup \sigma_2 = \{v_1, v_2, ...,$ $v_6\}$. The corresponding hypergraph is shown in Fig. 5.35.

Definition 5.39 Let $H_1 = (\sigma_1, \varepsilon_1)$ and $H_2 = (\sigma_2, \varepsilon_2)$ be two *m*-polar fuzzy directed hypergraphs. The vertex-wise multiplication of two *m*-polar fuzzy directed hypergraphs over a fixed set V is denoted by $H_1 \otimes H_2 = (\sigma_1 \otimes \sigma_2, \varepsilon_1 \otimes \varepsilon_2)$ and defined as

$$P_k o(\sigma_1 \otimes \sigma_2) = \inf\{P_k o\sigma_1(v_r), P_k o\sigma_2(v_r)\} \text{ if } v_r \in \sigma_1 \cap \sigma_2, \tag{5.3}$$

$$P_k o(\varepsilon_1 \otimes \varepsilon_2)(e_{rs})$$
$$= \inf\{P_k o\varepsilon_1(e_{ij}), P_k o\sigma_2(e_{pq})\} \text{ if } v_r = v_i = v_p \in \sigma_1 \cap \sigma_2, v_s = v_j = v_q \in \sigma_1 \cap \sigma_2. \tag{5.4}$$

Fig. 5.35 $H_1 \boxplus H_2$

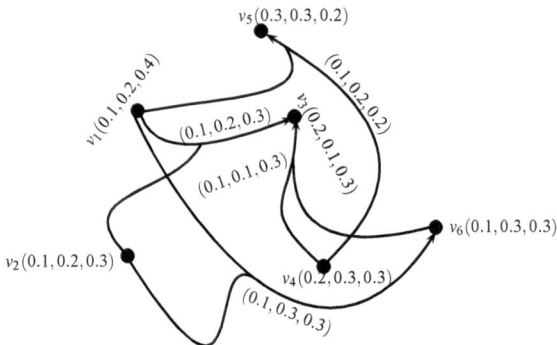

Table 5.19 Index matrix of $H_1 \otimes H_2$

$H_1 \otimes H_2$	v_1	v_2	v_3	v_4	v_5
v_1	0	0	0	0	0
v_2	0	0	0	0	0
v_3	0	0	0	0	0
v_4	0	0	0	0	0
v_5	$(0.1, 0.2, 0.2)$	$(0.1, 0.2, 0.3)$	0	0	0

Fig. 5.36 $H_1 \otimes H_2$

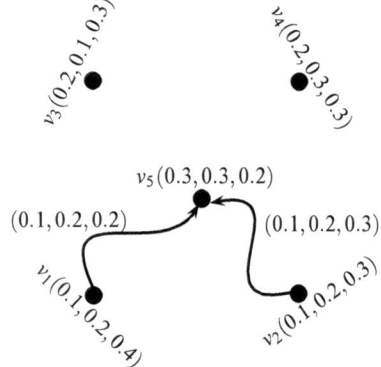

Example 5.22 Let $H_1 = (\sigma_1, \varepsilon_1)$ and $H_2 = (\sigma_2, \varepsilon_2)$ be two 3-polar fuzzy directed hypergraphs as shown in Figs. 5.33 and 5.34, respectively. The index matrix of $H_1 \otimes H_2$ is shown in Table 5.19, where $\sigma_1 \cap \sigma_2 = \{v_1, v_2, ..., v_5\}$.

The graph of $H_1 \otimes H_2$ is shown in the Fig. 5.36.

Definition 5.40 Let $H_1 = (\sigma_1, \varepsilon_1)$ and $H_2 = (\sigma_2, \varepsilon_2)$ be two m-polar fuzzy directed hypergraphs. The structural subtraction of two m-polar fuzzy directed hypergraphs over a fixed set V is denoted by $H_1 \boxminus H_2 = (\sigma_2 - \sigma_1, \varepsilon_2 - \varepsilon_1)$ and defined as

Table 5.20 Index matrix of
$H_1 \boxminus H_2$

$H_1 \boxminus H_2$	v_6
v_6	0

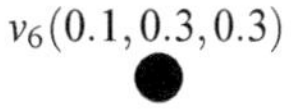

Fig. 5.37 $H_1 \boxminus H_2$ $v_6(0.1, 0.3, 0.3)$

$$P_k o(\sigma_2 - \sigma_1)(v_r) = \begin{cases} P_k o \sigma_1(v_r), & \text{if } v_r \in \sigma_1, \\ P_k o \sigma_2(v_r), & \text{if } v_r \in \sigma_2, \\ 0, & \text{otherwise.} \end{cases} \quad (5.5)$$

$$P_k o(\varepsilon_2 - \varepsilon_1)(e_{rs}) = P_k o \varepsilon_1(e_{ij}) \text{ if } v_r = v_i \in \sigma_2 - \sigma_1 \text{ and } v_s = v_j \in \sigma_2 - \sigma_1. \quad (5.6)$$

The graph $H_1 \boxminus H_2$ is empty when $\sigma_2 - \sigma_1 = \emptyset$.

Example 5.23 Let $H_1 = (\sigma_1, \varepsilon_1)$ and $H_2 = (\sigma_2, \varepsilon_2)$ be two 3-polar fuzzy directed hypergraphs as shown in Figs. 5.33 and 5.34, respectively. The index matrix of $H_1 \boxminus H_2$ is shown in Table 5.20, where $\sigma_2 - \sigma_1 = \{v_6\}$.

The graph $H_1 \boxminus H_2$ is shown in the following Fig. 5.37

Definition 5.41 Let $H_1 = (\sigma_1, \varepsilon_1)$ and $H_2 = (\sigma_2, \varepsilon_2)$ be two *m*-polar fuzzy directed hypergraphs. The multiplication of two *m*-polar fuzzy directed hypergraphs H_1 and H_2, denoted by $H_1 \odot H_2 = (\sigma_1 \odot \sigma_2, \varepsilon_1 \odot \varepsilon_2)$ is defined as

$$P_k o(\sigma_1 \odot \sigma_2)(v_r) = \begin{cases} P_k o \sigma_1(v_r), & \text{if } v_r \in \sigma_1, \\ P_k o \sigma_2(v_r), & \text{if } v_r \in \sigma_2, \\ \inf\{P_k o \sigma_1(v_r), P_k o \sigma_2(v_r)\}, & \text{if } v_r \in \sigma_1 \cap \sigma_2. \end{cases} \quad (5.7)$$

$P_k o(\varepsilon_1 \odot \varepsilon_2)(e_{rs})$

$$= \begin{cases} P_k o \varepsilon_1(e_{ij}), & \text{if } v_r = v_i \in \sigma_1 \text{ and } v_s = v_j \in \sigma_1 \setminus \sigma_2, \\ P_k o \varepsilon_2(e_{pq}), & \text{if } v_r = v_p \in \sigma_2 \text{ and } v_s = v_q \in \sigma_2 \setminus \sigma_1, \\ \sup_{i,q} \{\inf_{j,p}\{P_k o \varepsilon_1(e_{ij}), P_k o \varepsilon_2(e_{pq})\}\}, & \text{if } v_r = v_i \in \sigma_1 \cap \sigma_2 \text{ and } v_s = v_q \in \sigma_1 \cap \sigma_2. \end{cases} \quad (5.8)$$

Table 5.21 Index matrix of $H_1 \odot H_2$

$H_1 \odot H_2$	v_1	v_2	v_3	v_4	v_5	v_6
v_1	0	0	0	0	0	0
v_2	0	0	0	0	0	0
v_3	$(0.1, 0.2, 0.3)$	$(0.1, 0.2, 0.3)$	0	$(0.1, 0.2, 0.3)$	0	$(0.1, 0.1, 0.3)$
v_4	0	0	0	0	0	0
v_5	$(0.1, 0.2, 0.2)$	$(0.1, 0.2, 0.2)$	0	$(0.1, 0.2, 0.2)$	0	0
v_6	$(0.1, 0.3, 0.3)$	0	0	$(0.1, 0.3, 0.3)$	0	0

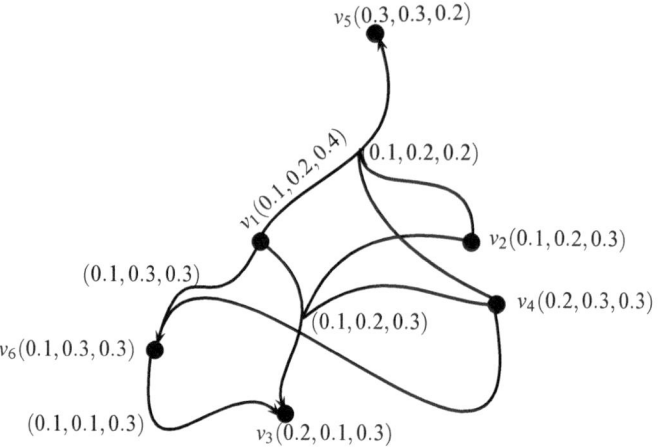

Fig. 5.38 $H_1 \odot H_2$

Example 5.24 The index matrix of graph $H_1 \odot H_2$ is shown in Table 5.21, where $\sigma_2 \cup (\sigma_1 - \sigma_2) = \{v_1, v_2, v_3, ..., v_6\}$ is given in Table 5.20.

The corresponding hypergraph is shown in Fig. 5.38.

5.5   Application of m-Polar Fuzzy Directed Hypergraphs

Decision-making is regarded as the intellectual process resulting in the selection of a belief or a course of action among several alternative possibilities. Every decision-making process produces a final choice, which may or may not prompt action. Decision-making is the process of identifying and choosing alternatives based on the values, preferences, and beliefs of the decision-maker. Problems in almost every credible discipline, including decision-making can be handled using graphical models.

5.5.1   Business Strategy Company

A business strategy is a registered plan on how an organization is setting out to fulfill their ambitions. A business strategy has a variety of successful key of principles that sketch how a company will go about achieving their dreams in business. It deals with competitors, look at their needs and expectations of customers and will examine the long-term growth and sustainability of their organization.

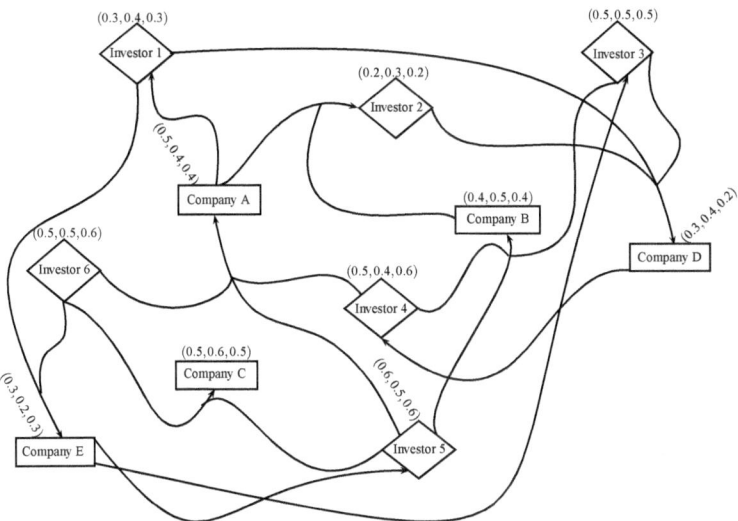

Fig. 5.39 3-polar fuzzy directed hypergraph model

Table 5.22 Collective interest of investors toward companies

Business strategy company	Positive effects of investors
Company A	(0.5, 0.4, 0.6)
Company B	(0.5, 0.4, 0.5)
Company C	(0.5, 0.5, 0.6)
Company D	(0.2, 0.3, 0.2)
Company E	(0.3, 0.4, 0.3)

In this fast running world where every investor is searching out a best business strategy company so that they invest their money on the company to promote the business and to compete their competitors. Then to select a good marketing business company which will achieve its goals, meet the expectations and sustain a competitive advantage in the marketplace, we develop a 3-polar fuzzy directed hypergraphical model that how an investor can choice the greatest salubrious company to promote the business by following a step by step procedure. A 3-polar fuzzy directed hypergraph demonstrating a group of investors as members of different business strategy companies is shown in Fig. 5.39.

If an investor wants to adopt the most suitable and powerful business company to which he works and get the progress in business, the following procedure can help the investors. Firstly, one should think about the cooperative contribution of investors toward the company, which can be found out by means of membership values of 3-polar fuzzy directed hypergraphs. The membership values given in Table 5.22 shows the collective interest of investors toward the company.

Table 5.23 Benefits of company on the investors

Business strategy company	Effects of company on investors
Company A	$(0.5, 0.4, 0.4)$
Company B	$(0.4, 0.5, 0.4)$
Company C	$(0.5, 0.6, 0.5)$
Company D	$(0.3, 0.4, 0.2)$
Company E	$(0.3, 0.2, 0.3)$

Table 5.24 In-degrees and out-degrees of companies

Business strategy company	In-degrees	out-degrees
Company A	$(0.5, 0.4, 0.4)$	$(0.5, 0.7, 0.5)$
Company B	$(0.4, 0.4, 0.4)$	$(0.2, 0.3, 0.2)$
Company C	$(0.5, 0.5, 0.5)$	$(0, 0, 0)$
Company D	$(0.2, 0.3, 0.2)$	$(0.3, 0.4, 0.3)$
Company E	$(0.3, 0.2, 0.3)$	$(0.6, 0.4, 0.6)$

The first membership value showing how much investors invest money on company, second showing the sharp-minded quality of investors to run the business and third showing how can strongly they make production by working with company. It can be noticed that the company C has strong collective interest in investors which is maximum among all other companies. Secondly, one should do his research on the powerful impacts of all under consideration companies on their investors. The membership degrees of all company nodes show their effects on their investors as given in Table 5.23.

The membership values showing three different positive effects of company on investor, first one shows how much a company is financially strong already, second showing its business growth in the market, and third one showing the strong competitive position of company. Note that, company C has the most benefits for investors. Thirdly, an investor can observe the influence of a company by calculating its in-degrees and out-degrees. In-degrees show the percentage of investors joining the company and out-degrees show the percentage of investors leaving that company. The in-degrees and out-degrees of all business strategy companies are given in Table 5.24.

Hence, a best business strategy company has maximum in-degrees and minimum out-degrees. However, in case when two companies have same minimum out-degrees, then we compare their in-degrees. Similarly, when in-degrees same, we compare out-degrees. From all the above discussion, we conclude that company C is the most appropriate company to fulfill the requirements of the investors because it is more financially strong, best in competitive position and business growth of this company is more suitable to run the business and compete with the competitors. The method of searching out the constructive and profitable business strategy company is explained in the following Algorithm 5.5.1.

Algorithm 5.5.1   To find out the constructive and profitable business strategy company
1. Input the membership values of all nodes(investors) $v_1, v_2, ..., v_n$.
2. Determine the augmentation of investors toward companies by calculating the membership values of all directed hyperedges as

$$P_k o \varepsilon_r \leq \inf\{P_k o v_1, P_k o v_2, ..., P_k o v_n\}, 1 \leq k \leq m.$$

3. Obtain the most suitable company as

$$\sup P_k o \varepsilon_r.$$

4. Find the company having strong and more benefits for investors as,

$$\sup P_k o v_r,$$

where all v_r here are vertices represent the different business strategy company.
5. Find the profitable influence of companies v_r on the investors by calculating the in-degrees $D^-(v_r)$ as

$$\sum_{v_r \in h(\varepsilon_r)} P_k o \varepsilon_r.$$

6. Find the profitless impact of companies v_k on the investors by calculating the out-degrees $D^+(v_r)$ as,

$$\sum_{v_r \in t(\varepsilon_r)} P_k o \varepsilon_r.$$

7. Obtain the most advantageous business strategy company as

$$(\sup D^-(v_r), \inf D^+(v_r)).$$

The algorithm runs linearly and its net time complexity is $\bigcirc(n)$, where n is the number of membership values of all nodes(investors).

References

1. Akram, M.: Bipolar fuzzy graphs. Inf. Sci. **181**, 5548–5564 (2011)
2. Akram, M.: Bipolar fuzzy graphs with applications. Knowl. Based Syst. **39**, 1–8 (2013)
3. Akram, M.: *m*-polar fuzzy graphs: theory, methods & applications. In: Studies in Fuzziness and Soft Computing, vol. 371, pp. 1-284. Springer (2019)

4. Akram, M., Dudek, W.A., Sarwar, S.: Properties of bipolar fuzzy hypergraphs. Ital. J. Pure Appl. Math. **31**, 141–160 (2013)
5. Akram, M., Sarwar, M.: Novel applications of m-polar fuzzy hypergraphs. J. Intell. Fuzzy Syst. **32**(3), 2747–2762 (2016)
6. Akram, M., Sarwar, M.: Transversals of m-polar fuzzy hypergraphs with applications. J. Intell. Fuzzy Syst. **33**(1), 351–364 (2017)
7. Akram, M., Shahzadi, G.: Hypergraphs in m-polar fuzzy environment. Mathematics **6**(2), 28 (2018). https://doi.org/10.3390/math6020028
8. Akram, M., Shahzadi, G.: Directed hypergraphs under m-polar fuzzy environment. J. Intell. Fuzzy Syst. **34**(6), 4127–4137 (2018)
9. Akram, M., Shahzadi, G., Shum, K.P.: Operations on m-polar fuzzy r-uniform hypergraphs. Southeast Asian Bull. Math. **44** (2020)
10. Berge, C.: Graphs and Hypergraphs. North-Holland, Amsterdam (1973)
11. Chen, S.M.: Interval-valued fuzzy hypergraph and fuzzy partition. IEEE Trans. Syst. Man Cybern. (Cybern.) **27**(4), 725–733 (1997)
12. Chen, J., Li, S., Ma, S., Wang, X.: m-polar fuzzy sets: An extension of bipolar fuzzy sets. Sci. World J. **8**, (2014). https://doi.org/10.1155/2014/416530
13. Gallo, G., Longo, G., Pallottino, S.: Directed hypergraphs and applications. Discret. Appl. Math. **42**, 177–201 (1993)
14. Kaufmann, A.: Introduction a la Thiorie des Sous-Ensemble Flous, 1. Masson, Paris (1977)
15. Lee-kwang, H., Lee, K.-M.: Fuzzy hypergraph and fuzzy partition. IEEE Trans. Syst. Man Cybern. **25**(1), 196–201 (1995)
16. Mordeson, J.N., Nair, P.S.: Fuzzy Graphs and Fuzzy Hypergraphs, 2nd edn. Physica Verlag, Heidelberg (2001)
17. Parvathi, R., Thilagavathi, S., Karunambigai, M.G.: Intuitionistic fuzzy hypergraphs. Cybern. Inf. Technol. **9**(2), 46–53 (2009)
18. Rosenfeld, A.: Fuzzy graphs. In: Zadeh, L.A., Fu, K.S., Shimura, M. (eds.) Fuzzy Sets and their Applications, pp. 77–95. Academic Press, New York (1975)
19. Zadeh, L.A.: Fuzzy sets. Inf. Control **8**(3), 338–353 (1965)
20. Zhang, W.R.: Bipolar fuzzy sets and relations: a computational framework forcognitive modeling and multiagent decision analysis. In: Proceedings of IEEE Conference, pp. 305–309 (1994)

Chapter 6
(Directed) Hypergraphs: q-Rung Orthopair Fuzzy Models and Beyond

A q-rung orthopair fuzzy set is a powerful tool for depicting fuzziness and uncertainty, as compared to the Pythagorean fuzzy model. In this chapter, we present concepts including q-rung orthopair fuzzy hypergraphs, (α, β)-level hypergraphs, and transversals and minimal transversals of q-rung orthopair fuzzy hypergraphs. We implement some interesting notions of q-rung orthopair fuzzy hypergraphs into decision-making. We describe additional concepts like q-rung orthopair fuzzy directed hypergraphs, dual directed hypergraphs, line graphs, and coloring of q-rung orthopair fuzzy directed hypergraphs. We also apply other interesting notions of q-rung orthopair fuzzy directed hypergraphs to real life problems. We introduce complex q-rung orthopair fuzzy graphs, complex Pythagorean fuzzy hypergraphs, and complex q-rung orthopair fuzzy hypergraphs. We study the transversals and minimal transversals of complex q-rung orthopair fuzzy hypergraphs. We present some algorithms to construct the minimal transversals and certain related concepts. Finally, we illustrate a collaboration network model through complex q-rung orthopair fuzzy hypergraphs to find the author having powerful collaboration skills using score and choice values. This chapter is basically due to [22–24, 35].

6.1 Introduction

Zadeh [37] proposed the notion of fuzzy sets in his monumental paper in 1965, to model uncertainty or vague ideas by nominating a degree of membership to each entity, ranging between 0 and 1. In 1983, intuitionistic fuzzy sets, primarily proposed by Atanassov [14], offered many significant advantages in representing human knowledge by denoting fuzzy membership not only with a single value but pairs of mutually orthogonal fuzzy sets called orthopairs, which allow the incorporation of uncertainty. Since intuitionistic fuzzy sets confine the selection of orthopairs to come only from a triangular region, as shown in Fig. 6.1, Pythagorean fuzzy sets, proposed

© Springer Nature Singapore Pte Ltd. 2020
M. Akram and A. Luqman, *Fuzzy Hypergraphs and Related Extensions*,
Studies in Fuzziness and Soft Computing 390,
https://doi.org/10.1007/978-981-15-2403-5_6

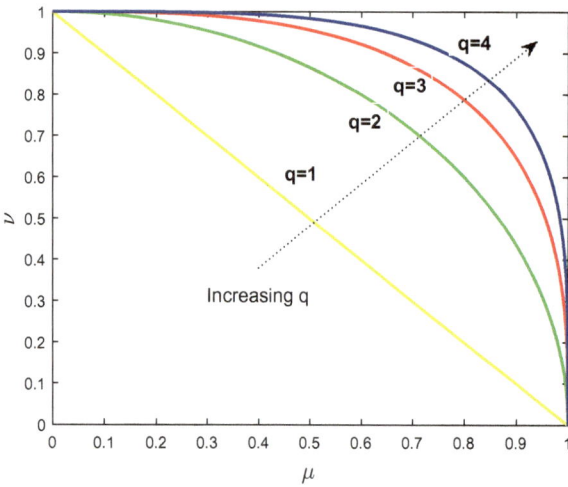

Fig. 6.1 Spaces of acceptable q-rung orthopairs

by Yager [32], as a new extension of intuitionistic fuzzy sets have emerged as an efficient tool for conducting uncertainty more properly in human analysis. Although both intuitionistic fuzzy sets and Pythagorean fuzzy sets make use of orthopairs to narrate assessment objects, they still have visible differences. The truth-membership function $T : X \rightarrow [0, 1]$ and falsity-membership function $F : X \rightarrow [0, 1]$ of intuitionistic fuzzy sets are required to satisfy the constraint condition $T(x) + F(x) \leq 1$. However, these two functions in Pythagorean fuzzy sets are needed to satisfy the condition $T(x)^2 + F(x)^2 \leq 1$, which shows that Pythagorean fuzzy sets have expanded space to assign orthopairs, as compared to intuitionistic fuzzy sets, displayed in Fig. 6.1.

A q-rung orthopair fuzzy set, originally proposed by Yager [35] in 2017, is a new generalization of orthopair fuzzy sets, which further relax the constraint of orthopair membership grades with $T(x)^q + F(x)^q \leq 1$ $(q \geq 1)$ [21]. As q increases, it is easy to see that the representation space of allowable orthopair membership grade increases. Figure 6.1 displays spaces of the most widely acceptable orthopairs for different q rungs. Ali [12] calculated the area of spaces with admissible orthopairs up to 10-rungs. Consider an example in the field of economics: in a market structure, a huge number of firms compete against each other with differentiated products with respect to branding or quality, which in nature are vague words. Since intuitionistic fuzzy sets have the capability to explore both aspects of ambiguous words, for example, it assigns an orthopair membership grade to "quality", i.e., support for quality and support for not-quality of an object with the condition that their sum is bounded by 1. This constraint clearly limits the selection of orthopairs.

The innovative concept of complex fuzzy sets was initiated by Ramot et al. [28] as an extension of fuzzy sets. Opposing to a fuzzy characteristic function, the range of complex fuzzy set's membership degrees is not restricted to [0, 1], but extends to the complex plane with the unit circle. Ramot et al. [29] discussed the union, intersection, and compliment of complex fuzzy sets with the help of illustrative examples. To generalize the concepts of intuitionistic fuzzy sets, complex intuitionistic fuzzy sets were introduced by Alkouri and Salleh [13]. As an extension of Pythagorean fuzzy sets and complex intuitionistic fuzzy sets, Ullah et al. [31] proposed complex Pythagorean fuzzy sets and discussed some applications. In complex Pythagorean fuzzy sets, membership $\mu = ue^{i\alpha}$ and nonmembership $v = ve^{i\beta}$ can take values in the unit circle subjected to the constraint $\mu^2 + v^2 \leq 1$. Complex Pythagorean fuzzy model, containing the phase term, is a more effective tool to capture the vague and uncertain data of periodic nature than the Pythagorean fuzzy model.

Definition 6.1 A *q-rung orthopair fuzzy set* Q in the universe X is an object having the representation

$$Q = (x, T_Q(x), F_Q(x)|x \in X),$$

where the function $T_Q : X \rightarrow [0, 1]$ defines the truth-membership and $F_Q : X \rightarrow [0, 1]$ defines the falsity-membership of the element $x \in X$ and for every $x \in X$, $0 \leq T_Q^q(x) + F_Q^q(x) \leq 1, q \geq 1$.

Furthermore, $\pi_Q(x) = \sqrt[q]{1 - T_Q^q(x) - F_Q^q(x)}$ is called a q-rung orthopair fuzzy index or indeterminacy degree of x to the set Q.

For convenience, Liu and Wang [21] called the pair $(T_Q^q(x), F_Q^q(x))$ as a q-rung orthopair fuzzy number, which is denoted as (T_Q^q, F_Q^q).

Definition 6.2 A *q-rung orthopair fuzzy relation* $\mathcal{R}$ in X is defined as $\mathcal{R} = \{x_1x_2, T_{\mathcal{R}}(x_1x_2), F_{\mathcal{R}}(x_1x_2)|x_1, x_2 \in X \times X\}$, where $T_{\mathcal{R}} : X \times X \rightarrow [0, 1]$ and $F_{\mathcal{R}} : X \times X \rightarrow [0, 1]$ represent the truth-membership and falsity-membership function of $\mathcal{R}$, respectively, such that $0 \leq T_{\mathcal{R}}^q(x_1x_2) + F_{\mathcal{R}}^q(x_1x_2) \leq 1$, for all $x_1x_2 \in X \times X$.

Example 6.1 Let $X = \{x_1, x_2, x_3\}$ be a non-empty set and $\mathcal{R}$ be a subset of $X \times X$ such that $\mathcal{R} = \{(x_1x_2, 0.9, 0.7),(x_1x_3, 0.7, 0.9), (x_2x_3, 0.6, 0.8)\}$. Note that, $0 \leq T_{\mathcal{R}}^5(x_1x_2) + F_{\mathcal{R}}^5(x_1x_2) \leq 1$, for all $x_1x_2 \in X \times X$. Hence, $\mathcal{R}$ is a 5-rung orthopair fuzzy relation on X.

For further terminologies and studies on Pythagorean fuzzy graphs and q-rung orthopair fuzzy graphs, readers are referred to [1–11, 15–20, 25–27, 30, 33, 34, 36].

6.2 q-Rung Orthopair Fuzzy Hypergraphs

Definition 6.3 A *q-rung orthopair fuzzy graph* on a non-empty set X is defined as an ordered pair $\mathcal{G} = (\mathcal{V}, \mathcal{E})$, where $\mathcal{V}$ is a q-rung orthopair fuzzy set on X and $\mathcal{E}$ is a q-rung orthopair fuzzy relation on X such that

$$T_{\mathcal{E}}(x_1x_2) \leq \min\{T_{\mathcal{V}}(x_1), T_{\mathcal{V}}(x_2)\}, F_{\mathcal{E}}(x_1x_2) \leq \max\{F_{\mathcal{V}}(x_1), F_{\mathcal{V}}(x_2)\},$$

and $0 \leq T_E^q(x_1 x_2) + F_E^q(x_1 x_2) \leq 1$, $q \geq 1$, for all $x_1, x_2 \in X$, where $T_{\mathscr{E}} : X \times X \to [0, 1]$ and $F_{\mathscr{E}} : X \times X \to [0, 1]$ represent the truth-membership and falsity-membership degrees of $\mathscr{E}$, respectively.

Remark 6.1

- When $q = 1$, 1-rung orthopair fuzzy graph is called an intuitionistic fuzzy graph.
- When $q = 2$, 2-rung orthopair fuzzy graph is called Pythagorean fuzzy graph.

Definition 6.4 The *support* of a q-rung orthopair fuzzy set $Q = (x, T_Q(x), F_Q(x)|$ $x \in X)$ is defined as $supp(Q) = \{x | T_Q(x) \neq 0, F_Q(x) \neq 1\}$.
The *height* of a q-rung orthopair fuzzy set $Q = (x, T_Q(x), F_Q(x)| x \in X)$ is defined as $h(Q) = (\max_{x \in X} T_Q(x), \min_{x \in X} F_Q(x))$.
If $h(Q) = (1, 0)$, then q-rung orthopair fuzzy set Q is called *normal*.

Example 6.2 Let $Q = \{(q_1, 1, 0), (q_2, 0, 1), (q_3, 0.5, 0.6), (q_4, 0.6, 0.7), (q_5, 0.9, 0.3)\}$ be a 4-rung orthopair fuzzy set on X. Then, the support and height of Q are given as, $supp(Q) = \{q_1, q_3, q_4, q_5\}$, $h(Q) = (1, 0)$, respectively. Note that Q is normal.

Definition 6.5 Let X be a non-empty set. A *q-rung orthopair fuzzy hypergraph* $\mathscr{H}$ on X is defined in the form of an ordered pair $\mathscr{H} = (\mathscr{Q}, \zeta)$, where $\mathscr{Q} = \{\mathscr{Q}_1, \mathscr{Q}_2, \mathscr{Q}_3, \ldots \mathscr{Q}_n\}$ is a finite collection of nontrivial q-rung orthopair fuzzy subsets on X and ζ is a q-rung orthopair fuzzy relation on q-rung orthopair fuzzy sets $\mathscr{Q}_i$'s such that

1. $T_\zeta(E_k) = T_\zeta(x_1, x_2, x_3, \ldots, x_m) \leq \min\{\mathscr{Q}_i(x_1), \mathscr{Q}_i(x_2), \mathscr{Q}_i(x_3), \ldots, \mathscr{Q}_i(x_m)\}$,
   $F_\zeta(E_k) = F_\zeta(x_1, x_2, x_3, \ldots, x_m) \leq \max\{\mathscr{Q}_i(x_1), \mathscr{Q}_i(x_2), \mathscr{Q}_i(x_3), \ldots, \mathscr{Q}_i(x_m)\}$,
   for all $x_1, x_2, x_3, \ldots, x_m \in X$,

2. $\bigcup_i supp(\mathscr{Q}_i) = X$, for all $\mathscr{Q}_i \in \mathscr{Q}$.

Definition 6.6 The *height* of a q-rung orthopair fuzzy hypergraph $\mathscr{H} = (\mathscr{Q}, \zeta)$ is defined as $h(\mathscr{H}) = \{\max(\zeta_l), \min(\zeta_m)\}$, where $\zeta_l = \max T_{\zeta_j}(x_i)$ and $\zeta_m = \min F_{\zeta_j}(x_i)$. Here, $T_{\zeta_j}(x_i)$ and $F_{\zeta_j}(x_i)$ denote the truth-membership degree and falsity-membership degree of vertex x_i to the hyperedge ζ_j, respectively.

Definition 6.7 Let $\mathscr{H} = (\mathscr{Q}, \zeta)$ be a q-rung orthopair fuzzy hypergraph. The *order* of $\mathscr{H}$, which is denoted by $O(\mathscr{H})$, and is defined as $O(\mathscr{H}) = \sum_{x \in X} \wedge \mathscr{Q}_i(x)$. The *size* of $\mathscr{H}$, which is denoted by $S(\mathscr{H})$, and is defined as $S(\mathscr{H}) = \sum_{x \in X} \vee \mathscr{Q}_i(x)$.

In a q-rung orthopair fuzzy hypergraph, adjacent vertices x_i and x_j are the vertices which are the part of the same q-rung orthopair fuzzy hyperedge. Two q-rung orthopair fuzzy hyperedges ζ_i and ζ_j are said to be adjacent hyperedges if they possess the non-empty intersection, i.e., $supp(\zeta_i) \cap supp(\zeta_i) \neq \emptyset$.
    We now define the adjacent level between two q-rung orthopair fuzzy vertices and q-rung orthopair fuzzy hyperedges.

Definition 6.8 The *adjacent level* between two vertices x_i and x_j is denoted by $\gamma(x_i, x_j)$ and is defined as $\gamma(x_i, x_j) = (\max_k \min[T_k(x_i), T_k(x_j)], \min_k \max[F_k(x_i), F_k(x_j)])$.

The *adjacent level* between two hyperedges ζ_i and ζ_j is denoted by $\sigma(\zeta_i, \zeta_j)$ and is defined as $\sigma(\zeta_i, \zeta_j) = (\max_j \min[T_j(x), T_k(x)], \min_j \max[F_j(x), F_k(x)])$.

Definition 6.9 A *simple* q-rung orthopair fuzzy hypergraph $\mathcal{H} = (\mathcal{Q}, \zeta)$ is defined as a hypergraph, which has no repeated hyperedges contained in it, i.e., if $\zeta_i, \zeta_j \in \zeta$ and $\zeta_i \subseteq \zeta_j$, then $\zeta_i = \zeta_j$.

A q-rung orthopair fuzzy hypergraph $\mathcal{H} = (\mathcal{Q}, \zeta)$ is *support simple* if $\zeta_i, \zeta_j \in \zeta$, $supp(\zeta_i) = supp(\zeta_j)$, and $\zeta_i \subseteq \zeta_j$, then $\zeta_i = \zeta_j$.

A q-rung orthopair fuzzy hypergraph $\mathcal{H} = (\mathcal{Q}, \zeta)$ is *strongly support simple* if $\zeta_i, \zeta_j \in \zeta$ and $supp(\zeta_i) = supp(\zeta_j)$, then $\zeta_i = \zeta_j$.

Definition 6.10 A q-rung orthopair fuzzy set $Q : X \to [0, 1]$ is called an *elementary* set if T_Q and F_Q are single-valued on the support of Q.
A q-rung orthopair fuzzy hypergraph $\mathcal{H} = (\mathcal{Q}, \zeta)$ is *elementary* if all it's hyperedges are elementary.

Proposition 6.1 *A q-rung orthopair fuzzy hypergraph $\mathcal{H} = (\mathcal{Q}, \zeta)$ is the generalization of fuzzy hypergraph and intuitionistic fuzzy hypergraph.*

An upper bound on the cardinality of hyperedges of a q-rung orthopair fuzzy hypergraph of order n can be achieved by using the following result.

Theorem 6.1 *Let $\mathcal{H} = (\mathcal{Q}, \zeta)$ be a simple q-rung orthopair fuzzy hypergraph of order n. Then, $|\zeta|$ acquires no upper bound.*

Proof Let $X = \{x_1, x_2\}$. Define $\zeta_N = \{\mathcal{Q}_j, j = 1, 2, 3, \ldots, N\}$, where

$$T_{\mathcal{Q}_j}(x_1) = \frac{1}{1+j}, F_{\mathcal{Q}_j}(x_1) = 1 - \frac{1}{1+j}$$

and

$$T_{\mathcal{Q}_j}(x_2) = \frac{1}{1+j}, F_{\mathcal{Q}_j}(x_2) = 1 - \frac{1}{1+j}.$$

Then, $\mathcal{H}_N = (\mathcal{Q}, \zeta_N)$ is a simple q-rung orthopair fuzzy hypergraph having N hyperedges.

Theorem 6.2 *Let $\mathcal{H} = (\mathcal{Q}, \zeta)$ be an elementary and simple q-rung orthopair fuzzy hypergraph on a non-empty set X having n elements. Then $|\zeta| \leq 2^n - 1$. The equality holds if and only if $\{supp(\zeta_j)|\zeta_j \in \zeta, \zeta \neq 0\} = P(X)\backslash\emptyset$.*

Proof Since $\mathcal{H}$ is elementary and simple then at most one $\zeta_i \in \zeta$ can have each nontrivial subset of X as its support, therefore, we have $|\zeta| \leq 2^n - 1$.

To prove that the relation satisfies the equality, consider a set of mappings $\zeta = \{(T_A, F_A)|A \subseteq X\}$ such that,

$$T_A(x) = \begin{cases} \frac{1}{|A|}, & \text{if } x \in A, \\ 0, & \text{otherwise.} \end{cases} , \quad F_A(x) = \begin{cases} \frac{1}{|A|}, & \text{if } x \in A, \\ 0, & \text{otherwise.} \end{cases}$$

Then each set containing single element has height $(1, 1)$ and the height of the set having two elements is $(0.5, 0.5)$ and so on. Hence, $\mathscr{H}$ is simple and elementary with $|\zeta| = 2^n - 1$.

Definition 6.11 The *cut level set* of a q-rung orthopair fuzzy set Q is defined to be a crisp set of the following form, $Q^{(\alpha,\beta)} = \{x \in X | T_Q(x) \geq \alpha, F_Q(x) \leq \beta\}$, where $\alpha, \beta \in [0, 1]$.

Definition 6.12 Let $\mathscr{H} = (\mathscr{Q}, \zeta)$ be a q-rung orthopair fuzzy hypergraph. The (α, β)-*level hypergraph* of $\mathscr{H}$ is defined as $\mathscr{H}^{(\alpha,\beta)} = (\mathscr{Q}^{(\alpha,\beta)}, \zeta^{(\alpha,\beta)})$, where

1. $\zeta^{(\alpha,\beta)} = \{\zeta_i^{(\alpha,\beta)} : \zeta_i \in \zeta\}$ and $\zeta_i^{(\alpha,\beta)} = \{x \in X | T_{\zeta_i}(x) \geq \alpha, F_{\zeta_i}(x) \leq \beta\}$,
2. $\mathscr{Q}^{(\alpha,\beta)} = \bigcup_{\zeta_i \in \zeta} \zeta_i^{(\alpha,\beta)}$.

Example 6.3 Let $\mathscr{H} = (\mathscr{Q}, \zeta)$ be a 4-rung orthopair fuzzy hypergraph as shown in Fig. 6.2, where $\zeta = \{\zeta_1, \zeta_2, \zeta_3, \zeta_4, \zeta_5\}$. Incidence matrix of $\mathscr{H}$ is given in Table 6.1.

By direct calculations, it can be seen that it is a 4-rung orthopair fuzzy hypergraph. All the above mentioned concepts can be well explained by considering this example. Here, $h(\mathscr{H}) = \{\max(\zeta_l), \min(\zeta_m)\} = (0.6, 0.2)$. Since, $\mathscr{H}$ does not contain repeated hyperedges, it is simple 4-rung orthopair fuzzy hypergraph. Also, $\mathscr{H}$ is support simple and strongly support simple, i.e., whenever $\zeta_i, \zeta_j \in \zeta$ and $supp(\zeta_i) = supp(\zeta_j)$, then $\zeta_i = \zeta_j$. Adjacency level between x_1, x_2 and between two hyperedges ζ_1, ζ_2 is given as follows:

Table 6.1 Incidence matrix of $\mathscr{H}$

I	ζ_1	ζ_2	ζ_3	ζ_4	ζ_5
x_1	$(0.1, 0.2)$	$(0.1, 0.2)$	$(0.1, 0.2)$	$(0, 1)$	$(0, 1)$
x_2	$(0.2, 0.3)$	$(0, 1)$	$(0, 1)$	$(0, 1)$	$(0, 1)$
x_3	$(0.3, 0.4)$	$(0, 1)$	$(0, 1)$	$(0, 1)$	$(0.3, 0.4)$
x_4	$(0, 1)$	$(0, 1)$	$(0.4, 0.5)$	$(0, 1)$	$(0, 1)$
x_5	$(0, 1)$	$(0.5, 0.6)$	$(0, 1)$	$(0, 1)$	$(0, 1)$
x_6	$(0, 1)$	$(0, 1)$	$(0, 1)$	$(0, 1)$	$(0.5, 0.4)$
x_7	$(0, 1)$	$(0, 1)$	$(0.4, 0.3)$	$(0.4, 0.3)$	$(0, 1)$
x_8	$(0, 1)$	$(0, 1)$	$(0, 1)$	$(0.6, 0.5)$	$(0, 1)$
x_9	$(0, 1)$	$(0, 1)$	$(0, 1)$	$(0.6, 0.7)$	$(0.6, 0.7)$

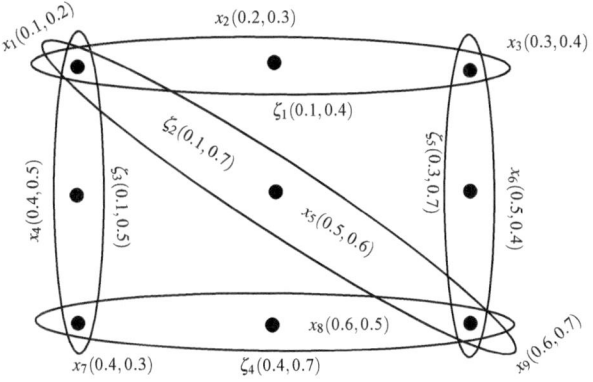

Fig. 6.2 4-rung orthopair fuzzy hypergraph

Fig. 6.3 (0.1, 0.4)-level hypergraph of $\mathcal{H}$

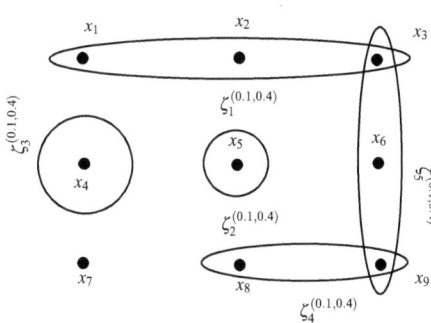

$$\gamma(x_1, x_2) = (\max_{k} \min[T_k(x_1), T_k(x_2)], \min_{k} \max[F_k(x_1), F_k(x_2)]), k = 1, 2, 3, 4, 5.$$
$$= (0.1, 0.3),$$
$$\sigma(\zeta_1, \zeta_2) = (\max \min[T_1(x), T_2(x)], \min \max[F_1(x), F_2(x)])$$
$$= (0.2, 0.6).$$

For $\alpha = 0.1$, $\beta = 0.4 \in [0, 1]$, (0.1, 0.4)-level hypergraph of $\mathcal{H}$ is $\mathcal{H}^{(0.1,0.4)} = (\mathcal{Q}^{(0.1,0.4)}, \zeta^{(0.1,0.4)})$, where

$$\zeta^{(0.1,0.4)} = \{\zeta_1^{(0.1,0.4)}, \zeta_2^{(0.1,0.4)}, \zeta_3^{(0.1,0.4)}, \zeta_4^{(0.1,0.4)}, \zeta_5^{(0.1,0.4)}\}$$
$$= \{\{x_1, x_2, x_3\}, \{x_5\}, \{x_4\}, \{x_8, x_9\}, \{x_3, x_6, x_9\}\},$$
$$\mathcal{Q}^{(0.1,0.4)} = \{x_1, x_2, x_3\} \cup \{x_5\} \cup \{x_4\} \cup \{x_8, x_9\} \cup \{x_3, x_6, x_9\}$$
$$= \{x_1, x_2, x_3, x_4, x_5, x_6, x_8, x_9\}.$$

Note that, (0.1, 0.4)-level hypergraph of $\mathcal{H}$ is a crisp hypergraph as shown in Fig. 6.3.

Remark 6.2 If $\alpha \geq \mu$ and $\beta \leq \nu$ and $\mathcal{Q}$ is a q-rung orthopair fuzzy set on X, then $\mathcal{Q}^{(\alpha,\beta)} \subseteq \mathcal{Q}^{(\mu,\nu)}$. Thus, we can have $\zeta^{(\alpha,\beta)} \subseteq \zeta^{(\mu,\nu)}$, for level hypergraphs of $\mathcal{H}$, i.e., if a q-rung orthopair fuzzy hypergraph has distinct hyperedges, its (α, β)-level hyperedges may be same and hence (α, β)-level hypergraphs of a simple q-rung orthopair fuzzy hypergraphs may have repeated edges.

Definition 6.13 Let $\mathcal{H} = (\mathcal{Q}, \zeta)$ be a q-rung orthopair fuzzy hypergraph and $\mathcal{H}^{(\alpha,\beta)}$ be the (α, β)-level hypergraph of $\mathcal{H}$. The sequence of real numbers $\rho_1 = (T_{\rho_1}, F_{\rho_1})$, $\rho_2 = (T_{\rho_2}, F_{\rho_2})$, $\rho_3 = (T_{\rho_3}, F_{\rho_3})$, ..., $\rho_n = (T_{\rho_n}, F_{\rho_n})$, $0 < T_{\rho_1} < T_{\rho_2} < T_{\rho_3} < \cdots < T_{\rho_n}, F_{\rho_1} > F_{\rho_2} > F_{\rho_3} > \cdots > F_{\rho_n} > 0$, where $(T_{\rho_n}, F_{\rho_n}) = h(\mathcal{H})$ such that

(i) if $\rho_{i-1} = (T_{\rho_{i-1}}, F_{\rho_{i-1}}) < \rho = (T_\rho, F_\rho) \leq \rho_i = (T_{\rho_i}, F_{\rho_i})$, then $\zeta^\rho = \zeta^{\rho_i}$,
(ii) $\zeta^{\rho_i} \subseteq \zeta^{\rho_{i+1}}$,

is called the *fundamental sequence* of $\mathcal{H}$, denoted by $f_S(H)$. The set of ρ_i-level hypergraphs $\{\mathcal{H}^{\rho_1}, \mathcal{H}^{\rho_2}, \mathcal{H}^{\rho_3}, \ldots, \mathcal{H}^{\rho_n}\}$ is called the *core hypergraphs* of $\mathcal{H}$ or simply the *core set* of $\mathcal{H}$ and is denoted by $c(\mathcal{H})$.

Definition 6.14 A q-rung orthopair fuzzy hypergraph $\mathcal{H}_1 = (\mathcal{Q}_1, \zeta_1)$ is called *partial hypergraph* of $\mathcal{H}_2 = (\mathcal{Q}_2, \zeta_2)$ if $\zeta_1 \subseteq \zeta_2$ and is denoted as $\mathcal{H}_1 \subseteq \mathcal{H}_2$.

Definition 6.15 Let $\mathcal{H} = (\mathcal{Q}, \zeta)$ be a q-rung orthopair fuzzy hypergraph having fundamental sequence $f_S(\mathcal{H}) = \{\rho_1, \rho_2, \rho_3, \ldots, \rho_n\}$ and let $\rho_{n+1} = 0$, if for all hyperedges $\zeta_k \in \zeta$, $k = 1, 2, 3, \ldots, n$, and for all $\rho \in (\rho_{i+1}, \rho_i]$, we have $\zeta_i^\rho = \zeta_i^{\rho_i}$ then $\mathcal{H}$ is called *sectionally elementary*.

Theorem 6.3 *Let $\mathcal{H} = (\mathcal{Q}, \zeta)$ be an elementary q-rung orthopair fuzzy hypergraph. Then the necessary and sufficient condition for $\mathcal{H} = (\mathcal{Q}, \zeta)$ to be strongly support simple is that $\mathcal{H}$ is support simple.*

Proof Suppose that $\mathcal{H}$ is support simple, elementary and $supp(\zeta_i) = supp(\zeta_j)$, for $\zeta_i, \zeta_j \in \zeta$. Let $h(\zeta_i) \leq h(\zeta_j)$. Since $\mathcal{H}$ is elementary, we have $\zeta_i \leq \zeta_j$ and since $\mathcal{H}$ is support simple, we have $\zeta_i = \zeta_j$. Hence, $\mathcal{H}$ is strongly support simple. On the same lines, the converse part may be proved.

Definition 6.16 A q-rung orthopair fuzzy hypergraph $\mathcal{H} = (\mathcal{Q}, \zeta)$ is said to be a $\mathcal{B} = (T_\mathcal{B}, F_\mathcal{B})$ *tempered* q-rung orthopair fuzzy hypergraph if for $H = (X, \xi)$, a crisp hypergraph, and a q-rung orthopair fuzzy set $\mathcal{B} = (T_\mathcal{B}, F_\mathcal{B}):X \to [0, 1]$ such that, $\zeta = \{D_A = (T_{D_A}, F_{D_A})|A \subset X\}$, where

$$T_{D_A}(x) = \begin{cases} \min(T_\mathcal{B}(y)) : y \in A, & \text{if} \quad x \in A, \\ 0, & \text{otherwise.} \end{cases}$$

$$F_{D_A}(x) = \begin{cases} \max(F_\mathcal{B}(y)) : y \in A, & \text{if} \quad x \in A, \\ 0, & \text{otherwise.} \end{cases}$$

Table 6.2 Incidence matrix of $\mathscr{H}$

I	ζ_1	ζ_2	ζ_3	ζ_4
x_1	(0.6, 0.7)	(0, 1)	(0.6, 0.7)	(0, 1)
x_2	(0, 1)	(0.7, 0.6)	(0, 1)	(0.7, 0.6)
x_3	(0.8, 0.7)	(0.8, 0.7)	(0, 1)	(0, 1)
x_4	(0, 1)	(0.6, 0.5)	(0.6, 0.7)	(0, 1)
x_5	(0.7, 0.8)	(0, 1)	(0, 1)	(0.7, 0.8)

Fig. 6.4 $\mathscr{B}$-tempered 3-rung orthopair fuzzy hypergraph

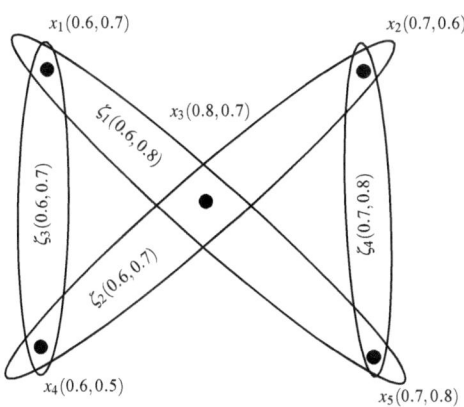

Example 6.4 Consider a 3-rung orthopair fuzzy hypergraph $\mathscr{H} = (\mathscr{Q}, \zeta)$ as shown in Fig. 6.4. Incidence matrix of $\mathscr{H} = (\mathscr{Q}, \zeta)$ is given in Table 6.2.

Define a 3-rung orthopair fuzzy set $\mathscr{B} = \{(x_1, 0.6, 0.7), (x_2, 0.7, 0.6), (x_3, 0.8, 0.7), (x_4, 0.6, 0.5), (x_5, 0.7, 0.8)\}$. By direct calculations, we have

$$T_{D_{\{x_1,x_3,x_5\}}}(x_1) = \min\{0.6, 0.8, 0.7\} = 0.6, \quad F_{D_{\{x_1,x_3,x_5\}}}(x_1) = \max\{0.7, 0.8, 0.7\} = 0.8,$$

$$T_{D_{\{x_2,x_3,x_4\}}}(x_2) = \min\{0.7, 0.8, 0.6\} = 0.6, \quad F_{D_{\{x_2,x_3,x_4\}}}(x_2) = \max\{0.6, 0.5, 0.7\} = 0.7,$$

$$T_{D_{\{x_1,x_4\}}}(x_4) = \min\{0.6, 0.6\} = 0.6, \quad F_{D_{\{x_1,x_4\}}}(x_4) = \max\{0.7, 0.7\} = 0.7,$$

$$T_{D_{\{x_2,x_5\}}}(x_5) = \min\{0.7, 0.7\} = 0.7, \quad F_{D_{\{x_2,x_5\}}}(x_5) = \max\{0.6, 0.8\} = 0.8.$$

Similarly, all other values can be calculated by using the same method. Thus, we have $\zeta_1 = (T_{D_{\{x_1,x_3,x_5\}}}, F_{D_{\{x_1,x_3,x_5\}}}), \zeta_2 = (T_{D_{\{x_2,x_3,x_4\}}}, F_{D_{\{x_2,x_3,x_4\}}}), \zeta_3 = (T_{D_{\{x_1,x_4\}}}, F_{D_{\{x_1,x_4\}}}), \zeta_4 = (T_{D_{\{x_2,x_5\}}}, F_{D_{\{x_2,x_5\}}}).$

Hence, $\mathscr{H}$ is $\mathscr{B}$-tempered 3-rung orthopair fuzzy hypergraph.

6.3   Transversals of q-Rung Orthopair Fuzzy Hypergraphs

Definition 6.17 Let $\mathcal{H} = (\mathcal{Q}, \zeta)$ be a q-rung orthopair fuzzy hypergraph on X. A q-rung orthopair fuzzy subset τ of X, which satisfies the condition $\tau^{h(\zeta_i)} \cap \zeta_i^{h(\zeta_i)} \neq \emptyset$, for all $\zeta_i \in \zeta$, is called a *q-rung orthopair fuzzy transversal* of $\mathcal{H}$.
τ is called *minimal transversal* of $\mathcal{H}$ if $\tau_1 \subset \tau$, τ_1 is not a q-rung orthopair fuzzy transversal. $t_r(\mathcal{H})$ denotes the collection of minimal transversals of $\mathcal{H}$.

We now discuss some results on q-rung orthopair fuzzy transversals.

Remark 6.3 Although τ can be regarded as a minimal transversal of $\mathcal{H}$, it is not necessary for $\tau^{(\alpha,\beta)}$ to be the minimal transversal of $\mathcal{H}^{(\alpha,\beta)}$, for all α, $\beta \in [0, 1]$. Also, it is not necessary for the family of minimal q-rung orthopair fuzzy hypergraphs to form a hypergraph on X. For those q-rung orthopair fuzzy transversals that satisfy the above property, we have the following definition.

Definition 6.18 A q-rung orthopair fuzzy transversal τ with the property that $\tau^{(\alpha,\beta)}$ is a minimal transversal of $\mathcal{H}^{(\alpha,\beta)}$, for α, $\beta \in [0, 1]$, is called *locally minimal q-rung orthopair fuzzy transversal* of $\mathcal{H}$. The collection of locally minimal q-rung orthopair fuzzy transversals of $\mathcal{H}$ is denoted by $t_r^*(\mathcal{H})$.

Lemma 6.1 *Let $f_S(\mathcal{H}) = \{\rho_1, \rho_2, \rho_3, \ldots, \rho_n\}$ be the fundamental sequence of a q-rung orthopair fuzzy hypergraph $\mathcal{H}$ and τ be the q-rung orthopair fuzzy transversal of $\mathcal{H}$. Then, $h(\tau) \geq h(\zeta_i)$, for each $\zeta_i \in \zeta$ and if τ is minimal, then $h(\tau) = \max\{h(\zeta_i)|\zeta_i \in \zeta\} = \rho_1$.*

Proof Since τ is a q-rung orthopair fuzzy transversal of $\mathcal{H}$ then $\tau^{h(\zeta_i)} \cap \zeta_i^{h(\zeta_i)} \neq \emptyset$. Consider an arbitrary element of $supp(\tau)$, then $\zeta_i(x) > h(\zeta_i)$ and we have $h(\tau) \geq h(\zeta_i)$. If τ is minimal transversal then $h(\zeta_i) = \{\max T_{\zeta_i}(x), \min F_{\zeta_i}(x)|x \in X$ and $\zeta_i \in \zeta\} = \rho_1$. Hence, $h(\tau) = \max\{h(\zeta_i)|\zeta_i \in \zeta\} = \rho_1$.

Theorem 6.4 *Let $\mathcal{H} = (\mathcal{Q}, \zeta)$ be a q-rung orthopair fuzzy hypergraph then the statements,*

(i) *τ is a q-rung orthopair fuzzy transversal of $\mathcal{H}$,*
(ii) *For all $\zeta_i \in \zeta$ and for each $\rho = \{T_\rho, F_\rho\} \in [0, 1]$ satisfying $0 < (T_\rho, F_\rho) < h(\zeta_i)$, $\tau^\rho \cap \zeta^\rho \neq \emptyset$,*
(iii) *τ^ρ is a transversal of $\mathcal{H}^\rho$, for all $\rho \in [0, 1]$, $0 < \rho < \rho_1$,*

are equivalent.

Proof $(i) \Rightarrow (ii)$. Suppose τ is a q-rung orthopair fuzzy transversal of $\mathcal{H}$. For any $\rho \in [0, 1]$, which satisfies $0 < (T_\rho, F_\rho) < h(\zeta_i)$, $\tau^\rho \supseteq \tau^{h(\zeta_i)}$ and $\zeta_i^\rho \supseteq \zeta_i^{h(\zeta_i)}$. Hence, $\tau^\rho \cap \zeta^\rho \supseteq \tau^{h(\zeta_i)} \cap \zeta_i^{h(\zeta_i)} \neq \emptyset$, because τ is a transversal.
$(ii) \Rightarrow (iii)$. Let $\tau^\rho \cap \zeta_i^\rho \neq \emptyset$, for all $\zeta_i \in \zeta$ and $0 < T_\rho < T_{\rho_1}, 0 > F_\rho < F_{\rho_1}$, which implies that τ^ρ is a transversal of $\mathcal{H}^\rho$.
$(iii) \Rightarrow (i)$. This part can be proved trivially.

Theorem 6.5 *Let $\mathcal{H} = (\mathcal{Q}, \zeta)$ be a q-rung orthopair fuzzy hypergraph. For each $x \in X$ such that $\tau(x) \in f_S(\mathcal{H})$ and for all $\tau \in t_r(\mathcal{H})$, the fundamental sequence of $t_r(\mathcal{H}) \subset f_S(\mathcal{H})$.*

Proof Let the fundamental sequence of $\mathcal{H}$ be $f_S(\mathcal{H}) = \{\rho_1, \rho_2, \rho_3, \ldots, \rho_n\}$ and $\tau \in t_r(\mathcal{H})$, for $\tau(x) \in (\rho_{i+1}, \rho_i]$. Consider a mapping ψ defined by

$$\psi(u) = \begin{cases} \rho_i, & \text{if } x = u, \\ \tau(u), & \text{otherwise.} \end{cases}$$

Thus, from the definition of ψ, we have $\psi^{\rho_i} = \tau^{\rho_i}$ and the Definition 6.13 implies that $\mathcal{H}^\rho = \mathcal{H}^{\rho_i}$, for all $\rho \in (\rho_{i+1}, \rho_i]$. Since τ is a q-rung orthopair fuzzy transversal of $\mathcal{H}$ and $\psi^\rho = \tau^\rho$, for all $\rho \notin (\rho_{i+1}, \rho_i]$, ψ is a q-rung orthopair fuzzy transversal. Now $\psi \leq \tau$ and minimality of τ both implies that $\psi = \tau$. Hence, $\tau(x) = \psi(x) = \rho_1$. Thus, $\tau(x) \in f_S(\mathcal{H})$, therefore we have $f_S(t_r(\mathcal{H})) \subseteq f_S(\mathcal{H})$.

Theorem 6.6 *The collection of all minimal transversals $t_r(\mathcal{H})$ is sectionally elementary.*

Proof Let the fundamental sequence of $t_r(\mathcal{H})$ be $f_S(t_r(\mathcal{H})) = \{\rho_1, \rho_2, \rho_3, \ldots, \rho_n\}$. Consider an element τ of $t_r(\mathcal{H})$ and some $\rho \in (\rho_{i+1}, \rho_i]$ such that $\tau^{\rho_i} \subset \tau^\rho$. In consideration of $[t_r(\mathcal{H})]^\rho = [t_r(\mathcal{H})]^{\rho_i}$, we have $\psi \in t_r(\mathcal{H})$ satisfying $\psi^\rho = \tau^{\rho_i}$. Then, the condition $\psi^\rho \supset \tau^{\rho_i}$ implies the existence of a q-rung orthopair fuzzy set $\mathcal{R}$ such that,

$$\mathcal{R}(x) = \begin{cases} \rho, & \text{if } x \in \psi^{\rho_i} \setminus \tau^{\rho_i}, \\ \psi(x), & \text{otherwise,} \end{cases}$$

is the q-rung orthopair fuzzy transversal of $\mathcal{H}$. Now, $\rho < \psi$ yields a contradiction to the minimality of ψ.

Lemma 6.2 *Let $\mathcal{H} = (\mathcal{Q}, \zeta)$ be a q-rung orthopair fuzzy hypergraph. Consider an element x of $supp(\tau)$, where $\tau \in t_r(\mathcal{H})$, then there exists a q-rung orthopair fuzzy hyperedge ζ of $\mathcal{H}$ such that,*

(i) $\tau(x) = h(\zeta) = \zeta(x) > 0$,
(ii) $\tau^{h(\zeta)} \cap \zeta^{h(\zeta)} = \{x\}$.

Proof (i) Let $\tau(x) > 0$ and Q denotes the set of all q-rung orthopair fuzzy hyperedges of $\mathcal{H}$ such that for each element ζ of Q, $\zeta(x) \geq \tau(x)$. Then this set is non-empty because $\tau^{\tau(x)}$ is a transversal of $\mathcal{H}^{\tau(x)}$ and $x \in \tau^{\tau(x)}$. Additionally, each element ζ of Q satisfies the inequality $h(\zeta) \geq \zeta(x) \geq \tau(x)$. Suppose on contrary, (i) is false then for each $\zeta \in Q$, $h(\zeta) > \tau(x)$ and we have an element $x^\zeta \neq x$, where $x^\zeta \in \zeta^{h(\zeta)} \cap \tau^{h(\zeta)}$. Here, we define a q-rung orthopair fuzzy set Q' as

$$Q'(v) = \begin{cases} \tau(v), & \text{if } x \neq v, \\ \max\{h(\zeta)|h(\zeta) < \tau(x)\}, & \text{if } x = v. \end{cases}$$

Note that, Q' is a q-rung orthopair fuzzy transversal of $\mathcal{H}$ and $Q' < \tau$, which is contradiction to the fact that τ is minimal. Hence, (i) holds for some ζ.

(ii) Suppose each element of Q satisfies (i) and also have an element $x^\zeta \neq x$, where $x^\zeta \in \zeta^{h(\zeta)} \cap \tau^{h(\zeta)}$. The same arguments as given above completes the proof.

Theorem 6.7 *Let $\mathcal{H} = (\mathcal{Q}, \zeta)$ be an ordered q-rung orthopair fuzzy hypergraph with $f_S(\mathcal{H}) = \{\rho_1, \rho_2, \rho_3, \ldots, \rho_n\}$ and $c(\mathcal{H}) = \{\mathcal{H}^{\rho_1}, \mathcal{H}^{\rho_2}, \mathcal{H}^{\rho_3}, \ldots, \mathcal{H}^{\rho_n}\}$. Then, $t_r^\star(\mathcal{H})$ is non-empty. Further, if τ_n is a minimal transversal of $\mathcal{H}^{\rho_n}$ then there exists $T \in t_r^\star(\mathcal{H})$ such that $supp(T) = \tau_n$.*

Proof Let τ_n be a minimal transversal of $\mathcal{H}^{\rho_n}$, $\mathcal{H}^{\rho_{n-1}}$ is a partial hypergraph of $\mathcal{H}^{\rho_n}$ because $\mathcal{H}$ is ordered and consequently τ_{n-1} is minimal transversal of $\mathcal{H}^{\rho_{n-1}}$ such that $\tau_{n-1} \subseteq \tau_n$. By continuing the same argument, we establish a nested sequence of minimal transversals $\tau_1 \subseteq \tau_2 \subseteq \tau_3 \subseteq \cdots \subseteq \tau_n$, where every τ_i is minimal transversal of $\mathcal{H}^{\rho_i}$. Let $\eta_j = \eta_j(\tau_j, \rho_j)$ is an elementary q-rung orthopair fuzzy set having height ρ_j and support τ_j. Then, $T = \max\{\eta_j | 1 \leq j \leq n\}$ is locally minimal transversal of $\mathcal{H}$ having support τ_n.

We now give an Algorithm 6.3.1 for finding $t_r(\mathcal{H})$.

Algorithm 6.3.1 Algorithm for finding $t_r(\mathcal{H})$

Let $\mathcal{H} = (\mathcal{Q}, \zeta)$ be a q-rung orthopair fuzzy hypergraph having the set of core hypergraphs $c(\mathcal{H}) = \{\mathcal{H}^{\rho_1}, \mathcal{H}^{\rho_2}, \mathcal{H}^{\rho_3}, \ldots, \mathcal{H}^{\rho_n}\}$. An iterative procedure to find the minimal transversal τ of $\mathcal{H}$ is as follows,

1. Find a crisp minimal transversal τ_1 of $\mathcal{H}^{\rho_1}$.
2. Find a minimal transversal τ_2 of $\mathcal{H}^{\rho_2}$, which satisfies $\tau_1 \subseteq \tau_2$, i.e., formulate a new hypergraph $\mathcal{H}_2$ having hyperedges ζ^{ρ_2} which is augmented having a loop at each $x \in \tau_1$. In accordance with, we can say that $\zeta(H_2) = \zeta^{\rho_2} \cup \{\{x\}|x \in \tau_1\}$. Let τ_2 be an arbitrary minimal transversal of $\mathcal{H}_2$.
3. By continuing the same procedure repeatedly, we have a sequence of minimal transversals $\tau_1 \subseteq \tau_2 \subseteq \tau_3 \subseteq \cdots \subseteq \tau_j$ such that τ_j be the minimal transversal of $\mathcal{H}^{\rho_j}$ with the property $\tau_{j-1} \subseteq \tau_j$.
4. Consider an elementary q-rung orthopair fuzzy set μ_j having the support τ_j and $h(\mu_j) = \rho_j$, $1 \leq j \leq n$. Then, $\tau = \bigcup_{j=1}^{n}\{\mu_j | 1 \leq j \leq n\}$ is a minimal q-rung orthopair fuzzy transversal of $\mathcal{H}$.

Example 6.5 Consider a 5-rung orthopair fuzzy hypergraph $\mathcal{H} = (\mathcal{Q}, \zeta)$, as shown in Fig. 6.5, where $\zeta = \{\zeta_1, \zeta_2, \zeta_3\}$. Incidence matrix of $\mathcal{H} = (\mathcal{Q}, \zeta)$ is given in Table 6.3.

By routine calculations, we have $h(\zeta_1) = (0.8, 0.6)$, $h(\zeta_2) = (0.8, 0.5)$, and $h(\zeta_3) = (0.8, 0.5)$. Consider a 5-rung orthopair fuzzy subset τ_1 of X such that

Fig. 6.5 5-rung orthopair
fuzzy hypergraph

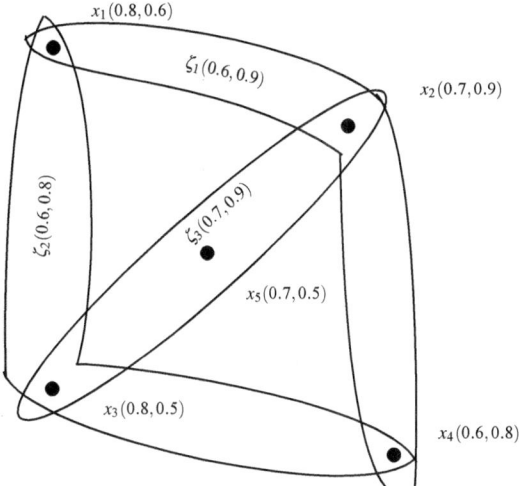

Table 6.3 Incidence matrix of $\mathcal{H}$

I	ζ_1	ζ_2	ζ_3
x_1	$(0.8, 0.6)$	$(0.8, 0.6)$	$(0, 1)$
x_2	$(0.7, 0.9)$	$(0, 1)$	$(0.7, 0.9)$
x_3	$(0, 1)$	$(0.8, 0.5)$	$(0.8, 0.5)$
x_4	$(0.6, 0.8)$	$(0.6, 0.8)$	$(0, 1)$
x_5	$(0, 1)$	$(0, 1)$	$(0.7, 0.5)$

$\tau_1 = \{(x_1, 0.8, 0.6), (x_2, 0.7, 0.9), (x_3, 0.8, 0.5)\}$. Note that, $\zeta_1^{h(\zeta_1)} = \{x_1\}$, $\zeta_2^{h(\zeta_2)} = \{x_3\}$ and $\zeta_3^{h(\zeta_3)} = \{x_3\}$. Also $\tau_1^{(0.8, 0.6)} = \{x_1\}$, $\tau_2^{(0.8, 0.5)} = \{x_3\}$ and $\tau_3^{(0.8, 0.5)} = \{x_3\}$. It can be seen that $\tau_1^{h(\zeta_i)} \cap \zeta_i^{h(\zeta_i)} \neq \emptyset$, for all $\zeta_i \in \zeta$. Thus, τ_1 is a 5-rung orthopair fuzzy transversal of $\mathcal{H}$. Similarly, $\tau_2 = \{(x_1, 0.8, 0.6), (x_3, 0.8, 0.5)\}$, $\tau_3 = \{(x_1, 0.8, 0.6), (x_3, 0.8, 0.5), (x_4, 0.6, 0.8)\}$, $\tau_4 = \{(x_1, 0.8, 0.6), (x_3, 0.8, 0.5), (x_5, 0.7, 0.5), \}$ are other transversals of $\mathcal{H}$. The minimal transversal is τ_2, i.e., whenever $\tau \subseteq \tau_2$, τ is not a 5-rung orthopair fuzzy transversal.

Let $\alpha = 0.8, \beta = 0.5$, then $\zeta_1^{(0.8, 0.5)} = \{\emptyset\}$, $\zeta_2^{(0.8, 0.5)} = \{x_3\}$, $\zeta_3^{(0.8, 0.5)} = \{x_3\}$ shows that $\tau_2^{(0.8, 0.5)}$ is not a minimal transversal of $\mathcal{H}^{(0.8, 0.5)}$.

Theorem 6.8 *Let $\mathcal{H} = (\mathcal{Q}, \zeta)$ be a q-rung orthopair fuzzy hypergraph and $x \in X$. Then, there exists an element τ of $t_r(\mathcal{H})$ such that $x \in supp(\tau)$ if and only if there is an hyperedge $\zeta_1 \in \zeta$ which satisfies,*

- *$\zeta_1(x) = h(\zeta')$,*
- *For every $\xi \in \zeta$ with $h(\xi) > h(\zeta_1)$, $\xi^{h(\zeta_i)} \not\subset \zeta_1^{h(\zeta_1)}$,*
- *$h(\zeta_1)$ level cut of ζ_1 is not a proper subset of any other hyperedge of $\mathcal{H}^{h(\zeta_1)}$.*

Proof Let us suppose that $\tau(x) > 0$ and τ is an element of $t_r(\mathscr{H})$, then first condition directly follows from Lemma 6.2.

To prove the second condition, suppose that for every ζ_1 which satisfies the first condition, there is $\xi \in \zeta$ such that $h(\xi) > h(\zeta_1)$ and $\xi^{h(\xi)} \subseteq \zeta_1^{h(\zeta_1)}$. Then there exists an element $v \neq x$, where $v \in \xi^{h(\xi)} \cap \tau^{h(\xi)} \subseteq \zeta_1^{h(\zeta_1)} \cap \tau^{h(\zeta_1)}$, which is a contradiction.

To prove that $h(\zeta_1)$ level cut of ζ_1 is not a proper subset of any other hyperedge of $\mathscr{H}^{h(\zeta_1)}$, suppose that for every ζ_1, which satisfies the above two conditions, there is $\xi \in \zeta$ with $\emptyset \subset \xi^{h(\xi)} \subset \zeta_1^{h(\zeta_1)}$, as $\xi^{h(\xi)} \neq \emptyset$ and from second condition, we have $h(\xi) = \zeta_1(x) = \tau(x)$. If $h(\xi) = \zeta_1(x)$, our supposition accommodates $\xi' \in \zeta$ such that $\emptyset \subset \xi'^{h(\zeta_1)} \subset \xi^{h(\zeta_1)} \subset \zeta_1^{h(\zeta_1)}$. This recursive procedure must end after a finite number of steps, so assume that $\xi(x) < h(\xi)$, which implies the existence of an element $v \neq x$, where $v \in \xi^{h(\zeta_1)} \cap \tau^{h(\zeta_1)} \subseteq \zeta_1^{h(\zeta_1)} \cap \tau^{h(\zeta_1)}$, which is again a contradiction.

The sufficient condition is proved by using the construction given in Algorithm 6.3.1. By using first condition, we have $h(\zeta_1) = \rho_1, \rho_1 \in f_s(\mathscr{H})$ and from other two conditions, we have $y_\xi \in \xi^{h(\xi)} \backslash \zeta_1^{h(\zeta_1)}$ such that $\xi \neq \zeta_1$ and $h(\xi) \geq h(\zeta_1)$. Then $Q \cap \zeta_1^{h(\zeta_1)}$, where Q is the collection of all such vertices. An initial sequence of transversals of is constructed in a way that $\tau_j \subseteq Q$, for $1 \leq j \leq n$ and $\tau_i \subseteq Q \cup \{x\}$. Continuing the construction given in Algorithm 6.3.1 will give a minimal q-rung orthopair fuzzy transversal with $\tau(x) = \zeta_1(x) = h(\zeta_1)$.

Definition 6.19 Let Q be a q-rung orthopair fuzzy set and $\alpha, \beta \in [0, 1]$. The *lower truncation* of Q at level α, β is a q-rung orthopair fuzzy set $Q_{\langle \alpha, \beta \rangle}$ given by

$$Q_{\langle \alpha, \beta \rangle}(x) = \begin{cases} Q(x), & \text{if } x \in Q^{(\alpha, \beta)}, \\ (0, 1), & \text{otherwise.} \end{cases}$$

The *upper truncation* of Q at level α, β is a q-rung orthopair fuzzy set $Q^{\langle \alpha, \beta \rangle}$ given by

$$Q^{\langle \alpha, \beta \rangle}(x) = \begin{cases} (\alpha, \beta), & \text{if } x \in Q^{(\alpha, \beta)}, \\ Q(x), & \text{otherwise.} \end{cases}$$

Definition 6.20 Let $\mathscr{E}$ be a collection of q-rung orthopair fuzzy sets of X and $\mathscr{E}^{\langle \alpha, \beta \rangle} = \{q^{\langle \alpha, \beta \rangle} | q \in \mathscr{E}\}$, $\mathscr{E}_{\langle \alpha, \beta \rangle} = \{q_{\langle \alpha, \beta \rangle} | q \in \mathscr{E}\}$. Then, the *upper* and *lower* truncations of a q-rung orthopair fuzzy hypergraph $\mathscr{H} = (\mathscr{Q}, \zeta)$ at α, β level are a pair of q-rung orthopair fuzzy hypergraphs, $\mathscr{H}^{\langle \alpha, \beta \rangle}$ and $\mathscr{H}_{\langle \alpha, \beta \rangle}$, defined by $\mathscr{H}^{\langle \alpha, \beta \rangle} = (X, \mathscr{E}^{\langle \alpha, \beta \rangle})$ and $\mathscr{H}_{\langle \alpha, \beta \rangle} = (X, \mathscr{E}_{\langle \alpha, \beta \rangle})$.

Definition 6.21 Let Q be a q-rung orthopair fuzzy set on X, then each $(\mu, v) \in (0, h(Q))$ for which $Q^{(\alpha, \beta)} \not\subseteq Q^{(\mu, v)}$, $(\mu, v) < (\alpha, \beta) \leq h(Q)$, is called the *transition level* of Q.

Definition 6.22 Let Q be a nontrivial q-rung orthopair fuzzy set of X. Then,

(i) the sequence $\mathscr{S}(Q) = \{t_1^Q, t_2^Q, t_3^Q, \ldots, t_n^Q\}$ is called the *basic sequence* determined by Q, where

- $t_1^Q > t_2^Q > t_3^Q > \cdots > t_n^Q > 0$,
- $t_1^Q = h(Q)$,
- $\{t_2^Q, t_3^Q, \ldots, t_n^Q\}$ is the set of transition levels of Q.

(ii) The set of cuts of Q, $\mathscr{C}(Q)$, is defined as $\mathscr{C}(Q) = \{Q^t | t \in \mathscr{S}(Q)\}$.

(iii) The join $\max\{\eta(Q^t, t) | t \in \mathscr{S}(Q)\}$ of basic elementary q-rung orthopair fuzzy sets $E(Q) = \{\eta(Q^t, t) | t \in \mathscr{S}(Q)\}$ is called the *basic elementary join* of Q.

Lemma 6.3 *Let $\mathscr{H}$ be a q-rung orthopair fuzzy hypergraph with $f_S(\mathscr{H}) = \{\rho_1, \rho_2, \rho_3, \ldots, \rho_n\}$. Then,*

(i) *If $t = (\mu, \nu)$ is a transition level of $\tau \in t_r(\mathscr{H})$, then there is an $\varepsilon > 0$ such that, $\forall \, (\alpha, \beta) \in (t, t + \varepsilon]$, $\tau^{(\mu, \nu)}$ is a minimal $\mathscr{H}^{(\mu, \nu)}$-transversal extension of $\tau^{(\alpha, \beta)}$, i.e., if $\tau^{(\alpha, \beta)} \subseteq \tau' \subseteq \tau^{(\mu, \nu)}$ then τ' is not a transversal of $\mathscr{H}^{(\mu, \nu)}$.*

(ii) *$t_r(\mathscr{H})$ is sectionally elementary.*

(iii) *$f_S(t_r(\mathscr{H}))$ is properly contained in $f_S(\mathscr{H})$.*

(iv) *$\tau^{(\alpha, \beta)}$ is a minimal transversal of $\mathscr{H}^{(\alpha, \beta)}$, for each $\tau \in t_r(\mathscr{H})$ and $\rho_2 < (\alpha, \beta) \leq \rho_1$.*

Proof (i) Let $\tilde{t} = (\mu, \nu)$ be a transition level of $\tau \in t_r(\mathscr{H})$. Then by definition, we have $\tau^{(\alpha, \beta)} \not\subseteq \tau^{(\mu, \nu)}$, $(\mu, \nu) < (\alpha, \beta) \leq h(\mathscr{H})$, for all α, β. Since, τ possesses a finite support, this implies the existence of an $\varepsilon > 0$ such that $\tau^{(\alpha, \beta)}$ is constant on $(\tilde{t}, \tilde{t} + \varepsilon]$. Assume that there is a transversal T of $\mathscr{H}^{(\mu, \nu)}$ such that $\tau^{(\alpha', \beta')} \subseteq T \subseteq \tau^{(\mu, \nu)}$, for $\alpha', \beta' \in (\tilde{t}, \tilde{t} + \varepsilon]$. We claim that this supposition is false. To demonstrate the existence of this claim, we suppose that assumption is true and consider the collection of basic elementary q-rung orthopair fuzzy sets $E(\tau) = \{\eta(\tau^t, t) | t \in S(\tau)\}$ of τ. Note that a nested sequence of X is formed by $c(\tau) \cup T$, where $c(\tau)$ is used to denote the basic cuts of τ. Since $\mathscr{H} = (\mathscr{Q}, \zeta)$ is defined on a finite set X and $\mathscr{Q}$ is a finite collection of q-rung orthopair fuzzy sets of X, then each $\rho \in (0, h(\mathscr{H}))$ corresponds a number $\varepsilon_\rho > 0$ such that

- $\mathscr{H}^{(\alpha, \beta)}$ is constant on $(\rho, \rho + \varepsilon_\rho]$,
- $\mathscr{H}^{(\alpha, \beta)}$ is constant on $(\rho - \varepsilon_\rho, \rho]$.

It follows from these considerations that level cuts of $\tau^{\star(\alpha, \beta)}$ of the join $\tau^\star = \max\{\max\{E(\tau) \backslash \eta(\tau^{\tilde{t}}, \tilde{t}), \eta(\tau^{\tilde{t}}, \tilde{t} - \varepsilon_{\tilde{t}}), \eta(T, \tilde{t})\}\}$ persuade

$$\tilde{\tau}^{(\alpha, \beta)} = \begin{cases} T, & \text{if } (\alpha, \beta) \in (\tilde{t} - \varepsilon_{\tilde{t}}, \tilde{t}), \\ \tau^{(\alpha, \beta)}, & \text{if } (\alpha, \beta) \in (0, h(\mathscr{H})) \backslash (\tilde{t}, \tilde{t} - \varepsilon_{\tilde{t}})]. \end{cases}$$

This relation is derived because of supposition that $\varepsilon_{\tilde{t}}$ is so small that the open interval $(\tilde{t} - \varepsilon_{\tilde{t}}, \tilde{t})$ does not contain any other transition level of τ.

Since, it is assumed that T is a transversal of $\mathcal{H}^{\tilde{t}}$, T is a transversal of $\mathcal{H}^{(\alpha,\beta)}$, for all $(\alpha,\beta) \in (\tilde{t} - \varepsilon_{\tilde{t}}, \tilde{t})$ and $\mathcal{H}^{(\alpha,\beta)}$ is constant on $(\tilde{t} - \varepsilon_{\tilde{t}}, \tilde{t})$. Note that, $\tau^{(\alpha,\beta)}$ is a transversal of $\mathcal{H}^{(\alpha,\beta)}$, for all $(\alpha,\beta) \in (0, h(\mathcal{H})]$, therefore, it follows that $\tilde{\tau}$ is a q-rung orthopair fuzzy transversal of $\mathcal{H}$, as $\tilde{\tau} < \tau$, implies that $\tau \notin t_r(\mathcal{H})$, which leads to a contradiction. Hence, the supposition is false and claim is satisfied.

(ii) Let $\tau \in t_r(\mathcal{H})$, then $\tau^{(\alpha,\beta)}$ is a transversal of $\mathcal{H}^{(\alpha,\beta)}$ for $0 < (\alpha,\beta) < h(\mathcal{H})$. Suppose that a transition level t of τ corresponds an interval $(t, t + \varepsilon], \varepsilon > 0$, on which $\tau^{(\alpha,\beta)}$ is constant. Then for $(\alpha',\beta') \in (t, t + \varepsilon], \tau^{(\alpha',\beta')}$ is not a transversal of $\mathcal{H}^t$, which implies that $\tau^{(\alpha',\beta')} \notin (t_r(\mathcal{H}))^t$, where $t_r(\mathcal{H}))^t$ denotes the t-cut of $t_r(\mathcal{H})$. However, the definition of fundamental sequence of $t_r(\mathcal{H})$ implies that $t \in f_S(t_r(\mathcal{H}))$.

(iii) To prove (iii), we suppose that if $t = (\mu, \nu)$ is a transition level of some $\tau \in t_r(\mathcal{H})$, then t belongs to $f_S(\mathcal{H})$. On contrary, suppose that the transition level t of some $\tau \in t_r(\mathcal{H})$ does not belong to $f_S(\mathcal{H})$. Then for some $\rho_j \in f_S(\mathcal{H})$, we have $\rho_{j+1} < t < \rho_j$, where $\rho_{n+1} = 0$, as $\mathcal{H}^{(\alpha,\beta)} = \mathcal{H}^{\rho_j}$, for all $(\alpha,\beta) \in (\rho_{j+1}, \rho_j]$, follows that τ^t is a transversal of $\mathcal{H}^t = \mathcal{H}^{\rho_j}$. Furthermore, there exists an $\varepsilon > 0$, such that $\tau^{(\alpha,\beta)}$ is constant on $(t, t + \varepsilon]$. Without loss of generality, we assume that $t + \varepsilon \leq \rho_j$ and $(\alpha',\beta') \in (t, t + \varepsilon]$. Since t is a transition level of τ then $\tau^{(\alpha',\beta')} \subsetneq \tau^t$ and $\tau^{(\alpha',\beta')}$ is not a transversal of $\mathcal{H}^t$ (from i), which is not possible, as $\mathcal{H}^{(\alpha',\beta')} = \mathcal{H}^{\rho_j} = \mathcal{H}^t$, this proves our claim. Along with this result and the fact that $h(\tau) = \rho_1 \in f_S(\mathcal{H})$, it follows that $f_S(t_r(\mathcal{H})) \subseteq f_S(\mathcal{H})$, for all $\tau \in t_r(\mathcal{H})$.

(iv) First, we will show that τ^{ρ_1} is a minimal transversal of $\mathcal{H}^{\rho_1}$. Suppose on contrary that there is a minimal transversal T of $\mathcal{H}^{\rho_1}$ such that $T \subseteq \tau^{\rho_1}$. Let $\tilde{\tau} = \max\{\tau^{\rho_2}, \eta_1\}$, where η_1 is the basic elementary q-rung orthopair fuzzy set having support T and height ρ_1. τ^{ρ_2} is considered as the upper truncation of τ at level ρ_2. It is obvious that $\tilde{\tau}$ is a transversal of $\mathcal{H}$ with $\tilde{\tau} < \tau$, which is contradiction to the fact that τ is minimal. From (ii) and (iii) parts, it is followed that $\tau^{(\alpha,\beta)} \in t_r(\mathcal{H})^{(\alpha,\beta)}$, for $\rho_2 < (\alpha,\beta) < \rho_1$.

Theorem 6.9 *At least one minimal q-rung orthopair fuzzy transversal is contained in every q-rung orthopair fuzzy transversal of a q-rung orthopair fuzzy hypergraph $\mathcal{H}$.*

Proof Let $f_S(\mathcal{H}) = \{\rho_1, \rho_2, \rho_3, \ldots, \rho_n\}$ be the fundamental sequence of $\mathcal{H}$ and suppose that ξ be a transversal of $\mathcal{H}$, which is not minimal. Let τ be a minimal transversal of $\mathcal{H}$, $\tau \leq \xi$, which is constructed in such a way, $\{q_i \in Q(X) | i = 0, 1, 2, \ldots, n\}$ satisfying $\tau = q_n \leq \cdots \leq q_1 \leq q_0 \leq \xi$, where $Q(X)$ is the collection of q-rung orthopair fuzzy sets on X. It can be noted that $h(\xi) \geq h(\mathcal{H}) = \rho_1$ and $\xi^{(\alpha,\beta)}$ is a transversal of $\mathcal{H}^{(\alpha,\beta)}$, for $0 < (\alpha,\beta) \leq \rho_1$. Therefore, the reduction process is started as $q_0 = \xi^{\langle\rho_1\rangle}$, where $\xi^{\langle\rho_1\rangle}$ represents the upper truncation level of ξ at ρ_1. Since the top level cut ξ^{ρ_1} of ρ_0 comprises a crisp minimal transversal T_1 of $\mathcal{H}^{\rho_1}$, we have $q_1 = \max\{\xi^{\langle\rho_2\rangle}, \lambda^{T_1}\}$, where λ^{T_1} is elementary q-rung orthopair fuzzy set having height ρ_1 and support T_1. Note that, $q_1 \leq q_2 \leq \xi$. The same procedure will determine the all other remaining members. For instance, we have $q_2 = \max\{\xi^{\langle \rho_3 \rangle}, \lambda^{T_1}, \lambda^{T_2}\}$, where λ^{T_2} is an elementary q-rung orthopair fuzzy set having height ρ_2 and support T_2, such that

$$T_2 = \begin{cases} T_1, & \text{if } T_1 \text{ is a transversal of } \mathscr{H}^{\rho_2}, \\ B_2, & \text{otherwise,} \end{cases}$$

where B_2 is the minimal transversal extension of T_1, i.e., if $T_1 \subseteq B \subseteq B_2$, then B_2 is not considered as a transversal of $\mathscr{H}^{\rho_2}$ and B_2 is contained in ρ-level of ξ because ξ^{ρ_2} contains a transversal of $\mathscr{H}^{\rho_2}$. Further, as $T_2 \subseteq \xi^{\rho_2}$, it is obvious that $q_2 \le q_1$. When this process is finished, we certainly have $q_n = \tau$ a q-rung orthopair fuzzy transversal of $\mathscr{H}$ and is included in ξ. We now claim that τ is a minimal transversal of $\mathscr{H}$, i.e., $\tau \in t_r(\mathscr{H})$. On contrary, suppose that τ_1 is a transversal of $\mathscr{H}$ such that $\tau_1 < \tau$. Then, we have

(i) $\tau_1^{(\alpha,\beta)} \subseteq \tau^{(\alpha,\beta)}$ for all $\alpha, \beta \in (0, h(\mathscr{H})]$,
(ii) $\tau_1^{(\alpha',\beta')} \subseteq \tau^{(\alpha',\beta')}$ for some $\alpha', \beta' \in (0, h(\mathscr{H})]$.

However, no such α', β' exist. To prove this, let $\alpha, \beta \in (\rho_2, \rho_1]$, then as $\tau_1^{(\alpha,\beta)} \subseteq \tau^{(\alpha,\beta)}$, $\tau_1^{(\alpha,\beta)}$ is a transversal of $\mathscr{H}^{(\alpha,\beta)} = \mathscr{H}^{\rho_1}$ and $\tau^{(\alpha,\beta)} \in t_r(\mathscr{H}^{\rho_1})$, which implies that $\tau_1^{(\alpha,\beta)} = \tau^{(\alpha,\beta)}$ on $(\rho_2, \rho_1]$. Moreover, suppose that $\alpha, \beta \in (\rho_3, \rho_2]$ then by using $\tau_1^{(\alpha,\beta)} = \tau^{(\alpha,\beta)}$, we have $\tau_1^{(\alpha,\beta)} \supseteq \tau^{\rho_1}$ on $(\rho_3, \rho_2]$ and if $T_2 = T_1 = \tau^{\rho_1}$, then by previous arguments $\tau_1^{(\alpha,\beta)} = \tau^{(\alpha,\beta)}$ on $(\rho_3, \rho_2]$. Furthermore, if $T_1 \subseteq T_2$ and $T_1 \subseteq \tau_1^{(\alpha,\beta)} \subsetneq T_2$ then $\tau_1^{(\alpha,\beta)}$ is not a transversal of $\mathscr{H}^{(\alpha,\beta)} = \mathscr{H}^{\rho_2}$, which is contradiction to the fact that τ_1 is a transversal of $\mathscr{H}$. Hence, we have $\tau_1^{(\alpha,\beta)} = \tau^{(\alpha,\beta)}$ on $(\rho_3, \rho_2]$. In general, we have $\tau_1^{(\alpha,\beta)} = \tau^{(\alpha,\beta)}$ on $(0, h(\mathscr{H})]$, which completes the proof.

6.4 Applications to Decision-Making

Decision-making is considered as the abstract technique, which results in the selection of an opinion or a strategy among a couple of elective potential results. Every decision-making procedure delivers a final decision, which may or may not be appropriate for our problem. We have to make hundreds of decisions everyday, some are easy but others may be complicated, confused and miscellaneous. That is the reason which leads to the process of decision-making. Decision-making is the foremost way to choose the most desirable alternative. It is essential in real-life problems when there are many possible choices. Thus, decision makers evaluate numerous merits and demerits of every choice and try to select the most fitting alternative.

6.4.1   Selection of Most Desirable Appliance

Here, we consider a decision-making problem of selecting the most appropriate product from different brands or organizations. Suppose that a person wants to purchase a product, which is available in many brands. Let he/she considers the following nine organizations or brands $O = \{O_1, O_2, O_3, \ldots, O_9\}$, of which product can be chosen to purchase. We will discuss that how the (α, β)-level cuts can be applied to q-rung orthopair fuzzy hypergraph to make a good decision. A 6-rung orthopair fuzzy hypergraph model depicting the problem is shown in Fig. 6.6.

The truth-membership degrees and falsity-membership degrees of vertices (which represent the organizations) depicts that how much that organization fulfills the costumer's requirements and up to which percentage the product is not suitable. The hyperedges of our graph represent the characteristics of those organizations which are (as vertices) contained in that hyperedge. It can be shown from Table 6.4.

The attributes, which we have considered as hyperedges $\{\zeta_1, \zeta_2, \zeta_3, \zeta_4, \zeta_5, \zeta_6\}$ to describe the characteristics of different organizations are {Delivery and service, Durability, Affordability, Quality, Functionality, Marketability}. Note that, if ζ_2 is considered as durability, then the membership degrees $(0.9, 0.5)$ of O_3 describes that the product manufactured by organization O_3 is 90% durable and 50% lacks the requirements of the customer. Similarly, O_4 is 60% durable and 40% lacks the condition. In the same way, we can describe the characteristics of all products manufactured by different organizations. Now to select the most appropriate product, we will find out the (α, β)-level cuts of all hyperedges. We choose the values of α and β in such manner that they will be fixed according to customer's demand. Let $\alpha = 0.7$ and $\beta = 0.4$, it means that customer will consider that product, which will satisfy

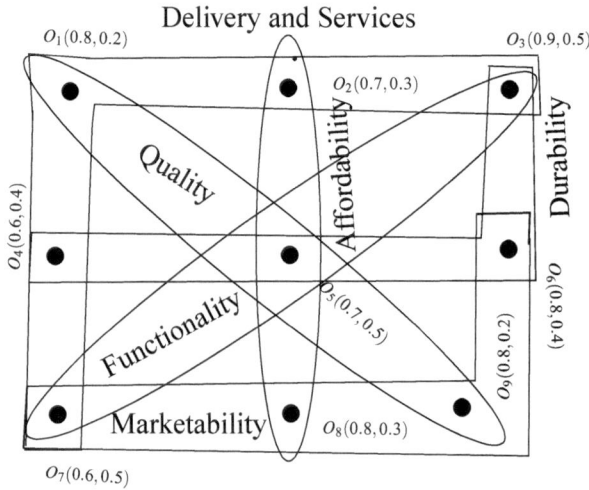

Fig. 6.6   6-rung orthopair fuzzy model for most appropriate appliance

Table 6.4 Incidence matrix

I	ζ_1	ζ_2	ζ_3	ζ_4	ζ_5	ζ_6
O_1	(0.8, 0.2)	(0, 1)	(0.8, 0.2)	(0, 1)	(0, 1)	(0, 1)
O_2	(0.7, 0.3)	(0, 1)	(0, 1)	(0.7, 0.3)	(0, 1)	(0, 1)
O_3	(0.9, 0.5)	(0.9, 0.5)	(0, 1)	(0, 1)	(0.9, 0.5)	(0, 1)
O_4	(0.6, 0.4)	(0.6, 0.4)	(0, 1)	(0, 1)	(0, 1)	(0, 1)
O_5	(0, 1)	(0.7, 0.5)	(0.7, 0.5)	(0.7, 0.5)	(0.7, 0.5)	(0, 1)
O_6	(0, 1)	(0.8, 0.4)	(0, 1)	(0, 1)	(0, 1)	(0.8, 0.4)
O_7	(0.6, 0.5)	(0, 1)	(0, 1)	(0, 1)	(0.6, 0.5)	(0.6, 0.5)
O_8	(0, 1)	(0, 1)	(0, 1)	(0.8, 0.3)	(0, 1)	(0.8, 0.3)
O_9	(0, 1)	(0, 1)	(0.8, 0.2)	(0, 1)	(0, 1)	(0.8, 0.2)

70% or more of the the characteristics mentioned above and will have deficiency less than or equal to 40%. The (α, β)-levels of all hyperedges are given as follows:

$$\zeta_1^{(0.7,0.4)} = \{O_1, O_2\}, \quad \zeta_2^{(0.7,0.4)} = \{O_6\}, \quad \zeta_3^{(0.7,0.4)} = \{O_1, O_5, O_9\},$$
$$\zeta_4^{(0.7,0.4)} = \{O_2, O_8\}, \quad \zeta_5^{(0.7,0.4)} = \{\emptyset\}, \quad \zeta_6^{(0.7,0.4)} = \{O_6, O_8, O_9\}.$$

Note that, $\zeta_1^{(0.7,0.4)}$ level set represents that O_1 and O_2 are the organizations that provide the best delivery services among all other organizations, $\zeta_2^{(0.7,0.4)}$ level set represents that O_6 is the organization, whose products are more durable as compared to all other organizations. Similarly, $\zeta_4^{(0.7,0.4)}$ indicates that the products proposed by O_2 and O_8 organizations, are more cheap and affordable in comparison to others. Thus, if a customer wants some specific speciality of product, for example he she wants to purchase a product with good marketablity, then the organizations O_6, O_8 and O_9 are more suitable. Similarly, if the satisfaction and dissatisfaction level of a customer are taken as $\alpha = 0.8$ and $\beta = 0.3$, respectively. Then, $(0.8, 0.3)$-level cuts are given as,

$$\zeta_1^{(0.8,0.3)} = \{O_1\}, \quad \zeta_2^{(0.8,0.3)} = \{\emptyset\}, \quad \zeta_3^{(0.8,0.3)} = \{O_1, O_9\},$$
$$\zeta_4^{(0.8,0.3)} = \{O_8\}, \quad \zeta_5^{(0.8,0.3)} = \{\emptyset\}, \quad \zeta_6^{(0.8,0.3)} = \{O_8, O_9\}.$$

Here, $\zeta_4^{(0.8,0.3)} = \{O_8\}$ indicates that the products proposed by organization O_8 satisfy the customer's requirement 80%, which is affordability and so on. For $\alpha = 0.7$ and $\beta = 0.3$, we have,

$$\zeta_1^{(0.7,0.3)} = \{O_1, O_2\}, \quad \zeta_2^{(0.7,0.3)} = \{\emptyset\}, \quad \zeta_3^{(0.7,0.3)} = \{O_1, O_9\},$$
$$\zeta_4^{(0.7,0.3)} = \{O_2, O_8\}, \quad \zeta_5^{(0.7,0.3)} = \{\emptyset\}, \quad \zeta_6^{(0.7,0.3)} = \{O_8, O_9\}.$$

Hence, by considering different (α, β)-levels corresponding to the satisfaction and dissatisfaction levels of customers, we can conclude that which organization fulfill the actual demands of a customer. The method adopted in this application is given in the following Algorithm 6.4.1.

Algorithm 6.4.1

Finding the most suitable organization

1. Input the degree of membership of all q−rung orthopair fuzzy vertices $O_1, O_2, O_3, \cdots, O_m$.
2. Calculate the membership degrees of q−rung orthopair fuzzy hyperedges using the formula,
$$T_\zeta(E_k) = T_\zeta(O_1, O_2, O_3, \ldots, O_m) \leq \min\{\mathscr{Q}_i(O_1), \mathscr{Q}_i(O_2), \mathscr{Q}_i(O_3), \ldots, \mathscr{Q}_i(O_m)\},$$
$$F_\zeta(E_k) = F_\zeta(O_1, O_2, O_3, \ldots, O_m) \leq \max\{\mathscr{Q}_i(O_1), \mathscr{Q}_i(O_2), \mathscr{Q}_i(O_3), \ldots, \mathscr{Q}_i(O_m)\},$$
for all $O_1, O_2, O_3, \cdots, O_m$ representing the organizations as vertices of hyperedge.
3. Calculate the (α, β)−levels of q−rung orthopair fuzzy hyperedges by using,
$$\zeta_i^{(\alpha,\beta)} = \{O_j \in O | T_{\zeta_i}(O_j) \geq \alpha, \ F_{\zeta_i}(O_j) \leq \beta\},$$
for $i = 1, 2, 3, \cdots, k, \ j = 1, 2, 3 \cdots, m$ and $\alpha, \beta \in [0, 1]$.
4. Crisp sets describe the most suitable organization according to the customer's satisfaction levels.

6.4.2 Adaptation of Most Alluring Residential Scheme

The essential factor for any purchase of property is the budget and location for a purchaser, particularly. However, it is a complicated procedure to select a residential area for buying a house. In addition to scrutinizing the further details such as the pricing, loan options, payments, and developer's credentials a customer must examine closely some other facilities which should be possessed by every housing colony. Now, to adopt a favorable housing scheme, an obvious initial step is to compare the differen societies. After analyzing the characteristics of different societies, one will be able to make a wise decision. We will investigate the problem of adopting the most alluring residential scheme using 7-rung orthopair fuzzy hypergraph. Let the set of vertices of 7-rung orthopair fuzzy hypergraph is taken as the representative of those attributes characteristics, which one has been considered to make a comparison between different housing societies. The hyperedges of 7-rung orthopair fuzzy hypergraph represents some housing schemes, which will be compared. The portrayal of our problem is illustrated in Fig. 6.7.

The description of hyperedges $\{\zeta_1, \zeta_2, \zeta_3, \zeta_4, \zeta_5, \zeta_6, \zeta_7\}$ and vertices $\{x_1, x_2, x_3, x_4, x_5, x_6, x_7, x_8, x_9, x_{10}\}$ of above hypergraph is given in Tables 6.5 and 6.6, respectively.

Note that, each hyperedge represents a distinct housing scheme and the vertices contained in hyperedges are those attributes, which will be provided by the societies represented through hyperedges. It means that Senate Avenue housing society provides 80% the basic facilities of life such as water, gas, and electricity and 20% deprives these facilities. Similarly, the same society 90% accommodates its residents being easy assessable and only 10% lacks the facility. In the same way, taking into account the truth-membership and falsity-membership degrees of all other attributes, we can identify the characteristics of all societies.

Now, to determine the overall comforts of each society, we will calculate the heights of all hyperedges and the society having the maximum truth-membership

Fig. 6.7 7-rung orthopair fuzzy hypergraph model

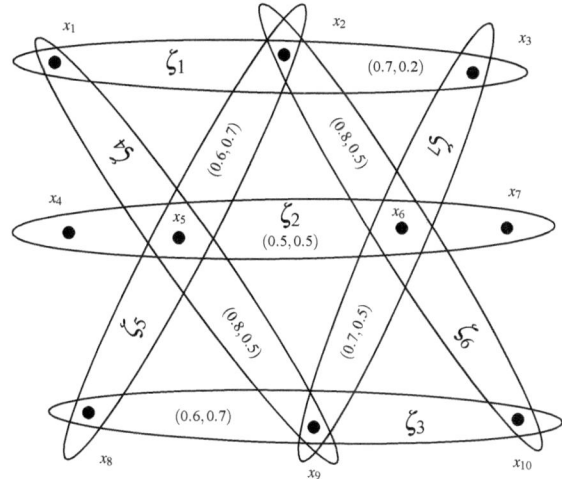

Table 6.5 Description of hyperedges

Set of hyperedges	Corresponding housing scheme	Provision of facilities (%)	Deprival of facilities (%)
ζ_1	Senate avenue	70	20
ζ_2	Soan gardens	50	50
ζ_3	CBR town	60	70
ζ_4	OPF housing scheme	80	50
ζ_5	Paradise city	60	70
ζ_6	RP corporation	80	50
ζ_7	Tele gardens housing scheme	70	50

Table 6.6 Description of attributes

Set of attributes	Depicting facility	Provision level of corresponding facility	Deprival of corresponding facility
x_1	Basic amenities of life	0.8	0.2
x_2	Easily assessable	0.9	0.1
x_3	Land ownership	0.7	0.2
x_4	The power back-up	0.6	0.3
x_5	Eco-friendly construction	0.9	0.4
x_6	Social infrastructure	0.8	0.5
x_7	Drainage system	0.5	0.6
x_8	Security	0.6	0.7
x_9	Regular sanitation	0.8	0.5
x_{10}	The parking area	0.9	0.3

Table 6.7 Heights of hyperedges

Heights of hyperedges	$(\max(\zeta_l), \min(\zeta_m))$
h (Senate Avenue)	$(0.9, 0.1)$
h (Soan Gardens)	$(0.9, 0.3)$
h (CBR Town)	$(0.9, 0.3)$
h (OPF Housing Scheme)	$(0.9, 0.2)$
h (Paradise City)	$(0.9, 0.1)$
h (RP Corporation)	$(0.9, 0.1)$
h (Tele Gardens Housing Scheme)	$(0.8, 0.2)$

and minimum falsity-membership will be considered as a most comfortable society to be live in. The calculated heights of all schemes are given in Table 6.7.

It can be noted from Table 6.7 that there are three societies which have the maximum membership and minimum nonmembership degrees, i.e., Senate Avenue, Paradise City, and RP Corporation are those housing societies which will provide 90% facilities to their habitants and only 10% amenities will be dispersed. Thus, it is more beneficial and substantial to select one of these three housing schemes.

The same problem can be speculated to a more extended idea that if some one wants to built a new housing scheme, which will carry out the facilities of all above societies. The concept of 7-rung orthopair fuzzy hypergraphs can be utilized to speculate such housing scheme. Consider a 7-rung orthopair fuzzy set of vertices given as follows,

$$\tau_1 = \{(x_1, 0.8, 0.2), (x_2, 0.9, 0.1), (x_5, 0.9, 0.3), (x_6, 0.8, 0.2), (x_{10}, 0.9, 0.3)\}.$$

By applying the definition of 7-rung orthopair fuzzy transversal, it can be seen that

$$\zeta_1^{(0.9,0.1)} \cap \tau_1^{(0.9,0.1)} = \{x_2\}, \qquad \zeta_2^{(0.9,0.3)} \cap \tau_1^{(0.9,0.3)} = \{x_5\},$$
$$\zeta_3^{(0.9,0.3)} \cap \tau_1^{(0.9,0.3)} = \{x_{10}\}, \qquad \zeta_4^{(0.9,0.2)} \cap \tau_1^{(0.9,0.2)} = \{x_5\},$$
$$\zeta_5^{(0.9,0.1)} \cap \tau_1^{(0.9,0.1)} = \{x_2\}, \qquad \zeta_6^{(0.9,0.1)} \cap \tau_1^{(0.9,0.1)} = \{x_2\},$$
$$\zeta_7^{(0.8,0.2)} \cap \tau_1^{(0.8,0.2)} = \{x_6\},$$

that is, the q-rung orthopair fuzzy subset τ_1 satisfies the condition of transversal and the housing society that will be represented through this hyperedge will contain at least one attribute of each scheme mentioned above. Similarly, some other societies can be figured out by following the same method. Hence, some other 7-rung orthopair fuzzy subsets are given as

$$\tau_2 = \{(x_1, 0.8, 0.2), (x_2, 0.9, 0.1), (x_3, 0.7, 0.2), (x_5, 0.9, 0.3), (x_6, 0.8, 0.2), (x_{10}, 0.9, 0.3)\},$$
$$\tau_3 = \{(x_2, 0.9, 0.1), (x_4, 0.6, 0.3), (x_5, 0.9, 0.3), (x_6, 0.8, 0.2), (x_{10}, 0.9, 0.3)\},$$
$$\tau_4 = \{(x_2, 0.9, 0.1), (x_5, 0.9, 0.3), (x_6, 0.8, 0.2), (x_{10}, 0.9, 0.3)\},$$
$$\tau_5 = \{(x_2, 0.9, 0.1), (x_5, 0.9, 0.3), (x_6, 0.8, 0.2), (x_7, 0.5, 0.5), (x_8, 0.6, 0.7), (x_{10}, 0.9, 0.3)\}.$$

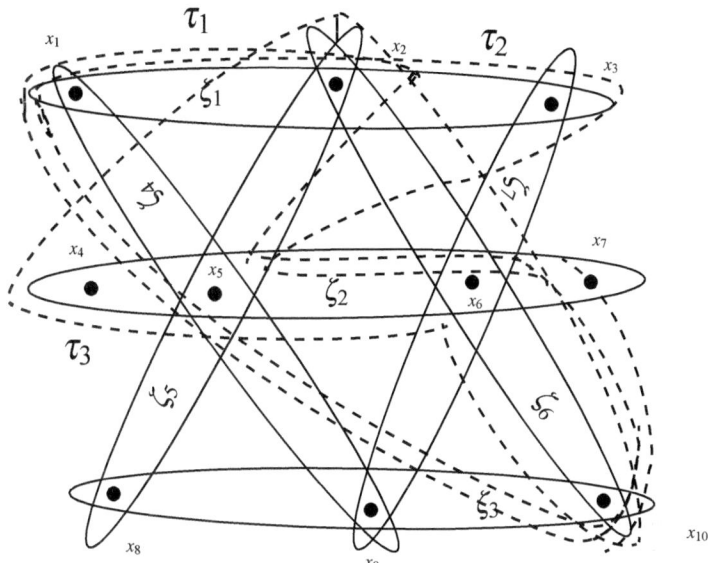

Fig. 6.8 7-rung orthopair fuzzy transversals

The graphical description of these schemes is displayed in Fig. 6.8 through dashed lines.

Thus, the schemes shown through dashed lines will contain the attributes of all other societies and may be more advantageous to their dwellers. The method adopted in our application is explained through the Algorithm 6.4.2.

Algorithm 6.4.2

Finding the more advantageous schemes

1. Input the degree of membership of all q−rung orthopair fuzzy vertices $x_1, x_2, x_3, \cdots, x_m$.
2. Calculate the membership degrees of q−rung orthopair fuzzy hyperedges using the formula,
$$T_\zeta(E_k) \le \min\{\mathcal{Q}_i(x_1), \mathcal{Q}_i(x_2), \ldots, \mathcal{Q}_i(x_m)\},$$
$$F_\zeta(E_k) \le \max\{\mathcal{Q}_i(x_1), \mathcal{Q}_i(x_2), \ldots, \mathcal{Q}_i(x_m)\},$$
for all $x_1, x_2, x_3, \cdots, x_m$ representing the attributes of housing societies.
3. Calculate the heights of all q−rung orthopair fuzzy hyperedges using,
$$h(\zeta_j) = (\max T_{\zeta_j}(x_i), \min F_{\zeta_j}(x_i)),$$
$$j = 1, 2, 3, \cdots, k \text{ and } i = 1, 2, 3, \cdots, m.$$
4. Maximum truth-membership and minimum falsity-membership will denote the most alluring residential area.
5. Input the different q−rung orthopair fuzzy subsets.
6. Determine the q−rung orthopair fuzzy transversals using the formula,
7. Find the more advantageous schemes, which will contain the attributes of all other societies.

6.5 q-Rung Orthopair Fuzzy Directed Hypergraphs

In this section, we define q-rung orthopair fuzzy digraphs and q-rung orthopair fuzzy directed hypergraphs. A q-rung orthopair fuzzy directed hypergraph generalizes the concept of an intuitionistic fuzzy directed hypergraph and broaden the space of orthopairs. We also define and construct the dual and line graphs of q-rung orthopair fuzzy directed hypergraphs. All these concepts are explained and justified through concrete examples.

Definition 6.23 A *q-rung orthopair fuzzy digraph* on a non-empty set X is a pair $\overrightarrow{D} = (\mathscr{A}, \overrightarrow{\mathscr{B}})$, where $\mathscr{A}$ is a q-rung orthopair fuzzy set on X and $\overrightarrow{\mathscr{B}}$ is a q-rung orthopair fuzzy relation on X such that

$$T_{\overrightarrow{\mathscr{B}}}(x_1x_2) \leq \min\{T_{\mathscr{A}}(x_1), T_{\mathscr{A}}(x_2)\}, \quad F_{\overrightarrow{\mathscr{B}}}(x_1x_2) \leq \max\{F_{\mathscr{A}}(x_1), F_{\mathscr{A}}(x_2)\},$$

and $0 \leq T_{\overrightarrow{\mathscr{B}}}^q(x_1x_2) + F_{\overrightarrow{\mathscr{B}}}^q(x_1x_2) \leq 1, q \geq 1$, for all $x_1, x_2 \in X$.

Remark 6.4 • When $q = 1$, 1-rung orthopair fuzzy digraph is called an intuitionistic fuzzy digraph.
• When $q = 2$, 2-rung orthopair fuzzy digraph is called Pythagorean fuzzy digraph.

Example 6.6 Let $X = \{x_1, x_2, x_3, x_4\}$ be the set of universe, $\mathscr{A} = \{(x_1, 0.7, 0.8), (x_2, 0.6, 0.9), (x_3, 0.5, 0.8), (x_4, 0.7, 0.8)\}$ be a 5-rung orthopair fuzzy set and $\overrightarrow{\mathscr{B}}$ be a 5-rung orthopair fuzzy relation on X such that, $0 \leq T_{\overrightarrow{\mathscr{B}}}^5(x_ix_j) + F_{\overrightarrow{\mathscr{B}}}^5(x_ix_j) \leq 1$, for all $x_i, x_j \in X$. The corresponding 5-rung orthopair fuzzy digraph $\overrightarrow{D} = (\mathscr{A}, \overrightarrow{\mathscr{B}})$ is shown in Fig. 6.9.

Definition 6.24 A *q-rung orthopair fuzzy directed hypergraph* $\mathscr{D}$ on X is defined as an ordered pair $\mathscr{D} = (Q, \xi)$, where Q is the collection of q-rung orthopair fuzzy subsets of X and ξ is a family of q-rung orthopair fuzzy directed hyperedges (or hyperarcs) such that,

Fig. 6.9 5-rung orthopair fuzzy digraph $\overrightarrow{D}$

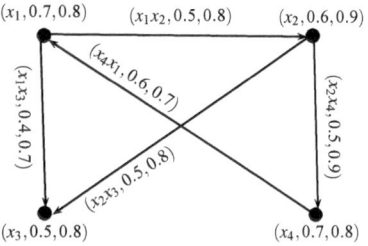

1.
$$T_\xi(E_k) = T_\xi(x_1, \ldots, x_m) \le \min\{T_{Q_i}(x_1), \ldots, T_{Q_i}(x_m)\},$$

$$F_\xi(E_k) = F_\xi(x_1, \ldots, x_m) \le \max\{F_{Q_i}(x_1), \ldots, F_{Q_i}(x_m)\},$$

for all $x_1, x_2, \ldots, x_m \in X$.
2. $\bigcup_i supp(Q_i) = X$, for all $Q_i \in Q$.

A q-rung orthopair fuzzy directed hyperedge $\xi_i \in \xi$ is defined as an ordered pair $(h(\xi_i), t(\xi_i))$, where $h(\xi_i)$ and $t(\xi_i) \in X - h(\xi_i)$, nontrivial subsets of X, are called the *head* of ξ_i and *tail* of ξ_i, respectively.
A *source vertex* v in ξ_i is defined as $h(\xi_i) \ne v$, for all $\xi_i \in \xi$ and a *destination vertex* v' in ξ_i is defined as $t(\xi_i) \ne v'$, for all $\xi_i \in \xi$.

Definition 6.25 A q-rung orthopair fuzzy directed hypergraph is called a *backward q-rung orthopair fuzzy directed hypergraph* if all of its hyperarcs are B-arcs, i.e., $\xi_i = (h(\xi_i), t(\xi_i))$ with $|h(\xi_i)| = 1$, for all $\xi_i \in \xi$.
A q-rung orthopair fuzzy directed hypergraph is called a *forward q-rung orthopair fuzzy directed hypergraph* if all of its hyperarcs are F-arcs, i.e., $\xi_i = (h(\xi_i), t(\xi_i))$ with $|t(\xi_i)| = 1$, for all $\xi_i \in \xi$.

Definition 6.26 The *height* of a q-rung orthopair fuzzy directed hypergraph $\mathscr{D} = (Q, \xi)$ is defined as $h^\star(\mathscr{D}) = \{\max(\xi_l), \min(\xi_m)\}$, where $\xi_l = \max T_{\xi_j}(x_i)$ and $\xi_m = \min F_{\xi_j}(x_i)$. Here, $T_{\xi_j}(x_i)$ and $F_{\xi_j}(x_i)$ denote the truth-membership and falsity-membership of vertex x_i to the directed hyperedge ξ_j, respectively.

Definition 6.27 Let $\mathscr{D} = (Q, \xi)$ be a q-rung orthopair fuzzy directed hypergraph. The *order* of $\mathscr{D}$, which is denoted by $O(\mathscr{D})$, and is defined as $O(\mathscr{D}) = \sum_{x \in X} \wedge \xi_i(x)$.
The *size* of $\mathscr{D}$, which is denoted by $S(\mathscr{D})$, and is defined as $S(\mathscr{D}) = \sum_{x \in X} \vee \xi_i(x)$.

Definition 6.28 A repeatedly occurring sequence $v_1, \xi_1, v_2, \xi_2, \ldots, v_{n-1}, \xi_{n-1}, v_n$ of definite vertices and directed hyperarcs such that,

- $0 < T_\xi(\xi_i) \le 1$ and $0 \le F_\xi(\xi_i) < 1$,
- $v_{i-1}, v_i \in \xi_i$, $i = 1, 2, 3, \ldots, n$,

is called a *q-rung orthopair fuzzy directed hyperpath* of length $n - 1$ from v_1 to v_n. If $v_1 = v_n$, then this q-rung orthopair fuzzy directed hyperpath is called a *q-rung orthopair fuzzy directed hypercycle*.

Definition 6.29 The *strength of q-rung orthopair fuzzy directed hyperpath* of length k, which connects the two vertices v_1 and v_2, is defined as $\lambda^k(v_1, v_2) = \{\min\{T_\xi(\xi_1), T_\xi(\xi_2), T_\xi(\xi_3), \ldots, T_\xi(\xi_k)\}, \max\{F_\xi(\xi_1), F_\xi(\xi_2), F_\xi(\xi_3), \ldots, F_\xi(\xi_k)\}\}, v_1 \in \xi_1, v_2 \in \xi_k$ and $\xi_1, \xi_2, \xi_3, \ldots, \xi_k$ are q-rung orthopair fuzzy directed hyperedges.
   The *strength of connectedness* between v_1 and v_2 is given as, $\lambda^\infty(v_1, v_2) = \{\max_k T(\lambda^k(v_1, v_2)), \min_k F(\lambda^k(v_1, v_2))\}$.

Fig. 6.10 A 5-rung orthopair fuzzy directed hypergraph

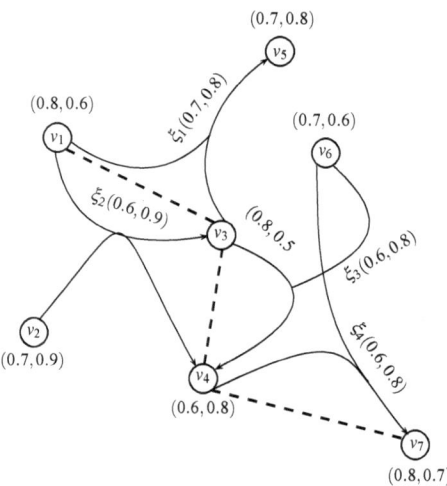

A *connected q-rung orthopair fuzzy directed hypergraph* is one in which we have at least one q-rung orthopair fuzzy directed hyperpath between each pair of vertices of $\mathscr{D}$.

We now illustrate the Definitions 6.24, 6.25, 6.26, 6.27, 6.28 and 6.29 through an example of 5-rung orthopair fuzzy directed hypergraph.

Example 6.7 Consider a 5-rung orthopair fuzzy directed hypergraph $\mathscr{D} = (Q, \xi)$, as shown in Fig. 6.10.

In this 5-rung orthopair fuzzy directed hypergraph, we have

$$\xi_1 = \{\{(v_1, 0.8, 0.6), (v_3, 0.8, 0.5)\}, \{(v_5, 0.7, 0.8)\}\} = \{t(\xi_1), h(\xi_1)\},$$
$$\xi_2 = \{\{(v_1, 0.8, 0.6), (v_2, 0.7, 0.9)\}, \{(v_3, 0.8, 0.5), (v_4, 0.6, 0.8)\}\} = \{t(\xi_2), h(\xi_2)\},$$
$$\xi_3 = \{\{(v_3, 0.8, 0.5), (v_6, 0.7, 0.6)\}, \{(v_4, 0.6, 0.8)\}\} = \{t(\xi_3), h(\xi_3)\},$$
$$\xi_4 = \{\{(v_4, 0.6, 0.8), (v_6, 0.7, 0.6)\}, \{(v_7, 0.8, 0.7)\}\} = \{t(\xi_4), h(\xi_4)\}.$$

A 5-rung orthopair fuzzy directed hyperpath from v_1 to v_7 of length 3 is shown through dashed lines and is given by an alternating sequence $v_1, \xi_2, v_3, \xi_3, v_4, \xi_4, v_7$ of distinct vertices and directed hyperarcs. The strength of this hyperpath is

$$\lambda^3(v_1, v_7) = \{\min\{T_\xi(\xi_2), T_\xi(\xi_3), T_\xi(\xi_4)\}, \max\{F_\xi(\xi_2), F_\xi(\xi_3), F_\xi(\xi_4)\}\}$$
$$= (0.6, 0.9),$$
$$\lambda^\infty(v_1, v_7) = (0.6, 0.9).$$

Note that, $\mathscr{D} = (Q, \xi)$ is not connected because we don't have a directed hyperpath between each pair of vertices, i.e., v_1 is not connected to v_6. A backward and forward 5-rung orthopair fuzzy directed hypergraph is shown in Fig. 6.11a, b, respectively.

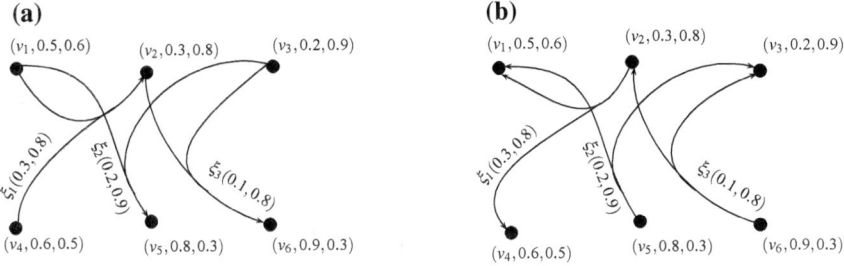

Fig. 6.11 Backward and forward 5-rung orthopair fuzzy directed hypergraphs

Definition 6.30 A *q*-rung orthopair fuzzy directed hypergraph $\mathscr{D} = (Q, \xi)$ is *linear* if every pair of *q*-rung orthopair fuzzy directed hyperedges $\xi_i, \xi_j \in \xi$ satisfies

- $supp(\xi_i) \subseteq supp(\xi_j) \Rightarrow i = j$,
- $|supp(\xi_i) \cap supp(\xi_j)| \leq 1$.

Example 6.8 Consider a 5-rung orthopair fuzzy directed hypergraph $\mathscr{D} = (Q, \xi)$, as shown in Fig. 6.10. In this 5-rung orthopair fuzzy directed hypergraph, we have $supp(\xi_1) = \{v_1, v_3, v_5\}$, $supp(\xi_2) = \{v_1, v_2, v_3, v_4\}$, $supp(\xi_3) = \{v_3, v_6, v_4\}$, $supp(\xi_4) = \{v_4, v_6, v_7\}$. Note that, $supp(\xi_i) \subseteq supp(\xi_j) \Rightarrow i = j$ and

$$|supp(\xi_1) \cap supp(\xi_2)| = |\{v_1, v_3\}| = 2,$$
$$|supp(\xi_1) \cap supp(\xi_3)| = |\{v_3\}| = 1,$$
$$|supp(\xi_1) \cap supp(\xi_4)| = |\{\emptyset\}| = 0,$$
$$|supp(\xi_2) \cap supp(\xi_3)| = |\{v_4, v_3\}| = 2,$$
$$|supp(\xi_2) \cap supp(\xi_4)| = |\{v_4\}| = 1,$$
$$|supp(\xi_3) \cap supp(\xi_4)| = |\{v_4, v_6\}| = 2.$$

That is, $|supp(\xi_i) \cap supp(\xi_j)| \nleq 1$, for all $\xi_i, \xi_j \in \xi$. Hence, $\mathscr{D} = (Q, \xi)$ is not linear.

Definition 6.31 Let $\mathscr{D} = (Q, \xi)$ be a *q*-rung orthopair fuzzy directed hypergraph. The *q*-*rung orthopair fuzzy line graph* of $\mathscr{D}$ is the graph $l(\mathscr{D}) = (X_l, \xi_l)$ such that,

1. $X_l = \xi$,
2. $\{\xi_i, \xi_j\} \in \xi_l \Leftrightarrow |supp(\xi_i) \cap supp(\xi_j)| \neq \emptyset$, for $i \neq j$.

The truth-membership and falsity-membership of vertices and edges of $l(\mathscr{D})$ are determined as follows:

- $X_l(\xi_i) = \xi(\xi_i)$,
- $T_{\xi_l}(\{\xi_i, \xi_j\}) = \min\{T_\xi(\xi_i), T_\xi(\xi_j)|\xi_i, \xi_j \in \xi\}$, $\qquad$ $F_{\xi_l}(\{\xi_i, \xi_j\}) = \max\{F_\xi(\xi_i),$ $F_\xi(\xi_j)|\xi_i, \xi_j \in \xi\}$.

Theorem 6.10 *Let $\mathscr{G} = (U, \varepsilon)$ be a simple q-rung orthopair fuzzy directed graph. Then $\mathscr{G}$ is the q-rung orthopair fuzzy line graph of a linear q-rung orthopair fuzzy directed hypergraph.*

Proof Let $\mathscr{G} = (U, \varepsilon)$ be a simple q-rung orthopair fuzzy directed graph. We suppose that $\mathscr{G} = (U, \varepsilon)$ is connected, with no loss of generality. A q-rung orthopair fuzzy directed hypergraph $\mathscr{D} = (Q, \xi)$ can be formulated from $\mathscr{G}$ as follows:

(i) The set of directed edges of $\mathscr{G}$ will be taken as vertices of $\mathscr{D}$, i.e., $\varepsilon = \{\varepsilon_1, \varepsilon_2, \varepsilon_3, \ldots, \varepsilon_n\}$ be the directed edges of $\mathscr{G}$ and hence the set of vertices of $\mathscr{D}$. Let $X = \{q_1, q_2, q_3, \ldots, q_k\}$ be the set of nontrivial q-rung orthopair fuzzy sets on U such that $q_i(\varepsilon_j) = (1, 0)$, $i = 1, 2, 3, \ldots, k$, $j = 1, 2, 3, \ldots, n$.

(ii) Let $U = \{u_1, u_2, u_3, \ldots, u_j\}$ then the directed hyperedges of $\mathscr{D}$ are $\xi = \{\xi_1, \xi_2, \xi_3, \ldots, \xi_n\}$, where ξ_i are those directed edges of $\mathscr{G}$, which contain the vertex u_i as their incidence vertex, i.e., $\xi_i = \{\varepsilon_j | u_i \in \varepsilon_j, j = 1, 2, 3, \ldots, n\}$. Moreover, $\xi(\xi_i) = U(u_i)$, $i = 1, 2, 3, \ldots, k$.

We now claim that $\mathscr{D} = (Q, \xi)$ is linear q-rung orthopair fuzzy directed hypergraph. Consider an arbitrary directed hyperedge $\xi_j = \{\varepsilon_1, \varepsilon_2, \varepsilon_3, \ldots, \varepsilon_r\}$ and from the defining relation of q-rung orthopair fuzzy directed hypergraph, we have

$$T_\xi(\xi_j) = \min\{T_{q_j}(\varepsilon_1), T_{q_j}(\varepsilon_2), \ldots, T_{q_j}(\varepsilon_r)\} = T_U(u_i) \leq 1,$$
$$F_\xi(\xi_j) = \max\{F_{q_j}(\varepsilon_1), F_{q_j}(\varepsilon_2), \ldots, F_{q_j}(\varepsilon_r)\} = F_U(u_i) \geq 0,$$

$i = 1, 2, 3, \ldots, k$ and $\bigcup_k supp(q_k) = X$, for all q_k.

We now prove that $\mathscr{D} = (Q, \xi)$ is linear.

1. By our supposition, membership degree of each vertex ε_i of $\mathscr{D}$ is $(1, 0)$. Thus, we have $supp(\xi_i) \subseteq supp(\xi_j)$ implies $i = j$.

2. Suppose on contrary that $|supp(\xi_i) \cap supp(\xi_j)| = \{\varepsilon_l, \varepsilon_m\}$, i.e., these edges have two incidence vertices in common, which is contradiction to the fact that $\mathscr{G}$ is simple. Hence, $|supp(\xi_i) \cap supp(\xi_j)| \leq 1$, for $1 \leq i, j \leq r$.

Theorem 6.11 *A necessary and sufficient condition for $l(\mathscr{D})$ to be connected is that $\mathscr{D}$ is connected.*

Proof Let $\mathscr{D} = (Q, \xi)$ be a connected q-rung orthopair fuzzy directed hypergraph and $l(\mathscr{D}) = (X_l, \xi_l)$ be the line graph of $\mathscr{D}$. Suppose that ξ_i and ξ_j be two vertices of $l(\mathscr{D})$ and $v_i \in \xi_i$, $v_j \in \xi_j$, for $v_i \neq v_j$. Since $\mathscr{D}$ is connected then there exists an alternating sequence $v_i, \xi_i, v_{i+1}, \xi_{i+1}, \ldots, \xi_j, v_j$, which connects v_i and v_j. From the definition of strength of connectedness between v_i and v_j, we have

$$\lambda^{\infty}(\xi_i, \xi_j) = \max_k T(\lambda^k(\xi_i, \xi_j)), \min_k F(\lambda^k(\xi_i, \xi_j))$$

$$= \{\max_k(T_{\xi_l}(\xi_i, \xi_{i+1}) \wedge T_{\xi_l}(\xi_{i+1}, \xi_{i+2}) \wedge \cdots \wedge T_{\xi_l}(\xi_{j-1}, \xi_j)),$$

$$\min_k(F_{\xi_l}(\xi_i, \xi_{i+1}) \vee F_{\xi_l}(\xi_{i+1}, \xi_{i+2}) \vee \cdots \vee F_{\xi_l}(\xi_{j-1}, \xi_j))\}, k = 1, 2, \ldots$$

$$= \{\max_k(T_{\xi_l}(\xi_i) \wedge T_{\xi_l}(\xi_{i+1}) \wedge T_{\xi_l}(\xi_{i+2}) \wedge \cdots \wedge T_{\xi_l}(\xi_{j-1}) \wedge T_{\xi_l}(\xi_j)),$$

$$\min_k(F_{\xi_l}(\xi_i) \vee F_{\xi_l}(\xi_{i+1}) \vee F_{\xi_l}(\xi_{i+2}) \vee \cdots \vee F_{\xi_l}(\xi_{j-1}) \vee F_{\xi_l}(\xi_j))\},$$

$$= \max T(\lambda^k(v_i, v_j)), \min F(\lambda^k(v_i, v_j))$$

$$= \lambda^{\infty}(v_i, v_j) > 0.$$

Hence, $l(\mathscr{D})$ is connected. By reversing the same procedure, we can easily prove that if $l(\mathscr{D})$ is connected then $\mathscr{D}$ is connected.

Let $\mathscr{D} = (Q, \xi)$ be a q-rung orthopair fuzzy directed hypergraph. The construction of a q-rung orthopair fuzzy directed line graph from a q-rung orthopair fuzzy directed hypergraph is illustrated in Algorithm 6.5.1.

Algorithm 6.5.1

Finding the q−rung orthopair fuzzy directed line graph
---
1. Input the number of directed hyperedges r of q−rung orthopair fuzzy directed hypergraph $\mathscr{D} = (Q, \xi)$.
2. Input the truth-membership and falsity membership of directed hyperedges $\xi_1, \xi_2, \xi_3, \cdots, \xi_r$.
3. Construct a q−rung orthopair fuzzy line graph $l(\mathscr{D}) = (X_l, \xi_l)$, whose vertices are taken as the directed hyperedges $\xi_1, \xi_2, \xi_3, \cdots, \xi_r$.
4. Calculate the degrees of membership of vertices $l(\mathscr{D}) = (X_l, \xi_l)$ as $X_l(\xi_j) = \xi(\xi_j)$.
5. Draw an edge between ξ_i and ξ_j in $l(\mathscr{D})$ if $|supp(\xi_i) \cap supp(\xi_j)| \geq 1$.
6. Calculate the degrees of membership of edges in $l(\mathscr{D})$ as,

$$\xi_l(\xi_i\xi_j) = (\min\{T_{\xi}(\xi_i), T_{\xi}(\xi_j)\}, \max\{F_{\xi}(\xi_i), F_{\xi}(\xi_j)\}).$$

Definition 6.32 The *2-section graph* of a q-rung orthopair fuzzy directed hypergraph $\mathscr{D} = (Q, \xi)$ is a q-rung orthopair fuzzy graph $[\mathscr{D}]_2 = (X', \mathscr{E})$ such that

(i) $X = X'$, i.e., the set of vertices of both graphs is same.
(ii) $\mathscr{E} = \{v_i v_j | v_i \neq v_j, v_i v_j \in \xi_k, k = 1, 2, 3, \ldots\}$, i.e., v_i and v_j are adjacent in $\mathscr{D}$.

We now justify the Definitions 6.31 and 6.32 through Example 6.9.

Example 6.9 Let $\mathscr{D} = (Q, \xi)$ be a 7-rung orthopair fuzzy directed hypergraph as shown in Fig. 6.12. By following the above Algorithm 6.5.1, it's line graph is constructed and shown by dashed lines.

The 2-section graph of 7-rung orthopair fuzzy directed hypergraph given in Fig. 6.12 is shown in Fig. 6.13.

Definition 6.33 Let $\mathscr{D} = (Q, \xi)$ be a q-rung orthopair fuzzy directed hypergraph. The *dual q−rung orthopair fuzzy directed hypergraph* $\mathscr{D}^d = (X^d, \xi^d)$ of $\mathscr{D} = (Q, \xi)$ is defined as,

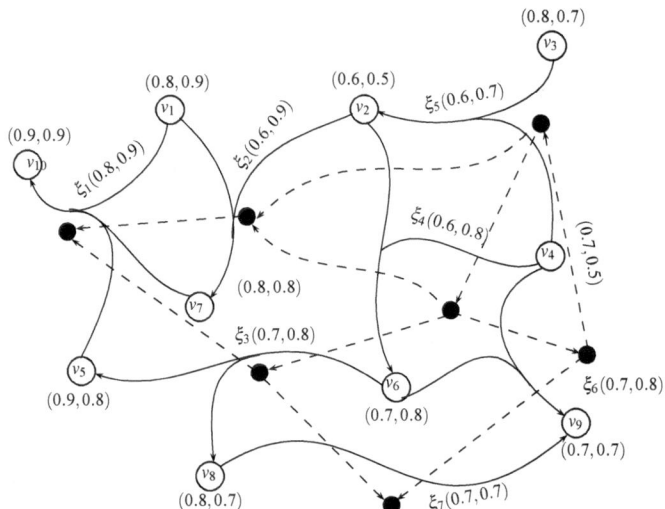

Fig. 6.12 A 7-rung orthopair fuzzy directed hypergraph and its line graph

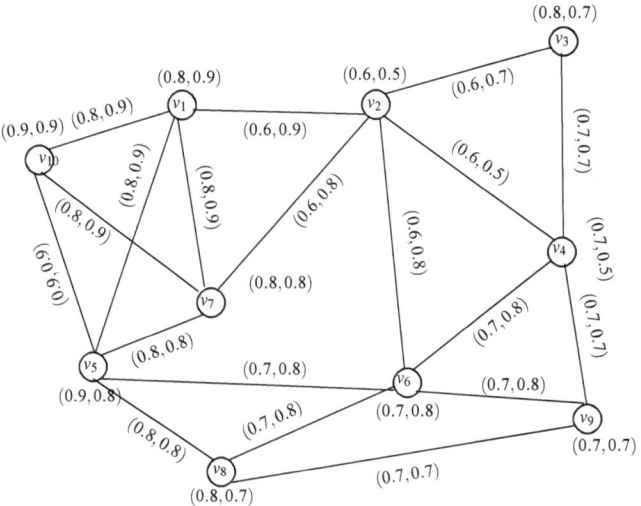

Fig. 6.13 The 2-section graph of 7-rung orthopair fuzzy directed hypergraph

(i) $X^d = \xi$ is the q-rung orthopair fuzzy set of vertices of $\mathscr{D}^d$.

(ii) If $|X| = n$, then ξ^d is q-rung orthopair fuzzy set on the set of directed hyperedges $\{X_1, X_2, X_3, \ldots, X_n\}$ such that $X_i = \{\xi_j | v_i \in \xi_j, \xi_j \in \xi\}$, i.e., X_i is the set of those directed hyperedges in which v_i is a common vertex.

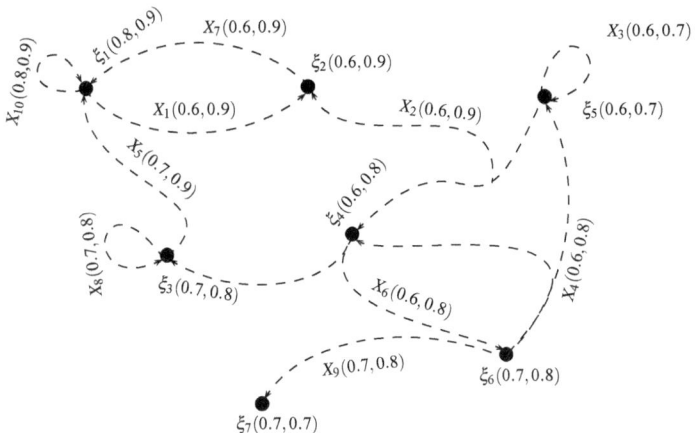

Fig. 6.14 Dual directed hypergraph of 7-rung orthopair fuzzy directed hypergraph

The membership degrees of X_i are defined as

$$T_{\xi^d}(X_i) = \min\{T_\xi(\xi_j)|v_i \in \xi_j\}, \quad F_{\xi^d}(X_i) = \max\{F_\xi(\xi_j)|v_i \in \xi_j\}.$$

The method of forming the dual of q-rung orthopair fuzzy directed hypergraph is described in Algorithm 6.5.2. We also explain this concept through an example.

Algorithm 6.5.2

The dual of q−rung orthopair fuzzy directed hypergraph

1. Input $\{v_1, v_2, v_3, \cdots, v_n\}$ the set of vertices and $\{\xi_1, \xi_2, \xi_3, \cdots, \xi_m\}$ the set of directed hyperedges of $\mathscr{D}$.
2. Formulate a q−rung orthopair fuzzy set of vertices of $\mathscr{D}^d$ as $X^d = \xi$.
3. Define a mapping $\psi : X \to \xi$, which maps the set of vertices to the directed hyperedges of $\mathscr{D}$, i.e.,
   if vertex v_i is contained in $\xi_l, \xi_{l+1}, \xi_{l+2}, \cdots, \xi_m$ then v_i is mapped onto $\xi_l, \xi_{l+1}, \xi_{l+2}, \cdots, \xi_m$.
4. Construct the directed hyperedges $\{X_1, X_2, X_3, \cdots, X_n\}$ of $\mathscr{D}^d$ such that $X_i = \{\xi_j|\psi(v_i) = \xi_j\}$.
5. Draw the q−rung orthopair fuzzy directed hyperedge, the vertex ξ_j of $\mathscr{D}^d$ is associated to $h(X_i)$ if
   and only if $v_i \in t(\xi_j)$ in $\mathscr{D}$ and viceversa.
6. Formulate the truth-membership and falsity-membership of directed hyperedges of $\mathscr{D}^d$ as,
   $$T_{\xi^d}(X_i) = \min\{T_\xi(\xi_j)|v_i \in \xi_j\}, \quad F_{\xi^d}(X_i) = \max\{F_\xi(\xi_j)|v_i \in \xi_j\}.$$

Example 6.10 Let $\mathscr{D} = (Q, \xi)$ be a 7-rung orthopair fuzzy directed hypergraph as shown in Fig. 6.12. The dual 7-rung orthopair fuzzy directed hypergraph $\mathscr{D}^d = (X^d, \xi^d)$ of $\mathscr{D} = (Q, \xi)$ is shown in Fig. 6.14, which is constructed by following the Algorithm 6.5.2.

Theorem 6.12 *The 2-section of dual of q-rung orthopair fuzzy directed hypergraph $[\mathscr{D}^d]_2$ is same as the line graph of $\mathscr{D}$, i.e., $[\mathscr{D}^d]_2 = l(\mathscr{D})$.*

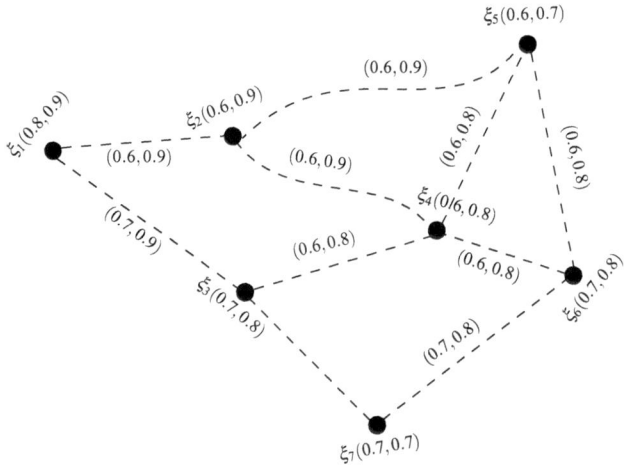

Fig. 6.15 $l(\mathscr{D})$

Proof Let $\mathscr{D} = (Q, \xi)$ be a q-rung orthopair fuzzy directed hypergraph having $\{v_1,$ $v_2, v_3, \ldots, v_n\}$ the set of vertices and $\{\xi_1, \xi_2, \xi_3, \ldots, \xi_m\}$ the set of directed hyperedges. Suppose that $l(\mathscr{D}) = (X_l, \xi_l)$, $\mathscr{D}^d = (X^d, \xi^d)$ and $[\mathscr{D}^d]_2 = (X^d, \mathscr{E})$ be the line graph, dual directed hypergraph, and 2-section of dual of $\mathscr{D}$, respectively. The 2-section $[\mathscr{D}^d]_2$ has the same vertex set as that of $l(\mathscr{D})$. Assume that the set of directed hyperedges of $\mathscr{D}^d$ be $\{X_1, X_2, X_3, \ldots, X_n\}$. Obviously $\{\xi_i \xi_j | \xi_i, \xi_j \in X_i\}$ are the edges of $[\mathscr{D}^d]_2$ and also the set of edges of $l(\mathscr{D})$. We now show that $\xi_l(\xi_i \xi_j) = \mathscr{E}(\xi_i \xi_j)$.

$$\xi_l(\xi_i \xi_j) = (\max\{T_\xi(\xi_i), T_\xi(\xi_i)\}, \min\{F_\xi(\xi_i), F_\xi(\xi_i)\}),$$
$$= (\max\{T_{\xi^d}(\xi_i), T_{\xi^d}(\xi_i)\}, \min\{F_{\xi^d}(\xi_i), F_{\xi^d}(\xi_i)\}),$$
$$= \mathscr{E}(\xi_i \xi_j),$$

which completes the proof.

We now justify the result of Theorem 6.12 through a concrete example.

Example 6.11 Let $\mathscr{D} = (Q, \xi)$ be a 7-rung orthopair fuzzy directed hypergraph as shown in Fig. 6.12. Its line graph is constructed and shown by dashed lines in Fig. 6.15.

The dual of $\mathscr{D}$ is shown in Fig. 6.14. We now determine the 2-section of $\mathscr{D}^d$, which is given in Fig. 6.16.

Thus, Figs. 6.15 and 6.16 show that $[\mathscr{D}^d]_2 = l(\mathscr{D})$.

Fig. 6.16 $[\mathscr{D}^d]_2$

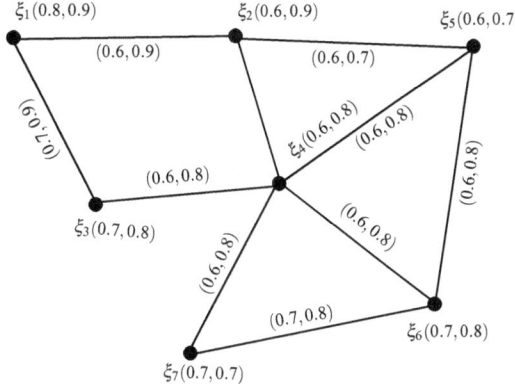

6.6 Coloring of q-Rung Orthopair Fuzzy Directed Hypergraphs

In this section, we define the (α, β)-level hypergraph of $\mathscr{D}$, which is a useful concept in the coloring of q-rung orthopair fuzzy directed hypergraphs. A sequence of real numbers, called the fundamental sequence of $\mathscr{D}$, is also defined using the (α, β)-level sets. The concept of the fundamental sequence is used to prove various results related to the coloring of q-rung orthopair fuzzy directed hypergraphs. Moreover, we define $\mathscr{L}$-coloring, chromatic number, and p-coloring of $\mathscr{D}$. We also prove some useful results, which simplify the complicated procedure of coloring and finding the chromatic number of q-rung orthopair fuzzy directed hypergraphs.

Definition 6.34 Let $\mathscr{D} = (X, \xi)$ be a q-rung orthopair fuzzy directed hypergraph. The (α, β)-*level hypergraph* of $\mathscr{D}$ is defined as $\mathscr{D}^{(\alpha,\beta)} = (X^{(\alpha,\beta)}, \xi^{(\alpha,\beta)})$, where

1. $\xi^{(\alpha,\beta)} = \{\xi_i^{(\alpha,\beta)} : \xi_i \in \xi\}$ and $\xi_i^{(\alpha,\beta)} = \{x \in X | T_{\xi_i}(x) \geq \alpha, F_{\xi_i}(x) \leq \beta\}$,
2. $X^{(\alpha,\beta)} = \bigcup\limits_{\xi_i \in \xi} \xi_i^{(\alpha,\beta)}$.

Definition 6.35 Let $\mathscr{D} = (X, \xi)$ be a q-rung orthopair fuzzy directed hypergraph and $\mathscr{D}^{(\alpha,\beta)}$ be the (α, β)-level hypergraph of $\mathscr{D}$. The sequence of real numbers $\rho_1 = (T_{\rho_1}, F_{\rho_1})$, $\rho_2 = (T_{\rho_2}, F_{\rho_2})$, $\rho_3 = (T_{\rho_3}, F_{\rho_3}), \ldots, \rho_n = (T_{\rho_n}, F_{\rho_n})$, $0 < T_{\rho_1} < T_{\rho_2} < T_{\rho_3} < \cdots < T_{\rho_n}$, $F_{\rho_1} > F_{\rho_2} > F_{\rho_3} > \cdots > F_{\rho_n} > 0$, where $(T_{\rho_n}, F_{\rho_n}) = h(\mathscr{H})$ such that,

(i) if $\rho_{i-1} = (T_{\rho_{i-1}}, F_{\rho_{i-1}}) < \rho = (T_\rho, F_\rho) \leq \rho_i = (T_{\rho_i}, F_{\rho_i})$ then $\xi^\rho = \xi^{\rho_i}$,
(ii) $\xi^{\rho_i} \subseteq \xi^{\rho_{i+1}}$,

is called the *fundamental sequence* of $\mathscr{D}$, denoted by $f_S(\mathscr{D})$. The set of ρ_i-level hypergraphs $\{\mathscr{D}^{\rho_1}, \mathscr{D}^{\rho_2}, \mathscr{D}^{\rho_3}, \ldots, \mathscr{D}^{\rho_n}\}$ is called the *core hypergraphs* of $\mathscr{D}$ or simply the *core set* of $\mathscr{D}$ and is denoted by $c(\mathscr{D})$.

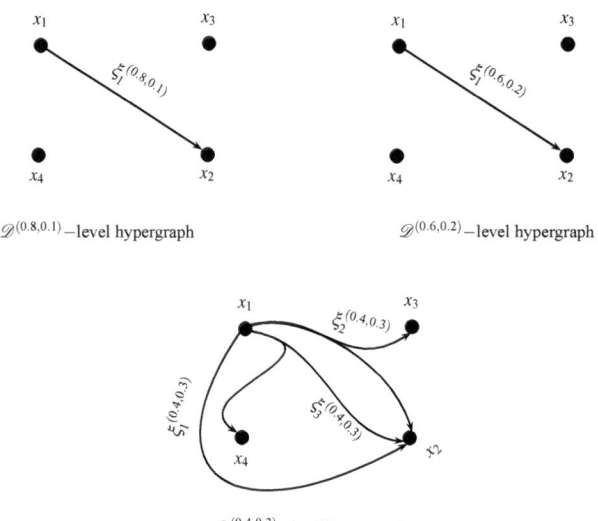

Fig. 6.17 Fundamental sequence of $\mathscr{D}$

Definition 6.36 A q-rung orthopair fuzzy directed hypergraph $\mathscr{D} = (X, \xi)$ is *ordered* if $c(\mathscr{D}) = \{\mathscr{D}^{\rho_1}, \mathscr{D}^{\rho_2}, \mathscr{D}^{\rho_3}, \ldots, \mathscr{D}^{\rho_n}\}$ is ordered, i.e., $\mathscr{D}^{\rho_1} < \mathscr{D}^{\rho_2} < \mathscr{D}^{\rho_3} < \cdots < \mathscr{D}^{\rho_n}$ and is *simply ordered* if $c(\mathscr{D})$ is simply ordered.

Example 6.12 Consider a 2-rung orthopair fuzzy directed hypergraph $\mathscr{D} = (X, \xi)$, where $X = \{x_1, x_2, x_3, x_4\}$ and $\xi = \{\xi_1, \xi_2, \xi_3\}$ such that $\xi_1 = \{(x_1, 0.8, 0.1), (x_2, 0.8, 0.1)\}, \xi_2 = \{(x_1, 0.6, 0.2), (x_2, 0.6, 0.2), (x_3, 0.4, 0.3)\}, \xi_3 = \{(x_1, 0.4, 0.3), (x_2, 0.4, 0.3), (x_4, 0.4, 0.3)\}$. By determining the (α, β)-level hypergraphs of $\mathscr{D}$, we have $\mathscr{D}^{(0.8, 0.1)} = \mathscr{D}^{(0.6, 0.2)}$ and $f_S(D) = \{(0.6, 0.2), (0.8, 0.1)\}$. Further, $\mathscr{D}^{(0.4, 0.3)} = \mathscr{D}^{(0.6, 0.2)}$. The corresponding sequence of level hypergraphs is shown in Fig. 6.17.

We now define the primitive k-coloring (or simply a p-coloring), $\mathscr{L}$-coloring, and chromatic number of q-rung orthopair fuzzy directed hypergraphs and illustrate these concepts by considering a concrete example.

Definition 6.37 Let $\mathscr{D} = (X, \xi)$ be a q-rung orthopair fuzzy directed hypergraph. A *primitive k-coloring C (or simply a p-coloring)* is defined as a partition of X in k subgroups, called colors, such that the elements from at least two colors of C are contained in the support of every q-rung orthopair fuzzy directed hyperedge of $\mathscr{D}$.

Definition 6.38 Let $\mathscr{D} = (X, \xi)$ be a q-rung orthopair fuzzy directed hypergraph and $c(\mathscr{D}) = \{\mathscr{D}^{\rho_1}, \mathscr{D}^{\rho_2}, \mathscr{D}^{\rho_3}, \ldots, \mathscr{D}^{\rho_n}\}$ be the set of core hypergraphs of $\mathscr{D}$. An $\mathscr{L}$-*coloring* is defined as a partition of X, with k components, into k subgroups $\{s_1, s_2, s_3, \ldots, s_k\}$ such that C persuades a coloring for each core hypergraph $\mathscr{D}^{\rho_i} = (X^{\rho_i}, \xi^{\rho_i})$.

Remark 6.5 Note that, an $\mathscr{L}$-coloring of $\mathscr{D}$ is a p-coloring, but in general, the converse does not hold. The preceding theorem states the condition under which an $\mathscr{L}$-coloring and p-coloring of $\mathscr{D}$ coincides.

Theorem 6.13 *Let $\mathscr{D} = (X, \xi)$ be an ordered q-rung orthopair fuzzy directed hypergraph and C is a p-coloring of $\mathscr{D}$ then $\mathscr{L}$-coloring of $\mathscr{D}$ is also C.*

Definition 6.39 Let $\mathscr{D} = (X, \xi)$ be a q-rung orthopair fuzzy directed hypergraph and let $k \geq 2$ be an integer then the *k-coloring of vertex set* is defined as a function $\kappa\colon X \to \{1, 2, 3, \ldots, k\}$ such that for all $\rho \in f_S(\mathscr{D})$ and for each hyperedge ξ^ρ, which is not a loop, κ is not a constant on ξ^ρ.
The minimum integer k, for which there exists a k-coloring of $\mathscr{D}$ is called *chromatic number* of $\mathscr{D}$, denoted by $\chi(\mathscr{D})$.

Example 6.13 Let $\mathscr{D} = (X, \xi)$ be a 1-rung orthopair fuzzy directed hypergraph, where $X = \{t_1, t_2, t_3, t_4, t_5, t_6, t_7\}$ and $\xi = \{\xi_1, \xi_2, \xi_3, \xi_4, \xi_5, \xi_6, \xi_7\}$ such that

$$\xi_1 = \{(t_1, 0.6, 0.3), (t_2, 0.6, 0.3), (t_4, 0.5, 0.2)\},$$
$$\xi_2 = \{(t_1, 0.6, 0.3), (t_3, 0.6, 0.3), (t_5, 0.3, 0.1), (t_7, 0.5, 0.2)\},$$
$$\xi_3 = \{(t_1, 0.6, 0.3), (t_3, 0.6, 0.3), (t_6, 0.2, 0.1), (t_7, 0.5, 0.2)\},$$
$$\xi_4 = \{(t_2, 0.6, 0.3), (t_3, 0.6, 0.3), (t_4, 0.5, 0.2)\},$$
$$\xi_5 = \{(t_2, 0.6, 0.3), (t_4, 0.5, 0.2), (t_5, 0.3, 0.1), (t_7, 0.5, 0.2)\},$$
$$\xi_6 = \{(t_2, 0.6, 0.3), (t_4, 0.5, 0.2), (t_6, 0.2, 0.1)\},$$
$$\xi_7 = \{(t_4, 0.5, 0.2), (t_5, 0.3, 0.1), (t_6, 0.2, 0.1)\},$$

Let $\rho_1 = (0.6, 0.3)$, $\rho_2 = (0.5, 0.2)$, $\rho_3 = (0.30.1)$ and $\rho_4 = (0.2, 0.1)$. The corresponding ρ_i-level hyperedges are given as follows:

$\xi^{\rho_1} = \{\{t_1, t_2\}, \{t_1, t_3\}, \{t_2, t_3\}\},$
$\xi^{\rho_2} = \{\{t_1, t_2, t_4\}, \{t_1, t_3, t_7\}, \{t_2, t_3, t_4\}, \{t_2, t_7, t_4\}\},$
$\xi^{\rho_3} = \{\{t_1, t_2, t_4\}, \{t_1, t_3, t_5, t_7\}, \{t_1, t_3, t_7\}, \{t_2, t_3, t_4\}, \{t_2, t_4, t_5\}, \{t_2, t_4\}, \{t_4, t_5\}\},$
$\xi^{\rho_3} = \{\{t_1, t_2, t_4\}, \{t_1, t_3, t_5, t_7\}, \{t_1, t_3, t_6, t_7\}, \{t_2, t_3, t_4\}, \{t_2, t_4, t_5, t_7\}, \{t_2, t_4, t_6\}, \{t_4, t_5, t_6\}\}.$

Suppose $\{C_1, C_2\}$ is a coloring of $\mathscr{D}^{\rho_1}$. Then, $\{t_1, t_2\} \cap \{C_1, C_2\} \neq \emptyset$, $\{t_1, t_3\} \cap \{C_1, C_2\} \neq \emptyset$ and $\{t_2, t_3\} \cap \{C_1, C_2\} \neq \emptyset$. Thus, $C_1 \cap C_2 \neq \emptyset$, which is a contradiction. Hence, $\chi(\mathscr{D}^{\rho_1}) = 3$. $\{\{t_1, t_2, t_3\}, \{t_4, t_5, t_6, t_7\}\}$ is the coloring of $\mathscr{D}^{\rho_2}$. Hence, $\chi(\mathscr{D}^{\rho_2}) = 2$. Similarly, $\chi(\mathscr{D}^{\rho_3}) = 3$ and $\chi(\mathscr{D}^{\rho_4}) = 3$.

Definition 6.40 Let $\mathscr{D} = (X, \xi)$ be a q-rung orthopair fuzzy directed hypergraph and $Q = \{q_1, q_2, q_3, \ldots, q_k\}$ be the collection of non trivial q-rung orthopair fuzzy sets on X then Q is a *q-rung orthopair fuzzy k-coloring* if Q satisfies the following:

- $\min\{q_i, q_j\} = (0, 1)$, if $i \neq j$,
- for every $(\alpha, \beta) \in (0, 1]$, $\bigcup_i q_i^{(\alpha, \beta)} = X$,

- for every $(\alpha, \beta) \in (0, 1]$, each hyperedge $\xi_j^{(\alpha,\beta)}$ possesses non-empty intersection with at least two color classes $q_i^{(\alpha,\beta)}$.

Observation 6.14 *Let $\mathscr{D} = (X, \xi)$ be a q-rung orthopair fuzzy directed hypergraph having the fundamental sequence $f_S(\mathscr{D}) = \{\rho_1, \rho_2, \rho_3, \ldots, \rho_n\}$. Then, the coloring of core hypergraph $\mathscr{D}^{\rho_i}$ can be enlarged to the coloring of $\mathscr{D}^{\rho_{i+1}}$ if and only if a single color class of κ does not contain any hyperedge of $\mathscr{D}^{\rho_{i+1}}$. Particularly, if $\mathscr{D}$ is simply ordered then any coloring κ of $\mathscr{D}^{\rho_i}$ maybe elongated to the coloring of $\mathscr{D}$.*

Theorem 6.14 *Let $\mathscr{D} = (X, \xi)$ be a q-rung orthopair fuzzy directed hypergraph having the fundamental sequence $f_S(\mathscr{D}) = \{\rho_1, \rho_2, \rho_3, \ldots, \rho_n\}$. Let $\widetilde{\mathscr{D}}^{\rho_n}$ be the core coloring of $\mathscr{D}^{\rho_n}$ then every coloring of $\mathscr{D}^{\rho_n}$ is a coloring of $\mathscr{D}$ if and only if for every $\rho \in f_S(\mathscr{D})$ there exists $A \in \widetilde{\mathscr{D}}^{\rho_n}$ such that $A \subseteq \xi_i^{\rho}$, for each $\xi_i \in \xi$ for which ξ_i^{ρ} is a non loop edge.*

Proof Suppose the existance of some $\rho \in f_S(\mathscr{D})$ and $\xi_i \in \xi$ such that $|\xi_i^{\rho}| \geq 2$ and $A \nsubseteq \xi_i^{\rho}$, for every $A \in \widetilde{\mathscr{D}}^{\rho_n}$. Let a color class is defined for the vertex set of ξ_i^{ρ}. Construct a sub-hypergraph $\mathscr{D}'$ of $\mathscr{D}$, which is constructed by removing ξ_i^{ρ} from the vertices of $\widetilde{\mathscr{D}}^{\rho_n}$. Thus, $\{A \backslash \xi_i^{\rho} | A \in \widetilde{\mathscr{D}}^{\rho_n}\}$ is the set of hyperedges of $\mathscr{D}'$. Since every $\xi_j^{\rho_n} \in \mathscr{D}^{\rho_n}$, which is not a loop and also including $\xi_i^{\rho_n}$, contains some $A \in \widetilde{\mathscr{D}}^{\rho_n}$ and this non loop edge $\xi_j^{\rho_n}$ has non empty intersection with the vertices of $\mathscr{D}'$. Let $\{q_2, q_3, \ldots, q_k\}$ be the coloring of $\mathscr{D}'$ then the coloring of $\mathscr{D}^{\rho_n}$ is $\{\xi^{\rho}, q_2, q_3, \ldots, q_k\}$, where ξ^{ρ} is contained in single color class. Hence, there exists a coloring of $\mathscr{D}^{\rho_n}$ which is not a coloring of $\mathscr{D}$.

Conversely, assume that there exists some $\rho \in f_S(\mathscr{D})$ and $\xi_i \in \xi$ such that $|\xi_i^{\rho}| \geq 2$ and $A \subseteq \xi_i^{\rho}$, for every $A \in \widetilde{\mathscr{D}}^{\rho_n}$. Suppose that ρ and ξ_i are taken as arbitrary but fixed and κ be the coloring of $\mathscr{D}^{\rho_n}$. Since κ is not a constant on A, it is also non constant on ξ_i^{ρ}, hence κ is a coloring of $\mathscr{D}$.

The coloring problem of $\mathscr{D}$ can be reduced to the correlated crisp coloring. It can be done by replacing $\mathscr{D}$ with a more simpler framework $\mathscr{D}^{\wedge}$, it will be noted that $\mathscr{D}^{\wedge}$ is ordered, simpler to color and every p-coloring of $\mathscr{D}^{\wedge}$ will generate the $\mathscr{L}$-coloring of $\mathscr{D}$.

Definition 6.41 A *spike reduction* of $\xi_i \in P(X)$, which is denoted by $\widetilde{\xi}_i$, is defined as

$$\widetilde{\xi}_i^{(\alpha,\beta)} = \begin{cases} \xi_i^{(\alpha,\beta)}, & \text{if } |\xi_i^{(\alpha,\beta)}| \geq 2, \\ \emptyset, & \text{if } |\xi_i^{(\alpha,\beta)}| \leq 1, \end{cases}$$

for $0 < \alpha, \beta \leq 1$. Particularly, if ξ_i is a loop then $\widetilde{\xi}_i = \emptyset$.

Definition 6.42 Given $\mathscr{D} = (Q, \xi)$ then $\widetilde{\mathscr{D}} = (\widetilde{X}, \widetilde{\xi})$, where $\widetilde{\xi} = \{\widetilde{\xi}_i | \xi_i \in \xi\}$.

Construction 6.2 Let $\mathscr{D} = (X, \xi)$ be a q-rung orthopair fuzzy directed hypergraph having the fundamental sequence $f_S(\mathscr{D}) = \{\rho_1, \rho_2, \rho_3, \ldots, \rho_n\}$ and $c(\mathscr{D}) = \{\mathscr{D}^{\rho_1}, \mathscr{D}^{\rho_2}, \mathscr{D}^{\rho_3}, \ldots, \mathscr{D}^{\rho_n}\}$. Then, the conversion of $\mathscr{D}$ into $\mathscr{D}^s$ is given in the following construction.

1. Obtain a partial hypergraph $\overline{\mathscr{D}}^{\rho_1}$ of $\mathscr{D}^{\rho_1}$ by abolishing all those directed hyperedges of $\mathscr{D}^{\rho_1}$ that properly accommodate any other hyperedge of $\mathscr{D}^{\rho_1}$.

2. Subsequently, obtain a partial hypergraph $\overline{\mathscr{D}}^{\rho_2}$ of $\mathscr{D}^{\rho_2}$ by abolishing all those directed hyperedges of $\mathscr{D}^{\rho_2}$ that properly accommodate any other hyperedge of $\mathscr{D}^{\rho_2}$ or (properly or improperly) contain a hyperedge of partial hypergraph $\overline{\mathscr{D}}^{\rho_1}$. (It may be possible that $\overline{\mathscr{D}}^{\rho_2}$ possesses no hyperedges, in such case existance of $\overline{\mathscr{D}}^{\rho_2}$ is ignored.)

3. By following the same procedure, obtain a partial hypergraph $\overline{\mathscr{D}}^{\rho_3}$ of $\mathscr{D}^{\rho_3}$ by abolishing all those directed hyperedges of $\mathscr{D}^{\rho_3}$ that properly accommodate any other hyperedge of $\mathscr{D}^{\rho_3}$ or (properly or improperly) contain a hyperedge of partial hypergraph either $\overline{\mathscr{D}}^{\rho_1}$ or $\overline{\mathscr{D}}^{\rho_2}$.

4. Following this iterative procedure, we obtain a subsequence of $f_S(\mathscr{D})$, $\rho_m^s \cdots < \rho_1^s = \rho_1$ and the set of partial hypergraphs corresponding to this subsequence is $c(\overline{\mathscr{D}}) = \{\overline{\mathscr{D}}^{\rho_1^s}, \overline{\mathscr{D}}^{\rho_2^s}, \overline{\mathscr{D}}^{\rho_3^s}, \ldots, \overline{\mathscr{D}}^{\rho_m^s}\}$ from the $c(\mathscr{D})$. It is obvious from above procedure that each $\overline{\mathscr{D}}^{\rho_i^s}$, $1 \le i \le m$, contain non-empty set of hyperedges because all those hypergraphs having empty set of hyperedges have been eliminated from the consideration.

5. Construct the elementary q-rung orthopair fuzzy directed hypergraph $\mathscr{D}^s = (X^s, \xi^s)$ satisfying the following conditions

   - $f_S(\mathscr{D}^s) = \{\rho_1^s, \rho_2^s, \rho_3^s, \ldots, \rho_m^s\}$,
   - if $\xi_j \in \xi^s$ then $h(\xi_j) \in \{\rho_1^s, \rho_2^s, \rho_3^s, \ldots, \rho_m^s\}$,
   - the family of hyperedges in ξ^s having heights ρ_k^s is the collection of elementary q-rung orthopair fuzzy sets $\{\eta(Q, \rho_k^s)|Q \in \overline{\mathscr{D}}^{\rho_k^s}\}$, for all k, $1 \le k \le m$.

Definition 6.43 Let $\mathscr{D}^\Lambda$ be a q-rung orthopair fuzzy directed hypergraph obtained from $\widetilde{\mathscr{D}}$ by the procedure described above, i.e., $\mathscr{D}^\Lambda = (\widetilde{\mathscr{D}})^s$.

Definition 6.44 Let $\mathscr{D} = (X, \xi)$ be a q-rung orthopair fuzzy directed hypergraph having the fundamental sequence $f_S(\mathscr{D}) = \{\rho_1, \rho_2, \rho_3, \ldots, \rho_n\}$ and $c(\mathscr{D}) = \{\mathscr{D}^{\rho_1}, \mathscr{D}^{\rho_2}, \mathscr{D}^{\rho_3}, \ldots, \mathscr{D}^{\rho_n}\}$ with $\mathscr{D}^{\rho_i} = (X_i, \mathscr{E}_i)$ and the elements of $f_S(\mathscr{D})$ are ordered then $\mathscr{D}$ is called *sequentially simple* if whenever $\mathscr{E} \in \mathscr{E}_i \backslash \mathscr{E}_{i-1}$ then $\mathscr{E} \not\subseteq X_{i-1}$, $i = 1, 2, 3, \ldots, n$.

Theorem 6.15 *Let $\mathscr{D} = (X, \xi)$ be a sequentially simple q-rung orthopair fuzzy directed hypergraph having core set $c(\mathscr{D}) = \{\mathscr{D}^{\rho_i} = (X_i, \mathscr{E}_i)|i = 1, 2, 3, \ldots, n\}$ and the elements of $f_S(\mathscr{D})$ are ordered. Suppose that $\mathscr{E} \in \mathscr{E}_{j+k} \backslash \mathscr{E}_j$, $j < n$ and $k \in \{1, 2, 3, \ldots, n - j\}$ then $\mathscr{E} \not\subseteq X_j$.*

Proof The general proof of this theorem is illustrated by considering an example. Assume that $\mathscr{E} \in \mathscr{E}_{j+3} \backslash \mathscr{E}_j$, then

(i) either $\mathscr{E} \in \mathscr{E}_{j+2}$ or $\mathscr{E} \not\in \mathscr{E}_{j+2}$. In the succeeding condition $\mathscr{E} \in \mathscr{E}_{j+3} \backslash \mathscr{E}_{j+2}$, which indicates that $\mathscr{E} \not\subseteq X_{j+2}$, thus $\mathscr{E} \not\subseteq X_j$ because $X_j \subseteq X_{j+2}$. Now suppose that $\mathscr{E} \in \mathscr{E}_{j+2}$. Then

(ii) either $\mathcal{E} \in \mathcal{E}_{j+1}$ or $\mathcal{E} \notin \mathcal{E}_{j+1}$. In the succeeding condition $\mathcal{E} \in \mathcal{E}_{j+2} \backslash \mathcal{E}_{j+1}$, which indicates that $\mathcal{E} \nsubseteq X_{j+1}$, thus $\mathcal{E} \nsubseteq X_j$ because $X_j \subseteq X_{j+1}$. Now suppose that $\mathcal{E} \in \mathcal{E}_{j+1}$. Then

(iii) since $\mathcal{E} \notin \mathcal{E}_j$, this implies that $\mathcal{E} \in \mathcal{E}_{j+1} \backslash \mathcal{E}_j$. Thus, $\mathcal{E} \nsubseteq X_j$. Hence it is clear that $\mathcal{E} \nsubseteq X_j$.

Theorem 6.16 *Let $\mathscr{D} = (X, \xi)$ be a sequentially simple q-rung orthopair fuzzy directed hypergraph the $\widetilde{\mathscr{D}}$, $\mathscr{D}^s$ and $\mathscr{D}^\Lambda$ are also sequentially simple q-rung orthopair fuzzy directed hypergraphs.*

Proof Since $\mathscr{D} = (X, \xi)$ is a sequentially simple q-rung orthopair fuzzy directed hypergraph. Since $\widetilde{\mathscr{D}}$ is obtained by removing all those hyperedges of $\mathscr{D}$, which are spikes(loops) and also by eliminating all terminal spikes from the directed hyperedges of $\mathscr{D}$. Certainly $\widetilde{\mathscr{D}}$ is a sequentially simple q-rung orthopair fuzzy directed hypergraph. Also the skeleton of $\mathscr{D}$, denoted by $\mathscr{D}^s$, is a sequentially simple q-rung orthopair fuzzy directed hypergraph. Therefore, $\mathscr{D}^\Lambda = (\widetilde{\mathscr{D}})^s$ is also sequentially simple q-rung orthopair fuzzy directed hypergraph.

6.7 Applications

6.7.1 The Most Proficient Arrangement for Hazardous Chemicals

Hazardous waste is a type of waste that is considered to have potential and substantial threats to the environment and human health. There are many human activities, including medical practice, industrial manufacturing procedures, and batteries that generate the hazardous waste in various categories, including solids, gases, liquids, and sludges. The improper arrangement of these hazardous wastes results in many serious tragedies. Serious health issues, including cancer, birth defects, and nerve damage may occur due to improper handling for those who ingest the contaminated air, water or food. Remediation and cleanup cost of these hazardous substances may amount to millions and billions of dollars. To ensure the well being of the population, protection of the surrounding environment, and to avoid any type of threat or hazard proper management of hazardous chemicals is extremely important. A q-rung orthopair fuzzy directed hypergraph can be used to well demonstrate the management system of hazardous elements. The 5-rung orthopair fuzzy directed hypergraph model of some compatible and incompatible elements is shown in Fig. 6.18.

The set of oval vertices $G = \{G_1, G_2, G_3, G_4, G_5\}$ of this directed hypergraph represents the types of those elements, which are adjacent to them. The description of these vertices is given in Table 6.8.

For the cost efficient and secure management of hazardous elements, it is imperative to fill the containers up to 75% and also the container's material should be

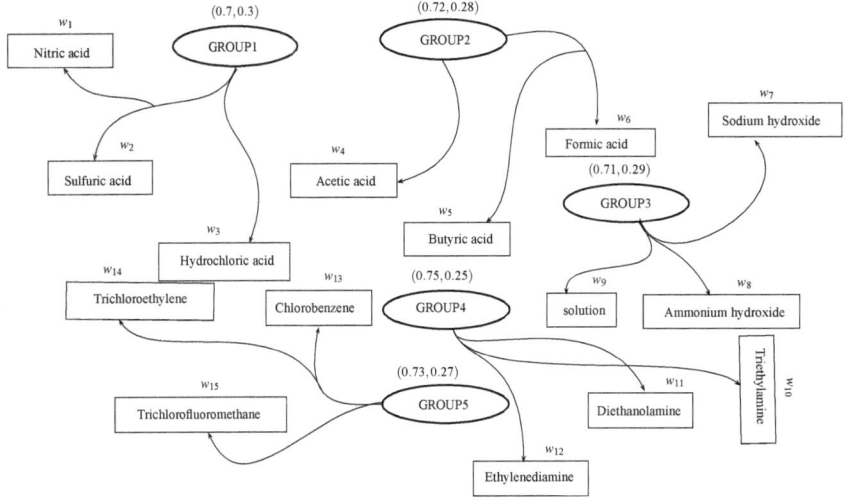

Fig. 6.18 5-rung orthopair fuzzy directed hypergraph model

Table 6.8 Description of oval vertices

	Category	Membership values	Proficiency (%)	Ineptness (%)
GROUP1	Inorganic acids	(0.7, 0.3)	70	30
GROUP2	Organic acids	(0.72, 0.28)	72	28
GROUP3	Caustics	(0.71, 0.29)	71	29
GROUP4	Amines and alkanolamines	(0.75, 0.25)	75	25
GROUP5	Halogenated compounds	(0.73, 0.27)	73	27

compatible to the elements stored in it. Only those chemical substances are connected through the same directed hyperedges, which are compatible to each other and are not dangerous when stored together. For a proficient management of such elements, one should know the characteristics of hazardous elements such as corrosivity, reactivity or toxicity of these elements. A 5-rung orthopair fuzzy set Q describes the corrosivity of these chemical substances.

$$Q = \{(w_1, 0.81, 0.23), (w_2, 0.81, 0.23), (w_3, 0.81, 0.23), (w_4, 0.90, 0.17),$$
$$(w_5, 0.90, 0.17), (w_6, 0.90, 0.17), (w_7, 0.87, 0.13), (w_8, 0.87, 0.13),$$
$$(w_9, 0.87, 0.13), (w_{10}, 0.75, 0.30), (w_{11}, 0.70, 0.20), (w_{12}, 0.85, 0.20),$$
$$(w_{13}, 0.70, 0.10), (w_{14}, 0.70, 0.10), (w_{15}, 0.90, 0.20)\}.$$

Table 6.9 describes the importance of defining this 5-rung orthopair fuzzy set.

Table 6.9 Corrosivity and fortifying level of square vertices

Square vertices	Corrosivity (%)	Vitriolicity (%)	Square vertices	Corrosivity (%)	Vitriolicity (%)
Nitric acid	81	23	Triethylamine	75	30
Sulfuric acid	81	23	Diethanolamine	70	20
Hydrochloric acid	81	23	Ethylenediamine	85	20
Acetic acid	90	17	Chlorobenzene	70	10
Butyric acid	90	17	Trichloroethylene	70	10
Formic acid	90	17	Trichlorofluoromethane	90	20
Sodium hydroxide	87	13	Solutions	87	13
Ammonium hydroxide	87	13			

Table 6.10 Compatibility and incompatibility levels of containers to chemicals

C	Inorganic	Organic	Caustics	Alkanolamines	Compounds
C_1	$(0.81, 0.23)$	$(0.001, 0.980)$	$(0.10, 0.75)$	$(0.001, 0.980)$	$(0.010, 0.908)$
C_2	$(0.10, 0.83)$	$(0.90, 0.17)$	$(0.75, 0.10)$	$(0.010, 0.908)$	$(0.75, 0.10)$
C_3	$(0.001, 0.980)$	$(0.81, 0.23)$	$(0.10, 0.83)$	$(0.10, 0.83)$	$(0.91, 0.23)$
C_4	$(0.10, 0.83)$	$(0.81, 0.23)$	$(0.90, 0.17)$	$(0.81, 0.23)$	$(0.81, 0.23)$
C_5	$(0.001, 0.980)$	$(0.71, 0.23)$	$(0.930, 0.200)$	$(0.001, 0.980)$	$(0.870, 0.210)$

The containers which are holding these chemicals should be in good condition, non-leaking and compatible and these wastes should not be kept in a container that is made of an incompatible material. For example, acids must not be stored in metal material, hydrofluoric acid should not be stored in glass and lightweight polyethylene containers should not be used to store or transfer solvents. Thus, one should make sure that containers possess a high-level of compatibility with chemicals. We now consider a set of containers/cabinets $C = \{C_1, C_2, C_3, C_4, C_5\}$ and define five 5-rung orthopair fuzzy sets on C according to their compatibility with these elements. For example, the membership degrees $C_1(G_1) = (0.001, 0.980)$ implies that C_1 container is made up of such material which is incompatible to store inorganic acids and suitable to store organic acids as $C_1(G_2) = (0.81, 0.23)$. Similarly, by taking the same assumptions, we define other 5-rung orthopair fuzzy sets as given in Table 6.10.

It can be noted from Table 6.10 that inorganic acids should be stored in C_1 container as this is highly compatible to inorganic acids, so this storage will be most secure and risk less. Note that, the material of C_2 is compatible with organic acids, caustics, and halogenated compounds but we will use this container to store organic acids because the truth-membership degree is greatest in this case. In the same way, we find that C_3 is good for halogenated compounds, C_4 is used to store amines and

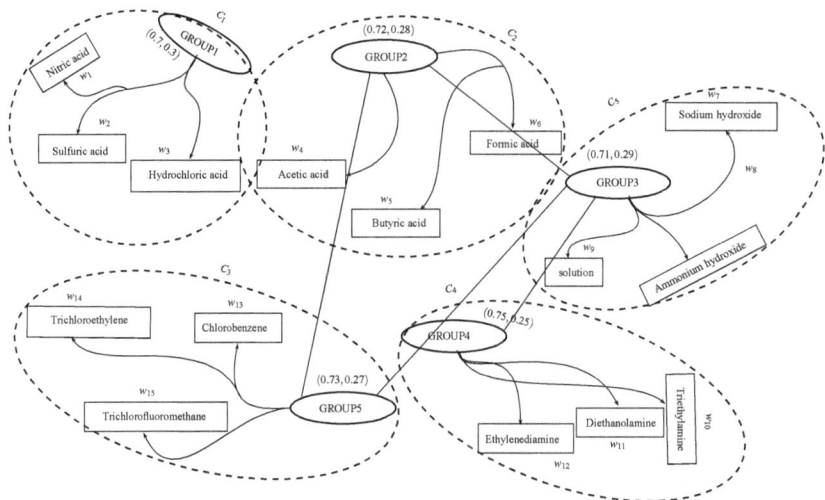

Fig. 6.19 Graphical representations of storages of chemical substances

alkanolamines and C_5 is suitable for storing caustics. The graphical representations of these storages are shown in Fig. 6.19.

Thus, by taking the above model under consideration, hazardous chemicals can be systemized in a more appropriate and acceptable manner to reduce the precarious risks to human health and environment.

6.7.2 Assessment of Collaborative Enterprise to Achieve a Particular Objective

Collaboration is the demonstration of working as a team of members to achieve some piece of work, including research projects. Many organizations are realizing the significance of collaboration as a key factor in innovations. The collaborative work provides more opportunities for studying team-work skills and improves personal and professional relationships. Here, we consider a few projects in chemical industry, which are assigned to different groups of trainees. A 7-rung orthopair fuzzy directed hypergraph model is used to well demonstrate this collaborative activity of different teams/groups.

6.7.2.1 The Project Possessesing the Powerful Collaboration

Consider the peculiar projects in the field of chemical industry, including *Zero Energy Homes, Heat Exchanger Network Retrofit, Genetic Algorithms for Process Optimization, Progressive Crude Distillation, Water Management* (for pollution prevention)

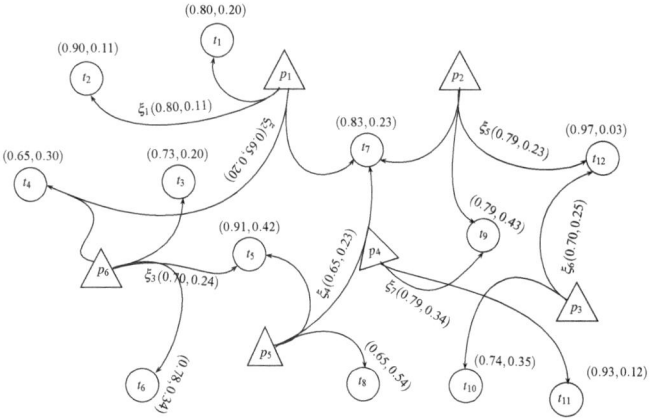

Fig. 6.20 7-rung orthopair fuzzy directed hypergraph model

Table 6.11 Collaboration capabilities of groups to projects

Assigned projects	Collaboration team	Collaborative competency (%)	Collaborative incompetency (%)
Zero energy homes	$\{t_1, t_2, t_4, t_7\}$	65	11
Heat exchanger network retrofit	$\{t_7, t_9, t_{12}\}$	79	23
Genetic algorithms for process optimization	$\{t_{10}, t_{12}\}$	70	25
Progressive crude distillation	$\{t_9, t_{11}\}$	79	34
Water management	$\{t_5, t_7, t_8\}$	65	23
Design of LNG facilities	$\{t_3, t_4, t_5, t_6\}$	70	23

and *Design of LNG Facilities*. The assignment of these projects to different groups is well explained through a 7-rung orthopair fuzzy directed hypergraph model as shown in Fig. 6.20.

Note that, the set of triangular vertices $\{p_1, p_2, p_3, p_4, p_5, p_6\}$ represents the projects that are considered to be worked on and the set of circular vertices $\{t_1, t_2, t_3, t_4, t_5, t_6, t_7, t_8, t_9, t_{10}, t_{11}, t_{12}\}$ represents the trainees, to whom these projects are assigned. Each directed hyperedge connects the corresponding project to it's allocated trainees. The projects assigned to different groups are illustrated through Table 6.11.

Note that, collaborative competency levels of different teams narrate that how much mutual understanding is there between the members of corresponding teams towards their projects. For example, the trainees of "*Zero Energy Homes*" project have 65% collaborative competency, i.e., they give respect to each other's ideas, contribution, and acknowledge the opinions of other trainees and their collective strength to achieve the goal is 65%. Incompetency degree shows that they have 11% conflicts of ideas and opinions. Similarly, the collaborative competency of all other

Table 6.12 Heights of all directed hyperedges

$h(\xi_1)$	$(0.90, 0.11)$	$h(\xi_2)$	$(0.83, 0.23)$
$h(\xi_3)$	$(0.91, 0.20)$	$h(\xi_4)$	$(0.91, 0.23)$
$h(\xi_5)$	$(0.97, 0.03)$	$h(\xi_6)$	$(0.97, 0.03)$
$h(\xi_7)$	$(0.93, 0.12)$		

teams can be studied through the table. Now, to evaluate the strength of determination and competent behavior of all teams towards their collaborative project, we calculate the heights of all directed hyperedges, which are given in Table 6.12.

The directed hyperedge having a maximum height, i.e., maximum truth-membership and minimum falsity-membership will correspond to the most efficient team working in collaboration. Note that, ξ_5 and ξ_6 have maximum heights showing that $\{t_7, t_9, t_{12}\}$ and $\{t_{10}, t_{12}\}$ share the most powerful collaborative characteristics. The method adopted in this part can be explained by a simple algorithm given in Table 6.13.

6.7.2.2 The Enduring Connection Between Projects:

Now, the line graph of the above 7-rung orthopair fuzzy directed hypergraph model can be used to determine the common trainees of distinct projects. The corresponding line graph is shown in Fig. 6.21.

The dashed lines between the projects demonstrate that they share some common trainees. The truth-membership and falsity-membership of these edges are given here.

$$(T_{p_1 p_2}, F_{p_1 p_2}) = (0.80, 0.11),$$
$$(T_{p_1 p_5}, F_{p_1 p_5}) = (0.80, 0.11),$$
$$(T_{p_1 p_6}, F_{p_1 p_6}) = (0.80, 0.11),$$
$$(T_{p_2 p_5}, F_{p_2 p_5}) = (0.79, 0.23),$$
$$(T_{p_2 p_3}, F_{p_2 p_3}) = (0.79, 0.23),$$
$$(T_{p_3 p_4}, F_{p_3 p_4}) = (0.79, 0.25).$$

The maximum truth-membership and minimum falsity-membership reveal the robust connection among the distinct projects. For instance, projects p_1 and p_5 are 80% connected to each other, i.e., the trainees of these projects can share their ideas, creative thinkings and motives among themselves to enhance the output of their projects. The method adopted in this section can be explained by a simple algorithm given in Table 6.14.

Table 6.13 Algorithm

Algorithm for powerful collaboration
1. m =input('enter the number of trainees');
2. T =input('enter the degrees of membership of vertices(trainees) as $m \times 2$');
3. r =input('enter the number of directed hyperedges');
4. Xi =input('enter the degrees of membership of directed hyperedges $r \times 2$');
5. Y =input('enter the set valued function that tells us how many vertices are contained in a hyperedge as $r \times m$');
6. $J=[\text{zeros}(r,1)\ \text{ones}(r,1)]$;
7.   **for** $i = 1 : r$
8.   **for** $k = 1 : m$
9.   **if** $Y(i, k) == 1$;
10. $J(i, 1) = \max(J(i, 1), T(k, 1))$;
11. $T(i, 2) = \min(J(i, 2), T(k, 2))$;
12.   **end**
13.   **end**
14.   **end**
15. $H = \max(J(:, 1))$;j=0;v=zeros(r,2); b=1;
16.   **for** $l = 1 : r$
17.   **if** J(1,1)==H
18.   j=j+1;v(l,1)=1;b=min(b,J(l,2));
19.   **end**
20.   **end**
21.   **if** j>1
22.   **for** $l = 1 : r$
23.   **if** J(1,2)==b
24.   k=k+1;v(l,2)=1;
25.   fprintf('you can choice (any of these) hyperedge(s) %d',l)
26.   **end**
27.   **end**
28.   **else**
29.   **for** $l = 1 : r$
30.   **if** J(1,1)==H
31.   fprintf('you can choice edge %d',l)
32.   **end**
33.   **end**
34.   **end**

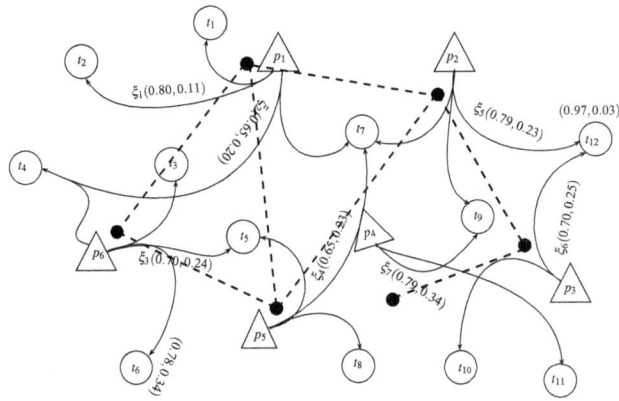

Fig. 6.21 Line graph of 7-rung orthopair fuzzy directed hypergraph

Table 6.14 Algorithm for the enduring connection between projects

1.	m =input('enter the number of vertices');
2.	T =input('enter the degrees of membership of vertices as $m \times 2$');
3.	r =input('enter the number of directed hyperedges');
4.	Xi =input('enter the degrees of membership of directed hyperedges $r \times 2$');
5.	s =input('enter the number of edges in line graph');
6.	P =input('enter the degrees of membership of edges $s \times 2$');
7.	$H = \max(P(:, 1))$;j=0;v=zeros(s,2); b=1;
8.	**for** n=1:s
9.	**if** P(n,1)==H
10.	j=j+1;v(n,1)=1;b=min(b,P(n,2));
11.	**end**
12.	**end**
13.	**if** j>1
14.	**for** n=1:s
15.	**if** P(n,2)==b
16.	k=k+1;v(n,2)=n;
17.	fprintf('you can choice (any of these) hyperedge(s) %d',n)
18.	**end**
19.	**end**
20.	**else**
21.	**for** n=1:s
22.	**if** P(n,1)==H
23.	fprintf('you can choice edge %d',n)
24.	**end**
25.	**end**
26.	**end**

6.8   Comparative Analysis

Orthopair fuzzy sets are defined as those fuzzy sets in which the membership degrees of an element is taken as the pair of values in the unit interval $[0, 1]$, given as $(T(x), F(x), T(x))$ indicates support for membership (truth-membership), and $F(x)$ indicates support against membership (falsity-membership) to the fuzzy set. Intuitionistic fuzzy sets and Pythagorean fuzzy sets are examples of orthopair fuzzy sets. Atanassov's [14] intuitionistic fuzzy set has been studied widely by various researchers, but the range of applicability of intuitionistic fuzzy set is limited because of its constraint that the sum of truth-membership and falsity-membership must be equal to or less than one. Under this condition, intuitionistic fuzzy sets cannot express some decision evaluation information effectively, because a decision maker may provide information for a particular attribute such that the sum of the degrees of truth-membership and the degrees of falsity-membership becomes greater than one. In order to solve such types of problems, Pythagorean fuzzy sets were defined by Yager [32], whose prominent characteristic is that the square sum of the truth-membership degree and the falsity-membership degree is less than or equal to one. Thus, a Pythagorean fuzzy set can solve a number of practical problems that cannot be handled using intuitionistic fuzzy set and is a generalization of intuitionistic fuzzy set. Due to the more complicated information in society and the development of theories, q-rung orthopair fuzzy sets were proposed by Yager [35]. A q-rung orthopair fuzzy set is characterized in such a way that the sum of the qth power of the truth-membership degree and the qth power of the degrees of falsity-membership is restricted to less than or equal to one. Note that, intuitionistic fuzzy sets and Pythagorean fuzzy sets are particular cases of q-rung orthopair fuzzy sets. The flexibility and effectiveness of a q-rung orthopair fuzzy model can be proven as follows: Suppose that (x, y) is an intuitionistic fuzzy grade, where $x \in [0, 1]$, $y \in [0, 1]$, and $0 \le x + y \le 1$, since $x^q \le x$, $y^q \le y$, $q \ge 1$, so we have $0 \le x^q + y^q \le 1$. Thus, every intuitionistic fuzzy grade is also a Pythagorean fuzzy grade, as well as a q-rung orthopair fuzzy grade. However, there are q-rung orthopair fuzzy grades that are not intuitionistic fuzzy nor Pythagorean fuzzy grades. For example, $(0.9, 0.8)$, here $(0.9)^5 + (0.8)^5 \le 1$, but $0.9 + 0.8 = 1.7 > 1$ and $(0.9)^2 + (0.8)^2 = 1.45 > 1$. This implies that the class of q-rung orthopair fuzzy sets extend the classes of intuitionistic fuzzy sets and Pythagorean fuzzy sets. It is worth noting that as the parameter q increases, the space of acceptable orthopairs also increases, and thus, the bounding constraint is satisfied by more orthopairs. Thus, a wider range of uncertain information can be expressed by using q-rung orthopair fuzzy sets. We can adjust the value of the parameter q to determine the expressed information range; thus, q-rung orthopair fuzzy sets are more effective and more practical for the uncertain environment. Based on these advantages of q-rung orthopair fuzzy sets, we proposed q-rung orthopair fuzzy hypergraphs and q-rung orthopair fuzzy directed hypergraphs to combine the benefits of both theories. A wider range of uncertain information can be expressed using the methods proposed in this paper, and they are closer to real decision-making. Our proposed models are more general as compared to the intuitionistic fuzzy and Pythagorean fuzzy models, as when $q = 1$, the model reduces to the intuitionistic fuzzy model, and when $q = 2$, it reduces to the Pythagorean fuzzy model. Hence, our approach is more flexible and generalized, and different values of q can be chosen by decision makers according to the different attitudes.

6.9 Complex Pythagorean Fuzzy Hypergraphs

A complex Pythagorean fuzzy set is an extension of a Pythagorean fuzzy set that is used to handle the vagueness with the degrees whose ranges are enlarged from real to complex subset with unit disc. For example, a clothing brand considers five locations to open new outlet regarding some particular criteria. If an expert assign membership 0.8 and nonmembership 0.6 to a location with respect to a criterion then intuitionistic fuzzy set fails to deal with this problem because $0.8 + 0.6 \geq 1$, but this problem can be effectively handled by Pythagorean fuzzy set as $0.8^2 + 0.6^2 \leq 1$. On the other hand, if we consider the maximum number of people visiting the outlet at a particular time then Pythagorean fuzzy set also fails because to handle time we have to introduce the periodic term. Now expert assign membership $0.8e^{\iota(1.4\pi)}$ and nonmembership $0.6e^{\iota(1.1\pi)}$ which satisfy the conditions of complex Pythagorean fuzzy set as $0.8^2 + 0.6^2 \leq 1$. Therefore, complex Pythagorean fuzzy set is proficient in dealing with data involving time period (periodic nature) due to complex membership and nonmembership grades along with the constraints.

Definition 6.45 A *complex Pythagorean fuzzy set* P on the universal set X is defined as, $P = \{(u, T_P(u)e^{i\phi_P(u)}, F_P(u)e^{i\psi_P(u)})|u \in X\}$, where $i = \sqrt{-1}$, $T_P(u), F_P(u) \in [0, 1], \phi_P(u), \psi_P(u) \in [0, 2\pi]$, and for every $u \in X, 0 \leq T_P^2(u) + F_P^2(u) \leq 1$. Here, $T_P(u), F_P(u)$ and $\phi_P(u), \psi_P(u)$ are called the amplitude terms and phase terms for truth membership and falsity membership grades, respectively.

Definition 6.46 A *complex Pythagorean fuzzy graph* on X is an ordered pair $G^* = (C, D)$, where C is a complex Pythagorean fuzzy set on X and D is complex Pythagorean fuzzy relation on X such that,

$$T_D(ab) \leq \min\{T_C(a), T_C(b)\},$$
$$F_D(ab) \leq \max\{F_C(a), F_C(b)\}, \text{ (for amplitude terms)}$$
$$\phi_D(ab) \leq \min\{\phi_C(a), \phi_C(b)\},$$
$$\psi_D(ab) \leq \max\{\psi_C(a), \psi_C(b)\}, \text{ (for phase terms)}$$

$0 \leq T_D^2(ab) + F_D^2(ab) \leq 1$, for all $a, b \in X$.

Definition 6.47 A *complex Pythagorean fuzzy hypergraph* on X is defined as an ordered pair $H^* = (\mathscr{C}^*, \mathscr{D}^*)$, where $\mathscr{C}^* = \{\beta_1, \beta_2, \ldots, \beta_k\}$ is a finite family of complex Pythagorean fuzzy sets on X and $\mathscr{D}^*$ is a complex Pythagorean fuzzy relation on complex Pythagorean fuzzy sets β_j's such that

(i)

$$T_{\mathscr{D}^*}(\{s_1, s_2, \ldots, s_l\}) \leq \min\{T_{\beta_j}(s_1), T_{\beta_j}(s_2), \ldots, T_{\beta_j}(s_l)\},$$
$$F_{\mathscr{D}^*}(\{s_1, s_2, \ldots, s_l\}) \leq \max\{F_{\beta_j}(s_1), F_{\beta_j}(s_2), \ldots, F_{\beta_j}(s_l)\}, \text{ (for amplitude terms)}$$
$$\phi_{\mathscr{D}^*}(\{s_1, s_2, \ldots, s_l\}) \leq \min\{\phi_{\beta_j}(s_1), \phi_{\beta_j}(s_2), \ldots, \phi_{\beta_j}(s_l)\},$$
$$\psi_{\mathscr{D}^*}(\{s_1, s_2, \ldots, s_l\}) \leq \max\{\psi_{\beta_j}(s_1), \psi_{\beta_j}(s_2), \ldots, \psi_{\beta_j}(s_l)\}, \text{ (for phase terms)}$$

$$0 \leq T_{\mathscr{D}^*}^2 + F_{\mathscr{D}^*}^2 \leq 1, \text{ for all } s_1, s_2, \ldots, s_l \in X.$$

(ii) $\bigcup_j supp(\beta_j) = X$, for all $\beta_j \in \mathscr{C}^*$.

Note that, $E_k = \{s_1, s_2, \ldots, s_l\}$ is the crisp hyperedge of $H^* = (\mathscr{C}^*, \mathscr{D}^*)$.

Example 6.15 Consider a complex Pythagorean fuzzy hypergraph $H^* = (\mathscr{C}^*, \mathscr{D}^*)$ on $X = \{s_1, s_2, s_3, s_4, s_5, s_6\}$. The complex Pythagorean fuzzy relation is defined as, $\mathscr{D}^*(s_1, s_2, s_3) = ((0.6e^{i(0.2)2\pi}, 0.5e^{i(0.9)2\pi}))$, $\mathscr{D}^*(s_4, s_5, s_6) = (0.6e^{i(0.4)2\pi}, 0.4e^{i(0.6)2\pi})$, $\mathscr{D}^*(s_3, s_6) = (0.6e^{i(0.6)2\pi}, 0.5e^{i(0.6)2\pi})$, $\mathscr{D}^*(s_2, s_5) = (0.6e^{i(0.4)2\pi}, 0.5e^{i(0.6)2\pi})$, and $\mathscr{D}^*(s_1, s_4) = (0.6e^{i(0.2)2\pi}, 0.9e^{i(0.9)2\pi})$. The corresponding complex Pythagorean fuzzy hypergraph is shown in Fig. 6.22.

Definition 6.48 A complex Pythagorean fuzzy hypergraph $H^* = (\mathscr{C}^*, \mathscr{D}^*)$ is *simple* if whenever $\mathscr{D}_j^*, \mathscr{D}_k^* \in \mathscr{D}^*$ and $\mathscr{D}_j^* \subseteq \mathscr{D}_k^*$, then $\mathscr{D}_j^* = \mathscr{D}_k^*$.
A complex Pythagorean fuzzy hypergraph $H^* = (\mathscr{C}^*, \mathscr{D}^*)$ is *support simple* if whenever $\mathscr{D}_j^*, \mathscr{D}_k^* \in \mathscr{D}^*$, $\mathscr{D}_j^* \subseteq \mathscr{D}_k^*$, and $supp(\mathscr{D}_j^*) = supp(\mathscr{D}_k^*)$, then $\mathscr{D}_j^* = \mathscr{D}_k^*$.

Definition 6.49 Let $H^* = (\mathscr{C}^*, \mathscr{D}^*)$ be a complex Pythagorean fuzzy hypergraph. Suppose that $\alpha_1, \beta_1 \in [0, 1]$ and $\theta, \varphi \in [0, 2\pi]$ such that $0 \leq \alpha_1^2 + \beta_1^2 \leq 1$. The

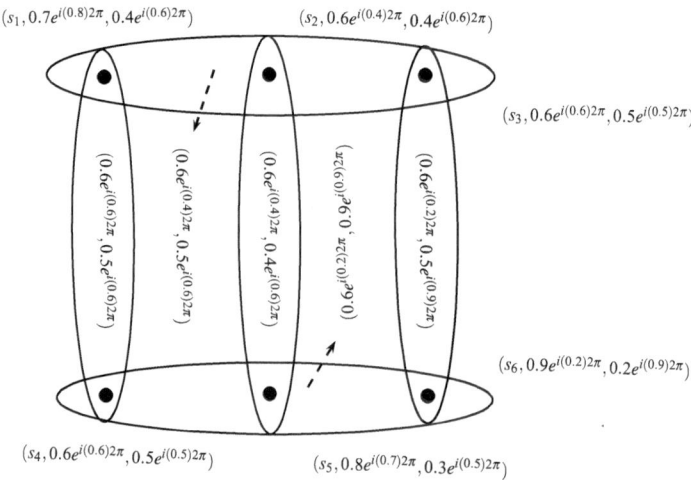

Fig. 6.22 Complex Pythagorean fuzzy hypergraph

Fig. 6.23 $(\alpha_1 e^{i\theta}, \beta_1 e^{i\varphi})$-level hypergraph of H^*

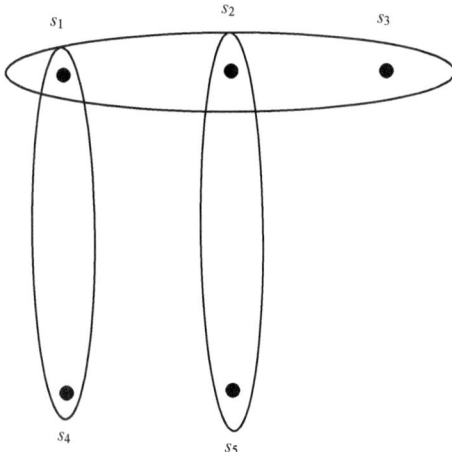

$(\alpha_1 e^{i\theta}, \beta_1 e^{i\varphi})$-*level hypergraph* of H^* is defined as an ordered pair $H^{*(\alpha_1 e^{i\theta}, \beta_1 e^{i\varphi})} = (\mathscr{C}^{*(\alpha_1 e^{i\theta}, \beta_1 e^{i\varphi})}, \mathscr{D}^{*(\alpha_1 e^{i\theta}, \beta_1 e^{i\varphi})})$, where

(i) $\mathscr{D}^{*(\alpha_1 e^{i\theta}, \beta_1 e^{i\varphi})} = \{D_j^{*(\alpha_1 e^{i\theta}, \beta_1 e^{i\varphi})} : D_j^* \in \mathscr{D}^*\}$ and $D_j^{*(\alpha_1 e^{i\theta}, \beta_1 e^{i\varphi})} = \{y \in X : T_{D_j^*}(y) \geq \alpha_1, \phi_{D_j^*}(y) \geq \theta$, and $F_{D_j^*}(y) \leq \beta_1, \psi_{D_j^*}(y) \leq \varphi\}$,

(ii) $\mathscr{C}^{*(\alpha_1 e^{i\theta}, \beta_1 e^{i\varphi})} = \bigcup_{D_j^* \in \mathscr{D}^*} D_j^{*(\alpha_1 e^{i\theta}, \beta_1 e^{i\varphi})}$.

Note that, $(\alpha_1 e^{i\theta}, \beta_1 e^{i\varphi})$-level hypergraph of H^* is a crisp hypergraph.

Example 6.16 Consider a complex Pythagorean fuzzy hypergraph $H^* = (\mathscr{C}^*, \mathscr{D}^*)$ as shown in Fig. 6.22. Let $\alpha_1 = 0.5$, $\beta_1 = 0.6$, $\theta = 0.3\pi$, and $\varphi = 0.7\pi$. Then, $(\alpha_1 e^{i\theta}, \beta_1 e^{i\varphi})$-level hypergraph of H^* is shown in Fig. 6.23.

Definition 6.50 Let $H^* = (\mathscr{C}^*, \mathscr{D}^*)$ be a complex Pythagorean fuzzy hypergraph. The *complex Pythagorean fuzzy line graph* of H^* is defined as an ordered pair $l(H^*) = (\mathscr{C}_l^*, \mathscr{D}_l^*)$, where $\mathscr{C}_l^* = \mathscr{D}^*$ and there exists an edge between two vertices in $l(H^*)$ if $|supp(D_j) \cap supp(D_k)| \geq 1$, for all $D_j, D_k \in \mathscr{D}^*$. The membership degrees of $l(H^*)$ are given as

(i) $\mathscr{C}_l^*(E_k) = \mathscr{D}^*(E_k)$,
(ii) $\mathscr{D}_l^*(E_j E_k) = (\min\{T_{\mathscr{D}^*}(E_j), T_{\mathscr{D}^*}(E_k)\}e^{i \min\{\phi_{\mathscr{D}^*}(E_j), \phi_{\mathscr{D}^*}(E_k)\}}, \max\{F_{\mathscr{D}^*}(E_j), F_{\mathscr{D}^*}(E_k)\}e^{i \max\{\psi_{\mathscr{D}^*}(E_j), \psi_{\mathscr{D}^*}(E_k)\}})$.

Definition 6.51 A complex Pythagorean fuzzy hypergraph $H^* = (\mathscr{C}^*, \mathscr{D}^*)$ is said to be *linear* if for every $D_j, D_k \in \mathscr{D}^*$,

(i) $supp(D_j) \subseteq supp(D_k) \Rightarrow j = k$,
(ii) $|supp(D_j) \cap supp(D_k)| \leq 1$.

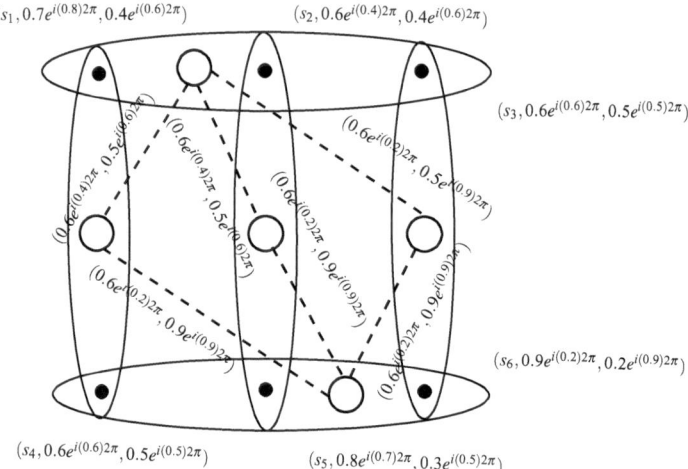

Fig. 6.24 Line graph of complex Pythagorean fuzzy hypergraph $H^* = (\mathscr{C}^*, \mathscr{D}^*)$

Example 6.17 Consider a complex Pythagorean fuzzy hypergraph $H^* = (\mathscr{C}^*, \mathscr{D}^*)$ as shown in Fig. 6.22. By direct calculations, we have

$$supp(\mathscr{D}_1) = \{s_1, s_2, s_3\}, \ supp(\mathscr{D}_2) = \{s_4, s_5, s_6\}, \ supp(\mathscr{D}_3) = \{s_1, s_4\},$$
$$supp(\mathscr{D}_4) = \{s_2, s_5\}, \ supp(\mathscr{D}_5) = \{s_3, s_6\}.$$

Note that, $supp(D_j) \subseteq supp(D_k) \Rightarrow j = k$ and $|supp(D_j) \cap supp(D_k)| \leq 1$. Hence, complex Pythagorean fuzzy hypergraph $H^* = (\mathscr{C}^*, \mathscr{D}^*)$ is linear. The corresponding complex Pythagorean fuzzy hypergraph $H^* = (\mathscr{C}^*, \mathscr{D}^*)$ and its line graph is shown in Fig. 6.24.

Theorem 6.17 *A simple strong complex Pythagorean fuzzy hypergraph is the complex Pythagorean fuzzy line graph of a linear complex Pythagorean fuzzy hypergraph.*

Definition 6.52 The *2-section* $H_2^* = (\mathscr{C}_2^*, \mathscr{D}_2^*)$ of a complex Pythagorean fuzzy hypergraph $H^* = (\mathscr{C}^*, \mathscr{D}^*)$ is a complex Pythagorean fuzzy graph having same set of vertices as that of H^*, $\mathscr{D}_2^*$ is a complex Pythagorean fuzzy set on $\{e = u_j u_k | u_j, u_k \in E_l, l = 1, 2, 3, \ldots\}$, and $\mathscr{D}_2^*(u_j u_k) = (\min\{\min T_{\beta_l}(u_j), \min T_{\beta_l}(u_k)\}$ $e^{i \min\{\min \phi_{\beta_l}(u_j), \min \phi_{\beta_l}(u_k)\}}, \max\{\max F_{\beta_l}(u_j), \max F_{\beta_l}(u_k)\} e^{i \max\{\max \psi_{\beta_l}(u_j), \max \psi_{\beta_l}(u_k)\}})$ such that $0 \leq T_{\mathscr{D}_2^*}^2(u_j u_k) + F_{\mathscr{D}_2^*}^2(u_j u_k) \leq 1$.

Example 6.18 An example of a complex Pythagorean fuzzy hypergraph is given in Fig. 6.25. The 2-section of H^* is presented with dashed lines.

Definition 6.53 Let $H^* = (\mathscr{C}^*, \mathscr{D}^*)$ be a complex Pythagorean fuzzy hypergraph. A *complex Pythagorean fuzzy transversal* τ is a complex Pythagorean fuzzy set of X

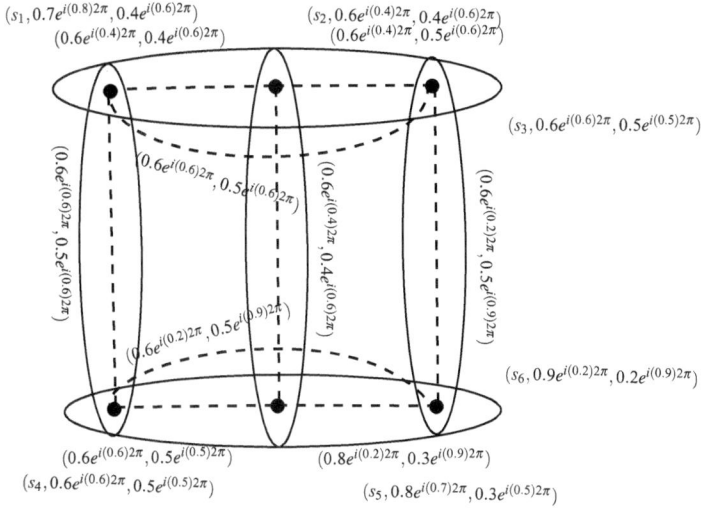

Fig. 6.25 2-section of complex Pythagorean fuzzy hypergraph H^*

satisfying the condition $\rho^{h(\rho)} \cap \tau^{h(\rho)} \neq \emptyset$, for all $\rho \in \mathscr{D}^*$, where $h(\rho)$ is the height of ρ.

A *minimal complex Pythagorean fuzzy transversal* t is the complex Pythagorean fuzzy transversal of H^* having the property that if $\tau \subset t$, then τ is not a complex Pythagorean fuzzy transversal of H^*.

6.10 Complex q-Rung Orthopair Fuzzy Hypergraphs

A complex q-rung orthopair fuzzy model provides more flexibility due to its most prominent feature that is the sum of the qth powers of the truth-membership, falsity-membership must be less than or equal to one, and the sum of qth powers of the corresponding phase angles should lie between 0 and 2π. A complex q-rung orthopair fuzzy hypergraph model proves to be more generalized framework to deal with vagueness in complex hypernetworks when the relationships are more generalized rather than the pairwise interactions. The generalization of our proposed model can be observed from the reduction of complex q-rung orthopair fuzzy model to complex intuitionistic fuzzy and complex Pythagorean fuzzy models for $q = 1$ and $q = 2$, respectively.

Definition 6.54 A *complex q-rung orthopair fuzzy set* S in the universal set X is given as

$$S = \{(u, T_S(u)e^{i\phi_S(u)}, F_S(u)e^{i\psi_S(u)})|u \in X\},$$

where $i = \sqrt{-1}$, $T_S(u)$, $F_S(u) \in [0, 1]$ are named as amplitude terms, $\phi_S(u)$, $\psi_S(u)$ $\in [0, 2\pi]$ are named as phase terms, and for every $u \in X$, $0 \leq T_S^q(u) + F_S^q(u) \leq 1$, $q \geq 1$.

Remark 6.6 • When $q = 1$, complex 1-rung orthopair fuzzy set is called a complex intuitionistic fuzzy set.
• When $q = 2$, complex 1-rung orthopair fuzzy set is called a complex Pythagorean fuzzy set.

Definition 6.55 Let $S_1 = \{(u, T_{S_1}(u)e^{i\phi_{S_1}(u)}, F_{S_1}(u)e^{i\psi_{S_1}(u)})|u \in X\}$ and $S_2 = \{(u, T_{S_2}(u)e^{i\phi_{S_2}(u)}, F_{S_2}(u)e^{i\psi_{S_2}(u)})|u \in X\}$ be two complex q-rung orthopair fuzzy sets in X, then

(i) $S_1 \subseteq S_2 \Leftrightarrow T_{S_1} \leq T_{S_2}(u)$, $F_{S_1}(u) \geq F_{S_2}(u)$, and $\phi_{S_1}(u) \leq \phi_{S_2}(u)$, $\psi_{S_1}(u) \geq \psi_{S_2}(u)$ for amplitudes and phase terms, respectively, for all $u \in X$.
(ii) $S_1 = S_2 \Leftrightarrow T_{S_1} = T_{S_2}(u)$, $F_{S_1}(u) = F_{S_2}(u)$, and $\phi_{S_1}(u) = \phi_{S_2}(u)$, $\psi_{S_1}(u) = \psi_{S_2}(u)$ for amplitudes and phase terms, respectively, for all $u \in X$.

Definition 6.56 Let $S_1 = \{(u, T_{S_1}(u)e^{i\phi_{S_1}(u)}, F_{S_1}(u)e^{i\psi_{S_1}(u)})|u \in X\}$ and $S_2 = \{(u, T_{S_2}(u)e^{i\phi_{S_2}(u)}, F_{S_2}(u)e^{i\psi_{S_2}(u)})|u \in X\}$ be two complex q-rung orthopair fuzzy sets in X, then

(i) $S_1 \cup S_2 = \{(u, \max\{T_{S_1}(u), T_{S_2}(u)\}e^{i\max\{\phi_{S_1}(u),\phi_{S_2}(u)\}}, \min\{F_{S_1}(u), F_{S_2}(u)\}$ $e^{i\min\{\psi_{S_1}(u),\psi_{S_2}(u)\}})|u \in X\}$.
(ii) $S_1 \cap S_2 = \{(u, \min\{T_{S_1}(u), T_{S_2}(u)\}e^{i\min\{\phi_{S_1}(u),\phi_{S_2}(u)\}}, \max\{F_{S_1}(u), F_{S_2}(u)\}$ $e^{i\max\{\psi_{S_1}(u),\psi_{S_2}(u)\}})|u \in X\}$.

Definition 6.57 A *complex q-rung orthopair fuzzy relation* is a complex q-rung orthopair fuzzy set on $X \times X$ given as

$$R = \{(rs, T_R(rs)e^{i\phi_R(rs)}, F_R(rs)e^{i\psi_R(rs)})|rs \in X \times X\},$$

where $i = \sqrt{-1}$, $T_R : X \times X \to [0, 1]$, $F_R : X \times X \to [0, 1]$ characterize the amplitudes of truth and falsity degrees of R, and $\phi_R(rs)$, $\psi_R(rs) \in [0, 2\pi]$ are called the phase terms such that for all $rs \in X \times X$, $0 \leq T_R^q(rs) + F_R^q(rs) \leq 1$, $q \geq 1$.

Example 6.19 Let $X = \{b_1, b_2, b_3\}$ be the universal set and $\{b_1b_2, b_2b_3, b_1b_3\}$ be the subset of $X \times X$. Then, the complex 5-rung orthopair fuzzy relation R is given as

$$R = \{(b_1b_2, 0.9e^{i(0.7)\pi}, 0.7e^{i(0.9)\pi}), (b_2b_3, 0.6e^{i(0.7)\pi}, 0.8e^{i(0.9)\pi}), (b_1b_3, 0.7e^{i(0.8)\pi}, 0.5e^{i(0.6)\pi})\}.$$

Note that, $0 \leq T_R^5(xy) + F_R^5(xy) \leq 1$, for all $xy \in X \times X$. Hence, R is a complex 5-rung orthopair fuzzy relation on X.

Definition 6.58 A *complex q-rung orthopair fuzzy graph* on X is an ordered pair $\mathscr{G} = (\mathscr{A}, \mathscr{B})$, where $\mathscr{A}$ is a complex q-rung orthopair fuzzy set on X and $\mathscr{B}$ is complex q-rung orthopair fuzzy relation on X such that

$$T_{\mathscr{B}}(ab) \leq \min\{T_{\mathscr{A}}(a), T_{\mathscr{A}}(b)\},$$
$$F_{\mathscr{B}}(ab) \leq \max\{F_{\mathscr{A}}(a), F_{\mathscr{A}}(b)\}, \text{ (for amplitude terms)}$$
$$\phi_{\mathscr{B}}(ab) \leq \min\{\phi_{\mathscr{A}}(a), \phi_{\mathscr{A}}(b)\},$$
$$\psi_{\mathscr{B}}(ab) \leq \max\{\psi_{\mathscr{A}}(a), \psi_{\mathscr{A}}(b)\}, \text{ (for phase terms)}$$

$$0 \leq T_{\mathscr{B}}^{q}(ab) + F_{\mathscr{B}}^{q}(ab) \leq 1, q \geq 1, \text{ for all } a, b \in X.$$

Remark 6.7 Note that,

- When $q = 1$, complex 1-rung orthopair fuzzy graph is called a complex intuition-istic fuzzy graph.
- When $q = 2$, complex 2-rung orthopair fuzzy graph is called a complex Pythagorean fuzzy graph.

Example 6.20 Let $\mathscr{G} = (\mathscr{A}, \mathscr{B})$ be a complex 6-rung orthopair fuzzy graph on $X = \{s_1, s_2, s_3, s_4\}$, where $\mathscr{A} = \{(s_1, 0.7e^{i(0.9)\pi}, 0.9e^{i(0.7)\pi}), (s_2, 0.5e^{i(0.6)\pi}, 0.6e^{i(0.5)\pi}), (s_3, 0.7e^{i(0.4)\pi}, 0.4e^{i(0.7)\pi}), (s_4, 0.8e^{i(0.5)\pi}, 0.5e^{i(0.8)\pi})\}$ and $\mathscr{B} = \{(s_1s_4, 0.7e^{i(0.7)\pi}, 0.8e^{i(0.8)\pi}), (s_2s_4, 0.5e^{i(0.5)\pi}, 0.6e^{i(0.8)\pi}), (s_3s_4, 0.7e^{i(0.4)\pi}, 0.5e^{i(0.8)\pi})\}$ are complex 6-rung orthopair fuzzy set and complex 6-rung orthopair fuzzy relation on X, respectively. The corresponding complex 6-rung orthopair fuzzy graph $\mathscr{G}$ is shown in Fig. 6.26.

We now define the more extended concept of complex q-rung orthopair fuzzy hypergraphs.

Definition 6.59 The *support* of a complex q-rung orthopair fuzzy set $S = \{(u, T_S(u)e^{i\phi_S(u)}, F_S(u)e^{i\psi_S(u)}) | u \in X\}$ is defined as $supp(S) = \{u | T_S(u) \neq 0, F_S(u) \neq 1, 0 < \phi_S(u), \psi_S(u) < 2\pi\}$.

Fig. 6.26 Complex 6-rung orthopair fuzzy graph

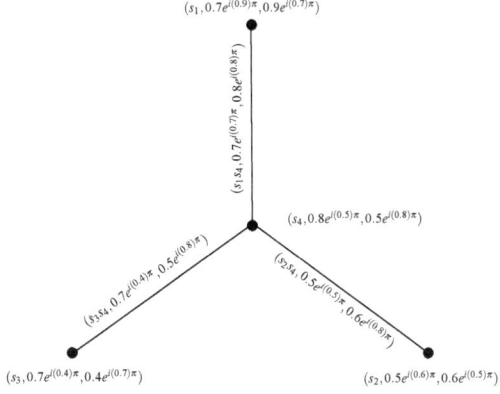

The *height* of a complex q-rung orthopair fuzzy set $S = \{(u, T_S(u)e^{i\phi_S(u)}, F_S(u)e^{i\psi_S(u)}) | u \in X\}$ is defined as

$$h(S) = \{\max_{u \in X} T_S(u)e^{i \max_{u \in X} \phi_S(u)}, \min_{u \in X} F_S(u)e^{i \min_{u \in X} \psi_S(u)}\}.$$

If $h(S) = (1e^{i2\pi}, 0e^{i0})$, then S is called *normal*.

Definition 6.60 Let X be a nontrivial set of universe. A *complex q-rung orthopair fuzzy hypergraph* is defined as an ordered pair $\mathscr{H} = (\mathscr{Q}, \eta)$, where $\mathscr{Q} = \{Q_1, Q_2, \ldots, Q_k\}$ is a finite family of complex q-rung orthopair fuzzy sets on X and η is a complex q-rung orthopair fuzzy relation on complex q-rung orthopair fuzzy sets Q_j's such that

(i)

$$T_\eta(\{a_1, a_2, \ldots, a_l\}) \leq \min\{T_{Q_j}(a_1), T_{Q_j}(a_2), \ldots, T_{Q_j}(a_l)\},$$
$$F_\eta(\{a_1, a_2, \ldots, a_l\}) \leq \max\{F_{Q_j}(a_1), F_{Q_j}(a_2), \ldots, F_{Q_j}(a_l)\}, \text{ (for amplitude terms)}$$
$$\phi_\eta(\{a_1, a_2, \ldots, a_l\}) \leq \min\{\phi_{Q_j}(a_1), \phi_{Q_j}(a_2), \ldots, \phi_{Q_j}(a_l)\},$$
$$\psi_\eta(\{a_1, a_2, \ldots, a_l\}) \leq \max\{\psi_{Q_j}(a_1), \psi_{Q_j}(a_2), \ldots, \psi_{Q_j}(a_l)\}, \text{ (for phase terms)}$$

$$0 \leq T_\eta^q + F_\eta^q \leq 1, q \geq 1, \text{ for all } a_1, a_2, \ldots, a_l \in X.$$
(ii) $\bigcup_j supp(Q_j) = X$, for all $Q_j \in \mathscr{Q}$.

Note that, $E_k = \{a_1, a_2, \ldots, a_l\}$ is the crisp hyperedge of $\mathscr{H} = (\mathscr{Q}, \eta)$.

Remark 6.8 Note that,

- When $q = 1$, complex 1-rung orthopair fuzzy hypergraph is a complex intuitionistic fuzzy hypergraph.
- When $q = 2$, complex 2-rung orthopair fuzzy hypergraph is a complex Pythagorean fuzzy hypergraph.

Definition 6.61 Let $\mathscr{H} = (\mathscr{Q}, \eta)$ be a complex q-rung orthopair fuzzy hypergraph. The *height* of $\mathscr{H}$, given as $h(\mathscr{H})$, is defined as $h(\mathscr{H}) = (\max \eta_l e^{i \max \phi}, \min \eta_m e^{i \min \psi})$, where $\eta_l = \max T_{\rho_j}(x_k)$, $\phi = \max \phi_{\rho_j}(x_k)$, $\eta_m = \min F_{\rho_j}(x_k)$, $\psi = \min \psi_{\rho_j}(x_k)$. Here, $T_{\rho_j}(x_k)$ and $F_{\rho_j}(x_k)$ denote the truth and falsity degrees of vertex x_k to hyperedge ρ_j, respectively.

Definition 6.62 Let $\mathscr{H} = (\mathscr{Q}, \eta)$ be a complex q-rung orthopair fuzzy hypergraph. Suppose that $\mu, \nu \in [0, 1]$ and $\theta, \varphi \in [0, 2\pi]$ such that $0 \leq \mu^q + \nu^q \leq 1$. The $(\mu e^{i\theta}, \nu e^{i\varphi})$-*level hypergraph* of $\mathscr{H}$ is defined as an ordered pair $\mathscr{H}^{(\mu e^{i\theta}, \nu e^{i\varphi})} = (\mathscr{Q}^{(\mu e^{i\theta}, \nu e^{i\varphi})}, \eta^{(\mu e^{i\theta}, \nu e^{i\varphi})})$, where

(i) $\eta^{(\mu e^{i\theta}, \nu e^{i\varphi})} = \{\rho_j^{(\mu e^{i\theta}, \nu e^{i\varphi})} : \rho_j \in \eta\}$ and $\rho_j^{(\mu e^{i\theta}, \nu e^{i\varphi})} = \{u \in X : T_{\rho_j}(u) \geq \mu, \phi_{\rho_j}(u) \geq \theta,$ and $F_{\rho_j}(u) \leq \nu, \psi_{\rho_j}(u) \leq \varphi\}$,
(ii) $\mathscr{Q}^{(\mu e^{i\theta}, \nu e^{i\varphi})} = \bigcup_{\rho_j \in \eta} \rho_j^{(\mu e^{i\theta}, \nu e^{i\varphi})}$.

Table 6.15 Incidence matrix of complex 6-rung orthopair fuzzy hypergraph $\mathscr{H}$

$u \in X$	η_1	η_2	η_3	η_4
u_1	$(0.8e^{i(0.8)\pi},$ $0.6e^{i(0.6)\pi})$	$(0,0)$	$(0,0)$	$(0.8e^{i(0.8)\pi},$ $0.6e^{i(0.6)\pi})$
u_2	$(0.7e^{i(0.7)\pi},$ $0.6e^{i(0.6)\pi})$	$(0,0)$	$(0,0)$	$(0,0)$
u_3	$(0.7e^{i(0.7)\pi},$ $0.8e^{i(0.8)\pi})$	$(0.7e^{i(0.7)\pi},$ $0.8e^{i(0.8)\pi})$	$(0,0)$	$(0,0)$
u_4	$(0,0)$	$(0.7e^{i(0.7)\pi},$ $0.8e^{i(0.8)\pi})$	$(0.7e^{i(0.7)\pi},$ $0.8e^{i(0.8)\pi})$	$(0,0)$
u_5	$(0,0)$	$(0.6e^{i(0.6)\pi},$ $0.8e^{i(0.8)\pi})$	$(0,0)$	$(0,0)$
u_6	$(0,0)$	$(0,0)$	$(0.9e^{i(0.9)\pi},$ $0.8e^{i(0.8)\pi})$	$(0.9e^{i(0.9)\pi},$ $0.8e^{i(0.8)\pi})$

Fig. 6.27 Complex 6-rung orthopair fuzzy hypergraph

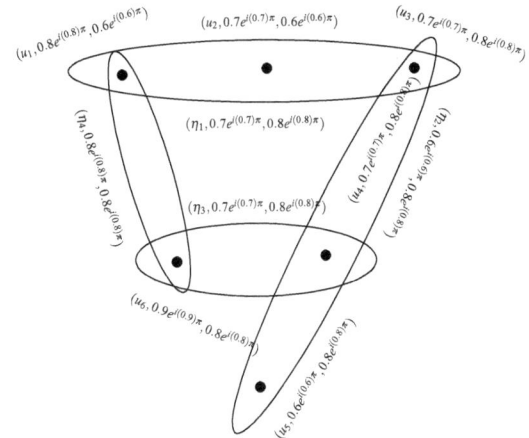

Note that, $(\mu e^{i\theta}, \nu e^{i\varphi})$-level hypergraph of $\mathscr{H}$ is a crisp hypergraph.

Example 6.21 Consider a complex 6-rung orthopair fuzzy hypergraph $\mathscr{H} = (\mathscr{Q}, \eta)$ on $X = \{u_1, u_2, u_3, u_4, u_5, u_6\}$. The complex 6-rung orthopair fuzzy relation η is given as, $\eta(u_1, u_2, u_3) = (0.7e^{i(0.7)\pi}, 0.8e^{i(0.8)\pi})$, $\eta(u_3, u_4, u_5) = (0.6e^{i(0.6)\pi}, 0.8e^{i(0.8)\pi})$, $\eta(u_1, u_6) = (0.8e^{i(0.8)\pi}, 0.8e^{i(0.8)\pi})$ and $\eta(u_4, u_6) = (0.7e^{i(0.7)\pi}, 0.8e^{i(0.8)\pi})$. The incidence matrix of $\mathscr{H}$ is given in Table 6.15.

The corresponding complex 6-rung orthopair fuzzy hypergraph $\mathscr{H} = (\mathscr{Q}, \eta)$ is shown in Fig. 6.27.

Let $\mu = 0.7$, $\nu = 0.6$, $\theta = 0.7\pi$, and $\varphi = 0.6\pi$, then $(0.7e^{i(0.7)\pi}, 0.6e^{i(0.6)\pi})$-level hypergraph of $\mathscr{H}$ is shown in Fig. 6.28.

Fig. 6.28 $(0.7e^{i(0.7)\pi}, 0.6e^{i(0.6)\pi})$-
level hypergraph of $\mathscr{H}$

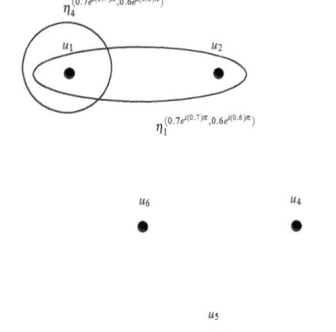

Note that,

$$\eta_1^{(0.7e^{i(0.7)\pi}, 0.6e^{i(0.6)\pi})} = \{u_1, u_2\}, \quad \eta_2^{(0.7e^{i(0.7)\pi}, 0.6e^{i(0.6)\pi})} = \{\emptyset\},$$

$$\eta_3^{(0.7e^{i(0.7)\pi}, 0.6e^{i(0.6)\pi})} = \{\emptyset\}, \quad \eta_4^{(0.7e^{i(0.7)\pi}, 0.6e^{i(0.6)\pi})} = \{u_1\}.$$

6.11   Transversals of Complex q-Rung Orthopair Fuzzy Hypergraphs

Definition 6.63 Let $\mathscr{H} = (\mathscr{Q}, \eta)$ be a complex q-rung orthopair fuzzy hypergraph and for $0 < \mu \leq T(h(\mathscr{H})), v \geq F(h(\mathscr{H})) > 0, 0 < \theta \leq \phi(h(\mathscr{H}))$, and $\varphi \geq \psi(h(\mathscr{H})) > 0$ let $\mathscr{H}^{(\mu e^{i\theta}, v e^{i\varphi})} = (\mathscr{Q}^{(\mu e^{i\theta}, v e^{i\varphi})}, \eta^{(\mu e^{i\theta}, v e^{i\varphi})})$ be the level hypergraph of $\mathscr{H}$. The sequence of complex numbers $\{(\mu_1 e^{i\theta_1}, v_1 e^{i\varphi_1}), (\mu_2 e^{i\theta_2}, v_2 e^{i\varphi_2}), \ldots, (\mu_n e^{i\theta_n}, v_n e^{i\varphi_n})\}$ such that $0 < \mu_1 < \mu_2 < \cdots < \mu_n = T(h(\mathscr{H})), v_1 > v_2 > \cdots > v_n = F(h(\mathscr{H})) > 0, \ 0 < \theta_1 < \theta_2 < \cdots < \theta_n = \phi(h(\mathscr{H})),$ and $\varphi_1 > \varphi_2 > \cdots > \varphi_n = \psi(h(\mathscr{H})) > 0$ satisfying the conditions

(i) if $\mu_{k+1} < \alpha \leq \mu_k, \ v_{k+1} > \beta \geq v_k, \ \theta_{k+1} < \phi \leq \theta_k, \ \varphi_{k+1} > \psi \geq \varphi_k,$ then $\eta^{(\alpha e^{i\phi}, \beta e^{i\psi})} = \eta^{(\mu_k e^{i\theta_k}, v_k e^{i\varphi_k})},$ and
(ii) $\eta^{(\mu_k e^{i\theta_k}, v_k e^{i\varphi_k})} \subset \eta^{(\mu_{k+1} e^{i\theta_{k+1}}, v_{k+1} e^{i\varphi_{k+1}})},$

is called the *fundamental sequence* of $\mathscr{H} = (\mathscr{Q}, \eta)$, denoted by $\mathscr{F}_s(\mathscr{H})$. The set of $(\mu_j e^{i\theta_j}, v_j e^{i\varphi_j})$-level hypergraphs $\{\mathscr{H}^{(\mu_1 e^{i\theta_1}, v_1 e^{i\varphi_1})}, \mathscr{H}^{(\mu_2 e^{i\theta_2}, v_2 e^{i\varphi_2})}, \ldots, \mathscr{H}^{(\mu_n e^{i\theta_n}, v_n e^{i\varphi_n})}\}$ is called the set of core hypergraphs or the *core set* of $\mathscr{H}$, denoted by $cor(\mathscr{H})$.

Definition 6.64 Let $\mathscr{H} = (\mathscr{Q}, \eta)$ be a complex q-rung orthopair fuzzy hypergraph. A *complex q-rung orthopair fuzzy transversal* τ is a complex q-rung orthopair fuzzy set of X satisfying the condition $\rho^{h(\rho)} \cap \tau^{h(\rho)} \neq \emptyset$, for all $\rho \in \eta$, where $h(\rho)$ is the height of ρ.

Table 6.16 Incidence matrix of complex 5-rung orthopair fuzzy hypergraph $\mathscr{H}$

$a \in X$	η_1	η_2	η_3
a_1	$(0.8e^{i(0.8)\pi}, 0.6e^{i(0.6)\pi})$	$(0.8e^{i(0.8)\pi}, 0.6e^{i(0.6)\pi})$	$(0,0)$
a_2	$(0.7e^{i(0.7)\pi}, 0.9e^{i(0.9)\pi})$	$(0,0)$	$(0.7e^{i(0.7)\pi}, 0.9e^{i(0.9)\pi})$
a_3	$(0,0)$	$(0.8e^{i(0.8)\pi}, 0.5e^{i(0.5)\pi})$	$(0.8e^{i(0.8)\pi}, 0.5e^{i(0.5)\pi})$
a_4	$(0.6e^{i(0.6)\pi}, 0.8e^{i(0.8)\pi})$	$(0.6e^{i(0.6)\pi}, 0.8e^{i(0.8)\pi})$	$(0,0)$
a_5	$(0,0)$	$(0,0)$	$(0.7e^{i(0.7)\pi}, 0.5e^{i(0.5)\pi})$

Fig. 6.29 Complex 5-rung orthopair fuzzy hypergraph

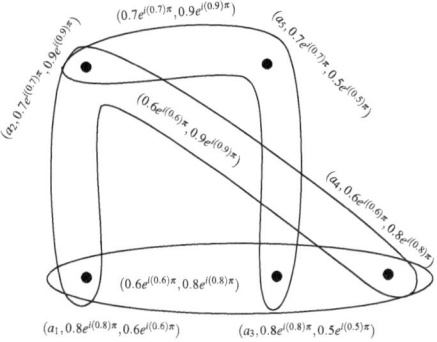

A *minimal complex q-rung orthopair fuzzy transversal t* is the complex q-rung orthopair fuzzy transversal of $\mathscr{H}$ having the property that if $\tau \subset t$, then τ is not a complex q-rung orthopair fuzzy transversal of $\mathscr{H}$.

Let us denote the family of minimal complex q-rung orthopair fuzzy transversals of $\mathscr{H}$ by $t_r(\mathscr{H})$.

Example 6.22 Consider a complex 5-rung orthopair fuzzy hypergraph $\mathscr{H} = (\mathscr{Q}, \eta)$ on $X = \{a_1, a_2, a_3, a_4, a_5\}$. The complex 5-rung orthopair fuzzy relation η is given as, $\eta(\{a_1 a_3, a_4\}) = (0.6e^{i(0.6)\pi}, 0.9e^{i(0.9)\pi})$, $\eta(\{a_2, a_3, a_5\}) = (0.7e^{i(0.7)\pi}, 0.9e^{i(0.9)\pi})$, and $\eta(\{a_1, a_2, a_4\}) = (0.6e^{i(0.6)\pi}, 0.9e^{i(0.9)\pi})$. The incidence matrix of $\mathscr{H}$ is given in Table 6.16.

The corresponding complex 5-rung orthopair fuzzy hypergraph is shown in Fig. 6.29.

By routine calculations, we have $h(\eta_1) = (0.8e^{i(0.8)\pi}, 0.6e^{i(0.6)\pi})$, $h(\eta_2) = (0.8e^{i(0.8)\pi}, 0.5e^{i(0.5)\pi})$, and $h(\eta_3) = (0.8e^{i(0.8)\pi}, 0.5e^{i(0.5)\pi})$. Consider a complex 5-rung orthopair fuzzy set τ_1 of X such that

$$\tau_1 = \{(a_1, 0.8e^{i(0.8)\pi}, 0.6e^{i(0.6)\pi}), (a_2, 0.7e^{i(0.7)\pi}, 0.9e^{i(0.9)\pi}), (a_3, 0.8e^{i(0.8)\pi}, 0.5e^{i(0.5)\pi})\}.$$

Note that,

$$\eta_1^{(0.8e^{i(0.8)\pi},0.6e^{i(0.6)\pi})} = \{a_1\}, \quad \eta_2^{(0.8e^{i(0.8)\pi},0.5e^{i(0.5)\pi})} = \{a_3\}, \quad \eta_3^{(0.8e^{i(0.8)\pi},0.5e^{i(0.5)\pi})} = \{a_3\},$$
$$\tau_1^{(0.8e^{i(0.8)\pi},0.6e^{i(0.6)\pi})} = \{a_1, a_3\}, \quad \tau_1^{(0.8e^{i(0.8)\pi},0.5e^{i(0.5)\pi})} = \{a_3\}, \quad \tau_1^{(0.8e^{i(0.8)\pi},0.5e^{i(0.5)\pi})} = \{a_3\}.$$

Thus, we have $\eta_j^{h(\eta_j)} \cap \tau_1^{h(\eta_j)} \neq \emptyset$, for all $\eta_j \in \eta$. Hence, τ_1 is a complex 5-rung orthopair fuzzy transversal of $\mathscr{H}$. Similarly,

$$\tau_2 = \{(a_1, 0.8e^{i(0.8)\pi}, 0.6e^{i(0.6)\pi}), (a_3, 0.8e^{i(0.8)\pi}, 0.5e^{i(0.5)\pi})\},$$
$$\tau_3 = \{(a_1, 0.8e^{i(0.8)\pi}, 0.6e^{i(0.6)\pi}), (a_3, 0.8e^{i(0.8)\pi}, 0.5e^{i(0.5)\pi}), (a_4, 0.6e^{i(0.6)\pi}, 0.8e^{i(0.8)\pi})\},$$
$$\tau_4 = \{(a_1, 0.8e^{i(0.8)\pi}, 0.6e^{i(0.6)\pi}), (a_3, 0.8e^{i(0.8)\pi}, 0.5e^{i(0.5)\pi}), (a_5, 0.7e^{i(0.7)\pi}, 0.5e^{i(0.5)\pi})\},$$

are complex 5-rung orthopair fuzzy transversals of $\mathscr{H}$.

Definition 6.65 A complex q-rung orthopair fuzzy hypergraph $\mathscr{H}_1 = (\mathscr{Q}_1, \eta_1)$ is a *partial* complex q-rung orthopair fuzzy hypergraph of $\mathscr{H}_2 = (\mathscr{Q}_2, \eta_2)$ if $\eta_1 \subseteq \eta_2$, denoted by $\mathscr{H}_1 \subseteq \mathscr{H}_2$.
A complex q-rung orthopair fuzzy hypergraph $\mathscr{H}_1 = (\mathscr{Q}_1, \eta_1)$ is *ordered* if the core set $cor(\mathscr{H}) = \{\mathscr{H}^{(\mu_1 e^{i\theta_1}, v_1 e^{i\varphi_1})}, \mathscr{H}^{(\mu_2 e^{i\theta_2}, v_2 e^{i\varphi_2})}, \ldots, \mathscr{H}^{(\mu_n e^{i\theta_n}, v_n e^{i\varphi_n})}\}$ is ordered, i.e., $\mathscr{H}^{(\mu_1 e^{i\theta_1}, v_1 e^{i\varphi_1})} \subseteq \mathscr{H}^{(\mu_2 e^{i\theta_2}, v_2 e^{i\varphi_2})} \subseteq \cdots \subseteq \mathscr{H}^{(\mu_n e^{i\theta_n}, v_n e^{i\varphi_n})}$. $\mathscr{H}$ is *simply ordered* if $\mathscr{H}$ is ordered and $\eta' \subset \eta^{(\mu_{l+1} e^{i\theta_{l+1}}, v_{l+1} e^{i\varphi_{l+1}})} \backslash \eta^{(\mu_l e^{i\theta_l}, v_l e^{i\varphi_l})} \Rightarrow \eta' \not\subseteq \mathscr{Q}^{(\mu_l e^{i\theta_l}, v_l e^{i\varphi_l})}$.

Definition 6.66 A complex q-rung orthopair fuzzy set S on X is *elementary* if S is single-valued on $supp(S)$. A complex q-rung orthopair fuzzy hypergraph $\mathscr{H} = (\mathscr{Q}, \eta)$ is *elementary* if every $Q_j \in \mathscr{Q}$ and η are elementary.

Proposition 6.2 *If τ is a complex q-rung orthopair fuzzy transversal of $\mathscr{H} = (\mathscr{Q}, \eta)$, then $h(\tau) \geq h(\rho)$, for all $\rho \in \eta$. Furthermore, if τ is minimal complex q-rung orthopair fuzzy transversal of $\mathscr{H} = (\mathscr{Q}, \eta)$, then $h(\tau) = \max\{h(\rho)|\rho \in \eta\} = h(\mathscr{H})$.*

Lemma 6.4 *Let $\mathscr{H}_1 = (\mathscr{Q}_1, \eta_1)$ be a partial complex q-rung orthopair fuzzy hypergraph of $\mathscr{H}_2 = (\mathscr{Q}_2, \eta_2)$. If τ_2 is minimal complex q-rung orthopair fuzzy transversal of $\mathscr{H}_2$, then there is a minimal complex q-rung orthopair fuzzy transversal of $\mathscr{H}_1$ such that $\tau_1 \subseteq \tau_2$.*

Proof Let S_1 be a complex q-rung orthopair fuzzy set on X, which is defined as $S_1 = \tau_2 \cap (\cup_{Q_{1j} \in \mathscr{Q}_1} Q_{1j})$. Then, S_1 is a complex q-rung orthopair fuzzy transversal of $\mathscr{H}_1 = (\mathscr{Q}_1, \eta_1)$. Thus, there exists a minimal complex q-rung orthopair fuzzy transversal of $\mathscr{H}_1$ such that $\tau_1 \subseteq S_1 \subseteq \tau_2$.

Lemma 6.5 *Let $\mathscr{H} = (\mathscr{Q}, \eta)$ be a complex q-rung orthopair fuzzy hypergraph then $f_s(t_r(\mathscr{H})) \subseteq f_s(\mathscr{H})$.*

Proof Let $f_s(\mathscr{H}) = \{(\mu_1 e^{i\theta_1}, v_1 e^{i\varphi_1}), (\mu_2 e^{i\theta_2}, v_2 e^{i\varphi_2}), \ldots, (\mu_n e^{i\theta_n}, v_n e^{i\varphi_n})\}$ and $\tau \in t_r(\mathscr{H})$. Suppose that for $u \in supp(\tau)$, $(T_\tau(u), F_\tau(u)) \in (\mu_{j+1}, \mu_j] \times (v_{j+1}, v_j]$, $\phi_\tau(u) \in (\theta_{j+1}, \theta_j]$, and $\psi_\tau(u) \in (\varphi_{j+1}, \varphi_j]$. Define a function λ by

$$T_\lambda(v)e^{i\phi} = \begin{cases} \mu_j e^{i\theta_j}, & \text{if } u = v, \\ T_\tau(u)e^{i\phi_\tau(u)}, & \text{otherwise.} \end{cases}, \quad F_\lambda(v)e^{i\psi} = \begin{cases} \mu_j e^{i\varphi_j}, & \text{if } u = v, \\ F_\tau(u)e^{i\psi_\tau(u)}, & \text{otherwise.} \end{cases}$$

From definition of λ, we have $\lambda^{(\mu_j e^{i\theta_j}, v_j e^{i\varphi_j})} = \tau^{(\mu_j e^{i\theta_j}, v_j e^{i\varphi_j})}$. Definition 6.63 implies that for every $t \in (\mu_{j+1}e^{i\theta_{j+1}}, \mu_j e^{i\theta_j}] \times (v_{j+1}e^{i\varphi_{j+1}}, v_j e^{i\varphi_j}]$, $\mathcal{H}^t = \mathcal{H}^{(\mu_1 e^{i\theta_1}, v_1 e^{i\varphi_1})}$. Thus, $\lambda^{(\mu_j e^{i\theta_j}, v_j e^{i\varphi_j})}$ is a complex q-rung orthopair fuzzy transversal of $\mathcal{H}^t$. Since, τ is minimal complex q-rung orthopair fuzzy transversal and $\lambda^t = \tau^t$, for all $t \notin (\mu_{j+1}e^{i\theta_{j+1}}, \mu_j e^{i\theta_j}] \times (v_{j+1}e^{i\varphi_{j+1}}, v_j e^{i\varphi_j}]$. This implies that λ is also a complex q-rung orthopair fuzzy transversal and $\lambda \leq \tau$ but the minimality of τ implies that $\lambda = \tau$. Hence, $\tau(u) = \lambda(u) = (\mu_j e^{i\theta_j}, v_j e^{i\varphi_j})$, which implies that for every complex q-rung orthopair fuzzy transversal $\tau \in t_r(\mathcal{H})$ and for each $u \in X$, $\tau(u) \in f_s(\mathcal{H})$ and so we have $f_s(t_r(\mathcal{H})) \subseteq f_s(\mathcal{H})$.

We now illustrate a recursive procedure to find $t_r(\mathcal{H})$ in Algorithm 6.11.1.

Algorithm 6.11.1 To find the family of minimal complex q-rung orthopair fuzzy transversals $t_r(\mathcal{H})$

Let $\mathcal{H} = (\mathcal{Q}, \eta)$ be a complex q-rung orthopair fuzzy hypergraph having the fundamental sequence $f_s(\mathcal{H}) = \{(\mu_1 e^{i\theta_1}, v_1 e^{i\varphi_1}), (\mu_2 e^{i\theta_2}, v_2 e^{i\varphi_2}), \ldots, (\mu_n e^{i\theta_n}, v_n e^{i\varphi_n})\}$ and core set $cor(\mathcal{H}) = \{\mathcal{H}^{(\mu_1 e^{i\theta_1}, v_1 e^{i\varphi_1})}, \mathcal{H}^{(\mu_2 e^{i\theta_2}, v_2 e^{i\varphi_2})}, \ldots, \mathcal{H}^{(\mu_n e^{i\theta_n}, v_n e^{i\varphi_n})}\}$. The minimal transversal of $\mathcal{H} = (\mathcal{Q}, \eta)$ is determined as follows:

1. Determine a crisp minimal transversal t_1 of $\mathcal{H}^{(\mu_1 e^{i\theta_1}, v_1 e^{i\varphi_1})}$.
2. Determine a crisp minimal transversal t_2 of $\mathcal{H}^{(\mu_2 e^{i\theta_2}, v_2 e^{i\varphi_2})}$ satisfying the condition $t_1 \subseteq t_2$, i.e., obtain an hypergraph H_2 having the hyperedges $\eta^{(\mu_2 e^{i\theta_2}, v_2 e^{i\varphi_2})}$ and a loop at every vertex $u \in t_1$. Thus, we have $\eta(H_2) = \eta(\mu_2 e^{i\theta_2}, v_2 e^{i\varphi_2}) \cup \{\{u \in t_1\}\}$.
3. Let t_2 be the minimal transversal of H_2.
4. Obtain a sequence of minimal transversals $t_1 \subseteq t_2 \subseteq \cdots \subseteq t_j$ such that t_j is the minimal transversal of $\mathcal{H}^{(\mu_j e^{i\theta_j}, v_j e^{i\varphi_j})}$ satisfying the condition $t_{j-1} \subseteq t_j$.
5. Define an elementary complex q-rung orthopair fuzzy set S_j having the support t_j and $h(S_j) = (\mu_j e^{i\theta_j}, v_j e^{i\varphi_j})$, $1 \leq j \leq n$.
6. Determine a minimal complex q-rung orthopair fuzzy transversal of $\mathcal{H}$ as $\tau = \bigcup_{j=1}^{n}\{S_j | 1 \leq j \leq n\}$.

Example 6.23 Consider a complex 5-rung orthopair fuzzy hypergraph $\mathcal{H} = (\mathcal{Q}, \eta)$ on $X = \{v_1, v_2, v_3, v_4, v_5, v_6\}$ as shown in Fig. 6.30. Let $(\mu_1 e^{i\theta_1}, v_1 e^{i\varphi_1}) = (0.9e^{i(0.9)2\pi}, 0.7e^{i(0.7)2\pi})$, $(\mu_2 e^{i\theta_2}, v_2 e^{i\varphi_2}) = (0.8e^{i(0.8)2\pi}, 0.5e^{i(0.5)2\pi})$, $(\mu_3 e^{i\theta_3}, v_3 e^{i\varphi_3}) = (0.6e^{i(0.6)2\pi}, 0.4e^{i(0.4)2\pi})$, and $(\mu_4 e^{i\theta_4}, v_4 e^{i\varphi_4}) = (0.3e^{i(0.3)2\pi}, 0.2e^{i(0.2)2\pi})$. Clearly, the sequence $\{(\mu_1 e^{i\theta_1}, v_1 e^{i\varphi_1}), (\mu_2 e^{i\theta_2}, v_2 e^{i\varphi_2}), (\mu_3 e^{i\theta_3}, v_3 e^{i\varphi_3}), (\mu_4 e^{i\theta_4}, v_4 e^{i\varphi_4})\}$ satisfies all the conditions of Definition 6.63. Hence, it is the fundamental sequence of $\mathcal{H}$.

Fig. 6.30 Complex 5-rung orthopair fuzzy hypergraph

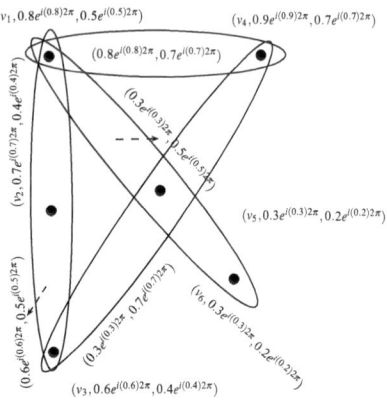

Note that, $t_1 = t_2 = \{v_4\}$ is the minimal transversal of $\mathscr{H}^{(\mu_1 e^{i\theta_1}, v_1 e^{i\varphi_1})}$ and $\mathscr{H}^{(\mu_2 e^{i\theta_2}, v_2 e^{i\varphi_2})}$, $t_3 = \{v_1\}$ is the minimal transversal of $\mathscr{H}^{(\mu_3 e^{i\theta_3}, v_3 e^{i\varphi_3})}$, and $t_4 = \{v_1, v_4\}$ is the minimal transversal of $\mathscr{H}^{(\mu_4 e^{i\theta_4}, v_4 e^{i\varphi_4})}$. Consider

$$S_1 = \{(v_4, 0.9e^{i(0.9)2\pi}, 0.7e^{i(0.7)2\pi})\} = S_2,$$
$$S_3 = \{(v_1, 0.8e^{i(0.8)2\pi}, 0.5e^{i(0.5)2\pi})\},$$
$$S_4 = \{(v_1, 0.8e^{i(0.8)2\pi}, 0.5e^{i(0.5)2\pi}), (v_4, 0.9e^{i(0.9)2\pi}, 0.7e^{i(0.7)2\pi})\}.$$

Hence, $\bigcup_{j=1}^{4} = \{(v_1, 0.8e^{i(0.8)2\pi}, 0.5e^{i(0.5)2\pi}), (v_4, 0.9e^{i(0.9)2\pi}, 0.7e^{i(0.7)2\pi})\}$ is a complex 5-rung orthopair fuzzy transversal of $\mathscr{H}$.

Lemma 6.6 *Let $\mathscr{H} = (\mathscr{Q}, \eta)$ be a complex q-rung orthopair fuzzy hypergraph with $f_s(\mathscr{H}) = \{(\mu_1 e^{i\theta_1}, v_1 e^{i\varphi_1}), (\mu_2 e^{i\theta_2}, v_2 e^{i\varphi_2}), \ldots, (\mu_n e^{i\theta_n}, v_n e^{i\varphi_n})\}$. If τ is a complex q-rung orthopair fuzzy transversal of $\mathscr{H}$, then $h(\tau) \geq h(Q_j)$, for every $Q_j \in \mathscr{Q}$. If $\tau \in t_r(\mathscr{H})$ then $h(\tau) = \max\{h(Q_j) | Q_j \in \mathscr{Q}\} = (\mu_1 e^{i\theta_1}, v_1 e^{i\varphi_1})$.*

Proof Since τ is a complex q-rung orthopair fuzzy transversal of $\mathscr{H}$, implies that $\tau^{h(Q_j)} \cap Q_j^{h(Q_j)} \neq \emptyset$. Let $a \in supp(\tau)$, then $T_\tau(a) \geq T(h(Q_j))$, $F_\tau(a) \leq F(h(Q_j))$, $\phi_\tau(a) \geq \phi(h(Q_j))$, and $\psi_\tau(a) \leq \psi(h(Q_j))$. This shows that $h(\tau) \geq h(Q_j)$. If $\tau \in t_r(\mathscr{H})$, i.e., τ is minimal complex q-rung orthopair fuzzy transversal then $h(Q_j) = (\max T_{Q_j}(a)e^{i \max \phi_{Q_j}(a)}, \min F_{Q_j}(a)e^{i \min \psi_{Q_j}(a)}) = (\mu_1 e^{i\theta_1}, v_1 e^{i\varphi_1})$. Thus, we have $h(\tau) = \max\{h(Q_j) | Q_j \in \mathscr{Q}\} = (\mu_1 e^{i\theta_1}, v_1 e^{i\varphi_1})$.

Lemma 6.7 *Let β be a complex q-rung orthopair fuzzy transversal of a complex q-rung orthopair fuzzy hypergraph $\mathscr{H}$. Then, there exists $\gamma \in t_r(\mathscr{H})$ such that $\gamma \leq \beta$.*

Proof Let $f_s(\mathscr{H}) = \{(\mu_1 e^{i\theta_1}, v_1 e^{i\varphi_1}), (\mu_2 e^{i\theta_2}, v_2 e^{i\varphi_2}), \ldots, (\mu_n e^{i\theta_n}, v_n e^{i\varphi_n})\}$. Suppose that $\lambda^{(\mu_k e^{i\theta_k}, v_k e^{i\varphi_k})}$ is a transversal of $\mathscr{H}^{(\mu_k e^{i\theta_k}, v_k e^{i\varphi_k})}$ and $\tau^{(\mu_k e^{i\theta_k}, v_k e^{i\varphi_k})} \in t_r(\mathscr{H}^{(\mu_k e^{i\theta_k}, v_k e^{i\varphi_k})})$, for $1 \leq k \leq n$ such that $\tau^{(\mu_k e^{i\theta_k}, v_k e^{i\varphi_k})} \subseteq \lambda^{(\mu_k e^{i\theta_k}, v_k e^{i\varphi_k})}$. Let β_k be an elementary complex q-rung orthopair fuzzy set having support λ_k and γ_k be an elementary complex q-rung orthopair fuzzy set having support τ_k, for $1 \le k \le n$. Then, Algorithm 6.11.1 implies that $\beta = \bigcup_{k=1}^{n} \beta_k$ is a complex q-rung orthopair fuzzy transversal of $\mathcal{H}$ and $\gamma = \bigcup_{k=1}^{n} \gamma_k$ is minimal complex q-rung orthopair fuzzy transversal of $\mathcal{H}$ such that $\gamma \le \beta$.

Theorem 6.18 *Let $\mathcal{H}_1 = (\mathcal{Q}_1, \eta_1)$ and $\mathcal{H}_2 = (\mathcal{Q}_2, \eta_2)$ be complex q-rung orthopair fuzzy hypergraphs. Then, $\mathcal{Q}_2 = t_r(\mathcal{H}_1) \Leftrightarrow \mathcal{H}_2$ is simple, $\mathcal{Q}_2 \subseteq \mathcal{Q}_1$, $h(\eta_k) = h(\mathcal{H}_1)$, for every $\rho_k \in \eta_2$, and for every complex q-rung orthopair fuzzy set $\xi \in \mathscr{P}(X)$, exactly one of the conditions must satisfy,*

(i) $\rho \le \xi$, for some $\rho \in \mathcal{Q}_2$ or
(ii) there is $Q_j \in \mathcal{Q}_1$ and $(\mu e^{i\theta}, v e^{i\varphi})$, where $(\mu, v) \in [0, T_{h(Q_j)}] \times [0, F_{h(Q_j)}]$, $\theta \in [0, \phi_{h(Q_j)}]$, $\varphi \in [0, \psi_{h(Q_j)}]$ such that $Q_j^{(\mu e^{i\theta}, v e^{i\varphi})} \cap \xi^{(\mu e^{i\theta}, v e^{i\varphi})} = \emptyset$, i.e., ξ is not a complex q-rung orthopair fuzzy transversal of $\mathcal{H}_1$.

Proof Let $\mathcal{Q}_2 = t_r(\mathcal{H}_1)$. Since, the family of all minimal complex q-rung orthopair fuzzy transversals form a simple complex q-rung orthopair fuzzy hypergraph on $X_1 \subseteq X_2$. Lemma 6.6 implies that every edge of $t_r(\mathcal{H}_1)$ has height $(\mu_1 e^{i\theta_1}, v_1 e^{i\varphi_1}) = h(\mathcal{H}_1)$. Let ξ be an arbitrary complex q-rung orthopair fuzzy set.

Case (i) If ξ is a complex q-rung orthopair fuzzy transversal of $\mathcal{H}_1$), then Lemma 6.7 implies the existence of a minimal complex q-rung orthopair fuzzy transversal ρ such that $\rho \le \xi$. Thus, the condition (i) holds and (ii) violates.

Case (ii) If ξ is not a complex q-rung orthopair fuzzy transversal of $\mathcal{H}_1$), then there is an edge $Q_j \in \mathcal{Q}_1$ such that $Q_j^{(\mu e^{i\theta}, v e^{i\varphi})} \cap \xi^{(\mu e^{i\theta}, v e^{i\varphi})} = \emptyset$. If condition (i) holds, $\rho \le \xi$ implies that $Q_j^{(\mu e^{i\theta}, v e^{i\varphi})} \cap \rho^{(\mu e^{i\theta}, v e^{i\varphi})} = \emptyset$, which is the contradiction against the fact that ρ is complex q-rung orthopair fuzzy transversal. Hence, condition (i) does not hold and (ii) is satisfied.

Conversely, suppose that $\mathcal{Q}_2$ satisfies all properties as mentioned above and $\rho \in \mathcal{Q}_2$. Let $\rho = \xi$, then we obtain $\rho \le \rho$ and conditions (ii) is not satisfied, so ρ is complex q-rung orthopair fuzzy transversal of $\mathcal{H}_1$. If t is minimal complex q-rung orthopair fuzzy transversal of $\mathcal{H}_1$ and $t \le \rho$, t does not satisfy (ii), this implies the existence of $\rho_2 \in \mathcal{Q}_2$ such that $\rho_2 \le t$, hence $\mathcal{Q}_2 \subseteq t_r(\mathcal{H}_1)$. Since, t is minimal complex q-rung orthopair fuzzy which implies that $\rho = t$, ρ and t were chosen arbitrarily therefore, we have $\mathcal{Q}_2 = t_r(\mathcal{H}_1)$.

The construction of fundamental subsequence and subcore of complex q-rung orthopair fuzzy hypergraph $\mathcal{H} = (\mathcal{Q}, \eta)$ is discussed in Algorithm 6.11.2.

Algorithm 6.11.2 Construction of fundamental subsequence and subcore
Let $\mathcal{H} = (\mathcal{Q}, \eta)$ be a complex q-rung orthopair fuzzy hypergraph and $\mathcal{H}_1 = (\mathcal{Q}_1, \eta_1)$ be a partial complex q-rung orthopair fuzzy hypergraph of $\mathcal{H}$. The fundamental subsequence $f_{ss}(\mathcal{H})$ is constructed as follows:

Let $f_s(\mathcal{H}) = \{(\mu_1 e^{i\theta_1}, \nu_1 e^{i\varphi_1}), (\mu_2 e^{i\theta_2}, \nu_2 e^{i\varphi_2}), \dots, (\mu_n e^{i\theta_n}, \nu_n e^{i\varphi_n})\}$ and $cor(\mathcal{H})$
$= \{\mathcal{H}^{(\mu_1 e^{i\theta_1}, \nu_1 e^{i\varphi_1})}, \mathcal{H}^{(\mu_2 e^{i\theta_2}, \nu_2 e^{i\varphi_2})}, \dots, \mathcal{H}^{(\mu_n e^{i\theta_n}, \nu_n e^{i\varphi_n})}\}$.

1. Construct $\widetilde{\mathcal{H}}^{(\mu_1 e^{i\theta_1}, \nu_1 e^{i\varphi_1})}$, a partial hypergraph of $\mathcal{H}^{(\mu_1 e^{i\theta_1}, \nu_1 e^{i\varphi_1})}$, by removing all hyperedges of $\mathcal{H}^{(\mu_1 e^{i\theta_1}, \nu_1 e^{i\varphi_1})}$, which contain properly any other hyperedge of $\mathcal{H}^{(\mu_1 e^{i\theta_1}, \nu_1 e^{i\varphi_1})}$.

2. In the same way, a partial hypergraph $\widetilde{\mathcal{H}}^{(\mu_2 e^{i\theta_2}, \nu_2 e^{i\varphi_2})}$ of $\mathcal{H}^{(\mu_2 e^{i\theta_2}, \nu_2 e^{i\varphi_2})}$ is constructed by removing all hyperedges of $\mathcal{H}^{(\mu_2 e^{i\theta_2}, \nu_2 e^{i\varphi_2})}$, which contain properly any other hyperedge of $\mathcal{H}^{(\mu_2 e^{i\theta_2}, \nu_2 e^{i\varphi_2})}$ or any other hyperedge of $\mathcal{H}^{(\mu_1 e^{i\theta_1}, \nu_1 e^{i\varphi_1})}$. $\widetilde{\mathcal{H}}^{(\mu_2 e^{i\theta_2}, \nu_2 e^{i\varphi_2})}$ is nontrivial iff there exists a complex q-rung orthopair fuzzy transversal $\tau \in t_r(\mathcal{H})$ and a vertex $u \in \mathcal{Q}^{(\mu_2 e^{i\theta_2}, \nu_2 e^{i\varphi_2})}$ such that $(T_\tau(u)e^{i\phi_\tau(u)}, F_\tau(u)e^{i\psi_\tau(u)}) = (\mu_2 e^{i\theta_2}, \nu_2 e^{i\varphi_2})$.

3. Continuing the same procedure, construct $\widetilde{\mathcal{H}}^{(\mu_k e^{i\theta_k}, \nu_k e^{i\varphi_k})}$, a partial hypergraph of $\mathcal{H}^{(\mu_k e^{i\theta_k}, \nu_k e^{i\varphi_k})}$, by removing all hyperedges of $\mathcal{H}^{(\mu_k e^{i\theta_k}, \nu_k e^{i\varphi_k})}$, which contain properly any other hyperedge of $\mathcal{H}^{(\mu_k e^{i\theta_k}, \nu_k e^{i\varphi_k})}$ or contain any other hyperedge of $\mathcal{H}^{(\mu_1 e^{i\theta_1}, \nu_1 e^{i\varphi_1})}, \mathcal{H}^{(\mu_2 e^{i\theta_2}, \nu_2 e^{i\varphi_2})}, \dots, \mathcal{H}^{(\mu_{k-1} e^{i\theta_{k-1}}, \nu_{k-1} e^{i\varphi_{k-1}})}$. $\widetilde{\mathcal{H}}^{(\mu_k e^{i\theta_k}, \nu_k e^{i\varphi_k})}$ is nontrivial if and only if there exists a complex q-rung orthopair fuzzy transversal $\tau \in t_r(\mathcal{H})$ and an element $u \in \mathcal{Q}^{(\mu_k e^{i\theta_k}, \nu_k e^{i\varphi_k})}$ such that $(T_\tau(u)e^{i\phi_\tau(u)}, F_\tau(u)e^{i\psi_\tau(u)}) = (\mu_k e^{i\theta_k}, \nu_k e^{i\varphi_k})$.

4. Let $\{(\tilde{\mu}_1 e^{i\tilde{\theta}_1}, \tilde{\nu}_1 e^{i\tilde{\varphi}_1}), (\tilde{\mu}_2 e^{i\tilde{\theta}_2}, \tilde{\nu}_2 e^{i\tilde{\varphi}_2}), \dots, (\tilde{\mu}_l e^{i\tilde{\theta}_l}, \tilde{\nu}_l e^{i\tilde{\varphi}_l})\}$ be the set of complex numbers such that the corresponding partial hypergraphs $\widetilde{\mathcal{H}}^{(\tilde{\mu}_1 e^{i\tilde{\theta}_1}, \tilde{\nu}_1 e^{i\tilde{\varphi}_1})}$, $\widetilde{\mathcal{H}}^{(\tilde{\mu}_2 e^{i\tilde{\theta}_2}, \tilde{\nu}_2 e^{i\tilde{\varphi}_2})}, \dots, \widetilde{\mathcal{H}}^{(\tilde{\mu}_l e^{i\tilde{\theta}_l}, \tilde{\nu}_l e^{i\tilde{\varphi}_l})}$ are non-empty.

5. Then, $f_{ss}(\mathcal{H}) = \{(\tilde{\mu}_1 e^{i\tilde{\theta}_1}, \tilde{\nu}_1 e^{i\tilde{\varphi}_1}), (\tilde{\mu}_2 e^{i\tilde{\theta}_2}, \tilde{\nu}_2 e^{i\tilde{\varphi}_2}), \dots, (\tilde{\mu}_l e^{i\tilde{\theta}_l}, \tilde{\nu}_l e^{i\tilde{\varphi}_l})\}$ and $\widetilde{cor}(\mathcal{H}) = \{\widetilde{\mathcal{H}}^{(\tilde{\mu}_1 e^{i\tilde{\theta}_1}, \tilde{\nu}_1 e^{i\tilde{\varphi}_1})}, \widetilde{\mathcal{H}}^{(\tilde{\mu}_2 e^{i\tilde{\theta}_2}, \tilde{\nu}_2 e^{i\tilde{\varphi}_2})}, \dots, \widetilde{\mathcal{H}}^{(\tilde{\mu}_l e^{i\tilde{\theta}_l}, \tilde{\nu}_l e^{i\tilde{\varphi}_l})}\}$ are subsequence and subcore set of $\mathcal{H}$, respectively.

Definition 6.67 Let $\mathcal{H} = (\mathcal{Q}, \eta)$ be a complex q-rung orthopair fuzzy hypergraph having fundamental subsequence $f_{ss}(\mathcal{H})$ and subcore $\widetilde{cor}(\mathcal{H})$ of $\mathcal{H}$. The *complex q-rung orthopair fuzzy transversal core* of $\mathcal{H}$ is defined as an elementary complex q-rung orthopair fuzzy hypergraph $\widehat{\mathcal{H}} = (\widehat{\mathcal{Q}}, \widehat{\eta})$ such that,

(i) $f_{ss}(\mathcal{H}) = f_{ss}(\widehat{\mathcal{H}})$, i.e., $f_{ss}(\mathcal{H})$ is also a fundamental subsequence of $\widehat{\mathcal{H}}$,

(ii) height of every $\widehat{Q}_j \in \widehat{\mathcal{Q}}$ is $(\tilde{\mu}_j e^{i\tilde{\theta}_j}, \tilde{\nu}_j e^{i\tilde{\varphi}_j}) \in f_{ss}(\mathcal{H})$ iff $supp(\widehat{Q}_j)$ is an hyperedge of $\widetilde{\mathcal{H}}^{(\tilde{\mu}_j e^{i\tilde{\theta}_j}, \tilde{\nu}_j e^{i\tilde{\varphi}_j})}$.

Theorem 6.19 *For every complex q-rung orthopair fuzzy hypergraph, we have* $t_r(\mathcal{H}) = t_r(\widehat{\mathcal{H}})$.

Proof Let $t \in t_r(\mathcal{H})$ and $\widehat{Q}_j \in \widehat{\mathcal{Q}}$. Definition 6.67 implies that $h(\widehat{Q}_j) = (\tilde{\mu}_j e^{i\tilde{\theta}_j}, \tilde{\nu}_j e^{i\tilde{\varphi}_j})$ and $\widehat{Q}_j^{(\tilde{\mu}_j e^{i\tilde{\theta}_j}, \tilde{\nu}_j e^{i\tilde{\varphi}_j})}$ is an hyperedge of $\widetilde{\mathcal{H}}^{(\tilde{\mu}_j e^{i\tilde{\theta}_j}, \tilde{\nu}_j e^{i\tilde{\varphi}_j})}$. Since $\widetilde{\mathcal{H}}^{(\tilde{\mu}_j e^{i\tilde{\theta}_j}, \tilde{\nu}_j e^{i\tilde{\varphi}_j})} \subseteq \mathcal{H}^{(\tilde{\mu}_j e^{i\tilde{\theta}_j}, \tilde{\nu}_j e^{i\tilde{\varphi}_j})}$ and $\tau^{(\mu_j e^{i\theta_j}, \nu_j e^{i\varphi_j})}$ is a transversal of $\mathcal{H}^{(\tilde{\mu}_j e^{i\tilde{\theta}_j}, \tilde{\nu}_j e^{i\tilde{\varphi}_j})}$ therefore $\widehat{Q}_j^{(\tilde{\mu}_j e^{i\tilde{\theta}_j}, \tilde{\nu}_j e^{i\tilde{\varphi}_j})} \cap \tau^{(\mu_j e^{i\theta_j}, \nu_j e^{i\varphi_j})} \neq \emptyset$. Thus, τ is acomplex q-rung orthopair fuzzy transversal of $\widehat{\mathcal{H}}$.

Let $\widehat{\tau} \in t_r(\widehat{\mathcal{H}})$ and $Q_j \in \mathcal{Q}$. Definition 6.63 implies that $Q_j^{h(Q_j)} \in \mathcal{H}^{(\mu_j e^{i\theta_j}, \nu_j e^{i\varphi_j})}$, for $h(Q_j) \leq (\mu_j e^{i\theta_j}, \nu_j e^{i\varphi_j}) \in f_s(\mathcal{H})$. Definition of subcore $\widetilde{cor}(\mathcal{H})$ implies the

Fig. 6.31 Complex 6-rung
orthopair fuzzy hypergraph

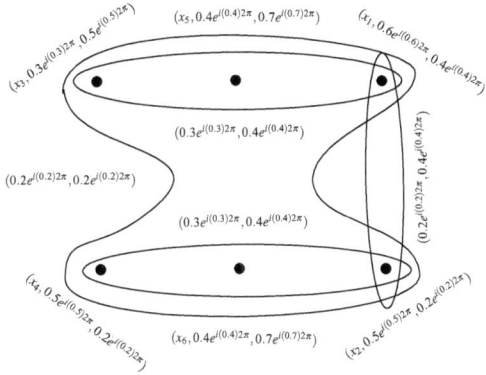

existence of an hyperedge $\widehat{Q}_j^{(\mu_j e^{i\theta_j}, \nu_j e^{i\varphi_j})}$ of $\widetilde{\mathscr{H}}^{(\mu_j e^{i\theta_j}, \nu_j e^{i\varphi_j})}$ such that $\widehat{Q}_j^{(\mu_j e^{i\theta_j}, \nu_j e^{i\varphi_j})} \subseteq Q_j^{h(Q_j)}$ and $(\mu_k e^{i\theta_k}, \nu_k e^{i\varphi_k}) \geq (\mu_j e^{i\theta_j}, \nu_j e^{i\varphi_j}) \geq h(Q_j)$. For $\widehat{\tau} \in t_r(\widetilde{\mathscr{H}})$, we have $u \in \widehat{Q}_j^{(\mu_j e^{i\theta_j}, \nu_j e^{i\varphi_j})} \cap \widehat{\tau}^{(\mu_j e^{i\theta_j}, \nu_j e^{i\varphi_j})} \subseteq \widehat{Q}_j^{h(Q_j)} \cap \widehat{\tau}^{(\mu_j e^{i\theta_j}, \nu_j e^{i\varphi_j})}$. Hence, $\widehat{\tau}$ is a complex q-rung orthopair fuzzy transversal of $\mathscr{H}$.

Let $\tau \in t_r(\mathscr{H}) \Rightarrow \tau$ is a complex q-rung orthopair fuzzy transversal of $\widetilde{\mathscr{H}}$. This implies that there is $\widehat{\tau}$ such that $\widehat{\tau} \subseteq \tau$. But $\widehat{\tau}$ is a complex q-rung orthopair fuzzy transversal of $\mathscr{H}$ and $\tau \in t_r(\mathscr{H})$ implies that $\widehat{\tau} = \tau$. Thus, $t_r(\mathscr{H}) \subseteq t_r(\widetilde{\mathscr{H}})$. Also $t_r(\widetilde{\mathscr{H}}) \subseteq t_r(\mathscr{H})$ implies that $t_r(\mathscr{H}) = t_r(\widetilde{\mathscr{H}})$.

Although τ can be taken as a minimal transversal of $\mathscr{H}$, it is not necessary for $\tau^{(\mu e^{i\theta}, \nu e^{i\varphi})}$ to be the minimal transversal of $\mathscr{H}^{(\mu e^{i\theta}, \nu e^{i\varphi})}$, for all $\mu, \nu \in [0, 1]$, and $\theta, \varphi \in [0, 2\pi]$. Furthermore, it is not necessary for the family of minimal complex q-rung orthopair fuzzy transversals to form a hypergraph on X. For those complex q-rung orthopair fuzzy transversals that satisfy the above property, we have

Definition 6.68 A complex q-rung orthopair fuzzy transversal τ having the property that $\tau^{(\mu e^{i\theta}, \nu e^{i\varphi})} \in t_r(\mathscr{H}^{(\mu e^{i\theta}, \nu e^{i\varphi})})$, for all $\mu, \nu \in [0, 1]$, and $\theta, \varphi \in [0, 2\pi]$ is called the *locally minimal complex q-rung orthopair fuzzy transversal* of $\mathscr{H}$. The collection of all locally minimal complex q-rung orthopair fuzzy transversals of $\mathscr{H}$ is represented by $t_r^*(\mathscr{H})$.

Note that, $t_r^*(\mathscr{H}) \subseteq t_r(\mathscr{H})$, but the converse is not generally true.

Example 6.24 Consider a complex 6-rung orthopair fuzzy hypergraph $\mathscr{H} = (\mathscr{Q}, \eta)$ as shown in Fig. 6.31. The complex 6-rung orthopair fuzzy set

$$\{(x_1, 0.6e^{i(0.6)2\pi}, 0.4e^{i(0.4)2\pi}), (x_5, 0.4e^{i(0.4)2\pi}, 0.7e^{i(0.7)2\pi}), (x_6, 0.4e^{i(0.4)2\pi}, 0.7e^{i(0.7)2\pi})\}$$

is a locally minimal complex 6-rung orthopair fuzzy transversal of $\mathscr{H}$.

Theorem 6.20 *Let $\mathscr{H} = (\mathscr{Q}, \eta)$ be an ordered complex q-rung orthopair fuzzy hypergraph with $f_s(\mathscr{H}) = \{(\mu_1 e^{i\theta_1}, \nu_1 e^{i\varphi_1}), (\mu_2 e^{i\theta_2}, \nu_2 e^{i\varphi_2}), \ldots, (\mu_n e^{i\theta_n}, \nu_n e^{i\varphi_n})\}$.*

If λ_k is a minimal transversal of $\mathscr{H}^{(\mu_k e^{i\theta_k}, \nu_k e^{i\varphi_k})}$, then there exists $\alpha \in t_r(\mathscr{H})$ such that $\alpha^{(\mu_k e^{i\theta_k}, \nu_k e^{i\varphi_k})} = \lambda_k$ and $\alpha^{(\mu_l e^{i\theta_l}, \nu_l e^{i\varphi_l})}$ is a minimal transversal of $\mathscr{H}^{(\mu_l e^{i\theta_l}, \nu_l e^{i\varphi_l})}$, for all $l < k$. In particular, if $\lambda_j \in t_r(\mathscr{H}^{(\mu_j e^{i\theta_j}, \nu_j e^{i\varphi_j})})$, then there exists a locally minimal complex q-rung orthopair fuzzy transversal $\alpha^{(\mu_j e^{i\theta_j}, \nu_j e^{i\varphi_j})} = \lambda_j$ and $t_r^*(\mathscr{H}) \neq \emptyset$.

Proof Let $\lambda_k \in t_r(\mathscr{H}^{(\mu_k e^{i\theta_k}, \nu_k e^{i\varphi_k})})$. Since, $\mathscr{H} = (\mathscr{Q}, \eta)$ is an ordered complex q-rung orthopair fuzzy hypergraph, therefore, $\mathscr{H}^{(\mu_{k-1} e^{i\theta_{k-1}}, \nu_{k-1} e^{i\varphi_{k-1}})} \subseteq \mathscr{H}^{(\mu_k e^{i\theta_k}, \nu_k e^{i\varphi_k})}$. Also, there exists $\lambda_{k-1} \in t_r(\mathscr{H}^{(\mu_{k-1} e^{i\theta_{k-1}}, \nu_{k-1} e^{i\varphi_{k-1}})})$ such that $\lambda_{k-1} \subseteq \lambda_k$. Following this iterative procedure, we have a nested sequence $\lambda_1 \subseteq \lambda_2 \subseteq \cdots \subseteq \lambda_{k-1} \subseteq \lambda_k$ of minimal transversals, where every $\lambda_l \in t_r(\mathscr{H}^{(\mu_l e^{i\theta_l}, \nu_l e^{i\varphi_l})})$. Let α_l be an elementary complex q-rung orthopair fuzzy set having height $(\mu_l e^{i\theta_l}, \nu_l e^{i\varphi_l})$ and support α_l. Let us define $\alpha(x)$ such that $\alpha(x) = \{(\max T_{\alpha_l}(x)e^{i \max \phi_{\alpha_l}(x)}, \min F_{\alpha_l}(x)e^{i \min \psi_{\alpha_l}(x)}) | 1 \leq l \leq n\}$, that generates the required minimal complex q-rung orthopair fuzzy transversal of $\mathscr{H}$. If $k = n$, α is locally minimal complex q-rung orthopair fuzzy transversal of $\mathscr{H}$. Hence, $t_r^*(\mathscr{H}) \neq \emptyset$.

Theorem 6.21 *Let $\mathscr{H} = (\mathscr{Q}, \eta)$ be a simply ordered complex q-rung orthopair fuzzy hypergraph with $f_s(\mathscr{H}) = \{(\mu_1 e^{i\theta_1}, \nu_1 e^{i\varphi_1}), (\mu_2 e^{i\theta_2}, \nu_2 e^{i\varphi_2}), \ldots, (\mu_n e^{i\theta_n}, \nu_n e^{i\varphi_n})\}$. If $\lambda_k \in t_r(\mathscr{H}^{(\mu_k e^{i\theta_k}, \nu_k e^{i\varphi_k})})$, then there exists $\alpha \in t_r^*(\mathscr{H})$ such that $\alpha^{(\mu_k e^{i\theta_k}, \nu_k e^{i\varphi_k})} = \lambda_k$.*

Proof Let $\lambda_k \in t_r(\mathscr{H}^{(\mu_k e^{i\theta_k}, \nu_k e^{i\varphi_k})})$ and $\mathscr{H} = (\mathscr{Q}, \eta)$ is a simply ordered q-rung orthopair fuzzy hypergraph. Theorem 6.20 implies that a nested sequence $\lambda_1 \subseteq \lambda_2 \subseteq \cdots \subseteq \lambda_{k-1} \subseteq \lambda_k$ of minimal transversals can be constructed. Let α_l be an elementary complex q-rung orthopair fuzzy set having height $(\mu_l e^{i\theta_l}, \nu_l e^{i\varphi_l})$ and support α_l such that $\alpha(x) = \{(\max T_{\alpha_l}(x)e^{i \max \phi_{\alpha_l}(x)}, \min F_{\alpha_l}(x)e^{i \min \psi_{\alpha_l}(x)}) | 1 \leq l \leq n\}$ generates the locally minimal complex q-rung orthopair fuzzy transversal of $\mathscr{H}$ with $\alpha^{(\mu_k e^{i\theta_k}, \nu_k e^{i\varphi_k})} = \lambda_k$.

6.12 Application of Complex q-Rung Orthopair Fuzzy Hypergraphs

Definition 6.69 Let $\mathscr{Q} = (Te^{i\phi}, Fe^{i\psi})$ be a complex q-rung orthopair fuzzy number. Then, *score function* of $\mathscr{Q}$ is defined as

$$s(\mathscr{Q}) = (T^q - F^q) + \frac{1}{2^q \pi^q}(\phi^q - \psi^q).$$

The *accuracy* of $\mathscr{Q}$ is defined as

$$a(\mathscr{Q}) = (T^q + F^q) + \frac{1}{2^q \pi^q}(\phi^q + \psi^q).$$

For two complex q-rung orthopair fuzzy numbers $\mathscr{Q}_1$ and $\mathscr{Q}_2$

1. if $s(\mathscr{Q}_1) > s(\mathscr{Q}_2)$, then $\mathscr{Q}_1 \succ \mathscr{Q}_2$,
2. if $s(\mathscr{Q}_1) = s(\mathscr{Q}_2)$, then

   - if $a(\mathscr{Q}_1) > a(\mathscr{Q}_2)$, then $\mathscr{Q}_1 \succ \mathscr{Q}_2$,
   - if $a(\mathscr{Q}_1) = a(\mathscr{Q}_2)$, then $\mathscr{Q}_1 \sim \mathscr{Q}_2$.

A complex 6-rung orthopair fuzzy hypergraph model of research collaboration network A collaboration network is a group of independent organizations or people that interact to complete a particular goal for achieving better collective results by the means of joint execution of task. The entities of a collaborative network may be geographically distributed and heterogenous in terms of their culture, goals, and operating environment but they collaborate to achieve compatible or common goals. For decades, science academies have been interested in research collaboration. The most common reasons of research collaboration are funding, more experts working on the same project imply the more chances for effectiveness, productivity, and innovativeness. Nowadays, most of the public research is based on collaboration of different types of expertise from different disciples and different economic sectors. In this section, we study a research collaboration network model through complex 6-rung orthopair fuzzy hypergraph. Consider a science academy wants to select an author among a group of researchers which has best collaborative skills. For this purpose, following are the characteristics that can be considered,

- Cooperative spirit
- Mutual respect
- Critical thinking
- Innovations
- Creativity
- Embrace diversity.

Consider a complex 6-rung orthopair fuzzy hypergraph $\mathscr{H} = (\mathscr{Q}, \eta)$ on $X = \{A_1, A_2, A_3, A_4, A_5, A_6, A_7, A_8, A_9, A_{10}\}$. The set of universe X represents the group of authors as the vertices of $\mathscr{H}$ and these authors are grouped through hyperedges if they have worked together on some projects. The truth-membership of each author represents the collaboration strength and falsity-membership describes the opposite behavior of corresponding author. Suppose that a team of experts assigns that the collaboration power of A_1 is 60% and non-collaborative behavior is 50% after carefully observing the different attributes. The corresponding phase terms illustrate the specific period of time in which the collaborative behavior of an author varies. We model this data as $(A_1, 0.6e^{i(0.5)2\pi}, 0.5e^{i(0.5)2\pi})$. The complex 6-rung orthopair fuzzy hypergraph $\mathscr{H} = (\mathscr{Q}, \eta)$ model of collaboration network is shown in Fig. 6.32.

The membership degrees of hyperedges represent the collective degrees of collaboration and non-collaboration of the corresponding authors combined through an hyperedge. The adjacency matrix of this network is given in Tables 6.17, 6.18, and 6.19.

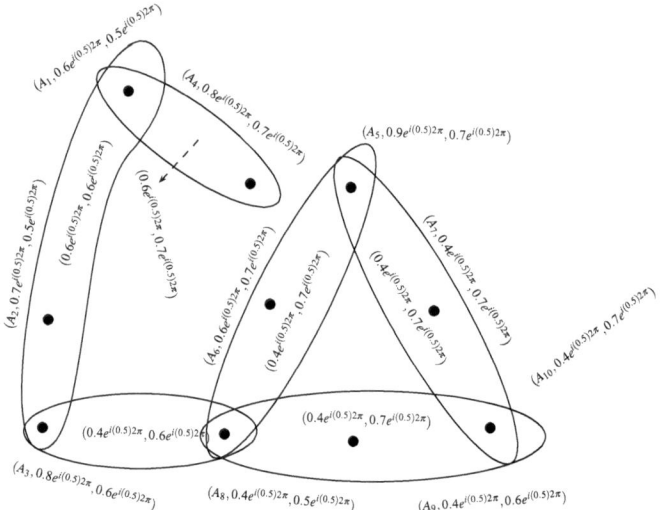

Fig. 6.32 Complex 6-rung orthopair fuzzy hypergraph model of collaboration network

Table 6.17 Adjacency matrix of collaboration network

η	A_1	A_2	A_3	A_4
A_1	$(0,0)$	$(0.6e^{i(0.5)2\pi},$ $0.6e^{i(0.5)2\pi})$	$(0.6e^{i(0.5)2\pi},$ $0.6e^{i(0.5)2\pi})$	$(0.6e^{i(0.5)2\pi},$ $0.6e^{i(0.5)2\pi})$
A_2	$(0.6e^{i(0.5)2\pi},$ $0.6e^{i(0.5)2\pi})$	$(0,0)$	$(0.6e^{i(0.5)2\pi},$ $0.6e^{i(0.5)2\pi})$	$(0,0)$
A_3	$(0.6e^{i(0.5)2\pi},$ $0.6e^{i(0.5)2\pi})$	$(0.6e^{i(0.5)2\pi},$ $0.6e^{i(0.5)2\pi})$	$(0,0)$	$(0,0)$
A_4	$(0.6e^{i(0.5)2\pi},$ $0.7e^{i(0.5)2\pi})$	$(0,0)$	$(0,0)$	$(0,0)$
A_5	$(0,0)$	$(0,0)$	$(0,0)$	$(0,0)$
A_6	$(0,0)$	$(0,0)$	$(0,0)$	$(0,0)$
A_7	$(0,0)$	$(0,0)$	$(0,0)$	$(0,0)$
A_8	$(0,0)$	$(0,0)$	$(0.4e^{i(0.5)2\pi},$ $0.6e^{i(0.5)2\pi})$	$(0,0)$
A_9	$(0,0)$	$(0,0)$	$(0,0)$	$(0,0)$
A_{10}	$(0,0)$	$(0,0)$	$(0,0)$	$(0,0)$

Table 6.18 Adjacency matrix of collaboration network

η	A_5	A_6	A_7	A_8
A_1	$(0,0)$	$(0,0)$	$(0,0)$	$(0,0)$
A_2	$(0,0)$	$(0,0)$	$(0,0)$	$(0,0)$
A_3	$(0,0)$	$(0,0)$	$(0,0)$	$(0.4e^{i(0.5)2\pi},$ $0.6e^{i(0.5)2\pi})$
A_4	$(0,0)$	$(0,0)$	$(0,0)$	$(0,0)$
A_5	$(0,0)$	$(0.4e^{i(0.5)2\pi},$ $0.7e^{i(0.5)2\pi})$	$(0.6e^{i(0.5)2\pi},$ $0.7e^{i(0.5)2\pi})$	$(0.4e^{i(0.5)2\pi},$ $0.7e^{i(0.5)2\pi})$
A_6	$(0.4e^{i(0.5)2\pi},$ $0.7e^{i(0.5)2\pi})$	$(0,0)$	$(0,0)$	$(0.4e^{i(0.5)2\pi},$ $0.7e^{i(0.5)2\pi})$
A_7	$(0.4e^{i(0.5)2\pi},$ $0.7e^{i(0.5)2\pi})$	$(0,0)$	$(0,0)$	$(0,0)$
A_8	$(0.4e^{i(0.5)2\pi},$ $0.7e^{i(0.5)2\pi})$	$(0.4e^{i(0.5)2\pi},$ $0.7e^{i(0.5)2\pi})$	$(0,0)$	$(0,0)$
A_9	$(0,0)$	$(0,0)$	$(0,0)$	$(0.4e^{i(0.5)2\pi},$ $0.7e^{i(0.5)2\pi})$
A_{10}	$(0.4e^{i(0.5)2\pi},$ $0.7e^{i(0.5)2\pi})$	$(0,0)$	$(0.4e^{i(0.5)2\pi},$ $0.7e^{i(0.5)2\pi})$	$(0.4e^{i(0.5)2\pi},$ $0.7e^{i(0.5)2\pi})$

Table 6.19 Adjacency matrix of collaboration network

η	A_9	A_{10}
A_1	$(0,0)$	$(0,0)$
A_2	$(0,0)$	$(0,0)$
A_3	$(0,0)$	$(0,0)$
A_4	$(0,0)$	$(0,0)$
A_5	$(0,0)$	$(0.6e^{i(0.5)2\pi}, 0.7e^{i(0.5)2\pi})$
A_6	$(0,0)$	$(0,0)$
A_7	$(0,0)$	$(0.4e^{i(0.5)2\pi}, 0.7e^{i(0.5)2\pi})$
A_8	$(0.4e^{i(0.5)2\pi}, 0.7e^{i(0.5)2\pi})$	$(0.4e^{i(0.5)2\pi}, 0.7e^{i(0.5)2\pi})$
A_9	$(0,0)$	$(0.4e^{i(0.5)2\pi}, 0.7e^{i(0.5)2\pi})$
A_{10}	$(0.4e^{i(0.5)2\pi}, 0.7e^{i(0.5)2\pi})$	$(0,0)$

The score values and choice values of a complex 6-rung orthopair fuzzy hypergraph $\mathscr{H} = (\mathscr{Q}, \eta)$ are calculated as follows:

$$s_{jk} = (T_{jk}^q + F_{jk}^q) + \frac{1}{2^q \pi^q}(\phi_{jk}^q + \psi_{jk}^q), \quad c_j = \sum_k s_{jk} + (T_j^q + F_j^q) + \frac{1}{2^q \pi^q}(\phi_j^q + \psi_j^q),$$

respectively. These values are given in Table 6.20.

The choice values of Table 6.20 show that A_5 is the author having maximum strength of collaboration and good collective skills among all the authors. Similarly,

Table 6.20 Score and choice values

s_{jk}	A_1	A_2	A_3	A_4	A_5	A_6	A_7	A_8	A_9	A_{10}	c_j
A_1	0	0.1245	0.1245	0.1245	0	0	0	0	0	0	0.88690
A_2	0.1245	0	0.1245	0	0	0	0	0	0	0	0.41377
A_3	0.1245	0.1245	0	0	0	0	0	0.0820	0	0	0.67105
A_4	0.1955	0	0	0	0	0	0	0	0	0	0.60654
A_5	0	0	0	0	0	0.1529	0.1955	0.1529	0	0.1955	1.37714
A_6	0	0	0	0	0.1529	0	0	0.1529	0	0	0.53480
A_7	0	0	0	0	0.1955	0	0	0	0	0.1529	0.50139
A_8	0	0	0.0820	0	0.1529	0.1529	0	0	0.1529	0.1529	0.74457
A_9	0	0	0	0	0	0	0	0.1529	0	0.1529	0.38780
A_{10}	0	0	0	0	0.1529	0	0.1529	0.1529	0.1529	0	0.76459

the choice values of all authors represent the strength of their respective collaboration skills in a specific period of time. The method adopted in our model to select the author having best collaboration skills is given in Algorithm 6.12.1.

Algorithm 6.12.1 Selection of author having maximum collaboration skills

1. Input the set of vertices (authors) $A_1, A_2, \ldots, A_j$.
2. Input the complex q-rung orthopair fuzzy set Q of vertices such that $Q(A_k) = (T_k e^{i\phi_k}, F_k e^{i\psi_k}), 1 \le k \le j, 0 \le T_k^q + F_k^q \le 1, q \ge 1$.
3. Input the adjacency matrix $\eta = [(T_{kl} e^{i\phi_{kl}}, F_{kl} e^{i\psi_{kl}})]_{j \times j}$ of vertices.
4.   **do** k from $1 \to j$
5. $c_k = 0$
6.   **do** l from $1 \to j$
7. $s_{jk} = (T_{kl}^q + F_{kl}^q) + \frac{1}{2^q \pi^q}(\phi_{kl}^q + \psi_{kl}^q)$
8. $c_k = c_k + s_{jk}$
9.   **end do**
10. $c_k = c_k + (T_k^q + F_k^q) + \frac{1}{2^q \pi^q}(\phi_k^q + \psi_k^q)$
11.   **do**
12. Select a vertex of $\mathscr{H} = (\mathscr{Q}, \eta)$ having maximum choice value as the author possessing strong collaboration powers.

6.13 Comparative Analysis

The proposed complex q-rung orthopair fuzzy model is more flexible and compatible to the system when the given data ranges over complex subset with a unit disk instead of the real subset with [0, 1]. We illustrate the flexibility of our proposed model by taking an example. Consider an educational institute that wants to establish its minimum branches in a particular city in order to facilitate the maximum number of

Table 6.21 Comparative analysis of three models

Methods	Score values	Ranking
Complex intuitionistic fuzzy model	0.4 1.0 0.6	$p_2 > p_3 > p_1$
Complex Pythagorean fuzzy model	0.4 0.9 0.42	$p_2 > p_3 > p_1$
Complex 3-rung orthopair fuzzy model	0.104 0.67 0.234	$p_2 > p_3 > p_1$

students according to some parameters such as transportation, suitable place, connectivity with the main branch, and expenditures. Suppose a team of three decision makers selects the different places. Let $X = \{p_1, p_2, p_3\}$ be the set of places where the team is interested to establish the new branches. After carefully observing the different attributes, the first decision makers assign the membership and nonmembership degrees to support the place p_1 as 60% and 40%, respectively. The phase terms represent the period of time for which the place p_1 can attract maximum number of students. This information is modeled using a complex intuitionistic fuzzy set as $(p_1, 0.6e^{i(0.6)2\pi}, 0.4e^{i(0.4)2\pi})$. Note that, $0 \leq 0.6 + 0.4 \leq 1$. Similarly, he models the other places as, $(p_2, 0.7e^{i(0.7)2\pi}, 0.2e^{i(0.2)2\pi})$, $(p_3, 0.5e^{i(0.5)2\pi}, 0.2e^{i(0.2)2\pi})$. We denote this complex intuitionistic fuzzy model as

$$I = \{(p_1, 0.6e^{i(0.6)2\pi}, 0.4e^{i(0.4)2\pi}), (p_2, 0.7e^{i(0.7)2\pi}, 0.2e^{i(0.2)2\pi}), (p_3, 0.5e^{i(0.5)2\pi}, 0.2e^{i(0.2)2\pi})\}.$$

Since, all complex intuitionistic fuzzy grades are complex Pythagorean fuzzy as well as complex q-rung orthopair fuzzy grades. We find the score functions of the above values using the formulas $s(p_j) = (T - F) + \frac{1}{2\pi}(\phi - \psi)$, $s(p_j) = (T^2 - F^2) + \frac{1}{2^2\pi^2}(\phi^2 - \psi^2)$, and $s(p_j) = (T^3 - F^3) + \frac{1}{2^3\pi^3}(\phi^3 - \psi^3)$. The results corresponding to these three approaches are given in Table 6.21.

Suppose that the second decision-maker assigns the membership values to these places as, $(p_1, 0.6e^{i(0.6)2\pi}, 0.4e^{i(0.4)2\pi})$, $(p_2, 0.7e^{i(0.7)2\pi}, 0.2e^{i(0.2)2\pi})$, $(p_3, 0.7e^{i(0.7)2\pi}, 0.5e^{i(0.5)2\pi})$. This information can not be modeled using complex intuitionistic fuzzy set as $0.7 + 0.5 = 1.2 > 1$. We model this information using a complex Pythagorean fuzzy set and the corresponding model is given as

$$P = \{(p_1, 0.6e^{i(0.6)2\pi}, 0.4e^{i(0.4)2\pi}), (p_2, 0.7e^{i(0.7)2\pi}, 0.2e^{i(0.2)2\pi}), (p_3, 0.7e^{i(0.7)2\pi}, 0.5e^{i(0.5)2\pi})\}.$$

Since, all complex Pythagorean fuzzy grades are also complex q-rung orthopair fuzzy grades. We find the score functions of the above values using the formulas $s(p_j) = (T^2 - F^2) + \frac{1}{2^2\pi^2}(\phi^2 - \psi^2)$ and $s(p_j) = (T^3 - F^3) + \frac{1}{2^3\pi^3}(\phi^3 - \psi^3)$. The results corresponding to these two approaches are given in Table 6.22.

Table 6.22 Comparative analysis of two models

Methods	Score values	Ranking
Complex Pythagorean fuzzy model	0.4 0.9 0.48	$p_2 > p_3 > p_1$
Complex 3-rung orthopair fuzzy model	0.104 0.67 0.436	$p_2 > p_3 > p_1$

We now suppose that the third decision maker assigns the membership values to these places as

$$(p_1, 0.6e^{i(0.6)2\pi}, 0.4e^{i(0.4)2\pi}), (p_2, 0.8e^{i(0.8)2\pi}, 0.7e^{i(0.7)2\pi}), (p_3, 0.7e^{i(0.7)2\pi}, 0.5e^{i(0.5)2\pi}).$$

This information cannot be modeled using complex intuitionistic fuzzy set and complex Pythagorean fuzzy set as $0.7 + 0.8 = 1.5 > 1$, $0.7^2 + 0.8^2 = 1.13 > 1$. We model this information using a complex 3-rung orthopair fuzzy set and the corresponding model is given as

$$Q = \{(p_1, 0.6e^{i(0.6)2\pi}, 0.4e^{i(0.4)2\pi}), (p_2, 0.8e^{i(0.8)2\pi}, 0.7e^{i(0.7)2\pi}), (p_3, 0.7e^{i(0.7)2\pi}, 0.5e^{i(0.5)2\pi})\}.$$

We find the score functions of the above values using the formula $s(p_j) = (T^3 - F^3) + \frac{1}{2^3\pi^3}(\phi^3 - \psi^3)$. The score values of complex 3-rung orthopair fuzzy information are given as

$$s(p_1) = 0.304, \quad s(p_2) = 0.438, \quad s(p_3) = 0.436.$$

Note that, p_2 is the best optimal choice to establish a new branch according to the given parameters. We see that every complex intuitionistic fuzzy grade is a complex Pythagorean fuzzy grade, as well as a complex q-rung orthopair fuzzy grade, however there are complex q-rung orthopair fuzzy grades that are not complex intuitionistic fuzzy nor complex Pythagorean fuzzy grades. This implies the generalization of complex q-rung orthopair fuzzy values. Thus, the proposed complex q-rung orthopair fuzzy model provides more flexibility due to its most prominent feature that is the adjustment of the range of demonstration of given information by changing the value of parameter q, $q \geq 1$. The generalization of our proposed model can also be observed from the reduction of complex q-rung orthopair fuzzy model to complex intuitionistic fuzzy and complex Pythagorean fuzzy models for $q = 1$ and $q = 2$, respectively.

References

1. Akram, A., Dar, J.M., Naz, S.: Certain graphs under Pythagorean fuzzy environment. Complex Intell. Syst. **5**(2), 127–144 (2019)
2. Akram, M., Ilyasa, F., Garg, H.: Multi-criteria group decision making based on ELECTRE I method in Pythagorean fuzzy information. Soft Comput. (2019). https://doi.org/10.1007/s00500-019-04105-0
3. Akram, M., Dar, J.M., Naz, S.: Pythagorean Dombi fuzzy graphs. Complex Intell. Syst. (2019). https://doi.org/10.1007/s40747-019-0109-0
4. Akram, M., Habib, A., Davvaz, B.: Direct sum of n Pythagorean fuzzy graphs with application to group decision-making. J. Mult.-Valued Log. Soft Comput. 1–41 (2019)
5. Akram, M., Naz, S.: A novel decision-making approach under complex Pythagorean fuzzy environment. Math. Comput. Appl. **24**(3), 73 (2019)
6. Akram, M., Naz, S., Davvaz, B.: Simplified interval-valued Pythagorean fuzzy graphs with application. Complex Intell. Syst. **5**(2), 229–253 (2019)
7. Akram, M., Ilyas, F., Saeid, A.B.: Certain notions of Pythagorean fuzzy graphs. J. Intell. Fuzzy Syst. **36**(6), 5857–5874 (2019)
8. Akram, M., Dudek, W.A., Ilyas, F.: Group decision making based on Pythagorean fuzzy TOPSIS method. Int. J. Intell. Syst. **34**(7), 1455–1475 (2019)
9. Akram, M, Habib , A., Koam , A.N.: A novel description on edge-regular q-rung picture fuzzy graphs with application. Symmetry **11**(4), 489 (2019). https://doi.org/10.3390/sym110
10. Akram, M., Habib, A., Ilyas, F., Dar, J.M.: Specific types of Pythagorean fuzzy graphs and application to decision-making. Math. Comput. Appl. **23**, 42 (2018)
11. Akram, M., Naz, S.: Energy of Pythagorean fuzzy graphs with applications. Mathematics **6**, 560 (2018). https://doi.org/10.3390/math6080136
12. Ali, M.I.: Another view on q-rung orthopair fuzzy sets. Int. J. Intell. Syst. **33**, 2139–2153 (2018)
13. Alkouri, A., Salleh, A.: Complex intuitionistic fuzzy sets. AIP Conf. Proc. **14**, 464–470 (2012)
14. Atanassov, K.T.: Intuitionistic fuzzy sets. Fuzzy Sets Syst. **20**(1), 87–96 (1986)
15. Berge, C.: Graphs and Hypergraphs. North-Holland, Amsterdam (1973)
16. Gallo, G., Longo, G., Pallottino, S.: Directed hypergraphs and applications. Discret. Appl. Math. **42**, 177–201 (1993)
17. Goetschel Jr., R.H., Craine, W.L., Voxman, W.: Fuzzy transversals of fuzzy hypergraphs. Fuzzy Sets Syst. **84**, 235–254 (1996)
18. Habib, A., Akram, M.M.: Farooq, q-rung orthopair fuzzy competition graphs with application in the soil ecosystem. Mathematics **7**(1), 91 (2019). https://doi.org/10.3390/math70100
19. Kaufmann, A.: Introduction a la Thiorie des Sous-Ensemble Flous, vol. 1. Masson, Paris (1977)
20. Li, L., Zhang, R., Wang, J., Shang, X., Bai, K.: A novel approach to multi-attribute group decision-making with q-rung picture linguistic information. Symmetry **10**(5), 172 (2018)
21. Liu, P.D., Wang, P.: Some q-rung orthopair fuzzy aggregation operators and their applications to multi-attribute decision making. Int. J. Intell. Syst. **33**, 259–280 (2018)
22. Luqman, A., Akram, M., Al-Kenani, A.N.: q-rung orthopair fuzzy hypergraphs with applications. Mathematics **7**, 260 (2019)
23. Luqman, A., Akram, M., Al-Kenani, A.N., Alcantud, J.C.R.: A study on hypergraph representations of complex fuzzy information. Symmetry **11**(11), 1381 (2019)
24. Luqman, A., Akram, M., Davvaz, B.: q-rung orthopair fuzzy directed hypergraphs: a new model with applications. J. Intell. Fuzzy Syst. **37**, 3777–3794 (2019)
25. Mordeson, J.N., Nair, P.S.: Fuzzy Graphs and Fuzzy Hypergraphs, 2nd edn. Physica Verlag, Heidelberg (2001)
26. Myithili, K.K., Parvathi, R.: Transversals of intuitionistic fuzzy directed hypergraphs. Notes Intuit. Fuzzy Sets **21**(3), 66–79 (2015)
27. Naz, S., Ashraf, S., Akram, M.: A novel approach to decision-making with Pythagorean fuzzy information. Mathematics **6**(6), 95 (2018). https://doi.org/10.3390/math6060095

28. Ramot, D., Milo, R., Friedman, M., Kandel, A.: Complex fuzzy sets. IEEE Trans. Fuzzy Syst. **10**(2), 171–186 (2002)
29. Ramot, D., Friedman, M., Langholz, G., Kandel, A.: Complex fuzzy logic. IEEE Trans. Fuzzy Syst. **11**(4), 450–461 (2003)
30. Thirunavukarasu, P., Suresh, R., Viswanathan, K.K.: Energy of a complex fuzzy graph. Int. J. Math. Sci. Eng. Appl. **10**, 243–248 (2016)
31. Ullah, K., Mahmood, T., Ali, Z., Jan, N.: On some distance measures of complex Pythagorean fuzzy sets and their applications in pattern recognition. Complex Intell. Syst. 1–13 (2019)
32. Yager, R.R.: Pythagorean fuzzy subsets. In: Proceedings of the Joint IFSA World Congress and NAIFPS Annual Meeting, Edmonton, Canada, pp. 57–61 (2013)
33. Yager, R.R., Abbasov, A.M.: Pythagorean membership grades, complex numbers and decision making. Int. J. Intell. Syst. **28**(5), 436–452 (2013)
34. Yager, R.R.: Pythagorean membership grades in multi-criteria decision making. IEEE Trans. Fuzzy Syst. **22**(4), 958–965 (2014)
35. Yager, R.R.: Generalized orthopair fuzzy sets. IEEE Trans. Fuzzy Syst. **25**, 1222–1230 (2017)
36. Yaqoob, N., Gulistan, M., Kadry, S., Wahab, H.: Complex intuitionistic fuzzy graphs with application in cellular network provider companies. Mathematics **7**(1), 35 (2019)
37. Zadeh, L.A.: Fuzzy sets. Inf. Control **8**(3), 338–353 (1965)

Chapter 7
Granular Computing Based on q-Rung Picture Fuzzy Hypergraphs

In this chapter, we present q-rung picture fuzzy hypergraphs and illustrate the formation of granular structures using q-rung picture fuzzy hypergraphs and level hypergraphs. Moreover, we define q-rung picture fuzzy equivalence relations and its' associated q-rung picture fuzzy hierarchical quotient space structures. We also present an arithmetic example in order to demonstrate the benefits and validity of this model. This chapter is due to [19, 24].

7.1 Introduction

Granular computing is defined as an identification of techniques, methodologies, tools, and theories that yields the advantages of clusters, groups or classes, i.e., the granules. The terminology was first introduced by Lin [20]. The fundamental concepts of granular computing are utilized in various fields, including rough set theory, cluster analysis, machine learning, and artificial intelligence. Different models have been proposed to study the numerous issues occurring in granular computing, including classification of the universe, illustration of granules, and the identification of relations among granules. For example, the procedure of problem-solving through granular computing can be considered as distinct descriptions of the problem at multilevels and these levels are linked together to construct a hierarchical space structure. Thus, this is a way of dealing with the formation of granules and the switching between different granularities. Here, the word "hierarchy" implies the methodology of hierarchical analysis in problem-solving and human activities. To understand this methodology, let us consider an example of national administration in which the complete nation is subdivided into various provinces. Further, divide every province into various divisions and similarly. The human activities and problem-solving involve the simplification of original complicated problem by ignoring some details rather than thinking about all points of the problem. This rationalize model is

© Springer Nature Singapore Pte Ltd. 2020
M. Akram and A. Luqman, *Fuzzy Hypergraphs and Related Extensions*,
Studies in Fuzziness and Soft Computing 390,
https://doi.org/10.1007/978-981-15-2403-5_7

Fig. 7.1 Comparison of spaces of q-rung picture fuzzy set

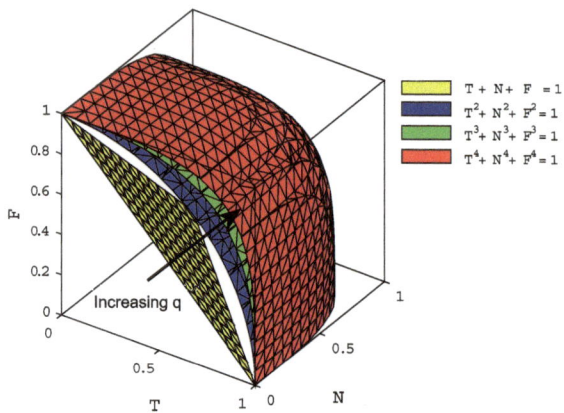

then further refined till the issue is completely solved. Thus, we resolve and interpret the complex problems from weaker grain to stronger one or from highest rank to lowest or from universal to particular, increasingly. This technique is called the hierarchical problem-solving. This is further acknowledged that hierarchical strategy is the only technique which is used by humans to deal with complicated problems and it enhances the competence and efficiency. This strategy is also known as the multi-granular computing.

Definition 7.1 A *picture fuzzy set* P is an object having the form, $P = \{(x, T_P(x), N_P(x), F_P(x)) | x \in X\}$, where the function $T_P : X \to [0, 1]$ represents the positive degree, $N_P : X \to [0, 1]$ represents the neutral degree, and $F_P : X \to [0, 1]$ represents the negative degree of the element $x \in X$ such that, $0 \le T_P(x) + N_P(x) + F_P(x) \le 1$, for all $x \in X$.

The refusal degree of x in P is defined as $(1 - T_P(x) - N_P(x) - F_P(x))$.

Definition 7.2 A *q-rung picture fuzzy set* R is an object having the form, $R = \{(x, T_R(x), N_R(x), F_R(x)) | x \in X\}$, where the function $T_R : X \to [0, 1]$ represents the positive degree, $N_R : X \to [0, 1]$ represents the neutral degree, and $F_R : X \to [0, 1]$ represents the negative degree of the element $x \in X$ such that, $0 \le T_R^q(x) + N_R^q(x) + F_R^q(x) \le 1, q \ge 1$, for all $x \in X$.

The refusal degree of x in R is defined as $\sqrt[q]{(1 - T_P^q(x) - N_P^q(x) - F_P^q(x))}$ (Fig. 7.1).

Picture fuzzy set proposed by Cuong and Kreinovich [16] in 2013 as an extension of intuitionistic fuzzy set, consists of four terms, degree of positive membership (T), degree of neutral membership (N), degree of negative membership (F), and degree of refusal membership in order to deal with real-life context more adequately, where degree of refusal membership (π) fully depends on preceding three terms as it is defined as $\pi = 1 - T - N - F$. The geometrical representation of picture fuzzy set in comparison with intuitionistic fuzzy set is first presented by Singh [26] in 2015. In 2017, Yager introduced a new generalization of intuitionistic fuzzy sets

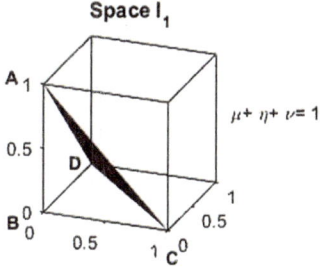

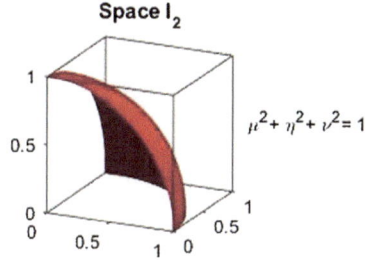

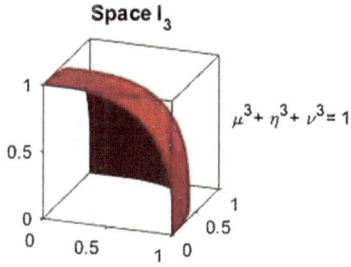

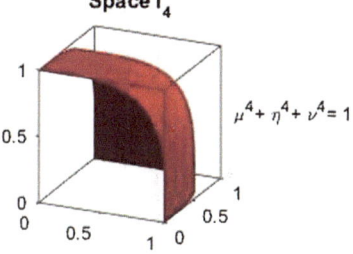

Fig. 7.2 Spaces of acceptable triplets upto 4-rungs

(orthopair fuzzy sets), called q-rung orthopair fuzzy sets which relax the constraint of orthopair membership grades with $T(x)^q + N(x)^q \leq 1$, $(q \geq 1)$. As q increases, the representation space of allowable orthopair membership grades increases. Motivated by the ideas of q-rung orthopair fuzzy sets and picture fuzzy sets, recently in 2018, Li et al. [19] proposed q-rung picture fuzzy set model which inherits the virtues of both q-rung orthopair fuzzy sets and picture fuzzy sets.

1. For $q = 1$, the q-rung picture fuzzy set reduces to picture fuzzy set (1-rung picture fuzzy set). The space I_1 in Fig. 7.2 bounded by surface $x + y + z = 1$ in first octant is equivalent to the volume occupied by a tetrahedron **ABCD**. Thus, any element belongs to 1-rung picture fuzzy set must lie within tetrahedron **ABCD**.

2. For $q = 2$, the q-rung picture fuzzy set reduces to spherical fuzzy set (2-rung picture fuzzy set) which covers more space of acceptable triplets than picture fuzzy set. The space I_2 in Fig. 7.2 bounded by surface $x^2 + y^2 + z^2 = 1$ in first octant is equivalent to the volume occupied by unit sphere in the first octant. Thus, any element belongs to 2-rung picture fuzzy set must lie within the unit sphere in the first octant.

3. Any element satisfying the constraint $x^3 + y^3 + z^3 \leq 1$ lies in the space I_3, representing a 3-rung picture fuzzy set.

4. The space I_4 of admissible triplets of 4-rung picture fuzzy set, displayed in Fig. 7.2, is vast than all preceding spaces.

The generalized structure of picture fuzzy set, called q-rung picture fuzzy set provides gradual increase in spaces bounded by surfaces $T(x)^q + N(x)^q + F(x)^q = 1$, obtained by varying q in first octant. Figure 7.2 displays spaces of most widely acceptable triplets for increasing q rungs.

Moreover, it is clear from the graphical structure of q-rung picture fuzzy sets that any object lying in tetrahedron, must lie not only within the unit sphere but also in all other spaces covered by surfaces $T^q + N^q + F^q = 1$ for $q > 2$. Since the volume occupied by a surface covers the volume occupied by all preceding surfaces (see in Fig. 7.2), therefore, it can be deduced that any element belongs to a particular q-rung picture fuzzy set, must qualify for all picture fuzzy sets of higher rungs (i.e., greater than q).

Definition 7.3 The *support* of a q-rung picture fuzzy set $R = \{(x, T_R(x), N_R(x), F_R(x))|x \in X\}$ is defined as

$$supp(R) = \{x|T_R(x) \neq 0, N_R(x) \neq 0, F_R(x) \neq 1\}.$$

The *height* of a q-rung picture fuzzy set $R = \{(x, T_R(x), N_R(x), F_R(x))|x \in X\}$ is defined as

$$h(R) = \{(\max\{T_R(x)\}, \min\{N_R(x)\}, \min\{F_R(x)\})|x \in X\}.$$

Definition 7.4 Let A and B be two q-rung picture fuzzy sets on X. The *union* and *intersection* of A and B are defined as

- $A \subseteq B \Leftrightarrow T_A(x) \leq T_B(x), N_A(x) \geq N_B(x), F_A(x) \geq F_B(x)$.
- $A \cup B = \{(x, \max(T_A(x), T_B(x)), \min(N_A(x), N_B(x)), \min(F_A(x), F_B(x)))| x \in X\}$.
- $A \cap B = \{(x, \min(T_A(x), T_B(x)), \min(N_A(x), N_B(x)), \max(F_A(x), F_B(x)))| x \in X\}$.

Definition 7.5 A *q-rung picture fuzzy graph* on X is defined as an ordered pair $G = (P, Q)$, where P is a q-rung picture fuzzy set on X and Q is a q-rung picture fuzzy relation on X such that

$$T_Q(x_1x_2) \leq \min\{T_P(x_1), T_P(x_2)\},$$

$$N_Q(x_1x_2) \leq \min\{N_P(x_1), N_P(x_2)\},$$

$$F_Q(x_1x_2) \leq \max\{F_P(x_1), F_P(x_2)\},$$

and $0 \leq T_Q^q(x_1x_2) + N_Q^q(x_1x_2) + F_Q^q(x_1x_2) \leq 1$, $q \geq 1$, for all $x_1, x_2 \in X$, where $T_Q : X \times X \to [0, 1]$, $N_Q : X \times X \to [0, 1]$ and $F_Q : X \times X \to [0, 1]$ represent the positive, neutral, and negative degrees of Q, respectively.

Definition 7.6 Let X be a set of universe. A *q-rung picture fuzzy hypergraph* on X is defined as an ordered pair $H = (\mathscr{R}, S)$, where $\mathscr{R} = \{R_1, R_2, R_3, \ldots, R_k\}$ is a collection of nontrivial q-rung picture fuzzy sets on X and S is a q-rung picture fuzzy relation on R_i such that

1. $T_S(E_i) = T_S(\{x_1, x_2, x_3, \ldots, x_r\}) \leq \min\{T_{R_i}(x_1), T_{R_i}(x_2), T_{R_i}(x_3), \ldots,$
   $T_{R_i}(x_r)\}$,
   $N_S(E_i) = N_S(\{x_1, x_2, x_3, \ldots, x_r\}) \leq \min\{N_{R_i}(x_1), N_{R_i}(x_2), N_{R_i}(x_3), \ldots,$
   $N_{R_i}(x_r)\}$,
   $F_S(E_i) = F_S(\{x_1, x_2, x_3, \ldots, x_r\}) \leq \max\{F_{R_i}(x_1), F_{R_i}(x_2), F_{R_i}(x_3), \ldots,$
   $F_{R_i}(x_r)\}$,
   for all $x_1, x_2, x_3, \ldots, x_r \in X$.
2. $X = \bigcup_k supp(R_k)$, for all $R_k \in \mathscr{R}$.

Here, $E = \{E_1, E_2, E_3, \ldots, E_l\}$ is the family of crisp hyperedges.

Definition 7.7 A q-rung picture fuzzy hypergraph $H = (\mathscr{R}, S)$ is *simple* if H does not contain repeated q-rung picture fuzzy hyperedges and whenever $E_i, E_j \in E$ and $E_i \subseteq E_j$ then $E_i = E_j$.

Definition 7.8 Let $H = (\mathscr{R}, S)$ be a q-rung picture fuzzy hypergraph and $E_i \in E$. Suppose $\alpha, \beta, \gamma \in [0, 1]$, the (α, β, γ)-cut of E_i, $E_i^{(\alpha,\beta,\gamma)}$, is defined as, $E_i^{(\alpha,\beta,\gamma)} = \{x \in X | T_{E_i}(x) \geq \alpha$, $N_{E_i}(x) \geq \beta$, $F_{E_i}(x) \leq \gamma\}$. Let

- $E^{(\alpha,\beta,\gamma)} = \{E_i^{(\alpha,\beta,\gamma)} | E_i \in E\}$,
- $X^{(\alpha,\beta,\gamma)} = \cup\{E_i^{(\alpha,\beta,\gamma)} | E_i \in E\}$.

If $E^{(\alpha,\beta,\gamma)}$ is non-empty, then the (α, β, γ)-*level hypergraph* of H is defined as a crisp hypergraph $H^{(\alpha,\beta,\gamma)} = (X^{(\alpha,\beta,\gamma)}, E^{(\alpha,\beta,\gamma)})$.

For further terminologies and studies on q-rung orthopair fuzzy sets and graphs, readers are referred to [1–13, 15, 17, 21–23, 25, 28–30, 33, 34, 36].

7.2 q-Rung Picture Fuzzy Hierarchical Quotient Space Structure

Different techniques have been proposed to deal with granular computing. Quotient space, fuzzy sets, and rough sets are three basic computing tools dealing with uncertainty. Based on fuzzy sets, fuzzy equivalence relation, as an extension of equivalence relation, was proposed by Zadeh [32]. The question of distinct membership degrees of same object from different scholars is arisen because of various ways of thinking about the interpretation of different functions dealing with the same problem. To resolve this issue, fuzzy set was structurally defined by Zhang and Zhang [37] which was based on quotient space theory and fuzzy equivalence relation. This definition provides some new intuitiveness regarding membership degree, called a hierarchical quotient space structure of a fuzzy equivalence relation. By following the same concept, we develop a hierarchical quotient space structure of a q-rung picture fuzzy equivalence relation.

Definition 7.9 Let X and Y be two finite non-empty sets then the Cartesian product between X and Y is $X \times Y$. Every q-rung picture fuzzy subset R of $X \times Y$ is defined as a *q-rung picture fuzzy binary relation* from X to Y. Let $X = \{x_1, x_2, x_3, \ldots, x_l\}$ and $Y = \{y_1, y_2, y_3, \ldots, y_m\}$, a q-rung picture fuzzy binary relation matrix $\tilde{M}_R$ is given as follows:

$$
\tilde{M}_R = \begin{bmatrix} d_R(x_1, y_1) & d_R(x_1, y_2) & \cdots & d_R(x_1, y_m) \\ d_R(x_2, y_1) & d_R(x_2, y_2) & \cdots & d_R(x_2, y_m) \\ \vdots & \vdots & \vdots & \vdots \\ d_R(x_l, y_1) & d_R(x_l, y_2) & \cdots & d_R(x_l, y_m) \end{bmatrix}.
$$

In general, $\tilde{M}_R$ is called *q-rung picture fuzzy relation matrix* of R, where $d_R(x, y) = (T_R(x, y), N_R(x, y), F_R(x, y))$ and $T_R : X \times Y \to [0, 1]$, $N_R : X \times Y \to [0, 1]$, $F_R : X \times Y \to [0, 1]$ represent the positive degree, neutral degree and negative degree of objects x and y satisfying the relation R, respectively, such that

$$
0 \leq T_R^q(x, y) + N_R^q(x, y) + F_R^q(x, y) \leq 1, \ q \geq 1,
$$

for all $(x, y) \in X \times Y$.

Definition 7.10 A q-rung picture fuzzy relation on a non-empty finite set X is called *q-rung picture fuzzy similarity relation* if it satisfies

1. $d_R(x, x) = (1, 0, 0)$, for all $x \in X$,
2. $d_R(x, y) = d_R(y, x)$, for all $x, y \in X$.

Definition 7.11 A q-rung picture fuzzy relation on a non-empty finite set X is called *q-rung picture fuzzy equivalence relation* if it satisfies the conditions

1. $d_R(x, x) = (1, 0, 0)$, for all $x \in X$,
2. $d_R(x, y) = d_R(y, x)$, for all $x, y \in X$,
3. for all $x, y, z \in X$

   - $T_R(x, z) = \sup_{y \in X} \{\min(T_R(x, y), T_R(y, z))\}$,
   - $N_R(x, z) = \sup_{y \in X} \{\min(N_R(x, y), N_R(y, z))\}$,
   - $F_R(x, z) = \inf_{y \in X} \{\max(F_R(x, y), F_R(y, z))\}$.

A *q-rung picture fuzzy quotient space* is denoted by a triplet $(X, \tilde{A}, R)$, where X is a finite domain, $\tilde{A}$ represents the attributes of X and R represents the q-rung picture fuzzy relationship between the objects of universe X, which is called the structure of the domain.

Definition 7.12 Let x_i and x_j be two objects in the universe X. The *similarity* between $x_i, x_j \in X$ having the attribute $\tilde{a}_k$ is defined as

$$T_R(x_i, x_j) = \frac{|\tilde{a}_{ik} \cap \tilde{a}_{jk}|}{|\tilde{a}_{ik} \cup \tilde{a}_{jk}|}, \quad N_R(x_i, x_j) = \frac{|\tilde{a}_{ik} \cap \tilde{a}_{jk}|}{|\tilde{a}_{ik} \cup \tilde{a}_{jk}|}, \quad F_R(x_i, x_j) = \frac{|\tilde{a}_{ik} \cup \tilde{a}_{jk}|}{|\tilde{a}_{ik} \cap \tilde{a}_{jk}|},$$

where $\tilde{a}_{ik}$ represents that object x_i possesses the attribute $\tilde{a}_k$ and $\tilde{a}_{jk}$ represents that object x_j possesses the attribute $\tilde{a}_k$.

It is noted that q-rung picture fuzzy relation matrix $\tilde{M}_R$ is symmetric and reflexive q-rung picture fuzzy relation but, in general, it does not satisfy the transitivity condition. In such cases, transitive closure is used to make a q-rung picture fuzzy equivalence relation from the given relation. The transitive closure $R^\star$ of a q-rung picture fuzzy relation R is defined as $R^\star = R \cup R^2 \cup R^3 \cdots R^{n-1}$.

Proposition 7.1 *Let R be a q-rung picture fuzzy relation on a finite domain X and $R^{(\alpha,\beta,\gamma)} = \{(x, y)|T_R(x, y) \geq \alpha, N_R(x, y) \geq \beta, F_R(x, y) \leq \gamma\}, (\alpha, \beta, \gamma) \in [0, 1]$. Then, $R^{(\alpha,\beta,\gamma)}$ is an equivalence relation on X and is said to be cut-equivalence relation of R.*

Proposition 7.1 depicts that $R^{(\alpha,\beta,\gamma)}$ is a crisp relation, which is equivalence on X and it's *knowledge space* is given as, $\tau_{R^{(\alpha,\beta,\gamma)}}(X) = X/R^{(\alpha,\beta,\gamma)}$. The *value domain* of an equivalence relation R on X is defined as $D = \{d_R(x, y)|x, y \in X\}$ such that

- $T_X(x) \wedge T_X(y) \wedge T_R(x, y) > 0$,
- $N_X(x) \wedge N_X(y) \wedge N_R(x, y) > 0$,
- $F_X(x) \vee F_X(y) \vee F_R(x, y) > 0$.

Definition 7.13 Let R be a q-rung picture fuzzy equivalence relation on a finite set X and D be the value domain of R. The set given by $\tau_X(R) = \{X/R^{(\alpha,\beta,\gamma)}|(\alpha, \beta, \gamma) \in D\}$ is called q-rung picture fuzzy hierarchical quotient space structure of R.

Example 7.1 Let $X = \{x_1, x_2, x_3, x_4, x_5\}$ and R be a 5-rung picture fuzzy equivalence relation on X, the corresponding relation matrix $\tilde{M}_R$ is given as follows:

$$\tilde{M}_R = \begin{bmatrix} (1, 0, 0) & (0.3, 0.4, 0.3) & (0.7, 0.5, 0.1) & (0.5, 0.5, 0.2) & (0.5, 0.5, 0.2) \\ (0.3, 0.4, 0.3) & (1, 0, 0) & (0.3, 0.4, 0.3) & (0.3, 0.4, 0.3) & (0.3, 0.4, 0.3) \\ (0.7, 0.5, 0.1) & (0.3, 0.4, 0.3) & (1, 0, 0) & (0.5, 0.5, 0.2) & (0.5, 0.5, 0.2) \\ (0.5, 0.5, 0.2) & (0.3, 0.4, 0.3) & (0.5, 0.5, 0.2) & (1, 0, 0) & (0.9, 0.5, 0.1) \\ (0.5, 0.5, 0.2) & (0.3, 0.4, 0.3) & (0.5, 0.5, 0.2) & (0.9, 0.5, 0.1) & (1, 0, 0) \end{bmatrix}.$$

It's corresponding hierarchical quotient space structure is given as

$X/R^{(\alpha_1,\beta_1,\gamma_1)} = \{\{x_1, x_2, x_3, x_4, x_5\}\}, \ 0 < \alpha_1 \leq 0.3, 0 < \beta_1 \leq 0.4, 0.3 > \gamma_1 \geq 0,$

$X/R^{(\alpha_2,\beta_2,\gamma_2)} = \{\{x_2\}, \{x_1, x_3, x_4, x_5\}\}, \ 0.3 < \alpha_2 \leq 0.5, 0.4 < \beta_2 \leq 0.5, 0.3 > \gamma_2 \geq 0.2,$

$X/R^{(\alpha_3,\beta_3,\gamma_3)} = \{\{x_1, x_3\}, \{x_2\}, \{x_4, x_5\}\}, \ 0.5 < \alpha_3 \leq 0.7, 0.5 < \beta_3 \leq 0.5, 0.2 > \gamma_3 \geq 0.1,$

$X/R^{(\alpha_4,\beta_4,\gamma_4)} = \{\{x_1\}, \{x_2\}, \{x_3\}, \{x_4, x_5\}\}, \ 0.7 < \alpha_4 \leq 0.9, 0.5 < \beta_4 \leq 0.5, 0.1 > \gamma_4 \geq 0.1,$

$X/R^{(\alpha_5,\beta_5,\gamma_5)} = \{\{x_1\}, \{x_2\}, \{x_3\}, \{x_4\}, \{x_5\}\}, \ 0.9 < \alpha_5 \leq 1, 0.5 < \beta_5 \leq 1, 0.1 > \gamma_4 \geq 0.$

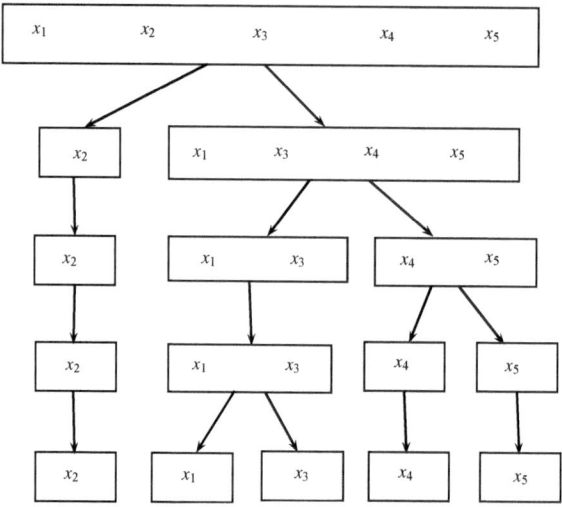

Fig. 7.3 The hierarchical quotient space structure of R

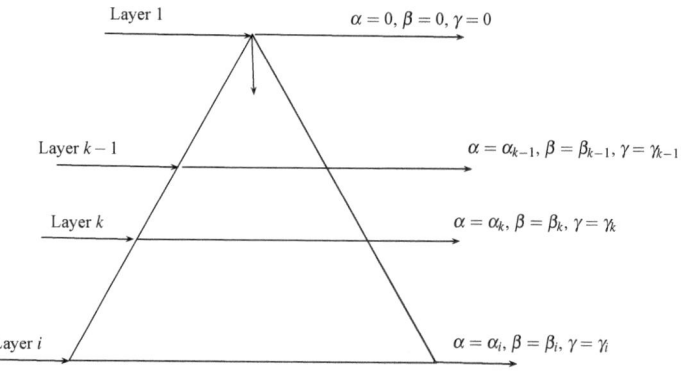

Fig. 7.4 The pyramid model of a hierarchical quotient space structure

Hence, a hierarchical quotient space structure induced by 5-rung picture fuzzy equivalence relation R is given as $\tau_X(R) = \{X/R^{(\alpha_1,\beta_1,\gamma_1)}, X/R^{(\alpha_2,\beta_2,\gamma_2)}, X/R^{(\alpha_3,\beta_3,\gamma_3)}, X/R^{(\alpha_4,\beta_4,\gamma_4)}, X/R^{(\alpha_5,\beta_5,\gamma_5)}\}$ and is shown in Fig. 7.3.

Furthermore, assuming the number of blocks in every distinct layer of this hierarchical quotient space structure, a pyramid model can also be constructed as shown in Fig. 7.4.

It is worth to note that the same hierarchical quotient space structure can be formed by different 5-rung picture fuzzy equivalence relations. For instance, the relation matrix $\tilde{M}_{R_1}$ of 5-rung picture fuzzy equivalence relation generates the same hierarchical quotient space structure as given by $\tilde{M}_R$. The relation matrix $\tilde{M}_{R_1}$ is given as

$$\tilde{M}_{R_1} = \begin{bmatrix} (1,0,0) & (0.2,0.3,0.2) & (0.9,0.5,0.1) & (0.6,0.6,0.3) & (0.6,0.6,0.3) \\ (0.2,0.3,0.2) & (1,0,0) & (0.2,0.3,0.2) & (0.2,0.3,0.2) & (0.2,0.3,0.2) \\ (0.9,0.5,0.1) & (0.2,0.3,0.2) & (1,0,0) & (0.6,0.6,0.3) & (0.6,0.6,0.3) \\ (0.6,0.6,0.3) & (0.2,0.3,0.2) & (0.6,0.6,0.3) & (1,0,0) & (0.7,0.5,0.1) \\ (0.6,0.6,0.3) & (0.2,0.3,0.2) & (0.6,0.6,0.3) & (0.7,0.5,0.1) & (1,0,0) \end{bmatrix}.$$

Definition 7.14 Let R be a q-rung picture fuzzy equivalence relation on X. Let $\tau_X(R) = \{X(\rho_1), X(\rho_2), X(\rho_3), \ldots, X(\rho_j)\}$ be its corresponding hierarchical quotient space structure, where $\rho_i = (\alpha_i, \beta_i, \gamma_i), i = 1, 2,\ldots,j$ and $X(\rho_j) < X(\rho_{j-1}) < \cdots < X(\rho_1)$. Then, the *partition sequence* of $\tau_X(R)$ is given as $P(\tau_X(R)) = \{P_1, P_2, P_3, \ldots, P_j\}$, where $P_i = |X(\rho_i)|, i = 1, 2, \ldots, j$ and $|.|$ denotes the number of elements in a set.

Definition 7.15 Let R be a q-rung picture fuzzy equivalence relation on X. Let $\tau_X(R) = \{X(\rho_1), X(\rho_2), X(\rho_3), \ldots, X(\rho_j)\}$ be its corresponding hierarchical quotient space structure, where $\rho_i = (\alpha_i, \beta_i, \gamma_i), i = 1, 2,\ldots,j$ and $X(\rho_j) < X(\rho_{j-1}) < \cdots < X(\rho_1)$, $P(\tau_X(R)) = \{P_1, P_2, P_3, \ldots, P_j\}$ be the partition sequence of $\tau_X(R)$. Assume that $X(\rho_i) = \{X_{i1}, X_{i2}, \ldots, X_{iP_i}\}$. The *information entropy* $I_X(\rho_i)$ is defined as $I_X(\rho_i) = -\sum_{t=1}^{P_i} \frac{|X_{it}|}{|X|} \ln(\frac{|X_{it}|}{|X|})$.

Theorem 7.1 *Let R be a q-rung picture fuzzy equivalence relation on X. Let $\tau_X(R) = \{X(\rho_1), X(\rho_2), X(\rho_3), \ldots, X(\rho_j)\}$ be its corresponding hierarchical quotient space structure, where $\rho_i = (\alpha_i, \beta_i, \gamma_i), i = 1, 2,\ldots,j$, then the entropy sequence $I(\tau_X(R)) = \{I_X(\rho_1), I_X(\rho_2), \ldots, I_X(\rho_j)\}$ increases monotonically and strictly.*

Proof The terminology of hierarchical quotient space structure implies that $X(\rho_j) < X(\rho_{j-1}) < \cdots < X(\rho_1)$, i.e., $X(\rho_{j-1})$ is a quotient subspace of $X(\rho_j)$. Suppose that $X(\rho_i) = \{X_{i1}, X_{i2}, \ldots, X_{iP_i}\}$ and $X(\rho_{i-1}) = \{X_{(i-1)1}, X_{(i-1)2}, \ldots, X_{(i-1)P_{(i-1)}}\}$, then every subblock of $X(\rho_{i-1})$ is an amalgam of sub blocks of $X(\rho_i)$. Without loss of generality, it is assumed that only one subblock $X_{i-1,j}$ in $X(\rho_{i-1})$ is formed by the combination of two subblocks X_{ir}, X_{is} in $X(\rho_i)$ and all other remaining blocks are equal in both sequences. Thus,

$$I_X(\rho_{j-1}) = -\sum_{t=1}^{P_{i-1}} \frac{|X_{i-1,t}|}{|X|} \ln(\frac{|X_{i-1,t}|}{|X|})$$

$$= -\sum_{t=1}^{P_{j-1}} \frac{|X_{i-1,t}|}{|X|} \ln(\frac{|X_{i-1,t}|}{|X|}) - \sum_{t=j+1}^{P_{i-1}} \frac{|X_{i-1,t}|}{|X|} \ln(\frac{|X_{i-1,t}|}{|X|}) - \frac{|X_{i-1,j}|}{|X|} \ln(\frac{|X_{i-1,j}|}{|X|})$$

$$= -\sum_{t=1}^{P_{j-1}} \frac{|X_{i,t}|}{|X|} \ln(\frac{|X_{i,t}|}{|X|}) - \sum_{t=j+1}^{P_i} \frac{|X_{i,t}|}{|X|} \ln(\frac{|X_{i,t}|}{|X|}) - \frac{|X_{i,r}| + |X_{i,s}|}{|X|} \ln(\frac{|X_{i,r}| + |X_{i,s}|}{|X|})$$

Since,

$$\frac{|X_{i,r}| + |X_{i,s}|}{|X|} \ln\left(\frac{|X_{i,r}| + |X_{i,s}|}{|X|}\right) = \frac{|X_{i,r}|}{|X|} \ln\left(\frac{|X_{i,r}| + |X_{i,s}|}{|X|}\right) + \frac{|X_{i,s}|}{|X|} \ln\left(\frac{|X_{i,r}| + |X_{i,s}|}{|X|}\right)$$

$$> \frac{|X_{i,r}|}{|X|} \ln\left(\frac{|X_{i,r}|}{|X|}\right) + \frac{|X_{i,s}|}{|X|} \ln\left(\frac{|X_{i,s}|}{|X|}\right)$$

Therefore, we have

$$I_X(\rho_{j-1}) < -\sum_{t=1}^{P_{j-1}} \frac{|X_{i,t}|}{|X|} \ln\left(\frac{|X_{i,t}|}{|X|}\right) - \sum_{t=j+1}^{P_i} \frac{|X_{i,t}|}{|X|} \ln\left(\frac{|X_{i,t}|}{|X|}\right) - \frac{|X_{i,r}|}{|X|} \ln\left(\frac{|X_{i,r}|}{|X|}\right) - \frac{|X_{i,s}|}{|X|} \ln\left(\frac{|X_{i,s}|}{|X|}\right)$$

$$= I_X(\rho_j), (2 \le j \le n).$$

Hence, $I_X(\rho_1) < I_X(\rho_2) < I_X(\rho_2) < \cdots < I_X(\rho_j)$.

Definition 7.16 Let $X = \{x_1, x_2, x_3, \ldots, x_n\}$ be a non-empty set of universe and let $\mathscr{P}_t(X) = \{X_1, X_2, X_3, \ldots, X_t\}$ be a partition space of X, where $|\mathscr{P}_t(X)| = t$ then $\mathscr{P}_t(X)$ is called t-order partition space on X.

Definition 7.17 Let X be a finite non-empty universe and let $\mathscr{P}_t(X) = \{X_1, X_2, X_3, \ldots, X_t\}$ be a t-order partition space on X. Let $|X_1| = b_1, |X_2| = b_2, \ldots, |X_t| = b_t$ and the sequence $\{b_1, b_2, \ldots, b_t\}$ is arranged in increasing order then we got a new sequence $H(t) = \{b'_1, b'_2, \ldots, b'_t\}$ which is also increasing and called a *subblock sequence* of $\mathscr{P}_t(X)$.

Note that, two different t-order partition spaces on X may possess the similar subblock sequence $H(t)$.

Definition 7.18 Let X be a finite non-empty universe and let $\mathscr{P}_t(X) = \{X_1, X_2, X_3, \ldots, X_t\}$ be a partition space of X. Suppose that $H_1(t) = \{b'_1, b'_2, \ldots, b'_t\}$ be a subblock sequence of $\mathscr{P}_t(X)$, then the ω-displacement of $H_1(t)$ is defined as increasing sequence $H_2(t) = \{b'_1, b'_2, \ldots, b'_r + 1, \ldots, b'_s - 1, \ldots, b'_t\}$, where $r < s$, $b'_r + 1 < b'_s - 1$.

An ω-displacement is obtained by subtracting 1 from some bigger term and adding 1 to some smaller element such that the sequence keeps its increasing property.

Theorem 7.2 *A single time ω-displacement $H_2(t)$ which is derived from $H_1(t)$ satisfies $I(H_1(t)) < I(H_2(t))$.*

Proof Let $H_1(t) = \{b'_1, b'_2, \ldots, b'_t\}$ and $H_2(t) = \{b'_1, b'_2, \ldots, b'_r + 1, \ldots, b'_s - 1, \ldots, b'_t\}$, $b'_1 + b'_2 + \cdots + b'_t = k$ then we have

$$I(H_2(t)) = -\sum_{j=1}^{t} \frac{b'_j}{k} \ln\frac{b'_j}{k} + \frac{b'_r}{k} \ln\frac{b'_r}{k} + \frac{b'_s}{k} \ln\frac{b'_s}{k} - \frac{b'_r + 1}{k} \ln\frac{b'_r + 1}{k} - \frac{b'_s - 1}{k} \ln\frac{b'_s - 1}{k}.$$

Let $\varphi(x) = -\frac{x}{k}\ln\frac{x}{k} - \frac{l-x}{k}\ln\frac{l-x}{k}$, where $l = b'_r + b'_s$ and $\varphi'(x) = \frac{1}{k}\ln\frac{l-x}{x}$. Suppose that $\varphi'(x) = 0$ then we obtain a solution, i.e., $x = \frac{l}{2}$. Furthermore,

$\varphi''(x) = \frac{-l}{k(l-x)x} < 0, 0 \leq x \leq \frac{l}{2}$ and $\varphi(x)$ is increasing monotonically. Let $x_1 = b'_r$ and $x_2 = b'_r + 1$, $b'_r + 1 < b'_s - 1$, i.e., $x_1 < x_2 \leq \frac{l}{2} = \frac{b'_r+b'_s}{2}$. Since, $\varphi(x)$ is monotone then $\varphi(x_2) - \varphi(x_1) > 0$. Thus,

$$\frac{b'_r}{k}\ln\frac{b'_r}{k} + \frac{b'_s}{k}\ln\frac{b'_s}{k} - \frac{b'_r+1}{k}\ln\frac{b'_r+1}{k} - \frac{b'_s-1}{k}\ln\frac{b'_s-1}{k} > 0.$$

Therefore, we have

$$I(H_2(t)) = -\sum_{j=1}^{t}\frac{b'_l}{k}\ln\frac{b'_l}{k} + \frac{b'_r}{k}\ln\frac{b'_r}{k} + \frac{b'_s}{k}\ln\frac{b'_s}{k} - \frac{b'_r+1}{k}\ln\frac{b'_r+1}{k} - \frac{b'_s-1}{k}\ln\frac{b'_s-1}{k}$$

$$> -(\frac{b'_r+1}{k}\ln\frac{b'_r+1}{k} + \frac{b'_s-1}{k}\ln\frac{b'_s-1}{k})$$

$$> -\sum_{j=1}^{t}\frac{b'_l}{k}\ln\frac{b'_l}{k}$$

$$= I(H_1(t)).$$

7.3   A q-Rung Picture Fuzzy Hypergraph Model of Granular Computing

In the following section, we construct a q-rung picture fuzzy hypergraph model of granular computing.

7.3.1   Model Construction

Granular computing may utilize frameworks in terms of levels, granules, and hierarchies which are based on multiple representations and multilevel [14, 27]. A *granule* is defined as a collection of objects or elements having same attributes or characteristics and can be treated as a single unit.

Definition 7.19 An *object space* is defined as a system (X, R), where X is a universe of objects or elements and $R = \{r_1, r_2, r_3, \ldots, r_k\}$, $k = |X|$ is a family of relations between the elements of X. For $n \leq k$, $r_n \in R$, $r_n \subseteq X \times X \times X \cdots \times X$, if $(x_1, x_2, \ldots, x_n) \subseteq r_n$, then there exists an n-array relation r_n on $(x_1, x_2, \ldots, x_n)$.

The set of those objects which have some relation $r_i \in R$ in an object space can be assumed as a granule. A single object is considered as a smallest granule and the set of all elements is said to be the largest granule in an object space.

We consider one vertex of a q-rung picture fuzzy hypergraph as a representation of an object in the object space and the set of objects having some relationship s_i

is represented by q-rung picture fuzzy hyperedge. The positive membership T of an element x_i refers to the belonging, neutral membership N refers to the unbiased behavior and negative membership F refers to the disconnection to the granule, where $T(x_i)$, $N(x_i)$, $F(x_i) \in [0, 1]$, satisfying the condition $T^q(x_i) + N^q(x_i) + F^q(x_i) \leq 1$, $q \geq 1$. Thus, we can establish a q-rung picture fuzzy hypergraph model of granular computing. An example of constructing such a model is given in Example 7.2.

Example 7.2 Let $X = \{x_1, x_2, x_3, x_4, x_5\}$ be the set of objects and $R = \{r_1, r_2, r_3, r_4, r_5, r_6\}$ be the set of relations. A hypergraph $H = (\mathcal{R}, S)$ representing the objects and the relations $r_i \in R (1 \leq i \leq 6)$ between them is shown in Fig. 7.5. By assigning the positive membership $T \in [0, 1]$, neutral membership $N \in [0, 1]$, and negative membership $F \in [0, 1]$ to each element x_i, we form a 7-rung picture fuzzy hypergraph. Let $X = \{x_1, x_2, x_3, x_4, x_5\}$ and $S = \{\mathcal{E}_1, \mathcal{E}_2, \mathcal{E}_3, \mathcal{E}_4\}$ and the corresponding incidence matrix is given in Table 7.1.

Let $\alpha = 0.5$, $\beta = 0.3$, $\gamma = 0.5$, then we have $r_1 = \{(x_3), (x_4)\}$, $r_2 = \{(x_2, x_4)\}$, $r_3, r_4, r_5 = \{\emptyset\}$. Hypergraph representation of granular in a level is given in Fig. 7.6.

The relationships between the vertices can be obtained by following the present situation, through computing the mathematical function or by figuring out the internal, external, and contextual properties of the granule [14]. After the relationship is being computed, the set of vertices possessing the relationship among them can be combined into one module. By taking in to account the actual condition, we

Fig. 7.5 Hypergraph representation of granules

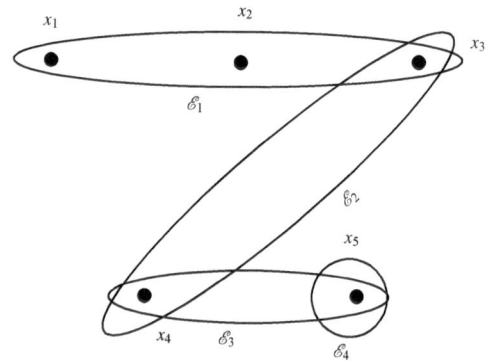

Table 7.1 Incidence matrix of H

I	$\mathcal{E}_1$	$\mathcal{E}_2$	$\mathcal{E}_3$	$\mathcal{E}_4$
x_1	$(0.7, 0.2, 0.1)$	$(0, 0, 1)$	$(0, 0, 1)$	$(0, 0, 1)$
x_2	$(0.8, 0.2, 0.3)$	$(0, 0, 1)$	$(0, 0, 1)$	$(0, 0, 1)$
x_3	$(0.5, 0.3, 0.4)$	$(0.5, 0.3, 0.1)$	$(0, 0, 1)$	$(0, 0, 1)$
x_4	$(0, 0, 1)$	$(0.6, 0.2, 0.1)$	$(0.5, 0.3, 0.1)$	$(0, 0, 1)$
x_5	$(0, 0, 1)$	$(0, 0, 1)$	$(0.6, 0.2, 0.1)$	$(0.5, 0.3, 0.1)$

Fig. 7.6 Representation of granular in a level through hypergraph

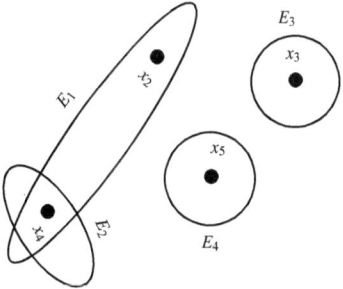

can calculate or assign positive, neutral, and negative degrees through the membership functions $T(x)$, $N(x)$, and $F(x)$, respectively. As a result, we have formed a granule which is also called the q-rung picture fuzzy hyperedge. The (α, β, γ)-cut are computed for this q-rung picture fuzzy hyperedge. When all under consideration vertices are to be integrated into a single unit and the membership degrees are assigned or computed of the unit, we have constructed a single-level model of granular computing.

The amalgamation of the elements or objects which are processed at the same time while resolving a problem is called a granule. It reflects the representation of the perceptions and the attributes of the integration. A granule may be a small part of another granule and can also be considered as the block to form larger granules. It can also be a collection of granules or may be thought as a whole unit. Thus, a granule may play two distinct roles.

A granule affiliates to a particular level. The whole view of granules at every level can be taken as a complete description of a particular problem at that level of granularity [14]. A q-rung picture fuzzy hypergraph formed by the set of relations R and membership functions $T(x)$, $N(x)$, $F(x)$ of objects in the space is considered as a specific level of granular computing model. All q-rung picture fuzzy hyperedges in that q-rung picture fuzzy hypergraph can be regarded as the complete granule in that particular level.

Definition 7.20 A *partition* of a set X established on the basis of relations between objects is defined as a collection of non-empty subsets which are pair-wise disjoint and whose union is whole of X. These subsets which form the partition of X are called *blocks*. Every partition of a finite set X contains the finite number of blocks. Corresponding to the q-rung picture fuzzy hypergraph, the constraints of partition $\psi = \{\mathscr{E}_i | 1 \leq i \leq n\}$ can be stated as follows:

- each $\mathscr{E}_i$ is non-empty,
- for $i \neq j$, $\mathscr{E}_i \cap \mathscr{E}_j = \emptyset$,
- $\cup \{supp(\mathscr{E}_i) | 1 \leq i \leq n\} = X$.

Definition 7.21 A *covering* of a set X is defined as a collection of non-empty subsets whose union is whole of X. The conditions for the covering $c = \{\mathscr{E}_i | 1 \leq i \leq n\}$ of X are stated as

- each $\mathscr{E}_i$ is non-empty,
- $\cup \{supp(\mathscr{E}_i) | 1 \leq i \leq n\} = X$.

The corresponding definitions in classical hypergraph theory are completely analogous to the above Definitions 7.20 and 7.21. In a crisp hypergraph, if the hyperedges E_i and E_j do not intersect each other, i.e., $E_i, E_j \in E$ and $E_i \cap E_j = \emptyset$ then these hyperedges form a *partition* of granules in this level. Furthermore, if $E_i, E_j \in E$ and $E_i \cap E_j \neq \emptyset$, i.e., the hyperedges E_i and E_j intersect each other, then these hyperedges form a *covering* in this level.

In a q-rung picture fuzzy hypergraph, if $\mathscr{E}_i, \mathscr{E}_j \in S$ and $\mathscr{E}_i \cap \mathscr{E}_j = \emptyset$, i.e., the hyperedges $\mathscr{E}_i$ and $\mathscr{E}_j$ do not intersect each other, then these hyperedges form a *partition* of granules. Furthermore, if $\mathscr{E}_i, \mathscr{E}_j \in S$ and $\mathscr{E}_i \cap \mathscr{E}_j \neq \emptyset$, i.e., the hyperedges $\mathscr{E}_i$ and $\mathscr{E}_j$ intersect each other, then these hyperedges form a *covering* of granules in this level. Note that, the definition of q-rung picture fuzzy hypergraph concludes that the q-rung picture fuzzy hypergraph forms a covering of set of universe X.

Example 7.3 Let $X = \{x_1, x_2, x_3, x_4, x_5, x_6, x_7, x_8, x_9, x_{10}\}$. The partition and covering of X are given in Figs. 7.7 and 7.8, respectively.

The fundamental properties which are possessed by granules are internal property, external property, and contextual property. The elements belonging to the granule determine the internal property and studies the relationships between objects. The internal property in this model incorporates the interaction of vertices. The external property reflects the relationships between granules. This property incorporates those vertices which belong to a q-rung picture fuzzy hyperedge under the membership functions T, N, and F. The existence of a granule is indicated by contextual property in an environment.

A set-theoretic way to study the granular computing model uses the following operators in a q-rung picture fuzzy hypergraph model.

Fig. 7.7 A partition of granules in a level

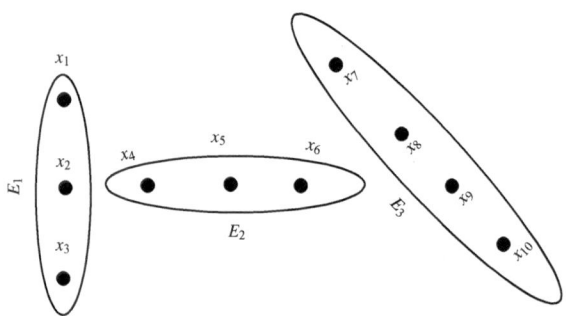

Fig. 7.8 A covering of
granules in a level

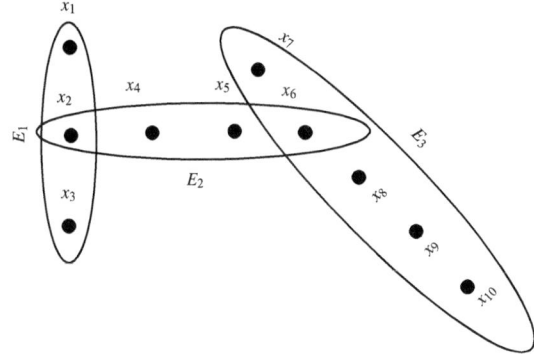

Definition 7.22 Let $\mathcal{G}_1$ and $\mathcal{G}_2$ be two granules in our model and the q-rung picture fuzzy hyperedges $\mathcal{E}_1$, $\mathcal{E}_2$ represent their external properties. The *union* of two granules $\mathcal{G}_1 \cup \mathcal{G}_2$ is defined as a larger q-rung picture fuzzy hyperedge that contains the vertices of both $\mathcal{E}_1$ and $\mathcal{E}_2$. If $x_i \in \mathcal{G}_1 \cup \mathcal{G}_2$, then the membership degrees of x_i in larger granule $\mathcal{G}_1 \cup \mathcal{G}_2$ are defined as follows:

$$T_{\mathcal{G}_1 \cup \mathcal{G}_2}(x_i) = \begin{cases} \max\{T_{\mathcal{E}_1}(x_i), T_{\mathcal{E}_2}(x_i)\}, & \text{if } x_i \in \mathcal{E}_1 \text{ and } x_i \in \mathcal{E}_2, \\ T_{\mathcal{E}_1}(x_i), & \text{if } x_i \in \mathcal{E}_1 \text{ and } x_i \notin \mathcal{E}_2, \\ T_{\mathcal{E}_2}(x_i), & \text{if } x_i \in \mathcal{E}_2 \text{ and } x_i \notin \mathcal{E}_1. \end{cases}$$

$$N_{\mathcal{G}_1 \cup \mathcal{G}_2}(x_i) = \begin{cases} \max\{N_{\mathcal{E}_1}(x_i), N_{\mathcal{E}_2}(x_i)\}, & \text{if } x_i \in \mathcal{E}_1 \text{ and } x_i \in \mathcal{E}_2, \\ N_{\mathcal{E}_1}(x_i), & \text{if } x_i \in \mathcal{E}_1 \text{ and } x_i \notin \mathcal{E}_2, \\ N_{\mathcal{E}_2}(x_i), & \text{if } x_i \in \mathcal{E}_2 \text{ and } x_i \notin \mathcal{E}_1. \end{cases}$$

$$F_{\mathcal{G}_1 \cup \mathcal{G}_2}(x_i) = \begin{cases} \min\{F_{\mathcal{E}_1}(x_i), F_{\mathcal{E}_2}(x_i)\}, & \text{if } x_i \in \mathcal{E}_1 \text{ and } x_i \in \mathcal{E}_2, \\ F_{\mathcal{E}_1}(x_i), & \text{if } x_i \in \mathcal{E}_1 \text{ and } x_i \notin \mathcal{E}_2, \\ F_{\mathcal{E}_2}(x_i), & \text{if } x_i \in \mathcal{E}_2 \text{ and } x_i \notin \mathcal{E}_1. \end{cases}$$

Definition 7.23 Let $\mathcal{G}_1$ and $\mathcal{G}_2$ be two granules in our model and the q-rung picture fuzzy hyperedges $\mathcal{E}_1$, $\mathcal{E}_2$ represent their external properties. The *intersection* of two granules $\mathcal{G}_1 \cap \mathcal{G}_2$ is defined as a smaller q-rung picture fuzzy hyperedge that contains those vertices belonging to both $\mathcal{E}_1$ and $\mathcal{E}_2$. If $x_i \in \mathcal{G}_1 \cap \mathcal{G}_2$, then the membership degrees of x_i in smaller granule $\mathcal{G}_1 \cap \mathcal{G}_2$ are defined as follows:

$$T_{\mathcal{G}_1 \cap \mathcal{G}_2}(x_i) = \begin{cases} \min\{T_{\mathcal{E}_1}(x_i), T_{\mathcal{E}_2}(x_i)\}, & \text{if } x_i \in \mathcal{E}_1 \text{ and } x_i \in \mathcal{E}_2, \\ T_{\mathcal{E}_1}(x_i), & \text{if } x_i \in \mathcal{E}_1 \text{ and } x_i \notin \mathcal{E}_2, \\ T_{\mathcal{E}_2}(x_i), & \text{if } x_i \in \mathcal{E}_2 \text{ and } x_i \notin \mathcal{E}_1. \end{cases}$$

$$N_{\mathscr{G}_1 \cap \mathscr{G}_2}(x_i) = \begin{cases} \min\{N_{\mathscr{E}_1}(x_i), N_{\mathscr{E}_2}(x_i)\}, & \text{if } x_i \in \mathscr{E}_1 \text{ and } x_i \in \mathscr{E}_2, \\ N_{\mathscr{E}_1}(x_i), & \text{if } x_i \in \mathscr{E}_1 \text{ and } x_i \notin \mathscr{E}_2, \\ N_{\mathscr{E}_2}(x_i), & \text{if } x_i \in \mathscr{E}_2 \text{ and } x_i \notin \mathscr{E}_1. \end{cases}$$

$$F_{\mathscr{G}_1 \cap \mathscr{G}_2}(x_i) = \begin{cases} \max\{F_{\mathscr{E}_1}(x_i), F_{\mathscr{E}_2}(x_i)\}, & \text{if } x_i \in \mathscr{E}_1 \text{ and } x_i \in \mathscr{E}_2, \\ F_{\mathscr{E}_1}(x_i), & \text{if } x_i \in \mathscr{E}_1 \text{ and } x_i \notin \mathscr{E}_2, \\ F_{\mathscr{E}_2}(x_i), & \text{if } x_i \in \mathscr{E}_2 \text{ and } x_i \notin \mathscr{E}_1. \end{cases}$$

Definition 7.24 Let $\mathscr{G}_1$ and $\mathscr{G}_2$ be two granules in our model and the q-rung picture fuzzy hyperedges $\mathscr{E}_1$, $\mathscr{E}_2$ represent their external properties. The *difference* between two granules $\mathscr{G}_1 - \mathscr{G}_2$ is defined as a smaller q-rung picture fuzzy hyperedge that contains those vertices belonging to $\mathscr{E}_1$ but not to $\mathscr{E}_2$.
Note that, if a vertex $x_i \in \mathscr{E}_1$ and $x_i \notin \mathscr{E}_2$, then $T_{\mathscr{E}_1}(x_i) > 0$, $N_{\mathscr{E}_1}(x_i) > 0$, $F_{\mathscr{E}_1}(x_i) < 1$ and $T_{\mathscr{E}_2}(x_i) = 0$, $N_{\mathscr{E}_2}(x_i) = 0$, $F_{\mathscr{E}_2}(x_i) = 1$.

Definition 7.25 A granule $\mathscr{G}_1$ is said to be the *sub-granule* of $\mathscr{G}_2$, if each vertex x_i of $\mathscr{E}_1$ also belongs to $\mathscr{E}_2$, i.e., $\mathscr{E}_1 \subseteq \mathscr{E}_2$. In such case, $\mathscr{G}_2$ is called the *super-granule* of $\mathscr{G}_1$.

Note that, if $\mathscr{E}(x_i) = \{0, 1\}$, then all the above-described operators are reduced to classical hypergraphs theory of granular computing.

7.3.2   The Construction of Hierarchical Structures

As earlier, we have constructed a granular structure of a specific level as a q-rung picture fuzzy hypergraph. In this way, we can interpret a problem in distinct levels of granularities. Hence, these granular structures at different levels produce a set of q-rung picture fuzzy hypergraphs. The upper set of these hypergraphs constructs a hierarchical structure in distinct levels. The relationships between granules are expressed by lower level, which represents the problem as a concrete example of granularity. The relationships between granule sets are expressed by higher level, which represents the problem as an abstract example of granularity. Thus, the single-level structures can be constructed and then can be subdivided into hierarchical structures using the relational mappings between different levels.

Definition 7.26 Let $H_1 = (\mathscr{R}_1, S_1)$ and $H_2 = (\mathscr{R}_2, S_2)$ be two q-rung picture fuzzy hypergraphs. In a hierarchy structure, their level cuts are $H_1^{(\alpha,\beta,\gamma)}$ and $H_2^{(\alpha,\beta,\gamma)}$, respectively. Let $(\alpha, \beta, \gamma) \in [0, 1]$ and $T_{\mathscr{E}_i^1} \geq \alpha$, $N_{\mathscr{E}_i^1} \geq \beta$, $F_{\mathscr{E}_i^1} \leq \gamma$, where $\mathscr{E}_i^1 \in S_1$, then a mapping $\phi : H_1^{(\alpha,\beta,\gamma)} \to H_2^{(\alpha,\beta,\gamma)}$ from $H_1^{(\alpha,\beta,\gamma)}$ to $H_2^{(\alpha,\beta,\gamma)}$ maps the $\mathscr{E}_i^{1(\alpha,\beta,\gamma)}$ in $H_1^{(\alpha,\beta,\gamma)}$ to a vertex x_i^2 in $H_2^{(\alpha,\beta,\gamma)}$. Furthermore, the mapping $\phi^{-1} : H_2^{(\alpha,\beta,\gamma)} \to H_1^{(\alpha,\beta,\gamma)}$ maps a vertex x_i^2 in $H_2^{(\alpha,\beta,\gamma)}$ to (α, β, γ)-cut of q-rung picture fuzzy hyperedge $\mathscr{E}_i^{1(\alpha,\beta,\gamma)}$ in $H_1^{(\alpha,\beta,\gamma)}$. It can be denoted as $\phi(\mathscr{E}_i^{1(\alpha,\beta,\gamma)}) = x_i^2$ or $\phi^{-1}(x_i^2) = \mathscr{E}_i^{1(\alpha,\beta,\gamma)}$, for $1 \leq i \leq n$.

The q-rung picture fuzzy hypergraph H_1 possesses the finer granularity than H_2, so H_1 is referred to as the finer granularity and H_2 as the coarser granularity.

In a q-rung picture fuzzy hypergraph model, the mappings are used to describe the relations among different levels of granularities. At each distinct level, the problem is interpreted w.r.t the q-rung picture fuzzy granularity of that level. The mapping associates the different descriptions of the same problem at distinct levels of granularities. There are two fundamental types to construct the method of hierarchical structures, the *top-down construction procedure* and *the bottom-up construction procedure* [31]. The possible method for *the bottom-up construction* is described in Algorithm 7.3.1.

Algorithm 7.3.1

The bottom-up construction

Input: An undirected q-rung picture fuzzy hypergraph H.
Output: The bottom-up construction of granular structures.
1. Combine the vertices having some relationship $r \in R$ into a unit U_j, according to the set
   of relations $R = \{r_1, r_2, r_3, \cdots, r_n\}$.
2. Compute the membership degrees through the functions $T(x)$, $N(x)$ and $F(x)$ of each unit U_j,
   which are q-rung picture fuzzy hyperedges and regarded as granules.
3. Now formulate the i-level of the model using the following steps.
4. Input the number of q-RPF hyperedges or granules k.
5. Fix parameters α, β and γ.
6.   **for** $\alpha, \beta, \gamma \in [0, 1]$
7.   **do** s from 1 to k
8.   **if** $(T_{U_s}(a) \geq \alpha, N_{U_s}(a) \geq \beta, F_{U_s}(a) \leq \gamma)$ **then**
9. $a \in U_s^{(\alpha, \beta, \gamma)}$
10.   **end if**
11.   **if** $(U_s^{(\alpha, \beta, \gamma)} \neq \emptyset)$ **then**
12. $U_s^{(\alpha, \beta, \gamma)} \in H_i^{(\alpha, \beta, \gamma)}$
13.   **end if**
14.   **print***, $H_i^{(\alpha, \beta, \gamma)}$ is the i-level of H.
15.   **end do**
16.   **end for**
17. A single level of bottom-up construction is constructed.
18. Perform the steps form 1 to 16 repeatedly to construct the $i + 1$-level of granularity.
19. Except the last level, each level is mapped to the next level using some operators.
20. Step 1– Step 16 are performed repeatedly until the complete set is formulated to a single granule.

The method of bottom-up construction is explained by the following example.

Example 7.4 Let $H = (\mathscr{R}, S)$ be a 7-rung picture fuzzy hypergraph as shown in Fig. 7.9. Let $X = \{x_1, x_2, x_3, x_4, x_5, x_6, x_7, x_8, x_9, x_{10}, x_{11}, x_{12}\}$ and $S = \{\mathscr{E}_1, \mathscr{E}_2, \mathscr{E}_3, \mathscr{E}_4, \mathscr{E}_5\}$.

For $\alpha = 0.5$, $\beta = 0.5$ and $\gamma = 0.6$, the $(0.5, 0.5, 0.6)$-level hypergraph of H is given in Fig. 7.10.

By considering the fixed α, β, γ and following the Algorithm 7.3.1, the bottom-up construction of this model is given in Fig. 7.11.

In granular computing, more than one hierarchical structure are considered to emphasize the multiple approaches. These different hierarchical structures can be formed

Fig. 7.9 A 7-rung picture
fuzzy hypergraph

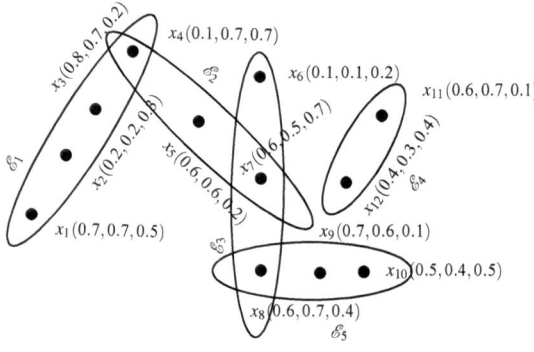

Fig. 7.10 $(0.5, 0.5, 0.6)$-
level hypergraph of H

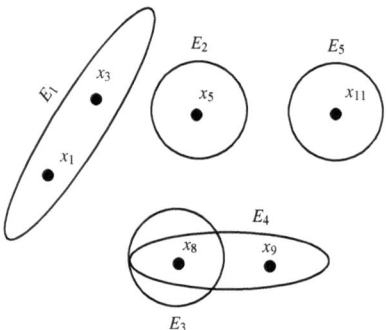

by considering the different interpretations of relations set R. Every hierarchical
structure is a distinct aspect of the problem. A multiple or a multilevel model of
granular computing based on hypergraph is formed by combining these different
hierarchical structures. In a hypergraph model, a series of correlated hypergraphs is
used to present every hierarchical structure. The mapping relates the distinct levels in
a hierarchical structure and each hypergraph denotes a particular level in that structure. Each specific level has multiple hyperedges and each hyperedge contains the
set of objects having similar attributes. An example to construct a model of granular
computing based on fuzzy hypergraph is illustrated in [27]. We now extend the same
example to construct a q-rung picture fuzzy hypergraph model of granular computing
to illustrate the validity and flexibility of our model.

Example 7.5 Consider an express hyper network, where the vertices represent the
express corporations. These vertices are combined together in one unit U according
to the relation set R and possessing some type of relation among them. To form a 6-
rung picture fuzzy hyperedge, which is also called a granule, we calculate and assign
the membership degrees to each unit. Each 6-rung picture fuzzy hyperedge denotes
a shop demanding express services and the vertices containing in that hyperedge are
the express corporations serving that shop. There are ten $\{x_1, x_2, x_3, x_4, x_5, x_6, x_7, x_8,$

Fig. 7.11   The bottom-up
construction procedure

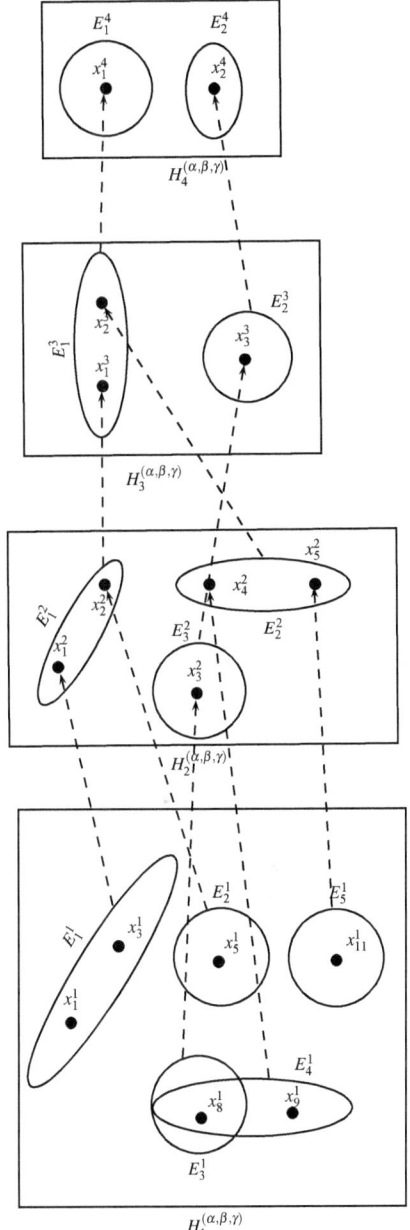

Table 7.2 Incidence matrix of express hypernetwork

	Shop1	Shop2	Shop3	Shop4	Shop5
x_1	$(0.8, 0.2, 0.1)$	$(0, 0, 1)$	$(0, 0, 1)$	$(0, 0, 1)$	$(0, 0, 1)$
x_2	$(0.9, 0.3, 0.5)$	$(0, 0, 1)$	$(1, 0, 0)$	$(0, 0, 1)$	$(0, 0, 1)$
x_3	$(0.3, 0.5, 0.9)$	$(0.6, 0.5, 0.2)$	$(0, 0, 1)$	$(0, 0, 1)$	$(0, 0, 1)$
x_4	$(0.8, 0.2, 0.1)$	$(0, 0, 1)$	$(0, 0, 1)$	$(0, 0, 1)$	$(0, 0, 1)$
x_5	$(0, 0, 1)$	$(0.3, 0.5, 0.9)$	$(0, 0, 1)$	$(0, 0, 1)$	$(0, 0, 1)$
x_6	$(0, 0, 1)$	$(0.7, 0.2, 0.3)$	$(0, 0, 1)$	$(0, 0, 1)$	$(0, 0, 1)$
x_7	$(0, 0, 1)$	$(0.5, 0.3, 0.6)$	$(0, 0, 1)$	$(0.1, 0.2, 0.8)$	$(0, 0, 1)$
x_8	$(0, 0, 1)$	$(0, 0, 1)$	$(0, 0, 1)$	$(0.6, 0.5, 0.2)$	$(0, 0, 1)$
x_9	$(0, 0, 1)$	$(0, 0, 1)$	$(0.6, 0.5, 0.2)$	$(0.7, 0.2, 0.3)$	$(0.4, 0.7, 0.5)$
x_{10}	$(0, 0, 1)$	$(0, 0, 1)$	$(0, 0, 1)$	$(0, 0, 1)$	$(0.5, 0.2, 0.7)$

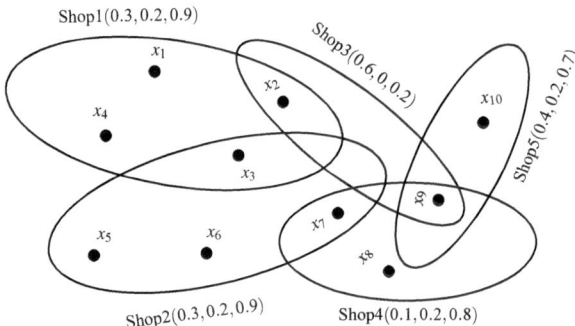

Fig. 7.12 A 6-rung picture fuzzy hypergraph model of express corporations

$x_9, x_{10}\}$ express corporations and the corresponding incidence matrix of this model is given in Table 7.2.

The corresponding 6-rung picture fuzzy hypergraph is shown in Fig. 7.12.

A 6-rung picture fuzzy hypergraph model of granular computing illustrates an uncertain set of objects with specific positive, neutral, and negative degrees, which incorporate the undetermined behavior of an element to its granule. In this example, there are five shops requiring the services of express corporation. The membership degrees of each corporation to distinct shops are different because every shop selects a corporation by considering different factors, including consignment, life span, limited liability, scale of that express corporation, the distance of receiving and mailing parcels, and so on. Note that, using a q-rung picture fuzzy hypergraph in granular computing is meaningful and more flexible as compared to fuzzy hypergraph because it also deals with negative membership and neutral behavior of objects toward their granules.

7.3.3 Zooming-In and Zooming-Out Operators

A formal discussion is provided to interpret a q-rung picture fuzzy hypergraph model in granular computing, which is more compatible to human thinking. Zhang and Zhang [35] highlighted that one of the most important and acceptable characteristics of human intelligence is that the same problem can be viewed and analyzed in different granularities. Their claim is that the problem cannot only be solved using various world of granularities but also can be switched easily and quickly. Hence, the procedure of solving a problem can be considered as the calculations in different hierarchies within that model.

A multilevel granularity of the problem is represented by a q-rung picture fuzzy hypergraph model, which allows the problem solvers to decompose it into various minor problems and transform it in other granularities. The transformation of problem in other granularities is performed by using two operators, i.e, zooming-in and zooming-out operators. The transformation from weaker level to finer level of granularity is done by zoom-in operator and the zoom-out operator deals with the shifting of problem from coarser to finer granularity.

Definition 7.27 Let $H_1 = (\mathscr{R}_1, S_1)$ and $H_2 = (\mathscr{R}_2, S_2)$ be two q-rung picture fuzzy hypergraphs, which are considered as two levels of hierarchical structures and H_2 owns the coarser granularity than H_1. Suppose $H_1^{(\alpha,\beta,\gamma)} = (X_1, E_1^{(\alpha,\beta,\gamma)})$ and $H_2^{(\alpha,\beta,\gamma)} = (X_2, E_2^{(\alpha,\beta,\gamma)})$ are the corresponding (α, β, γ)-level hypergraphs of H_1 and H_2, respectively. Let $e_i^1 \in E_1^{(\alpha,\beta,\gamma)}$, $x_j^1 \in X_1$, $e_j^2 \in E_2^{(\alpha,\beta,\gamma)}$, $x_l^2, x_m^2 \in X_2$, and $x_l^2, x_m^2 \in e_j^2$. If $\phi(e_i^1) = x_l^2$, then $r(x_j^1, x_m^2)$ is the relationship between x_j^1 and x_m^2 and is obtained by the characteristics of granules.

Definition 7.28 Let the hyperedge $\phi^{-1}(x_l)$ be a vertex in a new level and the relation between hyperedges in this level is same as that of relationship between vertices in previous level. This is called the *zoom-in operator* and transforms a weaker level to a stronger level. The function $r(x_j^1, x_m^2)$ defines the relation between vertices of original level as well as new level.

Let the vertex $\phi(e_i)$ be a hyperedge in a new level and the relation between vertices in this level is same as that of relationship between hyperedges in corresponding level. This is called the *zoom-out operator* and transforms a finer level to a coarser level.

By using these zoom-in and zoom-out operators, a problem can be viewed at multilevels of granularities. These operations allow us to solve the problem more appropriately and granularity can be switched easily at any level of problem-solving.

In a q-rung picture fuzzy hypergraph model of granular computing, the membership degrees of elements reflect the actual situation more efficiently and a wide variety of complicated problems in uncertain and vague environments can be presented by means of q-rung picture fuzzy hypergraphs. The previous analysis concludes that this model of granular computing generalizes the classical hypergraph model and fuzzy hypergraph model.

We now construct a hypergraph model of granular computing based on q-rung picture fuzzy equivalence relation.

7.4   A Level Hypergraph Partition Model

A fuzzy (or crisp) partition of universe set X is determined by fuzzy equivalence relation (or crisp equivalence relation) and generates a family of fuzzy equivalence classes (or crisp equivalence classes) [27, 31]. A partition model of granular computing using level hypergraph which is based on [27] is proposed. In this partition model, the blocks/subsets in a partition are represented by hyperedges and the objects having q-rung picture fuzzy equivalence relation R are contained in the same block. Let π_R be the partition of X induced by R then the q-rung picture fuzzy equivalence relation R_π is defined as, $x R_\pi y$ if and only if they are contained in the similar block/subset of partition π.

Definition 7.29 A system (X, R^ρ), $\rho = (\alpha, \beta, \gamma)$ is called an *object space*, where R^ρ is a non-empty collection of equivalence relations between the elements of X. $R^\rho = \{r_1^\rho, r_2^\rho, r_3^\rho, \ldots, r_n^\rho\}$, $n = |X|$. For $i \leq n$, $r_i^\rho \in R^\rho$, $r_i^\rho \subseteq X \times X \times X \cdots \times X$, if $(x_1, x_2, \ldots, x_i) \subseteq r_i^\rho$, then there exists an i-array relation r_i^ρ on $(x_1, x_2, \ldots, x_i)$.

Example 7.6 Let $X = \{x_1, x_2, x_3, x_4, x_5\}$ and $R^\rho = \{r_1^\rho, r_2^\rho, r_3^\rho, r_4^\rho, r_5^\rho\}$, for $0.7 < \alpha \leq 0.9, 0.5 < \beta \leq 0.5, 0.1 > \gamma \geq 0.1$, as shown in Example 7.2, $r_1^\rho = \{(x_1), (x_2), (x_3)\}$, $r_2^\rho = \{(x_4, x_5)\}$, $r_3^\rho, r_4^\rho, r_5^\rho = \{\emptyset\}$.

The q-rung picture fuzzy equivalence relations between the vertices are obtained from the actual situation and then the vertices having some relation are combined to a hyperedge. After the integration of all vertices to hyperedges is done, a single-level hypergraph model is formed. The construction of level hypergraph model is illustrated through the following example, which is the extension of example given in [27].

Example 7.7 Consider an express hypernetwork, as discussed in Example 7.5, we consider the same hypernetwork to illustrate the construction of level hypergraph model. Ten express corporations are considered and suppose that each corporation experiences six attributes. These attributes are *transit time, freight, tracking and circulation information, time window service, the distance of taking and mailing parcels and scale of the corresponding corporation* and are denoted by $A_1, A_2, A_3, A_4, A_5,$ and A_6, respectively. To indicate the values of attributes, 0 and 1 are used, as shown in Table 7.3.

The information about express corporations having these attributes is given in Table 7.4.

Table 7.3 Description of attributes

Attributes				
Transit time	Short	1	Long	0
Freight	Low	1	High	0
Tracking and circulation information	Existence	1	Non existence	0
Time window service	Existence	1	Non existence	0
The distance of mailing and taking parcels	Close	1	Away	0
Scale of the express corporation	Large	1	Small	0

Table 7.4 Corporations having attributes

X	Corporations	A_1	A_2	A_3	A_4	A_5	A_6
x_1	Express1	1	0	0	1	0	0
x_2	Express2	1	0	0	1	0	0
x_3	Express3	1	1	0	1	0	0
x_4	Express4	1	1	0	1	1	0
x_5	Express5	1	1	0	0	0	1
x_6	Express6	1	1	0	0	0	0
x_7	Express7	1	1	1	0	0	0
x_8	Express8	1	0	1	0	0	0
x_9	Express9	1	0	1	0	1	0
x_{10}	Express10	0	0	1	0	1	0

The positive membership function $T(x_i) \in [0, 1]$ represents the beneficial and helpful relationships between these corporations and this affirmative relationship is described through the following matrix.

$$
T_{\tilde{M}_R} = \begin{bmatrix}
1 & 1 & 0.67 & 0.55 & 0.25 & 0.33 & 0.25 & 0.33 & 0.25 & 0 \\
1 & 1 & 0.67 & 0.55 & 0.25 & 0.33 & 0.25 & 0.33 & 0.25 & 0 \\
0.67 & 0.67 & 1 & 0.75 & 0.55 & 0.66 & 0.55 & 0.25 & 0.20 & 0 \\
0.55 & 0.55 & 0.75 & 1 & 0.40 & 0.40 & 0.40 & 0.20 & 0.40 & 0.20 \\
0.25 & 0.25 & 0.55 & 0.40 & 1 & 0.67 & 0.55 & 0.25 & 0.20 & 0 \\
0.33 & 0.33 & 0.67 & 0.40 & 0.67 & 1 & 0.67 & 0.33 & 0.25 & 0 \\
0.25 & 0.25 & 0.55 & 0.40 & 0.55 & 0.67 & 1 & 0.67 & 0.55 & 0.25 \\
0.33 & 0.33 & 0.25 & 0.20 & 0.25 & 0.33 & 0.67 & 1 & 0.67 & 0.33 \\
0.25 & 0.25 & 0.20 & 0.40 & 0.20 & 0.25 & 0.55 & 0.67 & 1 & 0.67 \\
0 & 0 & 0 & 0.20 & 0 & 0 & 0.25 & 0.33 & 0.67 & 1
\end{bmatrix}.
$$

The neutral membership function $N(x_i) \in [0, 1]$ represents the unbiased and vague relationships between these corporations and this undeterminate behavior is described through the following matrix.

$$
N_{\tilde{M}_R} = \begin{bmatrix}
0 & 0 & 0.23 & 0.31 & 0.31 & 0.12 & 0.31 & 0.12 & 0.31 & 0 \\
0 & 0 & 0.23 & 0.31 & 0.31 & 0.12 & 0.31 & 0.12 & 0.31 & 0 \\
0.23 & 0.23 & 0 & 0.23 & 0.31 & 0.33 & 0.31 & 0.31 & 0.12 & 0 \\
0.31 & 0.31 & 0.23 & 0 & 0.20 & 0.20 & 0.20 & 0.12 & 0.20 & 0.12 \\
0.31 & 0.31 & 0.31 & 0.20 & 0 & 0.23 & 0.31 & 0.31 & 0.12 & 0 \\
0.12 & 0.12 & 0.23 & 0.20 & 0.23 & 0 & 0.23 & 0.12 & 0.31 & 0 \\
0.31 & 0.31 & 0.31 & 0.20 & 0.31 & 0.23 & 0 & 0.23 & 0.31 & 0.31 \\
0.12 & 0.12 & 0.31 & 0.12 & 0.31 & 0.12 & 0.23 & 0 & 0.23 & 0.12 \\
0.31 & 0.31 & 0.12 & 0.20 & 0.12 & 0.31 & 0.31 & 0.23 & 0 & 0.23 \\
0 & 0 & 0 & 0.12 & 0 & 0 & 0.31 & 0.12 & 0.23 & 0
\end{bmatrix}.
$$

The negative membership function $F(x_i) \in [0, 1]$ represents the contrary and competent relationships between these corporations and this opposing relationship is described through the following matrix.

$$
F_{\tilde{M}_R} = \begin{bmatrix}
0 & 0 & 0.32 & 0.45 & 0.65 & 0.65 & 0.65 & 0.65 & 0.65 & 1 \\
0 & 0 & 0.32 & 0.45 & 0.65 & 0.65 & 0.65 & 0.65 & 0.65 & 1 \\
0.32 & 0.32 & 0 & 0.25 & 0.45 & 0.23 & 0.45 & 0.65 & 0.76 & 1 \\
0.45 & 0.45 & 0.25 & 0 & 0.50 & 0.50 & 0.50 & 0.76 & 0.50 & 0.76 \\
0.65 & 0.65 & 0.45 & 0.50 & 0 & 0.32 & 0.45 & 0.65 & 0.76 & 1 \\
0.65 & 0.65 & 0.32 & 0.50 & 0.32 & 0 & 0.32 & 0.65 & 0.65 & 1 \\
0.65 & 0.65 & 0.45 & 0.50 & 0.45 & 0.32 & 0 & 0.32 & 0.45 & 0.65 \\
0.65 & 0.65 & 0.65 & 0.76 & 0.65 & 0.65 & 0.32 & 0 & 0.32 & 0.65 \\
0.65 & 0.65 & 0.76 & 0.50 & 0.76 & 0.65 & 0.45 & 0.32 & 0 & 0.32 \\
1 & 1 & 1 & 0.76 & 1 & 1 & 0.65 & 0.65 & 0.32 & 0
\end{bmatrix}.
$$

A 5-rung picture fuzzy relation matrix can be obtained by combining positive membership, neutral membership, and negative membership degrees as shown in the following matrix.

$$
\tilde{M}_R = \begin{pmatrix}
(1,0,0) & (1,0,0) & \cdots & (0,0,1) \\
(1,0,0) & (1,0,0) & \cdots & (0,0,1) \\
(0.67,0.23,0.32) & (0.67,0.23,0.32) & \cdots & (0,0,1) \\
\vdots & \vdots & \vdots & \vdots \\
(0,0,1) & (0,0,1) & \cdots & (1,0,0)
\end{pmatrix}.
$$

The positive $T(m_{ij})$, neutral $N(m_{ij})$, and negative $F(m_{ij})$ degrees of each m_{ij} entry of the above matrix describe the healthy, neutral, and rival relationships among the x_i and x_j corporations, respectively. The transitive closure of the above matrix is computed using the method described in [18] as shown in the matrix below. The transitive closure of 5-rung fuzzy equivalence relation is also an equivalence relation.

Fig. 7.13 A single level of hypergraph model

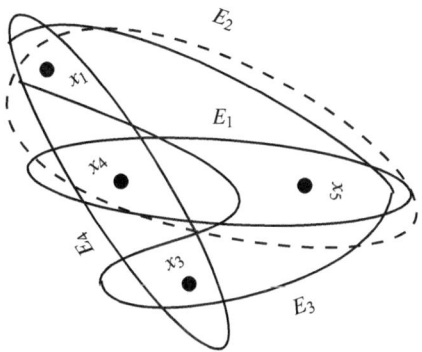

$$
\begin{pmatrix}
1 & (0.40,0.20,0.50) & (0.67,0.23,0.32) & (0.55,0.31,0.45) & (0.55,0.31,0.45) & 0 & 0 & 0 & 0 & 0 \\
(0.40,0.20,0.50) & 1 & (0.40,0.20,0.50) & (0.40,0.20,0.50) & (0.40,0.20,0.50) & 0 & 0 & 0 & 0 & 0 \\
(0.67,0.23,0.32) & (0.40,0.20,0.50) & 1 & (0.55,0.31,0.45) & (0.55,0.31,0.45) & 0 & 0 & 0 & 0 & 0 \\
(0.55,0.31,0.45) & (0.40,0.20,0.50) & (0.55,0.31,0.45) & 1 & (0.75,0.23,0.32) & 0 & 0 & 0 & 0 & 0 \\
(0.55,0.31,0.45) & (0.40,0.20,0.50) & (0.55,0.31,0.45) & (0.75,0.23,0.32) & 1 & 0 & 0 & 0 & 0 & 0 \\
0 & 0 & 0 & 0 & 0 & 1 & 0 & (0.33,0.12,0.65) & (0.20,0.12,0.76) & (0.20,0.12,0.76) \\
0 & 0 & 0 & 0 & 0 & 0 & 1 & 0 & 0 & 0 \\
0 & 0 & 0 & 0 & 0 & (0.33,0.12,0.65) & 0 & 1 & (0.20,0.12,0.76) & (0.20,0.12,0.76) \\
0 & 0 & 0 & 0 & 0 & (0.20,0.12,0.76) & 0 & (0.20,0.12,0.76) & 1 & (0.25,0.31,0.65) \\
0 & 0 & 0 & 0 & 0 & (0.20,0.12,0.76) & 0 & (0.20,0.12,0.76) & (0.25,0.31,0.65) & 1 \\
\end{pmatrix}
$$

Note that, here $\mathbf{0} = (0, 0, 1)$ and $\mathbf{1} = (1, 0, 0)$. The positive, neutral, and negative value domains are given as follows, $T_D = \{0.75, 0.67, 0.55, 0.40, 0.33, 0.25, 0.20\}$, $N_D = \{0.31, 0.23, 0.20, 0.12\}$, $F_D = \{0.76, 0.65, 0.50, 0.45, 0.32\}$. Let $0.20 < \alpha \leq 0.25$, $0.12 < \beta \leq 0.20$ and $0.32 < \gamma \leq 0.45$ and their corresponding hierarchical quotient space is given as follows, $X/R^{(\alpha,\beta,\gamma)} = \{(x_4, x_5), (x_1, x_4, x_5), (x_1, x_3, x_5), (x_1, x_3, x_4)\}$.

Thus, we can conclude that $r_1 = \{\emptyset\}$, $r_2 = \{(x_4, x_5)\}$, $r_3 = \{(x_1, x_4, x_5), (x_1, x_3, x_5), (x_1, x_3, x_4)\}$, $r_4 = r_5 = r_6 = r_7 = r_8 = r_9 = r_{10} = \{\emptyset\}$. We obtain four hyperedges $E_1 = \{x_4, x_5\}$, $E_2 = \{x_1, x_4, x_5\}$, $E_3 = \{x_1, x_3, x_5\}$, $E_4 = \{x_1, x_3, x_4\}$. Thus, we have constructed a single level of hypergraph model as shown in Fig. 7.13.

It is noted that a q-rung picture fuzzy equivalence relation determines the partition of domain set into different layers. All q-rung fuzzy equivalence relations, which are isomorphic, can also determine the same classification.

Definition 7.30 [14] Let H_1 and H_2 be two crisp hypergraphs. Suppose that H_1 owns the finer q-rung picture fuzzy granularity than H_2. A mapping from H_1 to H_2 $\psi : H_1 \to H_2$ maps a hyperedge of H_1 to the vertex of H_2 and the mapping $\psi^{-1} : H_2 \to H_1$ maps a vertex of H_2 to the hyperedge of H_1.

The procedure of bottom-up construction for level hypergraph model is illustrated in Algorithm 7.4.1.

Algorithm 7.4.1

The bottom-up construction for level hypergraphs

1. Input the number of vertices m of universal set X.
2. Input a q-rung picture fuzzy equivalence relation matrix $R=[r_{ij}]_{m \times m}$ of membership degrees.
3. Determine the value domain D of R as follows:
4. 　　　**do** i from 1 to m
5. 　　　　　**do** j from 1 to m
6. 　　　　　　　$x_i, y_j \in X$
7. 　　　　　　　**if** $(T_X(x_i) \wedge T_X(y_j) \wedge T_R(x_i, y_j) > 0,\ N_X(x_i) \wedge N_X(y_j) \wedge N_R(x_i, y_j) > 0,$
8. 　　　　　　　　$F_X(x_i) \vee F_X(y_j) \vee F_R(x_i, y_j) > 0)$ **then**
9. 　　　　　　　　$(x_i, y_j) \in D$
10. 　　　　　**end if**
11. 　　　　　　　**print***, D is the value domain of R.
12. Fix the parameters $\alpha,\ \beta,\ \gamma \in [0, 1]$, where $\alpha,\ \beta,\ \gamma \in D$.
13. 　　　　**if** $(T_R(x_i, y_j) \geq \alpha,\ N_R(x_i, y_j) \geq \beta,\ F_R(x_i, y_j) \leq \gamma)$ **then**
14. 　　　　　$(x_i, y_j) \in R^{(\alpha, \beta, \gamma)}$
15. 　　　**end if**
16. 　　　　　**print***, $\{X/R^{(\alpha, \beta, \gamma)}|(\alpha, \beta, \gamma) \in D\}$ is q-rung picture fuzzy hierarchical quotient
　　　　　　space structure of R.
17. 　　　　　**end do**
18. 　　　**end do**
19. 　　　**end for**
20. The subsets $\{X/R^{(\alpha, \beta, \gamma)}|(\alpha, \beta, \gamma) \in D\}$ of q-rung picture fuzzy hierarchical quotient space structure
　　　corresponds to the hyperedges or granules of level hypergraph.
21. The k-level of granularity is constructed.
22. These hyperedges or granules in k-level are mapped to $(k + 1)$-level.
23. Step 1 - Step 22 are repeated until the whole universe is formulated to a single granule.
24. Except the last level, each level is mapped to the next level of granularity through
　　　different operators.

Definition 7.31 Let R be a q-rung picture fuzzy equivalence relation on X. A *coarse gained universe* $X/R^{(\alpha, \beta, \gamma)}$ can be obtained by using q-rung picture fuzzy equivalence relation, where $[x_i]_{R^{\alpha, \beta, \gamma}} = \{x_j \in X | x_i R x_j\}$. This equivalence class $[x_i]_{R^{(\alpha, \beta, \gamma)}}$ is considered as an hyperedge in the level hypergraph.

Definition 7.32 Let $H_1 = (X_1, E_1)$ and $H_2 = (X_2, E_2)$ be level hypergraphs of q-rung picture fuzzy hypergraphs and H_2 has weaker granularity than H_1. Suppose that $e_i^1,\ e_j^2 \in E_1$ and $x_i^2,\ x_j^2 \in X_2$, $i, j = 1, 2, \ldots, n$. The *zoom-in operator* $\kappa : H_2 \rightarrow H_1$ is defined as $\kappa(x_i^2) = e_i^1,\ e_i^1 \in E_1$. The relations between the vertices of H_2 define the relationships among the hyperedges in new level. The zoom-in operator of two levels is shown in Fig. 7.14.

Remark 7.1 For all $X_2',\ X_2'' \subseteq X_2$, we have $\kappa(X_2') = \bigcup_{x_i^2 \in X_2'} \kappa(x_i^2)$ and $\kappa(X_2'') = \bigcup_{x_j^2 \in X_2''} \kappa(x_j^2)$.

Theorem 7.3 *Let $H_1 = (X_1, E_1)$ and $H_2 = (X_2, E_2)$ be two levels and $\kappa : H_2 \rightarrow H_1$ be the zoom-in operator. Then, for all $X_2',\ X_2'' \subseteq X_2$, the zoom-in operator satisfies*

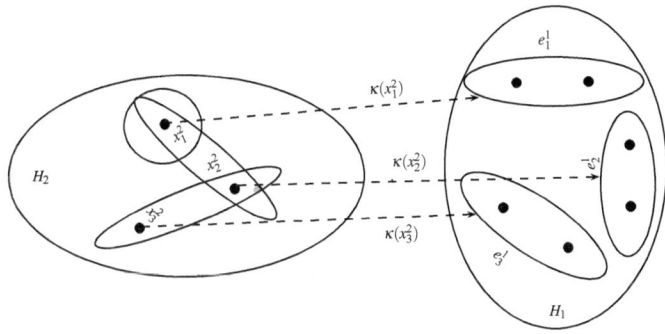

Fig. 7.14 The zoom-in operator

(i) κ maps the empty set to an empty set, i.e., $\kappa(\emptyset) = \emptyset$.

(ii) $\kappa(X_2) = E_1$.

(iii) $\kappa([X_2']^c) = [\kappa(X_2')]^c$.

(iv) $\kappa(X_2' \cap X_2'') = \kappa(X_2') \cap \kappa(X_2'')$.

(v) $\kappa(X_2' \cup X_2'') = \kappa(X_2') \cup \kappa(X_2'')$.

(vi) $X_2' \subseteq X_2''$ if and only if $\kappa(X_2') \subseteq \kappa(X_2'')$.

Proof (i) It is trivially satisfied that $\kappa(\emptyset) = \emptyset$.

(ii) As we know that for all $x_i^2 \in X_2$, we have $\kappa(X_2') = \bigcup_{x_i^2 \in X_2'} \kappa(x_i^2)$. Since $\kappa(x_i^2) =$

e_i^1, thus we have $\kappa(X_2') = \bigcup_{x_i^2 \in X_2'} \kappa(x_i^2) = \bigcup_{e_i^1 \in E_1} e_i^1 = E_1$.

(iii) Let $[X_2']^c = Z_2'$ and $[X_2'']^c = Z_2''$, then it is obvious that $Z_2' \cap X_2' = \emptyset$ and $Z_2' \cup X_2' = X_2$. It follows from (ii) that $\kappa(X_2) = E_1$ and we denote by W_1' that edge set of H_1 on which the vertex set Z_2' of H_2 is mapped under κ, i.e., $\kappa(Z_2') = W_1'$. Then $\kappa([X_2']^c) = \kappa(Z_2') = \bigcup_{x_i^2 \in Z_2'} \kappa(x_i^2) = \bigcup_{e_i^1 \in W_1'} e_i^1 = Z_1'$ and

$[\kappa(X_2')]^c = [\bigcup_{x_j^2 \in X_2'} \kappa(x_j^2)]^c = [\bigcup_{e_j^1 \in E_1'} e_j^1]^c = (E_1')^c$. Since, the relationship between

hyperedges in new level is same as that of relations among vertices in original level so we have $(E_1')^c = Z_1'$. Hence, we conclude that $\kappa([X_2']^c) = [\kappa(X_2')]^c$.

(iv) Assume that $X_2' \cap X_2'' = \tilde{X}_2$ then for all $x_i^2 \in \tilde{X}_2$ implies that $x_i^2 \in X_2'$ and $x_i^2 \in X_2''$. Further, we have $\kappa(X_2' \cap X_2'') = \kappa(\tilde{X}_2) = \bigcup_{x_i^2 \in \tilde{X}_2} \kappa(x_i^2) = \bigcup_{e_i^1 \in \tilde{E}_1} \kappa(e_i^1) = \tilde{E}_1$.

$\kappa(X_2') \cap \kappa(X_2'') = \{\bigcup_{x_i^2 \in X_2'} \kappa(x_i^2)\} \cap \{\bigcup_{x_j^2 \in X_2''} \kappa(x_j^2)\} = \bigcup_{e_i^1 \in E_1'} e_i^1 \cap \bigcup_{e_j^1 \in E_1''} e_j^1 = E_1' \cap E_1''$.

Since, the relationship between hyperedges in new level is same as that of relations among vertices in original level so we have $E_1' \cap E_1'' = \tilde{E}_1$. Hence, we conclude that $\kappa(X_2' \cap X_2'') = \kappa(X_2') \cap \kappa(X_2'')$.

(v) Assume that $X_2' \cup X_2'' = \bar{X}_2$. Then, we have $\kappa(X_2' \cup X_2'') = \kappa(\bar{X}_2) = \bigcup_{x_i^2 \in \bar{X}_2} \kappa(x_i^2)$

$$= \bigcup_{e_i^1 \in \bar{E}_1} \kappa(e_i^1) = \bar{E}_1.$$

$\kappa(X_2') \cup \kappa(X_2'') = \{ \bigcup_{x_i^2 \in X_2'} \kappa(x_i^2) \} \cup \{ \bigcup_{x_j^2 \in X_2''} \kappa(x_j^2) \} = \bigcup_{e_i^1 \in E_1'} e_i^1 \cup \bigcup_{e_j^1 \in E_1''} e_j^1 = E_1' \cup$

E_1''. Since the relationship between hyperedges in new level is same as that of relations among vertices in original level so we have $E_1' \cup E_1'' = \bar{E}_1$. Hence, we conclude that $\kappa(X_2' \cup X_2'') = \kappa(X_2') \cup \kappa(X_2'')$.

(vi) First we show that $X_2' \subseteq X_2''$ implies that $\kappa(X_2') \subseteq \kappa(X_2'')$. Since $X_2' \subseteq X_2''$, which implies that $X_2' \cap X_2'' = X_2'$ and $\kappa(X_2') = \bigcup_{x_i^2 \in X_2'} \kappa(x_i^2) = \bigcup_{e_i^1 \in E_1'} e_i^1 = E_1'$.

Also $\kappa(X_2'') = \bigcup_{x_j^2 \in X_2''} \kappa(x_j^2) = \bigcup_{e_j^1 \in E_1''} e_j^1 = E_1''$. Since, the relationship between

hyperedges in new level is same as that of relations among vertices in original level so we have $E_1' \subseteq E_1''$, i.e., $\kappa(X_2') \subseteq \kappa(X_2'')$. Hence, $X_2' \subseteq X_2''$ implies that $\kappa(X_2') \subseteq \kappa(X_2'')$.

We now prove that $\kappa(X_2') \subseteq \kappa(X_2'')$ implies that $X_2' \subseteq X_2''$. Suppose on contrary that whenever $\kappa(X_2') \subseteq \kappa(X_2'')$ then there is at least one vertex $x_i^2 \in X_2'$ but $x_i^2 \notin X_2''$, i.e., $X_2' \not\subseteq X_2''$. Since, $\kappa(x_i^2) = e_i^1$ and the relationship between hyperedges in new level is same as that of relations among vertices in original level so we have $e_i^1 \in E_1'$ but $e_i^1 \notin E_1''$, i.e., $E_1' \not\subseteq E_1''$, which is contradiction to the supposition. Thus, we have $\kappa(X_2') \subseteq \kappa(X_2'')$ implies that $X_2' \subseteq X_2''$. Hence, $X_2' \subseteq X_2''$ if and only if $\kappa(X_2') \subseteq \kappa(X_2'')$.

Definition 7.33 Let $H_1 = (X_1, E_1)$ and $H_2 = (X_2, E_2)$ be level hypergraphs of q-rung picture fuzzy hypergraphs and H_2 has weaker granularity than H_1. Suppose that $e_i^1, e_j^2 \in E_1$ and $x_i^2, x_j^2 \in X_2$, $i, j = 1, 2, \ldots, n$. The *zoom-out operator* $\sigma : H_1 \rightarrow H_2$ is defined as $\sigma(e_i^1) = x_i^2, x_i^2 \in X_2$. The zoom-out operator of two levels is shown in Fig. 7.15.

Theorem 7.4 *Let $\sigma : H_1 \rightarrow H_2$ be the zoom-out operator from $H_1 = (X_1, E_1)$ to $H_2 = (X_2, E_2)$ and let $E_1' \subseteq E_1$. Then, the zoom-out operator σ satisfies,*

(i) $\sigma(\emptyset) = \emptyset$.

(ii) σ maps the set of hyperedges of H_1 onto the set of vertices of H_2, i.e., $\sigma(E_1) = X_2$.

(iii) $\sigma([E_1']^c) = [\sigma(E_1')]^c$.

Proof (i) This part is trivially satisfied.

(ii) By applying the definition of σ, we have $\sigma(e_i^1) = x_i^2$. Since, the hyperedges define a partition of hypergraph so we have $E_1 = \{e_1^1, e_2^1, e_3^1, \ldots, e_n^1\} = \bigcup_{e_i^1 \in E_1} e_i^1$.

Then, $\sigma(E_1) = \sigma(\bigcup_{e_i^1 \in E_1} e_i^1) = \bigcup_{e_i^1 \in E_1} \sigma(e_i^1) = \bigcup_{x_i^2 \in X_2} x_i^2 = X_2$.

(iii) Assume that $[E_1']^c = V_1'$ then it is obvious that $E_1' \cap V_1' = \emptyset$ and $E_1' \cup V_1' = E_1$. Suppose on contrary that there exists at least one vertex $x_i^2 \in \sigma([E_1']^c)$ but $x_i^2 \notin$

$[\sigma(E_1')]^c$. $x_i^2 \in \sigma([E_1']^c)$ implies that $x_i^2 \in \sigma(V_1') \Rightarrow x_i^2 \in \bigcup_{e_i^1 \in V_1'} \sigma(e_i^1) \Rightarrow x_i^2 \in$

$\bigcup_{e_i^1 \in E_1 \setminus E_1'} \sigma(e_i^1)$. Since $x_i^2 \notin [\sigma(E_1')]^c \Rightarrow x_i^2 \in \sigma(E_1') \Rightarrow x_i^2 \in \bigcup_{e_i^1 \in E_1'} \sigma(e_i^1)$, which

is contradiction to our assumption. Hence, $\sigma([E_1']^c) = [\sigma(E_1')]^c$.

Definition 7.34 Let $H_1 = (X_1, E_1)$ and $H_2 = (X_2, E_2)$ be two levels of q-rung picture fuzzy hypergraphs and H_1 possesses the stronger granularity than H_2. Let $E_1' \subseteq E_1$ then $\hat{\sigma}(E_1') = \{e_i^2 | e_i^2 \in E_2, \kappa(e_i^2) \subseteq E_1'\}$ is called *internal zoom-out operator*.

The operator $\check{\sigma}(E_1') = \{e_i^2 | e_i^2 \in E_2, \kappa(e_i^2) \cap E_1' \neq \emptyset\}$ is called *external zoom-out operator*.

Example 7.8 Let $H_1 = (X_1, E_1)$ and $H_2 = (X_2, E_2)$ be two levels of q-rung picture fuzzy hypergraphs and H_1 possesses the stronger granularity than H_2, where

Fig. 7.15 The zoom-out operator

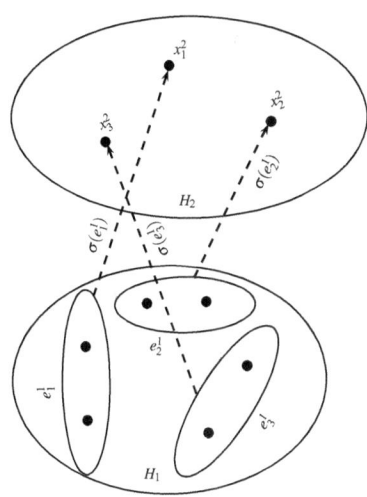

Fig. 7.16 The internal and external zoom-out operators

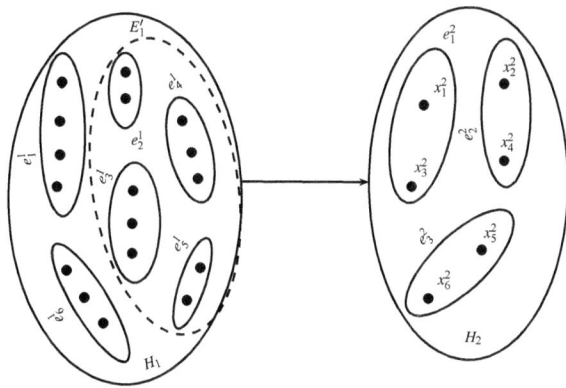

$E_1 = \{e_1^1, e_2^1, e_3^1, e_4^1, e_5^1, e_6^1\}$ and $E_2 = \{e_1^2, e_2^2, e_3^2\}$. Furthermore, $e_1^2 = \{x_1^2, x_3^2\}, e_2^2 = \{x_2^2, x_4^2\}, e_3^2 = \{x_5^2, x_6^2\}$ as shown in Fig. 7.16.

Let $E_1' = \{e_2^1, e_3^1, e_4^1, e_5^1\}$ be the subset of hyperedges of H_1 then we can not zoom-out to H_2 directly, thus by using the internal and external zoom-out operators we have the following relations.

$\hat{\sigma}(\{e_2^1, e_3^1, e_4^1, e_5^1\}) = \{e_2^2\},$
$\check{\sigma}(\{e_2^1, e_3^1, e_4^1, e_5^1\}) = \{e_1^2, e_2^2, e_3^2\}.$

References

1. Akram: Fuzzy Lie: algebras. Studies in Fuzziness and Soft Computing, vol. 9, pp. 1–302. Springer (2018)
2. Akram, A., Dar, J.M., Naz, S.: Certain graphs under Pythagorean fuzzy environment. Complex Intell. Syst. **5**(2), 127–144 (2019)
3. Akram, M., Ilyasa, F., Garg, H.: Multi-criteria group decision making based on ELECTRE I method in Pythagorean fuzzy information. Soft Comput. (2019). https://doi.org/10.1007/s00500-019-04105-0
4. Akram, M., Dar, J.M., Naz, S.: Pythagorean Dombi fuzzy graphs. Complex Intell. Syst. (2019). https://doi.org/10.1007/s40747-019-0109-0
5. Akram, M., Habib, A., Davvaz, B.: Direct sum of n Pythagorean fuzzy graphs with application to group decision-making. J. Mult.-Valued Log. Soft Comput. 1–41 (2019)
6. Akram, M., Naz, S., Davvaz, B.: Simplified interval-valued Pythagorean fuzzy graphs with application. Complex Intell. Syst. **5**(2), 229–253 (2019)
7. Akram, M., Ilyas, F., Saeid, A.B.: Certain notions of Pythagorean fuzzy graphs. J. Intell. Fuzzy Syst. **36**(6), 5857–5874 (2019)
8. Akram, M., Dudek, W.A., Ilyas, F.: Group decision making based on Pythagorean fuzzy TOPSIS method. Int. J. Intell. Syst. **34**(7), 1455–1475 (2019)
9. Akram, M., Habib, A., Koam, A.N.: A novel description on edge-regular q-rung picture fuzzy graphs with application. Symmetry **11**(4), 489 (2019). https://doi.org/10.3390/sym110
10. Akram, M., Habib, A.: q-rung picture fuzzy graphs: a creative view on regularity with applications. J. Appl. Math. Comput. (2019). https://doi.org/10.1007/s12190-019-01249-y
11. Ali, M.I.: Another view on q-rung orthopair fuzzy sets. Int. J. Intell. Syst. **33**, 2139–2153 (2018)
12. Atanassov. K.T.: Intuitionistic fuzzy sets. VII ITKR's Session, Sofia (deposed in Central Science-Technical Library of Bulgarian Academy of Science, 1697/84) (1983) (in Bulgarian)
13. Berge, C.: Graphs and Hypergraphs. North-Holland, Amsterdam (1973)
14. Chen, G., Zhong, N., Yao, Y.: A hypergraph model of granular computing. In: IEEE International Conference on Granular Computing, pp. 130–135 (2008)
15. Cuong, B.C.: Picture fuzzy sets. J. Comput. Sci. Cybern. **30**(4), 409 (2014)
16. Cuong, B.C., Kreinovich, V.: Picture fuzzy sets—a new concept for computational intelligence problems. In: 2013 Third World Congress on Information and Communication Technologies, pp. 1–6 (2013)
17. Habib, A., Akram, M.M.: Farooq, q-rung orthopair fuzzy competition graphs with application in the soil ecosystem. Mathematics **7**(1), 91 (2019). https://doi.org/10.3390/math70100
18. Lee, H.S.: An optimal algorithm for computing the maxmin transitive closure of a fuzzy similarity matrix. Fuzzy Sets Syst. **123**, 129–136 (2001)
19. Li, L., Zhang, R., Wang, J., Shang, X., Bai, K.: A novel approach to multi-attribute group decision-making with q-rung picture linguistic information. Symmetry **10**(5), 172 (2018)

20. Lin, T.Y.: Granular computing. In: Announcement of the BISC Special Interest Group on Granular Computing (1997)
21. Liu, Q., Jin, W.B., Wu, S.Y., Zhou, Y.H.: Clustering research using dynamic modeling based on granular computing. In: Proceeding of IEEE International Conference on Granular Computing, pp. 539–543 (2005)
22. Luqman, A., Akram, M., Al-Kenani, A.N.: q-rung orthopair fuzzy hypergraphs with applications. Mathematics **7**, 260 2019
23. Luqman, A., Akram, M., Davvaz, B.: q-rung orthopair fuzzy directed hypergraphs: a new model with applications. J. Intell. Fuzzy Syst. (2019)
24. Luqman, A., Akram, M., Koam, A.N.: Granulation of hypernetwork models under the q-rung picture fuzzy environment. Mathematics **7**(6), 496 (2019)
25. Mordeson, J.N., Nair, P.S.: Fuzzy Graphs and Fuzzy Hypergraphs, 2nd edn. Physica Verlag, Heidelberg (2001)
26. Singh, P.: Correlation coefficients for picture fuzzy sets. J. Intell. Fuzzy Syst. **28**, 591604 (2015)
27. Wang, Q., Gong, Z.: An application of fuzzy hypergraphs and hypergraphs in granular computing. Inf. Sci. **429**, 296–314 (2018)
28. Wong, S.K.M., Wu, D.: Automated mining of granular database scheme. In: Proceeding of IEEE International Conference on Fuzzy Systems, pp. 690–694 (2002)
29. Yang, J., Wang, G., Zhang, Q.: Knowledge distance measure in multigranulation spaces of fuzzy equivalence relation. Inf. Sci. **448**, 18–35 (2018)
30. Yager, R.R.: Generalized orthopair fuzzy sets. IEEE Trans. Fuzzy Syst. **25**, 1222–1230 (2017)
31. Yao, Y.Y.: A partition model of granular computing. In: LNCS vol. 3100, 232–253 (2004)
32. Zadeh, L.A.: Similarity relations and fuzzy orderings. Inf. Sci. **3**(2), 177–200 (1971)
33. Zadeh, L.A.: The concept of a linguistic and application to approximate reasoning-I. Inf. Sci. **8**, 199–249 (1975)
34. Zadeh, L.A.: Toward a generalized theory of uncertainty (GTU) an outline. Inf. Sci. **172**, 1–40 (2005)
35. Zhang, L., Zhang, B.: The structural analysis of fuzzy sets. J. Approx. Reason. **40**, 92–108 (2005)
36. Zhang, L., Zhang, B.: The Theory and Applications of Problem Solving-Quotient Space Based Granular Computing. Tsinghua University Press, Beijing (2007)
37. Zhang, L., Zhang, B.: Hierarchy and Multi-granular Computing, Quotient Space Based Problem Solving, pp. 45–103. Tsinghua University Press, Beijing (2014)

Chapter 8
Granular Computing Based on m-Polar Fuzzy Hypergraphs

An m-polar fuzzy model, as an extension of fuzzy and bipolar fuzzy models, plays a vital role in modeling of real-world problems that involve multi-attribute, multipolar information, and uncertainty. The m-polar fuzzy models give increasing precision and flexibility to the system as compared to the fuzzy and bipolar fuzzy models. An m-polar fuzzy set assigns the membership degree to an object belonging to $[0, 1]^m$ describing the m distinct attributes of that element. Granular computing deals with representing and processing information in the form of information granules. These information granules are collections of elements combined together due to their similarity and functional/physical adjacency. In this chapter, we illustrate the formation of granular structures using m-polar fuzzy hypergraphs and level hypergraphs. Further, we define m-polar fuzzy hierarchical quotient space structures. The mappings between the m-polar fuzzy hypergraphs depict the relationships among granules that occurred in different levels. The consequences reveal that the representation of partition of universal set is more efficient through m-polar fuzzy hypergraphs as compared to crisp hypergraphs. We also present some examples and a real-world problem to signify the validity of our proposed model. This chapter is due to [11, 12, 18].

8.1 Introduction

Granular computing is defined as an identification of techniques, methodologies, tools, and theories that yields the advantages of clusters, groups or classes, i.e., the granules. The terminology was first introduced by Lin [15]. The fundamental concepts of granular computing are utilized in various disciplines, including machine learning, rough set theory, cluster analysis, and artificial intelligence. Different models have been proposed to study the various issues occurring in granular computing, including classification of the universe, illustration of granules, and the identification of relations among granules. For example, the procedure of problem-solving through

© Springer Nature Singapore Pte Ltd. 2020

M. Akram and A. Luqman, *Fuzzy Hypergraphs and Related Extensions*,
Studies in Fuzziness and Soft Computing 390,
https://doi.org/10.1007/978-981-15-2403-5_8

granular computing can be considered as subdivisions of the problem at multilevels and these levels are linked together to construct a hierarchical space structure. Thus, this is a way of dealing with the formation of granules and the switching between different granularities. Here, the word "hierarchy" implies the methodology of hierarchical analysis in solving a problem and human activities [32]. To understand this methodology, let us consider an example of national administration in which the complete nation is subdivided into various provinces. Further, we divide every province into various divisions and so on. The human activities and problem-solving involve the simplification of original complicated problem by ignoring some details rather than thinking about all points of the problem. This rationalize model is then further refined till the issue is completely solved. Thus, we resolve and interpret the complex problems from weaker grain to stronger one or from highest rank to lowest or from universal to particular, etc. This technique is called the hierarchical problem-solving. This is further acknowledged that hierarchical strategy is the only technique which is used by humans to deal with complicated problems and it enhances the competence and efficiency. This strategy is also known as the multi-granular computing.

Hypergraphs, as an extension of classical graphs, experience various properties which appear very effective and useful as the basis of different techniques in many fields, including problem-solving, declustering, and databases [10]. The real-world problems which are represented and solved using hypergraphs have been achieved very good impacts. The formation of hypergraphs is same as that of granule structures and the relations between the vertices and hyperedges of hypergraphs can depict the relationships of granules and objects. A hyperedge can contain n vertices representing n-ary relations and hence can provide more effective analysis and description of granules. Many researchers have used hypergraph methods to study the clustering of complex documentation by means of granular computing and investigated the database techniques [16, 22]. Chen et al. [11] proposed a model of granular computing based on crisp hypergraph. They related a crisp hypergraph to a set of granules and represented the hierarchical structures using series of hypergraphs. They proved a hypergraph model as a visual description of granular computing.

Zadeh's [25] fuzzy set has been acquired greater attention by researchers in a wide range of scientific areas, including management sciences, robotics, decision theory, and many other disciplines. Zhang [29] generalized the idea of fuzzy sets to the concept of bipolar fuzzy sets whose membership degrees range over the interval $[-1, 1]$. An m-polar fuzzy set, as an extension of fuzzy set and bipolar fuzzy set, was proposed by Chen et al. [12] and it proved that 2-polar fuzzy sets and bipolar fuzzy sets are equivalent concepts in mathematics. An m-polar fuzzy set corresponds to the existence of "multipolar information" because there are many real-world problems which take data or information from n agents ($n \geq 2$). For example, in the case of telecommunication safety, the exact membership degree lies in the interval $[0, 1]^n$ ($n \approx 7 \times 10^9$) as the distinct members are monitored at different times. Similarly, there are many problems which are based on n logic implication operators ($n \geq 2$), including rough measures, ordering results of magazines and fuzziness measures, etc. To handle uncertainty in the representation of different objects or in the relationships between them, fuzzy graphs were defined by Rosenfeld [20]. m-polar fuzzy graphs and their interesting properties were discussed by Akram et al. [1] to deal with the network models possessing multi-attribute and multipolar data. As an extension of fuzzy graphs, Kaufmann [13] defined fuzzy hypergraphs. Although, many researchers have been explored the construction of granular structures using hypergraphs in various fields. However, there are many graph theoretic problems which may contain uncertainty and vagueness. To overcome the problems of uncertainty in models of granular computing, Wang and Gong [21] studied the construction of granular structures by means of fuzzy hypergraphs. They concluded that the representation of granules and partition is much efficient through the fuzzy hypergraphs. Novel applications and transversals of m-polar fuzzy hypergraphs were defined by Akram and Sarwar [5, 6]. Further, Akram and Shahzadi [7] studied various operations on m-polar fuzzy hypergraphs. Akram and Luqman [3, 4] introduced intuitionistic single-valued and bipolar neutrosophic hypergraphs. The basic purpose of this work is to develop an interpretation of granular structures using m-polar fuzzy hypergraphs. In the proposed model, the vertex of m-polar fuzzy hypergraph denotes an object and an m-polar fuzzy hyperedge represents a granule. The "refinement" and "coarsening" operators are defined to switch the different granularities from coarser to finer and vice versa, respectively.

For further terminologies and studies on m-polar fuzzy hypergraphs, readers are referred to [2, 8, 9, 14, 17, 19, 23, 26–28].

8.2 Fundamental Features of m-Polar Fuzzy Hypergraphs

Definition 8.1 An *m-polar fuzzy set* M on a universal set X is defined as a mapping $M:X \rightarrow [0, 1]^m$. The membership degree of each element $z \in X$ is represented by $M(z)=(\mathcal{P}_1 \circ M(z), \mathcal{P}_2 \circ M(z), \mathcal{P}_3 \circ M(z),\ldots, \mathcal{P}_m \circ M(z))$, where $\mathcal{P}_j \circ M(z) : [0, 1]^m \rightarrow [0, 1]$ is defined as j-th projection mapping.

Note that, the m-th power of $[0, 1]$ (i.e., $[0, 1]^m$) is regarded as a partially ordered set with the point-wise order $\leq$, where m is considered as an ordinal number ($m = n|n < m$ when $m > 0$), $\leq$ is defined as $z_1 \leq z_2$ if and only if $\mathcal{P}_j(z_1) \leq \mathcal{P}_j(z_2)$, for every $1 \leq j \leq m$. $\mathbf{0} = (0, 0, \ldots, 0)$ and $\mathbf{1} = (1, 1, \ldots, 1)$ are the smallest and largest values in $[0, 1]^m$, respectively.

Definition 8.2 Let M be an m-polar fuzzy set on X. An *m-polar fuzzy relation* $N = (\mathcal{P}_1 \circ N, \mathcal{P}_2 \circ N, \mathcal{P}_3 \circ N,\ldots, \mathcal{P}_m \circ N)$ on M is a mapping $N : M \rightarrow M$ such that $N(z_1z_2) \leq \inf\{M(z_1), M(z_2)\}$, for all $z_1, z_2 \in X$, i.e., for each $1 \leq j \leq m$, $\mathcal{P}_j \circ N(z_1z_2) \leq \inf\{\mathcal{P}_j \circ M(z_1), \mathcal{P}_j \circ M(z_2)\}$, where $\mathcal{P}_j \circ M(z)$ and $\mathcal{P}_j \circ N(z_1z_2)$ denote the j-th membership degree of an element $z \in X$ and the pair z_1z_2, respectively.

Definition 8.3 An *m-polar fuzzy graph* on X is defined as an ordered pair of functions $G = (C, D)$, where $C : X \rightarrow [0, 1]^m$ is an m-polar vertex set and $D : X \times X \rightarrow [0, 1]^m$ is an m-polar edge set of G such that $D(wz) \leq \inf\{C(w), C(z)\}$, i.e., $\mathcal{P}_j \circ D(wz) \leq \inf\{\mathcal{P}_j \circ C(w), \mathcal{P}_j \circ C(z)\}$, for all $w, z \in X$ and $1 \leq j \leq m$.

Definition 8.4 An *m-polar fuzzy hypergraph* on a non-empty set X is a pair $H = (A, B)$, where $A = \{M_1, M_2, \ldots, M_r\}$ is a finite family of m-polar fuzzy sets on X and B is an m-polar fuzzy relation on m-polar fuzzy sets M_k such that

- $B(E_k) = B(\{z_1, z_2, \ldots, z_l\}) \leq \inf\{M_k(z_1), M_k(z_2), \ldots, M_k(z_l)\}$,
- $\bigcup\limits_{k=1}^{r} supp(M_k) = X$, for all $M_k \in A$ and for all $z_1, z_2, \ldots, z_l \in X$.

Definition 8.5 Let $H = (A, B)$ be an m-polar fuzzy hypergraph and $\tau \in [0, 1]^m$. Then the τ-cut level set of an m-polar fuzzy set M is defined as $M_\tau = \{z | \mathscr{P}_j \circ M(z) \geq t_j, 1 \leq j \leq m\}, \tau = (t_1, t_2, \ldots, t_m)$.

$H_\tau = (A_\tau, B_\tau)$ is called a τ-*cut level hypergraph* of H, where $A_\tau = \bigcup\limits_{i=1}^{r} M_{i\tau}$.

8.2.1 Uncertainty Measures of m-Polar Fuzzy Hierarchical Quotient Space Structure

The question of distinct membership degrees of same object from different scholars is arisen because of various ways of thinking about the interpretation of different functions dealing with the same problem. To resolve this issue, fuzzy set was structurally defined by Zhang and Zhang [31] which was based on quotient space theory and fuzzy equivalence relation [30]. This definition provides some new initiatives regarding to membership degree, called a hierarchical quotient space structure of a fuzzy equivalence relation. By following the same concept, we develop a hierarchical quotient space structure of an m-polar fuzzy equivalence relation.

Definition 8.6 An m-polar fuzzy equivalence relation on a non-empty finite set X is called an *m-polar fuzzy similarity relation* if it satisfies,

1. $N(z, z) = (\mathscr{P}_1 \circ N(z, z), \mathscr{P}_2 \circ N(z, z), \ldots, \mathscr{P}_m \circ N(z, z)) = (1, 1, \ldots, 1)$, for all $z \in X$,
2. $N(u, w) = (\mathscr{P}_1 \circ N(u, w), \mathscr{P}_2 \circ N(u, w), \ldots, \mathscr{P}_m \circ N(u, w)) = (\mathscr{P}_1 \circ N(w, u), \mathscr{P}_2 \circ N(w, u), \ldots, \mathscr{P}_m \circ N(w, u)) = N(w, u)$, for all $u, w \in X$.

Definition 8.7 An m-polar fuzzy equivalence relation on a non-empty finite set X is called an *m-polar fuzzy equivalence relation* if it satisfies the conditions,

1. $N(z, z) = (\mathscr{P}_1 \circ N(z, z), \mathscr{P}_2 \circ N(z, z), \ldots, \mathscr{P}_m \circ N(z, z)) = (1, 1, \ldots, 1)$, for all $z \in X$,
2. $N(u, w) = (\mathscr{P}_1 \circ N(u, w), \mathscr{P}_2 \circ N(u, w), \ldots, \mathscr{P}_m \circ N(u, w)) = (\mathscr{P}_1 \circ N(w, u), \mathscr{P}_2 \circ N(w, u), \ldots, \mathscr{P}_m \circ N(w, u)) = N(w, u)$, for all $u, w \in X$,
3. for all $u, v, w \in X$, $N(u, w) = \sup\limits_{v \in X}\{\min(N(u, v), N(v, w))\}$, i.e., $\mathscr{P}_j \circ N(u, w) = \sup\limits_{v \in X}\{\min(\mathscr{P}_j \circ N(u, v), \mathscr{P}_j \circ N(v, w))\}, 1 \leq j \leq m$.

Definition 8.8 An *m-polar fuzzy quotient space* is denoted by a triplet $(X, \tilde{C}, N)$, where X is a finite domain, $\tilde{C}$ represents the attributes of X and N represents the m-polar fuzzy relationship between the objects of universe X, which is called the structure of the domain.

Definition 8.9 Let z_i and z_j be two objects in the universe X. The *similarity* between $z_i, z_j \in X$ having the attribute $\tilde{c}_k$ is defined as,

$$N(z_i, z_j) = \frac{|\tilde{c}_{ik} \cap \tilde{c}_{jk}|}{|\tilde{c}_{ik} \cup \tilde{c}_{jk}|},$$

where $\tilde{c}_{ik}$ represents that object z_i possesses the attribute $\tilde{c}_k$ and $\tilde{c}_{jk}$ represents that object z_j possesses the attribute $\tilde{c}_k$.

Proposition 8.1 *Let N be an m-polar fuzzy relation on a finite domain X and $N_\tau = \{(x, w) | \mathscr{P}_j \circ N(x, w) \geq t_j, 1 \leq j \leq m\}$, $\tau = (t_1, t_2, \ldots, t_j) \in [0, 1]$. Then, N_τ is an equivalence relation on X and is said to be cut-equivalence relation of N.*

Proposition 8.1 represents that N_τ is a crisp relation, which is equivalence on X and its *knowledge space* is given as $\xi_{N_\tau}(X) = X/N_\tau$.

The *value domain* of an equivalence relation N on X is defined as $D = \{N(w, y) | w, y \in X\}$ such that, $\mathscr{P}_j \circ X(w) \wedge \mathscr{P}_j \circ X(y) \wedge \mathscr{P}_j \circ N(x, y) > 0, 1 \leq j \leq m$.

Definition 8.10 Let N be an m-polar fuzzy equivalence relation on a finite set X and D be the value domain of N. The set given by $\xi_X(N) = \{X/N_\tau | \tau \in D\}$ is called *m-polar fuzzy hierarchical quotient space structure* of N.

Example 8.1 Let $X = \{w_1, w_2, w_3, w_4, w_5, w_6\}$ be a finite set of elements and N_1 be a 4-polar fuzzy equivalence relation on X, the relation matrix $\tilde{M}_{N_1}$ corresponding to N_1 is given as follows:

$$\tilde{M}_{N_1} = \begin{bmatrix} (1,1,1,1) & (0.4,0.4,0.5,0.5) & (0.5,0.5,0.4,0.4) & (0.5,0.5,0.4,0.4) & (0.5,0.5,0.4,0.4) & (0.5,0.5,0.4,0.4) \\ (0.4,0.4,0.5,0.5) & (1,1,1,1) & (0.8,0.8,0.9,0.9) & (0.8,0.8,0.6,0.6) & (0.8,0.8,0.6,0.6) & (0.6,0.6,0.5,0.5) \\ (0.5,0.5,0.4,0.4) & (0.8,0.8,0.9,0.9) & (1,1,1,1) & (0.6,0.6,0.7,0.7) & (0.6,0.6,0.7,0.7) & (0.6,0.6,0.5,0.5) \\ (0.5,0.5,0.4,0.4) & (0.8,0.8,0.6,0.6) & (0.6,0.6,0.7,0.7) & (1,1,1,1) & (0.7,0.8,0.7,0.8) & (0.6,0.6,0.5,0.5) \\ (0.5,0.5,0.4,0.4) & (0.8,0.8,0.6,0.6) & (0.6,0.6,0.7,0.7) & (0.7,0.8,0.7,0.8) & (1,1,1,1) & (0.6,0.6,0.5,0.5) \\ (0.5,0.5,0.4,0.4) & (0.6,0.6,0.5,0.5) & (0.6,0.6,0.5,0.5) & (0.6,0.6,0.5,0.5) & (0.6,0.6,0.5,0.5) & (1,1,1,1) \end{bmatrix}.$$

Its corresponding m-polar fuzzy hierarchical quotient space structure is given as

$$X/N_{1\tau_1} = X/N_{1(t_1,t_2,t_3,t_4)} = \{\{w_1, w_2, w_3, w_4, w_5, w_6\}\},$$
$$X/N_{1\tau_2} = X/N_{1(t_1',t_2',t_3',t_4')} = \{\{w_1\}, \{w_2, w_3, w_4, w_5, w_6\}\},$$
$$X/N_{1\tau_3} = X/N_{1(t_1'',t_2'',t_3'',t_4'')} = \{\{w_1\}, \{w_2, w_3, w_4, w_5\}, \{w_6\}\},$$
$$X/N_{1\tau_4} = X/N_{1(t_1''',t_2''',t_3''',t_4''')} = \{\{w_1\}, \{w_2, w_3\}, \{w_4, w_5\}, \{w_6\}\},$$
$$X/N_{1\tau_5} = X/N_{1(t_1'''',t_2'''',t_3'''',t_4'''')} = \{\{w_1\}, \{w_2, w_3\}, \{w_4\}, \{w_5\}, \{w_6\}\},$$
$$X/N_{1\tau_6} = X/N_{1(t_1''''',t_2''''',t_3''''',t_4''''')} = \{\{w_1\}, \{w_2\}, \{w_3\}, \{w_4\}, \{w_5\}, \{w_6\}\},$$

where

$$0 < \tau_1 = (t_1, t_2, t_3, t_4) \leq 0.4,$$
$$0.4 < \tau_2 = (t_1', t_2', t_3', t_4') \leq 0.5,$$
$$0.5 < \tau_3 = (t_1'', t_2'', t_3'', t_4'') \leq 0.6,$$
$$0.6 < \tau_4 = (t_1''', t_2''', t_3''', t_4''') \leq 0.7,$$
$$0.7 < \tau_5 = (t_1'''', t_2'''', t_3'''', t_4'''') \leq 0.8,$$
$$0.8 < \tau_6 = (t_1''''', t_2''''', t_3''''', t_4''''') \leq 1.$$

Hence, a 4-polar fuzzy hierarchical quotient space structure is given as $\xi_{X(N_1)} = \{X/N_{\tau_1}, X/N_{\tau_2}, X/N_{\tau_3}, X/N_{\tau_4}, X/N_{\tau_5}, X/N_{\tau_6}\}$ and is shown in Fig. 8.1.

It is worth to note that the same hierarchical quotient space structure can be formed by different 4-polar fuzzy equivalence relations. For instance, the relation matrix $\tilde{M}_{N_2}$ of 4-polar fuzzy equivalence relation generates the same hierarchical quotient space structure as given by $\tilde{M}_{N_1}$. The relation matrix $\tilde{M}_{N_2}$ is given as

$$\tilde{M}_{N_2} = \begin{bmatrix}
(1,1,1,1) & (0.2,0.2,0.5,0.5) & (0.6,0.6,0.4,0.4) & (0.6,0.6,0.4,0.4) & (0.6,0.6,0.4,0.4) & (0.6,0.6,0.4,0.4) \\
(0.2,0.2,0.5,0.5) & (1,1,1,1) & (0.2,0.2,0.5,0.5) & (0.2,0.2,0.5,0.5) & (0.2,0.2,0.5,0.5) & (0.2,0.2,0.5,0.5) \\
(0.6,0.6,0.4,0.4) & (0.2,0.2,0.5,0.5) & (1,1,1,1) & (0.7,0.7,0.7,0.7) & (0.7,0.7,0.7,0.7) & (0.7,0.7,0.7,0.7) \\
(0.6,0.6,0.4,0.4) & (0.2,0.2,0.5,0.5) & (0.7,0.7,0.7,0.7) & (1,1,1,1) & (0.8,0.8,0.7,0.8) & (0.6,0.6,0.5,0.5) \\
(0.6,0.6,0.4,0.4) & (0.2,0.2,0.5,0.5) & (0.7,0.7,0.7,0.7) & (0.8,0.8,0.7,0.8) & (1,1,1,1) & (0.6,0.6,0.5,0.5) \\
(0.6,0.6,0.4,0.4) & (0.2,0.2,0.5,0.5) & (0.7,0.7,0.7,0.7) & (0.6,0.6,0.5,0.5) & (0.6,0.6,0.5,0.5) & (1,1,1,1)
\end{bmatrix}.$$

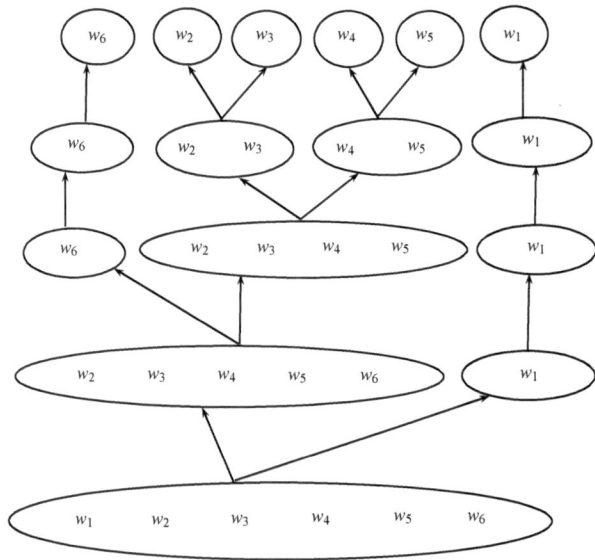

Fig. 8.1 A 4-polar fuzzy hierarchical quotient space structure

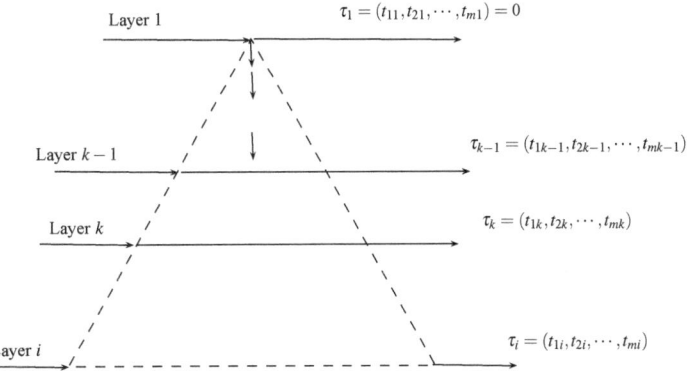

Fig. 8.2 Pyramid model of m-polar fuzzy hierarchical quotient space structure

Furthermore, assuming the number of blocks in every distinct layer of this hierarchical quotient space structure, a pyramid model can also be constructed as shown in Fig. 8.2.

8.2.2 Information Entropy of m-Polar Fuzzy Hierarchical Quotient Space Structure

Definition 8.11 Let N be an m-polar fuzzy equivalence relation on X. Let $\xi_X(N) = \{X(\tau_1), X(\tau_2), X(\tau_3), \ldots, X(\tau_j)\}$ be its corresponding hierarchical quotient space structure, where $\tau_i = (t_{1i}, t_{2i}, \ldots, t_{mi}), i = 1, 2, \ldots, j$ and $X(\tau_j) < X(\tau_{j-1}) < \cdots < X(\tau_1)$. Then, the *partition sequence* of $\xi_X(N)$ is given as $P(\xi_X(N)) = \{P_1, P_2, P_3, \ldots, P_j\}$, where $P_i = |X(\tau_i)|, i = 1, 2, \ldots, j$ and $|.|$ denotes the number of elements in a set.

Definition 8.12 Let N be an m-polar fuzzy equivalence relation on X. Let $\xi_X(N) = \{X(\tau_1), X(\tau_2), X(\tau_3), \ldots, X(\tau_j)\}$ be its corresponding hierarchical quotient space structure, where $\tau_i = (t_{1i}, t_{2i}, \ldots, t_{mi})$, $i = 1, 2, \ldots, j$ and $X(\tau_j) < X(\tau_{j-1}) < \cdots < X(\tau_1)$, $P(\xi_X(N)) = \{P_1, P_2, \ldots, P_j\}$ be the partition sequence of $\xi_X(N)$. Assume that $X(\tau_i) = \{X_{i1}, X_{i2}, \ldots, X_{iP_i}\}$. The *information entropy* $E_{X(\tau_i)}$ is defined as $E_{X(\tau_i)} = -\sum_{r=1}^{\mathscr{P}_i} \frac{|X_{ir}|}{|X|} ln(\frac{|X_{ir}|}{|X|})$.

Theorem 8.1 *Let N be an m-polar fuzzy equivalence relation on X. Let $\xi_X(N) = \{X(\tau_1), X(\tau_2), X(\tau_3), \ldots, X(\tau_j)\}$ be its corresponding hierarchical quotient space structure, where $\tau_i = (t_{1i}, t_{2i}, \ldots, t_{mi})$, $i = 1, 2, \ldots, j$, then the entropy sequence $E(\xi_X(N)) = \{E_{X(\tau_1)}, E_{X(\tau_2)}, \ldots, E_{X(\tau_j)}\}$ increases monotonically and strictly.*

Proof The terminology of hierarchical quotient space structure implies that $X(\tau_j) < X(\tau_{j-1}) < \cdots < X(\tau_1)$, i.e., $X(\tau_{j-1})$ is a quotient subspace of $X(\tau_j)$. Suppose that

$X(\tau_i) = \{X_{i1}, X_{i2}, \ldots, X_{iP_i}\}$ and $X(\tau_{i-1}) = \{X_{(i-1)1}, X_{(i-1)2}, \ldots, X_{(i-1)P_{(i-1)}}\}$, then every subblock of $X(\tau_{i-1})$ is an amalgam of subblocks of $X(\tau_i)$. Without loss of generality, it is assumed that only one subblock $X_{i-1,j}$ in $X(\tau_{i-1})$ is formed by the combination of two subblocks X_{ir}, X_{is} in $X(\tau_i)$ and all other remaining blocks are equal in both sequences. Thus,

$$
\begin{aligned}
E_{X(\tau_{j-1})} &= -\sum_{r=1}^{P_{i-1}} \frac{|X_{i-1,r}|}{|X|} \ln(\frac{|X_{i-1,r}|}{|X|}) \\
&= -\sum_{r=1}^{P_{j-1}} \frac{|X_{i-1,r}|}{|X|} \ln(\frac{|X_{i-1,r}|}{|X|}) - \sum_{r=j+1}^{P_{i-1}} \frac{|X_{i-1,r}|}{|X|} \ln(\frac{|X_{i-1,r}|}{|X|}) - \frac{|X_{i-1,j}|}{|X|} \ln(\frac{|X_{i-1,j}|}{|X|}) \\
&= -\sum_{r=1}^{P_{j-1}} \frac{|X_{i,r}|}{|X|} \ln(\frac{|X_{i,r}|}{|X|}) - \sum_{r=j+1}^{P_i} \frac{|X_{i,r}|}{|X|} \ln(\frac{|X_{i,r}|}{|X|}) - \frac{|X_{i,r}| + |X_{i,s}|}{|X|} \ln(\frac{|X_{i,r}| + |X_{i,s}|}{|X|}).
\end{aligned}
$$

Since,

$$
\begin{aligned}
\frac{|X_{i,r}| + |X_{i,s}|}{|X|} \ln(\frac{|X_{i,r}| + |X_{i,s}|}{|X|}) &= \frac{|X_{i,r}|}{|X|} \ln(\frac{|X_{i,r}| + |X_{i,s}|}{|X|}) + \frac{|X_{i,s}|}{|X|} \ln(\frac{|X_{i,r}| + |X_{i,s}|}{|X|}) \\
&> \frac{|X_{i,r}|}{|X|} \ln(\frac{|X_{i,r}|}{|X|}) + \frac{|X_{i,s}|}{|X|} \ln(\frac{|X_{i,s}|}{|X|}).
\end{aligned}
$$

Therefore, we have

$$
\begin{aligned}
E_{X(\tau_{j-1})} &< -\sum_{r=1}^{P_{j-1}} \frac{|X_{i,r}|}{|X|} \ln(\frac{|X_{i,r}|}{|X|}) - \sum_{r=j+1}^{P_i} \frac{|X_{i,r}|}{|X|} \ln(\frac{|X_{i,r}|}{|X|}) - \frac{|X_{i,r}|}{|X|} \ln(\frac{|X_{i,r}|}{|X|}) - \frac{|X_{i,s}|}{|X|} \ln(\frac{|X_{i,s}|}{|X|}), \\
&= E_{X(\tau_j)}, (2 \leq j \leq n).
\end{aligned}
$$

Hence, $E_{X(\tau_1)} < E_{X(\tau_2)} < E_{X(\tau_2)} < \cdots < E_{X(\tau_j)}$.

Definition 8.13 Let $X = \{s_1, s_2, s_3, \ldots, s_n\}$ be a non-empty set of universe and let $P_d(X) = \{X_1, X_2, X_3, \ldots, X_d\}$ be a partition space of X, where $|P_d(X)| = d$ then $P_d(X)$ is called *d-order partition space* on X.

Definition 8.14 Let X be a finite non-empty universe and let $P_d(X) = \{X_1, X_2, X_3, \ldots, X_d\}$ be a d-order partition space on X. Let $|X_1| = l_1, |X_2| = l_2, \ldots, |X_d| = l_d$ and the sequence $\{l_1, l_2, \ldots, l_d\}$ is arranged in increasing order then we got a new sequence $\chi(d) = \{l'_1, l'_2, \ldots, l'_d\}$ which is also increasing and called a *subblock sequence* of $P_d(X)$.

Note that, two different d-order partition spaces on X may possess the similar subblock sequence $\chi(d)$.

Definition 8.15 Let X be a finite non-empty universe and let $P_d(X) = \{X_1, X_2, X_3, \ldots, X_d\}$ be a partition space of X. Suppose that $\chi_1(d) = \{l'_1, l'_2, \ldots, l'_d\}$ be a subblock sequence of $P_d(X)$, then the *ω-displacement* of $\chi_1(d)$ is defined as an increasing sequence $\chi_2(d) = \{l'_1, l'_2, \ldots, l'_r + 1, \ldots, l'_s - 1, \ldots, l'_d\}$, where $r < s$, $l'_r + 1 < l'_s - 1$.

An ω-*displacement* is obtained by subtracting 1 from some bigger term and adding 1 to some smaller element such that the sequence keeps its increasing property.

Theorem 8.2 *A single time ω-displacement $\chi_2(d)$ which is derived from $\chi_1(d)$ satisfies $E(\chi_1(d)) < E(\chi_2(d))$.*

Proof Let $\chi_1(d) = \{l'_1, l'_2, \ldots, l'_d\}$ and $\chi_2(d) = \{l'_1, l'_2, \ldots, l'_r + 1, \ldots, l'_s - 1, \ldots, l'_d\}$, $l'_1 + l'_2 + \cdots + l'_d = k$ then we have

$$E(\chi_2(t)) = -\sum_{j=1}^{d} \frac{l'_l}{k}\ln\frac{l'_l}{k} + \frac{l'_r}{k}\ln\frac{l'_r}{k} + \frac{l'_s}{k}\ln\frac{l'_s}{k} - \frac{l'_r + 1}{k}\ln\frac{l'_r + 1}{k} - \frac{l'_s - 1}{k}\ln\frac{l'_s - 1}{k}.$$

Let $g(z) = -\frac{z}{k}\ln\frac{z}{k} - \frac{l-z}{k}\ln\frac{l-z}{k}$, where $l = l'_r + l'_s$ and $g'(z) = \frac{1}{k}\ln\frac{l-z}{z}$. Suppose that $g'(z) = 0$, then we obtain a solution, i.e., $z = \frac{l}{2}$. Furthermore, $g''(z) = \frac{-l}{k(l-z)z} < 0$, $0 \le z \le \frac{l}{2}$ and $g(z)$ is increasing monotonically. Let $z_1 = l'_r$ and $z_2 = l'_r + 1$, $l'_r + 1 < l'_s - 1$, i.e., $z_1 < z_2 \le \frac{l}{2} = \frac{l'_r + l'_s}{2}$. Since, $g(z)$ is monotone, then $g(z_2) - g(z_1) > 0$. Thus,

$$\frac{l'_r}{k}\ln\frac{l'_r}{k} + \frac{l'_s}{k}\ln\frac{l'_s}{k} - \frac{l'_r + 1}{k}\ln\frac{l'_r + 1}{k} - \frac{l'_s - 1}{k}\ln\frac{l'_s - 1}{k} > 0.$$

Hence,

$$E(\chi_2(d)) = -\sum_{j=1}^{d} \frac{l'_l}{k}\ln\frac{l'_l}{k} + \frac{l'_r}{k}\ln\frac{l'_r}{k} + \frac{l'_s}{k}\ln\frac{l'_s}{k} - \frac{l'_r + 1}{k}\ln\frac{l'_r + 1}{k} - \frac{l'_s - 1}{k}\ln\frac{l'_s - 1}{k}$$

$$> -(\frac{l'_r + 1}{k}\ln\frac{l'_r + 1}{k} + \frac{l'_s - 1}{k}\ln\frac{l'_s - 1}{k})$$

$$> -\sum_{j=1}^{t} \frac{l'_l}{k}\ln\frac{l'_l}{k}$$

$$= E(\chi_1(d)).$$

This completes the proof.

8.3 An m-Polar Fuzzy Hypergraph Model of Granular Computing

Definition 8.16 An *object space* is defined as a system (X, N), where X is a universe of objects or elements and $N = \{n_1, n_2, n_3, \ldots, n_k\}$, $k = |X|$ is a family of relations between the elements of X. For $r \le k$, $n_r \in N$, $n_r \subseteq X \times X \times \cdots \times X$, if $(z_1, z_2, \ldots, z_r) \subseteq n_r$, then there exists an r-array relation n_r on $(z_1, z_2, \ldots, z_n)$.

A granule affiliates to a particular level. The whole view of granules at every level can be taken as a complete description of a particular problem at that level of granularity [11]. An m-polar fuzzy hypergraph formed by the set of relations N and membership degrees $X(w) = \mathscr{P}_j \circ X(w)$, $1 \leq j \leq m$ of objects in the space is considered as a specific level of granular computing model. All m-polar fuzzy hyperedges in that m-polar fuzzy hypergraph can be regarded as the complete granule in that particular level.

Definition 8.17 A *partition* of a set X established on the basis of relations between objects is defined as a collection of non-empty subsets which are pair-wise disjoint and whose union is whole of X. These subsets which form the partition of X are called *blocks*. Every partition of a finite set X contains the finite number of blocks. Corresponding to the m-polar fuzzy hypergraph, the constraints of partition $\psi = \{\mathscr{E}_i | 1 \leq i \leq n\}$.

(i) each $\mathscr{E}_i$ is non-empty,
(ii) for $i \neq j$, $\mathscr{E}_i \cap \mathscr{E}_j = \emptyset$,
(iii) $\cup\{supp(\mathscr{E}_i) | 1 \leq i \leq n\} = X$.

Definition 8.18 A *covering* of a set X is defined as a collection of non-empty subsets whose union is whole of X. The conditions for the covering $c = \{\mathscr{E}_i | 1 \leq i \leq n\}$ of X are stated as

(i) each $\mathscr{E}_i$ is non-empty,
(ii) $\cup\{supp(\mathscr{E}_i) | 1 \leq i \leq n\} = X$.

The corresponding definitions in classical hypergraph theory are completely analogous to the above Definitions 8.17 and 8.18. In a crisp hypergraph, if the hyperedges E_i and E_j do not intersect each other, i.e., $E_i, E_j \in E$ and $E_i \cap E_j = \emptyset$ then these hyperedges form a *partition* of granules in this level. Furthermore, if $E_i, E_j \in E$ and $E_i \cap E_j \neq \emptyset$, i.e., the hyperedges E_i and E_j intersect each other, then these hyperedges form a *covering* in this level.

Example 8.2 Let $X = \{w_1, w_2, w_3, w_4, w_5, w_6, w_7, w_8, w_9, w_{10}\}$. The partition and covering of X are given in Figs. 8.3 and 8.4, respectively.

Fig. 8.3 A partition of granules in a level

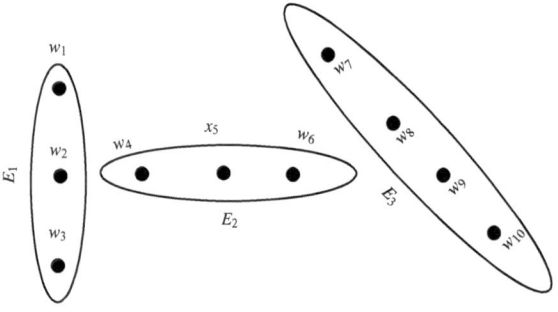

Fig. 8.4 A covering of granules in a level

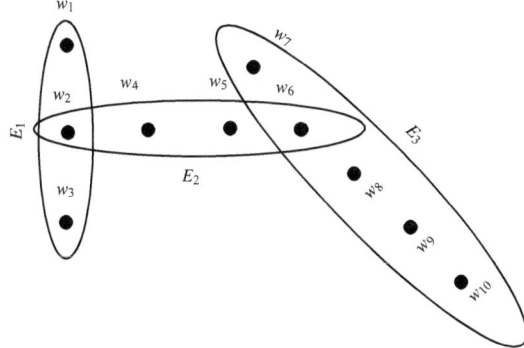

A set-theoretic way to study the granular computing model uses the following operators in an *m*-polar fuzzy hypergraph model.

Definition 8.19 Let $\mathscr{G}_1$ and $\mathscr{G}_2$ be two granules in our model and the *m*-polar fuzzy hyperedges $\mathscr{E}_1$, $\mathscr{E}_2$ represent their external properties. The *union* of two granules $\mathscr{G}_1 \cup \mathscr{G}_2$ is defined as a larger *m*-polar fuzzy hyperedge that contains the vertices of both $\mathscr{E}_1$ and $\mathscr{E}_2$. If $w_i \in \mathscr{G}_1 \cup \mathscr{G}_2$, then the membership degree $(\mathscr{G}_1 \cup \mathscr{G}_2)(w_i)$ of w_i in larger granule $\mathscr{G}_1 \cup \mathscr{G}_2$ is defined as follows:

$$\mathscr{P}_j \circ (\mathscr{G}_1 \cup \mathscr{G}_2)(w_i) = \begin{cases} \max\{\mathscr{P}_j \circ (\mathscr{E}_1)(w_i), \mathscr{P}_j \circ (\mathscr{E}_2)(w_i)\}, & \text{if } w_i \in \mathscr{E}_1 \text{ and } w_i \in \mathscr{E}_2, \\ \mathscr{P}_j \circ (\mathscr{E}_1)(w_i), & \text{if } w_i \in \mathscr{E}_1 \text{ and } w_i \notin \mathscr{E}_2, \\ \mathscr{P}_j \circ (\mathscr{E}_2)(w_i), & \text{if } w_i \in \mathscr{E}_2 \text{ and } w_i \notin \mathscr{E}_1, \end{cases}$$

$1 \le j \le m$.

Definition 8.20 Let $\mathscr{G}_1$ and $\mathscr{G}_2$ be two granules in our model and the *m*-polar fuzzy hyperedges $\mathscr{E}_1$, $\mathscr{E}_2$ represent their external properties. The *intersection* of two granules $\mathscr{G}_1 \cap \mathscr{G}_2$ is defined as a larger *m*-polar fuzzy hyperedge that contains the vertices of both $\mathscr{E}_1$ and $\mathscr{E}_2$. If $w_i \in \mathscr{G}_1 \cap \mathscr{G}_2$, then the membership degree $(\mathscr{G}_1 \cap \mathscr{G}_2)(w_i)$ of w_i in smaller granule $\mathscr{G}_1 \cap \mathscr{G}_2$ is defined as follows,

$$\mathscr{P}_j \circ (\mathscr{G}_1 \cap \mathscr{G}_2)(w_i) = \begin{cases} \min\{\mathscr{P}_j \circ (\mathscr{E}_1)(w_i), \mathscr{P}_j \circ (\mathscr{E}_2)(w_i)\}, & \text{if } w_i \in \mathscr{E}_1 \text{ and } w_i \in \mathscr{E}_2, \\ \mathscr{P}_j \circ (\mathscr{E}_1)(w_i), & \text{if } w_i \in \mathscr{E}_1 \text{ and } w_i \notin \mathscr{E}_2, \\ \mathscr{P}_j \circ (\mathscr{E}_2)(w_i), & \text{if } w_i \in \mathscr{E}_2 \text{ and } w_i \notin \mathscr{E}_1, \end{cases}$$

$1 \le j \le m$.

Definition 8.21 Let $\mathscr{G}_1$ and $\mathscr{G}_2$ be two granules in our model and the *m*-polar fuzzy hyperedges $\mathscr{E}_1$, $\mathscr{E}_2$ represent their external properties. The *difference* between two granules $\mathscr{G}_1 - \mathscr{G}_2$ is defined as a smaller *m*-polar fuzzy hyperedge that contains those vertices belonging to $\mathscr{E}_1$ but not to $\mathscr{E}_2$.

Note that, if a vertex $w_i \in \mathscr{E}_1$ and $w_i \notin \mathscr{E}_2$, then $\mathscr{P}_j \circ (\mathscr{E}_1)(w_i) > 0$ and $\mathscr{P}_j \circ (\mathscr{E}_2)(w_i) = 0$, $1 \le j \le m$.

Definition 8.22 A granule $\mathcal{G}_1$ is said to be the *sub-granule* of $\mathcal{G}_2$, if each vertex w_i of $\mathcal{E}_1$ also belongs to $\mathcal{E}_2$, i.e., $\mathcal{E}_1 \subseteq \mathcal{E}_2$. In such case, $\mathcal{G}_2$ is called the *super-granule* of $\mathcal{G}_1$.

Note that, if $\mathcal{E}(w_i) = \{0, 1\}$, then the all above described operators are reduced to classical hypergraphs theory of granular computing.

8.4 Formation of Hierarchical Structures

We can interpret a problem in distinct levels of granularities. These granular structures at different levels produce a set of m-polar fuzzy hypergraphs. The upper set of these hypergraphs constructs a hierarchical structure in distinct levels. The relationships between granules are expressed by lower level, which represents the problem as a concrete example of granularity. The relationships between granule sets are expressed by higher level, which represents the problem as an abstract example of granularity. Thus, the single-level structures can be constructed and then can be subdivided into hierarchical structures using the relational mappings between different levels.

Definition 8.23 Let $H^1 = (A^1, B^1)$ and $H^2 = (A^2, B^2)$ be two m-polar fuzzy hypergraphs. In an hierarchy structure, their level cuts are H_τ^1 and H_τ^2, respectively, where $\tau = (t_1, t_2, \ldots, t_m)$. Let $\tau \in [0, 1]$ and $\mathcal{P}_j \circ \mathcal{E}_i^1 \geq t_j, 1 \leq j \leq m$, where $\mathcal{E}_i^1 \in B^1$, then a mapping $\phi : H_\tau^1 \to H_\tau^2$ from H_τ^1 to H_τ^2 maps the $\mathcal{E}_{\tau_i}^1$ in H_τ^1 to a vertex w_i^2 in H_τ^2. Furthermore, the mapping $\phi^{-1} : H_\tau^2 \to H_\tau^1$ maps a vertex w_i^2 in H_τ^2 to τ-cut of m-polar fuzzy hyperedge $\mathcal{E}_\tau^1 i$ in $H_{\tau_i}^1$. It can be denoted as $\phi(\mathcal{E}_{\tau_i}^1) = w_i^2$ or $\phi^{-1}(w_i^2) = \mathcal{E}_{\tau_i}^1$, for $1 \leq i \leq n$.

In an m-polar fuzzy hypergraph model, the mappings are used to describe the relations among different levels of granularities. At each distinct level, the problem is interpreted w.r.t the m-PF granularity of that level. The mapping associates the different descriptions of the same problem at distinct levels of granularities. There are two fundamental types to construct the method of hierarchical structures, the *top-down construction procedure* and *the bottom-up construction procedure* [24].

A formal discussion is provided to interpret an m-polar fuzzy hypergraph model ingranular computing, which is more compatible to human thinking. Zhang and Zhang [30] highlighted that one of the most important and acceptable characteristic of human intelligence is that the same problem can be viewed and analyzed in different granularities. Their claim is that the problem can not only be solved using various world of granularities but also can be switched easily and quickly. Hence, the procedure of solving a problem can be considered as the calculations in different hierarchies within that model.

A multilevel granularity of the problem is represented by an m-polar fuzzy hypergraph model, which allows the problem solvers to decompose it into various minor problems and transform it in other granularities. The transformation of problem in other granularities is performed by using two operators, i.e., zooming-in and zooming-out operators. The transformation from weaker level to finer level of granularity is done by zoom-in operator and the zoom-out operator deals with the shifting of problem from coarser to finer granularity.

Definition 8.24 Let $H^1 = (A^1, B^1)$ and $H^2 = (A^2, B^2)$ be two m-polar fuzzy hypergraphs, which are considered as two levels of hierarchical structures and H^2 owns the coarser granularity than H^1. Suppose $H_\tau^1 = (X^1, E_\tau^1)$ and $H_\tau^2 = (X^2, E_\tau^2)$ are the corresponding τ-level hypergraphs of H^1 and H^2, respectively. Let $e_i^1 \in E_\tau^1$, $z_j^1 \in X^1$, $e_j^2 \in E_\tau^2$, $z_l^2, z_m^2 \in X^2$ and $z_l^2, z_m^2 \in e_j^2$. If $\phi(e_i^1) = z_l^2$, then $n(z_j^1, z_m^2)$ is the relationship between z_j^1 and z_m^2 and is obtained by the characteristics of granules.

Definition 8.25 Let the hyperedge $\phi^{-1}(z_l)$ be a vertex in a new level and the relation between hyperedges in this level is same as that of relationship between vertices in previous level. This is called the *zoom-in operator* and transforms a weaker level to a stronger level. The function $r(z_j^1, z_m^2)$ defines the relation between vertices of original level as well as new level.

   Let the vertex $\phi(e_i)$ be a hyperedge in a new level and the relation between vertices in this level is same as that of relationship between hyperedges in corresponding level. This is called the *zoom-out operator* and transforms a finer level to a coarser level.

By using these zoom-in and zoom-out operators, a problem can be viewed at multilevels of granularities. These operations allow us to solve the problem more appropriately and granularity can be switched easily at any level of problem-solving.

   In an m-polar fuzzy hypergraph model of granular computing, the membership degrees of elements reflect the actual situation more efficiently and a wide variety of complicated problems in uncertain and vague environments can be presented by means of m-polar fuzzy hypergraphs. The previous analysis conclude that this model of granular computing generalizes the classical hypergraph model and fuzzy hypergraph model.

Definition 8.26 Let H^1 and H^2 be two crisp hypergraphs. Suppose that H^1 owns the finer m-polar fuzzy granularity than H^2. A mapping from H^1 to H^2 $\psi : H^1 \to H^2$ maps a hyperedge of H^1 to the vertex of H^2 and the mapping $\psi^{-1} : H^2 \to H^1$ maps a vertex of H^2 to the hyperedge of H^1.

The procedure of bottom-up construction for level hypergraph model is illustrated in Algorithm 8.4.1.

Algorithm 8.4.1

The procedure of bottom-up construction for level hypergraph
1. Determine an $m-$polar fuzzy equivalence relation matrix according to the actual circumstances.
2. For fixed $\tau \in [0, 1]$, obtain the corresponding hierarchical quotient space structure.
3. Obtain the hyperedges through the hierarchical quotient space structure.
4. Granules in i-level are mapped to $(i + 1)-$level.
5. Calculate the $m-$polar fuzzy relationships between the vertices of $(i + 1)-$level and determine the $m-$polar fuzzy equivalence relation matrix.
6. Determine the corresponding hierarchical quotient space structure according to τ, which is fixed in Step 2.
7. Get the hyperedges in $(i + 1)-$level and $(i + 1)-$level of the model is constructed.
8. Step 1 - Step 5 are repeated until the whole universe is formulated to a single granule.

Definition 8.27 Let N be an m-polar fuzzy equivalence relation on X. A *coarse gained universe* X/N_τ can be obtained by using m-polar fuzzy equivalence relation, where $[w_i]_{N_\tau} = \{w_j \in X | w_i N w_j\}$. This equivalence class $[w_i]_{N_\tau}$ is considered as an hyperedge in the level hypergraph.

Definition 8.28 Let $H_1 = (X_1, E_1)$ and $H_2 = (X_2, E_2)$ be level hypergraphs of m-polar fuzzy hypergraphs and H_2 has weaker granularity than H_1. Suppose that e_i^1, $e_j^2 \in E_1$ and $w_i^2, w_j^2 \in X^2$, $i, j = 1, 2, \ldots, n$. The *zoom-in operator* $\omega : H_2 \to H_1$ is defined as $\omega(w_i^2) = e_i^1$, $e_i^1 \in E_1$. The relations between the vertices of H^2 define the relationships among the hyperedges in new level. The zoom-in operator of two levels is shown in Fig. 8.5.

Remark 1 For all X_2', $X_2'' \subseteq X_2$, we have $\omega(X_2') = \bigcup\limits_{w_i^2 \in X_2'} \omega(w_i^2)$ and $\omega(X_2'') = \bigcup\limits_{w_j^2 \in X_2''} \omega(w_j^2)$.

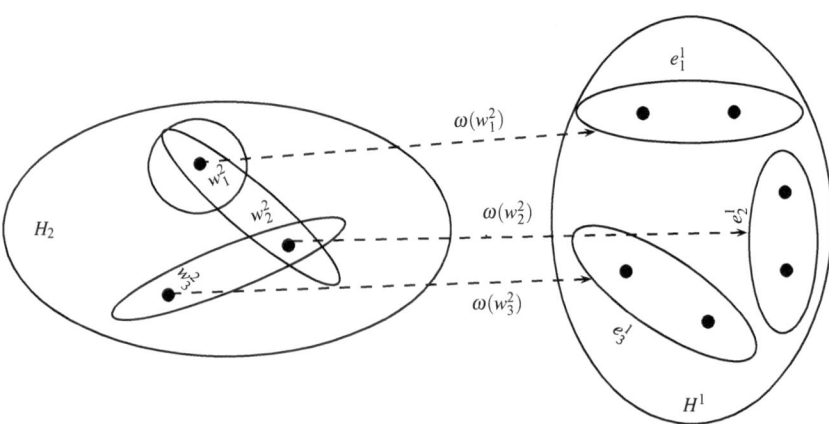

Fig. 8.5 Zoom-in operator

Theorem 8.3 *Let $H_1 = (X_1, E_1)$ and $H_2 = (X_2, E_2)$ be two levels and $\omega : H_2 \rightarrow H_1$ be the zoom-in operator. Then for all $X_2', X_2'' \subseteq X_2$, the zoom-in operator satisfies*

(i) ω maps the empty set to an empty set, i.e., $\omega(\emptyset) = \emptyset$,
(ii) $\omega(X_2) = E_1$,
(iii) $\omega([X_2']^c) = [\omega(X_2')]^c$,
(iv) $\omega(X_2' \cap X_2'') = \omega(X_2') \cap \omega(X_2'')$,
(v) $\omega(X_2' \cup X_2'') = \omega(X_2') \cup \omega(X_2'')$,
(vi) $X_2' \subseteq X_2''$ if and only if $\omega(X_2') \subseteq \omega(X_2'')$.

Proof (i) It is trivially satisfied that $\omega(\emptyset) = \emptyset$.

(ii) As we know that for all $w_i^2 \in X_2$, we have $\omega(X_2') = \bigcup\limits_{w_i^2 \in X_2'} \omega(w_i^2)$. Since $\omega(w_i^2) = e_i^1$, we have $\omega(X_2') = \bigcup\limits_{w_i^2 \in X_2'} \omega(w_i^2) = \bigcup\limits_{e_i^1 \in E_1} e_i^1 = E_1$.

(iii) Let $[X_2']^c = X_2'$ and $[X_2'']^c = X_2''$, then it is obvious that $X_2'' \cap X_2' = \emptyset$ and $X_2' \cup X_2' = X_2$. It follows from (ii) that $\omega(X_2) = E_1$ and we denote by W_1' that edge set of H_1 on which the vertex set X_2' of H_2 is mapped under ω, i.e., $\omega(X_2') = W_1'$. Then $\omega([X_2']^c) = \omega(X_2') = \bigcup\limits_{w_i^2 \in X_2'} \omega(w_i^2) = \bigcup\limits_{e_i^1 \in W_1'} e_i^1 = X_1'$

and $[\omega(X_2')]^c = [\bigcup\limits_{w_j^2 \in X_2'} \omega(w_j^2)]^c = [\bigcup\limits_{e_j^1 \in E_1'} e_j^1]^c = (E_1')^c$. Since, the relationship between hyperedges in new level is same as that of relations among vertices in original level so we have $(E_1')^c = X_1'$. Hence, we conclude that $\omega([X_2']^c) = [\omega(X_2')]^c$.

(iv) Assume that $X_2' \cap X_2'' = \tilde{X}_2$ then for all $w_i^2 \in \tilde{X}_2$ implies that $w_i^2 \in X_2'$ and $w_i^2 \in X_2''$. Further, we have $\omega(X_2' \cap X_2'') = \omega(\tilde{X}_2) = \bigcup\limits_{w_i^2 \in \tilde{X}_2} \omega(w_i^2) = \bigcup\limits_{e_i^1 \in \tilde{E}_1} \omega(e_i^1) = \tilde{E}_1$.

$\omega(X_2') \cap \omega(X_2'') = \{ \bigcup\limits_{w_i^2 \in X_2'} \omega(w_i^2) \} \cap \{ \bigcup\limits_{w_j^2 \in X_2''} \omega(w_j^2) \} = \bigcup\limits_{e_i^1 \in E_1'} e_i^1 \cap \bigcup\limits_{e_j^1 \in E_1''} e_j^1 = E_1' \cap E_1''$. Since, the relationship between hyperedges in new level is same as that of relations among vertices in original level so we have $E_1' \cap E_1'' = \tilde{E}_1$. Hence, we conclude that $\omega(X_2' \cap X_2'') = \omega(X_2') \cap \omega(X_2'')$.

(v) Assume that $X_2' \cup X_2'' = \tilde{X}_2$. Then we have $\omega(X_2' \cup X_2'') = \omega(\tilde{X}_2) = \bigcup\limits_{w_i^2 \in \tilde{X}_2} \omega(w_i^2) = \bigcup\limits_{e_i^1 \in \tilde{E}_1} \omega(e_i^1) = \tilde{E}_1$.

$\omega(X_2') \cup \omega(X_2'') = \{ \bigcup\limits_{w_i^2 \in X_2'} \omega(w_i^2) \} \cup \{ \bigcup\limits_{w_j^2 \in X_2''} \omega(w_j^2) \} = \bigcup\limits_{e_i^1 \in E_1'} e_i^1 \cup \bigcup\limits_{e_j^1 \in E_1''} e_j^1 = E_1' \cup E_1''$. Since, the relationship between hyperedges in new level is same as that of relations among vertices in original level so we have $E_1' \cup E_1'' = \tilde{E}_1$. Hence, we conclude that $\omega(X_2' \cup X_2'') = \omega(X_2') \cup \omega(X_2'')$.

(vi) First we show that $X_2' \subseteq X_2''$ implies that $\omega(X_2') \subseteq \omega(X_2'')$. Since, $X_2' \subseteq X_2''$, which implies that $X_2' \cap X_2'' = X_2'$ and $\omega(X_2') = \bigcup\limits_{w_i^2 \in X_2'} \omega(w_i^2) = \bigcup\limits_{e_i^1 \in E_1'} e_i^1 = E_1'$.

Also $\omega(X_2'') = \bigcup\limits_{w_j^2 \in X_2''} \omega(w_j^2) = \bigcup\limits_{e_j^1 \in E_1''} e_j^1 = E_1''$. Since, the relationship between

Fig. 8.6 Zoom-out operator

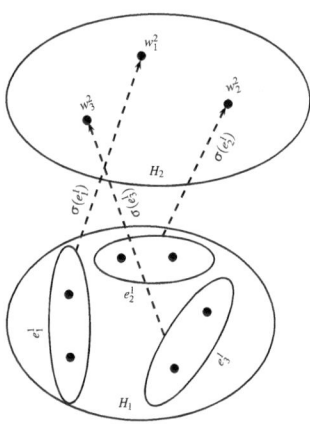

hyperedges in new level is same as that of relations among vertices in original level so we have $E'_1 \subseteq E''_1$, i.e., $\omega(X'_2) \subseteq \omega(X''_2)$. Hence, $X'_2 \subseteq X''_2$ implies that $\omega(X'_2) \subseteq \omega(X''_2)$.

We now prove that $\omega(X'_2) \subseteq \omega(X''_2)$ implies that $X'_2 \subseteq X''_2$. Suppose on contrary that whenever $\omega(X'_2) \subseteq \omega(X''_2)$ then there is at least one vertex $w_i^2 \in X'_2$ but $w_i^2 \notin X''_2$, i.e., $X'_2 \not\subseteq X''_2$. Since, $\omega(w_i^2) = e_i^1$ and the relationship between hyperedges in new level is same as that of relations among vertices in original level so we have $\qquad e_i^1 \in E'_1 \qquad$ but $\qquad e_i^1 \notin E''_1, \qquad$ i.e., $\qquad E'_1 \not\subseteq E''_1,$ which is contradiction to the supposition. Thus, we have $\omega(X'_2) \subseteq \omega(X''_2)$ implies that $X'_2 \subseteq X''_2$. Hence, $X'_2 \subseteq X''_2$ if and only if $\omega(X'_2) \subseteq \omega(X''_2)$.

Definition 8.29 Let $H_1 = (X_1, E_1)$ and $H_2 = (X_2, E_2)$ be level hypergraphs of m-polar fuzzy hypergraphs and H_2 has weaker granularity than H_1. Suppose that e_i^1, $e_j^2 \in E_1$ and $w_i^2, w_j^2 \in X_2$, $i, j = 1, 2, \ldots, n$. The *zoom-out operator* $\sigma : H_1 \to H_2$ is defined as $\sigma(e_i^1) = w_i^2$, $w_i^2 \in X_2$. The zoom-out operator of two levels is shown in Fig. 8.6.

Theorem 8.4 *Let* $\sigma : H_1 \to H_2$ *be the zoom-out operator from* $H_1 = (X_1, E_1)$ *to* $H_2 = (X_2, E_2)$ *and let* $E'_1 \subseteq E_1$. *Then, the zoom-out operator* σ *satisfies the following properties:*

  (i) $\sigma(\emptyset) = \emptyset$,
  (ii) σ *maps the set of hyperedges of* H_1 *onto the set of vertices of* H_2, *i.e.,* $\sigma(E_1) = X_2$,
  (iii) $\sigma([E'_1]^c) = [\sigma(E'_1)]^c$.

Fig. 8.7 Internal and external zoom-out operators

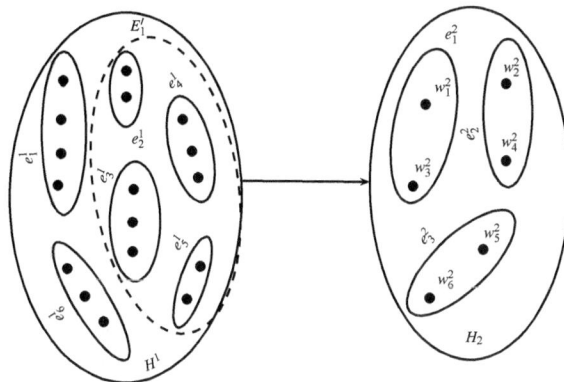

Proof (i) This part is trivially satisfied.

(ii) According to the definition of σ, we have $\sigma(e_i^1) = w_i^2$. Since, the hyperedges define a partition of hypergraph so we have $E_1 = \{e_1^1, e_2^1, e_3^1, \ldots, e_n^1\} = \bigcup_{e_i^1 \in E_1} e_i^1$.

Then
$$\sigma(E_1) = \sigma(\bigcup_{e_i^1 \in E_1} e_i^1) = \bigcup_{e_i^1 \in E_1} \sigma(e_i^1) = \bigcup_{w_i^2 \in X_2} w_i^2 = X_2.$$

(iii) Assume that $[E_1']^c = V_1'$ then it is obvious that $E_1' \cap V_1' = \emptyset$ and $E_1' \cup V_1' = E_1$. Suppose on contrary that there exists at least one vertex $w_i^2 \in \sigma([E_1']^c)$ but $w_i^2 \notin [\sigma(E_1')]^c$. $w_i^2 \in \sigma([E_1']^c)$ implies that $w_i^2 \in \sigma(V_1') \Rightarrow w_i^2 \in \bigcup_{e_i^1 \in V_1'} \sigma(e_i^1) \Rightarrow w_i^2 \in \bigcup_{e_i^1 \in E_1 \setminus E_1'} \sigma(e_i^1)$. Since, $w_i^2 \notin [\sigma(E_1')]^c \Rightarrow w_i^2 \in \sigma(E_1') \Rightarrow w_i^2 \in \bigcup_{e_i^1 \in E_1'} \sigma(e_i^1)$, which is contradiction to our assumption. Hence, $\sigma([E_1']^c) = [\sigma(E_1')]^c$.

Definition 8.30 Let $H_1 = (X_1, E_1)$ and $H_2 = (X_2, E_2)$ be two levels of m-polar fuzzy hypergraphs and H_1 possesses the stronger granularity than H_2. Let $E_1' \subseteq E_1$ then $\hat{\sigma}(E_1') = \{e_i^2 | e_i^2 \in E_2, \kappa(e_i^2) \subseteq E_1'\}$ is called *internal zoom-out operator*.

The operator $\check{\sigma}(E_1') = \{e_i^2 | e_i^2 \in E_2, \kappa(e_i^2) \cap E_1' \neq \emptyset\}$ is called *external zoom-out operator*.

Example 8.3 Let $H_1 = (X_1, E_1)$ and $H_2 = (X_2, E_2)$ be two levels of m-polar fuzzy hypergraphs and H_1 possesses the stronger granularity than H_2, where $E_1 = \{e_1^1, e_2^1, e_3^1, e_4^1, e_5^1, e_6^1\}$ and $E_2 = \{e_1^2, e_2^2, e_3^2\}$. Furthermore, $e_1^2 = \{w_1^2, w_3^2\}$, $e_2^2 = \{w_2^2, w_4^2\}$, $e_3^2 = \{w_5^2, w_6^2\}$ as shown in Fig. 8.7.

Let $E_1' = \{e_2^1, e_3^1, e_4^1, e_5^1\}$ be the subset of hyperedges of H_1 then we can not zoom-out to H_2 directly, thus by using the internal and external zoom-out operators we have the following relations.
$$\hat{\sigma}(\{e_2^1, e_3^1, e_4^1, e_5^1\}) = \{e_2^2\},$$
$$\check{\sigma}(\{e_2^1, e_3^1, e_4^1, e_5^1\}) = \{e_1^2, e_2^2, e_3^2\}.$$

8.5   A Granular Computing Model of Web Searching Engines

The most fertile way to direct a search on the Internet is through a search engine. A web search engine is defined as a system software which is designed to search for queries on World Wide Web. A user may utilize a number of search engines to gather information and similarly various searchers may make an effective use of same engine to fulfill their queries. In this section, we construct a granular computing model of web searching engines based on 4-polar fuzzy hypergraph. In a web searching hypernetwork, the vertices denote the various search engines. According to the relation set N, the vertices having some relationship are united together as an hyperedge, in which the search engines serve only one user. After assigning the membership degrees to that unit, a 4-polar fuzzy hyperedge is constructed, which is also considered as a granule. A 4-polar fuzzy hyperedge indicates a user who wants to gather some information and the vertices in that hyperedge represent those search engines which provide relevant data to the user. Let us consider there are ten search engines and the corresponding 4-polar fuzzy hypergraph $H = (A, B)$ is shown in Fig. 8.8. Note that, $A = \{e_1, e_2, e_3, \ldots, e_{10}\}$ and $B = \{U_1, U_2, U_3, U_4, U_5\}$.

The incidence matrix of 4-polar fuzzy hypergraph is given in Table 8.1.

Fig. 8.8 A 4-polar fuzzy hypergraph representation of web searching

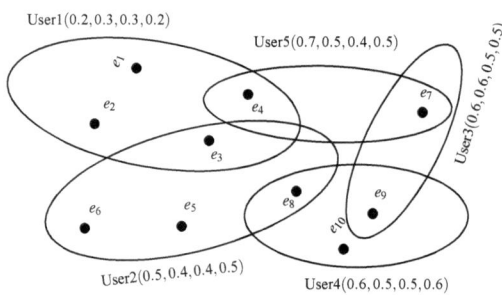

Table 8.1 Incidence matrix

X	$U1$	$U2$	$U3$	$U4$	$U5$
e_1	(0.2, 0.3, 0.3, 0.2)	(0, 0, 0, 0)	(0, 0, 0, 0)	(0, 0, 0, 0)	(0, 0, 0, 0)
e_2	(0.2, 0.3, 0.3, 0.2)	(0, 0, 0, 0)	(0, 0, 0, 0)	(0, 0, 0, 0)	(0, 0, 0, 0)
e_3	(0.2, 0.3, 0.3, 0.2)	(0.5, 0.4, 0.4, 0.5)	(0, 0, 0, 0)	(0, 0, 0, 0)	(0, 0, 0, 0)
e_4	(0.2, 0.3, 0.3, 0.2)	(0, 0, 0, 0)	(0, 0, 0, 0)	(0, 0, 0, 0)	(0.7, 0.5, 0.4, 0.5)
e_5	(0, 0, 0, 0)	(0.5, 0.4, 0.4, 0.5)	(0, 0, 0, 0)	(0, 0, 0, 0)	(0, 0, 0, 0)
e_6	(0, 0, 0, 0)	(0.5, 0.4, 0.4, 0.5)	(0, 0, 0, 0)	(0, 0, 0, 0)	(0, 0, 0, 0)
e_7	(0, 0, 0, 0)	(0, 0, 0, 0)	(0.6, 0.6, 0.5, 0.5)	(0, 0, 0, 0)	(0.7, 0.5, 0.4, 0.5)
e_8	(0, 0, 0, 0)	(0, 0, 0, 0)	(0, 0, 0, 0)	(0.6, 0.5, 0.5, 0.6)	(0, 0, 0, 0)
e_9	(0, 0, 0, 0)	(0, 0, 0, 0)	(0.6, 0.6, 0.5, 0.5)	(0.6, 0.5, 0.5, 0.6)	(0, 0, 0, 0)
e_{10}	(0, 0, 0, 0)	(0, 0, 0, 0)	(0, 0, 0, 0)	(0.6, 0.5, 0.5, 0.6)	(0, 0, 0, 0)

Table 8.2 The information table

X	Core technology	Scalability	Content processing	Query functionality
e_1	0.7	0.6	0.5	0.7
e_2	0.6	0.5	0.5	0.6
e_3	0.7	0.8	0.8	0.7
e_4	0.8	0.6	0.6	0.8
e_5	0.7	0.5	0.5	0.7
e_6	0.7	0.6	0.5	0.7
e_7	0.6	0.5	0.5	0.6
e_8	0.7	0.8	0.8	0.7
e_9	0.8	0.6	0.6	0.8
e_{10}	0.7	0.8	0.8	0.7

An m-polar fuzzy hypergraph model of granular computing illustrates a vague set having some membership degrees. In this model, there are five users need the search engines to gather information. Note that, the membership degrees of these engines are different to the users because whenever a user selects a search engine, he/she considers various factors or attributes. Hence, an m-polar fuzzy hypergraph in granular computing is more meaning full and effective.

Let us suppose that each search engine possesses four attributes which are *Core Technology, Scalability, Content Processing, Query Functionality*. The information table for various search engines having these attributes is given in Table 8.2.

The membership degrees of search engines reveal the percentage of attributes possessed by them, e.g., e_1 own 70% of *core technology*, 60% *scalability*, 50% provide *content processing* and *query functionality* of this engine is 70%. The 4-polar fuzzy equivalence relation matrix describes the similarities between these search engines and is given as follows:

$$\tilde{P}_N = \begin{bmatrix} 1 & 1 & 0.6 & 0.6 & 0.6 & 0.6 & 0.6 & 0.6 & 0 \\ 1 & 1 & 0.7 & 0.7 & 0.7 & 0.7 & 0.7 & 0.7 & 0 \\ 0.6 & 0.7 & 1 & 0.8 & 0.8 & 0.8 & 0.8 & 0.8 & 0.8 & 0 \\ 0.6 & 0.7 & 0.8 & 1 & 0.6 & 0.6 & 0.6 & 0.6 & 0.6 & 0.6 \\ 0.6 & 0.7 & 0.8 & 0.6 & 1 & 0.5 & 0.5 & 0.5 & 0.5 & 0 \\ 0.6 & 0.7 & 0.8 & 0.6 & 0.5 & 1 & 0.6 & 0.6 & 0.6 & 0 \\ 0.6 & 0.7 & 0.8 & 0.6 & 0.5 & 0.6 & 1 & 0.7 & 0.7 & 0.7 \\ 0.6 & 0.7 & 0.8 & 0.6 & 0.5 & 0.6 & 0.7 & 1 & 0.8 & 0.8 \\ 0.6 & 0.7 & 0.8 & 0.6 & 0.5 & 0.6 & 0.7 & 0.8 & 1 & 0.8 \\ 0 & 0 & 0 & 0.6 & 0 & 0 & 0.7 & 0.8 & 0.8 & 1 \end{bmatrix},$$

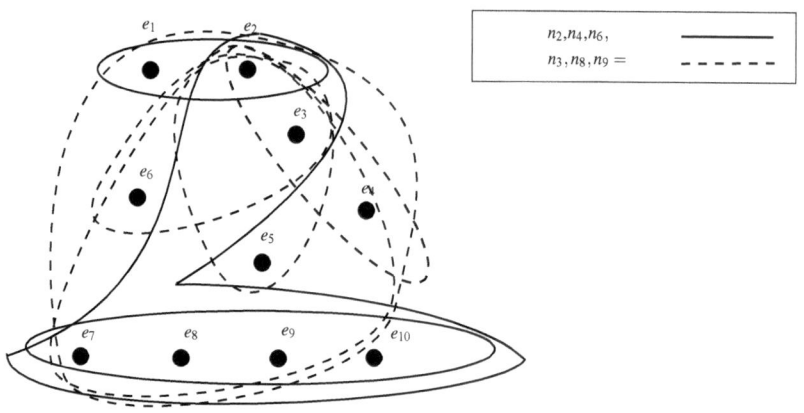

Fig. 8.9 A single-level model of 4-polar fuzzy hypergraph

where $\mathbf{1} = (1, 1, 1, 1)$, $\mathbf{0} = (0, 0, 0, 0)$, $\mathbf{0.5} = (0.5, 0.5, 0.5, 0.5)$, $\mathbf{0.6} = (0.6, 0.6, 0.6, 0.6)$, $\mathbf{0.7} = (0.7, 0.7, 0.7, 0.7)$ and $\mathbf{0.8} = (0.8, 0.8, 0.8, 0.8)$. Let $\tau = (t_1, t_2, t_3, t_4) = (0.7, 0.7, 0.7, 0.7)$, then its corresponding hierarchical quotient space structure is given as follows:

$$X/N_\tau = X/N_{(0.7, 0.7, 0.7, 0.7)} = \{\{e_1, e_2\}, \{e_1, e_2, e_3, e_4, e_5, e_6, e_7, e_8, e_9\}, \{e_2, e_3, e_5\},$$
$$\{e_2, e_3, e_4, e_5, e_6, e_7, e_8, e_9\}, \{e_2, e_3, e_4\}, \{e_2, e_3, e_6\},$$
$$\{e_2, e_3, e_7, e_8, e_9, e_{10}\}, \{e_7, e_8, e_9, e_{10}\}\}.$$

Note that, $n_1 = n_5 = n_7 = n_{10} = \{\emptyset\}$, $n_2 = \{(e_1, e_2)\}$, $n_3 = \{(e_2, e_3, e_4), (e_2, e_3, e_5), (e_2, e_3, e_6)\}$, $n_4 = \{(e_7, e_8, e_9, e_{10})\}$, $n_6 = \{(e_2, e_3, e_7, e_8, e_9, e_{10})\}$, $n_8 = \{(e_2, e_3, e_4, e_5, e_6, e_7, e_8, e_9)\}$, $n_9 = \{(e_1, e_2, e_3, e_4, e_5, e_6, e_7, e_8, e_9)\}$. Hence, a single level of 4-polar fuzzy hypergraph model is constructed and is shown in Fig. 8.9.

Thus, we can obtain eight hyperedges $E_1 = \{e_1, e_2\}$, $E_2 = \{e_2, e_3, e_4\}$, $E_3 = \{e_2, e_3, e_5\}$, $E_4 = \{e_2, e_3, e_6\}$, $E_5 = \{e_7, e_8, e_9, e_{10}\}$, $E_6 = \{e_2, e_3, e_7, e_8, e_9, e_{10}\}$, $E_7 = \{e_2, e_3, e_4, e_5, e_6, e_7, e_8, e_9\}$, $E_8 = \{e_1, e_2, e_3, e_4, e_5, e_6, e_7, e_8, e_9\}$. The procedure of constructing this single-level model is explained in the following flow chart Fig. 8.10.

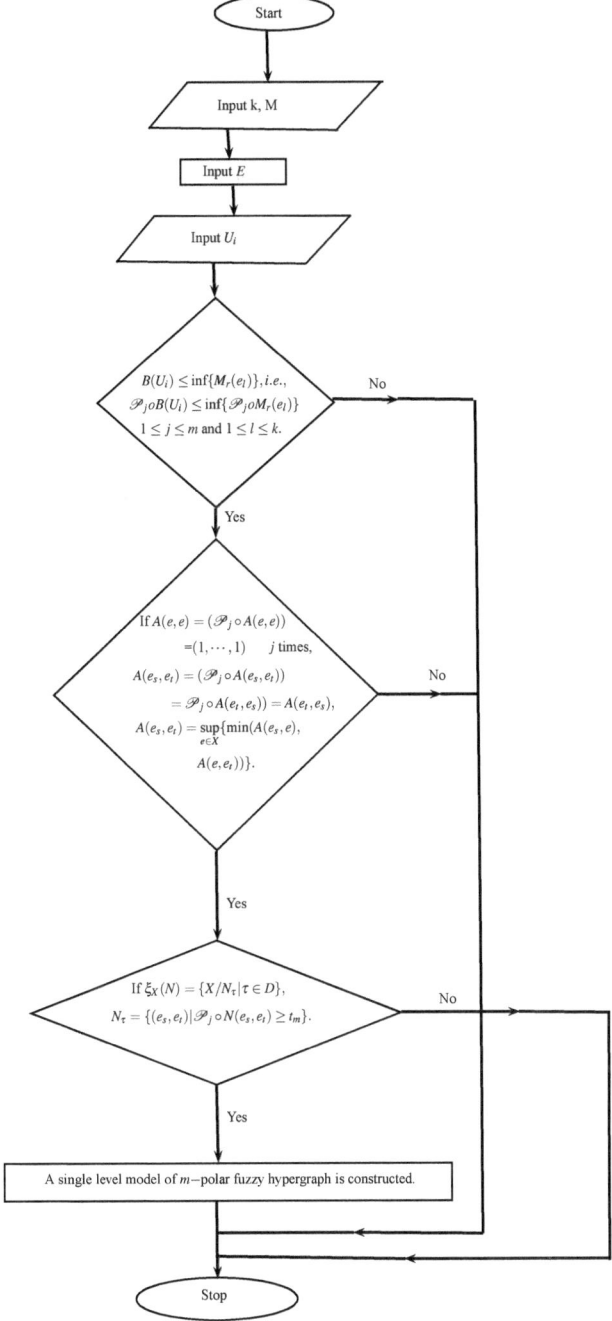

Fig. 8.10 Flow chart of single-level model of m-polar fuzzy hypergraph

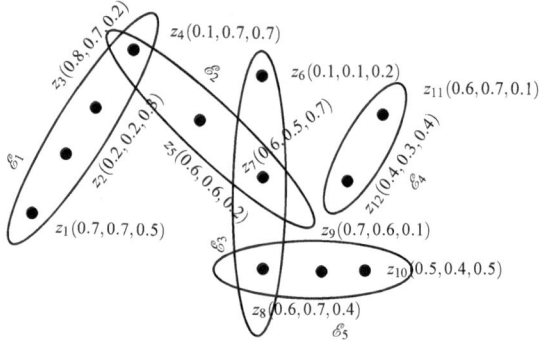

Fig. 8.11 A 3-polar fuzzy hypergraph

Fig. 8.12 $(0.5, 0.5, 0.6)$-
level hypergraph of H

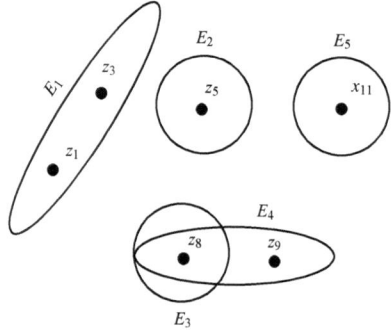

Example 8.4 Let $H = (A, B)$ be a 3-polar fuzzy hypergraph as shown in Fig. 8.11.
Let $X = \{z_1, z_2, z_3, z_4, z_5, z_6, z_7, z_8, z_9, z_{10}, z_{11}, z_{12}\}$ and $B = \{\mathscr{E}_1, \mathscr{E}_2, \mathscr{E}_3, \mathscr{E}_4, \mathscr{E}_5\}$.

For $t_1 = 0.5$, $t_2 = 0.5$ and $t_3 = 0.6$, the $(0.5, 0.5, 0.6)$-level hypergraph of H is
given in Fig. 8.12.

By considering the fixed t_1, t_2, t_3 and following the Algorithm 8.4.1, the
bottom-up construction of this model is given in Fig. 8.13.

The possible method for *the bottom-up construction* is described in Algorithm 8.5.1.

Algorithm 8.5.1

Algorithm for the method of the bottom-up construction

```
1.  clc
2.  𝒫_j ∘ z_i=input('𝒫_j ∘ z_i='); T=input('τ='); q=1;
3.      while q==1
4.      [r, m]=size(𝒫_j ∘ z_i);N=zeros(r, r);N=input('N='); [r_1, r]=size(N); D=ones(r_1, m)+1;
5.          for l=1:r_1
6.              if N(l,:)==zeros(1, r)
7.                  D(l,:)=zeros(1, m);
8.              else
9.                      for k=1:r
10.                         if N(l, k)==1
11.                             for j=1:m
12.                                 D(l, j)=min(D(l, j),𝒫_j ∘ z_i(k, j));
13.                             end
14.                         else
15.                             s=0;
16.                         end
17.                     end
18.             end
19.     end
20.     D
21.     𝒫_j ∘ ℰ_i=input('𝒫_j ∘ ℰ_i=');
22.     if size(𝒫_j ∘ ℰ_i)==[r_1, m]
23.         if 𝒫_j ∘ ℰ_i <=D
24.             if size(T)==[1, m]
25.                 S=zeros(r_1, r);s=zeros(r_1, 1);
26.                     for l=1:r_1
27.                         for k=1:r
28.                             if N(l, k)==1
29.                                 if 𝒫_j ∘ z_i(k,:)>=T(1,:)
30.                                     S(l, k)=1;
31.                                     s(l, 1)=s(l, 1)+1;
32.                                 else
33.                                     S(l, k)=0;
34.                                 end
35.                             end
36.                         end
37.                     end
38.                 S
39.                 if s==ones(r_1, 1)
40.                     q=2;
41.                 else
42.                     𝒫_j ∘ z_i = 𝒫_j ∘ ℰ_i;
43.                 end
44.             else
45.                 fprintf('error')
46.             end
47.         else
48.             fprintf('error')
49.         end
50.     else
51.         fprintf('error')
52.     end
53.  end
```

Fig. 8.13 Bottom-up
construction procedure

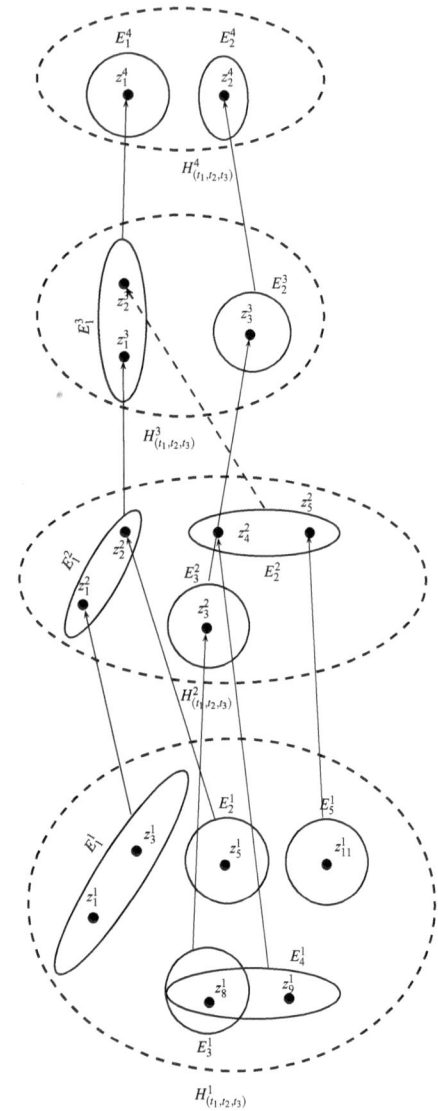

References

1. Akram, M.: m-polar fuzzy graphs: theory, methods & applications. Studies in Fuzziness and Soft Computing, vol. 371, pp. 1–284. Springer (2019)
2. Akram, M.: Fuzzy Lie algebras. Studies in Fuzziness and Soft Computing, vol. 9, pp. 1–302. Springer (2018)
3. Akram, M., Luqman, A.: Intuitionistic single-valued neutrosophic hypergraphs. OPSEARCH **54**(4), 799–815 (2017)
4. Akram, M., Luqman, A.: Bipolar neutrosophic hypergraphs with applications. J. Intell. Fuzzy Syst. **33**(3), 1699–1713 (2017)
5. Akram, M., Sarwar, M.: Novel applications of m-polar fuzzy hypergraphs. J. Intell. Fuzzy Syst. **32**(3), 2747–2762 (2016)
6. Akram, M., Sarwar, M.: Transversals of m-polar fuzzy hypergraphs with applications. J. Intell. Fuzzy Syst. **33**(1), 351–364 (2017)
7. Akram, M., Shahzadi, G.: Hypergraphs in m-polar fuzzy environment. Mathematics **6**(2), 28 (2018). https://doi.org/10.3390/math6020028
8. Akram, M., Shahzadi, G.: Directed hypergraphs under m-polar fuzzy environment. J. Intell. Fuzzy Syst. **34**(6), 4127–4137 (2018)
9. Akram, M., Shahzadi, G., Shum, K.P.: Operations on m-polar fuzzy r-uniform hypergraphs. Southeast Asian Bull. Math. (2019)
10. Berge, C.: Graphs and Hypergraphs. North-Holland, Amsterdam (1973)
11. Chen, G., Zhong, N., Yao, Y.: A hypergraph model of granular computing. In: IEEE International Conference on Granular Computing, pp. 130–135 (2008)
12. Chen, J., Li, S., Ma, S., Wang, X.: m-polar fuzzy sets: an extension of bipolar fuzzy sets. Sci. World J. **8** (2014). https://doi.org/10.1155/2014/416530
13. Kaufmann, A.: Introduction a la Thiorie des Sous-Ensemble Flous, vol. 1. Masson, Paris (1977)
14. Lee, H.S.: An optimal algorithm for computing the maxmin transitive closure of a fuzzy similarity matrix. Fuzzy Sets Syst. **123**, 129–136 (2001)
15. Lin, T.Y.: Granular computing. Announcement of the BISC Special Interest Group on Granular Computing (1997)
16. Liu, Q., Jin, W.B., Wu, S.Y., Zhou, Y.H.: Clustering research using dynamic modeling based on granular computing. In: Proceeding of IEEE International Conference on Granular Computing, pp. 539–543 (2005)
17. Luqman, A., Akram, M., Koam, A.N.: Granulation of hypernetwork models under the q-rung picture fuzzy environment. Mathematics **7**(6), 496 (2019)
18. Luqman, A., Akram, M., Koam, A.N.: An m-polar fuzzy hypergraph model of granular computing. Symmetry **11**, 483 (2019)
19. Mordeson, J.N., Nair, P.S.: Fuzzy Graphs and Fuzzy Hypergraphs, 2nd edn. Physica Verlag, Heidelberg (2001)
20. Rosenfeld, A.: Fuzzy graphs. In: Zadeh, L.A., Fu, K.S., Shimura, M. (eds.) Fuzzy Sets and Their Applications, pp. 77–95. Academic Press, New York (1975)
21. Wang, Q., Gong, Z.: An application of fuzzy hypergraphs and hypergraphs in granular computing. Inf. Sci. **429**, 296–314 (2018)
22. Wong, S.K.M., Wu, D.: Automated mining of granular database scheme. In: Proceeding of IEEE International Conference on Fuzzy Systems, pp. 690–694 (2002)
23. Yang, J., Wang, G., Zhang, Q.: Knowledge distance measure in multigranulation spaces of fuzzy equivalence relation. Inf. Sci. **448**, 18–35 (2018)
24. Yao, Y.Y.: A partition model of granular computing. In: LNCS, vol. 3100, 232–253 (2004)
25. Zadeh, L.A.: Fuzzy sets. Inf. Control **8**(3), 338–353 (1965)
26. Zadeh, L.A.: Similarity relations and fuzzy orderings. Inf. Sci. **3**(2), 177–200 (1971)
27. Zadeh, L.A.: The concept of a linguistic and application to approximate reasoning-I. Inf. Sci. **8**, 199–249 (1975)
28. Zadeh, L.A.: Toward a generalized theory of uncertainty (GTU) an outline. Inf. Sci. **172**, 1–40 (2005)

29. Zhang, W.R., Bipolar fuzzy sets and relations: a computational framework for cognitive modeling and multiagent decision analysis. Proc. IEEE Conf. 305–309 (1994)
30. Zhang, L., Zhang, B.: The structural analysis of fuzzy sets. J. Approx. Reason. **40**, 92–108 (2005)
31. Zhang, L., Zhang, B.: The Theory and Applications of Problem Solving-Quotient Space Based Granular Computing. Tsinghua University Press, Beijing (2007)
32. Zhang, L., Zhang, B.: Hierarchy and Multi-granular Computing, Quotient Space Based Problem Solving, pp. 45–103. Tsinghua University Press, Beijing (2014)

Chapter 9
Some Types of Hypergraphs for Single-Valued Neutrosophic Structures

In this chapter, we present concepts including single-valued neutrosophic hypergraphs, dual single-valued neutrosophic hypergraphs, and transversal single-valued neutrosophic hypergraphs. Additionally, we discuss the notions of intuitionistic single-valued neutrosophic hypergraphs and dual intuitionistic single-valued neutrosophic hypergraphs. We describe an application of intuitionistic single-valued neutrosophic hypergraphs in a clustering problem. Then, we present other related concepts like single-valued neutrosophic directed hypergraphs, single-valued neutrosophic line directed graphs, and dual single-valued neutrosophic directed hypergraphs. Finally, we describe applications of single-valued neutrosophic directed hypergraphs. We define complex neutrosophic hypergraphs and discuss their certain properties including, lower truncation, upper truncation, and transition levels. Further, we define T-related complex neutrosophic hypergraphs and properties of minimal transversals of complex neutrosophic hypergraphs. We represent the modeling of certain social networks having intersecting communities through the score functions and choice values of complex neutrosophic hypergraphs. This chapter is mainly due to [3–5, 14, 20, 22–24].

9.1 Introduction

Zadeh [28] introduced the degree of membership/truth (T) in 1965 and defined the fuzzy set. Atanassov [8] introduced the degree of nonmembership/falsehood (F) in 1983 and defined the intuitionistic fuzzy set. Smarandache [20] introduced the degree of indeterminacy/neutrality (I) as independent component in 1995 and defined the neutrosophic set on three components $(T, I, F) =$ (Truth, Indeterminacy, Falsity). Fuzzy set theory and intuitionistic fuzzy set theory are useful models for dealing with uncertainty and incomplete information. But they may not be sufficient in the modeling of indeterminate and inconsistent information encountered in the real world.

© Springer Nature Singapore Pte Ltd. 2020
M. Akram and A. Luqman, *Fuzzy Hypergraphs and Related Extensions*,
Studies in Fuzziness and Soft Computing 390,
https://doi.org/10.1007/978-981-15-2403-5_9

In order to cope with this issue, (The words "neutrosophy" and "neutrosophic" were invented by Smarandache in 1995. Neutrosophy is a new branch of philosophy that studies the origin, nature, and scope of neutralities, as well as their interactions with different ideational spectra. It is the base of neutrosophic logic, a multiple value logic that generalizes the fuzzy logic and deals with paradoxes, contradictions, antitheses, antinomies) neutrosophic set theory was proposed by Smarandache. However, since neutrosophic sets are identified by three functions called truth-membership (T), indeterminacy-membership (I), and falsity-membership (F), whose values are real standard or nonstandard subset of unit interval $]^-0, 1^+[$, where $^-0 = 0 - \varepsilon$, $1^+ = 1 + \varepsilon$, ε is an infinitesimal number. To apply neutrosophic set in real-life problems more conveniently, Smarandache [20] and Wang et al. [23] defined single-valued neutrosophic set, which takes the value from the subset of [0, 1]. Thus, a single-valued neutrosophic set is an instance of neutrosophic set, and can be used expediently to deal with real-world problems, especially in decision support. The innovative concept of complex fuzzy sets was initiated by Ramot et al. [17] as an extension of fuzzy sets. A complex fuzzy set is characterized by a membership function $\mu(x)$, whose range is not limited to [0, 1] but extends to the unit circle in the complex plane. Hence, $\mu(x)$ is a complex-valued function that assigns a grade of membership of the form $r(x)e^{i\alpha(x)}$, $i = \sqrt{-1}$ to any element x in the universe of discourse. Thus, the membership function $\mu(x)$ of complex fuzzy set consists of two terms, i.e., amplitude term $r(x)$ which lies in the unit interval [0, 1] and phase term (periodic term) $w(x)$ which lies in the interval $[0, 2\pi]$. This phase term distinguishes a complex fuzzy set model from all other models available in the literature. The potential of a complex fuzzy set for representing two-dimensional phenomena makes it superior to handle ambiguous and intuitive information that are prevalent in time-periodic phenomena. A systematic review of complex fuzzy sets was proposed by Yazdanbakhsh and Dick [27]. To generalize the concepts of intuitionistic fuzzy sets, complex intuitionistic fuzzy sets were introduced by Alkouri and Salleh [7] by adding nonmembership $v(x) = s(x)e^{i\beta(x)}$ to the complex fuzzy sets subjected to the constraint $r + s \leq 1$. To handle imprecise information having a periodic nature, complex neutrosophic sets were proposed by Ali and Smarandache [6]. As we see that uncertainty, indeterminacy, incompleteness, inconsistency, and falsity in data are periodic in nature, to handle these types of problems, the complex neutrosophic set plays an important role. A complex neutrosophic set is characterized by a complex-valued truth-membership function $T(x)$, complex-valued indeterminate membership function $I(x)$, and complex-valued false membership function $F(x)$, whose range is extended from [0, 1] to the unit disk in the complex plane.

Graphs are used to represent the pair-wise relationships between objects. However, in many real-world phenomena, sometimes relationships are much problematic that they cannot be perceived through simple graphs. By handling such complex relationships by pair-wise connections naively, one can face the loss of data which is considered to be worthwhile for learning errands. To overcome these difficulties, we take into account the generalization of simple graphs, named as hypergraphs, to personify the complex relationships. A hypergraph is an extension of a classical graph in this way that a hyperedge can combine two or more than two vertices.

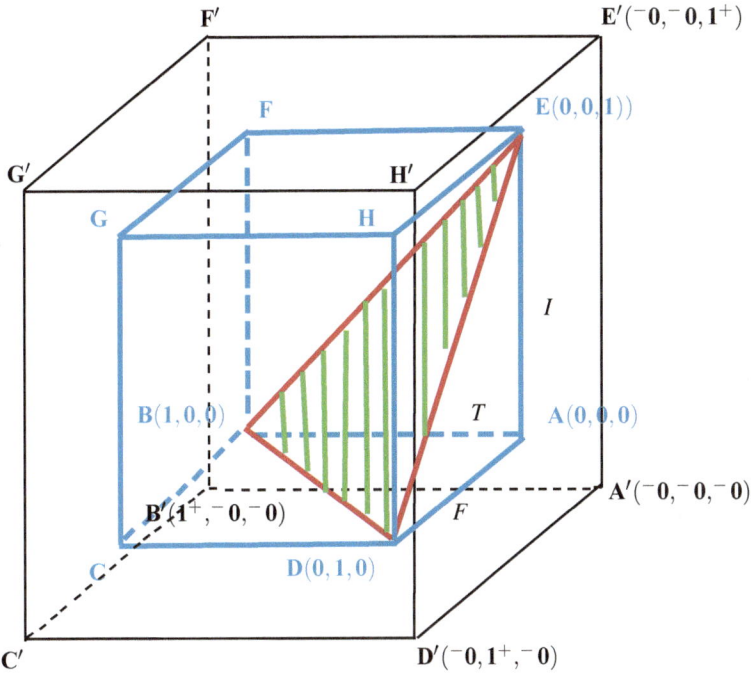

Fig. 9.1 A geometric interpretation of the neutrosophic set

Hypergraphs have many applications in various fields, including biological sciences, computer science, and natural sciences. To study the degree of dependence of an object to the other, Kaufmann [13] applied the concept of fuzzy sets to hypergraphs. Mordeson and Nair [15] presented fuzzy graphs and fuzzy hypergraphs. Parvathi et al. [16] established the notion of intuitionistic fuzzy hypergraph.

A Geometric Interpretation of the Neutrosophic Set

We describe a geometric interpretation of the neutrosophic set using the neutrosophic cube $A'B'C'D'E'F'G'H'$ as shown in Fig. 9.1. In technical applications, only the classical interval $[0, 1]$ is used as range for the neutrosophic parameters T, I, and F, we call the cube $ABCDEFGH$ the technical neutrosophic cube and its extension $A'B'C'D'E'F'G'H'$ the neutrosophic cube, used in the field where we need to differentiate between absolute and relative notions. Consider a 3-D Cartesian system of coordinates, where T is the truth axis with value range in $]^-0, 1^+[$, F is the false axis with value range in $]^-0, 1^+[$, and I is the indeterminate axis with value range in $]^-0, 1^+[$.

We now divide the technical neutrosophic cube $ABCDEFGH$ into three disjoint regions:

1. The equilateral triangle BDE, whose sides are equal to $\sqrt{2}$, which represents the geometrical locus of the points whose sum of the coordinates is 1. If a point Q is situated on the sides of the triangle BDE or inside of it, then $T_Q + I_Q + F_Q = 1$.
2. The pyramid $EABD$ situated in the right side of the $\triangle EBD$, including its faces $\triangle ABD$ (base), $\triangle EBA$, and $\triangle EDA$ (lateral faces), but excluding its faces $\triangle BDE$ is the locus of the points whose sum of their coordinates is less than 1. If $P \in EABD$, then $T_P + I_P + F_P < 1$.
3. In the left side of $\triangle BDE$ in the cube, there is the solid $EFGCDEBD$ (excluding $\triangle BDE$) which is the locus of points whose sum of their coordinates is greater than 1. If a point $R \in EFGCDEBD$, then $T_R + I_R + F_R > 1$.

It is possible to get the sum of coordinates strictly less than 1 or strictly greater than 1. For example,

(1) We have a source which is capable to find only the degree of membership of an element; but it is unable to find the degree of nonmembership.

(2) Another source which is capable to find only the degree of nonmembership of an element.

(3) Or a source which only computes the indeterminacy.

Thus, when we put the results together of these sources, it is possible that their sum is not 1, but smaller or greater.

On the other hand, in information fusion, when dealing with indeterminate models (i.e., elements of the fusion space which are indeterminate/unknown, such as intersections we do not know if they are empty or not since we do not have enough information, similarly for complements of indeterminate elements, etc.), if we compute the believe in that element (truth), the disbelieve in that element (falsehood), and the indeterminacy part of that element, then the sum of these three components is strictly less than 1 (the difference to 1 is the missing information).

Definition 9.1 Let X be a space of points (objects). A *single-valued neutrosophic set A* on a non-empty set X is characterized by a truth-membership function $T_A : X \rightarrow [0, 1]$, indeterminacy-membership function $I_A : X \rightarrow [0, 1]$, and a falsity-membership function $F_A : X \rightarrow [0, 1]$. Thus, $A = \{< x, T_A(x), I_A(x), F_A(x) > | x \in X\}$. There is no restriction on the sum of $T_A(x)$, $I_A(x)$ and $F_A(x)$ for all $x \in X$.

When X is continuous, a single-valued neutrosophic set A can be written as

$$A = \int_X \langle (T(x), I(x), F(x)) / x, x \in X \rangle.$$

When X is discrete, a single-valued neutrosophic set A can be written as

$$A = \sum_{i=1}^{n} \langle (T(x_i), I(x_i), F(x_i)) / x_i, x_i \in X \rangle.$$

Example 9.1 Assume that the universe of discourse $X = \{x_1, x_2, x_3\}$, where x_1 describes the capability, x_2 describes the trustworthiness, and x_3 describes the prices of the objects. It may be further assumed that the values of x_1, x_2, and x_3 are in [0, 1] and they are obtained from some questionnaires of some experts. The experts may impose their opinion in three components, namely, the degree of goodness, the degree of indeterminacy, and that of poorness to explain the characteristics of the objects. Suppose A is a single-valued neutrosophic set of X such that

$$A = \{< x_1, 0.3, 0.5, 0.6 >, < x_2, 0.3, 0.2, 0.3 >, < x_3, 0.3, 0.5, 0.6 >\},$$

where $< x_1, 0.3, 0.5, 0.6 >$ represents the degree of goodness of capability is 0.3, degree of indeterminacy of capability is 0.5, and degree of falsity of capability is 0.6.

Remark 9.1 When we consider that there are three different experts that are independent (i.e., they do not communicate with each other), so each one focuses on one attribute only (because each one is the best specialist in evaluating a single attribute). Therefore, each expert can assign 1 to his attribute value [for (1,1,1)], or each expert can assign 0 to his attribute value [for (0,0,0)], respectively.

When we consider a single expert for evaluating all three attributes, then he evaluates each attribute from a different point of view (using a different parameter) and arrives to (1,1,1) or (0,0,0), respectively.

For example, we examine a student "Muhammad" for his research in neutrosophic graphs he deserves 1, for his research in analytical mathematics he also deserves 1, and for his research in physics he deserves 1.

Yang et al. [24] introduced the concept of single-valued neutrosophic relations.

Definition 9.2 A *single-valued neutrosophic relation* on a non-empty set X is a single-valued neutrosophic subset of $X \times X$ of the form, $B = \{(yz, T_B(yz), I_B(yz), F_B(yz)) : yz \in X \times X\}$, where $T_B : X \times X \to [0, 1], I_B : X \times X \to [0, 1], F_B : X \times X \to [0, 1]$ denote the truth-membership, indeterminacy-membership, and falsity-membership functions of B, respectively.

Definition 9.3 A *single-valued neutrosophic graph* on a non-empty X is a pair $G = (A, B)$, where A is single-valued neutrosophic set in X and B single-valued neutrosophic relation on X such that

$$T_B(xy) \leq \min\{T_A(x), T_A(y)\},$$

$$I_B(xy) \leq \min\{I_A(x), I_A(y)\},$$

$$F_B(xy) \leq \max\{F_A(x), F_A(y)\},$$

for all $x, y \in X$. A is called *single-valued neutrosophic vertex set* of G and B is called *single-valued neutrosophic edge set* of G, respectively.

Remark 9.2 1. *B* is called symmetric single-valued neutrosophic relation on *A*.
2. If *B* is not a symmetric single-valued neutrosophic relation on *A*, then $G = (A, B)$ is called a *single-valued neutrosophic directed graph (digraph)*.
3. *X* and *E* are underlying vertex set and underlying edge set of *G*, respectively.

Definition 9.4 The *support* of a single-valued neutrosophic set $A = \{(x, T_A(x), I_A(x), F_A(x)): x \in X\}$ is denoted by supp(*A*), defined as $supp(A) = \{x \mid T_A(x) \neq 0, I_A(x) \neq 0, F_A(x) \neq 0\}$. The support of a single-valued neutrosophic set is a crisp set.

Definition 9.5 The *height* of a single-valued neutrosophic set $A = \{(x, T_A(x), I_A(x), F_A(x)) : x \in X\}$ is defined as $h(A) = (\sup_{x \in X} T_A(x), \sup_{x \in X} I_A(x), \inf_{x \in X} F_A(x))$. We call single-valued neutrosophic set *A* is *normal* if there exists at least one element $x \in X$ such that $T_A(x) = 1, I_A(x) = 1, F_A(x) = 0$.

Definition 9.6 Let $A = \{(x, T_A(x), I_A(x), F_A(x)) : x \in X\}$ be a single-valued neutrosophic set on *X* and let $\alpha, \beta, \gamma \in [0, 1]$ such that $\alpha + \beta + \gamma \leq 3$. Then, the set $A_{(\alpha, \beta, \gamma)} = \{x \mid T_A(x) \geq \alpha, I_A(x) \geq \beta, F_A(x) \leq \gamma\}$ is called (α, β, γ)-*level subset* of *A*. (α, β, γ)-level set is a crisp set.

For further terminologies and studies on single-valued neutrosophic theory, readers are referred to [1, 2, 9, 11, 12, 18, 19, 21, 25, 26].

9.2 Single-Valued Neutrosophic Hypergraphs

Definition 9.7 A *single-valued neutrosophic hypergraph* on a non-empty set *X* is a pair $\mathscr{H} = (X, \varepsilon)$, where *X* is a crisp set of vertices and $\varepsilon = \{E_1, E_2, \cdots, E_m\}$ be a finite family of nontrivial single-valued neutrosophic subsets of *X* such that

(i)
$$T_\varepsilon(\{x_1, x_2, \ldots, x_s\}) \leq \min\{T_{E_i}(x_1), T_{E_i}(x_2), \ldots, T_{E_i}(x_s)\},$$

$$I_\varepsilon(\{x_1, x_2, \ldots, x_s\}) \leq \min\{I_{E_i}(x_1), I_{E_i}(x_2), \ldots, I_{E_i}(x_s)\},$$

$$F_\varepsilon(\{x_1, x_2, \ldots, x_s\}) \leq \max\{F_{E_i}(x_1), F_{E_i}(x_2), \ldots, F_{E_i}(x_s)\},$$

for all $x_1, x_2, \ldots, x_s \in X$.
(ii) $\bigcup_i supp(E_i) = X$, for all $E_i \in \varepsilon$.

In single-valued neutrosophic hypergraph, two vertices *u* and *v* are *adjacent* if there exists an edge $E_i \in \varepsilon$ which contains the two vertices *v* and *u*, i.e., $u, v \in supp(E_i)$. In single-valued neutrosophic hypergraph $\mathscr{H}$, if two vertices *u* and *v* are connected, then there exists a sequence $u = u_0, u_1, u_2, \ldots, u_n = v$ of vertices of $\mathscr{H}$ such that u_{i-1} is adjacent u_i for $i = 1, 2, \ldots, n$. A *connected single-valued neutrosophic hypergraph* is a single-valued neutrosophic hypergraph in which every pair of vertices is connected.

In a single-valued neutrosophic hypergraph, two edges E_i and E_j are said to be *adjacent* if their intersection is non-empty, i.e., $supp(E_i) \cap supp(E_j) \neq \emptyset, i \neq j$. The *order* of a single-valued neutrosophic hypergraph is denoted by $|X|$ and *size* (number of edges) is denoted by $|\varepsilon|$. If $supp(E_i) = k$ for each $E_i \in \varepsilon$, then single-valued neutrosophic hypergraph $\mathscr{H} = (X, \varepsilon)$ is *k-uniform single-valued neutrosophic hypergraph*. The element a_{ij} of the single-valued neutrosophic matrix represents the truth-membership (participation) degree, indeterminacy-membership degree, and falsity-membership of v_i to E_j (that is $(T_{E_j}(v_i), I_{E_j}(v_i), F_{E_j}(v_i))$). Since, the diagram of single-valued neutrosophic hypergraph does not imply sufficiently the truth-membership degree, indeterminacy-membership degree, and falsity-membership degree of vertex to edges, we use incidence matrix $M_{\mathscr{H}}$ for the description of single-valued neutrosophic hyperedges.

Definition 9.8 The *height* of a single-valued neutrosophic hypergraph $\mathscr{H} = (X, \varepsilon)$, denoted by $h(\mathscr{H})$, is defined by $h(\mathscr{H}) = \max_i \{h(E_i)| E_i \in \varepsilon\}$.

Definition 9.9 Let $\mathscr{H} = (X, \varepsilon)$ be a single-valued neutrosophic hypergraph, the *cardinality* of a single-valued neutrosophic hyperedge is the sum of truth-membership, indeterminacy-membership, and falsity-membership values of the vertices connected to an hyperedge, it is denoted by $|E_i|$. The *degree of a single-valued neutrosophic hyperedge*, $E_i \in \varepsilon$ is its cardinality, that is $d_{\mathscr{H}}(E_i) = |E_i|$. The *rank* of a single-valued neutrosophic hypergraph is the maximum cardinality of any hyperedge in $\mathscr{H}$, i.e., $\max_{E_i \in \varepsilon} d_{\mathscr{H}}(E_i)$ and *anti rank* of a single-valued neutrosophic is the minimum cardinality of any hyperedge in $\mathscr{H}$, i.e., $\min_{E_i \in \varepsilon} d_{\mathscr{H}}(E_i)$.

Definition 9.10 A single-valued neutrosophic hypergraph is said to be *linear single-valued neutrosophic hypergraph* if every pair of distinct vertices of $\mathscr{H} = (X, \varepsilon)$ is in at most one edge of $\mathscr{H}$, i.e., $|supp(E_i) \cap supp(E_j)| \leq 1$ for all $E_i, E_j \in \varepsilon$. A 2-uniform linear single-valued neutrosophic hypergraph is a single-valued neutrosophic graph.

Example 9.2 Consider a single-valued neutrosophic hypergraph $\mathscr{H} = (X, \varepsilon)$ such that $X = \{v_1, v_2, v_3, v_4, v_5, v_6\}$, $\varepsilon = \{E_1, E_2, E_3, E_4, E_5, E_6\}$, where

$$E_1 = \{(v_1, 0.3, 0.4, 0.6), (v_3, 0.7, 0.4, 0.4)\},$$
$$E_2 = \{(v_1, 0.3, 0.4, 0.6), (v_2, 0.5, 0.7, 0.6)\},$$
$$E_3 = \{(v_2, 0.5, 0.7, 0.6), (v_4, 0.6, 0.4, 0.8)\},$$
$$E_4 = \{(v_3, 0.7, 0.4, 0.4), (v_6, 0.4, 0.2, 0.7)\},$$
$$E_5 = \{(v_3, 0.7, 0.4, 0.4), (v_5, 0.6, 0.7, 0.5)\},$$
$$E_6 = \{(v_5, 0.6, 0.7, 0.5), (v_6, 0.4, 0.2, 0.7)\},$$
$$E_7 = \{(v_4, 0.6, 0.4, 0.8), (v_6, 0.4, 0.2, 0.7)\}.$$

The corresponding single-valued neutrosophic hypergraph is shown in Fig. 9.2.
Its incidence matrix $M_{\mathscr{H}}$ is given in Table 9.1.

Fig. 9.2 Single-valued
neutrosophic hypergraph

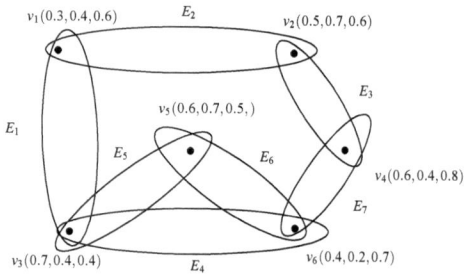

Table 9.1 Incidence matrix $M_{\mathcal{H}}$ of $\mathcal{H}$

$M_{\mathcal{H}}$	E_1	E_2	E_3	E_4	E_5	E_6	E_7
v_1	(0.3, 0.4, 0.6)	(0.3, 0.4, 0.6)	(0, 0, 0)	(0, 0, 0)	(0, 0, 0)	(0, 0, 0)	(0, 0, 0)
v_2	(0, 0, 0)	(0.5, 0.7, 0.6)	(0.5, 0.7, 0.6)	(0, 0., 0)	(0, 0, 0)	(0, 0, 0)	(0, 0, 0)
v_3	(0.7, 0.4, 0.4)	(0, 0, 0)	(0, 0, 0)	(0.7, 0.4, 0.4)	(0.7, 0.4, 0.4)	(0, 0, 0)	(0, 0, 0)
v_4	(0, 0, 0)	(0, 0, 0)	(0.6, 0.4, 0.8)	(0, 0, 0)	(0, 0, 0)	(0, 0, 0)	(0.6, 0.4, 0.8)
v_5	(0, 0, 0)	(0, 0, 0)	(0, 0, 0)	(0, 0, 0)	(0.6, 0.7, 0.5)	(0.6, 0.7, 0.5)	(0, 0, 0)
v_6	(0, 0, 0)	(0, 0, 0)	(0, 0, 0)	(0.4, 0.2, 0.7)	(0, 0, 0)	(0.4, 0.2, 0.7)	(0.4, 0.2, 0.7)

Definition 9.11 Let $\mathcal{H} = (X, \varepsilon)$ be a single-valued neutrosophic hypergraph, the
degree $d_{\mathcal{H}}(v)$ *of a vertex* v *in* $\mathcal{H}$ is $d_{\mathcal{H}}(v) = \sum_{v \in E_i} (T_{E_i}(v), I_{E_i}(v), F_{E_i}(v))$, where
E_i are the edges that contain the vertex v. The *maximum degree of a single-valued
neutrosophic hypergraph* is $\triangle(\mathcal{H}) = \max_{v \in X}(d_{\mathcal{H}}(v))$. A single-valued neutrosophic
hypergraph is said to be *regular single-valued neutrosophic hypergraph* in which all
the vertices have the same degree.

Proposition 9.1 *Let* $\mathcal{H} = (X, \varepsilon)$ *be a single-valued neutrosophic hypergraph, then*
$\sum_{v \in X} d_{\mathcal{H}}(v) = \sum_{E_i \in \varepsilon} d_{\mathcal{H}}(E_i).$

Proof Let $M_{\mathcal{H}}$ be the incidence matrix of single-valued neutrosophic hypergraph
$\mathcal{H}$, then the sum of the degrees of each vertex $v_i \in X$ and the sum of degrees of each
edge $E_i \in \varepsilon$ are equal. We obtain $\sum_{v \in X} d_{\mathcal{H}}(v) = \sum_{E_i \in \varepsilon} d_{\mathcal{H}}(E_i)$.

Definition 9.12 The *strength* η *of a single-valued neutrosophic hyperedge* E_i is the
minimum of truth-membership, indeterminacy-membership, and maximum falsity-
membership values in the edges E_i, i.e.,

$$\eta(E_i) = \{ \min_{v_j \in E_i}(T_{E_i}(v_j) \mid T_{E_i}(v_j) > 0), \ \min_{v_j \in E_i}(I_{E_i}(v_j) \mid I_{E_i}(v_j) > 0), \ \max_{v_j \in E_i}(F_{E_i}(v_j) \mid F_{E_i}(v_j) > 0) \}.$$

The strength of an edge in single-valued neutrosophic hypergraph interprets that the
edge E_i group elements are having participation degree at least $\eta(E_i)$.

Table 9.2 Incidence matrix $M_{\mathscr{H}}$

$M_{\mathscr{H}}$	E_1	E_2	E_3	E_4	E_5	E_6	E_7
v_1	(0.3, 0.4, 0.6)	(0.3, 0.4, 0.6)	(0, 0, 0)	(0, 0, 0)	(0, 0, 0)	(0, 0, 0)	(0, 0, 0)
v_2	(0, 0, 0)	(0.3, 0.4, 0.6)	(0.5, 0.4, 0.8)	(0, 0, 0)	(0, 0, 0)	(0, 0, 0)	(0, 0, 0)
v_3	(0.3, 0.4, 0.6)	(0, 0, 0)	(0, 0, 0)	(0.4, 0.2, 0.7)	(0.6, 0.4, 0.5)	(0, 0, 0)	(0, 0, 0)
v_4	(0, 0, 0)	(0, 0, 0)	(0.5, 0.4, 0.8)	(0, 0, 0)	(0, 0, 0)	(0, 0, 0)	(0.4, 0.2, 0.8)
v_5	(0, 0, 0)	(0, 0, 0)	(0, 0, 0)	(0, 0, 0)	(0.6, 0.4, 0.5)	(0.4, 0.2, 0.7)	(0, 0, 0)
v_6	(0, 0, 0)	(0, 0, 0)	(0, 0, 0)	(0.4, 0.2, 0.7)	(0, 0, 0)	(0.4, 0.2, 0.7)	(0.4, 0.2, 0.8)

Fig. 9.3 Single-valued neutrosophic graph

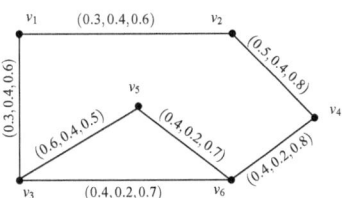

Example 9.3 Consider a single-valued neutrosophic hypergraph as shown in Fig. 9.2, the height of $\mathscr{H}$ is $h(\mathscr{H}) = (0.7, 0.7, 0.4)$, the strength of each edge is $\eta(E_1) = (0.3, 0.4, 0.6)$, $\eta(E_2) = (0.3, 0.4, 0.6)$, $\eta(E_3) = (0.5, 0.4, 0.8)$, $\eta(E_4) = (0.4, 0.2, 0.7)$, $\eta(E_5) = (0.6, 0.4, 0.5)$, $\eta(E_6) = (0.4, 0.2, 0.7)$, and $\eta(E_7) = (0.4, 0.2, 0.8)$, respectively. The edges with high strength are called the strong edges because the inter-relation (cohesion) in them is strong. Therefore, E_5 is stronger than each E_i, for $i = 1, 2, 3, 4, 6, 7$.

If we assign $\eta(E_i) = (T_{\eta(E_i)}, I_{\eta(E_i)}, F_{\eta(E_i)})$ to each clique in single-valued neutrosophic graph mapped to an edge E_i in single-valued neutrosophic hypergraph, we obtain a single-valued neutrosophic graph which represents subset with grouping strength(interrelationship). Consider the incidence matrix as shown in Table 9.2. We see that a single-valued neutrosophic graph can be associated with a single-valued neutrosophic hypergraph, a hyperedge with its strength η in the single-valued neutrosophic hypergraph is mapped to a clique in the single-valued neutrosophic graph, all edges in the clique have the same strength. Figure 9.3 shows corresponding single-valued neutrosophic graph to the single-valued neutrosophic hypergraph $\mathscr{H}$ shown in Fig. 9.2. In the corresponding single-valued neutrosophic graph, the numbers attached to the edges represent the truth-membership, indeterminacy-membership, and falsity-membership of the edges.

Proposition 9.2 *Single-valued neutrosophic graphs and single-valued neutrosophic digraphs are special cases of the single-valued neutrosophic hypergraphs.*

Definition 9.13 A single-valued neutrosophic set $A = \{(x, T_A(x)), I_A(x), F_A(x) \mid x \in X\}$ is an *elementary single-valued neutrosophic set* if A is single valued on $supp(A)$. An *elementary single-valued neutrosophic hypergraph* $\mathscr{H} = (X, \varepsilon)$ is a single-valued neutrosophic hypergraph in which each element of ε is elementary.

Definition 9.14 A single-valued neutrosophic hypergraph $\mathcal{H} = (X, \varepsilon)$ is called *simple* if ε has no repeated single-valued neutrosophic hyperedges and whenever $E_i, E_j \in \varepsilon$ and $T_{E_i} \leq T_{E_j}, I_{E_i} \leq I_{E_j}, F_{E_i} \geq F_{E_j}$, then $T_{E_i} = T_{E_j}, I_{E_i} = I_{E_j}, F_{E_i} = F_{E_j}$.

A single-valued neutrosophic hypergraph is called *support simple*, if whenever $E_i, E_j \in \varepsilon, E_i \subset E_j$ and $supp(E_i) = supp(E_j)$, then $E_i = E_j$.

A single-valued neutrosophic hypergraph is called *strongly support simple* if whenever $E_i, E_j \in \varepsilon$, and $supp(E_i) = supp(E_j)$, then $E_i = Ej$.

Definition 9.15 Let $\mathcal{H} = (X, \varepsilon)$ be a single-valued neutrosophic hypergraph. Suppose that $\alpha, \beta, \gamma \in [0, 1]$. Let $E^{(\alpha,\beta,\gamma)} = \{E_i^{(\alpha,\beta,\gamma)} \mid E_i \in \varepsilon\}$ and $X^{(\alpha,\beta,\gamma)} = \bigcup_{E_i \in \varepsilon} E_i^{(\alpha,\beta,\gamma)}$. $\mathcal{H}^{(\alpha,\beta,\gamma)} = (X^{(\alpha,\beta,\gamma)}, E^{(\alpha,\beta,\gamma)})$ is the (α, β, γ)-*level hypergraph* of $\mathcal{H} = (X, \varepsilon)$, where $E^{(\alpha,\beta,\gamma)} \neq \emptyset$. $\mathcal{H}^{(\alpha,\beta,\gamma)}$ is a crisp hypergraph.

Remark 9.3 1. A single-valued neutrosophic hypergraph $\mathcal{H} = (X, \varepsilon)$ is a single-valued neutrosophic graph (with loops) if and only if $\mathcal{H}$ is elementary, support simple, and each edge has two (or one) element support.

2. For a simple single-valued neutrosophic hypergraph $\mathcal{H} = (X, \varepsilon)$, (α, β, γ)-level hypergraph $\mathcal{H}^{(\alpha,\beta,\gamma)}$ may or may not be simple single-valued neutrosophic hypergraph. Clearly, it is possible that $E_i^{(\alpha,\beta,\gamma)} = E_j^{(\alpha,\beta,\gamma)}$ for $E_i \neq E_j$.

3. $\mathcal{H}$ and $\mathcal{H}'$ are two families of crisp sets (hypergraphs) produced by the (α, β, γ)-cuts of a single-valued neutrosophic hypergraph which share an important relationship with each other such that for each set $\mathcal{H} \in \mathcal{H}$ there is at least one set $\mathcal{H}' \in \mathcal{H}'$ which contains $\mathcal{H}$. We say that $\mathcal{H}'$ absorbs $\mathcal{H}$, i.e., $\mathcal{H} \subseteq \mathcal{H}'$. Since, it is possible $\mathcal{H}'$ absorbs $\mathcal{H}$ while $\mathcal{H}' \cap \mathcal{H} = \emptyset$, we have that $\mathcal{H} \subseteq \mathcal{H}'$ implies $\mathcal{H} \subseteq \mathcal{H}'$, but the converse is generally false, If $\mathcal{H} \subseteq \mathcal{H}'$ and $\mathcal{H} \neq \mathcal{H}'$, then $\mathcal{H} \subset \mathcal{H}'$.

Definition 9.16 Let $\mathcal{H} = (X, \varepsilon)$ be a single-valued neutrosophic hypergraph, and let $h(\mathcal{H}) = (r, s, t)$, $\mathcal{H}^{(r_i,s_i,t_i)} = (X^{(r_i,s_i,t_i)}, E^{(r_i,s_i,t_i)})$ be the (r_i, s_i, t_i)-level hypergraphs of $\mathcal{H}$. The sequence of real numbers $\{(r_1, s_1, t_1), (r_2, s_2, t_2), \cdots, (r_n, s_n, t_n)\}$, such that $0 < r_n < r_{n-1} < \cdots < r_1 = r$, $0 < s_n < s_{n-1} < \cdots < s_1 = s$, and $t_n > t_{n-1} > \cdots > t_1 = t > 0$, which satisfies the properties,

(i) if $r_{i+1} < r' < r_i, s_{i+1} < s' < s_i, t_{i+1} > t' > t_i (t_i < t' < t_{i+1})$, then $E^{(r',s',t')} = E^{(r_i,s_i,t_i)}$,

(ii) $E^{(r_i,s_i,t_i)} \subset E^{(r_i+1,s_i+1,t_i+1)}$,

is called the *fundamental sequence* of $\mathcal{H}$, and is denoted by $F(\mathcal{H})$ and the set of (r_i, s_i, t_i)-level hypergraphs $\{\mathcal{H}^{(r_1,s_1,t_1)}, \mathcal{H}^{(r_2,s_2,t_2)}, \ldots, \mathcal{H}^{(r_n,s_n,t_n)}\}$ is called the set of *core* hypergraphs of $\mathcal{H}$, and is denoted by $C(\mathcal{H})$.

If $r_1 < r \leq 1, s_1 < s \leq 1, 0 \leq t < t_1$, then $E^{(r,s,t)} = \{\emptyset\}$ and $\mathcal{H}^{(r,s,t)}$ does not exist.

Definition 9.17 Suppose $\mathcal{H} = (X, \varepsilon)$ is a single-valued neutrosophic hypergraph with $F(\mathcal{H}) = \{(r_1, s_1, t_1), (r_2, s_2, t_2), \cdots, (r_n, s_n, t_n)\}$ and $r_{n+1} = 0, s_{n+1} = 0, t_{n+1} = 0$. Then, $\mathcal{H}$ is called *sectionally elementary* if for each $E_i \in E$ and each $(r_i, s_i, t_i) \in F(\mathcal{H})$, $E_i^{(r_i,s_i,t_i)} = E_i^{(r,s,t)}$ for all $(r, s, t) \in \left((r_{i+1}, s_{i+1}, t_{i+1}), (r_i, s_i, t_i)\right)$.

Table 9.3 Incidence matrix

$M_{\mathscr{H}}$	E_1	E_2	E_3	E_4	E_5
v_1	$(0.7, 0.6, 0.5)$	$(0.9, 0.8, 0.1)$	$(0, 0, 0)$	$(0, 0, 0)$	$(0.4, 0.3, 0.3)$
v_2	$(0.7, 0.6, 0.5)$	$(0.9, 0.8, 0.1)$	$(0.9, 0.8, 0.1)$	$(0.7, 0.6, 0.5)$	$(0, 0, 0)$
v_3	$(0, 0, 0)$	$(0, 0, 0)$	$(0.9, 0.8, 0.1)$	$(0.7, 0.6, 0.5)$	$(0.4, 0.3, 0.3)$
v_4	$(0, 0, 0)$	$(0.4, 0.3, 0.3)$	$(0, 0, 0)$	$(0.4, 0.3, 0.3)$	$(0.4, 0.3, 0.3)$

Definition 9.18 Suppose that $\mathscr{H} = (X, \varepsilon)$ and $\mathscr{H}' = (X', \varepsilon')$ are single-valued neutrosophic hypergraphs. $\mathscr{H}$ is called a *partial single-valued neutrosophic hypergraph* of $\mathscr{H}'$ if $\varepsilon \subseteq \varepsilon'$. If $\mathscr{H}$ is partial single-valued neutrosophic hypergraph of $\mathscr{H}'$, we write $\mathscr{H} \subseteq \mathscr{H}'$. If $\mathscr{H}$ is partial single-valued neutrosophic hypergraph of $\mathscr{H}'$ and $\varepsilon \subset \varepsilon'$, then we denote as $\mathscr{H} \subset \mathscr{H}'$.

Example 9.4 Consider the single-valued neutrosophic hypergraph $\mathscr{H} = (X, \varepsilon)$, where $X = \{v_1, v_2, v_3, v_4\}$ and $\varepsilon = \{E_1, E_2, E_3, E_4, E_5\}$, which is represented by the following incidence matrix given in Table 9.3.

Clearly, $h(\mathscr{H}) = (0.9, 0.8, 0.1)$, $E_1^* = E^{(0.9, 0.8, 0.1)} = \{\{v_2, v_3\}\}$, $E_2^* = E^{(0.7, 0.6, 0.5)} = \{\{v_1, v_2\}\}$, and $E_3^* = E^{(0.4, 0.3, 0.3)} = \{\{v_1, v_2, v_4\}, \{v_2, v_3\}, \{v_2, v_3, v_4\}\}$. Therefore, fundamental sequence is $F(\mathscr{H}) = \{(r_1, s_1, t_1) = (0.9, 0.8, 0.1), (r_2, s_2, t_2) = (0.7, 0.6, 0.5), (r_3, s_3, t_3) = (0.4, 0.3, 0.3)\}$, and the set of core hypergraph is $C(\mathscr{H}) = \{\mathscr{H}^{(0.9, 0.8, 0.1)} = (X_1, E_1^*), \mathscr{H}^{(0.7, 0.6, 0.5)} = (X_2, E_2^*), \mathscr{H}^{(0.4, 0.3, 0.3)} = (X_3, E_3^*)\}$. Note that $E^{(0.9, 0.8, 0.1)} \sqsubseteq E^{(0.4, 0.3, 0.3)}$ and $E^{(0.9, 0.8, 0.1)} \neq E^{(0.4, 0.3, 0.3)}$. As $E_5 \subseteq E_2$, $\mathscr{H}$ is not simple single-valued neutrosophic hypergraph but $\mathscr{H}$ is support simple. In single-valued neutrosophic graph $\mathscr{H} = (X, \varepsilon)$, $E^{(r, s, t)} \neq E^{(0.9, 0.8, 0.1)}$ for $(r, s, t) = (0.7, 0.6, 0.5)$, $\mathscr{H}$ is not sectionally elementary.

The partial single-valued neutrosophic hypergraphs, $\mathscr{H}' = (X', E')$, where $E' = \{E_2, E_3, E_4, E_1\}$ is simple, $\mathscr{H}'' = (X'', E'')$, where $E'' = \{E_2, E_3, E_5\}$ is sectionally elementary, and $\mathscr{H}''' = (X''', E''')$, where $E''' = \{E_1, E_3, E_5\}$ is elementary.

Definition 9.19 A single-valued neutrosophic hypergraph $\mathscr{H}$ is said to be *ordered* if $C(\mathscr{H})$ is ordered. That is, if $C(\mathscr{H}) = \{\mathscr{H}^{(r_1, s_1, t_1)}, \mathscr{H}^{(r_2, s_2, t_2)}, \ldots, \mathscr{H}^{(r_n, s_n, t_n)}\}$, then $\mathscr{H}^{(r_1, s_1, t_1)} \subseteq \mathscr{H}^{(r_2, s_2, t_2)} \subseteq \cdots \subseteq \mathscr{H}^{(r_n, s_n, t_n)}$.

Proposition 9.3 *If $\mathscr{H} = (X, \varepsilon)$ is an elementary single-valued neutrosophic hypergraph, then $\mathscr{H}$ is ordered. Also, if $\mathscr{H} = (X, \varepsilon)$ is an ordered single-valued neutrosophic hypergraph with $C(\mathscr{H}) = \{\mathscr{H}^{(r_1, s_1, t_1)}, \mathscr{H}^{(r_2, s_2, t_2)}, \ldots, \mathscr{H}^{(r_n, s_n, t_n)}\}$ and if $\mathscr{H}^{(r_n, s_n, t_n)}$ is simple, then $\mathscr{H}$ is elementary.*

Definition 9.20 A single-valued neutrosophic hypergraph $\mathscr{H} = (X, \varepsilon)$ is called a E^t *tempered single-valued neutrosophic hypergraph* of $H = (X, E)$ if there is a crisp hypergraph $H = (X, E)$ and a single-valued neutrosophic set E^t is defined on X, where $T_{E^t} : X \to (0, 1]$, $I_{E^t} : X \to (0, 1]$, and $F_{E^t} : X \to (0, 1]$ such that $\varepsilon = \{C_E \mid E \in E^*\}$, where

Table 9.4 Incidence matrix of E^t tempered hypergraph

$M_{\mathscr{H}}$	E_1	E_2	E_3	E_4
v_1	$(0.3, 0.4, 0.6)$	$(0, 0, 0)$	$(0.1, 0.4, 0.5)$	$(0.3, 0.4, 0.5)$
v_2	$(0, 0, 0)$	$(0.1, 0.4, 0.3)$	$(0, 0, 0)$	$(0.3, 0.4, 0.5)$
v_3	$(0.3, 0.4, 0.6)$	$(0, 0, 0)$	$(0, 0, 0)$	$(0, 0, 0)$
v_4	$(0, 0, 0)$	$(0.1, 0.4, 0.3)$	$(0.1, 0.4, 0.5)$	$(0, 0, 0)$

$$T_{C_E}(x) = \begin{cases} \min\{T_{E^t}(y) \mid y \in E\}, & \text{if } x \in E; \\ 0, & \text{otherwise.} \end{cases}$$

$$I_{C_E}(x) = \begin{cases} \min\{I_{E^t}(y) \mid y \in E\}, & \text{if } x \in E; \\ 0, & \text{otherwise.} \end{cases}$$

$$F_{C_E}(x) = \begin{cases} \max\{F_{E^t}(y) \mid y \in E\}, & \text{if } x \in E; \\ 0, & \text{otherwise.} \end{cases}$$

We let $E^t \otimes \mathscr{H}$ denote the E^t tempered single-valued neutrosophic hypergraph of $\mathscr{H}$ determined by the crisp hypergraph $H = (X, E)$ and the single-valued neutrosophic set E^t.

Example 9.5 Consider the single-valued neutrosophic hypergraph $\mathscr{H} = (X, \varepsilon)$, where $X = \{v_1, v_2, v_3, v_4\}$ and $\varepsilon = \{E_1, E_2, E_3, E_4\}$, which is represented by the following incidence matrix in Table 9.4.

Define $E^t = \{(v_1, 0.3, 0.4, 0.5), (v_2, 0.6, 0.5, 0.2), (v_3, 0.5, 0.4, 0.6), (v_4, 0.1, 0.4, 0.3)\}$. Note that

$$T_{\{v_1,v_3\}}(v_1) = \min\{T_{E^T}(v_1), T_{E^T}(v_3)\} = 0.3, \quad I_{\{v_1,v_3\}}(v_1) = \min\{I_{E^t}(v_1), I_{E^t}(v_3)\} = 0.4,$$
$$F_{\{v_1,v_3\}}(v_1) = \max\{F_{E^T}(v_1), F_{E^T}(v_3)\} = 0.6, \quad T_{\{v_1,v_3\}}(v_3) = \min\{T_{E^T}(v_3), T_{E^T}(v_1)\} = 0.3,$$
$$I_{\{v_1,v_3\}}(v_3) = \min\{I_{E^t}(v_3), I_{E^t}(v_1)\} = 0.4, \quad F_{\{v_1,v_3\}}(v_3) = \max\{F_{E^T}(v_3), F_{E^T}(v_1)\} = 0.6.$$

Then, $C_{\{v_1,v_3\}} = E_1$, $C_{\{v_2,v_4\}} = E_2$, $C_{\{v_1,v_4\}} = E_3$, and $C_{\{v_1,v_2\}} = E_4$. Thus, $\mathscr{H}$ is E^t tempered.

Theorem 9.1 *A single-valued neutrosophic hypergraph $\mathscr{H} = (X, \varepsilon)$ is a E^t tempered single-valued neutrosophic hypergraph of some crisp hypergraph $\mathscr{H}$ if and only if $\mathscr{H}$ is elementary, support simple, and simple ordered.*

Proof Suppose $\mathscr{H} = (X, \varepsilon)$ is a E^t tempered single-valued neutrosophic hypergraph of some crisp hypergraph $\mathscr{H}$. Clearly, $\mathscr{H}$ is elementary and support simple. We show that $\mathscr{H}$ is simply ordered.

Let $C(\mathcal{H}) = \{\mathcal{H}^{(r_1,s_1,t_1)} = (X_1, E_1^*), \; \mathcal{H}^{(r_2,s_2,t_2)} = (X_2, E_2^*), \ldots, \mathcal{H}^{(r_n,s_n,t_n)} = (X_n, E_n^*)\}$. Since, $\mathcal{H}$ is elementary, it follows that $\mathcal{H}$ is ordered. To show that $\mathcal{H}$ is simply ordered, suppose there exists $E \in E_{i+1}^* \setminus E_i^*$. Then, there exist $v \in E$ such that $T_E(v) = r_{i+1}$, $I_E(v) = s_{i+1}$, $F_E(v) = t_{i+1}$. Since, $T_E(v) = r_{i+1} < r_i$, $I_E(v) = s_{i+1} < s_i$, $F_E(v) = t_{i+1} > t_i$, it follows that $v \notin X_i$ and $E \nsubseteq X_i$, hence $\mathcal{H}$ is simply ordered.

Conversely, suppose $\mathcal{H} = (X, \varepsilon)$ is elementary, support simple, and simply ordered. For $C(\mathcal{H}) = \{\mathcal{H}^{(r_1,s_1,t_1)} = (X_1, E_1^*), \; \mathcal{H}^{(r_2,s_2,t_2)} = (X_2, E_2^*), \ldots, \mathcal{H}^{(r_n,s_n,t_n)} = (X_n, E_n^*)\}$, fundamental sequence is $F(\mathcal{H}) = \{(r_1, s_1, t_1),(r_2, s_2, t_2), \cdots, (r_n, s_n, t_n)\}$ with $0 < r_n < r_{n-1} < \cdots < r_1$, $0 < s_n < s_{n-1} < \cdots < s_1$, and $0 < t_1 < t_2 < \cdots < t_n$. $\mathcal{H}^{(r_n,s_n,t_n)} = (X_n, E_n^*)$ and single-valued neutrosophic set E^t on X_n defined by

$$T_{E^t}(v) = \begin{cases} r_1, & \text{if } v \in X_1; \\ r_i, & \text{if } v \in X_i \setminus X_{i-1}, i = 2, 3, \ldots, n. \end{cases}$$

$$I_{E^t}(v) = \begin{cases} s_1, & \text{if } v \in X_1; \\ s_i, & \text{if } v \in X_i \setminus X_{i-1}, i = 2, 3, \ldots, n. \end{cases}$$

$$F_{E^t}(v) = \begin{cases} t_1, & \text{if } v \in X_1; \\ t_i, & \text{if } v \in X_i \setminus X_{i-1}, i = 2, 3, \ldots, n. \end{cases}$$

We show that $\varepsilon = \{C_E \mid E \in E_n^*\}$, where

$$T_{C_E}(x) = \begin{cases} \min\{T_{E^t}(y) \mid y \in E\}, & \text{if } x \in E; \\ 0, & \text{otherwise.} \end{cases}$$

$$I_{C_E}(x) = \begin{cases} \min\{I_{E^t}(y) \mid y \in E\}, & \text{if } x \in E; \\ 0, & \text{otherwise.} \end{cases}$$

$$F_{C_E}(x) = \begin{cases} \max\{F_{E^t}(y) \mid y \in E\}, & \text{if } x \in E; \\ 0, & \text{otherwise.} \end{cases}$$

Let $E \in E_n^*$. Since $\mathcal{H}$ is elementary and support simple, there is a unique single-valued neutrosophic hyperedge E_j in ε having support $E \in E_n^*$. We have to show that E^t tempered single-valued neutrosophic hypergraph $\mathcal{H} = (X, \varepsilon)$ is determined by the crisp graph $\mathcal{H}_n = (X_n, E_n^*)$, i.e., $C_{E \in E_n^*} = E_i$, $i = 1, 2, \ldots, m$. As all single-valued neutrosophic hyperedges are elementary and $\mathcal{H}$ is support simple, then different edges have different supports, that is $h(E_j)$ is equal to some member (r_i, s_i, t_i) of $F(\mathcal{H})$. Consequently, $E \subseteq X_i$ and if $i > 1$, then $E \in E_i^* \setminus E_{i-1}^*$, $T_E(v) \geq r_i$, $I_E(v) \geq s_i$, $F_E(v) \leq t_i$, for some $v \in E$. Since, $E \subseteq X_i$, we claim that $T_{E^t}(v) = r_i$, $I_{E^t}(v) = s_i$, $F_{E^t}(v) = t_i$, for some $v \in E$, if not then $T_{E^t}(v) \geq r_{i-1}$, $I_{E^t}(v) \geq s_{i-1}$, $F_{E^t}(v) \leq t_{i-1}$, for all $v \in E$, which implies $E \subseteq X_{i-1}$ and since $\mathcal{H}$ is simply ordered, $E \in E_i^* \setminus E_{i-1}^*$, then $E \nsubseteq X_{i-1}$, a contradiction. Thus, $C_E = E_i$, $i = 1, 2, \ldots, m$, by the definition of C_E. $\qquad \square$

Corollary 9.1 *Suppose $\mathcal{H} = (X, \varepsilon)$ is a simply ordered single-valued neutrosophic hypergraph and $F(\mathcal{H}) = \{(r_1, s_1, t_1), (r_2, s_2, t_2), \ldots, (r_n, s_n, t_n)\}$. If $\mathcal{H}^{(r_n,s_n,t_n)}$ is a*

simple hypergraph, then there is a partial single-valued neutrosophic hypergraph
$\mathscr{H}' = (X, \varepsilon')$ *of* $\mathscr{H}$ *such that following statements hold.*

(i) $\mathscr{H}'$ *is a* E^t *tempered single-valued neutrosophic hypergraph of* $\mathscr{H}^{(r_n, s_n, t_n)}$.
(ii) $F(\mathscr{H}') = (\mathscr{H})$ *and* $C(\mathscr{H}') = C(\mathscr{H})$.

Proof Since $\mathscr{H}$ is simple ordered, then $\mathscr{H}$ is an elementary single-valued neutrosophic hypergraph. We obtain the partial fuzzy hypergraph $\mathscr{H}' = (X, \varepsilon')$ of $\mathscr{H} = (X, \varepsilon)$ by removing all edges from ε that are properly contained in another edge of $\mathscr{H}$, where $\varepsilon' = \{E_i \in E \mid \text{if } E_i \subseteq E_j \text{ and } E_j \in E, \text{ then } E_i = E_j\}$. Since, $\mathscr{H}^{(r_n, s_n, t_n)}$ is simple and all edges are elementary, any edge in $\mathscr{H}$ contain another edge, then both have the same support. Hence, $F(\mathscr{H}') = \mathscr{H}$ and $C(\mathscr{H}') = C(\mathscr{H})$. By the definition of ε', $\mathscr{H}'$ is elementary, support simple. Thus, by the Theorem 9.1 $\mathscr{H}'$ is a E^t tempered single-valued neutrosophic graph.   □

Definition 9.21 Let $L(G^*) = (C, D)$ be a line graph of a crisp graph $G^* = (X, E)$, where $C = \{\{x\} \cup \{u_x, v_x\} \mid x \in E, u_x, v_x \in X, x = u_x v_x\}$, and $D = \{S_x S_y \mid S_x \cap S_y \neq \emptyset, x, y \in E, x \neq y\}$, and $S_x = \{\{x\} \cup \{u_x, v_x\}\}, x \in E$.

Let $G = (A_1, B_1)$ be a single-valued neutrosophic graph with underlying set X. The *single-valued neutrosophic line graph* of G is a single-valued neutrosophic graph $L(G) = (A_2, B_2)$ such that

(i)

$$T_{A_2}(S_x) = T_{B_1}(x) = T_{B_1}(u_x v_x),$$

$$I_{A_2}(S_x) = I_{B_1}(x) = I_{B_1}(u_x v_x),$$

$$F_{A_2}(S_x) = F_{B_1}(x) = F_{B_1}(u_x v_x),$$

(ii)

$$T_{B_2}(S_x S_y) = \min\{T_{B_1}(x), T_{B_1}(y)\},$$

$$I_{B_2}(S_x S_y) = \min\{I_{B_1}(x), I_{B_1}(y)\},$$

$$F_{B_2}(S_x S_y) = \max\{F_{B_1}(x), F_{B_1}(y)\},$$

for all $S_x S_y \in D$.

Proposition 9.4 $L(G) = (A_2, B_2)$ *is a single-valued neutrosophic line graph of some single-valued neutrosophic graph* $G = (A_1, B_1)$ *if and only if*

$$T_{B_2}(S_x S_y) = \min\{T_{A_2}(S_x), T_{A_2}(S_y)\},$$
$$I_{B_2}(S_x S_y) = \min\{I_{A_2}(S_x), T_{A_2}(S_y)\},$$
$$F_{B_2}(S_x S_y) = \max\{F_{A_2}(S_x), F_{A_2}(S_y)\},$$

for all $S_x S_y \in D$.

Definition 9.22 Let $\mathscr{H} = (X, \varepsilon)$ be a single-valued neutrosophic hypergraph of a simple graph $\mathscr{H} = (X, E)$, and $L(\mathscr{H}) = (X, \varepsilon)$ be a line graph of $\mathscr{H}$. The *single-valued neutrosophic line graph* $L(\mathscr{H})$ *of a single-valued neutrosophic hypergraph* $\mathscr{H}$ is defined to be a pair $L(\mathscr{H}) = (A, B)$, where A is the vertex set of $L(\mathscr{H})$ and B is the edge set of $L(\mathscr{H})$ defined as follows:

(i) A is a single-valued neutrosophic set of X such that $T_A(E_i) = \max\limits_{v \in E_i}(T_{E_i}(v))$, $I_A(E_i) = \max\limits_{v \in E_i}(I_{E_i}(v))$, and $F_A(E_i) = \min\limits_{v \in E_i}(F_{E_i}(v))$, for all $E_i \in \varepsilon$.

(ii) B is a single-valued neutrosophic set of ε such that $T_B(E_j E_k) = \min\limits_{i}\{\min(T_{E_j}(v_i),$ $T_{E_k}(v_i))\}$, $T_B(E_j E_k) = \min\limits_{i}\{\min(T_{E_j}(v_i), T_{E_k}(v_i))\}$,   and $T_B(E_j E_k) = \max\limits_{i}$ $\{\max(T_{E_j}(v_i), T_{E_k}(v_i))\}$, where $v_i \in E_i \cap E_j$, $j, k = 1, 2, 3, \ldots, n$.

Example 9.6 Consider a crisp hypergraph $H = (X, E)$, where $X = \{v_1, v_2, v_3, v_4,$ $v_5, v_6\}$ and $E = \{E_1, E_2, E_3, E_4, E_5, E_6\}$ such that $E_1 = \{v_1, v_3\}, E_2 = \{v_1, v_2\}, E_3 = \{v_2, v_4\}, E_4 = \{v_3, v_6\}, E_5 = \{v_3, v_5\}, E_6 = \{v_5, v_6\}, E_7 = \{(v_4, v_6\}.$ Let $\mathscr{H} = (X, \varepsilon)$ be a single-valued neutrosophic hypergraph, where $\varepsilon = \{E_1, E_2, E_3, E_4, E_5, E_6\}$, such that

$$E_1 = \{(v_1, 0.3, 0.4, 0.6), (v_3, 0.7, 0.4, 0.4)\},$$
$$E_2 = \{(v_1, 0.3, 0.4, 0.6), (v_2, 0.5, 0.7, 0.6)\},$$
$$E_3 = \{(v_2, 0.5, 0.7, 0.6), (v_4, 0.6, 0.4, 0.8)\},$$
$$E_4 = \{(v_3, 0.7, 0.4, 0.4), (v_6, 0.4, 0.2, 0.7)\},$$
$$E_5 = \{(v_3, 0.7, 0.4, 0.4), (v_5, 0.6, 0.7, 0.5)\},$$
$$E_6 = \{(v_5, 0.6, 0.7, 0.5), (v_6, 0.4, 0.2, 0.7)\},$$
$$E_7 = \{(v_4, 0.6, 0.4, 0.8), (v_6, 0.4, 0.2, 0.7)\}.$$

The single-valued neutrosophic line graph of $\mathscr{H}$ is shown in Fig. 9.4.

The line graph $L(\mathscr{H})$ of single-valued neutrosophic hypergraph $\mathscr{H}$ is $L(\mathscr{H}) = (A, B)$, where $A = \{(E_1, 0.7, 0.4, 0.4), (E_2, 0.5, 0.7, 0.6), (E_3, 0.6, 0.7, 0.6), (E_4, 0.7, 0.4, 0.4), (E_5, 0.7, 0.7, 0.4), (E_6, 0.6, 0.7, 0.5), (E_7, 0.6, 0.4, 0.7)\}$ is the vertex set and $B = \{(E_1 E_2, 0.3, 0.4, 0.6), (E_1 E_5, 0.7, 0.4, 0.4), (E_1 E_4, 0.7, 0.4, 0.4), (E_2 E_3, 0.5, 0.7, 0.6), (E_3 E_7, 0.6, 0.4, 0.8), (E_4 E_5, 0.7, 0.4, 0.4), (E_4 E_6, 0.4, 0.2, 0.7), (E_4 E_7, 0.4, 0.2, 0.7), (E_5 E_6, 0.6, 0.7, 0.5), (E_6 E_7, 0.4, 0.2, 0.7)\}$ is the edge set of the single-valued neutrosophic line graph of $\mathscr{H}$.

Proposition 9.5 *A single-valued neutrosophic hypergraph is connected if and only if line graph of a single-valued neutrosophic hypergraph is connected.*

Definition 9.23 The 2-*section* of a single-valued neutrosophic hypergraph $\mathscr{H} = (X, \varepsilon)$, denoted by $[\mathscr{H}]_2$, is a single-valued neutrosophic graph $[\mathscr{H}]_2 = (A^*, B^*)$,

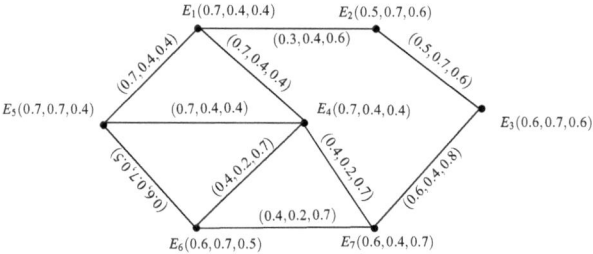

Fig. 9.4 Single-valued neutrosophic line graph $L(\mathscr{H})$ of $\mathscr{H}$

Fig. 9.5 Single-valued
neutrosophic hypergraph

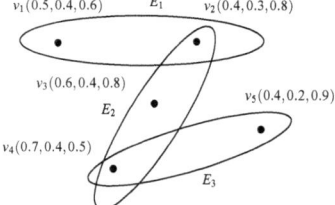

where A^* is the single-valued neutrosophic vertex of $[\mathscr{H}]_2$ and B^* is the single-valued neutrosophic edge set in which any two vertices form an edge if they are in the same single-valued neutrosophic hyperedge such that

$$T_{B^*}(e) = \min\{T_{E_k}(v_i), T_{E_k}(v_j)\},$$
$$I_{B^*}(e) = \min\{I_{E_k}(v_i), T_{E_k}(v_j)\},$$
$$F_{B^*}(e) = \max\{F_{E_k}(v_i), F_{E_k}(v_j)\},$$

for all $E_k \in \varepsilon, i \neq j, k = 1, 2, \ldots, m$.

We now introduce the concept of dual single-valued neutrosophic hypergraph for a single-valued neutrosophic hypergraph.

Definition 9.24 The *dual of a single-valued neutrosophic hypergraph* $\mathscr{H} = (X, \varepsilon)$ is a single-valued neutrosophic hypergraph $\mathscr{H}^d = (E, \mathbf{X})$, $E = \{e_1, e_2, \ldots, e_n\}$ set of vertices corresponding to $E_1, E_2, \ldots, E_n$, respectively, and $\mathbf{X} = \{X_1, X_2, \ldots, X_n\}$ set of hyperedges corresponding to $v_1, v_2, \ldots, v_n$, respectively.

Example 9.7 Consider a single-valued neutrosophic hypergraph $\mathscr{H} = (X, \varepsilon)$ as shown in Fig. 9.5 such that $X = \{v_1, v_2, v_3, v_4, v_5\}$ and $\varepsilon = \{E_1, E_2, E_3\}$, where

$$E_1 = \{(v_1, 0.5, 0.4, 0.6), (v_2, 0.4, 0.3, 0.8)\},$$
$$E_2 = \{(v_2, 0.4, 0.3, 0.8), (v_3, 0.6, 0.4, 0.8), (v_4, 0.7, 0.4, 0.5)\},$$
$$E_3 = \{(v_4, 0.7, 0.4, 0.5), (v_5, 0.4, 0.2, 0.9)\}.$$

Table 9.5 The incidence matrix of single-valued neutrosophic hypergraph

$M_{\mathscr{H}}$	E_1	E_2	E_3
v_1	$(0.5, 0.4, 0.6)$	$(0, 0, 0)$	$(0, 0, 0)$
v_2	$(0.4, 0.3, 0.8)$	$(0.4, 0.3, 0.8)$	$(0, 0, 0)$
v_3	$(0, 0, 0)$	$(0.6, 0.4, 0.8)$	$(0, 0, 0)$
v_4	$(0, 0, 0)$	$(0.7, 0.4, 0.5)$	$(0.7, 0.4, 0.5)$
v_5	$(0, 0, 0)$	$(0, 0, 0)$	$(0.4, 0.2, 0.9)$

Fig. 9.6 Dual single-valued neutrosophic hypergraph $\mathscr{H}$

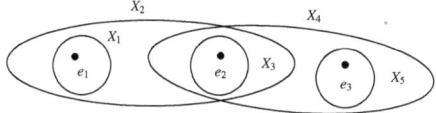

Table 9.6 Incidence matrix of $\mathscr{H}^d$

$M_{\mathscr{H}}$	X_1	X_2	X_3	X_4	X_5
e_1	$(0.5, 0.4, 0.6)$	$(0.4, 0.3, 0.8)$	$(0, 0, 0)$	$(0, 0, 0)$	$(0, 0, 0)$
e_2	$(0, 0, 0)$	$(0.4, 0.3, 0.8)$	$(0.6, 0.4, 0.8)$	$(0.7, 0.4, 0.5)$	$(0, 0, 0)$
e_3	$(0, 0, 0)$	$(0, 0, 0)$	$(0, 0, 0)$	$(0.7, 0.4, 0.5)$	$(0.4, 0.2, 0.9)$

The single-valued neutrosophic hypergraph can be represented by the following incidence matrix in Table 9.5.

Consider the dual single-valued neutrosophic hypergraph $\mathscr{H}^d = (E, \mathbf{X})$ of $\mathscr{H}$ such that $E = \{e_1, e_2, e_3\}$, $\mathbf{X} = \{X_1, X_2, X_3, X_4, X_5\}$, where

$$X_1 = \{(e_1, 0.5, 0.4, 0.6), (e_2, 0, 0, 0), (e_3, 0, 0, 0)\},$$
$$X_2 = \{(e_1, 0.4, 0.3, 0.8), (e_2, 0.4, 0.3, 0.8), (e_3, 0, 0, 0)\},$$
$$X_3 = \{(e_1, 0, 0, 0), (e_2, 0.6, 0.4, 0.8), (e_3, 0, 0, 0)\},$$
$$X_4 = \{(e_1, 0, 0, 0), (e_2, 0.7, 0.4, 0.5), (e_3, 0.7, 0.4, 0.5)\},$$
$$X_5 = \{(e_1, 0, 0, 0), (e_2, 0, 0, 0), (e_3, 0.4, 0.2, 0.9)\}.$$

The dual single-valued neutrosophic hypergraph is shown in Fig. 9.6 and its incidence matrix $M_{\mathscr{H}}$ is defined as in Table 9.6.

Remark 9.4 $\mathscr{H}^d$ is a single-valued neutrosophic hypergraph whose incidence matrix is the transpose of the incidence matrix of $\mathscr{H}$ and $\Delta(\mathscr{H}^d) = rank(\mathscr{H})$. The dual single-valued neutrosophic hypergraph $\mathscr{H}^d$ of a simple single-valued neutrosophic hypergraph $\mathscr{H}$ may or may not be simple.

Proposition 9.6 *The dual $\mathscr{H}^d$ of a linear single-valued neutrosophic hypergraph without isolated vertex is also a linear single-valued neutrosophic hypergraph.*

Proof Let $\mathscr{H}$ be a linear hypergraph. Assume that $\mathscr{H}^d$ is not linear. There are two distinct hyperedges X_i and X_j of $\mathscr{H}^d$ which intersect with at least two vertices e_1 and e_2. The definition of duality implies that v_i and v_j belong to E_1 and E_2 (the single-valued neutrosophic hyperedges of $\mathscr{H}$ standing for the vertices e_1, e_2 of $\mathscr{H}^d$, respectively) so $\mathscr{H}$ is not linear. Contradiction since $\mathscr{H}$ is linear. Hence, dual $\mathscr{H}^d$ of a linear single-valued neutrosophic hypergraph without isolated vertex is also linear single-valued neutrosophic hypergraph.   □

Definition 9.25 Let $\mathscr{H} = (X, \varepsilon)$ be a single-valued neutrosophic hypergraph. A *single-valued neutrosophic transversal* τ of $\mathscr{H}$ is a single-valued neutrosophic subset of X with the property that $\tau^{h(E)} \cap E^{h(E)} \neq \emptyset$ for each $E \in \varepsilon$, where $h(E)$ is the height of hyperedge E. A *minimal single-valued neutrosophic transversal* τ for $\mathscr{H}$ is a transversal of $\mathscr{H}$ with the property that if $\tau' \subset \tau$, then τ' is not a single-valued neutrosophic transversal of $\mathscr{H}$.

Proposition 9.7 *If τ is a single-valued neutrosophic transversal of a single-valued neutrosophic hypergraph $\mathscr{H} = (X, \varepsilon)$, then $h(\tau) > h(E)$ for each $E \in \varepsilon$. Moreover, if τ is a minimal single-valued neutrosophic transversal of $\mathscr{H}$, then $h(\tau) = \max\{h(E) \mid E \in \varepsilon\} = h(\mathscr{H})$.*

Theorem 9.2 *For a single-valued neutrosophic hypergraph $\mathscr{H}$, $Tr(\mathscr{H}) \neq \emptyset$, where $Tr(\mathscr{H})$ is the family of minimal single-valued neutrosophic transversal of $\mathscr{H}$.*

Proposition 9.8 *Let $\mathscr{H} = (X, E)$ be a single-valued neutrosophic hypergraph. The following statements are equivalent:*

(i) *τ is a single-valued neutrosophic transversal of $\mathscr{H}$.*
(ii) *For each $E \in \varepsilon$, $h(E) = (r', s', t')$, and each $0 < r \leq r', 0 < s \leq s', t \geq t'$, $\tau^{(r,s,t)} \cap E^{(r,s,t)} \neq \emptyset$.*

If the (r, s, t)-cut $\tau^{(r,s,t)}$ is a subset of the vertex set of $\mathscr{H}^{(r,s,t)}$ for each (r, s, t), $0 < r \leq r'$, $0 < s \leq s'$, $t \geq t'$, then

(iii) *For each (r, s, t), $0 < r \leq r'$, $0 < s \leq s'$, $t \geq t'$, $\tau^{(r,s,t)}$ is a transversal of $\mathscr{H}^{(r,s,t)}$.*
(iv) *Every single-valued neutrosophic transversal τ of $\mathscr{H}$ contains a single-valued neutrosophic transversal τ' for each (r, s, t), $0 < r \leq r'$, $0 < s \leq s'$, $t \geq t'$, $\tau'^{(r,s,t)}$ is a transversal of $\mathscr{H}^{(r,s,t)}$.*

Observation: If τ is a minimal transversal of single-valued neutrosophic graph $\mathscr{H}$, then $\tau^{(r,s,t)}$ not necessarily belongs to $Tr(\mathscr{H}^t)$ for each (r, s, t), satisfying $0 < r \leq r', 0 < s \leq s', t \geq t'$. Let $Tr^*(\mathscr{H})$ represents the collection of those minimal single-valued neutrosophic transversal, τ of $\mathscr{H}$, where $\tau^{(r,s,t)}$ is a minimal transversal of $\mathscr{H}^{(r,s,t)}$, for each (r, s, t), $0 < r \leq r', 0 < s \leq s', t \geq t'$, i.e., $Tr^* = \{\tau \in Tr(\mathscr{H}) \mid h(\tau) = h(\mathscr{H})$ and $\tau^{(r,s,t)} \in Tr(\mathscr{H}^{(r,s,t)})\}$.

Example 9.8 Consider the single-valued neutrosophic hypergraph $\mathscr{H} = (X, \varepsilon)$, where $X = \{v_1, v_2, v_3\}$ and $\varepsilon = \{E_1, E_2, E_3\}$, which is represented by the following incidence matrix given in Table 9.7.

Table 9.7 Incidence matrix

$M_{\mathcal{H}}$	E_1	E_2	E_3
v_1	$(0.9, 0.6, 0.1)$	$(0, 0, 0)$	$(0.4, 0.3, 0.2)$
v_2	$(0.4, 0.3, 0.2)$	$(0.4, 0.3, 0.2)$	$(0.4, 0.3, 0.2)$
v_3	$(0, 0, 0)$	$(0, 0, 0)$	$(0.4, 0.3, 0.2)$

Clearly, $h(\mathcal{H}) = (0.9, 0.6, 0.1)$, the only minimal transversal τ of single-valued neutrosophic hypergraph $\mathcal{H}$ is $\tau(\mathcal{H}) = \{(v_1, 0.9, 0.6, 0.1), (v_2, 0.4, 0.3, 0.2)\}$. The fundamental sequence of $\mathcal{H}$ is $F(\mathcal{H}) = \{(0.9, 0.6, 0.1), (0.4, 0.3, 0.2)\}$, $\tau^{(0.9,0.6,0.1)} = \{v_1\}$, $\tau^{(0.4,0.3,0.2)} = \{v_1, v_2\}$. Since, $\{v_2\}$ is the only minimal transversal of the $\mathcal{H}^{(0.4,0.3,0.2)}$, $E^{(0.4,0.3,0.2)} = \{\{v_1, v_2\}, \{v_2\}, \{v_1, v_2, v_3\}\}$, it follows that the only minimal transversal τ of $\mathcal{H}$ is not a member of $Tr^*(\mathcal{H})$. Hence, $Tr^*(\mathcal{H}) = \emptyset$.

9.3 Intuitionistic Single-Valued Neutrosophic Hypergraphs

Definition 9.26 Let X be a fixed set. A *generalized intuitionistic fuzzy set I* of X is an object having the form $I = \{(u, T_I(u), F_I(u))|u \in X\}$, where the functions $T_I(u) :\to$ $[0, 1]$ and $F_I(u) :\to [0, 1]$ define the truth-membership and falsity-membership of an element $u \in X$, respectively, such that

$$\min\{T_I(u), F_I(u)\} \le 0.5, \text{ for all } u \in X.$$

This condition is called the generalized intuitionistic condition.

Being motivated from this definition, Bhowmik and Pal [10] gave the idea of an intuitionistic single-valued neutrosophic set and discussed its certain properties.

Definition 9.27 An *intuitionistic single-valued neutrosophic set* on a universal set X can be stated as a set having the form $A = \{T_A(u), I_A(u), F_A(u) : u \in X\}$, where

$$\min\{T_A(u), I_A(u)\} \le 0.5,$$
$$\min\{F_A(u), I_A(u)\} \le 0.5,$$
$$\min\{T_A(u), F_A(u)\} \le 0.5,$$

and $0 \le T_A(u) + I_A(u) + F_A(u) \le 2$.

Definition 9.28 The *support* set of an intuitionistic single-valued neutrosophic set $A = \{(v, T_A(v), I_A(v), F_A(v)) : v \in X\}$ is defined as $supp(A) = \{v|T_A(v) \ne 0, I_A(v) \ne 0, F_A(v) \ne 0\}$. The *support* set of an intuitionistic single-valued neutrosophic set is a crisp set.

Definition 9.29 The *height* of an intuitionistic single-valued neutrosophic set $A = \{(v, T_A(v), I_A(v), F_A(v)) : v \in X\}$ is defined as $h(A) = (\max_{v \in X} T_A(v), \max_{v \in X} I_A(v), \min_{v \in X} F_A(v))$.

An intuitionistic single-valued neutrosophic set A is called *normal* if $h(A) = (1, 1, 0)$.

Definition 9.30 Let $A = \{(v, T_A(v), I_A(v), F_A(v)) : v \in X\}$ be an intuitionistic single-valued neutrosophic set on X. Let $\eta, \phi, \psi \in [0, 1]$ such that $\eta + \phi + \psi \leq 2$, then the (η, ϕ, ψ)-*level set* of A is defined as $A_{(\eta, \phi, \psi)} = \{v : T_A(v) \geq \eta, I_A(v) \geq \phi, F_A(v) \leq \psi\}$. Note that (η, ϕ, ψ)-level set is a crisp set.

We first define intuitionistic single-valued neutrosophic graph.

Definition 9.31 An *intuitionistic single-valued neutrosophic graph* with underlying set X is a pair $\mathscr{G} = (C, D)$, such that

(i) the degrees of truth-membership, indeterminacy-membership, and falsity-membership of the element $x_i \in X$ are defined by $T_C : X \to [0, 1]$, $I_C : X \to [0, 1]$, $F_C : X \to [0, 1]$, respectively, where

$$\min\{T_C(x_i), F_C(x_i)\} \leq 0.5,$$

$$\min\{T_C(x_i), I_C(x_i)\} \leq 0.5,$$

$$\min\{F_C(x_i), I_C(x_i)\} \leq 0.5,$$

for all $x_i \in X$, $i = 1, 2, 3, \ldots, m$, with the condition $0 \leq T_C(x_i) + F_C(x_i) + I_C(x_i) \leq 2$.

(ii) The mappings $T_D : D \subseteq X \times X \to [0, 1]$, $I_D : D \subseteq X \times X \to [0, 1]$, $F_D : D \subseteq X \times X \to [0, 1]$ defined by

$$T_D(x_i x_j) \leq \min\{T_C(x_i), T_C(x_j)\},$$
$$I_D(x_i x_j) \leq \min\{I_C(x_i), I_C(x_j)\},$$
$$F_D(x_i x_j) \leq \max\{F_C(x_i), F_C(x_j)\},$$

denote the degree of truth-membership, indeterminacy-membership, and falsity-membership of the edge $x_i x_j \in D$, respectively, where

$$\min\{T_D(x_i x_j), I_D(x_i x_j)\} \leq 0.5,$$
$$\min\{T_D(x_i x_j), F_D(x_i x_j)\} \leq 0.5,$$
$$\min\{F_D(x_i x_j), I_D(x_i x_j)\} \leq 0.5,$$

for all $x_i x_j \in D$, $i = 1, 2, 3, \ldots, m$, $j = 1, 2, 3, \ldots, m$, with the condition $0 \leq T_D(x_i x_j) + F_D(x_i x_j) + I_D(x_i x_j) \leq 2$.

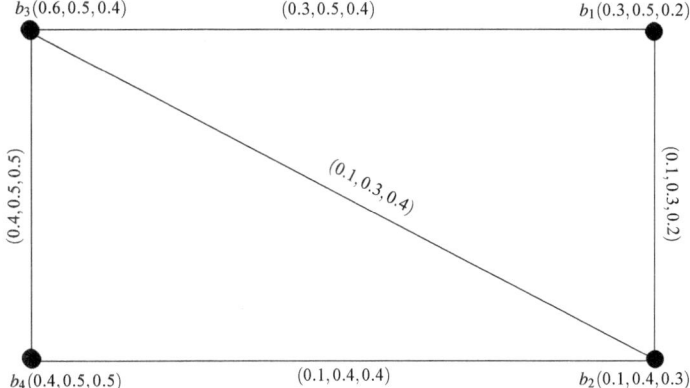

Fig. 9.7 Intuitionistic single-valued neutrosophic graph

Fig. 9.8 Intuitionistic single-valued neutrosophic hypergraph

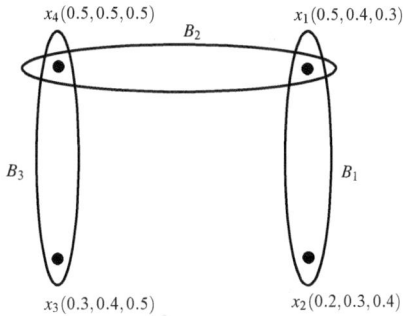

Example 9.9 An Intuitionistic single-valued neutrosophic graph is shown in Fig. 9.7.

We now define an intuitionistic single-valued neutrosophic hypergraph.

Definition 9.32 An *intuitionistic single-valued neutrosophic hypergraph* on a non-empty set X is a pair $\mathscr{H}^* = (X, \mathscr{B})$, where X is a crisp set of vertices and $\mathscr{B} = \{B_1, B_2, \cdots, B_m\}$ be a finite family of nontrivial intuitionistic single-valued neutrosophic subsets of X such that

(i)

$$\min\{T_{B_i}(v_j), I_{B_i}(v_j)\} \leq 0.5,$$

$$\min\{F_{B_i}(v_j), I_{B_i}(v_j)\} \leq 0.5,$$

$$\min\{T_{B_i}(v_j), F_{B_i}(v_j)\} \leq 0.5,$$

with the condition $0 \leq T_{B_i}(v_j) + F_{B_i}(v_j) + I_{B_j}(v_j) \leq 2$.

(ii) $\bigcup_i supp(B_i) = X$, for all $B_i \in \mathscr{B}$.

Table 9.8 Incidence matrix of $\mathscr{H}^*$

$I_{\mathscr{H}^*}$	B_1	B_2	B_3
x_1	$(0.5, 0.4, 0.3)$	$(0.5, 0.4, 0.3)$	$(0, 0, 0)$
x_2	$(0.2, 0.3, 0.4)$	$(0, 0, 0)$	$(0, 0, 0)$
x_3	$(0, 0, 0)$	$(0, 0, 0)$	$(0.3, 0.4, 0.5)$
x_4	$(0, 0, 0)$	$(0.5, 0.5, 0.5)$	$(0.5, 0.5, 0.5)$

Example 9.10 Consider an intuitionistic single-valued neutrosophic hypergraph $\mathscr{H}^* = (X, \mathscr{B})$ as shown in Fig. 9.8, such that $X = \{x_1, x_2, x_3, x_4\}$ and $\mathscr{B} = \{B_1, B_2, B_3\}$, which is represented by the following incidence matrix in Table 9.8.

Definition 9.33 In an intuitionistic single-valued neutrosophic hypergraph, two vertices x_1 and x_2 are said to be *adjacent* if there is a hyperedge $B_i \in \mathscr{B}$ which contains both x_1 and x_2, i.e., $x_1, x_2 \in supp(B_i)$.

Two hyperedges B_i and B_j are called *adjacent* edges if they have non-empty intersection, i.e., $supp(B_i) \cap supp(B_j) \neq \emptyset$, $i \neq j$. The number of elements in X, i.e., $|X|$ is called the *order* and $|\mathscr{B}|$ is called the *size* of an intuitionistic single-valued neutrosophic hypergraph.

An intuitionistic single-valued neutrosophic hypergraph is said to be *n-uniform* if $supp(B_i) = n$, for each $B_i \in \mathscr{B}$.

Definition 9.34 The *height* of an intuitionistic single-valued neutrosophic hypergraph $\mathscr{H}^* = (X, \mathscr{B})$ is defined as $h(\mathscr{H}) = \max\{h(B_i)|B_i \in \mathscr{B}\}$.

Definition 9.35 Consider an intuitionistic single-valued neutrosophic hypergraph $\mathscr{H}^* = (X, \mathscr{B})$, the *cardinality* of an intuitionistic single-valued neutrosophic hyperedge is the sum of truth-membership, indeterminacy-membership, and falsity-membership values of the vertices connected to a hyperedge, it is denoted by $|B_i|$.

The *degree* of an intuitionistic single-valued neutrosophic hyperedge, $B_i \in \mathscr{B}$ is its cardinality, i.e., $d_{\mathscr{H}}(B_i) = |B_i|$.

The *rank* of an intuitionistic single-valued neutrosophic hypergraph is the maximum cardinality of any hyperedge in $\mathscr{H}$, i.e., $\max_{B_i \in \mathscr{B}} d_{\mathscr{H}}(B_i)$ and *anti rank* is the minimum cardinality of any hyperedge in $\mathscr{H}$, i.e., $\min_{B_i \in \mathscr{B}} d_{\mathscr{H}}(B_i)$.

Remark 9.5 (i) If an intuitionistic single-valued neutrosophic hypergraph $\mathscr{H}^* = (X, \mathscr{B})$ is simple, then (η, ϕ, ψ)-level hypergraph $\mathscr{H}^*_{(\eta, \phi, \psi)}$ may or may not be simple. Also, it is possible $B_{i(\eta, \phi, \psi)} = B_{j(\eta, \phi, \psi)}$ for $B_i \neq B_j$, where $B_i, B_j \in \mathscr{B}$ are any two intuitionistic single-valued neutrosophic hyperedges.

(ii) An intuitionistic single-valued neutrosophic hypergraph $\mathscr{H}^* = (X, \mathscr{B})$ is an intuitionistic single-valued neutrosophic graph (with loops) if and only if $\mathscr{H}$ is elementary, support simple, and every hyperedge has two (or one) element support.

Definition 9.36 Let $\mathscr{H}^* = (X, \mathscr{B})$ be an intuitionistic single-valued neutrosophic hypergraph such that $h(\mathscr{H}^*) = (u, v, w)$. Let $\mathscr{H}^*_{(u_i, v_i, w_i)}$ be the (u_i, v_i, w_i)-level hypergraphs of $\mathscr{H}^*$. The sequence of real numbers $\{(u_1, v_1, w_1), (u_2, v_2, w_2), \ldots, (r_n, v_n, w_n)\}$, such that $0 < u_n < u_{n+1} <, \ldots, < u_1 = u, \, 0 < v_n < v_{n+1} <, \ldots, < v_1 = v$, and $w_n > w_{n+1} >, \ldots, >, w_1 = w > 0$, which satisfies the properties,

(i) if $u_{i+1} < u' < u_i$, $v_{i+1} < v' < v_i$, $w_{i+1} > w' > w_i (w_i < w' < w_{i+1})$, then
$$B_{j(u', v', w')} = B_{j(u_i, v_i, w_i)},$$
(ii) $B_{j(u_i, v_i, w_i)} \sqsubseteq B_{j(u_{i+1}, v_{i+1}, w_{i+1})},$

is called the *fundamental sequence* of intuitionistic single-valued neutrosophic hypergraph $\mathscr{H}^*$, denoted by $FS(\mathscr{H}^*)$, the set of (u_i, v_i, w_i)-level hypergraphs $\{\mathscr{H}^*_{(u_1, v_1, w_1)}, \mathscr{H}^*_{(u_2, v_2, w_2)}, \ldots, \mathscr{H}^*_{(u_n, v_n, w_n)}\}$ is known as *core hypergraphs* of intuitionistic single-valued neutrosophic hypergraph $\mathscr{H}$, and is denoted by $C(\mathscr{H}^*)$.

Definition 9.37 An intuitionistic single-valued neutrosophic hypergraph $\mathscr{H}^* = (X, \mathscr{B})$ is *simple* if $B_1, B_2 \in \mathscr{B}$ and $T_{B_1}(v_j) \leq T_{B_2}(v_j)$, $F_{B_1}(v_j) \geq F_{B_2}(v_j)$, $I_{B_1}(v_j) \leq I_{B_2}(v_j)$ imply $T_{B_1}(v_j) = T_{B_2}(v_j)$, $F_{B_1}(v_j) = F_{B_2}(v_j)$, $I_{B_1}(v_j) = I_{B_2}(v_j)$, $j = 1, 2, 3, \ldots, m$.

Definition 9.38 Let $\mathscr{H}^* = (X, \mathscr{B})$ be an intuitionistic single-valued neutrosophic hypergraph with $FS(\mathscr{H}^*) = \{(u_1, v_1, w_1), (u_2, v_2, w_2), \cdots, (u_m, v_m, w_m)\}$, $u_{m+1} = 0$, $v_{m+1} = 0$, $w_{m+1} = 0$. Then, $\mathscr{H}^*$ is *sectionally elementary* if for every hyperedge $B_i \in \mathscr{B}$ and each $(u_i, v_i, w_i) \in FS(\mathscr{H}^*)$, $B_{i(u_i, v_i, w_i)} = B_{i(u, v, w)}$, for all $(u, v, w) \in ((u_{i+1}, v_{i+1}, w_{i+1}), (u_i, v_i, w_i))$.

Definition 9.39 Let $\mathscr{H}_1^* = (X_1, \mathscr{B}_1)$ and $\mathscr{H}_2^* = (X_2, \mathscr{B}_2)$ be intuitionistic single-valued neutrosophic hypergraph and $\mathscr{B}_2 \subseteq \mathscr{B}_1$, then $\mathscr{H}_2^*$ is called a *partial intuitionistic single-valued neutrosophic hypergraph* of $\mathscr{H}_1^*$, denoted by, $\mathscr{H}_2^* \subseteq \mathscr{H}_1^*$. If $\mathscr{H}_2^*$ is a partial intuitionistic single-valued neutrosophic hypergraph of $\mathscr{H}_1^*$ and $\mathscr{B}_2 \subset \mathscr{B}_1$, then we write $\mathscr{H}_2^* \subset \mathscr{H}_1^*$.

Definition 9.40 An intuitionistic single-valued neutrosophic hypergraph $\mathscr{H}^* = (X, \mathscr{B})$ is *elementary*, whose hyperedges are elementary. An intuitionistic single-valued neutrosophic set $I = (T_I, I_I, F_I)$ is *elementary* if I is single-valued on $supp(I)$.

Example 9.11 Consider the intuitionistic single-valued neutrosophic hypergraph $\mathscr{H}^* = (X, \mathscr{B})$, where $X = \{x_1, x_2, x_3, x_4\}$ and $\mathscr{B} = \{B_1, B_2, B_3, B_4, B_5\}$. Then, the corresponding incidence matrix is given in Table 9.9.

Here, $h(\mathscr{H}^*) = (0.9, 0.5, 0.1)$. By calculating the (u_i, v_i, w_i)-level hypergraphs of $\mathscr{H}^*$, we have $\mathscr{B}_{(0.9, 0.5, 0.1)} = \{\{x_1, x_2\}, \{x_2, x_3\}\} = \mathscr{B}_{(0.7, 0.5, 0.5)}$, $\mathscr{B}_{(0.4, 0.3, 0.3)} = \{\{x_1, x_2, x_4\}, \{x_2, x_3\}, \{x_3, x_4\}\}$.
Note that $\mathscr{H}^*$ is not simple intuitionistic single-valued neutrosophic hypergraph and not support simple.

Further, $\mathscr{B}_{(0.9, 0.5, 0.1)} \neq \mathscr{B}_{(0.4, 0.3, 0.3)}$ and $\mathscr{B}_{(0.9, 0.5, 0.1)} \sqsubseteq \mathscr{B}_{(0.4, 0.3, 0.3)}$. So, the fundamental sequence is $FS(\mathscr{H}^*) = \{(u_1, v_1, w_1) = (0.9, 0.5, 0.1), (u_2, v_2, w_2) = (0.4, 0.3, 0.3)\}$ and the set of core hypergraphs is $C(\mathscr{H}^*) = \{\mathscr{H}^*_{(0.9, 0.5, 0.1)}, \mathscr{H}^*_{(0.4, 0.3, 0.3)}\}$. $\mathscr{H}^*$ is not sectionally elementary as $\mathscr{B}_{1(u, v, w)} \neq \mathscr{B}_{1(0.9, 0.5, 0.1)}$ for $(u, v, w) = (0.7, 0.5, 0.5)$.

Table 9.9 Incidence matrix of $\mathscr{H}^*$

$I_{\mathscr{H}^*}$	B_1	B_2	B_3	B_4	B_5
x_1	(0.7, 0.5, 0.5)	(0.9, 0.5, 0.1)	(0, 0, 0)	(0, 0, 0)	(0, 0, 0)
x_2	(0.7, 0.5, 0.5)	(0.9, 0.5, 0.1)	(0.9, 0.5, 0.1)	(0.7, 0.5, 0.5)	(0, 0, 0)
x_3	(0, 0, 0)	(0, 0, 0)	(0.9, 0.5, 0.1))	(0.7, 0.5, 0.5)	(0.4, 0.3, 0.3)
x_4	(0, 0, 0)	(0.4, 0.3, 0.3)	(0, 0, 0)	(0.4, 0.3, 0.3)	(0.4, 0.3, 0.3)

Definition 9.41 An *ordered* intuitionistic single-valued neutrosophic hypergraph is an intuitionistic single-valued neutrosophic hypergraph in which $C(\mathscr{H}^*)$ is ordered, i.e., if $C(\mathscr{H}^*) = \{\mathscr{H}^*_{(l_1,m_1,n_1)}, \mathscr{H}^*_{(l_2,m_2,n_2)}, \ldots, \mathscr{H}^*_{(l_n,m_n,n_n)}\}$, then $\mathscr{H}^*_{(l_1,m_1,n_1)} \subseteq \mathscr{H}^*_{(l_2,m_2,n_2)} \subseteq, \ldots, \subseteq \mathscr{H}^*_{(l_n,m_n,n_n)}$.

Definition 9.42 The *strength* of a hyperedge B_j, denoted by $\eta(B_j)$, is defined as $\eta(B_j) = (\min(T_{B_j}(v)), \min(I_{B_j}(v)), \max(F_{B_j}(v)))$, for every $T_{B_j}(v) > 0$, $F_{B_j}(v) > 0$, $I_{B_j}(v) > 0$.

Example 9.12 In Example 9.11, the strength of each hyperedge is given as $\eta(B_1) = (0.7, 0.5, 0.5)$, $\eta(B_2) = (0.4, 0.3, 0.3)$, $\eta(B_3) = (0.9, 0.5, 0.1)$, $\eta(B_4) = (0.4, 0.3, 0.5)$, $\eta(B_5) = (0.4, 0.3, 0.3)$. Thus, B_3 is the stronger edge than B_1, B_2, B_4, B_5.

We now define the D^t *tempered* intuitionistic single-valued neutrosophic hypergraph.

Definition 9.43 An intuitionistic single-valued neutrosophic hypergraph $\mathscr{H}^* = (X, \mathscr{B})$ is called a D^t *tempered* intuitionistic single-valued neutrosophic hypergraph if there is a crisp hypergraph $\mathscr{H}' = (X, D')$ and intuitionistic single-valued neutrosophic set D^t defined on X, such that $\mathscr{B} = \{C_D | D \in D'\}$, where

$$T_{C_D}(x) = \begin{cases} \min\{T_{D^T}(y)|y \in D\}, & \text{if } x \in D, \\ 0, & \text{otherwise.} \end{cases}$$

$$I_{C_D}(x) = \begin{cases} \min\{I_{D^T}(y)|y \in D\}, & \text{if } x \in D, \\ 0, & \text{otherwise.} \end{cases}$$

$$F_{C_D}(x) = \begin{cases} \max\{F_{D^T}(y)|y \in D\}, & \text{if } x \in D, \\ 0, & \text{otherwise.} \end{cases}$$

Theorem 9.3 *An intuitionistic single-valued neutrosophic hypergraph $\mathscr{H}^* = (X, \mathscr{B})$ is a D^t tempered intuitionistic single-valued neutrosophic hypergraph of $\mathscr{G}'$ if and only if $\mathscr{H}^*$ is elementary, support simple, and simply ordered.*

Proof Consider $\mathscr{H}^* = (X, \mathscr{B})$ is a D^t tempered intuitionistic single-valued neu-trosophic hypergraph of $\mathscr{G}'$. Clearly, $\mathscr{H}^*$ is elementary and support simple. We will prove that $\mathscr{H}^*$ is simply ordered. Let $C(\mathscr{H}^*) = \{\mathscr{H}^*_{(l_1,m_1,n_1)}, \mathscr{H}^*_{(l_2,m_2,n_2)}, \cdots, \mathscr{H}^*_{(l_n,m_n,n_n)}\}$. Since, $\mathscr{H}^*$ is elementary, then $\mathscr{H}^*$ is ordered. Suppose, there is $D \in D'_{j+1} \setminus D'_j$ and $d \in \mathscr{B}$ such that $T_{\mathscr{B}}(d) = l_{j+1}$, $I_{\mathscr{B}}(d) = m_{j+1}$, and $F_{\mathscr{B}}(d) = n_{j+1}$. Since, $T_{\mathscr{B}}(d) = l_{j+1} < l_j$, $I_{\mathscr{B}}(d) = m_{j+1} < m_j$, $F_{\mathscr{B}}(d) = n_{j+1} > n_j$, it follows that $d \notin X_j$ and $\mathscr{B} \nsubseteq X_j$, hence $\mathscr{H}^*$ is simply ordered.

Conversely, suppose that $\mathscr{H}^*$ is elementary, support simple, and simply ordered. For $C(\mathscr{H}^*) = \{\mathscr{H}^*_{(l_1,m_1,n_1)}, \mathscr{H}^*_{(l_2,m_2,n_2)}, \cdots, \mathscr{H}^*_{(l_n,m_n,n_n)}\}$, the fundamental sequence is $FS(\mathscr{H}^*) = \{(l_1, m_1, n_1), (l_2, m_2, n_2), \cdots, (l_n, m_n, n_n)\}$ with $0 < l_n < l_{n-1} <, \ldots, < l_1$, $0 < m_n < m_{n-1} <, \ldots, < m_1$, $n_n > n_{n-1} >, \ldots, > n_1 > 0$. $\mathscr{H}^*_{(l_n,m_n,n_n)}$ and intu-itionistic single-valued neutrosophic set D^t on X_n is defined as

$$T_{D^t}(d) = \begin{cases} l_1, & \text{if } d \in X_1, \\ l_j, & \text{if } d \in X_j \setminus X_{j-1}, j = 2, 3, \ldots, n. \end{cases}$$

$$I_{D^t}(d) = \begin{cases} m_1, & \text{if } d \in X_1, \\ m_j, & \text{if } d \in X_j \setminus X_{j-1}, j = 2, 3, \ldots, n. \end{cases}$$

$$F_{D^t}(d) = \begin{cases} n_1, & \text{if } d \in X_1, \\ n_j, & \text{if } d \in X_j \setminus X_{j-1}, j = 2, 3, \ldots, n. \end{cases}$$

Now we prove that $D = \{X_D | D \in D'_n\}$, where

$$T_{X_D}(a) = \begin{cases} \min\{T_{D^t}(y | y \in D)\}, & \text{if } a \in D, \\ 0, & \text{otherwise.} \end{cases}$$

$$I_{X_D}(a) = \begin{cases} \min\{I_{D^t}(y | y \in D)\}, & \text{if } a \in D, \\ 0, & \text{otherwise.} \end{cases}$$

$$F_{X_D}(a) = \begin{cases} \min\{F_{D^T}(y | y \in D)\}, & \text{if } a \in D, \\ 0, & \text{otherwise.} \end{cases}$$

Let $D \in D'_n$. As $\mathscr{H}^*$ is elementary and support simple, then there is a unique intu-itionistic single-valued neutrosophic hyperedge B_i in $\mathscr{B}$ having support $D \in D'_n$.

We will show that D^t tempered intuitionistic single-valued neutrosophic hyper-graph is determined by the crisp graph $G'_n = (X, D'_n)$, i.e., $X_{D \in D'_n} = B_i$, $i = 1, 2, 3, \ldots, n$. Since, all intuitionistic single-valued neutrosophic hyperedges are ele-mentary and $\mathscr{H}^*$ is support simple, then distinct edges have different supports, i.e., $h(B_i)$ is equal to some member (l_i, m_i, n_i) of $FS(\mathscr{H}^*)$. As a consequence, $\mathscr{B} \subseteq X_j$ and if $j > 1$, then $D \in D'_j \setminus D'_{j-1}$, $T_D(d) \geq l_i$, $I_D(d) \geq m_i$, $F_D(d) \leq n_i$, for some $d \in D$. Since, $\mathscr{B} \subseteq C_j$, we claim that $T_{D^t}(d) = l_i$, $I_{D^t}(d) = m_i$, $F_{D^t}(d) = n_i$, for some $d \in D$, if not then $T_{D^t}(d) \geq l_{i-1}$, $I_{D^t}(d) \geq m_{i-1}$, $F_{D^t}(d) \leq n_{i-1}$, for all $d \in D$ implies $\mathscr{B} \subseteq C_{i-1}$ and because $\mathscr{H}^*$ is simply ordered, $D \in D'_i \setminus D'_{i-1}$, then $D \nsubseteq X_{i-1}$, which is a contradiction. Thus, $C_D = B_i$, $i = 1, 2, \ldots, m$, by the definition of C_D. $\square$

Proposition 9.9 *Let* $\mathcal{H}^* = (X, \mathcal{B})$ *be a simply ordered intuitionistic single-valued neutrosophic hypergraph and* $FS(\mathcal{H}^*) = \{(l_1, m_1, n_1), (l_2, m_2, n_2), \cdots, (l_n, m_n, n_n)\}$. *For a crisp hypergraph* $G_{(l_n,m_n,n_n)}$, *there is a partial intuitionistic single-valued neutrosophic hypergraph* $G^* = (X, D^*)$ *of* $\mathcal{H}^* = (X, \mathcal{B})$ *such that the following conditions hold:*

(i) G^* *is a* D^t *tempered intuitionistic single-valued neutrosophic hypergraph of* $G_{(l_n,m_n,n_n)}$,
(ii) $FS(G^*) = FS(\mathcal{H}^*)$ *and* $C(G^*) = C(\mathcal{H}^*)$.

Proof Since, $\mathcal{H}^*$ is simply ordered, $\mathcal{H}^*$ is an elementary intuitionistic single-valued neutrosophic hypergraph. We take the partial intuitionistic single-valued neutrosophic hypergraph $G^* = (X, D^*)$ of $\mathcal{H}^* = (X, \mathcal{B})$ by removing all those edges of $\mathcal{B}$ which are properly contained in another edge, where $D^* = \{B_i \in D|$, if $B_i \subseteq B_j$, and $B_j \in D$, then $B_i = B_j\}$. Since, $G_{(l_n,m_n,n_n)}$ is simple and its all edges are elementary, if any hyperedge in $\mathcal{H}^*$ is subset of another hyperedge, then both edges have the same support. So $FS(G^*) = \mathcal{H}^*$ and $C(G^*) = C\mathcal{H}^*)$. From the definition of D^*, G^* is elementary and support simple. Thus, by Theorem 9.3, G^* is a D^t tempered intuitionistic single-valued neutrosophic hypergraph. $\qquad\square$

Example 9.13 Consider an intuitionistic single-valued neutrosophic hypergraph $\mathcal{H}^* = (X, \mathcal{B})$, where $X = \{x_1, x_2, x_3, x_4\}$ and $\mathcal{B} = \{B_1, B_2, B_3\}$ and the incidence matrix of $\mathcal{H}^*$ is given in Table 9.10.

Let $D^t = \{(x_1, 0.3, 0.4, 0.5), (x_2, 0.1, 0.2, 0.3), (x_3, 0.5, 0.4, 0.6), (x_4, 0.4, 0.3, 0.3)\}$ be an intuitionistic single-valued neutrosophic subset defined on X. Then, it can be seen that

$$T_{\{x_1,x_2,x_3\}}(x_1) = \min\{T_{D^t}(x_1), T_{D^t}(x_2), T_{D^t}(x_3)\} = 0.1,$$
$$I_{\{x_1,x_2,x_3\}}(x_1) = \min\{I_{D^t}(x_1), I_{D^t}(x_2), I_{D^t}(x_3)\} = 0.2,$$
$$F_{\{x_1,x_2,x_3\}}(x_1) = \max\{F_{D^t}(x_1), F_{D^t}(x_2), F_{D^t}(x_3)\} = 0.6,$$
$$T_{\{x_1,x_2,x_3\}}(x_2) = \min\{T_{D^t}(x_1), T_{D^t}(x_2), T_{D^t}(x_3)\} = 0.1,$$
$$I_{\{x_1,x_2,x_3\}}(x_2) = \min\{I_{D^t}(x_1), I_{D^t}(x_2), I_{D^t}(x_3)\} = 0.2,$$
$$F_{\{x_1,x_2,x_3\}}(x_2) = \max\{F_{D^t}(x_1), F_{D^t}(x_2), F_{D^t}(x_3)\} = 0.2,$$
$$T_{\{x_1,x_2,x_3\}}(x_3) = \min\{T_{D^t}(x_1), T_{D^t}(x_2), T_{D^t}(x_3)\} = 0.1,$$
$$I_{\{x_1,x_2,x_3\}}(x_3) = \min\{I_{D^t}(x_1), I_{D^t}(x_2), I_{D^t}(x_3)\} = 0.2,$$
$$F_{\{x_1,x_2,x_3\}}(x_3) = \max\{F_{D^t}(x_1), F_{D^t}(x_2), F_{D^t}(x_3)\} = 0.6.$$

That is, $C_{\{x_1,x_2,x_3\}} = B_1$. Also $C_{\{x_2,x_4\}} = B_2$, $C_{\{x_1,x_3\}} = B_3$. Thus, $\mathcal{H}^*$ is D^t tempered.

Definition 9.44 The *dual* of an intuitionistic single-valued neutrosophic hypergraph $\mathcal{H}^* = (X, \mathcal{B})$ is an intuitionistic single-valued neutrosophic hypergraph $\mathcal{G}^* = (D, X)$, where $D = \{e_1, e_2, \ldots, e_n\}$ is a set of vertices corresponding to $B_1, B_2, \ldots, B_n$, respectively, and $X = \{C_1, C_2, \ldots, C_n\}$ a set of hyperedges corresponding to $x_1, x_2, \ldots, x_n$, respectively.

Table 9.10 Incidence matrix of $\mathscr{H}^*$

$I_{\mathscr{H}^*}$	B_1	B_2	B_3
x_1	(0.1, 0.2, 0.6)	(0, 0, 0)	(0.3, 0.4, 0.6)
x_2	(0.1, 0.2, 0.6)	(0.1, 0.2, 0.3)	(0, 0, 0)
x_3	(0.1, 0.2, 0.6)	(0, 0, 0)	(0.3, 0.4, 0.6)
x_4	(0, 0, 0)	(0.1, 0.2, 0.3)	(0, 0, 0)

Table 9.11 Incidence matrix of $\mathscr{H}^*$

$M_{\mathscr{D}}$	B_1	B_2	B_3
x_1	(0.5, 0.4, 0.3)	(0.5, 0.4, 0.3)	(0, 0, 0)
x_2	(0.2, 0.3, 0.4)	(0, 0, 0)	(0, 0, 0)
x_3	(0, 0, 0)	(0, 0, 0)	(0.3, 0.4, 0.5)
x_4	(0, 0, 0)	(0.5, 0.5, 0.5)	(0.5, 0.5, 0.5)

Fig. 9.9 Intuitionistic single-valued neutrosophic hypergraph

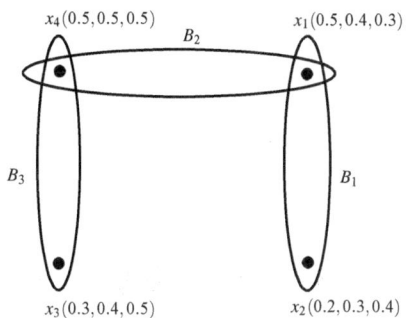

Example 9.14 Consider an intuitionistic single-valued neutrosophic hypergraph $\mathscr{H}^* = (X, \mathscr{B})$, where $X = \{x_1, x_2, x_3, x_4\}$ and $\mathscr{B} = \{B_1, B_2, B_3\}$ is represented by the incidence matrix in Table 9.11. The intuitionistic single-valued neutrosophic hypergraph and its dual are shown in Figs. 9.9 and 9.10, respectively.

The dual of intuitionistic single-valued neutrosophic hypergraph $\mathscr{H}^*$ is $\mathscr{D}^* = (D, X)$ such that $D = \{x_1, x_2, x_3\}$, $X = \{B_1, B_2, B_3, B_4\}$, where

$$B_1 = \{(x_1, 0.5, 0.4, 0.3), (x_2, 0.5, 0.4, 0.3), (x_3, 0, 0, 0)\},$$
$$B_2 = \{(x_1, 0.2, 0.3, 0.4), (x_2, 0, 0, 0), (x_3, 0, 0, 0)\},$$
$$B_3 = \{(x_1, 0, 0, 0), (x_2, 0, 0, 0), (x_3, 0.3, 0.4, 0.5)\},$$
$$B_4 = \{(x_1, 0, 0, 0), (x_2, 0.5, 0.5, 0.5), (x_3, 0.5, 0.5, 0.5)\}.$$

The incidence matrix of dual intuitionistic single-valued neutrosophic hypergraph is given in Table 9.12.

Fig. 9.10 Dual intuitionistic single-valued neutrosophic hypergraph $\mathscr{D}^\star$

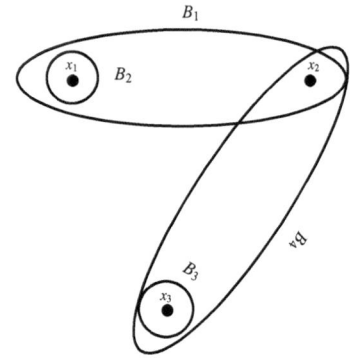

Table 9.12 Incidence matrix of dual intuitionistic single-valued neutrosophic hypergraph

$I_{\mathscr{D}^\star}$	B_1	B_2	B_3	B_4
x_1	(0.5, 0.4, 0.3)	(0.2, 0.3, 0.4)	(0, 0, 0)	(0, 0, 0)
x_2	(0.5, 0.4, 0.3)	(0, 0, 0)	(0, 0, 0)	(0.5, 0.5, 0.5)
x_3	(0, 0, 0)	(0, 0, 0)	(0.3, 0.4, 0.5)	(0.5, 0.5, 0.5)

Theorem 9.4 *If $\mathscr{H}^*$ is linear intuitionistic single-valued neutrosophic hypergraph, then its dual intuitionistic single-valued neutrosophic hypergraph $\mathscr{D}^*$ without isolated vertex is linear intuitionistic single-valued neutrosophic hypergraph.*

Proof Let $\mathscr{H}^*$ be a linear intuitionistic single-valued neutrosophic hypergraph. Suppose that $\mathscr{D}^*$ is not linear intuitionistic single-valued neutrosophic hypergraph, then there must be two distinct intuitionistic single-valued neutrosophic hyperedges X_i and X_j of $\mathscr{D}^*$ having at least two vertices e_1 and e_2 in common. By Definition 9.10 of dual intuitionistic single-valued neutrosophic hypergraph, v_i and v_j belongs to B_1 and B_2 (the intuitionistic single-valued neutrosophic hyperedges of $\mathscr{H}^*$ corresponds to the vertices e_1, e_2 of $\mathscr{D}^*$, respectively) so $\mathscr{H}^*$ is not linear intuitionistic single-valued neutrosophic hypergraph. A contradiction to the statement that $\mathscr{H}^*$ is linear intuitionistic single-valued neutrosophic hypergraph. Hence, dual $\mathscr{D}^*$ of a linear intuitionistic single-valued neutrosophic hypergraph without isolated vertex is also linear intuitionistic single-valued neutrosophic hypergraph. □

Remark 9.6 $\mathscr{D}^* = (\mathscr{H}^*)^t$, that is, incidence matrix of $\mathscr{D}^*$ is the transpose of the incidence matrix of $\mathscr{H}^*$. Also, the dual of a simple intuitionistic single-valued neutrosophic hypergraph may or may not be simple.

Now, we define the intuitionistic single-valued neutrosophic transversal of an intuitionistic single-valued neutrosophic hypergraph.

Definition 9.45 Let $\mathscr{H}^* = (X, \mathscr{B})$ be an intuitionistic single-valued neutrosophic hypergraph and $h(B_i)$ the height of intuitionistic single-valued neutrosophic hyperedge B_i. Then, the *intuitionistic single-valued neutrosophic transversal* τ of $\mathscr{H}^*$ is

Table 9.13 Incidence matrix of $\mathcal{H}^*$

$I_{\mathcal{H}^*}$	B_1	B_2	B_3
u_1	$(0, 0, 0)$	$(0, 0, 0)$	$(0.4, 0.3, 0.2)$
u_2	$(0.9, 0.3, 0.1)$	$(0.4, 0.3, 0.2)$	$(0.4, 0.3, 0.2)$
u_3	$(0.4, 0.3, 0.2)$	$(0, 0, 0)$	$(0.4, 0.3, 0.2)$

defined as an intuitionistic single-valued neutrosophic subset defined on X such that $\tau_{h(B_i)} \cap B_{h(B_i)} \neq \emptyset$ for all $B_i \in \mathcal{B}$.

If $\tau' \subset \tau$ and τ' is not an intuitionistic single-valued neutrosophic transversal of $\mathcal{H}^*$, then τ is called the *minimal transversal*.

Here, we state the following propositions without proof.

Proposition 9.10 *For an intuitionistic single-valued neutrosophic transversal of $\mathcal{H}^* = (X, \mathcal{B})$, we have $h(\tau) \geq h(B_i)$, for all $B_i \in \mathcal{B}$, and for a minimal transversal of $\mathcal{H}^*$, we have $h(\tau) = \max\{B_i | B_i \in \mathcal{B}\} = h(\mathcal{H}^*)$.*

Theorem 9.5 *Let $\mathcal{H}^*$ be an intuitionistic single-valued neutrosophic hypergraph and $Tr(\mathcal{H}^*)$ be the family of minimal intuitionistic single-valued neutrosophic transversals of $\mathcal{H}^*$, then $Tr(\mathcal{H}^*) \neq \emptyset$.*

Example 9.15 Consider the intuitionistic single-valued neutrosophic hypergraph $\mathcal{H}^* = (X, \mathcal{B})$, where $X = \{u_1, u_2, u_3\}$ and $\mathcal{B} = \{B_1, B_2, B_3\}$, which is represented by the following incidence matrix in Table 9.13.

Clearly, $h(\mathcal{H}^*) = (0.9, 0.3, 0.1)$, the intuitionistic single-valued neutrosophic transversals of $\mathcal{H}^*$ are $\tau_1(\mathcal{H}^*) = \{(u_2, 0.9, 0.3, 0.1), (u_3, 0.4, 0.3, 0.2)\}$, $\tau_2(\mathcal{H}^*) = \{(u_2, 0.9, 0.3, 0.1)\}$. Fundamental sequence of $\mathcal{H}^*$ is $FS(\mathcal{H}^*) = \{(0.9, 0.3, 0.1), (0.4, 0.3, 0.2)\}$. $\tau_{1(0.9, 0.3, 0.1)} = \{u_2\}$ and $\tau_{1(0.4, 0.3, 0.2)} = \{u_2, u_3\}$. $\tau_{2(0.9, 0.3, 0.1)} = \{u_2\}$ and $\tau_{2(0.4, 0.3, 0.2)} = \{u_2\}$. $B_{(0.9, 0.3, 0.1)} = \{u_2\}$, $B_{(0.4, 0.3, 0.2)} = \{\{u_2, u_3\}, \{u_2\}, \{u_1, u_2, u_3\}\}$. The minimal transversal of $\mathcal{H}^*$ is $\tau_2(\mathcal{G}) = \{(u_2, 0.9, 0.3, 0.1)\}$.

9.4 Clustering Problem

Clustering (or cluster analysis) involves the task of classifying data points into clusters or classes in such a way that the objects in the same class or cluster are similar and the objects belonging to different clusters are not much similar. The identification of clusters can be done by means of similarity measures. The connectivity and distance can be taken as the similarity measures. Similarity measures are chosen according to the choice of data or the application. The purpose of graph clustering is to group the vertices into classes according to the properties of the graph. So that the edges having high similarity are in the same group. In statistical data analysis, clustering analysis serves as a strong and significant tool, which can be widely used in various fields, like pattern recognition, banking sector, microbiology, document classification, and data mining, etc. In a computer cluster, a set of more than one connected computers work together. The benefit of such clustering of computers is that if any one computer of the cluster fails, another computer can manage the workload of failed computer.

Definition 9.46 Let X be a universal set. A collection of intuitionistic single-valued neutrosophic sets $\{A_1, A_2, A_3, \cdots, A_m\}$ is an intuitionistic single-valued neutrosophic *partition* if

(i) $\bigcup_j supp(A_j) = X, j = 1, 2, 3, ..., m,$

(ii) $\sum_{j=1}^{m} T_{A_j}(x) = 1$, for all $x \in X,$

(iii) $\sum_{j=1}^{m} I_{A_j}(x) = 1$, for all $x \in X,$

(iv) there is at most one j for which $F_{A_j}(x) = 0$, for all $x \in X$ (there is at most one intuitionistic single-valued neutrosophic set for which $T_{A_j}(x) + I_{A_j}(x) + F_{A_j}(x) = 2$, for all $x \in X$).

A family of intuitionistic single-valued neutrosophic subsets $\{A_1, A_2, A_3, \ldots, A_m\}$ is said to be an intuitionistic single-valued neutrosophic partition if it captivates the above conditions.

An intuitionistic single-valued neutrosophic matrix (a_{ij}) can be used to interpret an intuitionistic single-valued neutrosophic partition, where a_{ij} indicates the truth value, indeterminacy value, and falsity value of element x_i in class j. We see that the incidence matrix in intuitionistic single-valued neutrosophic hypergraph is as similar as this matrix. So that we can express an intuitionistic single-valued neutrosophic partition by an intuitionistic single-valued neutrosophic hypergraph $\mathscr{H}^* = (X, \mathscr{B})$ such that

(i) $X = \{x_1, x_2, x_3, \ldots, x_n\}$ is a set of elements $i = 1, 2, 3, \ldots, n,$

(ii) $\mathscr{B} = \{B_1, B_2, B_3, \ldots, B_m\}$ be a finite class of nontrivial intuitionistic single-valued neutrosophic sets,

(iii) $\bigcup_k supp(B_k) = X, k = 1, 2, 3, ..., n,$

(iv) $\sum_{k=1}^{m} T_{A_j}(x) = 1$, for all $x \in X,$

(v) $\sum_{k=1}^{m} I_{A_j}(x) = 1$, for all $x \in X,$

(vi) there is at most one j for which $F_{A_j}(x) = 0$, for all $x \in X$ (there is at most one intuitionistic single-valued neutrosophic set such that $T_{A_j}(x) + I_{A_j}(x) + F_{A_j}(x) = 2$ for all x).

It should be noted that the conditions $(iv)-(vi)$ are combined with the intuitionistic single-valued neutrosophic hypergraph for intuitionistic single-valued neutrosophic partition. Along with these three conditions, an intuitionistic single-valued neutrosophic covering can be represented as an intuitionistic single-valued neutrosophic

Table 9.14 Intuitionistic single-valued neutrosophic partition matrix

$\mathcal{H}^*$	A_t	B_h
a_1	(0.96, 0.50, 0.04)	(0.04, 0.50, 0.96)
a_2	(1, 0.50, 0)	(0, 0.50, 1)
a_3	(0.05, 0.50, 0.05)	(0.95, 0.50, 0.03)
a_4	(0.30, 0.50, 0.61)	(0.70, 0.50, 0.04)
a_5	(0.61, 0.50, 0.04)	(0.39, 0.50, 0.05)

hypergraph. Naturally, (η, ϕ, ψ)-level cut can be applied to intuitionistic single-valued neutrosophic partition (Table 9.14).

Example 9.16 Let us suppose the clustering problem as an illustrative example of an intuitionistic single-valued neutrosophic partition on the visual image processing. We take the five objects which are restricted into two classes:tank and house. To cluster these five objects $a_1, a_2, a_3, a_3, a_4, a_5$ into A_t (tank) and B_h (house), an intuitionistic single-valued neutrosophic partition matrix is given in table below in the form of incidence matrix of an intuitionistic single-valued neutrosophic hypergraph. By applying (η, ϕ, ψ)-cut to the hypergraph, we attain a hypergraph $\mathcal{H}^*_{(\eta,\phi,\psi)}$ which is not an intuitionistic single-valued neutrosophic hypergraph. We denote the edge in $\mathcal{H}^*_{(\eta,\phi,\psi)}$-cut hypergraph $\mathcal{H}^*_{(\eta,\phi,\psi)}$ as $B_{j(\eta,\phi,\psi)}$. This hypergraph $\mathcal{H}^*$ represents the covering because of conditions,

(iv) $\sum_{j=1}^{m} T_{A_j}(x) = 1$, for all $x \in X$,

(v) $\sum_{j=1}^{m} I_{A_j}(x) = 1$, for all $x \in X$,

(vi) there is at most one j, for which $F_{A_j}(x) = 0$, is not always guaranteed.

Incidence matrix of corresponding hypergraph is given in Table 9.15.

Incidence matrix of dual of the above hypergraph is given in Table 9.16. The clarifications for $\tilde{\eta}(B_{j(\eta,\phi,\psi)})$ are given as follows:

- The elements having at least $\tilde{\eta}$ truth value, indeterminacy value, and most falsity value are grouped as an edge in the partition hypergraph $\mathcal{H}^*_{(\eta,\phi,\psi)}$.

Table 9.15 Hypergraph $\mathcal{H}^*_{(0.60,0.50,0.04)}$

$\mathcal{H}^*_{(0.60,0.50,0.04)}$	$A_{t(0.60,0.50,0.04)}$	$B_{h(0.60,0.50,0.04)}$
a_1	1	0
a_2	1	0
a_3	0	1
a_4	0	1
a_5	1	0

Table 9.16 Dual of hypergraph

$\mathscr{G}^*_{(0.60,0.50,0.04)}$	X_1	X_2	X_3	X_4	X_5
A_t	1	1	0	0	1
B_h	0	0	1	1	0

- The strength of edge $\tilde{\eta}(B_{j(\eta,\phi,\psi)})$ in $\mathscr{H}^*_{(\eta,\phi,\psi)}$ is $\tilde{\eta}$. Thus, cohesion or strength of a class in a partition can be measured by the strength of edge. As an example, the strength of classes $A_{t(0.60,0.50,0.04)}$ and $B_{h(0.60,0.50,0.04)}$ at $\eta = 0.60$, $\phi = 0.50$, $\psi = 0.04$ are given as $\tilde{\eta}(A_{t(0.56,0.50,0.40)}) = (0.96, 0.50, 0.04)$ and $\tilde{\eta}(B_{h(0.60,0.50,0.04)}) = (0.70, 0.50, 0.04)$, respectively. Thus, we see that the class $\tilde{\eta}(A_{t(0.60,0.50,0.04)})$ is stronger than $\tilde{\eta}(B_{h(0.60,0.50,0.04)})$ because $\tilde{\eta}_T(A_{t(0.60,0.50,0.04)}) > \tilde{\eta}_T(B_{h(0.60,0.50,0.04)})$. Taking into account the above analysis on the hypergraph $\mathscr{H}^*_{(0.60,0.50,0.04)}$ and $\mathscr{H}^*_{(0.60,0.50,0.04)}$, we have,

  (i) An intuitionistic single-valued neutrosophic partition can be represented by intuitionistic single-valued neutrosophic hypergraph, visually. The (η, ϕ, ψ)-cut hypergraph also represents the (η, ϕ, ψ)-cut partition.

  (ii) The dual hypergraph $\mathscr{H}^*_{(0.60,0.50,0.04)}$ represents those elements X_i, which can be classified into same class $B_{j(\eta,\phi,\psi)}$. For example, the edges X_1, X_2, X_5 of the dual hypergraph represent that the elements a_1, a_2, a_5 can be grouped into A_t at level $(0.60, 0.50, 0.04)$.

  (iii) In an intuitionistic single-valued neutrosophic partition, we have $\sum_{j=1}^{m} T_{A_j}(x)$ $= 1$, $\sum_{j=1}^{m} I_{A_j}(x) = 1$, for all $x \in X$ and for all $x \in X$, there is at most one j such that $F_{A_j}(x) = 0$. If we take (η, ϕ, ψ)-cut at level $(\eta \geq 0.5$ or $\phi \geq 0.5$ or $\psi \leq 0.5)$, no element can be grouped into two classes at the same time. That is, if $\eta \geq 0.5$ or $\phi \geq 0.5$ or $\psi \leq 0.5$, distinct elements are contained in distinct classes in $\mathscr{H}^*_{(\eta,\phi,\psi)}$.

  (iv) At the $(\eta, \phi, \psi) = (0.60, 0.50, 0.04)$ level, $\eta(A_{t(0.60,0.50,0.04)})$ is strongest class as its strength is highest, i.e., $(0.96, 0.50, 0.04)$. It means that this class can be grouped independently from other parts. Thus, the class B_h can be removed from other classes and continue the clustering process. In this way, the elimination of weak classes from the others can allow us to decompose a clustering problem into smaller ones.

Following this strategy, we can reduce data in clustering problem.

9.5 Single-Valued Neutrosophic Directed Hypergraphs

Definition 9.47 A *single-valued neutrosophic directed hypergraph* on a non-empty set X is defined as an ordered pair $D = (V, H)$, where $V = \{A_1, A_2, A_3, \ldots, A_k\}$ is a family of nontrivial single-valued neutrosophic subsets on X and H is a single-valued neutrosophic relation on single-valued neutrosophic sets A_i such that

(i)
$$T_H(B_i) = T_H(\{v_1, v_2, v_3, \ldots, v_r\}) \leq \min\{T_{A_i}(v_1), T_{A_i}(v_2), T_{A_i}(v_3), \ldots, T_{A_i}(v_r)\},$$
$$I_H(B_i) = I_H(\{v_1, v_2, v_3, \ldots, v_r\}) \leq \min\{I_{A_i}(v_1), I_{A_i}(v_2), I_{A_i}(v_3), \ldots, I_{A_i}(v_r)\},$$
$$F_H(B_i) = F_H(\{v_1, v_2, v_3, \ldots, v_r\}) \leq \max\{F_{A_i}(v_1), F_{A_i}(v_2), F_{A_i}(v_3), \ldots, F_{A_i}(v_r)\},$$

for all $v_1, v_2, v_3, \ldots, v_r \in X$.

(ii) $X = \bigcup_k supp(A_k)$, for all $A_k \in V$.

Here $\{B_1, B_2, B_3, \ldots, B_r\}$ is the family of directed crisp hyperedges.

Definition 9.48 A single-valued neutrosophic *directed hyperedge* (or *hyperarc*) is defined as an ordered pair $A = (u, v)$, where u and v are disjoint subsets of nodes. u is taken as the *tail* of A and v is called its *head*. $t(A)$ and $h(A)$ are used to denote the *tail* and *head* of single-valued neutrosophic directed hyperarc, respectively.

In single-valued neutrosophic directed hypergraph $D = (V, H)$, any two vertices s and t are *adjacent vertices* if they both belong to the same directed hyperedge. A *source* vertex s is defined as a vertex in D if $h(x) \neq s$, for each $x \in H$. A *destination* vertex d is defined as a vertex if $t(x) \neq d$, for every $x \in H$.

Definition 9.49 A single-valued neutrosophic directed hypergraph $D = (V, H)$ can be represented by an incidence matrix. The *incidence matrix* of a single-valued neutrosophic directed hypergraph is defined by an $n \times m$ matrix $[b_{ij}]$ as

$$b_{ij} = \begin{cases} (T_{A_j}(a_i), I_{A_j}(a_i), F_{A_j}(a_i)), & \text{if } a_i \in A_j, \\ \mathbf{0}, & \text{otherwise.} \end{cases}$$

Note that $\mathbf{0} = (0, 0, 0)$.

We illustrate the concept of a single-valued neutrosophic directed hypergraph with an example.

Example 9.17 Consider a single-valued neutrosophic directed hypergraph $D = (V, H)$, such that $X = \{v_1, v_2, v_3, v_4, v_5, v_6, v_7\}$ and $H = \{H_1, H_2, H_3, H_4\}$, where

Table 9.17 Incidence matrix of single-valued neutrosophic directed hypergraph

I_D	H_1	H_2	H_3	H_4
v_1	$(0.2, 0.3, 0.4)$	$(0.2, 0.3, 0.4)$	0	0
v_2	0	$(0.1, 0.3, 0.5)$	0	0
v_3	$(0.2, 0.4, 0.2)$	$(0.2, 0.4, 0.2)$	$(0.2, 0.4, 0.2)$	0
v_4	0	$(0.1, 0.2, 0.3)$	$(0.1, 0.2, 0.3)$	0
v_5	$(0.5, 0.5, 0.3)$	0	0	$(0.5, 0.5, 0.3)$
v_6	0	0	0	$(0.4, 0.5, 0.6)$
v_7	0	0	$(0.1, 0.2, 0.3)$	$(0.1, 0.2, 0.3)$

Fig. 9.11 Single-valued neutrosophic directed hypergraph

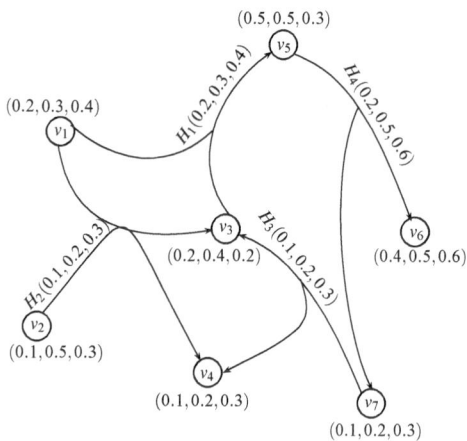

$H_1 = \{(v_1, 0.2, 0.3, 0.4), (v_3, 0.2, 0.4, 0.2), (v_5, 0.5, 0.5, 0.3)\}$,

$H_2 = \{(v_1, 0.2, 0.3, 0.4), (v_2, 0.1, 0.3, 0.5), (v_3, 0.2, 0.4, 0.2), (v_4, 0.1, 0.2, 0.3)\}$,

$H_3 = \{(v_3, 0.2, 0.4, 0.2), (v_4, 0.1, 0.2, 0.3), (v_7, 0.1, 0.2, 0.3)\}$,

$H_4 = \{(v_5, 0.5, 0.5, 0.3), (v_6, 0.4, 0.5, 0.6), (v_7, 0.1, 0.2, 0.3)\}$.

The incidence matrix of $D = (V, H)$ is given in Table 9.17.

The single-valued neutrosophic directed hypergraph is shown in Fig. 9.11.

Definition 9.50 The *height* of a single-valued neutrosophic directed hypergraph $D = (V, H)$ is defined as $h(D) = \{\max(H_k), \max(H_l), \min(H_m) \mid H_k, H_l, H_m \in H\}$, where $H_k = \max(T_{H_j}(v_i))$, $H_l = \max(I_{H_j}(v_i))$, $H_m = \min(F_{H_j}(v_i))$. Here, $T_{H_j}(v_i)$, $I_{H_j}(v_i)$, and $F_{H_j}(v_i)$ denote the truth-membership, indeterminacy, and falsity-membership values of vertex v_i to directed hyperedge H_j, respectively.

Definition 9.51 A single-valued neutrosophic set $S = \{(x, T_S(x), I_S(x), F_S(x)) : x \in X\}$ is called an *elementary* single-valued neutrosophic set if T_S, I_S, and F_S are single valued on the support of S.

A single-valued neutrosophic directed hypergraph $D = (V, H)$ is an *elementary single-valued neutrosophic directed hypergraph* if its all directed hyperedges are elementary.

Definition 9.52 The *strength* of a single-valued neutrosophic directed hyperedge is defined as

$$\eta(H_i) = \{\min_{v_j} \in H_i(T_{H_i}(v_j) : T_{H_i}(v_j) > 0), \ \min_{v_j} \in H_i(I_{H_i}(v_j) : I_{H_i}(v_j) > 0),$$

$$\max_{v_j} \in H_i(F_{H_i}(v_j) : F_{H_i}(v_j) > 0)\}.$$

The *strength* of directed hyperedge describes that the objects having the participation degree at least $\eta(H_i)$ are grouped in the hyperedge H_i.

Definition 9.53 A single-valued neutrosophic directed hypergraph $D = (V, H)$ is *simple* if $A_j, \ A_k \in H, A_j \le A_k$ imply $A_j = A_k$.

A single-valued neutrosophic directed hypergraph $D = (V, H)$ is called *support simple* if $A_j, \ A_k \in H, supp(A_j) = supp(A_k)$, and $A_j \le A_k$, then $A_j = A_k$.

A single-valued neutrosophic directed hypergraph $D = (V, H)$ is called *strongly support simple* if $A_j, \ A_k \in H$, and $supp(A_j) = supp(A_k)$ imply that $A_j = A_k$.

Theorem 9.6 *A single-valued neutrosophic directed hypergraph $D = (V, H)$ is single-valued neutrosophic directed graph (possibly with loops) if and only if D is support simple, elementary, and all the hyperedges have two(or one) element supports.*

Theorem 9.7 *Let $D = (V, H)$ be an elementary single-valued neutrosophic directed hypergraph. Then, D is support simple if and only if D is strongly support simple.*

Proof Suppose that $D = (V, H)$ is elementary, support simple, and $supp(A_j) = supp(A_k)$ for $A_j, A_k \in H$. We assume that $h(A_j) \le h(A_k)$. Since, D is elementary, we have $A_j \le A_k$, and since D is support simple we have $A_j = A_k$. Hence, D is strongly support simple. The converse part of the theorem can be proved trivially by using the definitions. $\square$

Theorem 9.8 *Let $D = (V, H)$ be a strongly support simple single-valued neutrosophic directed hypergraph of order n. Then, $|H| \le 2^n - 1$. The equality holds if and only if $\{supp(A_j)|A_j \in H\} = \mathscr{P}(V) \setminus \emptyset$.*

Proof Since every nontrivial subset of X can be the support set of at most one $A_j \in H$ so $|H| \le 2^n - 1$. The second part is trivial. $\square$

Theorem 9.9 *Let $D = (V, H)$ be a simple, elementary single-valued neutrosophic directed hypergraph of order n. Then, $|H| \le 2^n - 1$. The equality holds if and only if $\{supp(A_j)|A_j \in H\} = \mathscr{P}(V) \setminus \emptyset$.*

Proof Since D is simple and elementary, each nontrivial subset of X can be the support set of at most one $A_j \in H$. Hence, $|H| \leq 2^n - 1$.

We now prove that there exists an elementary, simple D having $|H| = 2^n - 1$. Let $A = \{(T_B(v), I_B(v), F_B(v)) | B \subseteq V\}$ be the set of mappings such that

$$T_B(v) = \begin{cases} \frac{1}{|B|}, & \text{if } v \in B, \\ 0, & \text{otherwise.} \end{cases}$$

$$I_B(v) = \begin{cases} \frac{1}{|B|}, & \text{if } v \in B, \\ 0, & \text{otherwise.} \end{cases}$$

$$F_B(v) = \begin{cases} \frac{1}{|B|}, & \text{if } v \in B, \\ 0, & \text{otherwise.} \end{cases}$$

Then, every set containing single element has height $(1, 1, 1)$, height of every set containing two elements is $(0.5, 0.5, 0.5)$, and so on. Hence, D is elementary, simple, and $|H| = 2^n - 1$. $\qquad\square$

Definition 9.54 Let $D = (V, H)$ be a single-valued neutrosophic directed hypergraph. Consider $\lambda \in [0, 1]$, $\mu \in [0, 1]$, and $\nu \in [0, 1]$ such that $0 \leq \lambda + \mu + \nu \leq 3$. Then, the (λ, μ, ν)-*level directed hypergraph* of D is defined as an ordered pair, $D^{(\lambda,\mu,\nu)} = (V^{(\lambda,\mu,\nu)}, H^{(\lambda,\mu,\nu)})$, where

(i) $H^{(\lambda,\mu,\nu)} = \{H_i^{(\lambda,\mu,\nu)} | H_i \in H\}$, $V^{(\lambda,\mu,\nu)} = \bigcup_{H_i \in H} H_i^{(\lambda,\mu,\nu)}$,

(ii) $H_i^{(\lambda,\mu,\nu)} = \{v_j \in X | T_{H_i}(v_j) \geq \lambda, I_{H_i}(v_j) \geq \mu, F_{H_i}(v_j) \leq \nu\}$.

Definition 9.55 Let $D = (V, H)$ be a single-valued neutrosophic directed hypergraph such that $h(D) = (u, v, w)$. Let $D^{(u_i,v_i,w_i)} = (V^{(u_i,v_i,w_i)}, H^{(u_i,v_i,w_i)})$ be the (u_i, v_i, w_i)-level hypergraphs of D. The sequence of real numbers $\{(u_1, v_1, w_1), (u_2, v_2, w_2), \ldots, (u_n, v_n, w_n)\}$, $0 < u_n < u_{n+1} <, \ldots, < u_1 = u$, $0 < v_n < v_{n+1} <, \ldots, < v_1 = v$, $w_n > w_{n+1} >, \ldots, > w_1 = w > 0$, which satisfies the properties,

(i) if $u_{i+1} < u' < u_i$, $v_{i+1} < v' < v_i$, $w_{i+1} > w' > w_i(w_i < w' < w_{i+1})$, then $H^{(u',v',w')} = H^{(u_i,v_i,w_i)}$,

(ii) $H^{(u_i,v_i,w_i)} \sqsubseteq H^{(u_{i+1},v_{i+1},w_{i+1})}$,

is called the *fundamental sequence* of single-valued neutrosophic directed hypergraph D, denoted by $FS(D)$. The set of (u_i, v_i, w_i)-level hypergraphs $\{D^{(u_1,v_1,w_1)}, D^{(u_2,v_2,w_2)}, \ldots, D^{(u_n,v_n,w_n)}\}$ is known as *core hypergraphs* of single-valued neutrosophic directed hypergraph D and is denoted by $c(D)$.

The corresponding sequence of (u_i, v_i, w_i)-level directed hypergraphs $\{D^{(u_1,v_1,w_1)} \subseteq D^{(u_2,v_2,w_2)} \subseteq \ldots \subseteq D^{(u_n,v_n,w_n)}\}$ is called the D *induced fundamental sequence*.

Example 9.18 Consider a single-valued neutrosophic directed hypergraph $D = (V, H)$, where $X = \{v_1, v_2, v_3, v_4, v_5\}$ and $H = \{H_1, H_2, H_3, H_4\}$. Incidence matrix of D is given in Table 9.18.

The corresponding graph is shown in Fig. 9.12.

Table 9.18 Incidence matrix of D

I_D	H_1	H_2	H_3	H_4
v_1	$(0.8, 0.7, 0.1)$	$(0.9, 0.8, 0.1)$	0	$(0.5, 0.4, 0.3)$
v_2	$(0.8, 0.7, 0.1)$	$(0.9, 0.8, 0.1)$	$(0.5, 0.4, 0.3)$	$(0.5, 0.4, 0.3)$
v_3	0	0	$(0.3, 0.3, 0.4)$	0
v_4	$(0.5, 0.4, 0.3)$	0	$(0.5, 0.4, 0.3)$	0
v_5	0	0	0	$(0.5, 0.4, 0.3)$

Fig. 9.12 Single-valued
neutrosophic directed
hypergraph

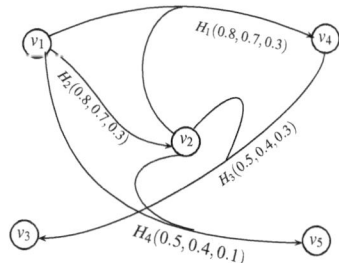

By routine calculations, we have $h(D) = (0.9, 0.8, 0.1)$, $H^{(0.9,0.8,0.1)} = \{\{v_1, v_2\}\}$, $H^{(0.8,0.7,0.1)} = \{\{v_1, v_2\}\}$, $H^{(0.5,0.4,0.3)} = \{\{v_1, v_2\}, \{v_1, v_2, v_5\}, \{v_1, v_2, v_4\}, \{v_2, v_4\}\}$. Therefore, the $FS(D)$ is $\{(0.9, 0.8, 0.1), (0.5, 0.4, 0.3)\}$. The set of core hypergraphs is $c(D) = \{D^{(0.9,0.8,0.1)} = (V_1, H_1), D^{(0.5,0.4,0.3)} = (V_2, H_2)\}$. Note that $H^{(0.9,0.8,0.1)} \subseteq H^{(0.5,0.4,0.3)}$, $H^{(0.9,0.8,0.1)} \neq H^{(0.5,0.4,0.3)}$, $H_i \not\subseteq H_j$, for all $H_i, H_j \in H$. Hence, D is simple. Further, it can be seen that $supp(H_i) = supp(H_j)$, for all $H_i, H_j \in H$ implies $H_i = H_j$. Thus, D is strongly support simple and support simple. The induced fundamental sequence of D is given in Fig. 9.13.

Definition 9.56 Let $D = (V, H)$ be a single-valued neutrosophic directed hypergraph and $FS(D) = \{(u_1, v_1, w_1), (u_2, v_2, w_2), \cdots, (u_n, v_n, w_n)\}$ be the fundamental sequence of D. If for each $H_i \in H$ and each $(l, m, n) \in ((u_{i+1}, v_{i+1}, w_{i+1}), (u_i, v_i, w_i))$, we have $H_i^{(l,m,n)} = H_i^{(u_i, v_i, w_i)}$ for all $(u_i, v_i, w_i) \in FS(D)$, then D is called *sectionally elementary* .

It can be noted that D is *sectionally elementary* if and only if $T_{H_i}(x)$, $I_{H_i}(x)$, $F_{H_i}(x) \in FS(D)$, for all $H_i \in H$ and for every $x \in V$.

Definition 9.57 Let $D = (V, H)$ be a single-valued neutrosophic directed hypergraph. The *partial single-valued neutrosophic directed hypergraph* of D is defined as an ordered pair $D' = (V', H')$, where $H' \subseteq H$ and $V' = \bigcup_i \{supp(H_i') | H_i' \in H\}$.

Then, D' is called *partial single-valued neutrosophic directed hypergraph* generated by H'.

Definition 9.58 A single-valued neutrosophic directed hypergraph $D = (V, H)$ is said to be *ordered* if $c(D)$ is ordered. That is, if $c(D) = \{D^{(u_1,v_1,w_1)}, D^{(u_2,v_2,w_2)}, \cdots, D^{(u_n,v_n,w_n)}\}$, then $D^{(u_1,v_1,w_1)} \subseteq D^{(u_2,v_2,w_2)} \subseteq \cdots \subseteq D^{(u_n,v_n,w_n)}$.

$D^{(0.9,0.8,0.1)}$−level single-valued neutrosophic directed hypergraph. $D^{(0.8,0.7,0.1)}$−level single-valued neutrosophic directed hypergraph.

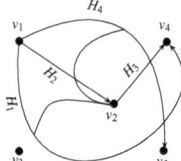

$D^{(0.5,0.4,0.3)}$−level single-valued neutrosophic directed hypergraph.

Fig. 9.13 Induced fundamental sequence of D

Table 9.19 Incidence matrix of D

I	H_1	H_2	H_3
v_1	$(0.7, 0.5, 0.1)$	0	$(0.5, 0.3, 0.1)$
v_2	$(0.7, 0.5, 0.1)$	$(0.5, 0.3, 0.1)$	0
v_3	0	$(0.5, 0.3, 0.1)$	$(0.5, 0.3, 0.1)$
v_4	0	0	$(0.5, 0.3, 0.1)$

The sequence is called *simply ordered* if it is *ordered* and if whenever $H^* \in H_{j+1}^* \setminus H_j^*$, then $H^* \not\subseteq V_j$. Thus, the single-valued neutrosophic directed hypergraph is also *simply ordered*.

Proposition 9.11 *Let $D = (V, H)$ be a single-valued neutrosophic directed hypergraph. If D is elementary, then it is ordered. Further, if D is an ordered single-valued neutrosophic directed hypergraph with simple support, then D is elementary.*

Example 9.19 Consider a single-valued neutrosophic directed hypergraph $D = (V, H)$, where $X = \{v_1, v_2, v_3, v_4\}$ and $H = \{H_1, H_2, H_3\}$, which is represented by the incidence matrix given in Table 9.19.

Here, $FS(D) = \{(0.7, 0.5, 0.1), (0.5, 0.3, 0.1)\}$. The single-valued neutrosophic directed hypergraph D is sectionally elementary. As for each $H_i \in H$ and for all $(l, m, n) \in ((0.5, 0.3, 0.1), (0.7, 0.5, 0.1)$, we have $H_i^{(l,m,n)} = H_i^{(0.7,0.5,0.1)}$. It can be seen that $H_1^{(0.6,0.35,0.1)} = \{v_1, v_2\} = H_1^{(0.7,0.5,0.1)}$ and so on. The corresponding single-valued neutrosophic directed hypergraph is shown in Fig. 9.14.

Fig. 9.14 Sectionally
elementary single-valued
neutrosophic directed
hypergraph

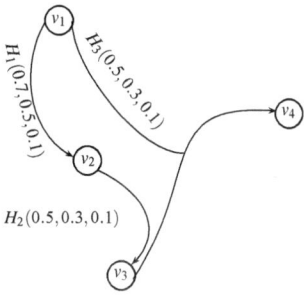

9.6 Single-Valued Neutrosophic Line Directed Hypergraphs

Definition 9.59 A *single-valued neutrosophic directed hyperpath* of length k in a single-valued neutrosophic directed hypergraph $D = (V, H)$ is defined as an alternating sequence $v_1, H_1, v_2, H_2, \ldots, v_k, H_k, v_{k+1}$ of distinct points and directed hyperedges such that

(i) $T_H(H_i) > 0$, $I_H(H_i) > 0$, and $F_H(H_i) > 0$,
(ii) $v_i, v_{i+1} \in H_i$, $i = 1, 2, 3, \ldots, k$.

A single-valued neutrosophic directed hyperpath is called a *single-valued neutrosophic directed hypercycle* if $v_1 = v_{k+1}$.

Definition 9.60 A single-valued neutrosophic directed hypergraph $D = (V, H)$ is *connected* if a single-valued neutrosophic directed hyperpath exists between each pair of distinct nodes.

Definition 9.61 Let any two vertices, say s and t, be connected through a single-valued neutrosophic directed hyperpath of length k in a single-valued neutrosophic directed hypergraph. Then, the *strength* of the single-valued neutrosophic directed hyperpath is defined as

$$\chi^k(s, t) = \{\min(T_H(H_1), T_H(H_2), \ldots, T_H(H_k)), \ \min(I_H(H_1), I_H(H_2), \ldots, I_H(H_k)),$$
$$\max(F_H(H_1), F_H(H_2), \ldots, F_H(H_k))\}.$$

$s \in H_1$, $t \in H_k$. $H_1, H_2, \ldots, H_k$ are directed hyperedges. The *strength of connectedness* between s and t is defined as $\chi^\infty(s, t) = \{\sup_k T(\chi^k(s, t)), \sup_k I(\chi^k(s, t)),$
$\inf_k F(\chi^k(s, t))\}$.

Theorem 9.10 *A single-valued neutrosophic directed hypergraph $D = (V, H)$ is connected if and only if $\chi^\infty(s, t) > 0$, for all $s, t \in V$.*

Proof Suppose that $D = (V, H)$ is connected single-valued neutrosophic directed hypergraph. Then, between each pair of distinct vertices there exists a single-valued neutrosophic directed hyperpath such that $\chi^k(s, t) > 0 \Rightarrow \{\sup_k T(\chi^k(s, t)),$

$\sup_k\{I(\chi^k(s, t)), \inf_k\{F(\chi^k(s, t))|k = 1, 2, \cdots\} > 0 \quad \Rightarrow \chi^\infty(s, t) > 0, \quad \text{for} \quad \text{all}$

$s, t \in V$.

Conversely, suppose that $\chi^\infty(s, t) > 0 \Rightarrow \{\sup_k T(\chi^k(s, t)), \sup_k\{I(\chi^k(s, t)),$

$\inf_k\{F(\chi^k(s, t))|k = 1, 2, \cdots\} > 0$. This shows that there exists at least one directed hyperpath between each pair of vertices. Hence, D is connected. □

Definition 9.62 A single-valued neutrosophic directed hypergraph $D = (V, H)$ is called *linear* if for every single-valued neutrosophic directed hyperedge $H_i, H_j \in H$,

(i) $supp(H_i) \subseteq supp(H_j)$ implies $i = j$,
(ii) $|supp(H_i) \cap supp(H_j)| \leq 1$.

We now define the dual single-valued neutrosophic directed hypergraphs.

Definition 9.63 Let $D = (V, H)$ be a single-valued neutrosophic directed hypergraph on a universal set V. The *dual single-valued neutrosophic directed hypergraph* of D is defined as an ordered pair $D^* = (V^*, H^*)$, where

(i) $V^* = H$ is single-valued neutrosophic set of vertices of D^*.
(ii) If $|V| = n$, then H^* is the single-valued neutrosophic set on the set of directed hyperedges $\{V_1, V_2, V_3, \cdots, V_n\}$ such that $V_i = \{H_j|v_i \in H_j, H_j \text{ is the directed hyperedge in } D\}$. This means that V_i is the set of those directed hyperedges which contain the vertex v_i as a common vertex.

The truth-membership, indeterminacy, and falsity-membership values of V_i are defined as

$$T_H^*(V_i) = \inf\{T_H(H_j) : v_i \in H_j\},$$
$$I_H^*(V_i) = \inf\{I_H(H_j) : v_i \in H_j\},$$
$$F_H^*(V_i) = \sup\{F_H(H_j) : v_i \in H_j\}.$$

We now describe the method of construction of dual single-valued neutrosophic directed hypergraph D^* of a single-valued neutrosophic directed hypergraph D as a simple procedure as given in Construction 9.3. We also describe an example.

Construction 9.3 *The construction of dual single-valued neutrosophic directed hypergraph D^**

Let $D = (V, H)$ be a single-valued neutrosophic directed hypergraph. The procedure of constructing the dual single-valued neutrosophic directed hypergraph contains the following steps:

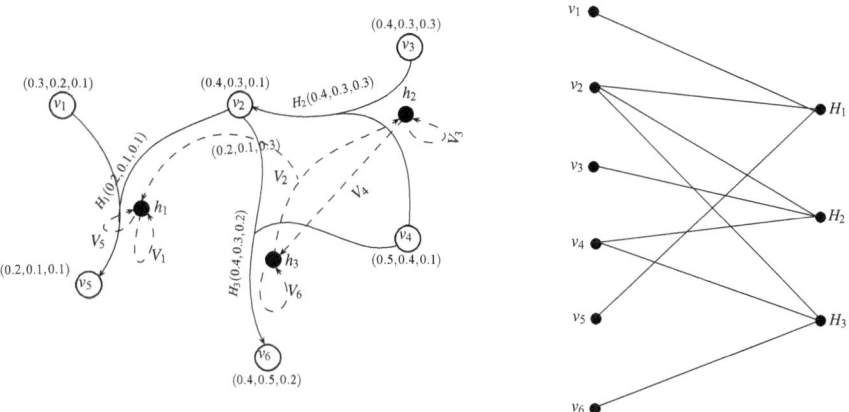

Fig. 9.15 Single-valued neutrosophic directed hypergraph and its D^*

1. Make the single-valued neutrosophic set of vertices of D^* as $V^* = H$.
2. Define a one to one function $f : V \rightarrow H$ from the set of vertices to the set of directed hyperedges of D in the way that if the directed hyperedges $H_s, H_{s+1}, H_{s+2}, \cdots, H_j$ contain the vertex v_i, then v_i is mapped onto $H_s, H_{s+1}, H_{s+2}, \cdots, H_j$ as shown in Fig. 9.15.
3. Draw the directed hyperedges $\{V_1, V_2, \cdots, V_n\}$ of D^* such that $V_i = \{H_j | f(v_i) = H_j\}$.
4. Make the directed hyperedges as the vertex H_j of D^* belongs to $h(V_i)$ if and only if $v_i \in t(H_j)$ in D and similarly H_j is in $t(V_i)$ if and only if $v_i \in h(H_j)$.
5. Calculate the truth-membership, indeterminacy, and falsity-membership values of directed hyperedges in D^* as

$$T_{H^*}(V_i) = \inf \{T_H(H_j) : v_i \in H_j\},$$
$$I_{H^*}(V_i) = \inf \{I_H(H_j) : v_i \in H_j\},$$
$$F_{H^*}(V_i) = \sup \{F_H(H_j) : v_i \in H_j\}.$$

Example 9.20 Consider a single-valued neutrosophic directed hypergraph $D = (V, H)$, where $X = \{v_1, v_2, v_3, v_4, v_5, v_6\}$ and $H = \{H_1, H_2, H_3\}$ as shown in Fig. 9.15. The dual single-valued neutrosophic directed hypergraph $D^* = (V^*, H^*)$ is shown with dashed lines such that $V^* = \{h_1, h_2, h_3\}$, $H^* = \{V_1, V_2, V_3, V_4, V_5, V_6\}$. The incidence matrix of D^* is given in Table 9.20.

Theorem 9.11 *Let D be a single-valued neutrosophic directed hypergraph. Then, $D^{**} = D$.*

Theorem 9.12 *The dual single-valued neutrosophic directed hypergraph of a linear single-valued neutrosophic hypergraph is also linear, that is, if D is linear, then D^* is also linear.*

Table 9.20 Incidence matrix of dual single-valued neutrosophic directed hypergraph

I_{D^*}	V_1	V_2	V_3	V_4	V_5	V_6
h_1	$(0.2, 0.1, 0.1)$	$(0.2, 0.1, 0.1)$	0	0	$(0.2, 0.1, 0.1)$	0
h_2	0	$(0.2, 0.1, 0.1)$	$(0.4, 0.3, 0.3)$	$(0.4, 0.3, 0.3)$	0	0
h_3	0	$(0.2, 0.1, 0.1)$	0	$(0.4, 0.3, 0.3)$	0	$(0.4, 0.3, 0.3)$

Proof Let $D = (V, H)$ be a linear single-valued neutrosophic directed hypergraph and $D^* = (V^*, H^*)$. Suppose on contrary that D^* is not linear, then there exists V_i and V_j such that $|supp(V_i) \bigcap supp(V_j)| = 2$. Let $|supp(V_i) \bigcap supp(V_j)| = \{H_l, H_m\}$. Then, the duality of D^* implies that $v_i, v_j \in H_l$ and $v_i, v_j \in H_m$, which is a contradiction to the statement that D is linear. Hence, D^* is linear.   □

Definition 9.64 Let $D = (V, H)$ be a single-valued neutrosophic directed hypergraph. The *single-valued line directed graph* of D is the directed graph $L(D) = (V_L, H_L)$, such that

(i) $V_L = H$,
(ii) $\{A_i, A_j\} \in H_L$ if and only if $|supp(A_i) \bigcap supp(A_j)| \neq \emptyset$ for $i \neq j$.

The truth-membership, indeterminacy and falsity-membership values of vertices and hyperedges of $L(D)$ are defined as

(i) $V_L(A_i) = H(A_i)$,
(ii) $T_{H_L}\{A_i, A_j\} = \min\{T_H(A_i), T_H(A_j) | A_i, A_j \in H\}$, $I_{H_L}\{A_i, A_j\} = \min\{I_H(A_i), I_H(A_j) | A_i, A_j \in H\}$, $F_{H_L}\{A_i, A_j\} = \max\{F_H(A_i), F_H(A_j) | A_i, A_j \in H\}$, respectively.

Example 9.21 Consider a single-valued neutrosophic directed hypergraph $D = (V, H)$ as given in Fig. 9.16. The single-valued neutrosophic line directed hypergraph $L(D) = (V_L, H_L)$ of D is shown with dashed hyperedges.

Theorem 9.13 Let $G = (U, W)$ be a simple single-valued neutrosophic digraph. Then, G is the single-valued neutrosophic line graph of a linear single-valued neutrosophic directed hypergraph.

Proof Let $G = (U, W)$ be a simple single-valued neutrosophic digraph on a set of universe Z. With no loss of generality, suppose that G is connected. A single-valued neutrosophic directed hypergraph $D = (V, H)$ can be formed from G as

1. Take the set of edges of G as the vertices of D. Let $W = \{w_1, w_2, w_3, \ldots, w_n\}$ be the directed edges of G and Z^D be the set of vertices of D, then $Z^D = W$. Let $V = \{\rho_1, \rho_2, \rho_3, \ldots, \rho_r\}$ be the collection of nontrivial single-valued neutrosophic sets on Z, such that $\rho_k(w_i) = 1, i = 1, 2, 3, \ldots, n$.

Fig. 9.16 Single-valued neutrosophic directed hypergraph and its $L(D)$

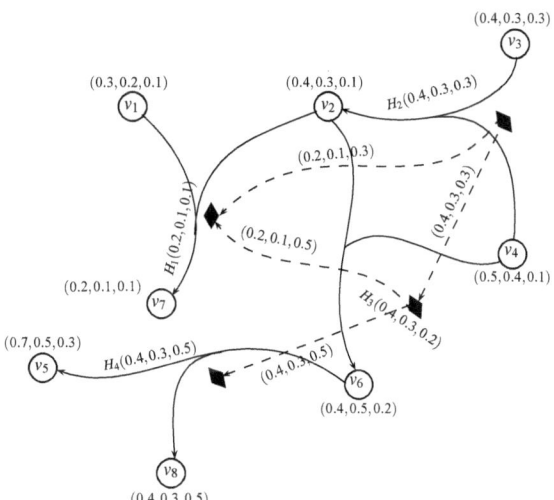

2. Let $Z = \{z_1, z_2, z_3, \ldots, z_m\}$, then the set of directed hyperedges of D is $H^D = \{H_1, H_2, H_3, \ldots, H_n\}$, where H_j are those edges of G in which z_i is the incidence vertex, that is, $H_i = \{w_j | z_i \in w_j, \; j = 1, 2, 3, \ldots, n.\}$. Further, $H(H_i) = U(z_i)$, $i = 1, 2, 3, \ldots, n$.

We claim that D is a linear single-valued neutrosophic directed hypergraph.

Consider the directed hyperedge $H_j = \{w_1, w_2, w_3, \ldots, w_k\}$. From the definition of single-valued neutrosophic directed hypergraph, we have

$$T_{H(H_i)} = \inf\{\wedge_j T_{\rho_j}(w_1), \quad \wedge_j T_{\rho_j}(w_2), \ldots, \wedge_j T_{\rho_j}(w_k)\} = T_U(z_i) \leq 1,$$

$$I_{H(H_i)} = \inf\{\wedge_j I_{\rho_j}(w_1), \quad \wedge_j I_{\rho_j}(w_2), \ldots, \wedge_j I_{\rho_j}(w_k)\} = I_U(z_i) \leq 1,$$

$$F_{H(H_i)} = \sup\{\vee_j F_{\rho_j}(w_1), \quad \vee_j F_{\rho_j}(w_2), \ldots, \vee_j F_{\rho_j}(w_k)\} = F_U(z_i) \leq 1,$$

$1 \leq i \leq n$, and $\bigcup_r(\rho_r) = Z^D$, for all $\rho_r \in V$.

Thus, D is single-valued neutrosophic directed hypergraph. We now prove that D is linear.

(i) Since, the truth-membership, indeterminacy, and falsity-membership values of all the vertices of D are same. Therefore, $supp(\rho_i) \subseteq supp(\rho_j)$ implies $i = j$, for each $1 \leq i, j \leq r$.

(ii) On contrary, suppose that $supp(\rho_i) \cap supp(\rho_j) = \{w_l, w_m\}$, that is, the both edges w_l, w_m have two incident vertices in G, which is a contradiction to the statement that G is simple. Hence, $|supp(\rho_i) \cap supp(\rho_j)| \leq 1$, $1 \leq i, j \leq r$.

□

Theorem 9.14 *A single-valued neutrosophic directed hypergraph $D = (V, H)$ is connected if and only if its line directed graph $L(D)$ is connected.*

Proof Suppose that $D = (V, H)$ is a connected single-valued neutrosophic directed hypergraph. Let $L(D) = (A, B)$ be the single-valued neutrosophic line directed graph of D and H_i, H_j be any two distinct vertices of $L(D)$. Consider $v_i \in H_i$ and $v_j \in H_j$. Since D is connected, there exists a single-valued neutrosophic directed hyperpath $v_i, H_i, v_{i+1}, H_{i+1}, \ldots, v_j, H_j$ between v_i and v_j. By definition of strength of connectedness, we have

$$\chi^\infty(H_i, H_j) = \{\sup_k T(\chi^k(H_i, H_j))\}, \sup_k\{I(\chi^k(H_i, H_j))\}, \inf_k\{F(\chi^k(H_i, H_j))\},$$

$$k = 1, 2, \cdots$$

$$= \sup_k\{T_B(H_i, H_{i+1}) \wedge T_B(H_{i+1}, H_{i+2}) \wedge \cdots \wedge T_B(H_{j-1}, H_j)\},$$

$$\sup_k\{I_B(H_i, H_{i+1}) \wedge I_B(H_{i+1}, H_{i+2}) \wedge \cdots \wedge I_B(H_{j-1}, H_j)\},$$

$$\inf_k\{F_B(H_i, H_{i+1}) \vee F_B(H_{i+1}, H_{i+2}) \vee \cdots \vee I_B(H_{j-1}, H_j)\},$$

$$= \sup\{T_H(H_i) \wedge T_H(H_{i+1}) \wedge \cdots \wedge T_H(H_{j-1}) \wedge T_H(H_j)\},$$

$$\sup\{I_H(H_i) \wedge I_H(H_{i+1}) \wedge \cdots \wedge T_H(H_{j-1}) \wedge I_H(H_j)\},$$

$$\inf\{F_H(H_i) \vee F_H(H_{i+1}) \vee \cdots \vee F_H(H_{j-1}) \vee F_H(H_j)\}, k = 1, 2, \ldots$$

$$= \sup\{T(\chi^k(v_i, v_j))\}, \sup\{I(\chi^k(v_i, v_j))\}, \inf\{F(\chi^k(v_i, v_j))\},$$

$$k = 1, 2, \cdots$$

$$= \chi^\infty(v_i, v_j) > 0.$$

Since, H_i and H_j were chosen arbitrarily. Hence, $L(D)$ is connected. The converse part of the theorem can be proved on the same lines. □

Definition 9.65 The 2-*section* $[D]_2$ of a single-valued neutrosophic directed hypergraph $D = (V, H)$ is the single-valued neutrosophic graph (V, E), where

(i) $V = V$, i.e., $[D]_2$ has the same set of vertices as D.
(ii) $E = \{h = v_iv_j | v_i \neq v_j, v_iv_j \in H_k, k = 1, 2, 3, \cdots\}$, i.e., two vertices v_i and v_j are adjacent in $[D]_2$ if they belong to the same directed hyperedge H_k in D and

$$T_E(v_iv_j) = \inf\{\wedge_k T_{H_k}(v_i), \wedge_k T_{H_k}(v_j)\},$$
$$I_E(v_iv_j) = \inf\{\wedge_k I_{H_k}(v_i), \wedge_k I_{H_k}(v_j)\},$$
$$F_E(v_iv_j) = \sup\{\vee_k F_{H_k}(v_i), \vee_k F_{H_k}(v_j)\}.$$

Example 9.22 A single-valued neutrosophic directed hypergraph $D = (V, H)$ and its 2-section $[D]_2$ is shown in Fig. 9.17.

Fig. 9.17 A single-valued neutrosophic directed hypergraph and its 2-section

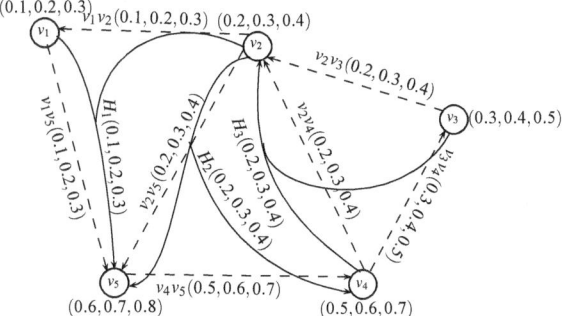

9.7 Applications of Neutrosophic Directed Hypergraphs

Graphs and hypergraphs can be used to describe the complex network systems. The complex network systems, including social networks, World Wide Web, neural networks are investigated by means of simple graphs and digraphs. The graphs take the nodes as a set of objects or people and the edges define the relations between them. In many cases, it is not possible to give full description of real-world systems using the simple graphs or digraphs. For example, if a collaboration network is represented through a simple graph. We only know that whether the two researchers are working together or not. We cannot know if three or more researchers, which are connected in the network, are coauthors of the same article or not. Further, in various situations, the given data contains the information of existence, indeterminacy, and nonexistence. To overcome such types of difficulties in complex networks, we use single-valued neutrosophic directed hypergraphs to describe the relationships between three or more elements and the networks are then called the *hypernetworks*.

9.7.1 Production and Manufacturing Networks

In a production system, there is a set of goods which can be produced using different technologies or devices. A single-valued neutrosophic directed hypergraph can be utilized more precisely to illustrate a production and manufacturing system. Consider a production system as given in Fig. 9.18, where the set of square vertices represents the products which are taken as input to produce the other products as given in elliptical vertices.

The set of directed hyperedges $\{d_1, d_2, d_3, d_4, d_5, d_6\}$ contains the devices or technologies which are used in our production system to design new products. Here, we use the devices *Silicon photovoltaic system, Electric hob, Ultrasonic shower, Electric heater shower, Harvesting system, Washing machine*, which are represented by directed hyperedges. A directed hyperarc $(t(d), h(d))$ represents that the goods in set $t(d)$ are required to manufacture the products in the set $h(d)$. The product nodes

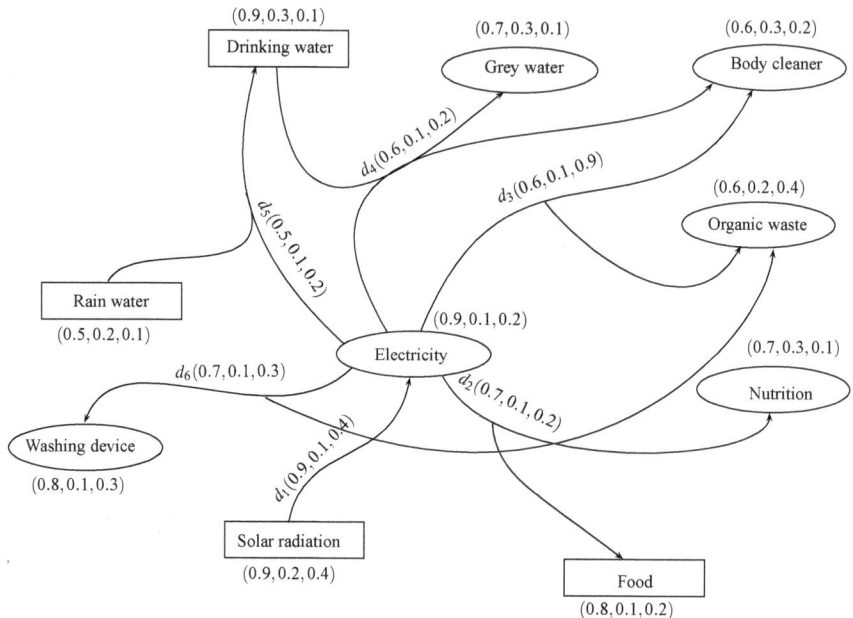

Fig. 9.18 Production system using a single-valued neutrosophic directed hypergraph

are taken as storage. The truth-membership and falsity-membership values of each product node interpret that how much of the product is available to supply and unavailable to fulfill the demand, respectively. The indeterminacy value contains the imprecise or inexact information about the product. The truth-membership degree of each directed hyperedge(or device) describes that how much this technology is appropriate to manufacture the product. For example, the directed hyperedge $d_2 = (\{\text{Electricity, Food}\}, \{\text{Nutrition}\})$ interprets that the electric hob uses electricity and food to produce nutrition. It is noted that more than one technology can be adapted to manufacture the same product using different or same inputs. The truth-membership degrees of each hyperedge evaluates the suitability of that device. For example, electric heater shower having membership degrees $(0.6, 0.1, 0.2)$ is a more useful device than an electric shower $(0.6, 0.1, 0.4)$ to a body cleaner, as its falsity-membership value is less than an electric shower.

9.7.2   Collaboration Network

We use a single-valued neutrosophic directed hypergraph as a directed hypernetwork to discuss the teamwork or joint efforts of researchers from different fields. Consider a single-valued neutrosophic directed hypergraph $D = (V, H)$ as a collaboration network. The vertices or nodes of the hypergraph are taken as the researchers.

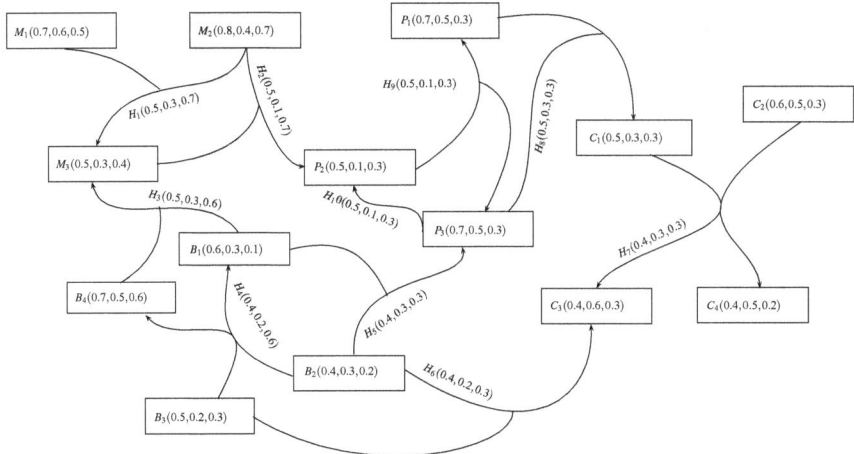

Fig. 9.19 Single-valued neutrosophic directed hypergraph for collaboration network

The set of vertices X is $\{M_1, M_2, M_3, P_1, P_2, P_3, C_1, C_2, C_3, C_4, B_1, B_2, B_3, B_4\}$, where the subset of vertices $\{M_1, M_2, M_3\}$ represents the group of researchers in field of Mathematics, $\{P_1, P_2, P_3\}$ represents the group of researchers in field of Physics, $\{C_1, C_2, C_3, C_4\}$ represents the group of researchers in field of Chemistry, and $\{B_1, B_2, B_3, B_4\}$ represents the group of researchers in field of Biology. The directed hyperedges of single-valued neutrosophic directed hypergraph interpret the group of members who are working together at the same project. The corresponding single-valued neutrosophic directed hypergraph is given in Fig. 9.19.

The truth-membership value of each researcher represents their published articles, indeterminacy shows their submitted articles that may be accepted or rejected and the falsity-membership value describes the rejected articles. For example, $(0.7, 0.6, 0.5)$ shows that the researcher M_1 has 70% publications, 60% submitted papers, and 50% of his research work is rejected. The value of a single-valued neutrosophic directed hyperedge depicts the joint work of the researchers which are connected through the hyperedge. For example, truth-membership, indeterminacy, and falsity-membership values $(0.5, 0.1, 0.3)$ of H_2 describe that the researchers M_2, M_3, P_2 from the field of Mathematics and Physics have 50% publications, 10% submitted papers, and 30% rejected papers, respectively, while working together. By calculating the strength of each single-valued neutrosophic directed hyperedge, we can conclude that which group of researchers has better work done as compared to others. By routine calculations, we have

Table 9.21 Algorithm for collaboration network

1. Input the degree of membership of all nodes(researchers) $v_1, v_2, \ldots, v_n$.

2. Input the number of directed hyperedges r.

3. Calculate the strength of single-valued neutrosophic directed hyperedge

$H_i = \{v_k, v_{k+1}, \ldots, v_l\}, 1 \le k \le n - 1, 2 \le l \le n$ as,

$\qquad S_i = \{\min_{v_j \in H_i} T_{H_i}(v_j)|T_{H_i}(v_j) > 0, \min_{v_j \in H_i} I_{H_i}(v_j)|I_{H_i}(v_j) > 0, \max_{v_j \in H_i} F_{H_i}(v_j)|F_{H_i}(v_j) > 0\},$

$1 \le i \le r.$

4. Find the strongest directed hyperedge using steps $5 - 14$.

5. **do** p from $1 \to r$

6. $\qquad J = (0, 0, 1)$

7. $\qquad$ **do** q from $1 \to r$

8. $\qquad\qquad$ **if**$(p \ne q)$ **then**

8. $\qquad\qquad\qquad J = (\max\{T(J), T(S_q)\}, \max\{I(J), I(S_q)\}, \min\{F(J), F(S_q)\})$

9. $\qquad\qquad$ **end if**

10. $\qquad$ **end do**

11. $\qquad$ **if**$(T(J) = T(S_p), I(J) = I(S_p)F(J) = F(S_p))$ **then**

12. $\qquad\qquad$ **print***, H_p is a strongest single-valued neutrosophic directed hyperedge.

13. $\qquad$ **end if**

14. **end do**

$$\eta(H_1) = (0.5, 0.3, 0.7), \quad \eta(H_2) = (0.5, 0.1, 0.7),$$
$$\eta(H_3) = (0.5, 0.3, 0.6), \quad \eta(H_4) = (0.4, 0.2, 0.6),$$
$$\eta(H_5) = (0.4, 0.3, 0.3), \quad \eta(H_6) = (0.4, 0.2, 0.3),$$
$$\eta(H_7) = (0.4, 0.3, 0.3), \quad \eta(H_8) = (0.5, 0.3, 0.3),$$
$$\eta(H_9) = (0.5, 0.1, 0.3), \quad \eta(H_{10}) = (0.5, 0.1, 0.3).$$

Thus, we have H_8 is the strongest edge among the all. So we conclude that the researchers P_1, P_3 from the field of Physics and C_1 from the field of Chemistry have done more joint work as compared to others, i.e., they have 50% publications, 30% of their research work is submitted, and 30% papers are rejected. The method adopted in our example can be explained by a simple algorithm given in Table 9.21.

9.7.3 Social Networking

A single-valued neutrosophic directed hypergraph can also be used to study and understand the social networks, using people as nodes (or vertices) and relationships between two or more than two people as single-valued neutrosophic directed hyperedges. Consider the representation of social clubs and its members as a single-valued neutrosophic directed hypergraph $D = (V, C)$, where $V = \{$Alen, Alex, Andy, Ben,

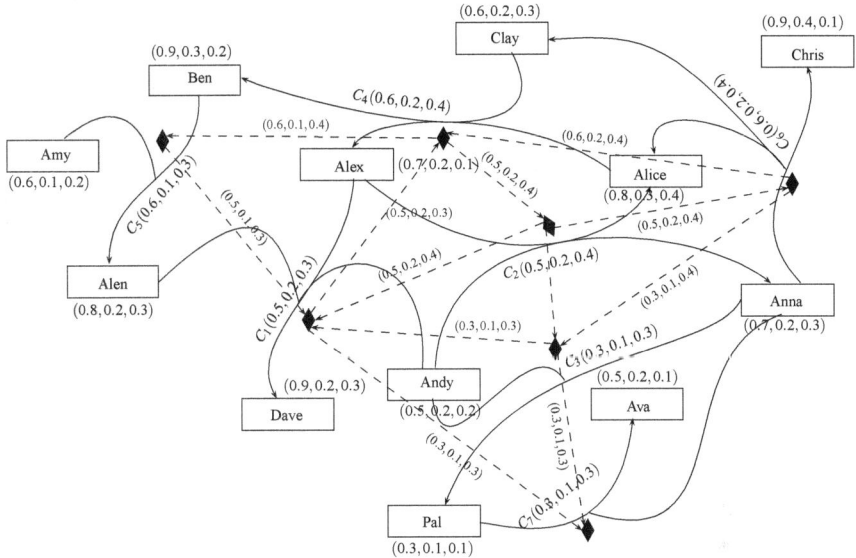

Fig. 9.20 Social network using single-valued neutrosophic directed hypergraph

Ava, Anna, Amy, Alice, Chris, Clay, Dave, Pal} interpret the members of different social clubs and the set of single-valued neutrosophic directed hyperedges $C = \{C_1, C_2, C_3, C_4, C_5, C_6, C_7, C_8, C_9, C_{10}\}$ represents the social clubs. Each directed hyperedge(or social club) connects the people having some common characteristics to each other. The social hypernetwork is shown in Fig. 9.20.

All the members of a social club connected through a single-valued neutrosophic directed hyperedge share some common characteristics, including emotional intelligence, good behavior, communication skills, and social sensitivity. For example, if the hyperedge C_1 describes the relation of *social sensitivity* (the capability to realize the emotions and thoughts of others) among the members of this club. Then, the truth-membership, indeterminacy, and falsity-membership values of each member indicate their sensitivity, unpredictable behavior, and insensitivity toward the other members of the club. The truth-membership, indeterminacy, and falsity-membership values (0.6, 0.1, 0.3) of a single-valued neutrosophic directed hyperedge C_6 interpret that 60% members have same characteristics, 30% have different, and 10% members of this club have unpredictable behavior. We use the concept of line directed graph to find out the common characteristics of different members of distinct clubs. By routine calculations, we have

Table 9.22 Algorithm for social networking

1. Input the number of directed hyperedges m of single-valued neutrosophic directed hypergraph $D = (V, H)$.
2. Input the degree of membership of all directed hyperedges $C_1, C_2, \ldots, C_m$.
3. Construct the single-valued neutrosophic line directed graph $L(D) = (V_L, H_L)$ by taking $\{C_1, C_2, C_3, \ldots, C_m\}$ as set of vertices such that $V_L(C_i) = D(C_i)$, $1 \le i \le m$.
4. Draw an edge between C_i and C_j if $|C_i \cap C_j| \neq \emptyset$ and
   $H_L(C_i C_j) = (\min\{T_H(C_i), T_H(C_j)\}, \min\{I_H(C_i), I_H(C_j)\}, \max\{F_H(C_i), F_H(C_j)\})$.
5. The edge $C_i C_j$ describes the common characteristics of members of various clubs.

$|supp(C_1) \bigcap supp(C_2)| = \{Alex, Andy\},$ $|supp(C_3) \bigcap supp(C_4)| = \{\emptyset\},$

$|supp(C_1) \bigcap supp(C_3)| = \{Andy\},$ $|supp(C_3) \bigcap supp(C_5)| = \{\emptyset\},$

$|supp(C_1) \bigcap supp(C_4)| = \{Alex\},$ $|supp(C_3) \bigcap supp(C_6)| = \{Anna\},$

$|supp(C_1) \bigcap supp(C_5)| = \{Alen\},$ $|supp(C_3) \bigcap supp(C_7)| = \{Pal\},$

$|supp(C_1) \bigcap supp(C_6)| = \{\emptyset\},$ $|supp(C_4) \bigcap supp(C_5)| = \{Ben\},$

$|supp(C_1) \bigcap supp(C_7)| = \{Dave\},$ $|supp(C_4) \bigcap supp(C_6)| = \{Alice, Clay\},$

$|supp(C_2) \bigcap supp(C_3)| = \{Andy\},$ $|supp(C_4) \bigcap supp(C_7)| = \{\emptyset\},$

$|supp(C_2) \bigcap supp(C_4)| = \{Alex, Alice\},$ $|supp(C_5) \bigcap supp(C_6)| = \{\emptyset\},$

$|supp(C_2) \bigcap supp(C_5)| = \{\emptyset\},$ $|supp(C_5) \bigcap supp(C_7)| = \{\emptyset\},$

$|supp(C_2) \bigcap supp(C_6)| = \{Alice, Anna\},$ $|supp(C_6) \bigcap supp(C_7)| = \{\emptyset\}.$

$|supp(C_2) \bigcap supp(C_7)| = \{\emptyset\},$

The line directed graph of social network single-valued neutrosophic directed hypergraph is given in Fig. 9.20 with dashed lines. Each common edge between two social clubs describes the common characteristics of members of different clubs. For example, the edge $C_1 C_2$ shows that the members of C_1 and C_2 have 50% common characteristics, 40% different to each other, and 20% they have unpredictable behavior. The procedure followed in our example can be explained by means of simple algorithm as given in Table 9.22.

9.8   Complex Neutrosophic Hypergraphs

The motivation behind this work is the existence of indeterminate information of periodic nature in hypernetwork models. A complex neutrosophic hypergraph model plays an important role in handling complicated behavior of indeterminacy and inconsistency with periodic nature. The proposed model generalizes the complex fuzzy model as well as complex intuitionistic fuzzy model. To prove the applicability of our proposed model, we consider two voting procedures. Suppose that 0.6 voters say "yes", 0.2 say "no", and 0.2 are "undecided" in the first voting procedure and 0.3 voters say "yes", 0.3 say "no", and 0.4 are "undecided" in the second voting procedure. We assume that these two procedures held at different days. It is clear that a complex fuzzy set cannot handle this situation as it only depicts the truth-membership 0.6 of voters but fails to represent the falsity and indeterminate degrees. Similarly, a complex intuitionistic fuzzy set represents the truth 0.6 and falsity 0.2 degrees of voters but it does not illustrate the 0.2 undecided voters. Now, if we set the amplitude terms as the membership degrees of first voting procedure and phase terms as the membership degrees of second voting procedure, then we can illustrate this information using a complex neutrosophic model as $\{0.6e^{\iota(0.3)2\pi}, 0.2e^{\iota(0.3)2\pi}, 0.2e^{\iota(0.4)2\pi}\}$. Thus, we apply the most generalized concept of complex neutrosophic sets to hypergraphs to deal periodic nature of inconsistent information existing in hypernetworks.

Complex neutrosophic sets are defined using single-valued neutrosophic sets.

Definition 9.66 A *complex neutrosophic set* $\mathcal{N}$ on the universal set X is defined as

$$\mathcal{N} = \{(u, T_{\mathcal{N}}(u)e^{i\phi_{\mathcal{N}}(u)}, I_{\mathcal{N}}(u)e^{i\varphi_{\mathcal{N}}(u)}, F_{\mathcal{N}}(u)e^{i\psi_{\mathcal{N}}(u)}) | u \in X\},$$

where $i = \sqrt{-1}$, $T_{\mathcal{N}}(u), I_{\mathcal{N}}(u), F_{\mathcal{N}}(u) \in [0, 1]$ are amplitude terms, $\phi_{\mathcal{N}}(u)$, $\varphi_{\mathcal{N}}(u), \psi_{\mathcal{N}}(u) \in [0, 2\pi]$ are phase terms for truth, indeterminacy, and falsity degrees, respectively, and for every $u \in X$, $0 \le T_{\mathcal{N}}(u) + I_{\mathcal{N}}(u) + F_{\mathcal{N}}(u) \le 3$.

Definition 9.67 A *complex neutrosophic relation* is a complex neutrosophic set on $X \times X$ given as

$$R = \{(rs, T_R(rs)e^{i\phi_R(rs)}, I_R(rs)e^{i\varphi_R(rs)}, F_R(rs)e^{i\psi_R(rs)}) | rs \in X \times X\},$$

where $i = \sqrt{-1}$, $T_R : X \times X \to [0, 1]$, $I_R : X \times X \to [0, 1]$, $F_R : X \times X \to [0, 1]$ characterize the amplitudes, and $\phi_R(rs), \varphi_R(rs), \psi_R(rs) \in [0, 2\pi]$ characterize the phase terms of truth, indeterminacy, and falsity degrees of R such that for all $rs \in X \times X$, $0 \le T_R(rs) + I_R(rs) + F_R(rs) \le 3$.

Definition 9.68 A *complex neutrosophic graph* on X is an ordered pair $G = (A, B)$, where A is a complex neutrosophic set on X and B is complex neutrosophic relation on X such that

$$T_B(ab) \le \min\{T_A(a), T_A(b)\},$$
$$I_B(ab) \le \min\{I_A(a), I_A(b)\},$$
$$F_B(ab) \le \max\{F_A(a), F_A(b)\}, \quad \text{(for amplitude terms)}$$
$$\phi_B(ab) \le \min\{\phi_A(a), \phi_A(b)\},$$
$$\varphi_B(ab) \le \min\{\varphi_A(a), \varphi_A(b)\},$$
$$\psi_B(ab) \le \max\{\psi_A(a), \psi_A(b)\}, \quad \text{(for phase terms)}$$

$0 \le T_B(ab) + I_B(ab) + F_B(ab) \le 3$, for all $a, b \in X$.

Fig. 9.21 Complex
neutrosophic graph

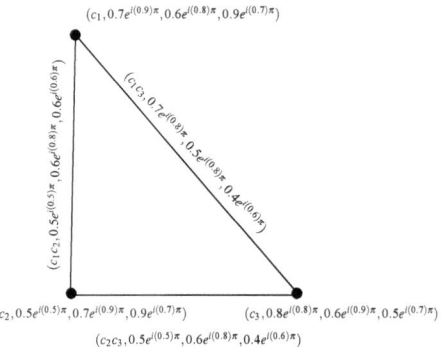

Example 9.23 Consider a complex neutrosophic graph $G = (A, B)$ on $X = \{c_1, c_2, c_3\}$, where $A = \{(c_1, 0.7e^{i(0.9)\pi}, 0.6e^{i(0.8)\pi}, 0.9e^{i(0.7)\pi}), (c_2, 0.5e^{i(0.5)\pi}, 0.7e^{i(0.9)\pi}, 0.9e^{i(0.7)\pi}), (c_3, 0.8e^{i(0.8)\pi}, 0.6e^{i(0.9)\pi}, 0.5e^{i(0.7)\pi})\}$ and $B = \{(c_1c_2, 0.5e^{i(0.5)\pi}, 0.6e^{i(0.8)\pi}, 0.6e^{i(0.6)\pi}), (c_2c_3, 0.5e^{i(0.5)\pi}, 0.6e^{i(0.8)\pi}, 0.4e^{i(0.6)\pi}), (c_1c_3, 0.7e^{i(0.8)\pi}, 0.5e^{i(0.8)\pi}, 0.4e^{i(0.6)\pi})\}$ are complex neutrosophic set and complex neutrosophic relation on X, respectively. The corresponding graph is shown in Fig. 9.21.

Definition 9.69 Let $N_1 = \{(u, T_{N_1}(u)e^{i\phi_{N_1}(u)}, I_{N_1}(u)e^{i\varphi_{N_1}(u)}, F_{N_1}(u)e^{i\psi_{N_1}(u)})|u \in X\}$ and $N_2 = \{(u, T_{N_2}(u)e^{i\phi_{N_2}(u)}, I_{N_2}(u)e^{i\varphi_{N_2}(u)}, F_{N_2}(u)e^{i\psi_{N_2}(u)})|u \in X\}$ be two complex neutrosophic sets in X, then

(i) $N_1 \subseteq N_2 \Leftrightarrow T_{N_1}(u) \leq T_{N_2}(u)$, $I_{N_1}(u) \leq I_{N_2}(u)$, $F_{N_1}(u) \geq F_{N_2}(u)$, and $\phi_{N_1}(u) \leq \phi_{N_2}(u)$, $\varphi_{N_1}(u) \leq \varphi_{N_2}(u)$, $\psi_{N_1}(u) \geq \psi_{N_2}(u)$ for amplitudes and phase terms, respectively, for all $u \in X$.

(ii) $N_1 = N_2 \Leftrightarrow T_{N_1}(u) = T_{N_2}(u)$, $I_{N_1}(u) = I_{N_2}(u)$, $F_{N_1}(u) = F_{N_2}(u)$, and $\phi_{N_1}(u) = \phi_{N_2}(u)$, $\varphi_{N_1}(u) = \varphi_{N_2}(u)$, $\psi_{N_1}(u) = \psi_{N_2}(u)$ for amplitudes and phase terms, respectively, for all $u \in X$.

(iii) $N_1 \cup N_2 = \{(u, \max\{T_{N_1}(u), T_{N_2}(u)\}e^{i \max\{\phi_{N_1}(u), \phi_{N_2}(u)\}}, \min\{I_{N_1}(u), I_{N_2}(u)\}e^{i \min\{\varphi_{N_1}(u), \varphi_{N_2}(u)\}}, \min\{F_{N_1}(u), F_{N_2}(u)\}e^{i \min\{\psi_{N_1}(u), \psi_{N_2}(u)\}})|u \in N_1 \cup N_2\}$.

(iv) $N_1 \cap N_2 = \{(u, \min\{T_{N_1}(u), T_{N_2}(u)\}e^{i \min\{\phi_{N_1}(u), \phi_{N_2}(u)\}}, \max\{I_{N_1}(u), I_{N_2}(u)\}e^{i \max\{\varphi_{N_1}(u), \varphi_{N_2}(u)\}}, \max\{F_{N_1}(u), F_{N_2}(u)\}e^{i \max\{\psi_{N_1}(u), \psi_{N_2}(u)\}})|u \in N_1 \cap N_2\}$.

Definition 9.70 The *support* of a complex neutrosophic set $N = \{(u, T_N(u)e^{i\phi_N(u)}, I_N(u)e^{i\varphi_N(u)}F_N(u)e^{i\psi_S(u)})|u \in X\}$ is defined as $supp(N) = \{u|T_N(u) \neq 0, I_N(u) \neq 0, F_N(u) \neq 1, 0 < \phi_N(u), \varphi_N(u), \psi_N(u) < 2\pi\}$. The *height* of a complex neutrosophic set $N = \{(u, T_N(u)e^{i\phi_N(u)}, I_N(u)e^{i\varphi_N(u)}F_N(u)e^{i\psi_S(u)})|u \in X\}$ is defined as

$$h(N) = \{\max_{u \in X} T_N(u)e^{i \max_{u \in X} \phi_N(u)}, \max_{u \in X} I_N(u)e^{i \max_{u \in X} \varphi_N(u)}, \min_{u \in X} F_N(u)e^{i \min_{u \in X} \psi_N(u)}\}.$$

Definition 9.71 A *complex neutrosophic hypergraph* on X is defined as an ordered pair $\mathcal{H} = (\mathcal{N}, \lambda)$, where $\mathcal{N} = \{N_1, N_2, \ldots, N_k\}$ is a finite family of complex neutrosophic sets on X and λ is a complex neutrosophic relation on complex neutrosophic sets N_j's such that

(i)

$$T_\lambda(\{r_1, r_2, \ldots, r_l\}) \leq \min\{T_{N_j}(r_1), T_{N_j}(r_2), \ldots, T_{N_j}(r_l)\},$$
$$I_\lambda(\{r_1, r_2, \ldots, r_l\}) \leq \min\{I_{N_j}(r_1), I_{N_j}(r_2), \ldots, I_{N_j}(r_l)\},$$
$$F_\lambda(\{r_1, r_2, \ldots, r_l\}) \leq \max\{F_{N_j}(r_1), F_{N_j}(r_2), \ldots, F_{N_j}(r_l)\}, \quad \text{(for amplitude terms)}$$
$$\phi_\lambda(\{r_1, r_2, \ldots, r_l\}) \leq \min\{\phi_{N_j}(r_1), \phi_{N_j}(r_2), \ldots, \phi_{N_j}(r_l)\},$$
$$\varphi_\lambda(\{r_1, r_2, \ldots, r_l\}) \leq \min\{\varphi_{N_j}(r_1), \varphi_{N_j}(r_2), \ldots, \varphi_{N_j}(r_l)\},$$
$$\psi_\lambda(\{r_1, r_2, \ldots, r_l\}) \leq \max\{\psi_{N_j}(r_1), \psi_{N_j}(r_2), \ldots, \psi_{N_j}(r_l)\}, \quad \text{(for phase terms)}$$

$0 \leq T_\lambda + I_\lambda + F_\lambda \leq 3$, for all $r_1, r_2, \ldots, r_l \in X$.

(ii) $\bigcup\limits_j supp(N_j) = X$, for all $N_j \in \mathcal{N}$.

Note that $E_k = \{r_1, r_2, \ldots, r_l\}$ is the crisp hyperedge of $\mathcal{H} = (\mathcal{N}, \lambda)$.

Definition 9.72 Let $\mathcal{H} = (\mathcal{N}, \lambda)$ be a complex neutrosophic hypergraph. The *height* of $\mathcal{H}$, denoted by $h(\mathcal{H})$, is defined as $h(\mathcal{H}) = (\max \lambda_l e^{i \max \phi}, \max \lambda_m e^{i \max \varphi}, \min \lambda_n e^{i \min \psi})$, where $\lambda_l = \max T_{\xi_j}(v_k)$, $\phi = \max \phi_{\xi_j}(v_k)$, $\lambda_m = \max I_{\xi_j}(v_k)$, $\varphi = \max \varphi_{\xi_j}(v_k)$, $\lambda_n = \min F_{\xi_j}(v_k)$, $\psi = \min \psi_{\xi_j}(v_k)$. Here, $T_{\xi_j}(v_k), I_{\xi_j}(v_k), F_{\xi_j}(v_k)$ denote the truth, indeterminacy, and falsity degrees of vertex v_k to hyperedge ξ_j, respectively.

Definition 9.73 Let $\mathcal{H} = (\mathcal{N}, \lambda)$ be a complex neutrosophic hypergraph. Suppose that $\alpha, \beta, \gamma \in [0, 1]$ and $\Theta, \Phi, \Psi \in [0, 2\pi]$ such that $0 \leq \alpha + \beta + \gamma \leq 3$. The $(\alpha e^{i\Theta}, \beta e^{i\Phi}, \gamma e^{i\Psi})$-*level hypergraph* of $\mathcal{H}$ is defined as an ordered pair $\mathcal{H}^{(\alpha e^{i\Theta}, \beta e^{i\Phi}, \gamma e^{i\Psi})} = (\mathcal{N}^{(\alpha e^{i\Theta}, \beta e^{i\Phi}, \gamma e^{i\Psi})}, \lambda^{(\alpha e^{i\Theta}, \beta e^{i\Phi}, \gamma e^{i\Psi})})$, where

(i) $\lambda^{(\alpha e^{i\Theta}, \beta e^{i\Phi}, \gamma e^{i\Psi})} = \{\lambda_j^{(\alpha e^{i\Theta}, \beta e^{i\Phi}, \gamma e^{i\Psi})} : \lambda_j \in \lambda\}$ and $\lambda_j^{(\alpha e^{i\Theta}, \beta e^{i\Phi}, \gamma e^{i\Psi})} = \{u \in X : T_{\lambda_j}(u) \geq \alpha, \phi_{\lambda_j}(u) \geq \Theta, I_{\lambda_j}(u) \geq \beta, \varphi_{\lambda_j}(u) \geq \Phi, \text{ and } F_{\lambda_j}(u) \leq \gamma, \psi_{\lambda_j}(u) \leq \Psi\}$,

(ii) $\mathcal{N}^{(\alpha e^{i\Theta}, \beta e^{i\Phi}, \gamma e^{i\Psi})} = \bigcup\limits_{\lambda_j \in \lambda} \lambda_j^{(\alpha e^{i\Theta}, \beta e^{i\Phi}, \gamma e^{i\Psi})}$.

Note that $(\alpha e^{i\Theta}, \beta e^{i\Phi}, \gamma e^{i\Psi})$-level hypergraph of $\mathcal{H}$ is a crisp hypergraph.

Definition 9.74 Let $\mathcal{H} = (\mathcal{N}, \lambda)$ be a complex neutrosophic hypergraph and for $0 < \alpha \leq T(h(\mathcal{H})), 0 < \beta \leq I(h(\mathcal{H})), \gamma \geq F(h(\mathcal{H})) > 0, 0 < \Theta \leq \phi(h(\mathcal{H})), 0 < \Phi \leq \varphi(h(\mathcal{H})),$ and $\Psi \geq \psi(h(\mathcal{H})) > 0$, let $\mathcal{H}^{(\alpha e^{i\Theta}, \beta e^{i\Phi}, \gamma e^{i\Psi})} = (\mathcal{N}^{(\alpha e^{i\Theta}, \beta e^{i\Phi}, \gamma e^{i\Psi})}, \lambda^{(\alpha e^{i\Theta}, \beta e^{i\Phi}, \gamma e^{i\Psi})})$ be the level hypergraph of $\mathcal{H}$. The sequence of complex numbers $\{(\alpha_1 e^{i\Theta_1}, \beta_1 e^{i\Phi_1}, \gamma_1 e^{i\Psi_1}), (\alpha_2 e^{i\Theta_2}, \beta_2 e^{i\Phi_2}, \gamma_2 e^{i\Psi_2}), \ldots, (\alpha_n e^{i\Theta_n}, \beta_n e^{i\Phi_n}, \gamma_n e^{i\Psi_n})\}$ such that $0 < \alpha_1 < \alpha_2 < \cdots < \alpha_n = T(h(\mathcal{H})), 0 < \beta_1 < \beta_2 < \cdots < \beta_n = I(h(\mathcal{H})), \gamma_1 > \gamma_2 > \cdots > \gamma_n = F(h(\mathcal{H})) > 0, 0 < \Theta_1 < \Theta_2 < \cdots < \Theta_n = \phi(h(\mathcal{H})), 0 < \Phi_1 < \Phi_2 < \cdots < \Phi_n = \varphi(h(\mathcal{H}))$, and $\Psi_1 > \Psi_2 > \cdots > \Psi_n = \psi(h(\mathcal{H})) > 0$ satisfying the conditions,

(i) if $\alpha_{k+1} < \alpha' \leq \alpha_k, \beta_{k+1} < \beta' \leq \beta_k, \gamma_{k+1} > \gamma' \geq \gamma_k, \Theta_{k+1} < \phi \leq \Theta_k, \Phi_{k+1} < \varphi \leq \Phi_k, \Psi_{k+1} > \psi \geq \Psi_k$, then $\lambda^{(\alpha' e^{i\phi}, \beta' e^{i\varphi}, \gamma' e^{i\psi})} = \lambda^{(\alpha_k e^{i\Theta_k}, \beta_k e^{i\Phi_k}, \gamma_k e^{i\psi_k})}$, and

(ii) $\lambda^{(\alpha_k e^{i\Theta_k}, \beta_k e^{i\Phi_k}, \gamma_k e^{i\psi_k})} \subset \lambda^{(\alpha_{k+1} e^{i\Theta_{k+1}}, \beta_{k+1} e^{i\Phi_{k+1}}, \gamma_{k+1} e^{i\psi_{k+1}})}$,

Fig. 9.22 Complex
neutrosophic hypergraph $\mathcal{H}$

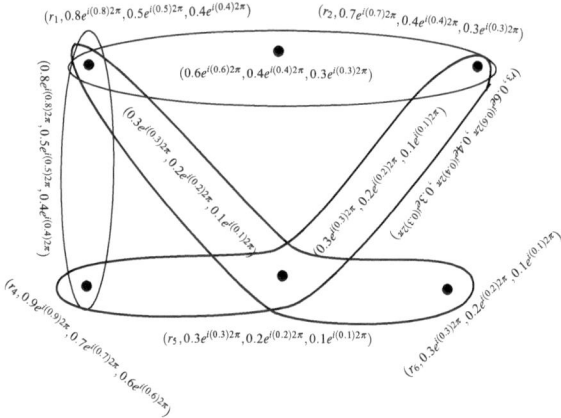

is called the *fundamental sequence* of $\mathcal{H} = (\mathcal{N}, \lambda)$, denoted by $\mathcal{F}_s(\mathcal{H})$. The set of $(\alpha_j e^{i\Theta_j}, \beta_j e^{i\Phi_j}, \gamma_j e^{i\Psi_j})$-level hypergraphs $\{\mathcal{H}^{(\alpha_1 e^{i\Theta_1}, \beta_1 e^{i\Phi_1}, \gamma_1 e^{i\Psi_1})},$ $\mathcal{H}^{(\alpha_2 e^{i\Theta_2}, \beta_2 e^{i\Phi_2}, \gamma_2 e^{i\Psi_2})}, \ldots, \mathcal{H}^{(\alpha_n e^{i\Theta_n}, \beta_n e^{i\Phi_n}, \gamma_n e^{i\Psi_n})}\}$ is called the *set of core hypergraphs* or the *core set* of $\mathcal{H}$, denoted by $c(\mathcal{H})$.

Example 9.24 Consider a complex neutrosophic hypergraph $\mathcal{H} = (\mathcal{N}, \lambda)$ on $X = \{r_1, r_2, r_3, r_4, r_5, r_6\}$. The complex neutrosophic relation λ is given as $\lambda(\{r_1, r_2, r_3\}) = (0.6e^{i(0.6)2\pi}, 0.4e^{i(0.4)2\pi}, 0.3e^{i(0.3)2\pi})$, $\lambda(\{r_1, r_4\}) = (0.8e^{i(0.8)2\pi}, 0.5e^{i(0.5)2\pi}, 0.4e^{i(0.4)2\pi})$, $\lambda(\{r_3, r_4, r_5\}) = (0.3e^{i(0.3)2\pi}, 0.2e^{i(0.2)2\pi}, 0.1e^{i(0.1)2\pi})$, and $\lambda(\{r_1, r_5, r_6\}) = (0.3e^{i(0.3)2\pi}, 0.2e^{i(0.2)2\pi}, 0.1e^{i(0.1)2\pi})$. The corresponding complex neutrosophic hypergraph is shown in Fig. 9.22. Let

$$(\alpha_1 e^{i\Theta_1}, \beta_1 e^{i\Phi_1}, \gamma_1 e^{i\Psi_1}) = (0.9e^{i(0.9)2\pi}, 0.7e^{i(0.7)2\pi}, 0.6e^{i(0.6)2\pi}),$$
$$(\alpha_2 e^{i\Theta_2}, \beta_2 e^{i\Phi_2}, \gamma_2 e^{i\Psi_2}) = (0.8e^{i(0.8)2\pi}, 0.5e^{i(0.5)2\pi}, 0.4e^{i(0.4)2\pi}),$$
$$(\alpha_3 e^{i\Theta_3}, \beta_3 e^{i\Phi_3}, \gamma_3 e^{i\Psi_3}) = (0.6e^{i(0.6)2\pi}, 0.4e^{i(0.4)2\pi}, 0.3e^{i(0.3)2\pi}),$$
$$(\alpha_4 e^{i\Theta_4}, \beta_4 e^{i\Phi_4}, \gamma_4 e^{i\Psi_4}) = (0.3e^{i(0.3)2\pi}, 0.2e^{i(0.2)2\pi}, 0.1e^{i(0.1)2\pi}).$$

Note that the sequence $\{(\alpha_1 e^{i\Theta_1}, \beta_1 e^{i\Phi_1}, \gamma_1 e^{i\Psi_1}), (\alpha_2 e^{i\Theta_2}, \beta_2 e^{i\Phi_2}, \gamma_2 e^{i\Psi_2}),$ $(\alpha_3 e^{i\Theta_3}, \beta_3 e^{i\Phi_3}, \gamma_3 e^{i\Psi_3}), (\alpha_4 e^{i\Theta_4}, \beta_4 e^{i\Phi_4}, \gamma_4 e^{i\Psi_4})\}$ satisfies all the conditions of Definition 9.74. Thus, it is a fundamental sequence of $\mathcal{H}$. The corresponding $(\alpha_j e^{i\Theta_j}, \beta_j e^{i\Phi_j}, \gamma_j e^{i\Psi_j})$-level hypergraphs are shown in Figs. 9.23, 9.24, and 9.25.

Definition 9.75 A complex neutrosophic hypergraph $\mathcal{H} = (\mathcal{N}, \lambda)$ is *ordered* if $c(\mathcal{H})$ is ordered, i.e., if $c(\mathcal{H}) = \{\mathcal{H}^{(\alpha_1 e^{i\Theta_1}, \beta_1 e^{i\Phi_1}, \gamma_1 e^{i\Psi_1})}, \mathcal{H}^{(\alpha_2 e^{i\Theta_2}, \beta_2 e^{i\Phi_2}, \gamma_2 e^{i\Psi_2})}, \ldots,$ $\mathcal{H}^{(\alpha_n e^{i\Theta_n}, \beta_n e^{i\Phi_n}, \gamma_n e^{i\Psi_n})}\}$, then $\{\mathcal{H}^{(\alpha_1 e^{i\Theta_1}, \beta_1 e^{i\Phi_1}, \gamma_1 e^{i\Psi_1})} \subset \mathcal{H}^{(\alpha_2 e^{i\Theta_2}, \beta_2 e^{i\Phi_2}, \gamma_2 e^{i\Psi_2})} \subset \cdots \subset \mathcal{H}^{(\alpha_n e^{i\Theta_n}, \beta_n e^{i\Phi_n}, \gamma_n e^{i\Psi_n})}\}$.

Fig. 9.23 $\mathscr{H}^{(\alpha_1 e^{i\Theta_1}, \beta_1 e^{i\Phi_1}, \gamma_1 e^{i\Psi_1})}$, $\mathscr{H}^{(\alpha_2 e^{i\Theta_2}, \beta_2 e^{i\Phi_2}, \gamma_2 e^{i\Psi_2})}$-level hypergraphs

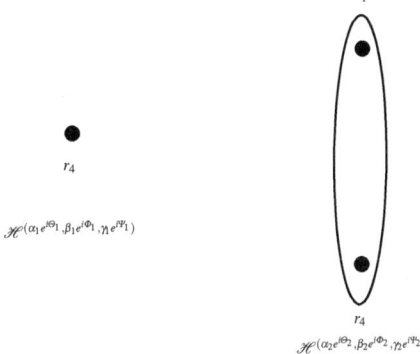

Fig. 9.24 $\mathscr{H}^{(\alpha_3 e^{i\Theta_3}, \beta_3 e^{i\Phi_3}, \gamma_3 e^{i\Psi_3})}$-level hypergraph

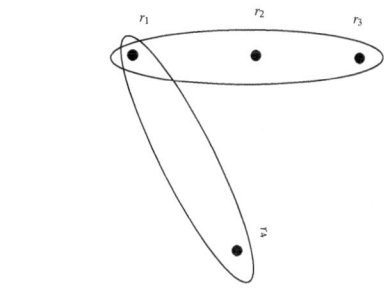

Fig. 9.25 $\mathscr{H}^{(\alpha_4 e^{i\Theta_4}, \beta_4 e^{i\Phi_4}, \gamma_4 e^{i\Psi_4})}$-level hypergraph

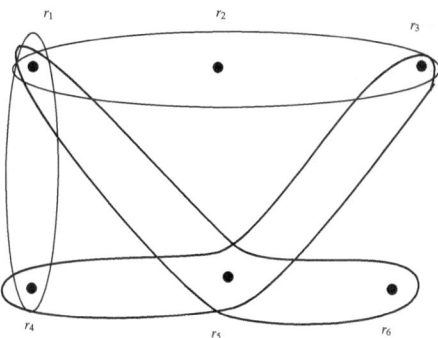

A complex neutrosophic hypergraph $\mathscr{H} = (\mathscr{N}, \lambda)$ is *simply ordered* if $c(\mathscr{H})$ is simply ordered, i.e., if $e \in E_{j+1} \setminus E_j$, then $e \nsubseteq X_j$.

Example 9.25 Consider a complex neutrosophic hypergraph $\mathscr{H} = (\mathscr{N}, \lambda)$ as shown in Fig. 9.22. The set of core hypergraphs is given as

$$c(\mathscr{H}) = \{\mathscr{H}^{(\alpha_1 e^{i\Theta_1}, \beta_1 e^{i\Phi_1}, \gamma_1 e^{i\Psi_1})}, \mathscr{H}^{(\alpha_2 e^{i\Theta_2}, \beta_2 e^{i\Phi_2}, \gamma_2 e^{i\Psi_2})},$$
$$\mathscr{H}^{(\alpha_3 e^{i\Theta_3}, \beta_3 e^{i\Phi_3}, \gamma_3 e^{i\Psi_3})}, \mathscr{H}^{(\alpha_4 e^{i\Theta_4}, \beta_4 e^{i\Phi_4}, \gamma_n e^{i\Psi_4})}\},$$

where

$$\mathcal{H}^{(\alpha_1 e^{i\Theta_1}, \beta_1 e^{i\Phi_1}, \gamma_1 e^{i\Psi_1})} = (X_1, E_1), \ X_1 = \{r_4\}, \ E_1 = \{\},$$

$$\mathcal{H}^{(\alpha_2 e^{i\Theta_2}, \beta_2 e^{i\Phi_2}, \gamma_2 e^{i\Psi_2})} = (X_2, E_2), \ X_2 = \{r_1, r_4\}, \ E_2 = \{\{r_1, r_4\}\},$$

$$\mathcal{H}^{(\alpha_3 e^{i\Theta_3}, \beta_3 e^{i\Phi_3}, \gamma_3 e^{i\Psi_3})} = (X_3, E_3), \ X_3 = \{r_1, r_2, r_3, r_4\}, \ E_3 = \{\{r_1, r_4\}, \{r_1, r_2, r_3\}\},$$

$$\mathcal{H}^{(\alpha_4 e^{i\Theta_4}, \beta_4 e^{i\Phi_4}, \gamma_4 e^{i\Psi_4})} = (X_4, E_4), \ X_4 = \{r_1, r_2, r_3, r_4, r_5, r_6\}, \ E_4 = \{\{r_1, r_4\}, \{r_1, r_2, r_3\},$$
$$\{r_1, r_5, r_6\}, \{r_3, r_4, r_5\}\}.$$

Note that

$$\mathcal{H}^{(\alpha_1 e^{i\Theta_1}, \beta_1 e^{i\Phi_1}, \gamma_1 e^{i\Psi_1})} \subseteq \mathcal{H}^{(\alpha_2 e^{i\Theta_2}, \beta_2 e^{i\Phi_2}, \gamma_2 e^{i\Psi_2})} \subseteq \mathcal{H}^{(\alpha_3 e^{i\Theta_3}, \beta_3 e^{i\Phi_3}, \gamma_3 e^{i\Psi_3})} \subseteq \mathcal{H}^{(\alpha_4 e^{i\Theta_4}, \beta_4 e^{i\Phi_4}, \gamma_4 e^{i\Psi_4})}.$$

Hence, $\mathcal{H} = (\mathcal{N}, \lambda)$ is an ordered complex neutrosophic hypergraph. Also, $\mathcal{H} = (\mathcal{N}, \lambda)$ is simply ordered.

Definition 9.76 A complex neutrosophic hypergraph $\mathcal{H} = (\mathcal{N}, \lambda)$ with $\mathcal{F}_s(\mathcal{H}) = \{(\alpha_1 e^{i\Theta_1}, \beta_1 e^{i\Phi_1}, \gamma_1 e^{i\Psi_1}), (\alpha_2 e^{i\Theta_2}, \beta_2 e^{i\Phi_2}, \gamma_2 e^{i\Psi_2}), \ldots, (\alpha_n e^{i\Theta_n}, \beta_n e^{i\Phi_n}, \gamma_n e^{i\Psi_n})\}$ is called *sectionally elementary* if for every $\lambda_j \in \lambda$ and for $k \in \{1, 2, \ldots, n\}$, $\lambda_j^{(\alpha e^{i\Theta}, \beta e^{i\Phi}, \gamma e^{i\Psi})} = \lambda_j^{(\alpha_k e^{i\Theta_k}, \beta_k e^{i\Phi_k}, \gamma_k e^{i\Psi_k})}$, for all $\alpha \in (\alpha_{k+1}, \alpha_k], \beta \in (\beta_{k+1}, \beta_k], \gamma \in (\gamma_{k+1}, \gamma_k], \Theta \in (\Theta_{k+1}, \Theta_k], \Phi \in (\Phi_{k+1}, \Phi_k]$, and $\Psi \in (\Psi_{k+1}, \Psi_k]$.

Definition 9.77 Let N be a complex neutrosophic set on X. The *lower truncation* of N at level $(\alpha e^{i\Theta}, \beta e^{i\Phi}, \gamma e^{i\Psi}), 0 < \alpha, \beta, \gamma \leq 1, 0 < \Theta, \Phi, \Psi \leq 2\pi$ is the complex neutrosophic set $N_{[(\alpha e^{i\Theta}, \beta e^{i\Phi}, \gamma e^{i\Psi})]}$ defined by

$$T_{N_{[(\alpha e^{i\Theta}, \beta e^{i\Phi}, \gamma e^{i\Psi})]}}(x) e^{i\phi_{N_{[(\alpha e^{i\Theta}, \beta e^{i\Phi}, \gamma e^{i\Psi})]}}(x)} = \begin{cases} T_N(x) e^{i\phi_N(x)}, & \text{if } x \in N^{(\alpha e^{i\Theta}, \beta e^{i\Phi}, \gamma e^{i\Psi})}, \\ 0, & \text{otherwise.} \end{cases}$$

$$I_{N_{[(\alpha e^{i\Theta}, \beta e^{i\Phi}, \gamma e^{i\Psi})]}}(x) e^{i\phi_{N_{[(\alpha e^{i\Theta}, \beta e^{i\Phi}, \gamma e^{i\Psi})]}}(x)} = \begin{cases} I_N(x) e^{i\varphi_N(x)}, & \text{if } x \in N^{(\alpha e^{i\Theta}, \beta e^{i\Phi}, \gamma e^{i\Psi})}, \\ 0, & \text{otherwise.} \end{cases}$$

$$F_{N_{[(\alpha e^{i\Theta}, \beta e^{i\Phi}, \gamma e^{i\Psi})]}}(x) e^{i\psi_{N_{[(\alpha e^{i\Theta}, \beta e^{i\Phi}, \gamma e^{i\Psi})]}}(x)} = \begin{cases} F_N(x) e^{i\psi_N(x)}, & \text{if } x \in N^{(\alpha e^{i\Theta}, \beta e^{i\Phi}, \gamma e^{i\Psi})}, \\ 0, & \text{otherwise.} \end{cases}$$

Definition 9.78 Let N be a complex neutrosophic set on X. The *upper truncation* of N at level $(\alpha e^{i\Theta}, \beta e^{i\Phi}, \gamma e^{i\Psi}), 0 < \alpha, \beta, \gamma \leq 1, 0 < \Theta, \Phi, \Psi \leq 2\pi$ is the complex neutrosophic set $N^{[(\alpha e^{i\Theta}, \beta e^{i\Phi}, \gamma e^{i\Psi})]}$ defined by

$$T_{N^{[(\alpha e^{i\Theta}, \beta e^{i\Phi}, \gamma e^{i\Psi})]}}(x) e^{i\phi_{N^{[(\alpha e^{i\Theta}, \beta e^{i\Phi}, \gamma e^{i\Psi})]}}(x)} = \begin{cases} \alpha e^{i\Theta}, & \text{if } x \in N^{(\alpha e^{i\Theta}, \beta e^{i\Phi}, \gamma e^{i\Psi})}, \\ T_N(x) e^{i\phi_N(x)}, & \text{otherwise.} \end{cases}$$

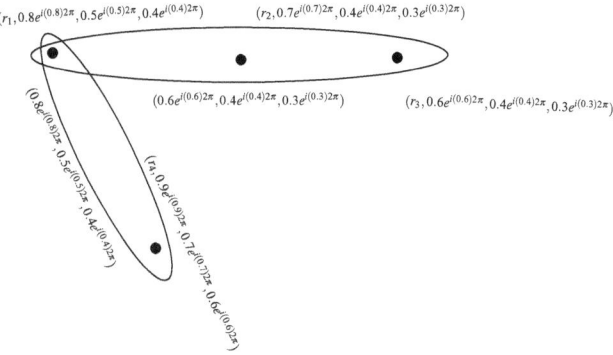

Fig. 9.26 Lower truncation of $\mathscr{H}$

$$I_{N^{[(\alpha e^{i\Theta}, \beta e^{i\Phi}, \gamma e^{i\Psi})]}}(x) e^{i\varphi_{N^{[(\alpha e^{i\Theta}, \beta e^{i\Phi}, \gamma e^{i\Psi})]}}}(x) = \begin{cases} \beta e^{i\Phi}, & \text{if } x \in N^{(\alpha e^{i\Theta}, \beta e^{i\Phi}, \gamma e^{i\Psi})}, \\ I_N(x) e^{i\varphi_N(x)}, & \text{otherwise.} \end{cases}$$

$$F_{N^{[(\alpha e^{i\Theta}, \beta e^{i\Phi}, \gamma e^{i\Psi})]}}(x) e^{i\psi_{N^{[(\alpha e^{i\Theta}, \beta e^{i\Phi}, \gamma e^{i\Psi})]}}}(x) = \begin{cases} \gamma e^{i\Psi}, & \text{if } x \in N^{(\alpha e^{i\Theta}, \beta e^{i\Phi}, \gamma e^{i\Psi})}, \\ F_N(x) e^{i\psi_N(x)}, & \text{otherwise.} \end{cases}$$

Definition 9.79 Let $\mathscr{H} = (\mathscr{N}, \lambda)$ be a complex neutrosophic hypergraph. The *lower truncation* $\mathscr{H}_{[(\alpha e^{i\Theta}, \beta e^{i\Phi}, \gamma e^{i\Psi})]}$ of $\mathscr{H}$ at level $(\alpha e^{i\Theta}, \beta e^{i\Phi}, \gamma e^{i\Psi})$ is defined as $\mathscr{H}_{[(\alpha e^{i\Theta}, \beta e^{i\Phi}, \gamma e^{i\Psi})]} = (\mathscr{N}_{[(\alpha e^{i\Theta}, \beta e^{i\Phi}, \gamma e^{i\Psi})]}, \lambda_{[(\alpha e^{i\Theta}, \beta e^{i\Phi}, \gamma e^{i\Psi})]}))$, where $\mathscr{N}_{[(\alpha e^{i\Theta}, \beta e^{i\Phi}, \gamma e^{i\Psi})]} = \{N_{[(\alpha e^{i\Theta}, \beta e^{i\Phi}, \gamma e^{i\Psi})]} | N \in \mathscr{N}\}$.

The *upper truncation* $\mathscr{H}^{[(\alpha e^{i\Theta}, \beta e^{i\Phi}, \gamma e^{i\Psi})]}$ of $\mathscr{H}$ at level $(\alpha e^{i\Theta}, \beta e^{i\Phi}, \gamma e^{i\Psi})$ is defined as $\mathscr{H}^{[(\alpha e^{i\Theta}, \beta e^{i\Phi}, \gamma e^{i\Psi})]} = (\mathscr{N}^{[(\alpha e^{i\Theta}, \beta e^{i\Phi}, \gamma e^{i\Psi})]}, \lambda^{[(\alpha e^{i\Theta}, \beta e^{i\Phi}, \gamma e^{i\Psi})]}))$, where $\mathscr{N}^{[(\alpha e^{i\Theta}, \beta e^{i\Phi}, \gamma e^{i\Psi})]} = \{N^{[(\alpha e^{i\Theta}, \beta e^{i\Phi}, \gamma e^{i\Psi})]} | N \in \mathscr{N}\}$.

Definition 9.80 Let N be a complex neutrosophic set on X. Then, each $(\alpha e^{i\Theta}, \beta e^{i\Phi}, \gamma e^{i\Psi})$, such that $\alpha \in (0, t(h(N)))$, $\beta \in (0, i(h(N)))$, $\gamma \in (0, f(h(N)))$, $\Theta \in (0, \phi(h(N)))$, $\Psi \in (0, \varphi(h(N)))$, and $\Psi \in (0, \psi(h(N)))$, for which $N^{(\alpha e^{i\Theta}, \beta e^{i\Phi}, \gamma e^{i\Psi})} \subset N^{(\alpha e^{i\Theta}, \beta e^{i\Phi}, \gamma e^{i\Psi})}$, is called a *transition level* of N.

Example 9.26 Consider a complex neutrosophic hypergraph $\mathscr{H} = (\mathscr{N}, \lambda)$ as shown in Fig. 9.22. The $(0.6 e^{i(0.6)2\pi}, 0.4 e^{i(0.4)2\pi}, 0.3 e^{i(0.3)2\pi})$-level hypergraph of $\mathscr{H}$ is shown in Fig. 9.24. Then, the lower truncation $\mathscr{H}_{[(0.6 e^{i(0.6)2\pi}, 0.4 e^{i(0.4)2\pi}, 0.3 e^{i(0.3)2\pi})]} = (\mathscr{N}_{[(0.6 e^{i(0.6)2\pi}, 0.4 e^{i(0.4)2\pi}, 0.3 e^{i(0.3)2\pi})]}, \lambda_{[(0.6 e^{i(0.6)2\pi}, 0.4 e^{i(0.4)2\pi}, 0.3 e^{i(0.3)2\pi})]})$ of $\mathscr{H}$ is a complex neutrosophic hypergraph on $X_1 = \{r_1, r_2, r_3, r_4\}$ as given in Fig. 9.26. Note that $X_1 = \bigcup_{N \in \mathscr{N}} N_{[(0.6 e^{i(0.6)2\pi}, 0.4 e^{i(0.4)2\pi}, 0.3 e^{i(0.3)2\pi})]}$. The upper truncation $\mathscr{H}^{[(0.6 e^{i(0.6)2\pi}, 0.4 e^{i(0.4)2\pi}, 0.3 e^{i(0.3)2\pi})]} = (\mathscr{N}^{[(0.6 e^{i(0.6)2\pi}, 0.4 e^{i(0.4)2\pi}, 0.3 e^{i(0.3)2\pi})]}, \lambda^{[(0.6 e^{i(0.6)2\pi}, 0.4 e^{i(0.4)2\pi}, 0.3 e^{i(0.3)2\pi})]})$ of $\mathscr{H}$ is a complex neutrosophic hypergraph on $X = \{r_1, r_2, r_3, r_4, r_5, r_6\}$ as given in Fig. 9.27.

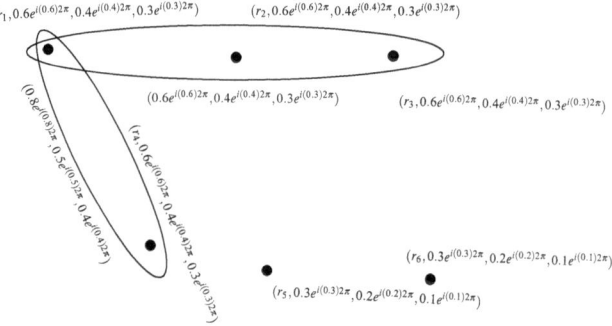

Fig. 9.27 Upper truncation of $\mathcal{H}$

Definition 9.81 Let $\mathcal{H} = (\mathcal{N}, \lambda)$ be a complex neutrosophic hypergraph. A *complex neutrosophic transversal* τ is a complex neutrosophic set of X satisfying the condition $\xi^{h(\xi)} \cap \tau^{h(\xi)} \neq \emptyset$, for all $\xi \in \lambda$, where $h(\xi)$ is the height of ξ.

A *minimal complex neutrosophic transversal* τ_1 is the complex neutrosophic transversal of $\mathcal{H}$ having the property that if $\tau \subset \tau_1$, then τ is not a complex neutrosophic transversal of $\mathcal{H}$.

Let us denote the family of minimal complex neutrosophic transversals of $\mathcal{H}$ by $T_r(\mathcal{H})$.

Definition 9.82 A complex neutrosophic transversal τ having the property that $\tau^{(\alpha e^{i\Theta}, \beta e^{i\Phi}, \gamma e^{i\Psi})} \in T_r(\mathcal{H}^{(\alpha e^{i\Theta}, \beta e^{i\Phi}, \gamma e^{i\Psi})})$, for all $\alpha, \beta, \gamma \in [0, 1]$, and $\Theta, \Phi, \Psi \in [0, 2\pi]$ is called the *locally minimal complex neutrosophic transversal* of $\mathcal{H}$. The collection of all locally minimal complex neutrosophic transversals of $\mathcal{H}$ is represented by $T_r^*(\mathcal{H})$.

Note that $T_r^*(\mathcal{H}) \subseteq T_r(\mathcal{H})$, but the converse is not generally true.

Definition 9.83 Let N be a complex neutrosophic set on X. Then, the *basic sequence* of N determined by N, denoted by $B_s(N)$, is defined as $\{(\alpha_1 e^{i\Theta_1}, \beta_1 e^{i\Phi_1}, \gamma_1 e^{i\Psi_1})^N, (\alpha_2 e^{i\Theta_2}, \beta_2 e^{i\Phi_2}, \gamma_2 e^{i\Psi_2})^N, \ldots, (\alpha_n e^{i\Theta_n}, \beta_n e^{i\Phi_n}, \gamma_n e^{i\Psi_n})^N\}$, where

(i) $\alpha_1 > \alpha_2 > \cdots > \alpha_n$, $\beta_1 > \beta_2 > \cdots > \beta_n$, $\gamma_1 < \gamma_2 < \cdots < \gamma_n$, $\Theta_1 > \Theta_2 > \cdots > \Theta_n$, $\Phi_1 > \Phi_2 > \cdots > \Phi_n$, $\Psi_1 < \Psi_2 < \cdots < \Psi_n$,
(ii) $(\alpha_1 e^{i\Theta_1}, \beta_1 e^{i\Phi_1}, \gamma_1 e^{i\Psi_1}) = h(N)$,
(iii) $\{(\alpha_2 e^{i\Theta_2}, \beta_2 e^{i\Phi_2}, \gamma_2 e^{i\Psi_2})^N, \ldots, (\alpha_n e^{i\Theta_n}, \beta_n e^{i\Phi_n}, \gamma_n e^{i\Psi_n})^N\}$ are the transition levels of N.

Definition 9.84 Let $B_s(N) = \{(\alpha_1 e^{i\Theta_1}, \beta_1 e^{i\Phi_1}, \gamma_1 e^{i\Psi_1})^N, (\alpha_2 e^{i\Theta_2}, \beta_2 e^{i\Phi_2}, \gamma_2 e^{i\Psi_2})^N, \ldots, (\alpha_n e^{i\Theta_n}, \beta_n e^{i\Phi_n}, \gamma_n e^{i\Psi_n})^N\}$ be the basic sequence of N. Then, the *set of basic cuts* $B_c(N)$ is defined as $B_c(N) = \{N^{(\alpha e^{i\Theta}, \beta e^{i\Phi}, \gamma e^{i\Psi})} | (\alpha e^{i\Theta}, \beta e^{i\Phi}, \gamma e^{i\Psi}) \in B_s(N)\}$.

Lemma 9.1 Let $\mathcal{H} = (\mathcal{N}, \lambda)$ be a complex neutrosophic hypergraph with $\mathcal{F}_s(\mathcal{H}) = \{(\alpha_1 e^{i\Theta_1}, \beta_1 e^{i\Phi_1}, \gamma_1 e^{i\Psi_1}), (\alpha_2 e^{i\Theta_2}, \beta_2 e^{i\Phi_2}, \gamma_2 e^{i\Psi_2}), \ldots, (\alpha_n e^{i\Theta_n}, \beta_n e^{i\Phi_n}, \gamma_n e^{i\Psi_n})\}$. Then,

(i) *If* $(\alpha e^{i\Theta}, \beta e^{i\Phi}, \gamma e^{i\Psi})$ *is a transition level of* $\tau \in T_r(\mathcal{H})$, *then there exists an* $\varepsilon > 0$ *such that for all* $\alpha_1 \in (\alpha, \alpha + \varepsilon]$, $\beta_1 \in (\beta, \beta + \varepsilon]$, $\gamma_1 \in (\gamma, \gamma + \varepsilon]$, $\Theta_1 \in (\Theta, \Theta + \varepsilon]$, $\Phi_1 \in (\Phi, \Phi + \varepsilon]$, $\Psi_1 \in (\Psi, \Psi + \varepsilon]$, $\tau^{(\alpha e^{i\Theta}, \beta e^{i\Phi}, \gamma e^{i\Psi})}$ *is a minimal* $\mathcal{H}^{(\alpha e^{i\Theta}, \beta e^{i\Phi}, \gamma e^{i\Psi})}$ *transversal extension of* $\tau^{(\alpha_1 e^{i\Theta_1}, \beta_1 e^{i\Phi_1}, \gamma_1 e^{i\Psi_1})}$, *i.e., if* $\tau^{(\alpha_1 e^{i\Theta_1}, \beta_1 e^{i\Phi_1}, \gamma_1 e^{i\Psi_1})} \subseteq C \subset \tau^{(\alpha e^{i\Theta}, \beta e^{i\Phi}, \gamma e^{i\Psi})}$, *then* C *is not a transversal of* $\mathcal{H}^{(\alpha e^{i\Theta}, \beta e^{i\Phi}, \gamma e^{i\Psi})}$.

(ii) $T_r(\mathcal{H})$, *i.e., the collection of minimal transversals of* $\mathcal{H}$ *is sectionally elementary.*

(iii) $\mathcal{F}_s(T_r(\mathcal{H}))$ *is properly contained in* $\mathcal{F}_s(\mathcal{H})$.

(iv) $\tau^{(\alpha e^{i\Theta}, \beta e^{i\Phi}, \gamma e^{i\Psi})} \in T_r(\mathcal{H}^{(\alpha e^{i\Theta}, \beta e^{i\Phi}, \gamma e^{i\Psi})})$, *for all* $\tau \in T_r(\mathcal{H})$ *and for every* $\alpha_2 < \alpha \leq \alpha_1, \beta_2 < \beta \leq \beta_1, \gamma_2 > \gamma \geq \gamma_1, \Theta_2 < \Theta \leq \Theta_1, \Phi_2 < \Phi \leq \Phi_1, \Psi_2 > \Psi \geq \Psi_1$.

Definition 9.85 Let $\mathcal{H} = (\mathcal{N}, \lambda)$ be a complex neutrosophic hypergraph. The *complex neutrosophic line graph* of $\mathcal{H}$ is defined as an ordered pair $l(\mathcal{H}) = (\mathcal{N}_l, \lambda_l)$, where $\mathcal{N}_l = \lambda$ and there exists an edge between two vertices in $l(\mathcal{H})$ if $|supp(\lambda_j) \cap supp(\lambda_k)| \geq 1$, for all $\lambda_j, \lambda_k \in \lambda$. The membership degrees of $l(\mathcal{H})$ are given as

(i) $\mathcal{N}_l(E_k) = \lambda(E_k)$,

(ii) $\lambda_l(E_j E_k) = (\min\{T_\lambda(E_j), T_\lambda(E_k)\}e^{i\min\{\phi_\lambda(E_j), \phi_\lambda(E_k)\}}, \min\{I_\lambda(E_j), I_\lambda(E_k)\}$ $e^{i\min\{\varphi_\lambda(E_j), \varphi_\lambda(E_k)\}}, \max\{F_\lambda(E_j), F_\lambda(E_k)\}e^{i\max\{\psi_\lambda(E_j), \psi_\lambda(E_k)\}})$.

9.9 T-Related Complex Neutrosophic Hypergraphs

Definition 9.86 A complex neutrosophic hypergraph $\mathcal{H} = (\mathcal{N}, \lambda)$ is N-tempered complex neutrosophic hypergraph of $H = (X, E)$ if there exists $H = (X, E)$, a crisp hypergraph, and a complex neutrosophic set N such that $\lambda = \{\delta_e | e \in E\}$, where

$$T_\delta(u)e^{i\phi_\delta(u)} = \begin{cases} \min\{T_N(x)e^{i\min\{\phi_N(x)\}}|x \in e\}, & \text{if } u \in e, \\ 0, & \text{otherwise.} \end{cases}$$

$$I_\delta(u)e^{i\varphi_\delta(u)} = \begin{cases} \min\{I_N(x)e^{i\min\{\varphi_N(x)\}}|x \in e\}, & \text{if } u \in e, \\ 0, & \text{otherwise.} \end{cases}$$

$$F_\delta(u)e^{i\psi_\delta(u)} = \begin{cases} \max\{F_N(x)e^{i\max\{\psi_N(x)\}}|x \in e\}, & \text{if } u \in e, \\ 0, & \text{otherwise} \end{cases}$$

An N-tempered complex neutrosophic hypergraph $\mathcal{H} = (\mathcal{N}, \lambda)$ determined by H and complex neutrosophic set N is denoted by $N \otimes H$.

Definition 9.87 A pair (G, J) of crisp hypergraphs is T-*related* if whenever g is a minimal transversal of G, k is any transversal of J, and $g \subseteq k$, then there exists a minimal transversal t of J such that $g \subseteq t \subseteq k$.

Definition 9.88 Let $\mathscr{H} = (\mathscr{N}, \lambda)$ be a complex neutrosophic hypergraph with $\mathscr{F}_s(\mathscr{H}) = \{(\alpha_1 e^{\iota\Theta_1}, \beta_1 e^{\iota\Phi_1}, \gamma_1 e^{\iota\Psi_1}), (\alpha_2 e^{\iota\Theta_2}, \beta_2 e^{\iota\Phi_2}, \gamma_2 e^{\iota\Psi_2}), \ldots, (\alpha_n e^{\iota\Theta_n}, \beta_n e^{\iota\Phi_n}, \gamma_n e^{\iota\Psi_n})\}$. Then, $\mathscr{H}$ is *T-related* if from the core set

$$c(\mathscr{H}) = \{\mathscr{H}^{(\alpha_1 e^{\iota\Theta_1}, \beta_1 e^{\iota\Phi_1}, \gamma_1 e^{\iota\Psi_1})}, \mathscr{H}^{(\alpha_2 e^{\iota\Theta_2}, \beta_2 e^{\iota\Phi_2}, \gamma_2 e^{\iota\Psi_2})}, \ldots, \mathscr{H}^{(\alpha_n e^{\iota\Theta_n}, \beta_n e^{\iota\Phi_n}, \gamma_n e^{\iota\Psi_n})}\}$$

of $\mathscr{H}$, every successive ordered pair $(\mathscr{H}^{(\alpha_j e^{\iota\Theta_j}, \beta_j e^{\iota\Phi_j}, \gamma_j e^{\iota\Psi_j})}, \mathscr{H}^{(\alpha_{j-1} e^{\iota\Theta_{j-1}}, \beta_{j-1} e^{\iota\Phi_{j-1}}, \gamma_{j-1} e^{\iota\Psi_{j-1}})})$ is *T*-related.

If $\mathscr{F}_s(\mathscr{H})$ contains only one element, $\mathscr{H}$ is considered to be trivially *T*-related.

Theorem 9.15 *Let $\mathscr{H} = (\mathscr{N}, \lambda)$ be a T-related complex neutrosophic hypergraph, then $T_r(\mathscr{H}) = T_r^*(\mathscr{H})$.*

Proof Let $\mathscr{H} = (\mathscr{N}, \lambda)$ be a *T*-related complex neutrosophic hypergraph with $\mathscr{F}_s(\mathscr{H}) = \{(\alpha_1 e^{\iota\Theta_1}, \beta_1 e^{\iota\Phi_1}, \gamma_1 e^{\iota\Psi_1}), (\alpha_2 e^{\iota\Theta_2}, \beta_2 e^{\iota\Phi_2}, \gamma_2 e^{\iota\Psi_2}), \ldots, (\alpha_1 e^{\iota\Theta_n}, \beta_n e^{\iota\Phi_n}, \gamma_n e^{\iota\Psi_n})\}$. Then, there arises two cases:

Case (i)　First we consider that $\mathscr{F}_s(\mathscr{H}) = \{(\alpha_1 e^{\iota\Theta_1}, \beta_1 e^{\iota\Phi_1}, \gamma_1 e^{\iota\Psi_1})\}$. Then, Lemma 9.1 implies that for each $\xi \in T_r(\mathscr{H})$, $\xi^{(\alpha e^{\iota\Theta}, \beta e^{\iota\Phi}, \gamma e^{\iota\Psi})} \in T_r(\mathscr{H}^{(\alpha e^{\iota\Theta}, \beta e^{\iota\Phi}, \gamma e^{\iota\Psi})})$, for all $0 < \alpha \le t(h(\mathscr{H}))$, $0 < \beta \le i(h(\mathscr{H}))$, $\gamma \ge f(h(\mathscr{H})) > 0$, $0 < \Theta \le \phi(h(\mathscr{H}))$, $0 < \Phi \le \varphi(h(\mathscr{H}))$, and $\Psi \ge \psi(h(\mathscr{H})) > 0$. Thus, $T_r(\mathscr{H}) = T_r^*(\mathscr{H})$.

Case (ii)　We now suppose that $|\mathscr{F}_s(\mathscr{H})| \ge 2$. Since, $T_r^*(\mathscr{H}) \subseteq T_r(\mathscr{H})$, we just have to prove that $T_r(\mathscr{H}) \subseteq T_r^*(\mathscr{H})$. Let $\xi \in T_r(\mathscr{H})$, and $\xi^{(\alpha_1 e^{\iota\Theta_1}, \beta_1 e^{\iota\Phi_1}, \gamma_1 e^{\iota\Psi_1})} \subset \xi^{(\alpha_2 e^{\iota\Theta_2}, \beta_2 e^{\iota\Phi_2}, \gamma_2 e^{\iota\Psi_2})}$.　As　$\xi^{(\alpha_1 e^{\iota\Theta_1}, \beta_1 e^{\iota\Phi_1}, \gamma_1 e^{\iota\Psi_1})} \in T_r(\mathscr{H}^{(\alpha_1 e^{\iota\Theta_1}, \beta_1 e^{\iota\Phi_1}, \gamma_1 e^{\iota\Psi_1})})$, $\xi^{(\alpha_2 e^{\iota\Theta_2}, \beta_2 e^{\iota\Phi_2}, \gamma_2 e^{\iota\Psi_2})} \in T_r(\mathscr{H}^{(\alpha_2 e^{\iota\Theta_2}, \beta_2 e^{\iota\Phi_2}, \gamma_2 e^{\iota\Psi_2})})$,　and　the　ordered　pair $(\mathscr{H}^{(\alpha_2 e^{\iota\Theta_2}, \beta_2 e^{\iota\Phi_2}, \gamma_2 e^{\iota\Psi_2})}, \mathscr{H}^{(\alpha_1 e^{\iota\Theta_1}, \beta_1 e^{\iota\Phi_1}, \gamma_1 e^{\iota\Psi_1})})$ is *T*-related. If $\xi^{(\alpha_2 e^{\iota\Theta_2}, \beta_2 e^{\iota\Phi_2}, \gamma_2 e^{\iota\Psi_2})} \notin T_r(\mathscr{H}^{(\alpha_2 e^{\iota\Theta_2}, \beta_2 e^{\iota\Phi_2}, \gamma_2 e^{\iota\Psi_2})})$, then there exists a minimal transversal τ of $\mathscr{H}^{(\alpha_2 e^{\iota\Theta_2}, \beta_2 e^{\iota\Phi_2}, \gamma_2 e^{\iota\Psi_2})}$ such that $\xi^{(\alpha_1 e^{\iota\Theta_1}, \beta_1 e^{\iota\Phi_1}, \gamma_1 e^{\iota\Psi_1})} \subseteq \tau_2 \subset \xi^{(\alpha_2 e^{\iota\Theta_2}, \beta_2 e^{\iota\Phi_2}, \gamma_2 e^{\iota\Psi_2})}$. Hence, we obtain a complex neutrosophic transversal δ of $\mathscr{H}$ such that $\delta \subset \xi$. Let $\xi^{(\alpha_1 e^{\iota\Theta_1}, \beta_1 e^{\iota\Phi_1}, \gamma_1 e^{\iota\Psi_1})} = \tau_1$ and $\delta = \xi^{(\alpha_3 e^{\iota\Theta_3}, \beta_3 e^{\iota\Phi_3}, \gamma_3 e^{\iota\Psi_3})} \cup \rho_2 \cap \rho_1$, where ρ_k is an elementary complex neutrosophic set having support τ_k and height $(\alpha_k e^{\iota\Theta_k}, \beta_k e^{\iota\Phi_k}, \gamma_k e^{\iota\Psi_k}), k = 1, 2$. This contradiction shows that $\xi^{(\alpha_2 e^{\iota\Theta_2}, \beta_2 e^{\iota\Phi_2}, \gamma_2 e^{\iota\Psi_2})} \in T_r(\mathscr{H}^{(\alpha_2 e^{\iota\Theta_2}, \beta_2 e^{\iota\Phi_2}, \gamma_2 e^{\iota\Psi_2})})$. Then, Lemma 9.1 implies that $\xi^{(\alpha e^{\iota\Theta}, \beta e^{\iota\Phi}, \gamma e^{\iota\Psi})} \in T_r(\mathscr{H}^{(\alpha e^{\iota\Theta}, \beta e^{\iota\Phi}, \gamma e^{\iota\Psi})})$, for $\alpha \in (\alpha_3, \alpha_1], \beta \in (\beta_3, \beta_1], \gamma \in (\gamma_3, \gamma_1], \Theta \in (\Theta_3, \Theta_1], \Phi \in (\Phi_3, \Phi_1], \Psi \in (\Psi_3, \Psi_1]$. Continuing the same recursive procedure, we show that $\xi^{(\alpha e^{\iota\Theta}, \beta e^{\iota\Phi}, \gamma e^{\iota\Psi})} \in T_r(\mathscr{H}^{(\alpha e^{\iota\Theta}, \beta e^{\iota\Phi}, \gamma e^{\iota\Psi})})$, for each $\alpha \in (0, \alpha_1], \beta \in (0, \beta_1], \gamma \in (0, \gamma_1], \Theta \in (0, \Theta_1], \Phi \in (0, \Phi_1], \Psi \in (0, \Psi_1]$.

Example 9.27 Let $\mathscr{H} = (\mathscr{N}, \lambda)$ be a complex neutrosophic hypergraph represented by the incidence matrix as given in Table 9.23. Note that

$$\lambda^{(0.9 e^{\iota(0.9)2\pi}, 0.9 e^{\iota(0.9)2\pi}, 0.9 e^{\iota(0.9)2\pi})} = \{\{j_1, j_2\}, \{j_1, j_3\}, \{j_2, j_3\}\},$$

$$\lambda^{(0.6 e^{\iota(0.6)2\pi}, 0.6 e^{\iota(0.6)2\pi}, 0.6 e^{\iota(0.6)2\pi})} = \{\{j_1, j_2, j_4\}, \{j_1, j_3, j_4\}, \{j_2, j_3, j_5\}\},$$

$$\lambda^{(0.3 e^{\iota(0.3)2\pi}, 0.3 e^{\iota(0.3)2\pi}, 0.3 e^{\iota(0.3)2\pi})} = \{\{j_1, j_2, j_4, j_5\}, \{j_1, j_3, j_4, j_5\}, \{j_2, j_3, j_4, j_5\}\}.$$

Table 9.23 Incidence matrix of complex neutrosophic hypergraph $\mathcal{H} = (\mathcal{N}, \lambda)$

I	λ_1	λ_2	λ_3
j_1	$(0.9e^{\iota(0.9)2\pi}, 0.9e^{\iota(0.9)2\pi},$ $0.9e^{\iota(0.9)2\pi})$	$(0.9e^{\iota(0.9)2\pi}, 0.9e^{\iota(0.9)2\pi},$ $0.9e^{\iota(0.9)2\pi})$	$(0, 0, 1)$
j_2	$(0.9e^{\iota(0.9)2\pi}, 0.9e^{\iota(0.9)2\pi},$ $0.9e^{\iota(0.9)2\pi})$	$(0, 0, 1)$	$(0.9e^{\iota(0.9)2\pi}, 0.9e^{\iota(0.9)2\pi},$ $0.9e^{\iota(0.9)2\pi})$
j_3	$(0, 0, 1)$	$(0.9e^{\iota(0.9)2\pi}, 0.9e^{\iota(0.9)2\pi},$ $0.9e^{\iota(0.9)2\pi})$	$(0.9e^{\iota(0.9)2\pi}, 0.9e^{\iota(0.9)2\pi},$ $0.9e^{\iota(0.9)2\pi})$
j_4	$(0.6e^{\iota(0.6)2\pi}, 0.6e^{\iota(0.6)2\pi},$ $0.6e^{\iota(0.6)2\pi})$	$(0.6e^{\iota(0.6)2\pi}, 0.6e^{\iota(0.6)2\pi},$ $0.6e^{\iota(0.6)2\pi})$	$(0.3e^{\iota(0.3)2\pi}, 0.3e^{\iota(0.3)2\pi},$ $0.3e^{\iota(0.3)2\pi})$
j_5	$(0.3e^{\iota(0.3)2\pi}, 0.3e^{\iota(0.3)2\pi},$ $0.3e^{\iota(0.3)2\pi})$	$(0.3e^{\iota(0.3)2\pi}, 0.3e^{\iota(0.3)2\pi},$ $0.3e^{\iota(0.3)2\pi})$	$(0.6e^{\iota(0.6)2\pi}, 0.6e^{\iota(0.6)2\pi},$ $0.6e^{\iota(0.6)2\pi})$

Clearly, $\mathcal{F}_s(\mathcal{H}) = \{(0.9e^{\iota(0.9)2\pi}, 0.9e^{\iota(0.9)2\pi}, 0.9e^{\iota(0.9)2\pi}), (0.6e^{\iota(0.6)2\pi}, 0.6e^{\iota(0.6)2\pi},$ $0.6e^{\iota(0.6)2\pi}), (0.3e^{\iota(0.3)2\pi}, 0.3e^{\iota(0.3)2\pi}, 0.3e^{\iota(0.3)2\pi})\}$. Also, $T_r(\mathcal{H}) = \{\tau_1, \tau_2, \tau_3\} = T_r^*(\mathcal{H})$, where

$$\tau_1 = \{(j_1, 0.9e^{\iota(0.9)2\pi}, 0.9e^{\iota(0.9)2\pi}, 0.9e^{\iota(0.9)2\pi}), (j_2, 0.9e^{\iota(0.9)2\pi}, 0.9e^{\iota(0.9)2\pi}, 0.9e^{\iota(0.9)2\pi})\},$$
$$\tau_2 = \{(j_1, 0.9e^{\iota(0.9)2\pi}, 0.9e^{\iota(0.9)2\pi}, 0.9e^{\iota(0.9)2\pi}), (j_3, 0.9e^{\iota(0.9)2\pi}, 0.9e^{\iota(0.9)2\pi}, 0.9e^{\iota(0.9)2\pi})\},$$
$$\tau_3 = \{(j_2, 0.9e^{\iota(0.9)2\pi}, 0.9e^{\iota(0.9)2\pi}, 0.9e^{\iota(0.9)2\pi}), (j_3, 0.9e^{\iota(0.9)2\pi}, 0.9e^{\iota(0.9)2\pi}, 0.9e^{\iota(0.9)2\pi})\}.$$

Since, $\{j_4, j_5\} \in T_r(\mathcal{H}^{(0.6e^{\iota(0.6)2\pi}, 0.6e^{\iota(0.6)2\pi}, 0.6e^{\iota(0.6)2\pi})})$ and $\{j_4\} \in T_r(\mathcal{H}^{(0.3e^{\iota(0.3)2\pi}, 0.3e^{\iota(0.3)2\pi}, 0.3e^{\iota(0.3)2\pi})})$, i.e., no minimal transversal of $\mathcal{H}^{(0.3e^{\iota(0.3)2\pi}, 0.3e^{\iota(0.3)2\pi}, 0.3e^{\iota(0.3)2\pi})}$ contains $\{j_4, j_5\}$. Thus, $(\mathcal{H}^{(0.6e^{\iota(0.6)2\pi}, 0.6e^{\iota(0.6)2\pi}, 0.6e^{\iota(0.6)2\pi})}, \mathcal{H}^{(0.3e^{\iota(0.3)2\pi}, 0.3e^{\iota(0.3)2\pi}, 0.3e^{\iota(0.3)2\pi})})$ is not *T*-related, therefore, $\mathcal{H}$ is not *T*-related.

Theorem 9.16 *Let $\mathcal{H} = (\mathcal{N}, \lambda)$ be an ordered complex neutrosophic hypergraph, then $T_r(\mathcal{H}) = T_r^*(\mathcal{H}) \Leftrightarrow \mathcal{H}$ is T-related.*

Proof In view of Theorem 9.15, this is enough to prove that $T_r(\mathcal{H}) = T_r^*(\mathcal{H})$ implies $\mathcal{H}$ is *T*-related. Suppose that $\mathcal{F}_s(\mathcal{H}) = \{(\alpha_1 e^{\iota\Theta_1}, \beta_1 e^{\iota\Phi_1}, \gamma_1 e^{\iota\Psi_1}),$ $(\alpha_2 e^{\iota\Theta_2}, \beta_2 e^{\iota\Phi_2}, \gamma_2 e^{\iota\Psi_2}), \ldots, (\alpha_n e^{\iota\Theta_n}, \beta_n e^{\iota\Phi_n}, \gamma_n e^{\iota\Psi_n})\}$ and $\mathcal{H}$ is not *T*-related. Here, we obtain $\xi \in T_r(\mathcal{H})$ such that $\xi \notin T_r^*(\mathcal{H})$. Assume that the ordered pair $(\mathcal{H}^{(\alpha_j e^{\iota\Theta_j}, \beta_j e^{\iota\Phi_j}, \gamma_j e^{\iota\Psi_j})}, \mathcal{H}^{(\alpha_{j+1} e^{\iota\Theta_{j+1}}, \beta_{j+1} e^{\iota\Phi_{j+1}}, \gamma_{j+1} e^{\iota\Psi_{j+1}})})$ is not *T*-related and $c(\mathcal{H}) = \{\mathcal{H}^{(\alpha_1 e^{\iota\Theta_1}, \beta_1 e^{\iota\Phi_1}, \gamma_1 e^{\iota\Psi_1})}, \mathcal{H}^{(\alpha_2 e^{\iota\Theta_2}, \beta_2 e^{\iota\Phi_2}, \gamma_2 e^{\iota\Psi_2})}, \ldots, \mathcal{H}^{(\alpha_n e^{\iota\Theta_n}, \beta_n e^{\iota\Phi_n}, \gamma_n e^{\iota\Psi_n})}\}$. Then, there exists a complex neutrosophic transversal τ_k such that $\tau_k \in T_r(\mathcal{H}^{(\alpha_k e^{\iota\Theta_k}, \beta_k e^{\iota\Phi_k}, \gamma_k e^{\iota\Psi_k})})$ and $\tau_k \subset \tau_{k+1}$, where

$$\tau_{k+1} \in T_r(\mathcal{H}^{(\alpha_{k+1} e^{\iota\Theta_{k+1}}, \beta_{k+1} e^{\iota\Phi_{k+1}}, \gamma_{k+1} e^{\iota\Psi_{k+1}})})$$

satisfying the condition that N is not a minimal transversal of $\mathcal{H}^{(\alpha_{k+1} e^{\iota\Theta_{k+1}}, \beta_{k+1} e^{\iota\Phi_{k+1}}, \gamma_{k+1} e^{\iota\Psi_{k+1}})}$, for every N, $\tau_k \subseteq N \subseteq \tau_{k+1}$. Since, $\mathcal{H} = (\mathcal{N}, \lambda)$ is an ordered complex neutrosophic hypergraph, then $\mathcal{H}^{(\alpha_k e^{\iota\Theta_k}, \beta_k e^{\iota\Phi_k}, \gamma_k e^{\iota\Psi_k})} \subseteq \mathcal{H}^{(\alpha_{k+1} e^{\iota\Theta_{k+1}}, \beta_{k+1} e^{\iota\Phi_{k+1}}, \gamma_{k+1} e^{\iota\Psi_{k+1}})}$, therefore, τ_k is not a transversal of $\mathcal{H}^{(\alpha_{k+1} e^{\iota\Theta_{k+1}}, \beta_{k+1} e^{\iota\Phi_{k+1}}, \gamma_{k+1} e^{\iota\Psi_{k+1}})}$, for otherwise $\tau_k \in T_r(\mathcal{H}^{(\alpha_{k+1} e^{\iota\Theta_{k+1}}, \beta_{k+1} e^{\iota\Phi_{k+1}}, \gamma_{k+1} e^{\iota\Psi_{k+1}})})$, which is a contradiction to our assumption. Let δ be an arbitrary CNT of $\mathscr{H}^{(\alpha_{k+1}e^{\iota\Theta_{k+1}},\beta_{k+1}e^{\iota\Phi_{k+1}},\gamma_{k+1}e^{\iota\Psi_{k+1}})}$ such that $\tau_k \subseteq \delta \subseteq \tau_{k+1}$. Now, if $\tau_k \subseteq Q \subset \delta$, then Q is not a crisp transversal of $\mathscr{H}^{(\alpha_{k+1}e^{\iota\Theta_{k+1}},\beta_{k+1}e^{\iota\Phi_{k+1}},\gamma_{k+1}e^{\iota\Psi_{k+1}})}$. As we know that $\delta \notin T_r(\mathscr{H}^{(\alpha_{k+1}e^{\iota\Theta_{k+1}},\beta_{k+1}e^{\iota\Phi_{k+1}},\gamma_{k+1}e^{\iota\Psi_{k+1}})})$ and $\tau_k \subset \delta$. Thus, we can obtain a minimal CNT ξ of $\mathscr{H}$ such that $\xi \notin T_r^*(\mathscr{H})$. First, we compute a minimal CNT ξ_1 of $\mathscr{H}^{(\alpha_k e^{\iota\Theta_k},\beta_k e^{\iota\Phi_k},\gamma_k e^{\iota\Psi_k})}$, where τ_k is the top level cut of ξ_1 at level $(\alpha_k e^{\iota\Theta_k}, \beta_k e^{\iota\Phi_k}, \gamma_k e^{\iota\Psi_k})$ and satisfies $\xi_1^{(\alpha_{k+1}e^{\iota\Theta_{k+1}},\beta_{k+1}e^{\iota\Phi_{k+1}},\gamma_{k+1}e^{\iota\Psi_{k+1}})} \subseteq \tau_{k+1}$. Then, Lemma 9.1 implies that the $(\alpha_{k+1}e^{\iota\Theta_{k+1}}, \beta_{k+1}e^{\iota\Phi_{k+1}}, \gamma_{k+1}e^{\iota\Psi_{k+1}})$-cut, $\xi_1^{(\alpha_{k+1}e^{\iota\Theta_{k+1}},\beta_{k+1}e^{\iota\Phi_{k+1}},\gamma_{k+1}e^{\iota\Psi_{k+1}})}$ of ξ_1 should equal to some δ that satisfies $\tau_k \subseteq \delta \subseteq \tau_{k+1}$ and $\tau_k \subseteq Q \subset \delta$, then Q is not a crisp transversal of $\mathscr{H}^{(\alpha_{k+1}e^{\iota\Theta_{k+1}},\beta_{k+1}e^{\iota\Phi_{k+1}},\gamma_{k+1}e^{\iota\Psi_{k+1}})}$. Thus, we obtain $\xi_1 \in T_r(\mathscr{H}^{(\alpha_k e^{\iota\Theta_k},\beta_k e^{\iota\Phi_k},\gamma_k e^{\iota\Psi_k})}) \setminus T_r^*(\mathscr{H}^{(\alpha_k e^{\iota\Theta_k},\beta_k e^{\iota\Phi_k},\gamma_k e^{\iota\Psi_k})})$.

We now assume that $(\alpha_k e^{\iota\Theta_k}, \beta_k e^{\iota\Phi_k}, \gamma_k e^{\iota\Psi_k}) \subset (\alpha_1 e^{\iota\Theta_1}, \beta_1 e^{\iota\Phi_1}, \gamma_1 e^{\iota\Psi_1})$. Since, $\mathscr{H}$ is ordered, then there exists an ordered sequence $t_k \supseteq t_{k-1} \supset \cdots \supseteq t_1$ of crisp minimal transversals of $\mathscr{H}^{(\alpha_k e^{\iota\Theta_k},\beta_k e^{\iota\Phi_k},\gamma_k e^{\iota\Psi_k})}$, $\mathscr{H}^{(\alpha_{k-1}e^{\iota\Theta_{k-1}},\beta_{k-1}e^{\iota\Phi_{k-1}},\gamma_{k-1}e^{\iota\Psi_{k-1}})}$, $\ldots$, $\mathscr{H}^{(\alpha_1 e^{\iota\Theta_1},\beta_1 e^{\iota\Phi_1},\gamma_1 e^{\iota\Psi_1})}$, respectively. Let ρ_l be an elementary CNSS having support t_l and height ξ_l. Then, $\xi = \rho_1 \cup \cdots \cup \rho_{l-1} \cup \delta$ such that $\xi \in T_r(\mathscr{H})$ and $\xi \notin T_r^*(\mathscr{H})$.

Corollary 9.2 *Let* $\mathscr{H} = (\mathcal{N}, \lambda)$ *be an ordered complex neutrosophic hypergraph with* $\mathscr{F}_s(\mathscr{H}) = \{(\alpha_1 e^{\iota\Theta_1}, \beta_1 e^{\iota\Phi_1}, \gamma_1 e^{\iota\Psi_1}), (\alpha_2 e^{\iota\Theta_2}, \beta_2 e^{\iota\Phi_2}, \gamma_2 e^{\iota\Psi_2}), \ldots, (\alpha_n e^{\iota\Theta_n}, \beta_n e^{\iota\Phi_n}, \gamma_n e^{\iota\Psi_n})\}$ *and*

$$c(\mathscr{H}) = \{\mathscr{H}^{(\alpha_1 e^{\iota\Theta_1},\beta_1 e^{\iota\Phi_1},\gamma_1 e^{\iota\Psi_1})}, \mathscr{H}^{(\alpha_2 e^{\iota\Theta_2},\beta_2 e^{\iota\Phi_2},\gamma_2 e^{\iota\Psi_2})}, \ldots, \mathscr{H}^{(\alpha_n e^{\iota\Theta_n},\beta_n e^{\iota\Phi_n},\gamma_n e^{\iota\Psi_n})}\}.$$

If an ordered pair $(\mathscr{H}^{(\alpha_j e^{\iota\Theta_j},\beta_j e^{\iota\Phi_j},\gamma_j e^{\iota\Psi_j})}, \mathscr{H}^{(\alpha_{j+1}e^{\iota\Theta_{j+1}},\beta_{j+1}e^{\iota\Phi_{j+1}},\gamma_{j+1}e^{\iota\Psi_{j+1}})})$ *is not T-related, then*

(i) $(\alpha_{j+1}e^{\iota\Theta_{j+1}}, \beta_{j+1}e^{\iota\Phi_{j+1}}, \gamma_{j+1}e^{\iota\Psi_{j+1}}) \in \mathscr{F}_s(T_r(\mathscr{H}))$.

(ii) $(\alpha_{j+1}e^{\iota\Theta_{j+1}}, \beta_{j+1}e^{\iota\Phi_{j+1}}, \gamma_{j+1}e^{\iota\Psi_{j+1}})$ *is a transition level for* $\xi \in T_r(\mathscr{H}) \setminus T_r^*(\mathscr{H})$.

Example 9.28 Let $N = \{(u, t_N(u)e^{\iota\phi_N(u)}, i_N(u)e^{\iota\varphi_N(u)}, f_N(u)e^{\iota\psi_N(u)})|u \in X\}$ be a complex neutrosophic set on $X = \{a_1, a_2, a_3, a_4, a_5, a_6, a_7\}$ such that $N(a_7) = (0.4e^{\iota(0.4)2\pi}, 0.4e^{\iota(0.4)2\pi}, 0.4e^{\iota(0.4)2\pi})$ and $N(a) = (0.9e^{\iota(0.9)2\pi}, 0.9 \ e^{\iota(0.9)2\pi}, 0.9e^{\iota(0.9)2\pi})$, for all $a \in X \setminus \{a_7\}$. Let $H = (X, E)$ be a crisp hypergraph on X, where $E_1 = \{a_1, a_2, a_4\}, E_2 = \{a_1, a_3, a_4\}, E_3 = \{a_4, a_5, a_6\}, E_4 = \{a_1, a_5\}$, and $E_5 = \{a_5, a_7\}$. Then, N-tempered complex neutrosophic hypergraph $\mathscr{H} = (\mathcal{N}, \lambda)$ is given by the incidence matrix as shown in Table 9.24.

Here, $\mathbf{0} = (0, 0, 1)$, $\mathbf{0.9e^{\iota(0.9)2\pi}} = (0.9e^{\iota(0.9)2\pi}, 0.9e^{\iota(0.9)2\pi}, 0.9e^{\iota(0.9)2\pi})$, and $\mathbf{0.4e^{\iota(0.4)2\pi}} = (0.4e^{\iota(0.4)2\pi}, 0.4 \ e^{\iota(0.4)2\pi}, 0.4e^{\iota(0.4)2\pi})$. Note that $\mathscr{F}_s(\mathscr{H}) = \{(0.9e^{\iota(0.9)2\pi}, 0.9e^{\iota(0.9)2\pi}, 0.9e^{\iota(0.9)2\pi}), (0.4e^{\iota(0.4)2\pi}, 0.4e^{\iota(0.4)2\pi}, 0.4e^{\iota(0.4)2\pi})\}$ and $c(\mathscr{H}) = \{\mathscr{H}^{(0.9e^{\iota(0.9)2\pi},0.9e^{\iota(0.9)2\pi},0.9e^{\iota(0.9)2\pi})}, \mathscr{H}^{(0.4e^{\iota(0.4)2\pi},0.4e^{\iota(0.4)2\pi},0.4e^{\iota(0.4)2\pi})}\}$, where

$$\mathscr{H}^{(0.9e^{\iota(0.9)2\pi},0.9e^{\iota(0.9)2\pi},0.9e^{\iota(0.9)2\pi})} = (X_1\}, \mathscr{E}_1), \ X_1 = \{a_1, a_2, a_3, a_4, a_5, a_6\},$$

$$\mathscr{E}_1 = \{\{a_1, a_2, a_4\}, \{a_1, a_3, a_4\}, \{a_4, a_5, a_6\}, \{a_1, a_5\}\},$$

$$\mathscr{H}^{(0.4e^{\iota(0.4)2\pi},0.4e^{\iota(0.4)2\pi},0.4e^{\iota(0.4)2\pi})} = (X_2, \mathscr{E}_2), \ X_2 = \{a_1, a_2, a_3, a_4, a_5, a_6, a_7\},$$

$$\mathscr{E}_2 = \{\{a_1, a_2, a_4\}, \{a_1, a_3, a_4\}, \{a_4, a_5, a_6\}, \{a_1, a_5\}\{a_5, a_7\}\}.$$

Table 9.24 Incidence matrix of N-tempered complex neutrosophic hypergraph $\mathscr{H}$

$\mathscr{H}$	λ_1	λ_2	λ_3	λ_4	λ_5
a_1	$0.9e^{\iota(0.9)2\pi}$	$0.9e^{\iota(0.9)2\pi}$	0	$0.9e^{\iota(0.9)2\pi}$	0
a_2	$0.9e^{\iota(0.9)2\pi}$	0	0	0	0
a_3	0	$0.9e^{\iota(0.9)2\pi}$	0	0	0
a_4	$0.9e^{\iota(0.9)2\pi}$	$0.9e^{\iota(0.9)2\pi}$	$0.9e^{\iota(0.9)2\pi}$	0	0
a_5	0	0	$0.9e^{\iota(0.9)2\pi}$	$0.9e^{\iota(0.9)2\pi}$	$0.4e^{\iota(0.4)2\pi}$
a_6	0	0	$0.9e^{\iota(0.9)2\pi}$	0	0
a_7	0	0	0	0	$0.4e^{\iota(0.4)2\pi}$

Note that

$$\{a_1, a_4\} \in T_r(\mathscr{H}^{(0.9e^{\iota(0.9)2\pi}, 0.9e^{\iota(0.9)2\pi}, 0.9e^{\iota(0.9)2\pi})}), \{a_1, a_4\} \notin T_r(\mathscr{H}^{(0.4e^{\iota(0.4)2\pi}, 0.4e^{\iota(0.4)2\pi}, 0.4e^{\iota(0.4)2\pi})}),$$

i.e., $\{a_1, a_4, a_5\}$ is a transversal of $\mathscr{H}^{(0.4e^{\iota(0.4)2\pi}, 0.4e^{\iota(0.4)2\pi}, 0.4e^{\iota(0.4)2\pi})}$ but not a minimal transversal. Therefore, the ordered pair $(\mathscr{H}^{(0.9e^{\iota(0.9)2\pi}, 0.9e^{\iota(0.9)2\pi}, 0.9e^{\iota(0.9)2\pi})},$ $\mathscr{H}^{(0.4e^{\iota(0.4)2\pi}, 0.4e^{\iota(0.4)2\pi}, 0.4e^{\iota(0.4)2\pi})})$ as well as $\mathscr{H}$ is not *T*-related.

Remark 9.7 • Example 9.28 shows that there exists some ordered complex neutrosophic hypergraphs that are not *T*-related.
• Every simply ordered complex neutrosophic hypergraph $\mathscr{H} = (\mathscr{N}, \lambda)$ satisfies $(T_r^*(\mathscr{H})^{(\alpha e^{\iota\Theta}, \beta e^{\iota\Phi}, \gamma e^{\iota\Psi})} = T_r(\mathscr{H}^{(\alpha e^{\iota\Theta}, \beta e^{\iota\Phi}, \gamma e^{\iota\Psi})})$, for all $\alpha \in (0, t(h(\mathscr{H}))]$, $\beta \in (0, i(h(\mathscr{H}))]$, $\gamma \in (0, f(h(\mathscr{H}))]$, $\Theta \in (0, \phi(h(\mathscr{H}))]$, $\Phi \in (0, \varphi(h(\mathscr{H}))]$, $\Psi \in (0, \psi(h(\mathscr{H}))]$.

Lemma 9.2 *Let $H = (X, E)$ be a crisp hypergraph and j be an arbitrary vertex of H. Then, $j \in \mathscr{E} \in T_r(H) \Leftrightarrow j \in E_k \in E$ such that for any hyperedge $E_l \neq E_k$ of H, $E_l \nsubseteq E_k$.*

Proposition 9.12 *Let $H_1 = (X_1, E_1)$ be a crisp partial hypergraph of $H = (X, E)$ that is obtained by removing those hyperedges of $H = (X, E)$ that contain any other edges properly. Then,*

(i) $T_r(H_1) = T_r(H)$,
(ii) $\cup T_r(H) = X_1$.

Definition 9.89 Let $\mathscr{H} = (\mathscr{N}, \lambda)$ be a complex neutrosophic hypergraph. The *join* of $\mathscr{H}$, denoted by $J(\mathscr{H})$, is defined as $J(\mathscr{H}) = \bigcup_{\rho \in \lambda} \rho$, where λ is the complex neutrosophic hyperedge set of $\mathscr{H}$.

For every $\alpha \in (0, t(h(\mathscr{H}))]$, $\beta \in (0, i(h(\mathscr{H}))]$, $\gamma \in (0, f(h(\mathscr{H}))]$, $\Theta \in (0, \phi(h(\mathscr{H}))]$, $\Phi \in (0, \varphi(h(\mathscr{H}))]$, $\Psi \in (0, \psi(h(\mathscr{H}))]$, the $(\alpha e^{\iota\Theta}, \beta e^{\iota\Phi}, \gamma e^{\iota\Psi})$-level cut of $J(\mathscr{H})$, i.e., $(J(\mathscr{H}))^{(\alpha e^{\iota\Theta}, \beta e^{\iota\Phi}, \gamma e^{\iota\Psi})}$ is the set of vertices of $(\alpha e^{\iota\Theta}, \beta e^{\iota\Phi}, \gamma e^{\iota\Psi})$-level hypergraph of $\mathscr{H}$, i.e., $(J(\mathscr{H}))^{(\alpha e^{\iota\Theta}, \beta e^{\iota\Phi}, \gamma e^{\iota\Psi})} = X(\mathscr{H}^{(\alpha e^{\iota\Theta}, \beta e^{\iota\Phi}, \gamma e^{\iota\Psi})})$.

Lemma 9.3 *Let $\mathscr{H} = (\mathscr{N}, \lambda)$ be a complex neutrosophic hypergraph and $\xi \in T_r(\mathscr{H})$. If $j \in supp(\xi)$, then there exists a complex neutrosophic hyperedge ρ of $\mathscr{H}$ such that*

(i) $\rho(j) = h(\rho) = \xi(j) > 0$,
(ii) $\xi^{h(\rho)} \cap \rho^{h(\rho)} = \{j\}$.

Proof Let $j_0 \in supp(\xi)$ such that $\xi \in T_r(\mathscr{H})$ and $\xi(j_0) = (\alpha_0 e^{i\phi_0}, \beta_0 e^{i\varphi_0}, \gamma_0 e^{i\psi_0})$. Since every ξ_1 that is a transversal of $\mathscr{H}$ contains a transversal ξ such that $\xi \subseteq j(\mathscr{H})$. This implies that $j_0 \in \mathscr{N}^{(\alpha_0 e^{i\phi_0}, \beta_0 e^{i\varphi_0}, \gamma_0 e^{i\psi_0})} = X(\mathscr{H}^{(\alpha_0 e^{i\phi_0}, \beta_0 e^{i\varphi_0}, \gamma_0 e^{i\psi_0})})$. Therefore, there exists at least one hyperedge ρ of $\mathscr{H}$ such that $\rho(j_0) \geq (\alpha_0 e^{i\phi_0}, \beta_0 e^{i\varphi_0}, \gamma_0 e^{i\psi_0})$. Let $\lambda = \{\lambda_1, \lambda_2, \ldots, \lambda_m\}$ be the set of hyperedges of $\mathscr{H}$ and $\rho(j_0) \geq (\alpha_0 e^{i\phi_0}, \beta_0 e^{i\varphi_0}, \gamma_0 e^{i\psi_0})$. We now prove that there exists at least one $\lambda_k \in \lambda$ such that $h(\lambda_j) = (\alpha_0 e^{i\phi_0}, \beta_0 e^{i\varphi_0}, \gamma_0 e^{i\psi_0})$. For otherwise, we have $h(\lambda_k) = (\alpha_k e^{i\phi_k}, \beta_k e^{i\varphi_k}, \gamma_k e^{i\psi_k}) \geq (\alpha_0 e^{i\phi_0}, \beta_0 e^{i\varphi_0}, \gamma_0 e^{i\psi_0})$, for all $\lambda_k \in \lambda$, $k = 1, 2, \ldots, m$. This implies that for every $\lambda_k \in \lambda$, there exists an element $u_k \in supp(\xi)$ such that $u_k \in (\lambda_k)^{(\alpha_k e^{i\phi_k}, \beta_k e^{i\varphi_k}, \gamma_k e^{i\psi_k})} \cap \xi^{(\alpha_k e^{i\phi_k}, \beta_k e^{i\varphi_k}, \gamma_k e^{i\psi_k})}$, for $(\alpha_k e^{i\phi_k}, \beta_k e^{i\varphi_k}, \gamma_k e^{i\psi_k}) \geq (\alpha_0 e^{i\phi_0}, \beta_0 e^{i\varphi_0}, \gamma_0 e^{i\psi_0})$. Since, $\xi(j_0) = (\alpha_0 e^{i\phi_0}, \beta_0 e^{i\varphi_0}, \gamma_0 e^{i\psi_0})$, then $h(\lambda_k) = (\alpha_k e^{i\phi_k}, \beta_k e^{i\varphi_k}, \gamma_k e^{i\psi_k}) \geq (\alpha_0 e^{i\phi_0}, \beta_0 e^{i\varphi_0}, \gamma_0 e^{i\psi_0})$ and $u_k \in (\lambda_k)^{(\alpha_k e^{i\phi_k}, \beta_k e^{i\varphi_k}, \gamma_k e^{i\psi_k})} \cap \xi^{(\alpha_k e^{i\phi_k}, \beta_k e^{i\varphi_k}, \gamma_k e^{i\psi_k})}$ imply that $u_k \neq j_0, k = 1, 2, \ldots, m$. If these hold, it could be shown that $\xi \notin T_r(\mathscr{H})$ by computing a complex neutrosophic transversal δ of $\mathscr{H}$ that satisfies $\delta \subset \xi$. This argument follows form the fact that X and λ are finite, there exist intervals $(\alpha_0 - \varepsilon, \alpha_0]$, $(\beta_0 - \varepsilon, \beta_0]$, $(\gamma_0 - \varepsilon, \gamma_0]$, $(\phi_0 - 2\pi\varepsilon, \phi_0]$, $(\varphi_0 - 2\pi\varepsilon, \varphi_0]$, and $(\psi_0 - 2\pi\varepsilon, \psi_0]$ such that $\mathscr{H}^{(\alpha e^{i\phi}, \beta e^{i\varphi}, \gamma e^{i\psi})} = \mathscr{H}^{(\alpha_0 e^{i\phi_0}, \beta_0 e^{i\varphi_0}, \gamma_0 e^{i\psi_0})}$ on $(\alpha_0 - \varepsilon, \alpha_0]$, $(\beta_0 - \varepsilon, \beta_0]$, $(\gamma_0 - \varepsilon, \gamma_0]$, $(\phi_0 - 2\pi\varepsilon, \phi_0]$, $(\varphi_0 - 2\pi\varepsilon, \varphi_0]$, and $(\psi_0 - 2\pi\varepsilon, \psi_0]$. Define $\delta(u)$ as

$$T_\delta(u) = \begin{cases} T_\xi(u), & \text{if } u \neq j_0, \\ \alpha_0 - \varepsilon, & \text{if } u = j_0. \end{cases}, \quad I_\delta(u) = \begin{cases} I_\xi(u), & \text{if } u \neq j_0, \\ \beta_0 - \varepsilon, & \text{if } u = j_0. \end{cases},$$

$$F_\delta(u) = \begin{cases} F_\xi(u), & \text{if } u \neq j_0, \\ \gamma_0 - \varepsilon, & \text{if } u = j_0. \end{cases}, \quad \phi_\delta(u) = \begin{cases} \phi_\xi(u), & \text{if } u \neq j_0, \\ \phi_0 - 2\pi\varepsilon, & \text{if } u = j_0. \end{cases},$$

$$\varphi_\delta(u) = \begin{cases} \varphi_\xi(u), & \text{if } u \neq j_0, \\ \varphi_0 - 2\pi\varepsilon, & \text{if } u = j_0. \end{cases}, \quad \psi_\delta(u) = \begin{cases} \psi_\xi(u), & \text{if } u \neq j_0, \\ \psi_0 - 2\pi\varepsilon, & \text{if } u = j_0. \end{cases}.$$

Clearly $\delta \subset \xi$ and δ is a transversal of $\mathscr{H}$. Also, $\xi^{(\alpha_0 e^{i\phi_0}, \beta_0 e^{i\varphi_0}, \gamma_0 e^{i\psi_0})} \setminus \{j_0\}$ contains $\{u_k | k = 1, 2, \ldots, m\}$. Therefore, $\xi^{(\alpha_0 e^{i\phi_0}, \beta_0 e^{i\varphi_0}, \gamma_0 e^{i\psi_0})} \setminus \{j_0\}$ is a transversal of $\mathscr{H}^{(\alpha_0 e^{i\phi_0}, \beta_0 e^{i\varphi_0}, \gamma_0 e^{i\psi_0})}$. The same argument holds for every $\mathscr{H}^{(\alpha e^{i\phi}, \beta e^{i\varphi}, \gamma e^{i\psi})}$, where $\alpha \in (\alpha_0 - \varepsilon, \alpha_0]$, $\beta \in (\beta_0 - \varepsilon, \beta_0]$, $\gamma \in (\gamma_0 - \varepsilon, \gamma_0]$, $\phi \in (\phi_0 - 2\pi\varepsilon, \phi_0]$, $\varphi \in (\varphi_0 - 2\pi\varepsilon, \varphi_0]$, $\psi \in (\psi_0 - 2\pi\varepsilon, \psi_0]$. Since $\delta^{(\alpha e^{i\phi}, \beta e^{i\varphi}, \gamma e^{i\psi})} = \xi^{(\alpha e^{i\phi}, \beta e^{i\varphi}, \gamma e^{i\psi})}$, for all $\alpha \in (0, t(h(\mathscr{H}))] \setminus (\alpha_0 - \varepsilon, \alpha_0]$, $\beta \in (0, i(h(\mathscr{H}))] \setminus (\beta_0 - \varepsilon, \beta_0]$, $\gamma \in (0, f(h(\mathscr{H}))] \setminus (\gamma_0 - \varepsilon, \gamma_0]$, $\phi \in (0, \phi(h(\mathscr{H}))] \setminus (\phi_0 - 2\pi\varepsilon, \phi_0]$, $\varphi \in (0, \varphi(h(\mathscr{H}))] \setminus (\varphi_0$

Table 9.25 Incidence matrix of $\mathcal{H}$

$I_{\mathcal{H}}$	λ_1	λ_2	λ_3	λ_4	λ_5
u_1	$\mathbf{0.7e^{\iota(0.7)2\pi}}$	$\mathbf{0.9e^{\iota(0.9)2\pi}}$	$(0,0,1)$	$(0,0,1)$	$\mathbf{0.4e^{\iota(0.4)2\pi}}$
u_2	$\mathbf{0.7e^{\iota(0.7)2\pi}}$	$\mathbf{0.9e^{\iota(0.9)2\pi}}$	$\mathbf{0.9e^{\iota(0.9)2\pi}}$	$\mathbf{0.7e^{\iota(0.7)2\pi}}$	$(0,0,1)$
u_3	$(0,0,1)$	$(0,0,1)$	$\mathbf{0.9e^{\iota(0.9)2\pi}}$	$\mathbf{0.7e^{\iota(0.7)2\pi}}$	$\mathbf{0.4e^{\iota(0.4)2\pi}}$
u_4	$(0,0,1)$	$\mathbf{0.4e^{\iota(0.4)2\pi}}$	$(0,0,1)$	$\mathbf{0.4e^{\iota(0.4)2\pi}}$	$\mathbf{0.4e^{\iota(0.4)2\pi}}$

$-2\pi\varepsilon, \varphi_0], \psi \in (0, \psi(h(\mathcal{H}))] \setminus (\psi_0 - 2\pi\varepsilon, \psi_0]$. This establishes the existence of $\rho \in \mathcal{H}$ for which $\rho(j_0) = h(\rho) = \xi(j_0) > 0$.

We now suppose that every hyperedge from the set $\lambda = \{\lambda_1, \lambda_2, \ldots, \lambda_m\}$ having height $\xi(j_0)$ contain two or more than two elements of $\xi^{(\alpha_0 e^{\iota\phi_0}, \beta_0 e^{\iota\varphi_0}, \gamma_0 e^{\iota\psi_0})} \setminus \{j_0\}$. By repeating the above procedure, we can establish that $\xi \notin T_r(\mathcal{H})$, which is a contradiction.

Example 9.29 Consider a complex neutrosophic hypergraph $\mathcal{H} = (\mathcal{N}, \lambda)$ on $X = \{u_1, u_2, u_3, u_4\}$ as represented by incidence matrix given in Table 9.25.

Here, $\mathbf{0.7e^{\iota(0.7)2\pi}} = (0.7e^{\iota(0.7)2\pi}, 0.7e^{\iota(0.7)2\pi}, 0.7e^{\iota(0.7)2\pi})$, $\mathbf{0.9e^{\iota(0.9)2\pi}} = (0.9e^{\iota(0.9)2\pi}, 0.9e^{\iota(0.9)2\pi}, 0.9e^{\iota(0.9)2\pi})$, $\mathbf{0.4e^{\iota(0.4)2\pi}} = (0.4e^{\iota(0.4)2\pi}, 0.4e^{\iota(0.4)2\pi}, 0.4e^{\iota(0.4)2\pi})$. Then, we see that λ_1, λ_3, and λ_5 have no transitions levels and $(0.4e^{\iota(0.4)2\pi}, 0.4e^{\iota(0.4)2\pi}, 0.4e^{\iota(0.4)2\pi})$ is the transition level of λ_2 and λ_4. The basic sequences are given as

$$B_s(\lambda_1) = \{(0.7e^{\iota(0.7)2\pi}, 0.7e^{\iota(0.7)2\pi}, 0.7e^{\iota(0.7)2\pi})\},$$
$$B_s(\lambda_2) = \{(0.9e^{\iota(0.9)2\pi}, 0.9e^{\iota(0.9)2\pi}, 0.9e^{\iota(0.9)2\pi}), (0.4e^{\iota(0.4)2\pi}, 0.4e^{\iota(0.4)2\pi}, 0.4e^{\iota(0.4)2\pi})\},$$
$$B_s(\lambda_3) = \{(0.9e^{\iota(0.9)2\pi}, 0.9e^{\iota(0.9)2\pi}, 0.9e^{\iota(0.9)2\pi})\},$$
$$B_s(\lambda_4) = \{(0.7e^{\iota(0.7)2\pi}, 0.7e^{\iota(0.7)2\pi}, 0.7e^{\iota(0.7)2\pi}), (0.4e^{\iota(0.4)2\pi}, 0.4e^{\iota(0.4)2\pi}, 0.4e^{\iota(0.4)2\pi})\},$$
$$B_s(\lambda_5) = \{(0.4e^{\iota(0.4)2\pi}, 0.4e^{\iota(0.4)2\pi}, 0.4e^{\iota(0.4)2\pi})\}.$$

Thus,

$$B_c(\lambda_1) = \{\lambda_1^{(0.7e^{\iota(0.7)2\pi}, 0.7e^{\iota(0.7)2\pi}, 0.7e^{\iota(0.7)2\pi})}\},$$
$$B_c(\lambda_2) = \{\lambda_2^{(0.9e^{\iota(0.9)2\pi}, 0.9e^{\iota(0.9)2\pi}, 0.9e^{\iota(0.9)2\pi})}, \lambda_2^{(0.4e^{\iota(0.4)2\pi}, 0.4e^{\iota(0.4)2\pi}, 0.4e^{\iota(0.4)2\pi})}\},$$
$$B_c(\lambda_3) = \{\lambda_3^{(0.9e^{\iota(0.9)2\pi}, 0.9e^{\iota(0.9)2\pi}, 0.9e^{\iota(0.9)2\pi})}\},$$
$$B_c(\lambda_4) = \{\lambda_4^{(0.7e^{\iota(0.7)2\pi}, 0.7e^{\iota(0.7)2\pi}, 0.7e^{\iota(0.7)2\pi})}, \lambda_4^{(0.4e^{\iota(0.4)2\pi}, 0.4e^{\iota(0.4)2\pi}, 0.4e^{\iota(0.4)2\pi})}\},$$
$$B_c(\lambda_5) = \{\lambda_5^{(0.4e^{\iota(0.4)2\pi}, 0.4e^{\iota(0.4)2\pi}, 0.4e^{\iota(0.4)2\pi})}\}.$$

Also, we have $\mathscr{F}_s(\mathcal{H}) = \{(0.9e^{\iota(0.9)2\pi}, 0.9e^{\iota(0.9)2\pi}, 0.9e^{\iota(0.9)2\pi}), (0.4e^{\iota(0.4)2\pi}, 0.4e^{\iota(0.4)2\pi}, 0.4e^{\iota(0.4)2\pi})\}$ and $c(\mathcal{H}) = \{\mathcal{H}^{(0.9e^{\iota(0.9)2\pi}, 0.9e^{\iota(0.9)2\pi}, 0.9e^{\iota(0.9)2\pi})}, \mathcal{H}^{(0.4e^{\iota(0.4)2\pi}, 0.4e^{\iota(0.4)2\pi}, 0.4e^{\iota(0.4)2\pi})}\}$, where

$$\lambda^{(0.9e^{\iota(0.9)2\pi},0.9e^{\iota(0.9)2\pi},0.9e^{\iota(0.9)2\pi})} = \{\{u_1,u_2,u_3\},\{u_1,u_2\},\{u_2,u_3\}\},$$

$$\lambda^{(0.4e^{\iota(0.4)2\pi},0.4e^{\iota(0.4)2\pi},0.4e^{\iota(0.4)2\pi})} = \{\{u_1,u_2,u_3,u_4\},\{u_1,u_2\},\{u_1,u_2,u_4\},\{u_2,u_3\},\{u_2,u_3,u_4\}\}.$$

We now determine $T_r(\mathscr{H})$ and $T_r^*(\mathscr{H})$. If $\tau \in T_r(\mathscr{H})$, then $\tau^{h(\lambda_1)} \cap \{u_1,u_2\} \neq \emptyset$, $\tau^{h(\lambda_2)} \cap \{u_1,u_2\} \neq \emptyset$, $\tau^{h(\lambda_3)} \cap \{u_2,u_3\} \neq \emptyset$, $\tau^{h(\lambda_4)} \cap \{u_2,u_3\} \neq \emptyset$, and $\tau^{h(\lambda_5)} \cap \{u_1,u_3,u_4\} \neq \emptyset$. Note that $T_r(\mathscr{H}) = \{\tau_1,\tau_2,\tau_3,\tau_4\}$, where

$$\tau_1 = \{(u_1,0.9e^{\iota(0.9)2\pi},0.9e^{\iota(0.9)2\pi},0.9e^{\iota(0.9)2\pi}),(u_3,0.9e^{\iota(0.9)2\pi},0.9e^{\iota(0.9)2\pi},0.9e^{\iota(0.9)2\pi})\},$$

$$\tau_2 = \{(u_2,0.9e^{\iota(0.9)2\pi},0.9e^{\iota(0.9)2\pi},0.9e^{\iota(0.9)2\pi}),(u_3,0.4e^{\iota(0.4)2\pi},0.4e^{\iota(0.4)2\pi},0.4e^{\iota(0.4)2\pi}\},$$

$$\tau_3 = \{(u_2,0.9e^{\iota(0.9)2\pi},0.9e^{\iota(0.9)2\pi},0.9e^{\iota(0.9)2\pi}),(u_4,0.4e^{\iota(0.4)2\pi},0.4e^{\iota(0.4)2\pi},0.4e^{\iota(0.4)2\pi}\},$$

$$\tau_4 = \{(u_2,0.9e^{\iota(0.9)2\pi},0.9e^{\iota(0.9)2\pi},0.9e^{\iota(0.9)2\pi}),(u_1,0.4e^{\iota(0.4)2\pi},0.4e^{\iota(0.4)2\pi},0.4e^{\iota(0.4)2\pi}\}.$$

Now $T_r(\mathscr{H}^{(0.9e^{\iota(0.9)2\pi},0.9e^{\iota(0.9)2\pi},0.9e^{\iota(0.9)2\pi})}) = \{\{u_2\},\{u_1,u_3\}\}$ and $T_r(\mathscr{H}^{(0.4e^{\iota(0.4)2\pi},0.4e^{\iota(0.4)2\pi},0.4e^{\iota(0.4)2\pi})}) = \{\{u_1,u_3\},\{u_2,u_3\},\{u_2,u_4\},\{u_1,u_2\},\{u_1,u_3,u_4\}\}$ and $\tau_k^{(\alpha e^{\iota\Theta},\beta e^{\iota\Phi},\gamma e^{\iota\Psi})} \in T_r(\mathscr{H}^{(\alpha e^{\iota\Theta},\beta e^{\iota\Phi},\gamma e^{\iota\Psi})})$, for all $\alpha \in (0,t(h(\mathscr{H}))]$, $\beta \in (0,i(h(\mathscr{H}))]$, $\gamma \in (0,f(h(\mathscr{H}))]$, $\Theta \in (0,\phi(h(\mathscr{H}))]$, $\Phi \in (0,\varphi(h(\mathscr{H}))]$, $\Psi \in (0,\psi(h(\mathscr{H}))]$. Hence, $T_r^*(\mathscr{H}) = \{\tau_1\}$.

We now illustrate Lemma 9.3 through the above example.

$$\lambda_2(u_1) = h(\lambda_2) = \tau_1(u_1) = (0.9e^{\iota(0.9)2\pi},0.9e^{\iota(0.9)2\pi},0.9e^{\iota(0.9)2\pi}),$$

$$\lambda_3(u_3) = h(\lambda_3) = \tau_1(u_3) = (0.9e^{\iota(0.9)2\pi},0.9e^{\iota(0.9)2\pi},0.9e^{\iota(0.9)2\pi}),$$

$$\lambda_2(u_2) = h(\lambda_2) = \tau_2(u_2) = (0.9e^{\iota(0.9)2\pi},0.9e^{\iota(0.9)2\pi},0.9e^{\iota(0.9)2\pi}),$$

$$\lambda_5(u_3) = h(\lambda_5) = \tau_2(u_3) = (0.4e^{\iota(0.4)2\pi},0.4e^{\iota(0.4)2\pi},0.4e^{\iota(0.4)2\pi}),$$

$$\lambda_3(u_2) = h(\lambda_3) = \tau_3(u_2) = (0.9e^{\iota(0.9)2\pi},0.9e^{\iota(0.9)2\pi},0.9e^{\iota(0.9)2\pi}),$$

$$\lambda_5(u_4) = h(\lambda_5) = \tau_3(u_4) = (0.4e^{\iota(0.4)2\pi},0.4e^{\iota(0.4)2\pi},0.4e^{\iota(0.4)2\pi}),$$

$$\lambda_5(u_1) = h(\lambda_5) = \tau_4(u_2) = (0.4e^{\iota(0.4)2\pi},0.4e^{\iota(0.4)2\pi},0.4e^{\iota(0.4)2\pi}),$$

$$\lambda_3(u_2) = h(\lambda_3) = \tau_4(u_2) = (0.9e^{\iota(0.9)2\pi},0.9e^{\iota(0.9)2\pi},0.9e^{\iota(0.9)2\pi}).$$

Also note that

$$\tau_1^{(0.9e^{\iota(0.9)2\pi},0.9e^{\iota(0.9)2\pi},0.9e^{\iota(0.9)2\pi})} \cap \lambda_2^{(0.9e^{\iota(0.9)2\pi},0.9e^{\iota(0.9)2\pi},0.9e^{\iota(0.9)2\pi})} = \{u_1\},$$

$$\tau_1^{(0.9e^{\iota(0.9)2\pi},0.9e^{\iota(0.9)2\pi},0.9e^{\iota(0.9)2\pi})} \cap \lambda_3^{(0.9e^{\iota(0.9)2\pi},0.9e^{\iota(0.9)2\pi},0.9e^{\iota(0.9)2\pi})} = \{u_3\},$$

$$\tau_2^{(0.9e^{\iota(0.9)2\pi},0.9e^{\iota(0.9)2\pi},0.9e^{\iota(0.9)2\pi})} \cap \lambda_2^{(0.9e^{\iota(0.9)2\pi},0.9e^{\iota(0.9)2\pi},0.9e^{\iota(0.9)2\pi})} = \{u_2\},$$

$$\tau_2^{(0.4e^{\iota(0.4)2\pi},0.4e^{\iota(0.4)2\pi},0.4e^{\iota(0.4)2\pi})} \cap \lambda_5^{(0.4e^{\iota(0.4)2\pi},0.4e^{\iota(0.4)2\pi},0.4e^{\iota(0.4)2\pi})} = \{u_3\},$$

$$\tau_3^{(0.9e^{\iota(0.9)2\pi},0.9e^{\iota(0.9)2\pi},0.9e^{\iota(0.9)2\pi})} \cap \lambda_3^{(0.9e^{\iota(0.9)2\pi},0.9e^{\iota(0.9)2\pi},0.9e^{\iota(0.9)2\pi})} = \{u_2\},$$

$$\tau_3^{(0.4e^{\iota(0.4)2\pi},0.4e^{\iota(0.4)2\pi},0.4e^{\iota(0.4)2\pi})} \cap \lambda_5^{(0.4e^{\iota(0.4)2\pi},0.4e^{\iota(0.4)2\pi},0.4e^{\iota(0.4)2\pi})} = \{u_4\},$$

$$\tau_4^{(0.4e^{\iota(0.4)2\pi},0.4e^{\iota(0.4)2\pi},0.4e^{\iota(0.4)2\pi})} \cap \lambda_5^{(0.4e^{\iota(0.4)2\pi},0.4e^{\iota(0.4)2\pi},0.4e^{\iota(0.4)2\pi})} = \{u_1\},$$

$$\tau_4^{(0.9e^{\iota(0.9)2\pi},0.9e^{\iota(0.9)2\pi},0.9e^{\iota(0.9)2\pi})} \cap \lambda_3^{(0.9e^{\iota(0.9)2\pi},0.9e^{\iota(0.9)2\pi},0.9e^{\iota(0.9)2\pi})} = \{u_2\}.$$

Hence, $(T_r(\mathscr{H}))^{(0.9e^{i(0.9)2\pi},0.9e^{i(0.9)2\pi},0.9e^{i(0.9)2\pi})} = \{\tau_1^{(0.9e^{i(0.9)2\pi},0.9e^{i(0.9)2\pi},0.9e^{i(0.9)2\pi})},$
$\tau_2^{(0.9e^{i(0.9)2\pi},0.9e^{i(0.9)2\pi},0.9e^{i(0.9)2\pi})}, \quad \tau_3^{(0.9e^{i(0.9)2\pi},0.9e^{i(0.9)2\pi},0.9e^{i(0.9)2\pi})}, \quad \tau_4^{(0.9e^{i(0.9)2\pi},0.9e^{i(0.9)2\pi},0.9e^{i(0.9)2\pi})}\}$

$= \{\{u_1,u_3\},\{u_2\},\{u_2\},\{u_2\}\} = \{\{u_1,u_3\},\{u_2\}\} = T_r(\mathscr{H}^{(0.9e^{i(0.9)2\pi},0.9e^{i(0.9)2\pi},0.9e^{i(0.9)2\pi})}))$.

Theorem 9.17 *Let $\mathscr{H} = (\mathscr{N}, \lambda)$ be a complex neutrosophic hypergraph and $j \in X$. If $\xi \in T_r(\mathscr{H})$ with $j \in supp(\xi)$, then there exists an hyperedge $\rho \in \lambda$ such that*

(i) $\rho(j) = h(\rho)$,
(ii) For $\lambda_1 \in \lambda$ such that $h(\lambda_1) \geq h(\rho)$, $\lambda_1^{h(\lambda_1)} \not\subseteq \rho^{h(\rho)}$,
(iii) $\mathscr{E}_k \not\subseteq \rho^{h(\rho)}$, where $\mathscr{E}_k$ is an arbitrary hyperedge of $\mathscr{H}^{h(\rho)}$,
(iv) $\xi(j) = \rho(j)$.

Corollary 9.3 *Let $\mathscr{H} = (\mathscr{N}, \lambda)$ be a complex neutrosophic hypergraph. If $\lambda_1 \in \lambda$ satisfies $h(\lambda_1) \geq h(\rho)$, $\lambda_1^{h(\lambda_1)} \not\subseteq \rho^{h(\rho)}$, then $h(\lambda_1) \in \mathscr{F}_s(\mathscr{H})$.*

9.10   Applications of Complex Neutrosophic Hypergraphs

In this section, we propose the modeling of overlapping communities that exist in different social networks through complex neutrosophic hypergraphs. These communities intersect each other when one person belongs to multiple communities at the same time. The vertices of the complex neutrosophic hypergraphs are used to represent different communities and the hyperlinks of individuals who participate in more than one community are illustrated using hyperedges of complex neutrosophic hypergraphs. Here, we define a score function for ranking complex neutrosophic sets by considering the truth, indeterminacy, and falsity degrees.

Definition 9.90 Let $N = (Te^{i\phi}, Ie^{i\varphi}, Fe^{i\psi})$ be a complex neutrosophic number, the score function S of N is defined as

$$S(N) = \frac{1 + T - 2I - F}{2} + \frac{2\pi + \phi - 2\varphi - \psi}{4\pi},$$

where $S(N) \in [-2, 2]$.

9.10.1   Modeling of Intersecting Research Communities

Research scholars have different fields of interest and these multiple research interests make researchers parts of different research communities at the same time. For example, Mathematics, Physics, and Computer Science may be the fields of interest for one researcher at the same time. That is how overlapping communities occur in research fields. We use a complex neutrosophic hypergraph to model intersecting communities that emerge in different research fields. The vertices of a complex

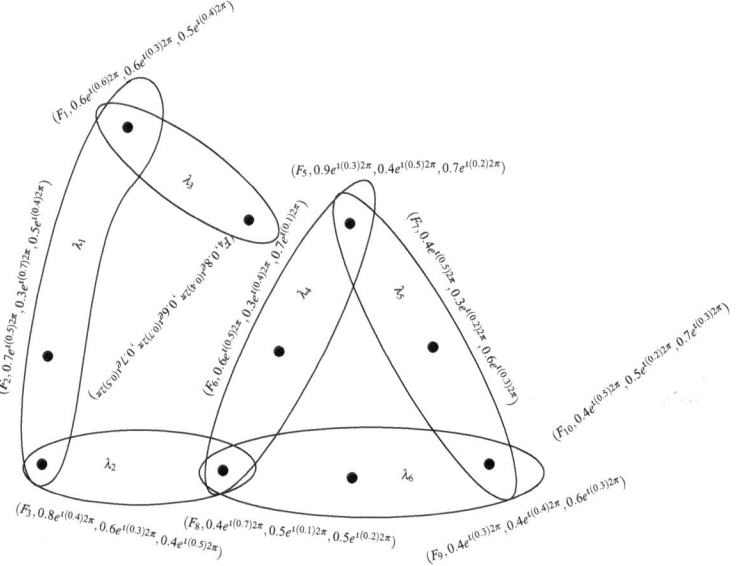

Fig. 9.28 Intersecting research communities

Table 9.26 Periodic behavior of research communities

Research fields	Accepted articles	Submitted articles	Rejected articles
F_1	$0.6e^{\iota(0.6)2\pi}$	$0.6e^{\iota(0.3)2\pi}$	$0.5e^{\iota(0.4)2\pi}$
F_2	$0.7e^{\iota(0.5)2\pi}$	$0.3e^{\iota(0.7)2\pi}$	$0.5e^{\iota(0.4)2\pi}$
F_3	$0.8e^{\iota(0.4)2\pi}$	$0.6e^{\iota(0.3)2\pi}$	$0.4e^{\iota(0.5)2\pi}$
F_4	$0.8e^{\iota(0.4)2\pi}$	$0.6e^{\iota(0.7)2\pi}$	$0.7e^{\iota(0.5)2\pi}$
F_5	$0.9e^{\iota(0.3)2\pi}$	$0.4e^{\iota(0.5)2\pi}$	$0.7e^{\iota(0.2)2\pi}$
F_6	$0.6e^{\iota(0.5)2\pi}$	$0.3e^{\iota(0.4)2\pi}$	$0.7e^{\iota(0.1)2\pi}$
F_7	$0.4e^{\iota(0.5)2\pi}$	$0.3e^{\iota(0.2)2\pi}$	$0.6e^{\iota(0.3)2\pi}$
F_8	$0.4e^{\iota(0.7)2\pi}$	$0.5e^{\iota(0.1)2\pi}$	$0.5e^{\iota(0.2)2\pi}$
F_9	$0.4e^{\iota(0.3)2\pi}$	$0.4e^{\iota(0.4)2\pi}$	$0.6e^{\iota(0.3)2\pi}$
F_{10}	$0.4e^{\iota(0.5)2\pi}$	$0.5e^{\iota(0.2)2\pi}$	$0.7e^{\iota(0.3)2\pi}$

neutrosophic hypergraph represent the different research fields and these fields are connected through an hyperedge that represents a research scholar who works in the corresponding fields. The corresponding model of intersecting research communities is shown in Fig. 9.28.

Here, the truth, indeterminacy, and falsity degrees of each vertex represent the accepted, submitted, and rejected articles of that community in a specific period of time that is represented by the phase terms. This inconsistent information having periodic nature is given in Table 9.26.

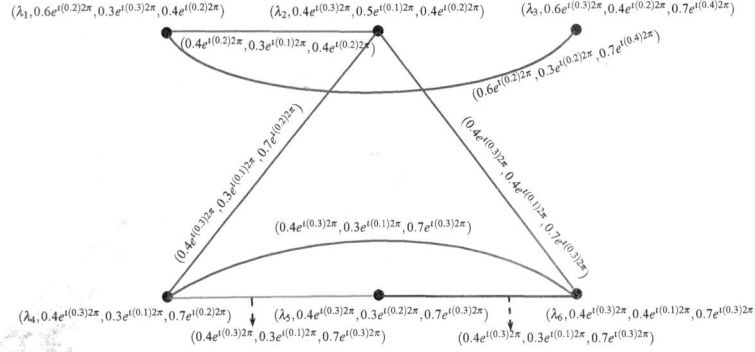

Fig. 9.29 Line graph of intersecting research communities

Note that number of accepted, submitted, and rejected articles of community F_1 are $0.6, 0.6$, and 0.5, and the corresponding behaviors repeat after $(0.6)2\pi$, $(0.3)2\pi$, and $(0.4)2\pi$ periods of time, respectively, and so on. The research scholar λ_1 belongs to communities F_1, F_2, and F_3 as he shares these three fields of interest. Similarly, λ_2 belongs to F_3 and F_8 and the communities overlap with each other. The indeterminate information about a researcher is calculated using complex neutrosophic relations given as

$$\lambda_1(\{F_1, F_2, F_3\}) = (0.6e^{i(0.2)2\pi}, 0.3e^{i(0.3)2\pi}, 0.4e^{i(0.2)2\pi}),$$
$$\lambda_2(\{F_3, F_8\}) = (0.4e^{i(0.3)2\pi}, 0.5e^{i(0.1)2\pi}, 0.4e^{i(0.2)2\pi}),$$
$$\lambda_3(\{F_1, F_4\}) = (0.6e^{i(0.3)2\pi}, 0.4e^{i(0.2)2\pi}, 0.7e^{i(0.4)2\pi}),$$
$$\lambda_4(\{F_5, F_8, F_6\}) = (0.4e^{i(0.3)2\pi}, 0.3e^{i(0.1)2\pi}, 0.7e^{i(0.2)2\pi}),$$
$$\lambda_5(\{F_5, F_7, F_{10}\}) = (0.4e^{i(0.3)2\pi}, 0.3e^{i(0.2)2\pi}, 0.7e^{i(0.3)2\pi}),$$
$$\lambda_6(\{F_8, F_9, F_{10}\}) = (0.4e^{i(0.3)2\pi}, 0.4e^{i(0.1)2\pi}, 0.7e^{i(0.3)2\pi}).$$

It shows the researcher represented by λ_1 has 0.6 accepted, 0.3 submitted, and 0.4 rejected articles within some specific periods of time. The line graph of intersecting communities as given in Fig. 9.28 is shown in Fig. 9.29. Here, the nodes represent the individuals and the communities are described by the links of same color.

This line graph represents the relationships between researchers. The researchers that belong to the community F_3 are connected through pink edge, members of F_1 are linked by red edge, members of F_{10} are connected by purple links, cyan and blue edges are used to represent the relation between the members of F_5 and F_8, respectively. The absence of F_2, F_4, F_6, F_7, and F_9 in the above graph shows that these communities share no common researchers as their members. The membership degrees of each edge of this line graph represent the collective work of corresponding researchers. The score functions and choice values of a complex neutrosophic graph are given as

Table 9.27 Score and choice values of complex neutrosophic line graph

S_{jk}	λ_1	λ_2	λ_3	λ_4	λ_5	λ_6	C_j
λ_1	0	0.600	0.350	0	0	0	0.450
λ_2	0.600	0	0	0.500	0	0.350	0.900
λ_3	0.350	0	0	0	0	0	−0.350
λ_4	0	0.500	0	0	0.450	0.450	0.900
λ_5	0	0	0	0.450	0	0.450	0.200
λ_6	0	0.350	0	0	0.450	0	−0.050

$$S_{jk} = \frac{1}{2}[1 + T_{jk} - 2I_{jk} - F_{jk}] + \frac{1}{4\pi}[2\pi + \phi_{jk} - 2\varphi_{jk} - \psi_{jk}],$$

$$C_j = \sum_k S_{jk} + \frac{1}{2}[1 + T_j - 2I_j - F_j] + \frac{1}{4\pi}[2\pi + \phi_j - 2\varphi_j - \psi_j],$$

respectively. The score functions and choice values of researchers represented by the line graph given in Fig. 9.29 are calculated in Table 9.27.

The choice values of Table 9.27 show that λ_2 and λ_4 are the most active and efficient participants of these research communities. Also, the score values show that λ_1 and λ_2 are the members having the strongest interactions between them and can share the most fruitful ideas relevant to their corresponding research fields being the participants of intersecting research communities. The procedure adopted in our application is described in Algorithm 9.10.1.

Algorithm 9.10.1 *Selection of a systematic member from intersecting research communities*

1. Input the set of vertices (research communities) $F_1, F_2, \ldots, F_j$.
2. Input the complex neutrosophic set N of vertices such that $N(F_k) = (T_k e^{\iota\phi_k}, I_k e^{\iota\varphi_k}, F_k e^{\iota\psi_k})$, $1 \le k \le j$, $0 \le T_k + I_k + F_k \le 3$.
3. Input the number of hyperedges (researchers) r of a complex neutrosophic hypergraph $\mathcal{H} = (\mathcal{N}, \lambda)$.
4. Input the membership degrees of the hyperedges $E_1, E_2, \ldots, E_r$.
5. Construct a complex neutrosophic line graph $l(\mathcal{H}) = (\mathcal{N}_l, \lambda_l)$ whose vertices are the r hyperedges $E_1, E_2, \cdots, E_n$ such that $\mathcal{N}_l(E_n) = \lambda(E_n)$.
6. If $|supp(\lambda_j) \cap supp(\lambda_k)| \ge 1$, then draw an edge between E_j and E_k and $\lambda_l(E_j E_k) = (\min\{T_\lambda(E_j), T_\lambda(E_k)\} \; e^{\iota \min\{\phi_\lambda(E_j), \phi_\lambda(E_k)\}}, \min\{I_\lambda(E_j), I_\lambda(E_k)\} e^{\iota \min\{\varphi_\lambda(E_j), \varphi_\lambda(E_k)\}}, \max\{F_\lambda(E_j), F_\lambda(E_k)\} e^{\iota \max\{\psi_\lambda(E_j), \psi_\lambda(E_k)\}})$.
7. Input the adjacency matrix $I = [(T_{mn}, I_{mn}, F_{mn})]_{r \times r}$ of vertices of complex neutrosophic line graph $l(\mathcal{H})$.
8. **do** m from $1 \to r$
9. 　　　$C_m = 0$
10. 　　**do** n from $1 \to r$
11. 　　　$S_{mn} = \frac{1}{2}[1 + T_{mn} - 2I_{mn} - F_{mn}] + \frac{1}{4\pi}[2\pi + \phi_{mn} - 2\varphi_{mn} - \psi_{mn}]$
12. 　　　$C_m = C_m + S_{mn}$

13. **end do**
14. $\quad C_m = C_m + \frac{1}{2}[1 + T_m - 2I_m - F_m] + \frac{1}{4\pi}[2\pi + \phi_m - 2\varphi_m - \psi_m]$
15. **end do**
16. The vertex with highest choice value in $l(\mathscr{H})$ is the most effective researcher among all the participants.

9.10.2 Influence of Modern Teaching Strategies on Educational Institutes

Teaching strategies are defined as the methods, techniques, and procedures that an educational institute utilizes to improve its performance. An educational institute can be judged according to its inputs and outputs that are highly effected through the teaching techniques adopted by that institute. Traditional teaching methods mainly depend on textbooks and emphasizes on basic skills while the modern techniques are based on technical approach and emphasize on creative ideas. Thus, modern teaching is very important and most effective in this technological era. Nowadays, educational institutes are modified through modern teaching strategies to enhance their outputs and these modern techniques play a vital role for teachers to explain the concepts in a more effective and radiant manner. Here, we consider a complex neutrosophic hypergraph model $\mathscr{H} = (\mathscr{N}, \lambda)$ to study the influence of modern teaching methods on a specific group of institutes in a time frame of 12 months. The vertices of a complex neutrosophic hypergraph represent the different teaching strategies and these techniques are grouped through an hyperedge if they are applied in the same institute. Since more than one institute can adopt the same strategy so the intersecting communities occur in this case. Each strategy is different from the other in terms of its positive, neutral, and negative impacts on students. The truth, indeterminacy, and falsity degrees of each strategy represent the positive, neutral, and negative effects of the corresponding technique on some institute during the time period of 12 months. The indeterminate information about modern teaching strategies having periodic nature is given in Table 9.28.

Table 9.28 Impacts of modern teaching strategies

Teaching strategy	Positive effects	Neutral behavior	Negative effects
Brain storming	$0.8e^{\iota(10/12)2\pi}$	$0.7e^{\iota(7/12)2\pi}$	$0.1e^{\iota(1/12)2\pi}$
Micro technique	$0.6e^{\iota(4/12)2\pi}$	$0.6e^{\iota(3/12)2\pi}$	$0.1e^{\iota(1/12)2\pi}$
Mind map	$0.6e^{\iota(6/12)2\pi}$	$0.3e^{\iota(5/12)2\pi}$	$0.7e^{\iota(7/12)2\pi}$
Cooperative learning	$0.8e^{\iota(10/12)2\pi}$	$0.7e^{\iota(7/12)2\pi}$	$0.1e^{\iota(1/12)2\pi}$
Dramatization	$0.5e^{\iota(3/12)2\pi}$	$0.3e^{\iota(3/12)2\pi}$	$0.2e^{\iota(2/12)2\pi}$
Educational software	$0.8e^{\iota(10/12)2\pi}$	$0.3e^{\iota(3/12)2\pi}$	$0.2e^{\iota(1/12)2\pi}$

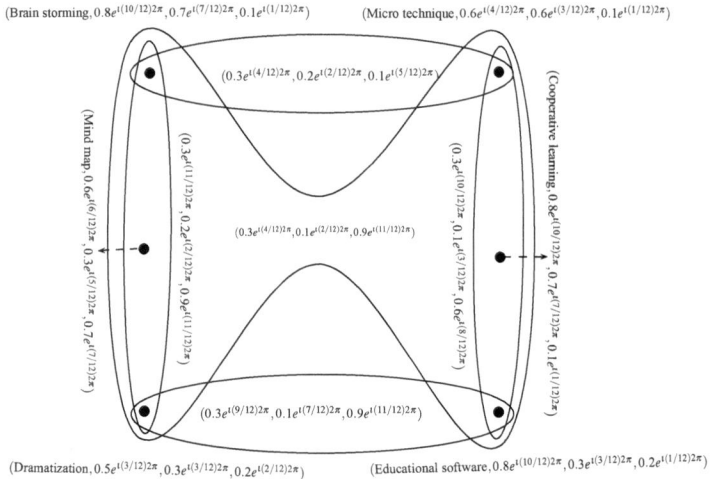

Fig. 9.30 Complex neutrosophic hypergraph model of modern teaching strategies

Note that the membership values $(0.8e^{\iota(10/12)2\pi}, 0.7e^{\iota(7/12)2\pi}, 0.1e^{\iota(1/12)2\pi})$ of brain storming show that this teaching technique has positive influence with degree 0.8 and this effect spreads over 10 months, the indeterminacy value represents the neutral effect or indeterminate behavior having degree 0.7 with time interval of 7 months, and the falsity degree 0.1 illustrates some negative effects of this strategy that spreads over 1 month. Similarly, the effects of all other strategies can be seen form Table 9.28 along with their time periods. An hyperedge of a complex neutrosophic hypergraph represents some institute in which the corresponding techniques are applied. The model of complex neutrosophic hypergraph grouping these strategies is shown in Fig. 9.30.

Here, each hyperedge represents an institute which groups the strategies adopted by that institute and the membership degrees of each hyperedge represent the combined effects of teaching strategies on corresponding institute. We now want to find a strategy or a collection of those techniques which are easy to apply, less in cost, and have higher positive effects on the performance of educational institutes. To find such methods, we calculate the minimal transversal of complex neutrosophic hypergraph given in Fig. 9.30 using Algorithm 9.10.2.

Algorithm 9.10.2 *Find a minimal complex neutrosophic transversal*

1. Input the complex neutrosophic sets $\lambda_1, \lambda_2, \ldots, \lambda_r$ of hyperedges.
2. Input the membership degrees of hyperedges.
3.   **do** j from $1 \rightarrow r$
4. $S_j = \lambda_j^{h(\lambda_j)}$
5. $S = S \cup S_j$
6.   **end do**
7. Take τ as the complex neutrosophic set having support S.

By following the Algorithm 9.10.2, we construct a minimal complex neutrosophic transversal of $\mathscr{H} = (\mathscr{N}, \lambda)$.

We have five hyperedges E_1, E_2, E_3, E_4, E_5 of $\mathscr{H}$. The heights of all complex neutrosophic hyperedges are given as

$h(\lambda_1) = (0.8e^{\iota(10/12)2\pi}, 0.7e^{\iota(7/12)2\pi}, 0.1e^{\iota(1/12)2\pi}),\ \lambda_1^{h(\lambda_1)} = \{\text{Brain storming}\},$

$h(\lambda_2) = (0.7e^{\iota(10/12)2\pi}, 0.6e^{\iota(7/12)2\pi}, 0.1e^{\iota(1/12)2\pi}),\ \lambda_2^{h(\lambda_2)} = \{\text{Brain storming}\},$

$h(\lambda_3) = (0.8e^{\iota(10/12)2\pi}, 0.3e^{\iota(3/12)2\pi}, 0.2e^{\iota(1/12)2\pi}),\ \lambda_3^{h(\lambda_3)} = \{\text{Educational software}\},$

$h(\lambda_4) = (0.8e^{\iota(10/12)2\pi}, 0.7e^{\iota(7/12)2\pi}, 0.1e^{\iota(1/12)2\pi}),\ \lambda_4^{h(\lambda_4)} = \{\text{Cooperative learning}\},$

$h(\lambda_5) = (0.8e^{\iota(10/12)2\pi}, 0.7e^{\iota(7/12)2\pi}, 0.1e^{\iota(1/12)2\pi}),\ \lambda_5^{h(\lambda_5)} = \{\text{Brain storming, Cooperative learn.}\}.$

$S = \lambda_1^{h(\lambda_1)} \cup \lambda_2^{h(\lambda_2)} \cup \lambda_3^{h(\lambda_3)} \cup \lambda_4^{h(\lambda_4)} \cup \lambda_5^{h(\lambda_5)}\{\text{Brain storming, Cooperative learning, Educational software}\}.$

The complex neutrosophic set having support S is given as

$\{(\text{Brain storming}, 0.8e^{\iota(10/12)2\pi}, 0.7e^{\iota(7/12)2\pi}, 0.1e^{\iota(1/12)2\pi}), (\text{Cooperative learning}, 0.8e^{\iota(10/12)2\pi},$
$0.7e^{\iota(7/12)2\pi}, 0.1e^{\iota(1/12)2\pi}), (\text{Educational software}, 0.8e^{\iota(10/12)2\pi}, 0.3e^{\iota(3/12)2\pi}, 0.2e^{\iota(1/12)2\pi})\},$

which is the minimal complex neutrosophic transversal of $\mathscr{H} = (\mathscr{N}, \lambda)$ and it shows that brain storming, cooperative learning, and educational software are the most influential teaching strategies for the given period of time. Thus, for some certain period of time, an influential and effective collection of modern teaching techniques can be determined.

9.11 Comparative Analysis

A complex neutrosophic set is characterized by truth, indeterminacy, and falsity degrees which are the combination of real-valued amplitude terms and complex-valued phase terms. To prove the flexibility and generalization of our proposed model complex neutrosophic hypergraphs, we propose the modeling of social networks through complex neutrosophic graphs, complex neutrosophic hypergraphs, and complex intuitionistic fuzzy hypergraphs. Consider a part of the social network as described in Sect. 9.10.2. Here, we consider only three modern techniques that are brain storming, cooperative learning, and educational software. A complex fuzzy hypergraph model of these techniques is given in Fig. 9.31.

Note that a complex fuzzy hypergraph model of intersecting techniques just illustrates the positive effects of these methods during a specific time interval. We see that a complex fuzzy hypergraph model fails to describe the negative effects of teaching strategies. To describe the positive as well as negative effects of these strategies, we utilize a complex intuitionistic fuzzy hypergraph model as shown in Fig. 9.32.

This shows that a complex intuitionistic fuzzy hypergraph model can well describe the positive and negative impacts of modern techniques on educational institutes but

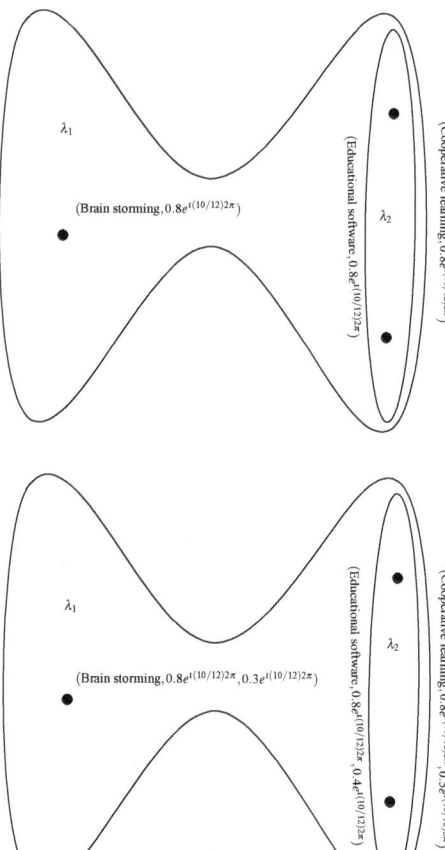

Fig. 9.31 Complex fuzzy hypergraph model of teaching techniques

Fig. 9.32 Complex intuitionistic fuzzy hypergraph model of teaching techniques

it cannot handle the situations when there is no effect during some time interval or there is indeterminate behavior. To handle such type of situations, we use a complex neutrosophic model as shown in Fig. 9.33.

Note that a complex neutrosophic graph model describes the truth, indeterminacy, and falsity degrees of impacts of teaching methods for some specific interval of time and proves to be a more generalized model as compared to complex fuzzy and complex intuitionistic fuzzy models. Figure 9.33 shows that λ_1 institute adopts the modern methods such as educational software and cooperative learning. Now, if an institute wants to utilize more than two strategies, then this model fails to model the required situation. For example, λ_1 wants to adopt all three modern teaching techniques. Then, we cannot model this social network using a simple graph. To handle such types of difficulties, i.e., for the modeling of indeterminate information with periodic nature existing in social hyeprnetworks, we have proposed complex

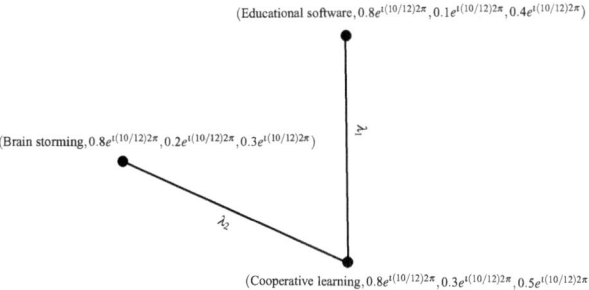

Fig. 9.33 Complex neutrosophic graph model of modern techniques

Table 9.29 Comparative analysis

Models	Edges	Hyperedge containing three strategies	Effects of modern techniques		
			Positive effects	Neutral behavior	Negative effects
Complex fuzzy hypergraph model	λ_1	{Brain storming,	$0.8e^{\iota(10/12)2\pi}$	–	–
		Cooperative learning,	$0.8e^{\iota(10/12)2\pi}$	–	–
		Educational software}	$0.8e^{\iota(10/12)2\pi}$	–	–
Complex intuitionistic fuzzy hypergraph model	λ_1	{Brain storming,	$0.8e^{\iota(10/12)2\pi}$	–	$0.3e^{\iota(10/12)2\pi}$
		Cooperative learning,	$0.8e^{\iota(10/12)2\pi}$	–	$0.5e^{\iota(10/12)2\pi}$
		Educational software}	$0.8e^{\iota(10/12)2\pi}$	–	$0.4e^{\iota(10/12)2\pi}$
Complex neutrosophic model	Cannot combine more than two elements	–	$0.8e^{\iota(10/12)2\pi}$	$0.2e^{\iota(10/12)2\pi}$	$0.3e^{\iota(10/12)2\pi}$
		–	$0.8e^{\iota(10/12)2\pi}$	$0.3e^{\iota(10/12)2\pi}$	$0.5e^{\iota(10/12)2\pi}$
		–	$0.8e^{\iota(10/12)2\pi}$	$0.1e^{\iota(10/12)2\pi}$	$0.4e^{\iota(10/12)2\pi}$
Complex neutrosophic hypergraph model	λ_1	{Brain storming,	$0.8e^{\iota(10/12)2\pi}$	$0.2e^{\iota(10/12)2\pi}$	$0.3e^{\iota(10/12)2\pi}$
		Cooperative learning,	$0.8e^{\iota(10/12)2\pi}$	$0.3e^{\iota(10/12)2\pi}$	$0.5e^{\iota(10/12)2\pi}$
		Educational software}	$0.8e^{\iota(10/12)2\pi}$	$0.1e^{\iota(10/12)2\pi}$	$0.4e^{\iota(10/12)2\pi}$

neutrosophic hypergraphs. The applicability and flexibility of our proposed model can be seen from Table 9.29.

It can be seen clearly from Table 9.29 that all existing models, including complex neutrosophic graphs, complex fuzzy hypergraphs, and complex intuitionistic fuzzy hypergraphs lack some information in order to handle the periodic and indeterminate data in case of hypernetworks. Thus, our proposed model is more flexible and applicable as it does not only deal with the reductant nature of imprecise information but also includes the benefits of hypergraphs. Hence, a complex neutrosophic hypergraph model combines the fruitful effects of complex neutrosophic sets and hypergraph theory.

References

1. Akram, M.: Single-valued neutrosophic planar graphs. Int. J. Algeb. Stat. **5**(2), 157–167 (2016)
2. Akram, M.: Single-valued neutrosophic graphs. Infosys Science Foundation Series in Mathematical Sciences, pp. 1–373. Springer (2018)
3. Akram, M., Luqman, A.: Intuitionistic single-valued neutrosophic hypergraphs. OPSEARCH **54**(4), 799–815 (2017)
4. Akram, M., Luqman, A.: Certain network models using single-valued neutrosophic directed hypergraphs. J. Intell. Fuzzy Syst. **33**(1), 575–588 (2017)
5. Akram, M., Shahzadi, S., Borumand Saeid, A.: Single-valued neutrosophic hypergraphs. TWMS J. Appl. Eng. Math. **8**(1), 122–135 (2018)
6. Ali, M., Smarandache, F.: Complex neutrosophic set. Neural Comput. Appl. **28**(7), 1817–1834 (2017)
7. Alkouri, A., Salleh, A.: Complex intuitionistic fuzzy sets, AIP Conference Proceedings, vol. 14, pp. 464–470 (2012)
8. Atanassov. K.T., Intuitionistic fuzzy sets. In: VII ITKR's Session, Sofia (deposed in Central Science-Technical Library of Bulgarian Academy of Science, 1697/84), (1983) (in Bulgarian)
9. Berge, C.: Graphs and Hypergraphs. North-Holland, Amsterdam (1973)
10. Bhowmik, M., Pal, M.: Intuitionistic neutrosophic set. J. Inf. Comput. Sci. **4**(2), 142–152 (2009)
11. Bhowmik, M., Pal, M.: Intuitionistic neutrosophic set relations and some of its properties. J. Inf. Comput. Sci. **5**(3), 183–192 (2010)
12. Gallo, G., Longo, G., Pallottino, S.: Directed hypergraphs and applications. Discret. Appl. Math. **42**, 177–201 (1993)
13. Kaufmann, A.: Introduction a la Thiorie des Sous-Ensemble Flous, 1. Masson, Paris (1977)
14. Luqman, A., Akram, M. and Smarandache, F.: Complex neutrosophic hypergraphs: new social network models. Algorithms **12**(11), 234 (2019)
15. Mordeson, J.N., Nair, P.S.: Fuzzy Graphs and Fuzzy Hypergraphs, 2nd edn. Physica Verlag, Heidelberg (2001)
16. Parvathi, R., Thilagavathi, S., Karunambigai, M.G.: Intuitionistic fuzzy hypergraphs. Cybern. Inf. Technol. **9**(2), 46–53 (2009)
17. Ramot, D., Milo, R., Friedman, M., Kandel, A.: Complex fuzzy sets. IEEE Trans. Fuzzy Syst. **10**(2), 171–186 (2002)
18. Ramot, D., Friedman, M., Langholz, G., Kandel, A.: Complex fuzzy logic. IEEE Trans. Fuzzy Syst. **11**(4), 450–461 (2003)
19. Rosenfeld, A.: Fuzzy graphs. In: Zadeh, L.A., Fu, K.S., Shimura, M. (eds.) Fuzzy Sets and their Applications, pp. 77–95. Academic Press, New York (1975)
20. Smarandache, F.: Neutrosophy: Neutrosophic Probability, Set and Logic. American Research Press, Rehoboth (1998)

21. Smarandache, F.: A Unifying Field in Logics Neutrosophy: Neutrosophic Probability, Set and Logic. American Research Press, Rehoboth (1999)
22. Smarandache, F.: Neutrosophic set-a generalization of the intuitionistic fuzzy set. Int. J. Pure Appl. Math. **24**(3), 287
23. Wang, H., Smarandache, F., Zhang, Y.Q., Sunderraman, R.: Single valued neutrosophic sets. Multispace and Multistructure **4**, 410–413 (2010)
24. Yang, H.L., Guo, Z.L., Liao, X.: On single-valued neutrosophic relations. J. Intell. Fuzzy Syst. **30**(2), 1045–1056 (2016)
25. Yaqoob, N., Akram, M.: Complex neutrosophic graphs. Bull. Comput. Appl. Math. **6**(2), 85–109 (2018)
26. Yaqoob, N., Gulistan, M., Kadry, S., Wahab, H.: Complex intuitionistic fuzzy graphs with application in cellular network provider companies. Mathematics **7**(1), 35 (2019)
27. Yazdanbakhsh, O., Dick, S.: A systematic review of complex fuzzy sets and logic. Fuzzy Sets Syst. **338**, 1–22 (2018)
28. Zadeh, L.A.: Fuzzy sets. Inf. Control **8**(3), 338–353 (1965)

Chapter 10
(Directed) Hypergraphs for Bipolar Neutrosophic Structures

In this chapter, we present bipolar neutrosophic hypergraphs and B-tempered bipolar neutrosophic hypergraphs. We describe the concepts of transversals, minimal transversals and locally minimal transversals of bipolar neutrosophic hypergraphs. Furthermore, we put forward some applications of bipolar neutrosophic hypergraphs in marketing and biology. We also introduce bipolar neutrosophic directed hypergraphs, regular bipolar neutrosophic directed hypergraphs, homomorphism, and isomorphism on bipolar neutrosophic directed hypergraphs. To conclude, we describe an efficient algorithm to solve decision-making problems. This chapter is due to [7, 10, 16].

10.1 Introduction

The notion of bipolar fuzzy sets (YinYang bipolar fuzzy sets) was introduced by Zhang [35, 36] in the space $\{\forall\ (x, y) \mid (x, y) \in [-1, 0] \times [0, 1]\}$. In Chinese medicine, *Yin* and *Yang* are the two sides. *Yin* is the negative side of a system and *Yang* is the positive side of a system. Although bipolar fuzzy sets and intuitionistic fuzzy sets look similar to each other, they are essentially different sets [21, 22]. Bipolar fuzzy sets are extension of fuzzy sets whose membership degree ranges $[-1, 1]$. In a bipolar fuzzy set, if the degree of membership is zero then we say the element is unrelated to the corresponding property, membership degree $(0, 1]$ indicates that the object satisfies a certain property, whereas the membership degree $[-1, 0)$ indicates that the element satisfies the implicit counter property. Positive information represents what is considered to be possible and negative information represents what is granted to be impossible. Actually, a variety of decision-making problems are based on two-sided bipolar judgements on a positive side and a negative side. Smarandache [29] incorporated indeterminacy membership function as independent component and defined neutrosophic set on three components truth, indeterminacy,

© Springer Nature Singapore Pte Ltd. 2020
M. Akram and A. Luqman, *Fuzzy Hypergraphs and Related Extensions*,
Studies in Fuzziness and Soft Computing 390,
https://doi.org/10.1007/978-981-15-2403-5_10

and falsehood. However, from practical point of view, Smarandache [29] and Wang et al. [32] defined single-valued neutrosophic sets where degree of truth membership, indeterminacy membership, and falsity membership belong to [0, 1]. Deli et al. [16] extended the ideas of bipolar fuzzy sets and neutrosophic sets to bipolar neutrosophic sets (bipolar single-valued neutrosophic sets) and studied its operations and applications in decision-making problems.

Definition 10.1 A *bipolar single-valued neutrosophic set* on a non-empty set X is an object of the form,

$$N = \{(y, T_N^+(y), I_N^+(y), F_N^+(y), T_N^-(y), I_N^-(y), F_N^-(y)) : y \in X\},$$

where $T_N^+, I_N^+, F_N^+ : X \to [0, 1]$ and $T_N^-, I_N^-, F_N^- : X \to [-1, 0]$ are mappings. The positive values $T_N^+(y), I_N^+(y), F_N^+(y)$ denote respectively the truth, indeterminacy and false membership degrees of an element $y \in X$, whereas $T_N^-(y), I_N^-(y), F_N^-(y)$ denote the implicit counter property of the truth, indeterminacy and false membership degrees of the element $y \in X$ corresponding to the bipolar neutrosophic set N.

Definition 10.2 A *bipolar single-valued neutrosophic relation* on a non-empty set X is a bipolar neutrosophic subset of $X \times X$ of the form,

$$D = \{(yz, T_D^+(yz), I_D^+(yz), F_D^+(yz), T_D^-(yz), I_D^-(yz), F_D^-(yz)) : yz \in X \times X\},$$

where $T_D^+, I_D^+, F_D^+, T_D^-, I_D^-, F_D^-$ are defined by the mappings $T_D^+, I_D^+, F_D^+ : X \times X \to [0, 1]$ and $T_D^-, I_D^-, F_D^- : X \times X \to [-1, 0]$.

Definition 10.3 A *bipolar single-valued neutrosophic graph* on a non-empty set A is a pair $G = (C, D)$, where C is a bipolar single-valued neutrosophic set on A and D is a bipolar single-valued neutrosophic relation in A such that

$$T_D^+(yz) \leq T_C^+(y) \wedge T_C^+(z), \quad I_D^+(yz) \leq I_C^+(y) \wedge I_C^+(z), \quad F_D^+(yz) \leq F_C^+(y) \vee F_C^+(z),$$
$$T_D^-(yz) \geq T_C^-(y) \vee T_C^-(z), \quad I_D^-(yz) \geq I_C^-(y) \vee I_C^-(z), \quad F_D^-(yz) \geq F_C^-(y) \wedge F_C^-(z)$$

for all $y, z \in X$. Note that $D(yz) = (0, 0, 1, 0, 0, -1)$ for all $yz \in X \times X \backslash E$.

Definition 10.4 The *support* of a bipolar neutrosophic set $N = \{(x, T_N^+(x), I_N^+(x), F_N^+(x), T_N^-(x), I_N^-(x), F_N^-(x)) | x \in X\}$ is defined as $supp(N) = supp^+(N) \cup supp^-(N)$, where

$$supp^+(N) = \{x \in X | T_N^+(x) \neq 0, I_N^+(x) \neq 0, F_N^+(x) \neq 0\},$$

$$supp^-(N) = \{x \in X | T_N^-(x) \neq 0, I_N^-(x) \neq 0, F_N^-(x) \neq 0\}.$$

$supp^+(N)$ is called the *positive support* and $supp^-(N)$ is called the *negative support* of N.

For further terminologies and studies on fuzzy sets, bipolar fuzzy sets, neutro-sophic sets, and bipolar neutrosophic sets and graphs, readers are referred to [1–9, 11–15, 17–20, 23–28, 30, 31, 33, 34].

10.2 Bipolar Neutrosophic Hypergraphs

Definition 10.5 A *bipolar neutrosophic graph* on a non-empty set X is a pair $G = (A, B)$, where A is a bipolar neutrosophic set on X and B is a bipolar neutrosophic relation in X such that

$$T_B^+(xy) \leq \min\{T_A^+(x), T_A^+(y)\}, \ T_B^-(xy) \geq \max\{T_A^-(x), T_A^-(y)\},$$
$$I_B^+(xy) \leq \min\{I_A^+(x), I_A^+(y)\}, \quad I_B^-(xy) \geq \max\{I_A^-(x), I_A^-(y)\},$$
$$F_B^+(xy) \leq \max\{F_A^+(x), F_A^+(y)\}, \ F_B^-(xy) \geq \min\{F_A^-(x), F_A^-(y)\},$$

for all $x, y \in X$. Note that, $D(xy) = (0, 0, 1, 0, 0, -1) = \mathbf{0}$, for all $xy \in X \times X \backslash E$.

Definition 10.6 Let X be a non-empty set. A *bipolar neutrosophic hypergraph* H on X is defined as an ordered pair $H = (\mu, \rho)$, where $\mu = \{\mu_1, \mu_2, \mu_3, \dots, \mu_n\}$ is a finite collection of bipolar neutrosophic subsets on X and ρ is a bipolar neutrosophic relation on bipolar neutrosophic subsets μ_i such that

1.

$$T_\rho^+(E_k) = T_\rho^+(\{x_1, x_2, \dots, x_m\}) \leq \min\{T_{\mu_i}^+(x_1), T_{\mu_i}^+(x_2), \dots, T_{\mu_i}^+(x_m)\},$$
$$I_\rho^+(E_k) = I_\rho^+(\{x_1, x_2, \dots, x_m\}) \leq \min\{I_{\mu_i}^+(x_1), I_{\mu_i}^+(x_2), \dots, I_{\mu_i}^+(x_m)\},$$
$$F_\rho^+(E_k) = F_\rho^+(\{x_1, x_2, \dots, x_m\}) \leq \max\{F_{\mu_i}^+(x_1), F_{\mu_i}^+(x_2), \dots, F_{\mu_i}^+(x_m)\},$$
$$T_\rho^-(E_k) = T_\rho^-(\{x_1, x_2, \dots, x_m\}) \geq \max\{T_{\mu_i}^-(x_1), T_{\mu_i}^-(x_2), \dots, T_{\mu_i}^-(x_m)\},$$
$$I_\rho^-(E_k) = I_\rho^-(\{x_1, x_2, \dots, x_m\}) \geq \max\{I_{\mu_i}^-(x_1), I_{\mu_i}^-(x_2), \dots, I_{\mu_i}^-(x_m)\},$$
$$F_\rho^-(E_k) = F_\rho^-(\{x_1, x_2, \dots, x_m\}) \geq \min\{F_{\mu_i}^-(x_1), F_{\mu_i}^-(x_2), \dots, F_{\mu_i}^-(x_m)\},$$

for all $x_1, x_2, x_3, \dots, x_m \in X$.
2. $\bigcup_i supp(\mu_i(x)) = X$, for all $\mu_i \in \mu$.

Definition 10.7 The *height* of a bipolar neutrosophic hypergraph $H = (\mu, \rho)$, denoted by $h(H)$, is defined as, $h(H) = \{(\max(\rho_l), \max(\rho_m), \min(\rho_n)) | \rho_l, \rho_m, \rho_n \in \rho\}$, where $\rho_l = \max T_{\rho_j}^+(x_i)$, $\rho_m = \max I_{\rho_j}^+(x_i)$, $\rho_n = \min F_{\rho_j}^+(x_i)$.

The *depth* of a bipolar neutrosophic hypergraph $H = (\mu, \rho)$, denoted by $d(H)$, is defined as $d(H) = \{(\min(\rho_l), \min(\rho_m), \max(\rho_n)) | \rho_l, \rho_m, \rho_n \in \rho\}$, where $\rho_l = \min T_{\rho_j}^-(x_i)$, $\rho_m = \min I_{\rho_j}^-(x_i)$, $\rho_n = \max F_{\rho_j}^-(x_i)$.

Here, $T_{\rho_j}^+(x_i)$, $I_{\rho_j}^+(x_i)$ and $F_{\rho_j}^+(x_i)$ denote the positive truth, indeterminacy and falsity membership values of vertex x_i to the hyperedge ρ_j, respectively, $T_{\rho_j}^-(x_i)$,

$I_{\rho_j}^-(x_i)$ and $F_{\rho_j}^-(x_i)$ denote the negative truth, indeterminacy, and falsity membership values of vertex x_i to the hyperedge ρ_j, respectively.

Definition 10.8 Let $H = (\mu, \rho)$ be a bipolar neutrosophic hypergraph. The *order* of H, denoted by $o(H)$, is defined as $o(H) = (o^+(H), o^-(H))$, where

$$o^+(H) = \sum_{x \in X} \min \mu_i^+(x), \quad o^-(H) = \sum_{x \in X} \max \mu_i^-(x).$$

The *size* of H, denoted by $s(H)$, is defined as $s(H) = (s^+(H), s^-(H))$, where

$$s^+(H) = \sum_{E_k \subset X} \rho^+(E_k), \quad s^-(H) = \sum_{E_k \subset X} \rho^-(E_k).$$

In a bipolar neutrosophic hypergraph, two vertices x_i and x_j are adjacent vertices if they both belong to the same bipolar neutrosophic hyperedge. Two bipolar neutrosophic hyperedges ρ_i and ρ_j are called adjacent if their intersection is non-empty, i.e., $supp(\rho_i) \cap supp(\rho_j) \neq \emptyset, i \neq j$.

We now define the adjacent level between two vertices and between two hyperedges.

Definition 10.9 The *adjacent level* between two vertices x_i and x_j, denoted by $\gamma(x_i, x_j)$, is defined as $\gamma(x_i, x_j) = (\gamma^+, \gamma^-)$, where

$$\gamma^+(x_i, x_j) = \max_k \min[T_k^+(x_i), T_k^+(x_j)], \ \max_k \min[I_k^+(x_i), I_k^+(x_j)], \ \min_k \max[F_k^+(x_i), F_k^+(x_j)],$$

$$\gamma^-(x_i, x_j) = \min_k \max[T_k^-(x_i), T_k^-(x_j)], \ \min_k \max[I_k^-(x_i), I_k^-(x_j)], \ \max_k \min[F_k^-(x_i), F_k^-(x_j)].$$

The *adjacent level* between two bipolar neutrosophic hyperedges ρ_i and ρ_j, denoted by $\sigma(\rho_i, \rho_j)$, is defined as $\sigma(\rho_i, \rho_j) = (\sigma^+, \sigma^-)$, where

$$\sigma^+(\rho_i, \rho_j) = \max_j \min[T_j^+(x), T_k^+(x)], \ \max_j \min[I_j^+(x), I_k^+(x)], \ \min_j \max[F_j^+(x), F_k^+(x)],$$

$$\sigma^-(\rho_i, \rho_j) = \min_j \max[T_j^-(x), T_k^-(x)], \ \min_j \max[I_j^-(x), I_k^-(x)], \ \max_j \min[F_j^-(x), F_k^-(x)].$$

Definition 10.10 A bipolar neutrosophic hypergraph $H = (\mu, \rho)$ is *simple* if it contains no repeated hyperedges, i.e., if $\rho_j, \rho_k \in \rho$ and $\rho_j \subseteq \rho_k$, then $\rho_j = \rho_k$.

A bipolar neutrosophic hypergraph $H = (\mu, \rho)$ is *support simple* if $\rho_j, \rho_k \in \rho$, $supp(\rho_j) = supp(\rho_k)$ and $\rho_j \subseteq \rho_k$, then $\rho_j = \rho_k$.

A bipolar neutrosophic hypergraph $H = (\mu, \rho)$ is *strongly support simple* if ρ_j, $\rho_k \in \rho$ and $supp(\rho_j) = supp(\rho_k)$, then $\rho_j = \rho_k$.

Remark 10.1 All these concepts imply that there are no multiple hyperedges. Simple bipolar neutrosophic hypergraphs are support simple and strongly support simple bipolar neutrosophic hypergraphs are support simple. These two concepts, simple and support simple, are independent o each other.

Definition 10.11 A bipolar neutrosophic set $N : X \to [0, 1] \times [-1, 0]$ is an *elementary* set if $T_N^+, I_N^+, F_N^+, T_N^-, I_N^-, F_N^-$ are all single valued on the support of N.

A bipolar neutrosophic hypergraph is an *elementary* if its all hyperedges are elementary.

Proposition 10.1 *A bipolar neutrosophic hypergraph is the generalization of fuzzy hypergraph, bipolar fuzzy hypergraph, and neutrosophic hypergraph.*

Theorem 10.1 *Let $H = (\mu, \rho)$ be a simple, elementary bipolar neutrosophic hypergraph on a non-empty set X having order n. Then, $|\rho| \leq 2^n - 1$. The equality holds if and only if $\{supp(\rho_j)|\rho_j \in \rho, \rho \neq 0\} = \mathscr{P}(X)\backslash\emptyset$.*

Proof Since H is simple and elementary, each nontrivial subset of X can be the support of at most one $\rho_j \in \rho$, therefore $|\rho| \leq 2^n - 1$. To prove that the equality holds, let $B = \{T_A^+(x), I_A^+(x), F_A^+(x), T_A^-(x), I_A^-(x), F_A^-(x)|A \subseteq X\}$ be the set of mappings such that

$$T_A^+(x) = \begin{cases} \frac{1}{|A|}, & \text{if } x \in A, \\ 0, & \text{otherwise.} \end{cases} \quad I_A^+(x) = \begin{cases} \frac{1}{|A|}, & \text{if } x \in A, \\ 0, & \text{otherwise.} \end{cases}$$

$$F_A^+(x) = \begin{cases} \frac{1}{|A|}, & \text{if } x \in A, \\ 0, & \text{otherwise.} \end{cases} \quad T_A^-(x) = \begin{cases} \frac{-1}{|A|}, & \text{if } x \in A, \\ 0, & \text{otherwise.} \end{cases}$$

$$I_A^-(x) = \begin{cases} \frac{-1}{|A|}, & \text{if } x \in A, \\ 0, & \text{otherwise.} \end{cases} \quad F_A^-(x) = \begin{cases} \frac{-1}{|A|}, & \text{if } x \in A, \\ 0, & \text{otherwise.} \end{cases}$$

Then, each set having single element has height $(1, 1, 1, -1, -1, -1)$, the set having two elements has height $(0.5, 0.5, 0.5, -0.5, -0.5, -0.5)$ and so on. Hence, H is elementary, simple, and $|\rho| = 2^n - 1$. $\square$

Definition 10.12 Let $H = (\mu, \rho)$ be a bipolar neutrosophic hypergraph. Suppose that $\alpha, \beta, \gamma \in [0, 1]$ and $\eta, \theta, \phi \in [-1, 0]$. Let

1. $\rho^{(\alpha,\beta,\gamma,\eta,\theta,\phi)} = \{\rho_i^{(\alpha,\beta,\gamma,\eta,\theta,\phi)} : \rho_i \in \rho\}$ and $\rho_i^{(\alpha,\beta,\gamma,\eta,\theta,\phi)} = \{x \in X | T_{\rho_i}^+(x) \geq \alpha,$ $I_{\rho_i}^+(x) \geq \beta, F_{\rho_i}^+(x) \leq \gamma$ and $T_{\rho_i}^-(x) \leq \eta, I_{\rho_i}^-(x) \leq \theta, F_{\rho_i}^-(x) \geq \phi\}$,
2. $\mu^{(\alpha,\beta,\gamma,\eta,\theta,\phi)} = \bigcup_{\rho_i \in \rho} \rho_i^{(\alpha,\beta,\gamma,\eta,\theta,\phi)}$.

Then, the ordered pair $H^{(\alpha,\beta,\gamma,\eta,\theta,\phi)} = (\mu^{(\alpha,\beta,\gamma,\eta,\theta,\phi)}, \rho^{(\alpha,\beta,\gamma,\eta,\theta,\phi)})$ is called the $(\alpha, \beta, \gamma, \eta, \theta, \phi)$–*level hypergraph* of H. Note that $H^{(\alpha,\beta,\gamma,\eta,\theta,\phi)}$ is a crisp hypergraph.

Remark 10.2 If $\alpha \geq \omega, \beta \geq \nu, \gamma \leq \sigma, \eta \leq \pi, \theta \leq \varepsilon, \phi \geq \varepsilon$ and μ is a bipolar neutrosophic subset on X, then $\mu^{(\alpha,\beta,\gamma,\eta,\theta,\phi)} \subseteq \mu^{(\omega,\nu,\sigma,\pi,\varepsilon,\varepsilon)}$. Thus, for level hypergraphs of H, we have $\rho^{(\alpha,\beta,\gamma,\eta,\theta,\phi)} \subseteq \rho^{(\omega,\nu,\sigma,\pi,\varepsilon,\varepsilon)}$. Thus the $(\alpha, \beta, \gamma, \eta, \theta, \phi)$-level hyperedges of distinct bipolar neutrosophic hyperedges of ρ can be same and hence the $(\alpha, \beta, \gamma, \eta, \theta, \phi)$-level hypergraphs $H^{(\alpha,\beta,\gamma,\eta,\theta,\phi)}$ of a simple bipolar neutrosophic hypergraph H could be multiedged.

For any bipolar neutrosophic hypergraph $H = (\mu, \rho)$, we can associate a finite sequence of real numbers, called the *fundamental sequence* of H.

Definition 10.13 Let $H = (\mu, \rho)$ be a bipolar neutrosophic hypergraph and $H^{(\alpha, \beta, \gamma, \eta, \theta, \phi)}$ be the $(\alpha, \beta, \gamma, \eta, \theta, \phi)$-level hypergraph of H. The sequence of real numbers, $\{\alpha_1 = (T_{\alpha_1}^+, I_{\alpha_1}^+, F_{\alpha_1}^+, T_{\alpha_1}^-, I_{\alpha_1}^-, F_{\alpha_1}^-), \alpha_2 = (T_{\alpha_2}^+, I_{\alpha_2}^+, F_{\alpha_2}^+, T_{\alpha_2}^-, I_{\alpha_2}^-, F_{\alpha_2}^-), \ldots, \alpha_n = (T_{\alpha_n}^+, I_{\alpha_n}^+, F_{\alpha_n}^+, T_{\alpha_n}^-, I_{\alpha_n}^-, F_{\alpha_n}^-)\}, 0 < T_{\alpha_1}^+, I_{\alpha_1}^+ < T_{\alpha_2}^+, I_{\alpha_2}^+ < \cdots T_{\alpha_n}^+, I_{\alpha_n}^+, F_{\alpha_1}^+ > F_{\alpha_2}^+ > \cdots > F_{\alpha_n}^+ > 0, \quad 0 > T_{\alpha_1}^-, I_{\alpha_1}^- > T_{\alpha_2}^-, I_{\alpha_2}^- > \cdots T_{\alpha_n}^-, I_{\alpha_n}^-, F_{\alpha_1}^- < F_{\alpha_2}^- < \cdots < F_{\alpha_n}^- < 0$, where $(T_{\alpha_n}^+, I_{\alpha_n}^+, F_{\alpha_n}^+) = h(H)$ and $(T_{\alpha_n}^-, I_{\alpha_n}^-, F_{\alpha_n}^-) = d(H)$, which satisfies the following properties:

(i) if $\alpha_{i-1} = (T_{\alpha_{i-1}}^+, I_{\alpha_{i-1}}^+, F_{\alpha_{i-1}}^+, T_{\alpha_{i-1}}^-, I_{\alpha_{i-1}}^-, F_{\alpha_{i-1}}^-) < \alpha = (T_{\alpha}^+, I_{\alpha}^+, F_{\alpha}^+, T_{\alpha}^-, I_{\alpha}^-, F_{\alpha}^-) \leq \alpha_i = (T_{\alpha_i}^+, I_{\alpha_i}^+, F_{\alpha_i}^+, T_{\alpha_i}^-, I_{\alpha_i}^-, F_{\alpha_i}^-)$, then $\rho^{\alpha} = \rho^{\alpha_i}$,

(ii) $\rho^{\alpha_i} \subseteq \rho^{\alpha_{i+1}}$,

is called the *fundamental sequence* of H, denoted by $F_S(H)$. The set of α_i-level hypergraphs $\{H^{\alpha_1}, H^{\alpha_2}, \ldots, H^{\alpha_n}\}$ is called the *core hypergraphs* of H or simply the *core set* of H and is denoted by $c(H)$.

Definition 10.14 A bipolar neutrosophic hypergraph $H = (\mu, \rho)$ is *ordered* if the core set $c(H) = \{H^{\alpha_1}, H^{\alpha_2}, \ldots, H^{\alpha_n}\}$ of H is ordered, i.e., $\{H^{\alpha_1} \subseteq H^{\alpha_2} \subseteq \cdots \subseteq H^{\alpha_n}\}$. H is *simply ordered* if H is ordered and whenever $\rho' \subset \rho^{\alpha_{i+1}} \backslash \rho^{\alpha_i}$ then $\rho' \not\subseteq \mu^{\alpha_i}$.

Definition 10.15 A bipolar neutrosophic hypergraph $H_1 = (\mu_1, \rho_1)$ is called a *partial* bipolar neutrosophic hypergraph of $H_2 = (\mu_2, \rho_2)$ if $\rho_1 \subseteq \rho_2$. We denote it as $H_1 \subseteq H_2$.

Definition 10.16 Let $H = (\mu, \rho)$ be a bipolar neutrosophic hypergraph on X having $F_S(H) = \{\alpha_1, \alpha_2, \alpha_3, \ldots, \alpha_n\}$, suppose $\alpha_{n+1} = 0$. Then, H is called *sectionally elementary* if for each hyperedge $\rho_j \in \rho$, $j \in \{1, 2, 3, \ldots, n\}$ and for all $\alpha \in (\alpha_{i+1}, \alpha_i]$, we have $\rho_i^{\alpha} = \rho_i^{\alpha_i}$.

Note that H is *sectionally elementary* if and only if $\rho_j(x) \in F_S(H)$ for each $\rho_j \in \rho$ and $x \in X$.

Example 10.1 Let $H = (\mu, \rho)$ be a bipolar neutrosophic hypergraph as shown in Fig. 10.1, where $\rho = \{\rho_1, \rho_2, \rho_3, \rho_4\}$. Incidence matrix of H is given in Table 10.1.

By routine calculations, we have $h(H) = (0.9, 0.8, 0.1)$ and $d(H) = (-0.9, -0.8, -0.1)$, $H^{(0.9, 0.8, 0.1, -0.9, -0.8, -0.1)} = \{x_1, x_2\}$, $H^{(0.8, 0.7, 0.1, -0.8, -0.7, -0.1)} = \{x_1, x_2\}$, $H^{(0.5, 0.4, 0.3, -0.5, -0.4, -0.3)} = \{\{x_1, x_2, x_4\}, \{x_1, x_2\}, \{x_4\}, \{x_1, x_5\}\}$. Therefore, $F_S(H) = \{(0.9, 0.8, 0.1, -0.9, -0.8, -0.1), (0.8, 0.7, 0.1, -0.8, -0.7, -0.1), (0.5, 0.4, 0.3, -0.5, -0.4, -0.3)\}$. The set of core hypergraphs is $c(H) = \{H^{(0.9, 0.8, 0.1, -0.9, -0.8, -0.1)} = (X_1, \rho_1^*), H^{(0.8, 0.7, 0.1, -0.8, -0.7, -0.1)} = (X_2, \rho_2^*), H^{(0.5, 0.4, 0.3, -0.5, -0.4, -0.3)} = (X_3, \rho_3^*)\}$.

Note that, $supp(\rho_i) = supp(\rho_j)$ for all $\rho_i, \rho_j \in \rho$ implies $\rho_i = \rho_j$. Thus, H is strongly support simple and support simple. The induced fundamental sequence of H is given in Fig. 10.2.

Table 10.1 Incidence matrix of H

I	ρ_1	ρ_2	ρ_3	ρ_4
x_1	$(0.8, 0.7, 0.1,$ $-0.8, -0.7, -0.1)$	$(0.9, 0.8, 0.1,$ $-0.9, -0.8, -0.1)$	0	$(0.5, 0.4, 0.3,$ $-0.5, -0.4, -0.3)$
x_2	$(0.8, 0.7, 0.1,$ $-0.8, -0.7, -0.1)$	$(0.9, 0.8, 0.1,$ $-0.9, -0.8, -0.1)$	0	0
x_3	0	0	$(0.3, 0.3, 0.4,$ $-0.3, -0.3, -0.4)$	0
x_4	$(0.5, 0.4, 0.3,$ $-0.5, -0.4, -0.3)$	0	$(0.5, 0.4, 0.3,$ $-0.5, -0.4, -0.3)$	0
x_5	0	0	0	$(0.5, 0.4, 0.3,$ $-0.5, -0.4, -0.3)$

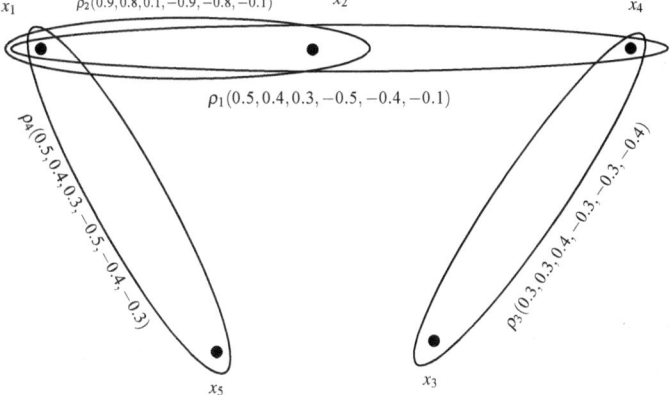

Fig. 10.1 Bipolar neutrosophic hypergraph

Theorem 10.2 *Let $H = (\mu, \rho)$ be an elementary bipolar neutrosophic hypergraph. Then, H is strongly support simple if and only if H is support simple.*

Proof Suppose that H is elementary, support simple, and $supp(\rho_i) = supp(\rho_j)$, for $\rho_i, \rho_j \in \rho$. Let $h(\rho_i) \leq h(\rho_j)$, $d(\rho_i) \geq d(\rho_j)$. Since, H is elementary, then we have $\rho_i \leq \rho_j$ and since H is support simple, we have $\rho_i = \rho_j$. Hence, H is strongly support simple. The converse part can be proved trivially. □

Theorem 10.3 *A bipolar neutrosophic hypergraph $H = (\mu, \rho)$ is single-valued bipolar neutrosophic graph (possibly with loops) if and only if H is elementary, support simple, and all the hyperedges have two(or one) element support.*

Proof Let $H = (\mu, \rho)$ be an elementary, support simple, and each hyperedge has two(or one) element support. Since, H is elementary, each hyperedge is single valued on $supp(\mu)$. Each hyperedge has two(or one) element support, i.e., each hyperedge contains exactly two(or one) elements in it. Hence, H is single-valued bipolar neutrosophic graph(possibly with loops). Converse part can be proved on the same lines. □

Fig. 10.2 Induced
fundamental sequence of H

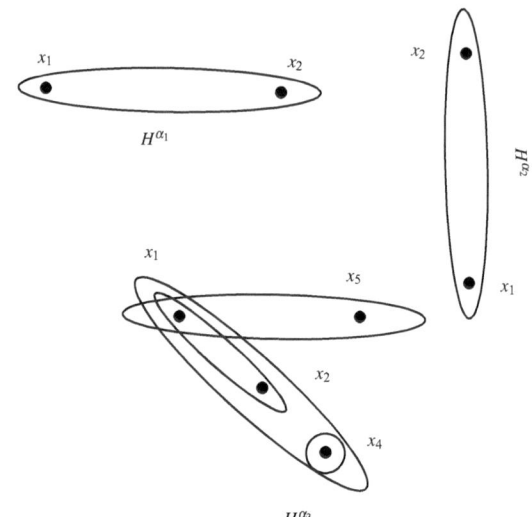

Definition 10.17 A bipolar neutrosophic hypergraph $H = (\mu, \rho)$ is called a $B = (T_B^+, I_B^+, F_B^+, T_B^-, I_B^-, F_B^-)$-*tempered* bipolar neutrosophic hypergraph if there exists a crisp hypergraph $H^* = (X, E)$ and a bipolar neutrosophic set $B = (T_B^+, I_B^+, F_B^+, T_B^-, I_B^-, F_B^-) : X \to [0, 1] \times [-1, 0]$ such that $\rho = \{D_F = (T_{D_F}^+, I_{D_F}^+, F_{D_F}^+, T_{D_F}^-, I_{D_F}^-, F_{D_F}^-) | F \in E\}$, where

$$T_{D_F}^+ (x) = \begin{cases} \min(T_B^+(y)) : y \in F, & \text{if } x \in F, \\ 0, & \text{otherwise.} \end{cases}$$

$$I_{D_F}^+ (x) = \begin{cases} \min(I_B^+(y)) : y \in F, & \text{if } x \in F, \\ 0, & \text{otherwise.} \end{cases}$$

$$F_{D_F}^+ (x) = \begin{cases} \max(F_B^+(y)) : y \in F, & \text{if } x \in F, \\ 0, & \text{otherwise.} \end{cases}$$

$$T_{D_F}^- (x) = \begin{cases} \max(T_B^-(y)) : y \in F, & \text{if } x \in F, \\ 0, & \text{otherwise.} \end{cases}$$

$$I_{D_F}^- (x) = \begin{cases} \max(I_B^-(y)) : y \in F, & \text{if } x \in F, \\ 0, & \text{otherwise.} \end{cases}$$

$$F_{D_F}^- (x) = \begin{cases} \min(F_B^-(y)) : y \in F, & \text{if } x \in F, \\ 0, & \text{otherwise.} \end{cases}$$

Example 10.2 Consider a bipolar neutrosophic hypergraph $H = (\mu, \rho)$ given in Fig. 10.3, where $\rho = \{\rho_1, \rho_2, \rho_3\}$. Incidence matrix of H is given in Table 10.2.

Define a bipolar neutrosophic set $B = (T_B^+, I_B^+, F_B^+, T_B^-, I_B^-, F_B^-) : X \to [0, 1] \times [-1, 0]$ as

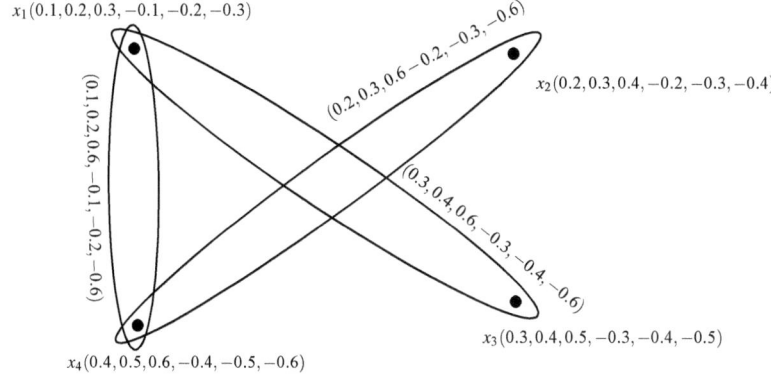

Fig. 10.3 B-tempered bipolar neutrosophic hypergraph

Table 10.2 B-tempered bipolar neutrosophic hypergraph

I	ρ_1	ρ_2	ρ_3
x_1	(0.1, 0.2, 0.6, $-0.1, -0.2, -0.6$)	0	0
x_2	0	(0.2, 0.3, 0.6, $-0.2, -0.3, -0.6$)	0
x_3	0	0	(0.3, 0.4, 0.6, $-0.3, -0.4, -0.6$)
x_4	(0.1, 0.2, 0.6, $-0.1, -0.2, -0.6$)	(0.2, 0.3, 0.6, $-0.2, -0.3, -0.6$)	(0.3, 0.4, 0.6, $-0.3, -0.4, -0.6$)

$$B = \{(x_1, 0.1, 0.2, 0.3, -0.1, -0.2, -0.3), (x_2, 0.2, 0.3, 0.4, -0.2, -0.3, -0.4),$$
$$(x_3, 0.3, 0.4, 0.5, -0.3, -0.4, -0.5), (x_4, 0.4, 0.5, 0.6, -0.4, -0.5, -0.6)\}.$$

By routine calculations, we have

$$T^+_{D_{\{x_1, x_4\}}}(x_1) = \min\{T^+_B(x_1), T^+_B(x_4)\} = \min\{0.1, 0.4\} = 0.1,$$
$$I^+_{D_{\{x_1, x_4\}}}(x_1) = \min\{I^+_B(x_1), I^+_B(x_4)\} = \min\{0.2, 0.5\} = 0.2,$$
$$F^+_{D_{\{x_1, x_4\}}}(x_1) = \max\{F^+_B(x_1), F^+_B(x_4)\} = \max\{0.3, 0.6\} = 0.6.$$

Similarly,

$$T^-_{D_{\{x_1, x_4\}}}(x_1) = -0.1, \, I^-_{D_{\{x_1, x_4\}}}(x_1) = -0.2, \, F^-_{D_{\{x_1, x_4\}}}(x_1) = -0.6.$$

The remaining values can be calculated using the same technique. Thus, we have

$$\rho_1 = (T^+_{D_{\{x_1,x_4\}}}, I^+_{D_{\{x_1,x_4\}}}, F^+_{D_{\{x_1,x_4\}}}, T^-_{D_{\{x_1,x_4\}}}, I^-_{D_{\{x_1,x_4\}}}, F^-_{D_{\{x_1,x_4\}}}),$$

$$\rho_2 = (T^+_{D_{\{x_2,x_4\}}}, I^+_{D_{\{x_2,x_4\}}}, F^+_{D_{\{x_2,x_4\}}}, T^-_{D_{\{x_2,x_4\}}}, I^-_{D_{\{x_2,x_4\}}}, F^-_{D_{\{x_2,x_4\}}}),$$

$$\rho_3 = (T^+_{D_{\{x_3,x_4\}}}, I^+_{D_{\{x_3,x_4\}}}, F^+_{D_{\{x_3,x_4\}}}, T^-_{D_{\{x_3,x_4\}}}, I^-_{D_{\{x_3,x_4\}}}, F^-_{D_{\{x_3,x_4\}}}).$$

Hence, H is B-tempered bipolar neutrosophic hypergraph.

Theorem 10.4 *A bipolar neutrosophic hypergraph* $H = (\mu, \rho)$ *is* $B = (T^+_B, I^+_B, F^+_B, T^-_B, I^-_B, F^-_B)$*-tempered bipolar neutrosophic hypergraph if and only if* H *is simply ordered, elementary, and support simple.*

Proof Suppose that H is a $B = (T^+_B, I^+_B, F^+_B, T^-_B, I^-_B, F^-_B)$-tempered bipolar neutrosophic hypergraph of a crisp hypergraph H^*. Then clearly H is elementary and hence also will be support simple. To prove H to be simply ordered, let $c(H) = \{(X_1, \rho^*_1), (X_2, \rho^*_2), (X_3, \rho^*_3), \ldots, (X_n, \rho^*_n)\}$ be the core hypergraphs of H. Since H is elementary, then H will be ordered. To show that H is simply ordered, let us suppose the existence of a crisp hyperedge $E \in \rho^*_{i+1} \setminus \rho^*_i$. Then there will be an element $x^* \in E$ satisfying $B^+(x^*) = \alpha^+_{i+1}$, $B^-(x^*) = \alpha^-_{i+1}$. Since $B^+(x^*) = \alpha^+_{i+1} < \alpha_i$, $B^-(x^*) = \alpha^-_{i+1} > \alpha_i$, it is followed that $x^* \notin X_i$ and $E \nsubseteq X_i$, hence H is simply ordered.

Conversely, suppose that H is simply ordered, elementary, and support simple. Let $c(H) = \{(X_1, \rho^*_1), (X_2, \rho^*_2), (X_3, \rho^*_3), \ldots, (X_n, \rho^*_n)\}$. Define $B = (T^+_B, I^+_B, F^+_B, T^-_B, I^-_B, F^-_B) : X_n \to [0, 1] \times [-1, 0]$ as

$$B^+(x) = \begin{cases} \alpha^+_1, & \text{if } x \in X_1, \\ \alpha^+_i, & \text{if } x \in X_i \setminus X_{i-1}, i = 1, 2, 3, \ldots, n \end{cases}$$

$$B^-(x) = \begin{cases} \beta^-_1, & \text{if } x \in X_1, \\ \beta^-_i, & \text{if } x \in X_i \setminus X_{i-1}, i = 1, 2, 3, \ldots, n \end{cases}$$

We prove that $\rho = \{D_F = (T^+_{D_F}, I^+_{D_F}, F^+_{D_F}, T^-_{D_F}, I^-_{D_F}, F^-_{D_F}) : F \in \rho^*\}$, where

$$T^+_{D_F}(x) = \begin{cases} \min(T^+_B(y)) : y \in F, & \text{if } x \in F, \\ 0, & \text{otherwise.} \end{cases}$$

$$I^+_{D_F}(x) = \begin{cases} \min(I^+_B(y)) : y \in F, & \text{if } x \in F, \\ 0, & \text{otherwise.} \end{cases}$$

$$F^+_{D_F}(x) = \begin{cases} \max(F^+_B(y)) : y \in F, & \text{if } x \in F, \\ 0, & \text{otherwise.} \end{cases}$$

$$T^-_{D_F}(x) = \begin{cases} \max(T^-_B(y)) : y \in F, & \text{if } x \in F, \\ 0, & \text{otherwise.} \end{cases}$$

$$I^-_{D_F}(x) = \begin{cases} \max(I^-_B(y)) : y \in F, & \text{if } x \in F, \\ 0, & \text{otherwise.} \end{cases}$$

$$F^-_{D_F}(x) = \begin{cases} \min(F^-_B(y)) : y \in F, & \text{if } x \in F, \\ 0, & \text{otherwise.} \end{cases}$$

Let $F \in \rho_n^*$. Because H is support simple and elementary, there is only one bipolar neutrosophic hyperedge $C_F = (T_{C_F}^+, (I_{C_F}^+, (F_{C_F}^+, T_{C_F}^-, I_{C_F}^-, F_{C_F}^-)$ in ρ whose support is ρ^*. Also the supports of different hyperedges in ρ are different and lies in ρ_n^*. Thus, to prove that $\rho = \{D_F = (T_{D_F}^+, I_{D_F}^+, F_{D_F}^+, T_{D_F}^-, I_{D_F}^-, F_{D_F}^-) : F \in \rho^*\}$, it is enough to show that

$$T_{C_F}^+ = T_{D_F}^+, \quad I_{C_F}^+ = I_{D_F}^+, \quad F_{C_F}^+ = F_{D_F}^+,$$
$$T_{C_F}^- = T_{D_F}^-, \quad I_{C_F}^- = I_{D_F}^-, \quad F_{C_F}^- = F_{D_F}^-,$$

for all $F \in \rho_n^*$. Since, H is support simple, that is, all the hyperedges have different support and all hyperedges are elementary, the definition of $F_S(H)$ implies that $h(C_F)$ is equal to some α_i. Therefore, $\rho^* \subseteq X_i$. If $i > 1$, then $F \in \rho^* \backslash \rho_{i-1}^*$. Since $F \subseteq X_i$, the definition of $B = (T_B^+, I_B^+, F_B^+, T_B^-, I_B^-, F_B^-)$ implies that for each $x \in F$, $B^+(x) \geq \alpha_i^+$ and $B^-(x) \leq \alpha_i^-$. Our claim is that $B^+(x) = \alpha_i^+$ and $B^-(x) = \alpha_i^-$, for some $x \in F$. If not, then the definition of $B = (T_B^+, I_B^+, F_B^+, T_B^-, I_B^-, F_B^-)$ implies that $B^+(x) \geq \alpha_i^+$, $B^-(x) \leq \alpha_i^-$ for all $x \in F$, which follows that $F \subseteq X_{i-1}$ and therefore $F \in \rho^* \backslash \rho_{i-1}^*$ and because H is simply ordered $F \not\subseteq X_{i-1}$, which is a contradiction. Hence it is followed from the definition of D_F that $D_F = C_F$. Hence, H is B-tempered bipolar neutrosophic hypergraph. This completes the proof. $\qquad\square$

10.3 Transversals of Bipolar Neutrosophic Hypergraphs

A transversal for a crisp hypergraph $G = (X, E)$ is an arbitrary subset t of X such that for all $\varepsilon_i \in E, t \cap \varepsilon_i \neq \emptyset$. A transversal t is called the minimal transversal if no any proper subset of t is a transversal of G. We now define the concepts of transversals and minimal transversals of bipolar neutrosophic hypergraphs.

Definition 10.18 Let $H = (\mu, \rho)$ be a bipolar neutrosophic hypergraph on X. A *bipolar neutrosophic transversal* $T = (\tau^+, \tau^-)$ of H is a bipolar neutrosophic subset of X having the property $T^{(h(\rho_i), d(\rho_i))} \cap \rho_i^{(h(\rho_i), d(\rho_i))} \neq \emptyset$, for all $\rho_i \in \rho$.

$\quad$ T is called the *minimal bipolar neutrosophic transversal* for H if whenever $T_1 \subset T$, T_1 is not a bipolar neutrosophic transversal of H. The collection of all minimal bipolar neutrosophic transversals(and the bipolar neutrosophic hypergraph formed by this set) is denoted by $T_r(H)$.

Theorem 10.5 *Let $H = (\mu, \rho)$ be a bipolar neutrosophic hypergraph then the following statements are equivalent:*

(i) T *is a bipolar neutrosophic transversal of H.*

(ii) *For all $\rho_i \in \rho$ and for each $\alpha = (T_\alpha^+, I_\alpha^+, F_\alpha^+, T_\alpha^-, I_\alpha^-, F_\alpha^-) \in [0, 1] \times [-1, 0]$, satisfying the conditions $0 < (T_\alpha^+, I_\alpha^+, F_\alpha^+) < h(\rho_i)$, $0 > (T_\alpha^-, I_\alpha^-, F_\alpha^-) > d(\rho_i)$, $T^\alpha \cap \rho^\alpha \neq \emptyset$.*

(iii) T^α *is a transversal of H^α, for all $\alpha \in [0, 1] \times [-1, 0]$ with $0 < \alpha^+ < \alpha_1^+$, $0 > \alpha^- > \alpha_1^-$.*

Proof $(i) \Rightarrow (ii)$. Since T is a bipolar neutrosophic transversal of H therefore for any $\alpha \in [0, 1] \times [-1, 0]$ satisfying the conditions, $0 < (T_\alpha^+, I_\alpha^+, F_\alpha^+) < h(\rho_i)$, $0 > (T_\alpha^-, I_\alpha^-, F_\alpha^-) > d(\rho_i)$, $T^{(\alpha^+, \alpha^-)} \supseteq T^{(h(\rho_i), d(\rho_i))}$ $\rho_i^{(\alpha^+, \alpha^-)} \supseteq \rho_i^{(h(\rho_i), d(\rho_i))}$. Hence, $T^{(\alpha^+, \alpha^-)} \cap \rho_i^{(\alpha^+, \alpha^-)} \supseteq T^{(h(\rho_i), d(\rho_i))} \cap \rho_i^{(h(\rho_i), d(\rho_i))} \neq \emptyset$. Since T is transversal.
$(ii) \Rightarrow (iii)$. Since $T^\alpha \cap \rho_i^\alpha \neq \emptyset$, for all $\rho_i \in \rho$ and $0 < \alpha^+ < \alpha_1^+, 0 > \alpha^- > \alpha_1^-$, therefore T^α is a transversal of H^α.
$(iii) \Rightarrow (i)$. The proof is trivial.   □

Remark 10.3 If T is a minimal bipolar neutrosophic transversal of H, then it is not necessary that T^α will be a minimal transversal of H^α for all $\alpha \in [0, 1] \times [-1, 0]$. Further, the set of all minimal transversals of H may not form a hypergraph on X.

Definition 10.19 Let T be a bipolar neutrosophic set satisfying the property that T^α is a minimal transversal for H^α for all $\alpha \in [0, 1] \times [-1, 0]$, then T is called the *locally minimal bipolar neutrosophic transversal* of H. The set of all *locally minimal bipolar neutrosophic transversals* of H is denoted by $T_r^*(H)$.

Lemma 10.1 Let $H = (\mu, \rho)$ be a bipolar neutrosophic hypergraph with $F_S(H) = \{\alpha_1, \alpha_2, \alpha_3, \ldots, \alpha_n\}$. If T is a bipolar neutrosophic transversal of H, then $h(T) > h(\rho_j)$, $d(T) < d(\rho_j)$, for all $\rho_j \in \rho$. If T is a minimal transversal then,

$$h(T) = \max\{h(\rho_j) : \rho_j \in \rho\} = \alpha_1^+,$$
$$d(T) = \min\{d(\rho_j) : \rho_j \in \rho\} = \alpha_1^-.$$

Proof Since T is a bipolar neutrosophic transversal of H, therefore, $T^{(h(\rho_j), d(\rho_j))} \cap \rho_j^{(h(\rho_j), d(\rho_j))} \neq \emptyset$, for all $\rho_j \in \rho$. Let x be a generic element of $supp(T)$ then, $\tau^+(x) \geq h(\rho_j), \tau^-(x) \leq d(\rho_j)$, which implies that $h(T) \geq \tau^+(x) \geq h(\rho_j), d(T) \leq \tau^-(x) \leq d(\rho_j)$. If T is minimal and

$$h(\rho_j) = \max\{\rho_j(x) : \text{ for all } x \in X, \rho_j \in \rho\} = \alpha_1^+,$$
$$d(\rho_j) = \min\{\rho_j(x) : \text{ for all } x \in X, \rho_j \in \rho\} = \alpha_1^-,$$

which implies that

$$h(T) = \max\{h(\rho_j) : \rho_j \in \rho\} = \alpha_1^+,$$
$$d(T) = \min\{d(\rho_j) : \rho_j \in \rho\} = \alpha_1^-.$$

□

Theorem 10.6 Let $H = (\mu, \rho)$ be a bipolar neutrosophic hypergraph on X. For all $T \in T_r(H)$ and for each $x \in X$ such that $T(x) \in F_S(H)$. The fundamental sequence of $T_r(H)$ is a subset of $F_S(H)$.

Proof Let $F_S(H) = \{\alpha_1, \alpha_2, \alpha_3, \ldots, \alpha_n\}$ be the fundamental sequence of H. Suppose $T = (\tau^+, \tau^-) \in T_r(H)$ and $T(x) \in (\alpha_{i+1}, \alpha_i]$. Define a mapping $\phi = (\phi^+, \phi^-)$ such that

$$\phi^+(v) = \begin{cases} \alpha_i^+, & \text{if } x = v, \\ \tau^+(v), & \text{otherwise.} \end{cases} \quad \phi^-(v) = \begin{cases} \alpha_i^-, & \text{if } x = v, \\ \tau^-(v), & \text{otherwise.} \end{cases}$$

From the definition of ϕ, we have $\phi^{\alpha_i} = T^{\alpha_i}$. The definition of $F_S(H)$ gives $H^\alpha = H^{\alpha_i}$, for all $\alpha \in (\alpha_{i+1}, \alpha_i]$. Therefore, ϕ^{α_i} is a transversal of H^α, for all $\alpha \in (\alpha_{i+1}, \alpha_i]$. Since $T = (\tau^+, \tau^-)$ is a bipolar neutrosophic transversal and $\phi^\alpha = T^\alpha$, for all $\alpha \notin (\alpha_{i+1}, \alpha_i]$, $\phi = (\phi^+, \phi^-)$ is a bipolar neutrosophic transversal.

Now $\phi = (\phi^+, \phi^-) \leq T = (\tau^+, \tau^-)$ and the minimality of T implies that $\phi = T$. Hence, $T(x) = \phi(x) = \alpha_1$. Thus, for each $T \in T_r(H)$ and for all $x \in X$, we have $T(x) \in F_S(H)$. Therefore, $F_S(T_r(H)) \subseteq F_S(H)$. □

Theorem 10.7 *The set $T_r(H)$ of all minimal transversals of H is sectionally elementary.*

Proof Let $F_S(T_r(H)) = \{\alpha_1, \alpha_2, \alpha_3, \ldots, \alpha_n\}$. Suppose that there is some $T \in T_r(H)$ and some $\alpha \in (\alpha_{i+1}, \alpha_i]$ such that T^{α_i} is a proper subset of T^α. Since $[T_r(H)]^\alpha = [T_r(H)]^{\alpha_i}$, there exists some $\phi \in T_r(H)$ such that $\phi^{\alpha_i} = T^\alpha$. Then $T^{\alpha_i} \subset \phi^{\alpha_i}$ implies the bipolar neutrosophic set β defined as

$$\beta(x) = \begin{cases} \alpha, & \text{if } x \in \phi^{\alpha_i} \setminus T^{\alpha_i}, \\ \phi(x), & \text{otherwise.} \end{cases}$$

is a bipolar neutrosophic transversal of H. Now $\alpha < \phi$, which is a contradiction to the minimality of ϕ. □

Theorem 10.8 *Let $F_S(H) = \{\alpha_1, \alpha_2, \alpha_3, \ldots, \alpha_n\}$. The top level cut of T, T^{α_1} is a minimal transversal of H^{α_1}, for each $T \in T_r(H)$.*

Proof On contrary, suppose that there is a minimal transversal T of H^{α_i} such that $T \subset T^{\alpha_i}$. Define a bipolar neutrosophic set β such that

$$\beta(x) = \begin{cases} \alpha_2, & \text{if } x \in T^{\alpha_1} \setminus T, \\ T(x), & \text{otherwise.} \end{cases}$$

Then β is a bipolar neutrosophic transversal of H and $\beta < T$, a contradiction to the minimality of T. □

Although finding $T_r(H)$ for bipolar neutrosophic hypergraphs is more complicated. We now give an algorithm for finding $T_r(H)$ in the following Construction 10.4.

Construction 10.4 *Algorithm for finding $T_r(H)$* Let $H = (\mu, \rho)$ be a bipolar neutrosophic hypergraph with the set of core hypergraphs $c(H) = \{H^{\alpha_1}, H^{\alpha_2}, H^{\alpha_3}, \ldots, H^{\alpha_n}\}$. An iterative procedure to find the minimal bipolar neutrosophic transversal T of H is as follows.

1. Find a crisp minimal transversal t_1 of H^{α_1}.
2. Find a minimal transversal t_2 of H^{α_2} satisfying the property $t_1 \subseteq t_2$, i.e., construct a new hypergraph H_2 with hyperedges ρ^{α_2} which is augmented having a loop at each $x \in t_1$. Equivalently, we can say that $\rho(H_2) = \rho^{\alpha_2} \cup \{\{x\}|x \in t_1\}$. Let t_2 be an arbitrary minimal transversal of H_2.
3. Continue the procedure iteratively, we get a sequence of minimal transversals $t_1 \subseteq t_2 \subseteq t_3 \subseteq \cdots \subseteq t_j$ such that t_j be the minimal transversal of H^{α_j} satisfying the property $t_{j-1} \subseteq t_j$.
4. Let μ_j be the elementary bipolar neutrosophic set having the support T_j and

$$(h(\mu_j), d(\mu_j)) = (\alpha_j^+, \alpha_j^-) = \alpha_j, \ 1 \le j \le n. \ \text{Then,} \ T = \bigcup_{j=1}^{n} \{\mu_j | 1 \le j \le n\} \ \text{is}$$

a minimal bipolar neutrosophic transversal of H.

To see the validity of this algorithm, note that T is a bipolar neutrosophic transversal of H. If $T_1 < T$, then there is an element $x \in X$ such that

$$T_{T_1}^+(x) < T_T^+(x), \quad I_{T_1}^+(x) < I_T^+(x), \quad F_{T_1}^+(x) > F_T^+(x),$$
$$T_{T_1}^-(x) > T_T^-(x), \quad I_{T_1}^-(x) > I_T^-(x), \quad F_{T_1}^-(x) < F_T^-(x).$$

But then x is not an element of $T(x)$-level cut of T_1. So $T_1^{T(x)} \subset T^{T(x)} = t^{T(x)}$, therefore $T_1^{T(x)}$ is not a transversal of $H^{T(x)}$. Hence T is a minimal bipolar neutrosophic transversal of H. The above construction clearly shows that T is also a locally minimal transversal, that is, $T \in T_r^*(H)$.

Example 10.3 Consider a bipolar neutrosophic hypergraph $H = (\mu, \rho)$ as shown in Fig. 10.1. The fundamental sequence of H is $F_S(H) = \{\alpha_1 = (0.9, 0.8, 0.1, -0.9, -0.8, -0.1), \ \alpha_2 = (0.8, 0.7, 0.1, -0.8, -0.7, -0.1), \ \alpha_3 = (0.5, 0.4, 0.3, -0.5, -0.4, -0.3)\}$. Clearly, $t_1 = t_2 = \{x_1\}$ is a minimal(crisp) transversal of H^{α_1} and H^{α_2} and $t_3 = \{x_1, x_2, x_4\}$ is a minimal transversal of H^{α_3}. Note that $t_1 \subseteq t_2 \subseteq t_3$.

Let $\mu_1 = \{(x_1, 0.9, 0.8, 0.1, -0.9, -0.8, -0.1)\}$, $\mu_2 = \{(x_1, 0.8, 0.7, 0.1, -0.8, -0.7, -0.1)\}$, $\mu_3 = \{(x_1, 0.5, 0.4, 0.3, -0.5, -0.4, -0.3), (x_2, 0.5, 0.4, 0.3, -0.5, -0.4, -0.3), (x_4, 0.5, 0.4, 0.3, -0.5, -0.4, -0.3)\}$. Hence, $T = \bigcup_{j=1}^{n} \mu_j = \{(x_1, 0.9, 0.8, 0.1, -0.9, -0.8, -0.1), (x_4, 0.5, 0.4, 0.3, -0.5, -0.4, -0.3)\}$, is a minimal bipolar neutrosophic transversal of H.

Theorem 10.9 *Let T be a bipolar neutrosophic transversal of bipolar neutrosophic hypergraph H. Then, there is a minimal bipolar neutrosophic transversal T_1 of H such that $T_1 \le T$.*

Proof Since T^{α_1} is a minimal transversal of H^{α_1}. Define recursively t_j as a transversal of H^{α_j}, then t_j is minimal transversal because of the property $t_{j-1} \subseteq t_j \subseteq T^{\alpha_j}$. Let T_j be the elementary bipolar neutrosophic set having support t_j, height α_j^+ and depth α_j^-. Then, $T_1 = \bigcup_{j=1}^{n} \{T_j | 1 \le j \le n\}$ is a minimal bipolar neutrosophic transversal of H satisfying the property $T_1 \le T$. $\square$

Theorem 10.10 *Let $H = (\mu, \rho)$ be a bipolar neutrosophic hypergraph and $T \in T_r(H)$. If $x \in supp(T)$ then there exists a bipolar neutrosophic hyperedge $\rho_i \in \rho$ of H for which the following conditions are satisfied.*

(i) $\rho_i^+(x) = h(\rho_i) = T^+(x) > 0$, $\rho_i^-(x) = d(\rho_i) = T^-(x) < 0$,

(ii) $T^{(h(\rho_i),d(\rho_i))} \cap \rho_i^{(h(\rho_i),d(\rho_i))} = \{x\}$.

Proof (i) Suppose that $T^+(x) > 0$ and $T^-(x) < 0$. Let B be the set of all bipolar neutrosophic hyperedges of H in which for all $\beta \in B$, $\beta^+(x) \geq T^+(x)$, $\beta^-(x) \leq T^-(x)$. Since, $T^{T(x)}$ is a transversal of $H^{T(x)}$ and $x \in T^{T(x)}$, then the set is non-empty. Further, for each $\beta \in B$ we have

$$h(\beta) \geq \beta^+(x) \geq T^+(x)$$
$$d(\beta) \leq \beta^-(x) \leq T^-(x).$$

On contrary suppose that (i) does not hold then for each $\beta \in B$ we have $h(\beta) > T^+(x)$, $d(\beta) < T^-(x)$ and there is $x_\beta \neq x$ such that $x_\beta \in \beta^{(h(\beta),d(\beta))} \cap T^{(h(\beta),d(\beta))}$ (using the definition of transversals). Define a bipolar neutrosophic set ψ as

$$\psi^+(v) = \begin{cases} T^+(v), & \text{if } x \neq v \\ \max\{h(\rho_i)|h(\rho_i) < T^+(x)\}, & \text{if } x = v \end{cases}$$

$$\psi^-(v) = \begin{cases} T^-(v), & \text{if } x \neq v \\ \min\{d(\rho_i)|d(\rho_i) < T^-(x)\}, & \text{if } x = v \end{cases}$$

Then, ψ is a transversal of H and $\psi < T$. A contradiction to the minimality of T. Hence (i) holds for some $\rho_i \in \rho$.

(ii) Suppose each $\beta \in B$ satisfies the condition (i) and contrary suppose contains an element $x_\beta \neq x$ such that $x_\beta \in \beta^{(h(\beta),d(\beta))} \cap T^{(h(\beta),d(\beta))}$. Then there exists a bipolar neutrosophic transversal ψ of H such that $\psi < T$, which is a contradiction. $\square$

Theorem 10.11 *Let $H = (\mu, \rho)$ be a bipolar neutrosophic hypergraph and let $x \in X$. Then there exists a minimal bipolar neutrosophic transversal $T \in T_r(H)$ if and only if there is a $\rho_i \in \rho$ satisfying the following conditions:*

(i) $\rho_i^+(x) = h(\rho_i)$ and $\rho_i^-(x) = d(\rho_i)$.

(ii) For all $\rho_i^* \in \rho$ with $h(\rho_i^*) > h(\rho_i)$ and $d(\rho_i^*) < d(\rho_i)$, $\rho_i^{*(h(\rho_i^*),d(\rho_i^*))} \not\subseteq \rho_i^{(h(\rho_i),d(\rho_i))}$.

(iii) Any hyperedge of $H^{(h(\rho_i),d(\rho_i))}$ is not a proper subset of $\rho_i^{(h(\rho_i),d(\rho_i))}$.

Proof (i) Suppose $T \in T_r(H)$ and $T^+(x) > 0$, $T^-(x) < 0$ then (i) follows from Theorem 10.10.

(ii) Suppose on contrary, for some $\rho_i' \in \rho$ having $h(\rho_i') > h(\rho_i)$ and $d(\rho_i') < d(\rho_i)$, we have $\rho_i'^{(h(\rho_i),d(\rho_i))} \subseteq \rho_i^{(h(\rho_i),d(\rho_i))}$. Then, there exists an element $v \neq x$ satisfying the condition $v \in \rho_i'^{(h(\rho_i'),d(\rho_i'))} \cap T^{(h(\rho_i'),d(\rho_i'))} \subseteq \rho_i^{(h(\rho_i),d(\rho_i))} \cap T^{(h(\rho_i),d(\rho_i))}$, which is contradiction to Theorem 10.10. Hence (ii) holds.

(iii) On contrary, suppose that for each $\rho_i \in \rho$ satisfying the conditions (i) and (ii), there exists $\rho_i^* \in \rho$ such that $\emptyset \subset \rho_i^{*(h(\rho_i),d(\rho_i))} \subset \rho_i^{(h(\rho_i),d(\rho_i))}$. Since, $\rho_i^{*(h(\rho_i),d(\rho_i))}$ is non-empty then we have, $\rho_i^{*+}(x) = h(\rho_i) = h(\rho_i^*)$, $\rho_i^{*-}(x) = d(\rho_i) = h(\rho_i^*)$. If $(\rho_i^{*+}(x), \rho_i^{*-}(x)) = (h(\rho_i^*), d(\rho_i^*))$, our supposition implies that there exists $\emptyset \subset \rho_i' \in \rho$ such that

$$\emptyset \subset \rho_i'^{(h(\rho_i),d(\rho_i))} \subset \rho_i^{*(h(\rho_i),d(\rho_i))} \subset \rho_i^{(h(\rho_i),d(\rho_i))}.$$

This recursive process must end after finitely many iterations, thus we suppose that $\rho_i^{*+}(x) < h(\rho_i^*)$, $\rho_i^{*-}(x) > d(\rho_i^*)$. Then, there exists some $v \neq x$ such that

$$x \in \rho_i^{*(h(\rho_i),d(\rho_i))} \cap T^{*(h(\rho_i),d(\rho_i))} \subseteq \rho_i^{(h(\rho_i),d(\rho_i))} \cap T^{*(h(\rho_i),d(\rho_i))},$$

which is a contradiction to Theorem 10.10. Hence, our supposition is false and (iii) holds.

Conversely, we assume that for any vertex $x \in X$ and hyperedges $\rho_i, \rho_i^* \in \rho$ all three conditions hold. Suppose that $h(\rho_i) \geq \alpha_i^+$, $d(\rho_i) \leq \alpha_i^-$, for some $\alpha_i \in F_S(H)$. From conditions (ii) and (iii), there exists $y \in \rho_i^{*(h(\rho_i^*),d(\rho_i^*))} \setminus \rho_i^{(h(\rho_i),d(\rho_i))}$, for each $\rho_i^* \in \rho$ and $h(\rho_i^*) \geq h(\rho_i), d(\rho_i^*) \leq d(\rho_i)$. Let Z be the set of all such vertices, thus $Z \cap \rho_i^{(h(\rho_i),d(\rho_i))} = \emptyset$. Let $T_1, T_2, T_3, \ldots, T_n$ be the sequence of transversals such that $T_j \subseteq Z$, for all $1 \leq j < i$ and $T_i \subseteq Z \cup \{x\}$. Then for each i, $x \in T_i$. Let μ_i be a bipolar neutrosophic set corresponding to T_i, then $T = \bigcup_{i=1}^{n} \mu_i$ is the minimal bipolar neutrosophic transversal and $x \in supp(T)$.   □

10.4   Applications of Bipolar Neutrosophic Hypergraphs

In recent years, research in anthropology (study of human and their culture) has been involved with evaluation of social networks. Such type of networks is developed by defining one or more relations on the set of individuals. The relations can be taken from effective relationships, aspects of some organizations and from a wide variety of other resources. Network models, represented as simple graphs, lack some information necessary for super-dyadic relationships between the vertices. Natural existence of hyperedges can be found in co-citation, e-mail networks, co-authorship, weblog networks and social networks, etc. Representation of these models as hypergraphs maintains the dyadic relationships.

10.4.1 Super-Dyadic Managements in Marketing Channels

Dyadic communication management has been a primary tool in marketing channels. Marketing researchers and managers have recognized mutual commitment in
marketing channels as central key to successful marketing and to produce benefits
for associations. Bipolar neutrosophic hypergraphs consist of marketing managers
as vertices and hyperedges represent their dyadic communications involving their
correlative thoughts, ideas, plans, projects, and objectives. Strong dyadic communications can improve the marketing strategies and the production of an organization. A bipolar neutrosophic network model representing the dyadic communications
among the marketing managers of an organization is given in Fig. 10.4. The positive
membership degrees of each person represent the percentage of its positive dyadic
behavior and negative membership degrees describe its negative dyadic behavior
toward the other members of the same dyad group. Adjacent level between any pair
of vertices depicts that how much their dyadic relationship is competent. By routine
calculations, we have adjacent levels as given in Table 10.3.

It can be noted that the most competent dyadic pair is (Abel, Ansel). Bipolar neutrosophic hyperedges are taken as the different digital marketing strategies adopted
by the different dyadic groups of the same organization. The main objective of this
model is to figure out the most effective dyad of digital marketing techniques. The
marketing managers are divided into six different groups and the digital marketing
strategies adopted by these six groups are represented by hyperedges, i.e., the bipolar
neutrosophic hyperedges $\{S_1, S_2, S_3, S_4, S_5, S_6\}$ represent the following strategies
{Spread the wealth, Align strategies, Analyze the competition, Provide value, Build
relationships, Promotions}, respectively. The individual effects of positive membership and negative membership degrees of each marketing strategy toward the
achievements of an organization are given in Tables 10.4 and 10.5, respectively.

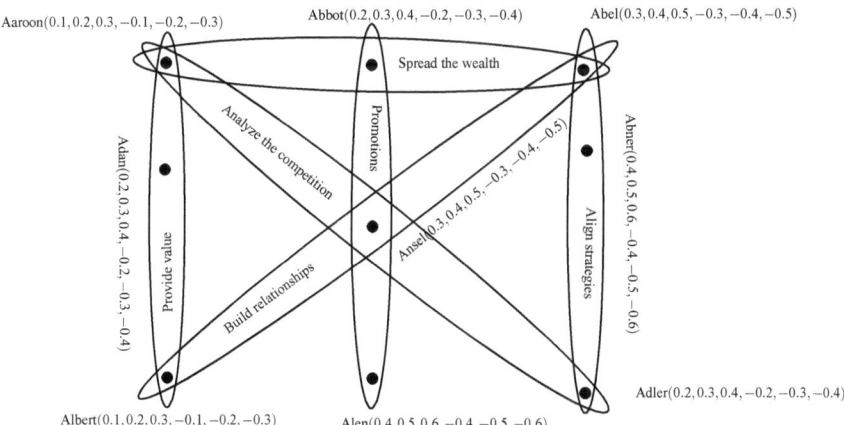

Fig. 10.4 Super-dyadic managements in marketing channels

Table 10.3 Adjacent levels

Dyad airs	Adjacent level	Dyad pairs	Adjacent level
γ(Aaroon, Abel)	$(0.1, 0.2, 0.5, -0.1, -0.2, -0.5)$	γ(Abel, Abner)	$(0.3, 0.4, 0.6, -0.3, -0.4, -0.6)$
γ(Abel, Adler)	$(0.2, 0.3, 0.5, -0.2, -0.3, -0.5)$	γ(Abner, Adler)	$(0.2, 0.3, 0.6, -0.2, -0.3, -0.6)$
γ(Abbot, Abel)	$(0.2, 0.3, 0.5 - 0.2, -0.3, -0.5)$	γ(Abel, Ansel)	$(0.3, 0.4, 0.5, -0.3, -0.4, -0.5)$
γ(Abel, Albert)	$(0.1, 0.2, 0.5, -0.1, -0.2, -0.5)$	γ(Ansel, Albert)	$(0.1, 0.2, 0.5, -0.1, -0.2, -0.5)$
γ(Abbot, Ansel)	$(0.2, 0.3, 0.5, -0.2, -0.3, -0.5)$	γ(Abbot, Alen)	$(0.2, 0.3, 0.6 - 0.2, -0.3, -0.6)$
γ(Alen, Alsen)	$(0.3, 0.4, 0.6, -0.3, -0.4, -0.6)$	γ(Aaroon, Ansel)	$(0.1, 0.2, 0.5, -0.1, -0.2, -0.5)$
γ(Aaroon, Adler)	$(0.1, 0.2, 0.4, -0.1, -0.2, -0.4)$	γ(Ansel, Adler)	$(0.2, 0.3, 0.5, -0.2, -0.3, -0.5)$
γ(Aaroon, Adan)	$(0.1, 0.2, 0.4, -0.1, -0.2, -0.4)$	γ(Aaroon, Albert)	$(0.1, 0.2, 0.3, -0.1, -0.2, -0.3)$
γ(Albert, Adan)	$(0.1, 0.2, 0.4, -01, -0.2, -0.4)$	γ(Aaroon, Abbot)	$(0.1, 0.2, 0.4, -0.1, -0.2, -0.4)$

Table 10.4 Positive effects of marketing strategies

Marketing strategy	Increase in earnings	Indeterminate factor	Break-even point
Spread the wealth	0.1	0.2	0.5
Align strategy	0.2	0.3	0.6
Analyze the competition	0.1	0.2	0.5
Provide value	0.1	0.2	0.4
Build the relationships	0.1	0.2	0.5
Promotions	0.2	0.3	0.6

Table 10.5 Negative effects of marketing strategies

Marketing strategy	Decrease in earnings	Indeterminate factor of loss	Critical point
Spread the wealth	−0.1	−0.2	−0.5
Align strategy	−0.2	−0.3	−0.6
Analyze the competition	−0.1	−0.2	−0.5
Provide value	−0.1	−0.2	−0.4
Build the relationships	−0.1	−0.2	−0.5
Promotions	−0.2	−0.3	−0.6

Table 10.6 Positive and negative effects of dyadic marketing

Dyadic strategies	Positive effects	Negative effects
σ(Spread the wealth, align strategy)	(0.1, 0.2, 0.6)	(−0.1, −0.2, −0.6)
σ(Spread the wealth, build the relationships)	(0.1, 0.2, 0.5)	(−0.1, −0.2, −0.5)
σ(Spread the wealth, promotions)	(0.1, 0.2, 0.6)	(−0.1, −0.2, −0.6)
σ(Spread the wealth, analyze the competition)	(0.1, 0.2, 0.5)	(−0.1, −0.2, −0.5)
σ(Spread the wealth, provide value)	(0.1, 0.2, 0.5)	(−0.1, −0.2 − 0.5)
σ(Align strategy, analyze the competition)	(0.1, 0.2, 0.6)	(−0.1, −0.2, −0.6)
σ(Align strategy, provide value)	(0.1, 0.2, 0.6)	(−0.1, −0.2, −0.6)
σ(Align strategy, build the relationships)	(0.1, 0.2, 0.6)	(−0.1, −0.2, −0.6)
σ(Align strategy, promotions)	(0.2, 0.3, 0.6)	(−0.2, −0.3, −0.6)
σ(Analyze the competition, build relationships)	(0.1, 0.2, 0.5)	(−0.1, −0.2, −0.6)
σ(Analyze the competition, promotions)	(0, 1, 0.2, 0.6)	(−0.1, −0.2, −0.5)
σ(Provide value, build relationships)	(0.1, 0.2, 0.5)	(−0.1, −0.2, −0.5)
σ(Provide value, promotions)	(0.1, 0.2, 0.6)	(−0.1, −0.2, −0.6)
σ(Build the relationships, promotions)	(0.1, 0.2, 0.6)	(−0.1, −0.2, −0.6)

Effective dyads of marketing strategies enhance the performance of an organization and perceive the better techniques to be adopted. Positive and negative adjacency of all dyadic communication managements is given in Table 10.6.

The most powerful and competent marketing strategies adopted mutually are Align strategy and Promotions. Thus, to improve the efficiency of an organization, dyadic managements should align their marketing strategies with their business strategies and should develop the strategies for the marketing areas that align with the marketing strategy. Then, they should use the promotion techniques to allure customers to purchase their products. The membership degrees of this dyad is $(0.2, 0.3, 0.6, −0.2, −0.3, −0.6)$ which shows that the combine effect of this dyad will increase the earnings of an organization up to 20% and loss will be reduced up to 20%, chances of break-even point will be 60%. Thus, we conclude that super dyad marketing communications are more effective to improve the performance of an organization. The method of searching out the most effective dyads is described in the following algorithm is given in Table 10.7.

10.4.2 Production of New Alleles Using Mutations

When a DNA gene is changed in such a way that the genetic message carried by this gene is altered is called mutation. Randomly genetic alterations caused by evaluation have different effects. Most genetic changes are neutral, some other are injurious and very rare of them have positive effects. If the germ cells (ova and spermatozoa) are affected by the mutations then the alteration will be passed to all cells and will

Table 10.7 Algorithm for super-dyadic managements

Algorithm for searching out the most effective dyads

1. Input the degrees of membership of all vertices(marketing managers) $x_1, x_2, \ldots, x_n$.

2. Input the degrees of membership of all hyperedges $S_1, S_2, \ldots, S_m$.

3. Calculate the adjacency level between vertices x_i and x_j,

$$\gamma^+(x_i, x_j) = \max_k \min[T_k^+(x_i), T_k^+(x_j)], \quad \max_k \min[I_k^+(x_i), I_k^+(x_j)], \quad \min_k \max[F_k^+(x_i), F_k^+(x_j)],$$
$$\gamma^-(x_i, x_j) = \min_k \max[T_k^-(x_i), T_k^-(x_j)], \quad \min_k \max[I_k^-(x_i), I_k^-(x_j)], \quad \max_k \min[F_k^-(x_i), F_k^-(x_j)].$$

4. Find the most competent dyadic pair as

$$\max T_\gamma^+(x_i, x_j), \quad \max I_\gamma^+(x_i, x_j), \quad \min F_\gamma^+(x_i x_j),$$
$$\min T_\gamma^-(x_i, x_j), \quad \min I_\gamma^-(x_i, x_j), \quad \max F_\gamma^-(x_i, x_j).$$

5. Calculate the adjacency level between hyperedges S_k and S_j as,

$$\sigma^+(S_k, S_j) = \max_k \min[T_k^+(x), T_j^+(x)], \max_k \min[I_k^+(x), I_j^+(x)], \min_k \max[F_k^+(x), F_j^+(x)],$$
$$\sigma^-(S_k, S_j) = \min_k \max[T_k^-(x), T_j^-(x)], \min_k \max[I_k^-(x), I_j^-(x)], \max_k \min[F_k^-(x), F_j^-(x)].$$

6. Find the most effective super dyad management as,

$$\max T_\sigma^+(S_k, S_j), \quad \max I_\sigma^+(S_k, x_j), \quad \min F_\sigma^+(S_k, S_j),$$
$$\min T_\sigma^-(S_k, S_j), \quad \min I_\sigma^-(S_k, x_j), \quad \max F_\sigma^-(S_k, S_j).$$

---

affect the future generations. Such type of mutations is called "germline mutations" and can cause the inherited properties. Understanding their multiple effects, some mutations can provide auspicious improvements in newly productive genes as their resistance to many diseases can be increased by inheriting the beneficial mutations from other genes. We will take the vertices of a bipolar neutrosophic hypergraph representing the peoples having genes of some beneficial mutations existing in human beings. Groups of families having the same mutant properties are shown by bipolar neutrosophic hyperedges $\{M_1, M_2, M_3, M_4, M_5\}$. Positive membership degrees and negative membership degrees contain the percentage of effects of mutations. To create new families from these genes, new alleles, we will apply the minimal property of transversals of bipolar neutrosophic hypergraphs. In this case, DNA mutations of all genes are copied into every new cell of the new growing embryo. Bipolar neutrosophic model is shown in Fig. 10.5.

The positive membership degrees and negative membership degrees of each mutation reveal the percentage of positive effects and negative effects of corresponding mutation, respectively. It can be noted from Table 10.8, that the families with Apo-AIM mutation have 90% lower risk of heart disease, 20% have chances to being neutral and there are 10% chances of failure of this mutation. Negative membership degrees show that the possibility of occurrence of bad effects is -90% and possibility of non occurrence is -10%. Consider a bipolar neutrosophic set $T = \{(g_1, 0.8, 0.2, 0.1, -0.8, -0.2, -0.1), (g_8, 0.6, 0.3, 0.2, -0.6, -0.3, -0.2), (g_{11}, 0.9, 0.2, 0.1, -0.9, -0.2, -0.1)\}$. By routine calculations, we have

Fig. 10.5 Bipolar
neutrosophic hypergraph
model

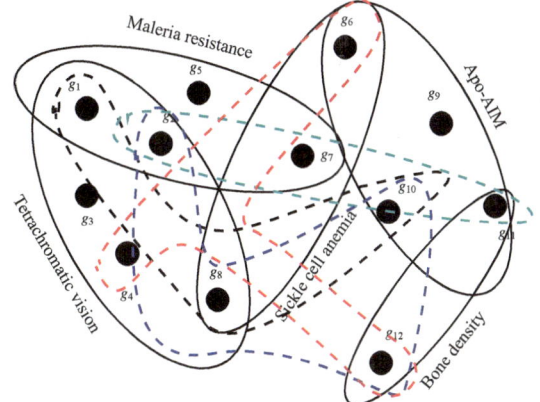

$$T^{(0.9,0.2,0.1,-0.9,-0.2,-0.1)} \cap M_1^{(0.9,0.2,0.1,-0.9,-0.2,-0.1)} = \{g_{11}\} \cap \{g_6, g_9, g_{10}, g_{11}\} = g_{11},$$

$$T^{(0.5,0.1,0.1,-0.5,-0.1,-0.1)} \cap M_2^{(0.5,0.1,0.1,-0.5,-0.1,-0.1)} = \{g_{11}, g_1\} \cap \{g_{12}, g_{11}\} = g_{11},$$

$$T^{(0.6,0.3,0.2,-0.6,-0.3,-0.2)} \cap M_3^{(0.6,0.3,0.2,-0.6,-0.3,-0.2)} = \{g_8\} \cap \{g_7, g_8\} = g_8,$$

$$T^{(0.5,0.1,0.1,-0.5,-0.1,-0.1)} \cap M_4^{(0.5,0.1,0.1,-0.5,-0.1,-0.1)} = \{g_1, g_{11}\} \cap \{g_1, g_2, g_5, g_7\} = g_1,$$

$$T^{(0.8,0.2,0.1,-0.8,-0.2,-0.1)} \cap M_5^{(0.8,0.2,0.1,-0.8,-0.2,-0.1)} = \{g_1, g_{11}\} \cap \{g_1, g_2, g_3, g_4, g_8\} = g_1.$$

Thus, T is a bipolar neutrosophic transversal. It can be seen that the intersection
of bipolar neutrosophic set is non-empty with all bipolar neutrosophic hyperedges.
Thus, this family will contain the maximum mutations among all the hyperedges.
By following the same technique, we can determine different families which inherit
the mutual properties of all the above mutations and thus are more strong, healthy,
and sharp. Hence,

$$T = \{(g_{11}, 0.8, 0.2, 0.1, -0.8, -0.2, -0.1), (g_7, 0.6, 0.3, 0.2, -0.6, -0.3, -0.2),$$
$$(g_2, 0.9, 0.2, 0.1, -0.9, -0.2, -0.1)\},$$
$$T = \{(g_{10}, 0.9, 0.2, 0.1, -0.9, -0.2, -0.1), (g_{12}, 0.9, 0.2, 0.1, -0.9, -0.2, -0.1),$$
$$(g_8, 0.6, 0.3, 0.2, -0.6, -0.3, -0.2)\},$$
$$T = \{(g_7, 0.5, 0.1, 0.1, -0.5, -0.2, -0.1), (g_4, 0.8, 0.2, 0.1, -0.8, -0.2, -0.1),$$
$$(g_6, 0.9, 0.2, 0.1, -0.9, -0.2, -0.1), (g_{12}, 0.5, 0.1, 0.1, -0.5, -0.1, -0.1)\},$$

are the required families and are shown by dashed lines in Fig. 10.5. The procedure
adopted in our application is given in following Table 10.9.

Table 10.8 Effects of mutations

Apo-AIM	$(0.9, 0.2, 0.1, -0.9, -0.2, -0.1)$
Bone density	$(0.5, 0.1, 0.1, -0.5, -0.1, -0.1)$
Sickle cell anemia	$(0.6, 0.3, 0.2, -0.6, -0.3, -0.2)$
Malaria resistance	$(0.5, 0.1, 0.1, -0.5, -0.1, -0.1)$
Tetrachromatic vision	$(0.8, 0.2, 0.1, -0.8, -0.2, -0.1)$

Table 10.9 Algorithm for new alleles

To find different families which inherit the mutual properties of all above mutations

1. Input the degree of membership of all bipolar neutrosophic hyperedges $M_1, M_2, \ldots, M_n$.

2. Calculate the height of bipolar neutrosophic hyperedge $M_j = \{x_l, x_{l+1}, \ldots, x_k\}$, $1 \leq l \leq n-1$, $2 \leq k \leq n$ as,

$$h(M_j) = \{\max T^+_{M_j}(x_i), \max I^+_{M_j}(x_i), \min F^+_{M_j}(x_i)\}.$$

3. Calculate the depth of bipolar neutrosophic hyperedge $M_j = \{x_l, x_{l+1}, \ldots, x_k\}$, $1 \leq l \leq n-1$, $2 \leq k \leq n$ as,

$$d(M_j) = \{\min T^-_{M_j}(x_i), \min I^-_{M_j}(x_i), \max F^-_{M_j}(x_i)\}.$$

4. Find a minimal bipolar neutrosophic transversal T such that $T^{(h(M_j), d(M_j))} \cap M_j^{(h(M_j), d(M_j))} \neq \emptyset$ using the following steps.

5. Find a crisp minimal transversal t_1 of H^{α_1}.

6. Find a minimal transversal t_2 of H^{α_2} satisfying the property $t_1 \subseteq t_2$, i.e.,

$$\rho(H_2) = \rho^{\alpha_2} \cup \{\{x\} | x \in t_1\}.$$

Let t_2 be an arbitrary minimal transversal of H_2.

7. Continue the procedure iteratively, obtain a sequence of minimal transversals $t_1 \subseteq t_2 \subseteq t_3 \subseteq \cdots \subseteq t_j$ such that t_j be the minimal transversal of H^{α_j} satisfying the property $t_{j-1} \subseteq t_j$.

8. Let μ_j be the elementary bipolar neutrosophic set having the support T_j and $(h(\mu_j), d(\mu_j)) = (\alpha^+_j, \alpha^-_j) = \alpha_j$, $1 \leq j \leq n$. Then $T = \bigcup_{j=1}^{n} \{\mu_j | 1 \leq j \leq n\}$ is a minimal bipolar neutrosophic transversal of H.

9. Family of minimal bipolar neutrosophic transversals describes the most powerful mutations.

10.5   Bipolar Neutrosophic Directed Hypergraphs

Definition 10.20 A *bipolar neutrosophic directed hypergraph* with underlying set X is an ordered pair $G = (\sigma, \varepsilon)$, where σ is non-empty set of vertices and ε is a set of bipolar neutrosophic directed hyperarcs(or hyperedges).

A *bipolar neutrosophic directed hyperarc(or hyperedge)* $\varepsilon_i \in \varepsilon$ is an ordered pair $(T(\varepsilon_i), H(\varepsilon_i))$, such that $T(\varepsilon_i) \subset X$, $T(\varepsilon_i) \neq \emptyset$, is called its *tail* and $H(\varepsilon_i) \neq T(\varepsilon_i)$ is its *head*.

Definition 10.21 Let $G = (\sigma, \varepsilon)$ be a bipolar neutrosophic directed hypergraph. The order of G, denoted by $O(G)$, is defined as $O(G) = (O^+(G), O^-(G))$, where

$$O^+(G) = \sum_{x \in X} \min \sigma^+_i(x) \quad O^-(G) = \sum_{x \in X} \max \sigma^-_i(x).$$

The *size* of G, denoted by $S(G)$, is defined as $S(G) = (S^+(G), S^-(G))$, where

$$S^+(G) = \sum_{E_k \subset X} \varepsilon^+(E_k), \quad S^-(G) = \sum_{E_k \subset X} \varepsilon^-(E_k).$$

In a bipolar neutrosophic directed hypergraph, the vertices u_i and u_j are *adjacent* vertices if they both belong to the same bipolar neutrosophic directed hyperedge. Two bipolar neutrosophic directed hyperedges ε_i and ε_j are called *adjacent* if they have non-empty intersection, i.e., $supp(\varepsilon_i) \cap supp(\varepsilon_j) \neq \emptyset, i \neq j$.

Definition 10.22 A bipolar neutrosophic directed hypergraph $G = (\sigma, \varepsilon)$ is *simple* if it contains no repeated directed hyperedges, i.e., if $\varepsilon_j, \varepsilon_k \in \varepsilon$ and $\varepsilon_j \subseteq \varepsilon_k$, then $\varepsilon_j = \varepsilon_k$.

A bipolar neutrosophic directed hypergraph $G = (\sigma, \varepsilon)$ is said to be *support simple* if $\varepsilon_j, \varepsilon_k \in \varepsilon$, $supp(\varepsilon_j) = supp(\varepsilon_k)$ and $\varepsilon_j \subseteq \varepsilon_k$, then $\varepsilon_j = \varepsilon_k$.

A bipolar neutrosophic directed hypergraph $G = (\sigma, \varepsilon)$ is *strongly support simple* if $\varepsilon_j, \varepsilon_k \in \varepsilon$ and $supp(\varepsilon_j) = supp(\varepsilon_k)$, then $\varepsilon_j = \varepsilon_k$.

Example 10.4 Consider a bipolar neutrosophic directed hypergraph $G = (\sigma, \varepsilon)$, where $\sigma = \{\sigma_1, \sigma_2, \sigma_3, \sigma_4\}$ be the family of bipolar neutrosophic subsets on $X = \{v_1, v_2, v_3, v_4, v_5, v_6\}$, as shown in Fig. 10.6, such that

$$\sigma_1 = \{(v_1, 0.1, 0.2, 0.3, -0.1, -0.2, -0.3), (v_2, 0.2, 0.2, 0.3, -0.2, -0.2, -0.3),$$
$$(v_5, 0.3, 0.2, 0.3, -0.3, -0.2, -0.3),$$
$$\sigma_2 = \{(v_1, 0.1, 0.2, 0.3, -0.1, -0.2, -0.3), (v_3, 0.4, 0.2, 0.3, -0.4, -0.2, -0.3),$$
$$(v_6, 0.3, 0.2, 0.3, -0.3, -0.2, -0.3)\},$$
$$\sigma_3 = \{(v_6, 0.1, 0.2, 0.3, -0.1, -0.2, -0.3), (v_4, 0.2, 0.2, 0.3, -0.2, -0.2, -0.3),$$
$$(v_5, 0.3, 0.2, 0.3, -0.3, -0.2, -0.3)\},$$
$$\sigma_4 = \{(v_6, 0.1, 0.2, 0.3, -0.1, -0.2, -0.3), (v_2, 0.2, 0.2, 0.3, -0.2, -0.2, -0.3),$$
$$(v_5, 0.3, 0.2, 0.3, -0.3, -0.2, -0.3)\}.$$

Bipolar neutrosophic relation ε is defined as

$$\varepsilon(v_1, v_2, v_5) = (0.1, 0.2, 0.3, -0.1, -0.2, -0.3),$$
$$\varepsilon(v_1, v_3, v_6) = (0.1, 0.2, 0.3, -0.1, -0.2, -0.3),$$
$$\varepsilon(v_6, v_4, v_5) = (0.1, 0.2, 0.3, -0.1, -0.2, -0.3),$$
$$\varepsilon(v_6, v_2, v_5) = (0.1, 0.2, 0.3, -0.1, -0.2, -0.3).$$

Note that, G is simple, strongly support simple and support simple, that is, it contains no repeated directed hyperedges and if whenever $\varepsilon_j, \varepsilon_k \in \varepsilon$ and $supp(\varepsilon_j) = supp(\varepsilon_k)$, then $\varepsilon_j = \varepsilon_k$. Further,

$$o(G) = (1.1, 1.2, 1.8, -1.1, -1.2, -1.8),$$
$$s(G) = (0.4, 0.6, 1.4, -0.4, -0.6, -1.4).$$

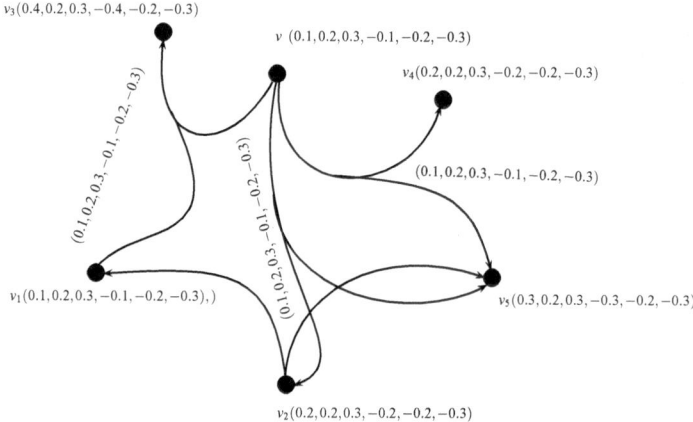

Fig. 10.6 Bipolar neutrosophic directed hypergraph

Definition 10.23 The *height* of a bipolar neutrosophic directed hypergraph $G = (\sigma, \varepsilon)$, denoted by $h(G)$, is defined as, $h(G) = \{\max(\varepsilon_l), \max(\varepsilon_m), \min(\varepsilon_n)|\varepsilon_l, \varepsilon_m, \varepsilon_n \in \varepsilon\}$, where $\varepsilon_l = \max T_{\varepsilon_j}^+(x_i)$, $\varepsilon_m = \max I_{\varepsilon_j}^+(x_i)$, $\varepsilon_n = \min F_{\varepsilon_j}^+(x_i)$.

The *depth* of a bipolar neutrosophic directed hypergraph $G = (\sigma, \varepsilon)$, denoted by $d(G)$, is defined as, $d(G) = \{\min(\varepsilon_l), \min(\varepsilon_m), \max(\varepsilon_n)|\varepsilon_l, \varepsilon_m, \varepsilon_n \in \varepsilon\}$, where $\varepsilon_l = \min T_{\varepsilon_j}^-(x_i)$, $\varepsilon_m = \min I_{\varepsilon_j}^-(x_i)$, $\varepsilon_n = \max F_{\varepsilon_j}^-(x_i)$.

The functions $T_{\varepsilon_j}^+(x_i)$, $I_{\varepsilon_j}^+(x_i)$ and $F_{\varepsilon_j}^+(x_i)$ denote the positive truth, indeterminacy and falsity membership values of vertex x_i to the hyperedge ε_j, respectively, $t_{\varepsilon_j}^-(x_i)$, $I_{\varepsilon_j}^-(x_i)$, and $F_{\varepsilon_j}^-(x_i)$ denote the negative truth, indeterminacy, and falsity membership values of vertex x_i to the hyperedge ε_j, respectively.

Definition 10.24 Let $\varepsilon = (\varepsilon^-, \varepsilon^+)$ be a directed hyperedge in a bipolar neutrosophic directed hypergraph. Then the vertex set ε^- is called the *in-set* and the vertex set ε^+ is called the *out-set* of the directed hyperedge ε. It is not necessary that the sets ε^-, ε^+ will be disjoint. The hyperedge ε is called the join of the vertices of ε^- and ε^+.

Definition 10.25 The *in-degree* $D_G^-(v)$ of a vertex v is defined as the sum of membership degrees of all those directed hyperedges such that v is contained in their out-set, i.e.,

$$D_G^-(v) = (\sum_{v \in H(E_k)} \varepsilon^+(E_k), \sum_{v \in H(E_k)} \varepsilon^-(E_k)).$$

The *out-degree* $D_G^+(v)$ of a vertex v is defined as the sum of membership degrees of all those directed hyperedges such that v is contained in their in-set, i.e.,

$$D_G^+(v) = (\sum_{v \in T(E_k)} \varepsilon^+(E_k), \sum_{v \in T(E_k)} \varepsilon^-(E_k)).$$

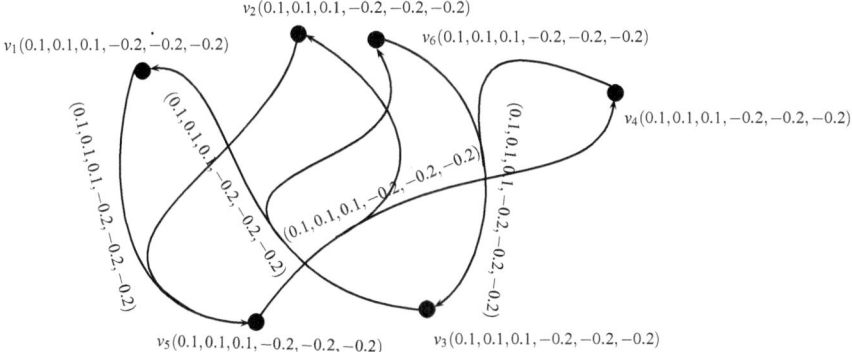

Fig. 10.7 Regular bipolar neutrosophic directed hypergraph

Definition 10.26 A bipolar neutrosophic directed hypergraph $G = (\sigma, \varepsilon)$ is called *k-regular* if in-degrees and out-degrees of all the vertices in G are same.

Example 10.5 Consider a bipolar neutrosophic directed hypergraph $G = (\sigma, \varepsilon)$ as shown in Fig. 10.7, where $\sigma = \{\sigma_1, \sigma_2, \sigma_3, \sigma_4\}$ is the family of bipolar neutrosophic subsets on $X = \{v_1, v_2, v_3, v_4, v_5, v_6\}$ and

$$\sigma_1 = \{(v_1, 0.1, 0.1, 0.1, -0.2, -0.2, -0.2), (v_2, 0.1, 0.1, 0.1, -0.2, -0.2, -0.2),$$
$$(v_5, 0.1, 0.1, 0.1, -0.2, -0.2, -0.2)\},$$
$$\sigma_2 = \{(v_1, 0.1, 0.1, 0.1, -0.2, -0.2, -0.2), (v_3, 0.1, 0.1, 0.1, -0.2, -0.2, -0.2),$$
$$(v_6, 0.1, 0.1, 0.1, -0.2, -0.2, -0.2)\},$$
$$\sigma_3 = \{(v_6, 0.1, 0.1, 0.1, -0.2, -0.2, -0.2), (v_4, 0.1, 0.1, 0.1, -0.2, -0.2, -0.2),$$
$$(v_3, 0.1, 0.1, 0.1, -0.2, -0.2, -0.2)\},$$
$$\sigma_4 = \{(v_2, 0.1, 0.1, 0.1, -0.2, -0.2, -0.2), (v_4, 0.1, 0.1, 0.1, -0.2, -0.2, -0.2),$$
$$(v_5, 0.1, 0.1, 0.1, -0.2, -0.2, -0.2)\}.$$

The bipolar neutrosophic relation ε is defined as

$$\varepsilon(v_1, v_2, v_5) = (0.1, 0.1, 0.1, -0.2, -0.2, -0.2),$$
$$\varepsilon(v_1, v_3, v_6) = (0.1, 0.1, 0.1, -0.2, -0.2, -0.2),$$
$$\varepsilon(v_6, v_4, v_3) = (0.1, 0.1, 0.1, -0.2, -0.2, -0.2),$$
$$\varepsilon(v_2, v_4, v_5) = (0.1, 0.1, 0.1, -0.2, -0.2, -0.2).$$

By routine calculations, we see that the bipolar neutrosophic directed hypergraph is regular.

Note that

$$D_G^-(v_1) = (0.1, 0.1, 0.1, -0.2, -0.2, -0.2) = D_G^+(v_1)$$
$$D_G^-(v_2) = (0.1, 0.1, 0.1, -0.2, -0.2, -0.2) = D_G^+(v_2).$$

Similarly, $D_G^-(v_3) = D_G^+(v_3), D_G^-(v_4) = D_G^+(v_4), D_G^-(v_5) = D_G^+(v_5), D_G^-(v_6) = D_G^+(v_6)$. Hence, G is regular bipolar neutrosophic directed hypergraph.

We now discuss the basic properties of isomorphism on bipolar neutrosophic directed hypergraphs.

Definition 10.27 Let $G = (\sigma, \varepsilon)$ and $G' = (\sigma', \varepsilon')$ be two bipolar neutrosophic directed hypergraphs, where $\sigma = \{\sigma_1, \sigma_2, \sigma_3, \ldots, \sigma_k\}$ and $\sigma' = \{\sigma_1', \sigma_2', \sigma_3', \ldots, \sigma_k'\}$. A *homomorphism* of bipolar neutrosophic directed hypergraphs $\chi : G \to G'$ is a mapping $\chi : X \to X'$ which satisfies

1.

$$\min T_{\sigma_i}^+(u) \le \min T_{\sigma_i'}^+(\chi(u)), \min I_{\sigma_i}^+(u) \le \min I_{\sigma_i'}^+(\chi(u)), \max F_{\sigma_i}^+(u) \ge \max F_{\sigma_i'}^+(\chi(u)),$$

$$\max T_{\sigma_i}^-(u) \ge \max T_{\sigma_i'}^-(\chi(u)), \max I_{\sigma_i}^-(u) \ge \max I_{\sigma_i'}^-(\chi(u)), \min F_{\sigma_i}^-(u) \le \min F_{\sigma_i'}^-(\chi(u)),$$

for all $u \in X$.

2.

$$T_\varepsilon^+(u_1, u_2, u_3, \ldots, u_k) \le T_{\varepsilon'}^+(\chi(u_1), \chi(u_2), \chi(u_3), \ldots, \chi(u_k)),$$

$$I_\varepsilon^+(u_1, u_2, u_3, \ldots, u_k) \le I_{\varepsilon'}^+(\chi(u_1), \chi(u_2), \chi(u_3), \ldots, \chi(u_k)),$$

$$F_\varepsilon^+(u_1, u_2, u_3, \ldots, u_k) \ge F_{\varepsilon'}^+(\chi(u_1), \chi(u_2), \chi(u_3), \ldots, \chi(u_k)),$$

$$t_\varepsilon^-(u_1, u_2, u_3, \ldots, u_k) \ge T_{\varepsilon'}^-(\chi(u_1), \chi(u_2), \chi(u_3), \ldots, \chi(u_k)),$$

$$I_\varepsilon^-(u_1, u_2, u_3, \ldots, u_k) \ge I_{\varepsilon'}^-(\chi(u_1), \chi(u_2), \chi(u_3), \ldots, \chi(u_k)),$$

$$F_\varepsilon^-(u_1, u_2, u_3, \ldots, u_k) \le F_{\varepsilon'}^-(\chi(u_1), \chi(u_2), \chi(u_3), \ldots, \chi(u_k)),$$

for all $\{u_1, u_2, u_3, \ldots, u_k\} = E_i \subset X$.

Note that, for a homomorphism $\chi : G \to G'$, $\chi(\varepsilon) = (T(\chi(\varepsilon)), H(\chi(\varepsilon)))$ is an hyperarc in G' if $\varepsilon = (T(\varepsilon), H(\varepsilon))$ is an hyperarc in G.

Definition 10.28 A *weak isomorphism* $\chi : G \to G'$ is a mapping $\chi : X \to X'$ which is a bijective homomorphism and satisfies

$$\min T_{\sigma_i}^+(v) = \min T_{\sigma_i'}^+(\chi(v)), \min I_{\sigma_i}^+(v) = \min I_{\sigma_i'}^+(\chi(v)), \max F_{\sigma_i}^+(v) = \max F_{\sigma_i'}^+(\chi(v)),$$

$$\max T_{\sigma_i}^-(v) = \max T_{\sigma_i'}^-(\chi(v)), \max I_{\sigma_i}^-(v) = \max I_{\sigma_i'}^-(\chi(v)), \min F_{\sigma_i}^-(v) = \min F_{\sigma_i'}^-(\chi(v)),$$

for all $v \in X$.

Definition 10.29 A *co-weak isomorphism* $\chi : G \to G'$ is a mapping $\chi : X \to X'$ which is a bijective homomorphism and satisfies

$$T_{\varepsilon}^+(x_1, x_2, x_3, \ldots, x_k) = T_{\varepsilon'}^+(\chi(x_1), \chi(x_2), \chi(x_3), \ldots, \chi(x_k)),$$

$$I_{\varepsilon}^+(x_1, x_2, x_3, \ldots, x_k) = I_{\varepsilon'}^+(\chi(x_1), \chi(x_2), \chi(x_3), \ldots, \chi(x_k)),$$

$$F_{\varepsilon}^+(x_1, x_2, x_3, \ldots, x_k) = F_{\varepsilon'}^+(\chi(x_1), \chi(x_2), \chi(x_3), \ldots, \chi(x_k)),$$

$$t_{\varepsilon}^-(x_1, x_2, x_3, \ldots, x_k) = T_{\varepsilon'}^-(\chi(x_1), \chi(x_2), \chi(x_3), \ldots, \chi(x_k)),$$

$$I_{\varepsilon}^-(x_1, x_2, x_3, \ldots, x_k) = I_{\varepsilon'}^-(\chi(x_1), \chi(x_2), \chi(x_3), \ldots, \chi(x_k)),$$

$$F_{\varepsilon}^-(x_1, x_2, x_3, \ldots, x_k) = F_{\varepsilon'}^-(\chi(x_1), \chi(x_2), \chi(x_3), \ldots, \chi(x_k)),$$

for all $\{x_1, x_2, x_3, \ldots, x_k\} = E_i \subset X$.

Definition 10.30 An *isomorphism* of bipolar neutrosophic directed hypergraphs $\chi : G \to G'$ is a mapping $\chi : X \to X'$ which is bijective homomorphism and satisfies

1.

$$\min T_{\sigma_i}^+(u) = \min T_{\sigma_i'}^+(\chi(u)), \min I_{\sigma_i}^+(u) = \min I_{\sigma_i'}^+(\chi(u)), \max F_{\sigma_i}^+(u) = \max F_{\sigma_i'}^+(\chi(u)),$$

$$\max T_{\sigma_i}^-(u) = \max T_{\sigma_i'}^-(\chi(u)), \max I_{\sigma_i}^-(u) = \max I_{\sigma_i'}^-(\chi(u)), \min F_{\sigma_i}^-(u) = \min F_{\sigma_i'}^-(\chi(u)),$$

for all $u \in X$.

2.

$$T_{\varepsilon}^+(u_1, u_2, u_3, \ldots, u_k) = T_{\varepsilon'}^+(\chi(u_1), \chi(u_2), \chi(u_3), \ldots, \chi(u_k)),$$

$$I_{\varepsilon}^+(u_1, u_2, u_3, \ldots, u_k) = I_{\varepsilon'}^+(\chi(u_1), \chi(u_2), \chi(u_3), \ldots, \chi(u_k)),$$

$$F_{\varepsilon}^+(u_1, u_2, u_3, \ldots, u_k) = F_{\varepsilon'}^+(\chi(u_1), \chi(u_2), \chi(u_3), \ldots, \chi(u_k)),$$

$$t_{\varepsilon}^-(u_1, u_2, u_3, \ldots, u_k) = T_{\varepsilon'}^-(\chi(u_1), \chi(u_2), \chi(u_3), \ldots, \chi(u_k)),$$

$$I_\varepsilon^-(u_1, u_2, u_3, \ldots, u_k) = I_{\varepsilon'}^-(\chi(u_1), \chi(u_2), \chi(u_3), \ldots, \chi(u_k)),$$

$$F_\varepsilon^-(u_1, u_2, u_3, \ldots, u_k) = F_{\varepsilon'}^-(\chi(u_1), \chi(u_2), \chi(u_3), \ldots, \chi(u_k)),$$

for all $\{u_1, u_2, u_3, \ldots, u_k\} = E_i \subset X$.

If two bipolar neutrosophic directed hypergraphs G and G' are isomorphic, we denote it as $G \cong G'$.

Example 10.6 Let $\sigma = \{\sigma_1, \sigma_2, \sigma_3\}$ and $\sigma' = \{\sigma_1', \sigma_2', \sigma_3'\}$ be the families of bipolar neutrosophic subsets on $X = \{v_1, v_2, v_3, v_4, v_5, v_6\}$ and $X' = \{v_1', v_2', v_3', v_4', v_5', v_6'\}$, respectively, as

$$
\begin{aligned}
\sigma_1 =\ & \{(v_1, 0.1, 0.2, 0.3, -0.1, -0.2, -0.3),\ (v_2, 0.2, 0.2, 0.3, -0.2, -0.2, -0.3), \\
& (v_5, 0.3, 0.2, 0.3, -0.3, -0.2, -0.3),\ (v_6, 0.4, 0.2, 0.5, -0.4, -0.2, -0.5)\}, \\
\sigma_2 =\ & \{(v_1, 0.1, 0.2, 0.3, -0.1, -0.2, -0.3),\ (v_3, 0.4, 0.2, 0.3, -0.4, -0.2, -0.3), \\
& (v_6, 0.3, 0.2, 0.3, -0.3, -0.2, -0.3)\}, \\
\sigma_3 =\ & \{(v_6, 0.1, 0.2, 0.3, -0.1, -0.2, -0.3),\ (v_4, 0.2, 0.2, 0.3, -0.2, -0.2, -0.3), \\
& (v_5, 0.3, 0.2, 0.3, -0.3, -0.2, -0.3)\}, \\
\sigma_1' =\ & \{(v_4', 0.1, 0.2, 0.3, -0.1, -0.2, -0.3),\ (v_3', 0.2, 0.2, 0.3, -0.2, -0.2, -0.3), \\
& (v_6', 0.3, 0.2, 0.3, -0.3, -0.2, -0.3),\ (v_5', 0.4, 0.2, 0.5, -0.4, -0.2, -0.5)\}, \\
\sigma_2' =\ & \{(v_4', 0.1, 0.2, 0.3, -0.1, -0.2, -0.3),\ (v_2', 0.4, 0.2, 0.3, -0.4, -0.2, -0.3), \\
& (v_5', 0.3, 0.2, 0.3, -0.3, -0.2, -0.3)\}, \\
\sigma_3' =\ & \{(v_5', 0.1, 0.2, 0.3, -0.1, -0.2, -0.3),\ (v_1', 0.2, 0.2, 0.3, -0.2, -0.2, -0.3), \\
& (v_6', 0.3, 0.2, 0.3, -0.3, -0.2, -0.3)\}.
\end{aligned}
$$

The bipolar neutrosophic relations ε and ε' are defined as

$$
\begin{aligned}
\varepsilon(v_1, v_2, v_5, v_6) &= (0.1, 0.2, 0.5, -0.1, -0.2, -0.5), \\
\varepsilon(v_1, v_3, v_6) &= (0.1, 0.2, 0.3, -0.1, -0.2, -0.3), \\
\varepsilon(v_6, v_4, v_5) &= (0.1, 0.2, 0.3, -0.1, -0.2, -0.3), \\
\varepsilon'(v_4', v_3', v_6', v_5') &= (0.1, 0.2, 0.5, -0.1, -0.2, -0.5), \\
\varepsilon'(v_4', v_2', v_5') &= (0.1, 0.2, 0.3, -0.1, -0.2, -0.3), \\
\varepsilon'(v_5', v_1', v_6') &= (0.1, 0.2, 0.3, -0.1, -0.2, -0.3).
\end{aligned}
$$

Define a mapping $\chi : X \to X'$ as

$$
\begin{aligned}
\chi(v_1) &= v_4',\ \chi(v_2) = v_3', \\
\chi(v_3) &= v_2',\ \chi(v_4) = v_1', \\
\chi(v_5) &= v_6',\ \chi(v_6) = v_5'.
\end{aligned}
$$

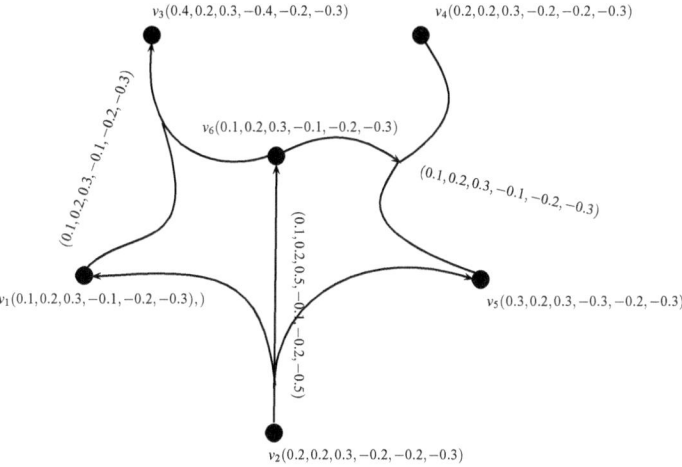

Fig. 10.8 Bipolar neutrosophic hypergraph G

Note that

$$\sigma_1(v_1) = (0.1, 0.2, 0.3, -0.1, -0.2, -0.3) = \sigma_1'(v_4') = \sigma_1'(\chi(v_1)),$$
$$\sigma_1(v_2) = (0.2, 0.2, 0.3, -0.2, -0.2, -0.3) = \sigma_1'(v_3') = \sigma_1'(\chi(v_2)),$$
$$\sigma_2(v_3) = (0.4, 0.2, 0.3, -0.4, -0.2, -0.3) = \sigma_2'(v_2') = \sigma_2'(\chi(v_3)),$$
$$\sigma_3(v_4) = (0.2, 0.2, 0.3, -0.2, -0.2, -0.3) = \sigma_3'(v_1') = \sigma_3'(\chi(v_4)),$$
$$\sigma_3(v_5) = (0.3, 0.2, 0.3, -0.3, -0.2, -0.3) = \sigma_3'(v_6') = \sigma_3'(\chi(v_5)).$$

Similarly, $\sigma(v) = \sigma'(\chi(v))$, for all $v \in X$, $\varepsilon(\{v_1, v_2, v_3, \ldots, v_k\}) = \varepsilon'(\{\chi(v_1), \chi(v_2), \chi(v_3), \ldots, \chi(v_k)\})$, for all $v_k \in X$. Hence, G and G' are isomorphic and the corresponding bipolar neutrosophic directed hypergraphs are shown in Figs. 10.8 and 10.9, respectively.

Note that, $\chi(\varepsilon) = (T(\chi(\varepsilon)), H(\chi(\varepsilon)))$ is an hyperarc in G' if $\varepsilon = (T(\varepsilon), H(\varepsilon))$ is an hyperarc in G.

Remark 10.4 A weak isomorphism of bipolar neutrosophic directed hypergraphs preserves the membership degrees of vertices but not necessarily the membership degrees of directed hyperedges. A co-weak isomorphism of bipolar neutrosophic directed hypergraphs preserves the membership degrees of directed hyperedges but not necessarily the membership degrees of vertices.

In isomorphism of crisp hypergraphs, isomorphic hypergraphs have same degree as well as the order. The same also holds in bipolar neutrosophic directed hypergraphs.

Theorem 10.12 *Let G and G' be two isomorphic bipolar neutrosophic directed hypergraphs. Then, they both have the same order and size.*

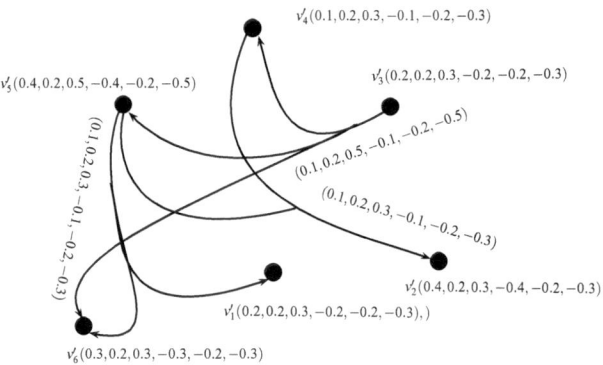

Fig. 10.9 Bipolar neutrosophic hypergraph G'

Proof Let $G = (\sigma, \varepsilon)$ and $G' = (\sigma', \varepsilon')$ be two bipolar neutrosophic directed hypergraphs, where $\sigma = \{\sigma_1, \sigma_2, \sigma_3, \ldots, \sigma_k\}$ and $\sigma' = \{\sigma_1', \sigma_2', \sigma_3', \ldots, \sigma_k'\}$ be the family of bipolar neutrosophic sets defined on X and X', respectively. Let $\chi : X \rightarrow X'$ be an isomorphism between G and G' then

$$\min \sigma_i^+(z) = \min \sigma_i'^+(\chi(z)), \quad \max \sigma_i^-(z) = \max \sigma_i'^-(\chi(z)), \quad \text{for all } z \in X$$

$$\varepsilon(z_1, z_2, z_3, \ldots, z_k) = \varepsilon'(\chi(z_1), \chi(z_2), \chi(z_3), \ldots, \chi(z_k)),$$

$$\mathcal{O}^+(G) = \sum_{z \in X} \min \sigma_i^+(z) = \sum_{z \in X} \min \sigma_i'^+(\chi(z)) = \sum_{z' \in X'} \min \sigma_i'^+(z') = \mathcal{O}^+(G'),$$

$$\mathcal{O}^-(G) = \sum_{z \in X} \max \sigma_i^-(z) = \sum_{z \in X} \max \sigma_i'^-(\chi(z)) = \sum_{z' \in X'} \max \sigma_i'^+(z') = \mathcal{O}^-(G'),$$

$$\mathscr{S}^+(G) = \sum_{E_k \subset X} \varepsilon^+(E_k) = \sum_{E_k \subset X} \varepsilon'^+(\chi(E_k)) = \sum_{E_k' \subset X'} \varepsilon'^+(E_k') = \mathscr{S}^+(G'),$$

$$\mathscr{S}^-(G) = \sum_{E_k \subset X} \varepsilon^-(E_k) = \sum_{E_k \subset X} \varepsilon'^-(\chi(E_k)) = \sum_{E_k' \subset X'} \varepsilon'^-(E_k') = \mathscr{S}^-(G'),$$

for all $\{z_1, z_2, z_3, \ldots, z_k\} = E_i \subset X$. This completes the proof. $\square$

Theorem 10.13 *Isomorphism between bipolar neutrosophic hypergraphs is an equivalence relation.*

Proof Let $G = (X, \sigma, \varepsilon)$, $G' = (X', \sigma', \varepsilon')$ and $G'' = (X'', \sigma'', \varepsilon'')$ be bipolar neutrosophic directed hypergraphs having underlying sets X, X' and X'', respectively.

(i) Reflexive: Consider the identity mapping $\chi : X \rightarrow X$, such that $\chi(z) = z$, for all $z \in X$. Then, χ is bijective homomorphism and satisfies $\min \sigma_i^+(z) = \min \sigma_i'^+(\chi(z))$, $\max \sigma_i^-(z) = \max \sigma_i'^-(\chi(z))$, for all $z \in X$ and

$$\varepsilon^+(z_1, z_2, z_3, \ldots, z_k) = \varepsilon'^+(\chi(z_1), \chi(z_2), \chi(z_3), \ldots, \chi(z_k)),$$

$$\varepsilon^-(z_1, z_2, z_3, \ldots, z_k) = \varepsilon'^-(\chi(z_1), \chi(z_2), \chi(z_3), \ldots, \chi(z_k)),$$

for all $\{z_1, z_2, z_3, \ldots, z_k\} = E_k \subset X$. Hence, χ is an isomorphism of bipolar neutrosophic directed hypergraphs to itself. Thus, reflexive relation is satisfied.

(ii) Symmetric: Let $\chi : X \to X'$ be an isomorphism between G and G', then χ is bijective mapping and $\chi(z) = z'$, for all $z \in X$. From the isomorphism of χ, we have $\min \sigma_i^+(z) = \min \sigma_i'^+(\chi(z)), \max \sigma_i^-(z) = \max \sigma_i'^-(\chi(z))$, for all $z \in X$ and

$$\varepsilon^+(z_1, z_2, z_3, \ldots, z_k) = \varepsilon'^+(\chi(z_1), \chi(z_2), \chi(z_3), \ldots, \chi(z_k)),$$

$$\varepsilon^-(z_1, z_2, z_3, \ldots, z_k) = \varepsilon'^-(\chi(z_1), \chi(z_2), \chi(z_3), \ldots, \chi(z_k)),$$

for all $z_1, z_2, z_3, \ldots, z_k\} = E_k \subset X$. Since, χ is bijective, $\chi^{-1} : X' \to X$ exists and $\chi^{-1}(z') = z$, for all $z' \in X'$. Then, $\min \sigma_i^+ \chi^{-1}(z') = \min \sigma_i'^+(z')$, $\max \sigma_i^- \chi^{-1}(z') = \max \sigma_i'^-(z')$, for all $z' \in X'$, and

$$\varepsilon^+(\chi^{-1}(z_1'), \chi^{-1}(z_2'), \chi^{-1}(z_3'), \ldots, \chi^{-1}(z_k')) = \varepsilon'^+(z_1', z_2', z_3', \ldots, z_k'),$$

$$\varepsilon^-(\chi^{-1}(z_1'), \chi^{-1}(z_2'), \chi^{-1}(z_3'), \ldots, \chi^{-1}(z_k')) = \varepsilon'^-(z_1', z_2', z_3', \ldots, z_k'),$$

for all $\{z_1', z_2', z_3', \ldots, z_k'\} = E_k' \subset X'$. Hence, we get a bijective mapping $\chi^{-1} : X' \to X$, which is isomorphism from G' to G, i.e., $G \cong G' \Rightarrow G' \cong G$.

(iii) Transitive: Let $\chi : X \to X'$ and Let $\lambda : X' \to X''$ be an isomorphism of bipolar neutrosophic directed hypergraphs G to G' and G' to G'', respectively, defined by $\chi(z) = z'$ and $\lambda(z') = z''$. Then, $\lambda \circ \chi : X \to X''$ is a bijective mapping from G to G'' such that $(\lambda \circ \chi)(z) = \lambda(\chi(z))$, for all $z \in X$. Since, $\chi : X \to X'$ is an isomorphism, we have

$$\min \sigma_i^+(z) = \min \sigma_i'^+(\chi(z)), \quad \max \sigma_i^-(z) = \max \sigma_i'^-(\chi(z)),$$

$$\varepsilon(z_1, z_2, z_3, \ldots, z_k) = \varepsilon'(\chi(z_1), \chi(z_2), \chi(z_3), \ldots, \chi(z_k)),$$

for all $z \in X$ and for all $\{z_1, z_2, z_3, \ldots, z_k\} = E_k \subset X$. Since, $\lambda : X' \to X''$ is an isomorphism, we have

$$\min \sigma_i'^+(z') = \min \sigma_i''^+(\lambda(z')), \max \sigma_i'^-(z') = \max \sigma_i''^-(\lambda(z')),$$

$$\varepsilon'(z_1', z_2', z_3', \ldots, z_k') = \varepsilon''(\lambda(z_1'), \lambda(z_2'), \lambda(z_3'), \ldots, \lambda(z_k')),$$

for all $z' \in X'$ and $\{z_1', z_2', z_3', \ldots, z_k'\} = E_k' \subset X'$. Thus, we have

$$\min \sigma_i^+(z) = \min \sigma_i'^+(\chi(z)) = \min \sigma_i'^+(z') = \min \sigma_i''^+(\lambda(z')) = \min \sigma_i''^+(\lambda(\chi(z))),$$

$$\varepsilon(z_1, z_2, z_3, \ldots, z_k) = \varepsilon'(\chi(z_1), \chi(z_2), \chi(z_3), \ldots, \chi(z_k)) = \varepsilon'(z_1', z_2', z_3', \ldots, z_k')$$
$$= \varepsilon''(\lambda(z_1'), \lambda(z_2'), \lambda(z_3'), \ldots, \lambda(z_k'))$$
$$= \varepsilon''(\lambda(\chi(z_1)), \lambda(\chi(z_2)), \lambda(\chi(z_3)), \ldots, \lambda(\chi(z_k))),$$

for all $z \in X$, $z' \in X'$, $z'' \in X''$ $\{z_1, z_2, z_3, \ldots, z_k\} = E_k \subset X$ and $\{z_1', z_2', z_3', \ldots, z_k'\} = E_k' \subset X'$.

Clearly, $\lambda \circ \chi$ is an isomorphism from G to G''. Hence, isomorphism of bipolar neutrosophic directed hypergraphs is an equivalence relation. □

Theorem 10.14 *A weak isomorphism between bipolar neutrosophic directed hypergraphs is a partial order relation.*

Proof Let $G = (X, \sigma, \varepsilon)$, $G' = (X', \sigma', \varepsilon')$ and $G'' = (X'', \sigma'', \varepsilon'')$ be bipolar neutrosophic directed hypergraphs having underlying sets X, X' and X'', respectively.

(i) Reflexive: Consider the identity mapping $\chi : X \to X$, such that $\chi(z) = z$, for all $z \in X$. Then, χ is bijective homomorphism and satisfies

$$\min \sigma_i^+(z) = \min \sigma_i'^+(\chi(z)), \quad \max \sigma_i^-(z) = \max \sigma_i'^-(\chi(z)),$$

for all $z \in X$,

$$T_\varepsilon^+(z_1, z_2, z_3, \ldots, z_k) \leq T_{\varepsilon'}^+(\chi(z_1), \chi(z_2), \chi(z_3), \ldots, \chi(z_k)),$$
$$I_\varepsilon^+(z_1, z_2, z_3, \ldots, z_k) \leq I_{\varepsilon'}^+(\chi(z_1), \chi(z_2), \chi(z_3), \ldots, \chi(z_k)),$$
$$F_\varepsilon^+(z_1, z_2, z_3, \ldots, z_k) \geq F_{\varepsilon'}^+(\chi(z_1), \chi(z_2), \chi(z_3), \ldots, \chi(z_k)),$$
$$T_\varepsilon^-(z_1, z_2, z_3, \ldots, z_k) \geq T_{\varepsilon'}^-(\chi(z_1), \chi(z_2), \chi(z_3), \ldots, \chi(z_k)),$$
$$I_\varepsilon^-(z_1, z_2, z_3, \ldots, z_k) \geq I_{\varepsilon'}^-(\chi(z_1), \chi(z_2), \chi(z_3), \ldots, \chi(z_k)),$$
$$F_\varepsilon^-(z_1, z_2, z_3, \ldots, z_k) \leq F_{\varepsilon'}^-(\chi(z_1), \chi(z_2), \chi(z_3), \ldots, \chi(z_k)).$$

Hence, χ is a weak isomorphism of bipolar neutrosophic directed hypergraphs to itself. Thus, reflexive relation is satisfied.

(ii) Anti symmetric: Let $\chi : X \to X'$ be a weak isomorphism between G and G' and $\lambda : X' \to X$ be a weak isomorphism between G' and G. Then χ is a bijective mapping $\chi(z) = z'$, satisfying

$$\min \sigma_i^+(z) = \min \sigma_i'^+(\chi(z)), \quad \max \sigma_i^-(z) = \max \sigma_i'^-(\chi(z)),$$

$$T_\varepsilon^+(z_1, z_2, z_3, \ldots, z_k) \leq T_{\varepsilon'}^+(\chi(z_1), \chi(z_2), \chi(z_3), \ldots, \chi(z_k)),$$
$$I_\varepsilon^+(z_1, z_2, z_3, \ldots, z_k) \leq I_{\varepsilon'}^+(\chi(z_1), \chi(z_2), \chi(z_3), \ldots, \chi(z_k)),$$
$$F_\varepsilon^+(z_1, z_2, z_3, \ldots, z_k) \geq F_{\varepsilon'}^+(\chi(z_1), \chi(z_2), \chi(z_3), \ldots, \chi(z_k)),$$
$$T_\varepsilon^-(z_1, z_2, z_3, \ldots, z_k) \geq T_{\varepsilon'}^-(\chi(z_1), \chi(z_2), \chi(z_3), \ldots, \chi(z_k)),$$

$$I_{\varepsilon}^{-}(z_1, z_2, z_3, \ldots, z_k) \geq I_{\varepsilon'}^{-}(\chi(z_1), \chi(z_2), \chi(z_3), \ldots, \chi(z_k)),$$
$$F_{\varepsilon}^{-}(z_1, z_2, z_3, \ldots, z_k) \leq F_{\varepsilon'}^{-}(\chi(z_1), \chi(z_2), \chi(z_3), \ldots, \chi(z_k)), \qquad (10.1)$$

for all $z \in X$, and for all $\{z_1, z_2, z_3, \ldots, z_k\} = E_i \subset X$. Since, λ is a bijective mapping $\lambda(z') = z$, satisfying

$$\min \sigma_i^{+}(z') = \min \sigma_i'^{+}(\lambda(z')), \quad \max \sigma_i^{-}(z') = \max \sigma_i'^{-}(\lambda(z')),$$

$$T_{\varepsilon}^{+}(z'_1, z'_2, z'_3, \ldots, z'_k) \leq T_{\varepsilon'}^{+}(\lambda(z'_1), \lambda(z'_2), \lambda(z'_3), \ldots, \lambda(z'_k)),$$
$$I_{\varepsilon}^{+}(z'_1, z'_2, z'_3, \ldots, z'_k) \leq I_{\varepsilon'}^{+}(\lambda(z'_1), \lambda(z'_2), \lambda(z'_3), \ldots, \lambda(z'_k)),$$
$$F_{\varepsilon}^{+}(z'_1, z'_2, z'_3, \ldots, z'_k) \geq F_{\varepsilon'}^{+}(\lambda(z'_1), \lambda(z'_2), \lambda(z'_3), \ldots, \lambda(z'_k)),$$
$$T_{\varepsilon}^{-}(z'_1, z'_2, z'_3, \ldots, z'_k) \geq T_{\varepsilon'}^{-}(\lambda(z'_1), \lambda(z'_2), \lambda(z'_3), \ldots, \lambda(z'_k)),$$
$$I_{\varepsilon}^{-}(z'_1, z'_2, z'_3, \ldots, z'_k) \geq I_{\varepsilon'}^{-}(\lambda(z'_1), \lambda(z'_2), \lambda(z'_3), \ldots, \lambda(z'_k)),$$
$$F_{\varepsilon}^{-}(z'_1, z'_2, z'_3, \ldots, z'_k) \leq F_{\varepsilon'}^{-}(\lambda(z'_1), \lambda(z'_2), \lambda(z'_3), \ldots, \lambda(z'_k)),$$

for all $z' \in X'$ and for all $\{z'_1, z'_2, z'_3, \ldots, z'_k\} = E'_i \subset X'$. The inequalities (1) and (2) hold true only if G and G' contain the same directed hyperedges having same membership degrees. Hence G and G' are equivalent.

(iii) Transitive: Let $\chi : X \to X'$ and let $\lambda : X' \to X''$ be weak isomorphism of bipolar neutrosophic directed hypergraphs G to G' and G' to G'', respectively, defined by $\chi(z) = z'$ and $\lambda(z') = z''$. Then $\lambda \circ \chi : X \to X''$ is a bijective mapping from G to G'' such that $(\lambda \circ \chi)(z) = \lambda(\chi(z))$, for all $z \in X$. Since $\chi : X \to X'$ is a weak isomorphism, we have $\min \sigma_i^{+}(z) = \min \sigma_i'^{+}(\chi(z))$, $\max \sigma_i^{-}(z) = \max \sigma_i'^{-}(\chi(z))$, for all $z \in X$ and

$$T_{\varepsilon}^{+}(z_1, z_2, z_3, \ldots, z_k) \leq T_{\varepsilon'}^{+}(\chi(z_1), \chi(z_2), \chi(z_3), \ldots, \chi(z_k)),$$

$$I_{\varepsilon}^{+}(z_1, z_2, z_3, \ldots, z_k) \leq I_{\varepsilon'}^{+}(\chi(z_1), \chi(z_2), \chi(z_3), \ldots, \chi(z_k)),$$

$$F_{\varepsilon}^{+}(z_1, z_2, z_3, \ldots, z_k) \geq F_{\varepsilon'}^{+}(\chi(z_1), \chi(z_2), \chi(z_3), \ldots, \chi(z_k)),$$

$$t_{\varepsilon}^{-}(z_1, z_2, z_3, \ldots, z_k) \geq T_{\varepsilon'}^{-}(\chi(z_1), \chi(z_2), \chi(z_3), \ldots, \chi(z_k)),$$

$$I_{\varepsilon}^{-}(z_1, z_2, z_3, \ldots, z_k) \geq I_{\varepsilon'}^{-}(\chi(z_1), \chi(z_2), \chi(z_3), \ldots, \chi(z_k)),$$

$$F_{\varepsilon}^{-}(z_1, z_2, z_3, \ldots, z_k) \leq F_{\varepsilon'}^{-}(\chi(z_1), \chi(z_2), \chi(z_3), \ldots, \chi(z_k)),$$

for all $\{z_1, z_2, z_3, \ldots, z_k\} = E_k \subset X$. Similarly λ is a weak isomorphism, we have $\min \sigma_i'^{+}(z') = \min \sigma_i''^{+}(\lambda(z'))$, $\max \sigma_i'^{-}(z') = \max \sigma_i''^{-}(\lambda(z'))$, for all $z' \in X'$ and

$$T_{\varepsilon'}^{+}(z'_1, z'_2, z'_3, \ldots, z'_k) \leq T_{\varepsilon''}^{+}(\lambda(z'_1), \lambda(z'_2), \lambda(z'_3), \ldots, \lambda(z'_k)),$$

$$I_{\varepsilon'}^{+}(z_1', z_2', z_3', \ldots, z_k') \le I_{\varepsilon''}^{+}(\lambda(z_1'), \lambda(z_2'), \lambda(z_3'), \ldots, \lambda(z_k')),$$

$$F_{\varepsilon'}^{+}(z_1', z_2', z_3', \ldots, z_k') \ge F_{\varepsilon''}^{+}(\lambda(z_1'), \lambda(z_2'), \lambda(z_3'), \ldots, \lambda(z_k')),$$

$$T_{\varepsilon'}^{-}(z_1', z_2', z_3', \ldots, z_k') \ge T_{\varepsilon''}^{-}(\lambda(z_1'), \lambda(z_2'), \lambda(z_3'), \ldots, \lambda(z_k')),$$

$$I_{\varepsilon'}^{-}(z_1', z_2', z_3', \ldots, z_k') \ge I_{\varepsilon''}^{-}(\lambda(z_1'), \lambda(z_2'), \lambda(z_3'), \ldots, \lambda(z_k')),$$

$$F_{\varepsilon'}^{-}(z_1', z_2', z_3', \ldots, z_k') \le F_{\varepsilon''}^{-}(\lambda(z_1'), \lambda(z_2'), \lambda(z_3'), \ldots, \lambda(z_k')),$$

for all $\{z_1', z_2', z_3', \ldots, z_k'\} = E_k' \subset X'$. From the above conditions, we have

$$\min \sigma_i^{+}(z) = \min \sigma_i'^{+}(\chi(z)) = \min \sigma_i'^{+}(z') = \min \sigma_i''^{+}(\lambda(z')) = \min \sigma_i''^{+}(\lambda(\chi(z))),$$

$$\max \sigma_i^{-}(z) = \max \sigma_i'^{-}(\chi(z)) = \max \sigma_i'^{-}(z') = \max \sigma_i'^{-}(z') \quad = \max \sigma_i''^{-}(\lambda(\chi(z))),$$

for all $z \in X$ and

$$\begin{aligned}
T_\varepsilon^{+}(z_1, z_2, z_3, \ldots, z_k) &\le T_{\varepsilon'}^{+}(\chi(z_1), \chi(z_2), \chi(z_3), \ldots, \chi(z_k)) = T_{\varepsilon'}^{+}(z_1', z_2', z_3', \ldots, z_k') \\
&\le T_{\varepsilon''}^{+}(\lambda(z_1'), \lambda(z_2'), \lambda(z_3'), \ldots, \lambda(z_k')) \\
&= T_{\varepsilon''}^{+}(\lambda(\chi(z_1)), \lambda(\chi(z_2)), \lambda(\chi(z_3)), \ldots, \lambda(\chi(z_k))), \\
I_\varepsilon^{+}(z_1, z_2, z_3, \ldots, z_k) &\le I_{\varepsilon'}^{+}(\chi(z_1), \chi(z_2), \chi(z_3), \ldots, \chi(z_k)) = I_{\varepsilon'}^{+}(z_1', z_2', z_3', \ldots, z_k') \\
&\le I_{\varepsilon''}^{+}(\lambda(z_1'), \lambda(z_2'), \lambda(z_3'), \ldots, \lambda(z_k')) \\
&= I_{\varepsilon''}^{+}(\lambda(\chi(z_1)), \lambda(\chi(z_2)), \lambda(\chi(z_3)), \ldots, \lambda(\chi(z_k))), \\
F_\varepsilon^{+}(z_1, z_2, z_3, \ldots, z_k) &\ge F_{\varepsilon'}^{+}(\chi(z_1), \chi(z_2), \chi(z_3), \ldots, \chi(z_k)) = F_{\varepsilon'}^{+}(z_1', z_2', z_3', \ldots, z_k') \\
&\ge F_{\varepsilon''}^{+}(\lambda(z_1'), \lambda(z_2'), \lambda(z_3'), \ldots, \lambda(z_k')) \\
&= F_{\varepsilon''}^{+}(\lambda(\chi(z_1)), \lambda(\chi(z_2)), \lambda(\chi(z_3)), \ldots, \lambda(\chi(z_k))).
\end{aligned}$$

Similarly, we have

$$\begin{aligned}
T_\varepsilon^{-}(z_1, z_2, z_3, \ldots, z_k) &\ge T_{\varepsilon'}^{-}(\chi(z_1), \chi(z_2), \chi(z_3), \ldots, \chi(z_k)) = t_{\varepsilon'}^{-}(z_1', z_2', z_3', \ldots, z_k') \\
&\ge T_{\varepsilon''}^{-}(\lambda(z_1'), \lambda(z_2'), \lambda(z_3'), \ldots, \lambda(z_k')) \\
&= t_{\varepsilon''}^{-}(\lambda(\chi(z_1)), \lambda(\chi(z_2)), \lambda(\chi(z_3)), \ldots, \lambda(\chi(z_k))), \\
I_\varepsilon^{-}(z_1, z_2, z_3, \ldots, z_k) &\ge I_{\varepsilon'}^{-}(\chi(z_1), \chi(z_2), \chi(z_3), \ldots, \chi(z_k)) = I_{\varepsilon'}^{-}(z_1', z_2', z_3', \ldots, z_k') \\
&\ge I_{\varepsilon''}^{-}(\lambda(z_1'), \lambda(z_2'), \lambda(z_3'), \ldots, \lambda(z_k')) \\
&= I_{\varepsilon''}^{-}(\lambda(\chi(z_1)), \lambda(\chi(z_2)), \lambda(\chi(z_3)), \ldots, \lambda(\chi(z_k))), \\
F_\varepsilon^{-}(z_1, z_2, z_3, \ldots, z_k) &\le F_{\varepsilon'}^{-}(\chi(z_1), \chi(z_2), \chi(z_3), \ldots, \chi(z_k)) = F_{\varepsilon'}^{-}(z_1', z_2', z_3', \ldots, z_k') \\
&\le F_{\varepsilon''}^{-}(\lambda(z_1'), \lambda(z_2'), \lambda(z_3'), \ldots, \lambda(z_k')) \\
&= F_{\varepsilon''}^{-}(\lambda(\chi(z_1)), \lambda(\chi(z_2)), \lambda(\chi(z_3)), \ldots, \lambda(\chi(z_k))).
\end{aligned}$$

for all $\{z_1, z_2, z_3, \ldots, z_k\} = E_k \subset X$, for all $\{z_1', z_2', z_3', \ldots, z_k'\} = E_k' \subset X'$.

Clearly, $\lambda \circ \chi$ is a weak isomorphism from G to G''. Hence weak isomorphism of bipolar neutrosophic directed hypergraphs is a partial order relation. $\qquad\square$

Remark 10.5 If G and G' are isomorphic bipolar neutrosophic directed hypergraphs, then their vertices preserve degrees but the converse is not true, that is, if degrees are preserved then bipolar neutrosophic directed hypergraphs may or may not be isomorphic.

To check whether the two bipolar neutrosophic directed hypergraphs are isomorphic or not, it is mandatory that they have same number of vertices having same degrees and same number of directed hyperedges.

Remark 10.6 • If two bipolar neutrosophic directed hypergraphs are weak isomorphic then they have same orders but converse may or may not be true.
• If two bipolar neutrosophic directed hypergraphs are co-weak isomorphic then they are of same size but the same size of bipolar neutrosophic directed hypergraphs does not imply to the co-weak isomorphism.
• Any two isomorphic bipolar neutrosophic directed hypergraphs have same order and size but the converse may or may not be true.

Definition 10.31 A *bipolar neutrosophic directed hyperpath* of length k in a bipolar neutrosophic directed hypergraph is defined as a sequence $x_1, E_1, x_2, E_2, \ldots, E_k, x_{k+1}$ of distinct vertices and directed hyperedges such that

1. $\varepsilon(E_i) > 0, i = 1, 2, 3, \ldots, k$,
2. $x_i, x_{i+1} \in E_i$.

The consecutive pairs (x_i, x_{i+1}) are called the directed arcs of the directed hyperpath.

Definition 10.32 Let s and t be any two arbitrary vertices in a bipolar neutrosophic directed hypergraph and they are connected through a directed hyperpath of length k then the *strength* of that directed hyperpath is $\eta_k(s, t) = (\eta_k^+(s, t), \eta_k^-(s, t))$, where the *positive strength* is defined as

$$
\begin{aligned}
\eta_k^+(s, t) = (&\min\{T_\varepsilon^+(E_1), T_\varepsilon^+(E_2), T_\varepsilon^+(E_3), \ldots, T_\varepsilon^+(E_k)\}, \\
&\min\{I_\varepsilon^+(E_1), I_\varepsilon^+(E_2), I_\varepsilon^+(E_3), \ldots, I_\varepsilon^+(E_k)\}, \\
&\max\{F_\varepsilon^+(E_1), F_\varepsilon^+(E_2), F_\varepsilon^+(E_3), \ldots, F_\varepsilon^+(E_k)\})
\end{aligned}
$$

and the *negative strength* is defined as

$$
\begin{aligned}
\eta_k^-(s, t) = (&\max\{T_\varepsilon^-(E_1), T_\varepsilon^-(E_2), T_\varepsilon^-(E_3), \ldots, T_\varepsilon^-(E_k)\}, \\
&\max\{I_\varepsilon^-(E_1), I_\varepsilon^-(E_2), I_\varepsilon^-(E_3), \ldots, I_\varepsilon^-(E_k)\}, \\
&\min\{F_\varepsilon^-(E_1), F_\varepsilon^-(E_2), T_\varepsilon^-(E_3), \ldots, T_\varepsilon^-(E_k)\}),
\end{aligned}
$$

$x \in E_1, y \in E_k$, where $E_1, E_2, E_3, \ldots, E_k$ are directed hyperedges.

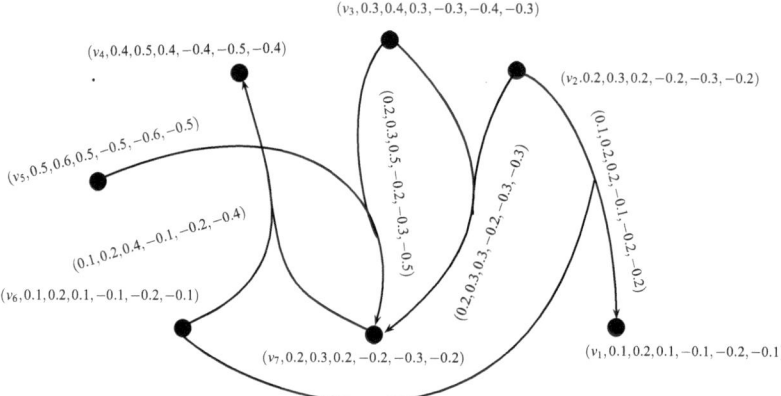

Fig. 10.10 Strong bipolar neutrosophic directed hypergraph

The *strength of connectedness* between x and y is defined as

$$\eta^\infty(s, t) = \{\sup t(\eta_k^+(s, t)), \sup I(\eta_k^+(s, t)), \inf f(\eta_k^+(s, t)), \inf t(\eta_k^-(s, t)),$$
$$\inf I(\eta_k^-(s, t)), \sup f(\eta_k^-(s, t)) | k = 1, 2, 3, \ldots\}.$$

Definition 10.33 A *strong* arc in a bipolar neutrosophic directed hypergraph is defined as $\eta(s, t) \geq \eta^\infty(s, t)$.

Definition 10.34 A bipolar neutrosophic directed hypergraph is said to be *connected* if $\eta^\infty(s, t) > 0$, for all $s, t \in X$, that is, there exists a bipolar neutrosophic directed hyperpath between each pair of vertices.

Definition 10.35 A *strong* or *effective* bipolar neutrosophic directed hypergraph is defined as

$$T_\varepsilon^+(E_k) = T_\varepsilon^+(\{x_1, x_2, x_3, \ldots, x_m\}) = \min\{T_{\sigma_i}^+(x_1), T_{\sigma_i}^+(x_2), \ldots, T_{\sigma_i}^+(x_m)\},$$
$$I_\varepsilon^+(E_k) = I_\varepsilon^+(\{x_1, x_2, x_3, \ldots, x_m\}) = \min\{I_{\sigma_i}^+(x_1), I_{\sigma_i}^+(x_2), \ldots, I_{\sigma_i}^+(x_m)\},$$
$$F_\varepsilon^+(E_k) = F_\varepsilon^+(\{x_1, x_2, x_3, \ldots, x_m\}) = \max\{F_{\sigma_i}^+(x_1), F_{\sigma_i}^+(x_2), \ldots, F_{\sigma_i}^+(x_m)\},$$
$$T_\varepsilon^-(E_k) = t_\varepsilon^-(\{x_1, x_2, x_3, \ldots, x_m\}) = \max\{t_{\sigma_i}^-(x_1), T_{\sigma_i}^-(x_2), \ldots, T_{\sigma_i}^-(x_m)\},$$
$$I_\varepsilon^-(E_k) = I_\varepsilon^-(\{x_1, x_2, x_3, \ldots, x_m\}) = \max\{I_{\sigma_i}^-(x_1), I_{\sigma_i}^-(x_2), \ldots, I_{\sigma_i}^-(x_m)\},$$
$$F_\varepsilon^-(E_k) = F_\varepsilon^-(\{x_1, x_2, x_3, \ldots, x_m\}) = \min\{F_{\sigma_i}^-(x_1), F_{\sigma_i}^-(x_2), \ldots, F_{\sigma_i}^-(x_m)\},$$

for all $\{x_1, x_2, x_3, \ldots, x_m\} = E_k \subset X$.

Example 10.7 Consider a bipolar neutrosophic directed hypergraph $G = (\sigma, \varepsilon)$, as shown in Fig. 10.10.

Note that

$$T_{\varepsilon}^{+}(E_1) = T_{\varepsilon}^{+}(\{v_1, v_2, v_6\}) = \min\{T_{\sigma_i}^{+}(v_1), T_{\sigma_i}^{+}(v_2), T_{\sigma_i}^{+}(x_6)\},$$

$$I_{\varepsilon}^{+}(E_1) = I_{\varepsilon}^{+}(\{v_1, v_2, v_6\}) = \min\{I_{\sigma_i}^{+}(v_1), I_{\sigma_i}^{+}(v_2), I_{\sigma_i}^{+}(x_6)\},$$

$$F_{\varepsilon}^{+}(E_1) = F_{\varepsilon}^{+}(\{v_1, v_2, v_6\}) = \max\{F_{\sigma_i}^{+}(v_1), F_{\sigma_i}^{+}(v_2), F_{\sigma_i}^{+}(x_6)\},$$

$$t_{\varepsilon}^{-}(E_1) = t_{\varepsilon}^{-}(\{v_1, v_2, v_6\}) = \max\{t_{\sigma_i}^{-}(v_1), T_{\sigma_i}^{-}(v_2), t_{\sigma_i}^{-}(x_6)\},$$

$$I_{\varepsilon}^{-}(E_1) = I_{\varepsilon}^{-}(\{v_1, v_2, v_6\}) = \max\{I_{\sigma_i}^{-}(v_1), I_{\sigma_i}^{-}(v_2), I_{\sigma_i}^{-}(x_6)\},$$

$$F_{\varepsilon}^{-}(E_1) = F_{\varepsilon}^{-}(\{v_1, v_2, v_6\}) = \min\{F_{\sigma_i}^{-}(v_1), F_{\sigma_i}^{-}(v_2), F_{\sigma_i}^{-}(x_6)\}.$$

Similarly, for all $\{x_1, x_2, x_3, \ldots, x_k\} = E_k \subset X$, we have

$$T_{\varepsilon}^{+}(E_k) = T_{\varepsilon}^{+}(\{v_1, v_2, \ldots, v_k\}) = \min\{T_{\sigma_i}^{+}(v_1), T_{\sigma_i}^{+}(v_2), \ldots, T_{\sigma_i}^{+}(x_k)\},$$

$$I_{\varepsilon}^{+}(E_k) = I_{\varepsilon}^{+}(\{v_1, v_2, \ldots, v_k\}) = \min\{I_{\sigma_i}^{+}(v_1), I_{\sigma_i}^{+}(v_2), \ldots, T_{\sigma_i}^{+}(x_k)\},$$

$$F_{\varepsilon}^{+}(E_k) = F_{\varepsilon}^{+}(\{v_1, v_2, \ldots, v_k\}) = \max\{F_{\sigma_i}^{+}(v_1), F_{\sigma_i}^{+}(v_2), \ldots, T_{\sigma_i}^{+}(x_k)\},$$

$$t_{\varepsilon}^{-}(E_k) = t_{\varepsilon}^{-}(\{v_1, v_2, \ldots, v_k\}) = \max\{t_{\sigma_i}^{-}(v_1), T_{\sigma_i}^{-}(v_2), \ldots, T_{\sigma_i}^{+}(x_k)\},$$

$$I_{\varepsilon}^{-}(E_k) = I_{\varepsilon}^{-}(\{v_1, v_2, \ldots, v_k\}) = \max\{I_{\sigma_i}^{-}(v_1), I_{\sigma_i}^{-}(v_2), \ldots, T_{\sigma_i}^{+}(x_k)\},$$

$$F_{\varepsilon}^{-}(E_k) = F_{\varepsilon}^{-}(\{v_1, v_2, \ldots, v_k\}) = \min\{F_{\sigma_i}^{-}(v_1), F_{\sigma_i}^{-}(v_2), \ldots, T_{\sigma_i}^{+}(x_k)\}.$$

Hence, G is strong.

Theorem 10.15 *Let G and G' be isomorphic bipolar neutrosophic directed hypergraphs, then G is connected if and only if G' is connected.*

Proof Let $G = (X, \sigma, \varepsilon)$ and $G' = (X', \sigma', \varepsilon')$ be two bipolar neutrosophic directed hypergraphs, where $\varepsilon = \{\varepsilon_1, \varepsilon_2, \varepsilon_3, \ldots, \varepsilon_k\}$ and $\varepsilon' = \{\varepsilon'_1, \varepsilon'_2, \varepsilon'_3, \ldots, \varepsilon'_k\}$ are directed hyperedges of G and G'. Let $\chi : G \to G'$ be an isomorphism between G and G'. Suppose that G is connected such that

$$0 < \eta^\infty(s, t) = \{\sup t(\eta_k^+(s, t)), \sup I(\eta_k^+(s, t)), \inf f(\eta_k^+(s, t)), \inf t(\eta_k^-(s, t)),$$

$$\inf I(\eta_k^-(s, t)), \sup f(\eta_k^-(s, t)) | k = 1, 2, 3, \ldots\}$$

$$= \{\sup \min_{i=1}^{k} T_\varepsilon^+(E_i), \sup \min_{i=1}^{k} I_\varepsilon^+(E_i), \inf \max_{i=1}^{k} F_\varepsilon^+(E_i),$$

$$\inf \max_{i=1}^{k} T_\varepsilon^-(E_i), \inf \max_{i=1}^{k} I_\varepsilon^-(E_i), \sup \min_{i=1}^{k} F_\varepsilon^-(E_i) | k = 1, 2, 3, \ldots\}$$

$$= \{\sup \min_{i=1}^{k} T_{\varepsilon'}^+(\chi(E_i)), \sup \min_{i=1}^{k} I_{\varepsilon'}^+(\chi(E_i)), \inf \max_{i=1}^{k} F_{\varepsilon'}^+(\chi(E_i)),$$

$$\inf \max_{i=1}^{k} T_{\varepsilon'}^-(\chi(E_i)), \inf \max_{i=1}^{k} I_{\varepsilon'}^-(\chi(E_i)), \sup \min_{i=1}^{k} F_{\varepsilon'}^-(\chi(E_i))$$

$$| k = 1, 2, 3, \ldots\}$$

$$= \{\sup t(\eta_k'^+(\chi(s), \chi(t))), \sup I(\eta_k'^+(\chi(s), \chi(t))),$$

$$\inf f(\eta_k'^+(\chi(s), \chi(t))), \inf t(\eta_k'^-(\chi(s), \chi(t))), \inf I(\eta_k'^-(\chi(s), \chi(t))),$$

$$\sup f(\eta_k'^-(\chi(s), \chi(t))) | k = 1, 2, 3, \ldots\}$$

$$= \eta'^\infty(\chi(s), \chi(t)) > 0$$

Hence, G' is connected. The converse part can be proved by following the same procedure. □

Theorem 10.16 *Let $G = (X, \sigma, \varepsilon)$ and $G' = (X', \sigma', \varepsilon')$ be two isomorphic bipolar neutrosophic directed hypergraphs. The arcs in G are strong if and only if their image arcs in G' are strong.*

Proof Let (s, t) be a strong arc in G such that $\eta(s, t) \geq \eta^\infty(s, t)$. Since G and G' are isomorphic, then there is a bijective mapping $\chi : G \to G'$ such that $\eta^\infty(s, t) \leq \eta(s, t) = \eta'(\chi(s), \chi(t)) \Rightarrow \eta'(\chi(s), \chi(t)) \geq \eta^\infty(s, t) = \eta'^\infty(\chi(s), \chi(t))$, which implies that $(\chi(s), \chi(t))$ is a strong arc in G'.

Converse part is trivial. □

Theorem 10.17 *Let G be a strong connected bipolar neutrosophic directed hypergraph, then every arc of G is a strong arc.*

Proof Let G be a strong strong bipolar neutrosophic directed hypergraph such that

$$T_\varepsilon^+(E_k) = T_\varepsilon^+(\{x_1, x_2, x_3, \ldots, x_m\}) = \min\{T_{\sigma_i}^+(x_1), T_{\sigma_i}^+(x_2), \ldots, T_{\sigma_i}^+(x_m)\},$$

$$I_\varepsilon^+(E_k) = I_\varepsilon^+(\{x_1, x_2, x_3, \ldots, x_m\}) = \min\{I_{\sigma_i}^+(x_1), I_{\sigma_i}^+(x_2), \ldots, I_{\sigma_i}^+(x_m)\},$$

$$F_\varepsilon^+(E_k) = F_\varepsilon^+(\{x_1, x_2, x_3, \ldots, x_m\}) = \max\{F_{\sigma_i}^+(x_1), F_{\sigma_i}^+(x_2), \ldots, F_{\sigma_i}^+(x_m)\},$$

$$t_\varepsilon^-(E_k) = t_\varepsilon^-(\{x_1, x_2, x_3, \ldots, x_m\}) = \max\{t_{\sigma_i}^-(x_1), T_{\sigma_i}^-(x_2), \ldots, T_{\sigma_i}^-(x_m)\},$$

$$I_\varepsilon^-(E_k) = I_\varepsilon^-(\{x_1, x_2, x_3, \ldots, x_m\}) = \max\{I_{\sigma_i}^-(x_1), I_{\sigma_i}^-(x_2), \ldots, I_{\sigma_i}^-(x_m)\},$$

$$F_\varepsilon^-(E_k) = F_\varepsilon^-(\{x_1, x_2, x_3, \ldots, x_m\}) = \min\{F_{\sigma_i}^-(x_1), F_{\sigma_i}^-(x_2), \ldots, F_{\sigma_i}^-(x_m)\},$$

for all $\{x_1, x_2, x_3, \ldots, x_m\} = E_k \subset X$ in ε.

There are following two cases:

1. If s and t are connected through only one directed hyperarc (s, t), then $\eta(s, t) = \eta^\infty(s, t)$.
2. If s and t are connected by two or more than two hyperpaths, then consider an arbitrary directed hyperpath $s = x_1, E_1, x_2, E_2, \ldots, E_k, x_{k+1} = t$. The strength of the path is

$$\eta(s, t) = \{\min_{i=1}^k T_\varepsilon^+(E_i), \min_{i=1}^k I_\varepsilon^+(E_i), \max_{i=1}^k F_\varepsilon^+(E_i), \max_{i=1}^k T_\varepsilon^-(E_i), \max_{i=1}^k I_\varepsilon^-(E_i), \min_{i=1}^k F_\varepsilon^-(E_i)\}$$

$$= \{\min_{i=1}^k(\min_{j=1}^m T_{\sigma_i}^+(x_j)), \min_{i=1}^k(\min_{j=1}^m I_{\sigma_i}^+(x_j)), \max_{i=1}^k(\max_{j=1}^m F_{\sigma_i}^+(x_j)),$$

$$\max_{i=1}^k(\max_{j=1}^m T_{\sigma_i}^-(x_j)), \max_{i=1}^k(\max_{j=1}^m I_{\sigma_i}^-(x_j)), \min_{i=1}^k(\min_{j=1}^m F_{\sigma_i}^-(x_j))\}$$

$$= \min T_{\sigma_i}^+(x_j), \min I_{\sigma_i}^+(x_j), \max F_{\sigma_i}^+(x_j), \max T_{\sigma_i}^-(x_j), \max I_{\sigma_i}^-(x_j), \min F_{\sigma_i}^-(x_j)$$

$$\leq [\min T_{\sigma_i}^+(s)] \min[\min T_{\sigma_i}^+(t)], \leq [\min I_{\sigma_i}^+(s)] \min[\min I_{\sigma_i}^+(t)],$$

$$\geq [\max F_{\sigma_i}^+(s)] \max[\max F_{\sigma_i}^+(t)], \geq [\max T_{\sigma_i}^-(s)] \max[\max T_{\sigma_i}^-(t)],$$

$$\geq [\max I_{\sigma_i}^-(s))] \max[\max I_{\sigma_i}^-(t)], \leq [\min F_{\sigma_i}^-(s)] \min[\min F_{\sigma_i}^-(t)]. \tag{10.2}$$

$$\eta^\infty(s, t) = \{\sup t(\eta_k^+(s, t)), \sup I(\eta_k^+(s, t)), \inf f(\eta_k^+(s, t)), \inf t(\eta_k^-(s, t)),$$

$$\inf I(\eta_k^-(s, t)), \sup f(\eta_k^-(s, t))\}$$

$$= \{\sup(\max t_\varepsilon^+(E_i)), \sup(\max I_\varepsilon^+(E_i)), \inf(\min f_\varepsilon^+(E_i)), \inf(\min t_\varepsilon^-(E_i)),$$

$$\inf(\max I_\varepsilon^-(E_i)), \sup(\max f_\varepsilon^-(E_i))\}$$

$$\leq [\min T_{\sigma_i}^+(s)] \min[\min T_{\sigma_i}^+(t)], \leq [\min I_{\sigma_i}^+(s)] \min[\min I_{\sigma_i}^+(t)],$$

$$\geq [\max F_{\sigma_i}^+(s)] \max[\max F_{\sigma_i}^+(t)], \geq [\max T_{\sigma_i}^-(s)] \max[\max T_{\sigma_i}^-(t)],$$

$$\geq [\max I_{\sigma_i}^-(s))] \max[\max I_{\sigma_i}^-(t)], \leq [\min F_{\sigma_i}^-(s)] \min[\min F_{\sigma_i}^-(t)]$$

$$= \eta(s, t) \qquad \text{(by using Eq. 10.2)}$$

$$\eta^\infty(s, t) \leq \eta(s, t).$$

Hence, every hyperarc in G is strong. $\qquad \square$

Theorem 10.18 *Let $G = (\sigma, \varepsilon)$ and $G' = (\sigma', \varepsilon')$ be isomorphic bipolar neutrosophic directed hypergraphs, then G is strong if and only if G' is strong.*

Proof Let $\chi : G \to G'$ be the isomorphism between G and G', such that

$$\min T_{\sigma_i}^+(w) = \min T_{\sigma_i'}^+(\chi(w)), \min I_{\sigma_i}^+(w) = \min I_{\sigma_i'}^+(\chi(w)),$$

$$\max F_{\sigma_i}^+(w) = \max F_{\sigma_i'}^+(\chi(w)), \max T_{\sigma_i}^-(w) = \max T_{\sigma_i'}^-(\chi(w)),$$

$$\max I_{\sigma_i}^-(w) = \max I_{\sigma_i'}^-(\chi(w)), \min F_{\sigma_i}^-(w) = \min F_{\sigma_i'}^-(\chi(w)),$$

for all $w \in X$.

$$T_\varepsilon^+(E_i) = T_\varepsilon^+(w_1, w_2, w_3, \ldots, w_k) = T_{\varepsilon'}^+(\chi(w_1), \chi(w_2), \chi(w_3), \ldots, \chi(w_k)),$$
$$I_\varepsilon^+(E_i) = I_\varepsilon^+(w_1, w_2, w_3, \ldots, w_k) = I_{\varepsilon'}^+(\chi(w_1), \chi(w_2), \chi(w_3), \ldots, \chi(w_k)),$$
$$F_\varepsilon^+(E_i) = F_\varepsilon^+(w_1, w_2, w_3, \ldots, w_k) = F_{\varepsilon'}^+(\chi(w_1), \chi(w_2), \chi(w_3), \ldots, \chi(w_k)),$$
$$T_\varepsilon^-(E_i) = t_\varepsilon^-(w_1, w_2, w_3, \ldots, w_k) = T_{\varepsilon'}^-(\chi(w_1), \chi(w_2), \chi(w_3), \ldots, \chi(w_k)),$$
$$I_\varepsilon^-(E_i) = I_\varepsilon^-(w_1, w_2, w_3, \ldots, w_k) = I_{\varepsilon'}^-(\chi(w_1), \chi(w_2), \chi(w_3), \ldots, \chi(w_k)),$$
$$F_\varepsilon^-(E_i) = F_\varepsilon^-(w_1, w_2, w_3, \ldots, w_k) = F_{\varepsilon'}^-(\chi(w_1), \chi(w_2), \chi(w_3), \ldots, \chi(w_k)),$$

for all $\{w_1, w_2, w_3, \ldots, w_k\} = E_i \subset X$. Let G be a strong bipolar neutrosophic directed hypergraph and

$$T_{\varepsilon'}^+(E_i') = T_{\varepsilon'}^+(\chi(E_i)) = T_\varepsilon^+(E_i) = \min T_{\sigma_i}^+(w_i) = \min T_{\sigma_i'}^+(w_i'),$$
$$I_{\varepsilon'}^+(E_i') = I_{\varepsilon'}^+(\chi(E_i)) = I_\varepsilon^+(E_i) = \min I_{\sigma_i}^+(w_i) = \min I_{\sigma_i'}^+(w_i'),$$
$$F_{\varepsilon'}^+(E_i') = F_{\varepsilon'}^+(\chi(E_i)) = F_\varepsilon^+(E_i) = \max T_{\sigma_i}^+(w_i) = \max T_{\sigma_i'}^+(w_i'),$$
$$T_{\varepsilon'}^-(E_i') = t_{\varepsilon'}^-(\chi(E_i)) = t_\varepsilon^-(E_i) = \max T_{\sigma_i}^-(w_i) = \max T_{\sigma_i'}^-(w_i'),$$
$$I_{\varepsilon'}^-(E_i') = I_{\varepsilon'}^-(\chi(E_i)) = I_\varepsilon^-(E_i) = \max I_{\sigma_i}^-(w_i) = \max I_{\sigma_i'}^-(w_i'),$$
$$F_{\varepsilon'}^-(E_i') = F_{\varepsilon'}^-(\chi(E_i)) = F_\varepsilon^-(E_i) = \min F_{\sigma_i}^-(w_i) = \min F_{\sigma_i'}^-(w_i').$$

Hence, G' is a strong bipolar neutrosophic directed hypergraph. The converse part is obvious.   □

Theorem 10.19 *Let $\chi : G \to G'$ be a co-weak isomorphism between G and G' and G' is strong. Then G is a strong bipolar neutrosophic directed hypergraph.*

Proof Let $\chi : G \to G'$ be a co-weak isomorphism between G and G', which satisfies

$$\min T_{\sigma_i}^+(x) \le \min T_{\sigma_i'}^+(\chi(x)), \quad \min I_{\sigma_i}^+(x) \le \min I_{\sigma_i'}^+(\chi(x)),$$

$$\max F_{\sigma_i}^+(x) \ge \max F_{\sigma_i'}^+(\chi(x)), \quad \max T_{\sigma_i}^-(x) \ge \max T_{\sigma_i'}^-(\chi(x)),$$

$$\max I_{\sigma_i}^-(x) \ge \max I_{\sigma_i'}^-(\chi(x)), \quad \min F_{\sigma_i}^-(x) \le \min F_{\sigma_i'}^-(\chi(x)),$$

for all $x \in X$.

$$T_\varepsilon^+(x_1, x_2, x_3, \ldots, x_k) = T_{\varepsilon'}^+(\chi(x_1), \chi(x_2), \chi(x_3), \ldots, \chi(x_k)),$$

$$I_\varepsilon^+(x_1, x_2, x_3, \ldots, x_k) = I_{\varepsilon'}^+(\chi(x_1), \chi(x_2), \chi(x_3), \ldots, \chi(x_k)),$$

$$F_{\varepsilon}^{+}(x_1, x_2, x_3, \ldots, x_k) = F_{\varepsilon'}^{+}(\chi(x_1), \chi(x_2), \chi(x_3), \ldots, \chi(x_k)),$$

$$t_{\varepsilon}^{-}(x_1, x_2, x_3, \ldots, x_k) = T_{\varepsilon'}^{-}(\chi(x_1), \chi(x_2), \chi(x_3), \ldots, \chi(x_k)),$$

$$I_{\varepsilon}^{-}(x_1, x_2, x_3, \ldots, x_k) = I_{\varepsilon'}^{-}(\chi(x_1), \chi(x_2), \chi(x_3), \ldots, \chi(x_k)),$$

$$F_{\varepsilon}^{-}(x_1, x_2, x_3, \ldots, x_k) = F_{\varepsilon'}^{-}(\chi(x_1), \chi(x_2), \chi(x_3), \ldots, \chi(x_k)),$$

for all $\{x_1, x_2, x_3, \ldots, x_k\} = E_i \subset X$. Since, G' is strong, then

$$
\begin{aligned}
T_{\varepsilon'}^{+}(\chi(x_1), \chi(x_2), \chi(x_3), \ldots, \chi(x_k)) &= \min\{T_{\sigma_i'}^{+}(\chi(x_1), T_{\sigma_i'}^{+}(\chi(x_2), T_{\sigma_i'}^{+}(\chi(x_3), \ldots, T_{\sigma_i'}^{+}(\chi(x_k))\} \\
&= T_{\varepsilon}^{+}(x_1, x_2, x_3, \ldots, x_k) \\
&\leq \min\{T_{\sigma_i}^{+}(x_1), T_{\sigma_i}^{+}(x_2), T_{\sigma_i}^{+}(x_3), \ldots, T_{\sigma_i}^{+}(x_k)\} \\
&\leq \min\{T_{\sigma_i'}^{+}(\chi(x_1), T_{\sigma_i'}^{+}(\chi(x_2), T_{\sigma_i'}^{+}(\chi(x_3), \ldots, T_{\sigma_i'}^{+}(\chi(x_k))\} \\
T_{\varepsilon}^{+}(x_1, x_2, x_3, \ldots, x_k) &= \min\{T_{\sigma_i}^{+}(x_1), T_{\sigma_i}^{+}(x_2), T_{\sigma_i}^{+}(x_3), \ldots, T_{\sigma_i}^{+}(x_k)\} \\
I_{\varepsilon'}^{+}(\chi(x_1), \chi(x_2), \chi(x_3), \ldots, \chi(x_k)) &= \min\{I_{\sigma_i'}^{+}(\chi(x_1), I_{\sigma_i'}^{+}(\chi(x_2), I_{\sigma_i'}^{+}(\chi(x_3), \ldots, I_{\sigma_i'}^{+}(\chi(x_k))\} \\
&= I_{\varepsilon}^{+}(x_1, x_2, x_3, \ldots, x_k) \\
&\leq \min\{I_{\sigma_i}^{+}(x_1), I_{\sigma_i}^{+}(x_2), I_{\sigma_i}^{+}(x_3), \ldots, I_{\sigma_i}^{+}(x_k)\} \\
&\leq \min\{I_{\sigma_i'}^{+}(\chi(x_1), I_{\sigma_i'}^{+}(\chi(x_2), I_{\sigma_i'}^{+}(\chi(x_3), \ldots, I_{\sigma_i'}^{+}(\chi(x_k))\} \\
I_{\varepsilon}^{+}(x_1, x_2, x_3, \ldots, x_k) &= \min\{I_{\sigma_i}^{+}(x_1), I_{\sigma_i}^{+}(x_2), I_{\sigma_i}^{+}(x_3), \ldots, I_{\sigma_i}^{+}(x_k)\}
\end{aligned}
$$

Similarly,

$$
\begin{aligned}
F_{\varepsilon}^{+}(x_1, x_2, x_3, \ldots, x_k) &= \max\{F_{\sigma_i}^{+}(x_1), F_{\sigma_i}^{+}(x_2), F_{\sigma_i}^{+}(x_3), \ldots, F_{\sigma_i}^{+}(x_k)\} \\
T_{\varepsilon}^{-}(x_1, x_2, x_3, \ldots, x_k) &= \max\{t_{\sigma_i}^{-}(x_1), T_{\sigma_i}^{-}(x_2), t_{\sigma_i}^{-}(x_3), \ldots, t_{\sigma_i}^{-}(x_k)\} \\
I_{\varepsilon}^{-}(x_1, x_2, x_3, \ldots, x_k) &= \max\{I_{\sigma_i}^{-}(x_1), I_{\sigma_i}^{-}(x_2), I_{\sigma_i}^{-}(x_3), \ldots, I_{\sigma_i}^{-}(x_k)\} \\
F_{\varepsilon}^{-}(x_1, x_2, x_3, \ldots, x_k) &= \min\{F_{\sigma_i}^{-}(x_1), F_{\sigma_i}^{-}(x_2), F_{\sigma_i}^{-}(x_3), \ldots, F_{\sigma_i}^{-}(x_k)\}
\end{aligned}
$$

Hence, G is strong bipolar neutrosophic directed hypergraph. $\qquad\square$

10.6 Applications of Bipolar Neutrosophic Directed Hypergraphs

Decision-making acts as a vital feature of current administration. Decisions are considered very important in this way that they determine both organizational and managerial actions. A decision can be defined as "a series of action which is consciously chosen from among a set of alternatives to achieve a desired result." It is appeared as a balanced commitment to action and a well-organized judgment. Problems in almost every conceivable discipline, including decision-making can be solved using graphical models.

10.6.1 Affiliation with an Apprenticeship Group

A social group is a unity of two or more humans, sharing similar activities and characteristics, who interact with one another. Social interactions can also occur on the Internet in online communities and these relationships preclude the face-to-face interactions. Different social groups are created on the basis of typical features, including education, apprenticeship, entertainment, tourism, ethics, and religion. It is bit difficult for an anonymous user to choose a social group that fulfills his desires and objectives appropriately. We develop a bipolar neutrosophic directed hypergraphical model depicting that how a user can join the most beneficial apprenticeship group by following a step-by-step procedure. A bipolar neutrosophic directed hypergraph illustrating a group of users as members of different apprenticeship groups is shown in Fig. 10.11.

If a user wants to select the most appropriate educational group, that is, the most effective one to promote and encourage a specific behavior or outcome, the following procedure can help him. Firstly, one should think about the collective contribution of members toward the group, which can be found out by means of membership degrees of bipolar neutrosophic directed hyperedges. The positive and negative contributions of users toward a specific apprenticeship group are given in Tables 10.10 and 10.11, respectively.

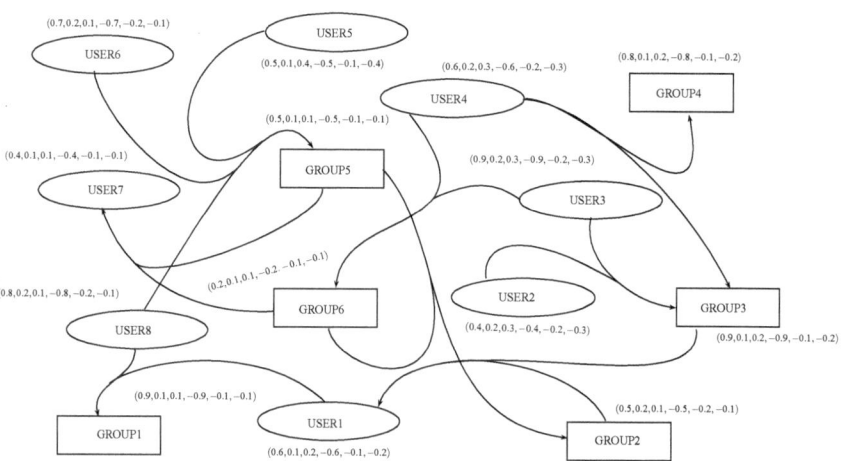

Fig. 10.11 A bipolar neutrosophic directed hypergraph illustrating the affiliations of users

Table 10.10 Didactical behavior of users toward apprenticeship groups

Apprenticeship groups	Didactical behavior	Indeterminate behavior	Irrelevant to didactics
GROUP1	0.6	0.1	0.2
GROUP2	0.5	0.1	0.1
GROUP3	0.4	0.2	0.3
GROUP4	0.6	0.2	0.3
GROUP5	0.5	0.1	0.4
GROUP6	0.6	0.2	0.3

Table 10.11 Prenicious behavior of users towards apprenticeship groups

Apprenticeship groups	Prenicious behavior	Indeterminate behavior	Extraneous behavior
GROUP1	−0.6	−0.1	−0.2
GROUP2	−0.5	−0.1	−0.1
GROUP3	−0.4	−0.2	−0.3
GROUP4	−0.6	−0.2	−0.3
GROUP5	−0.5	−0.1	−0.4
GROUP6	−0.6	−0.2	−0.3

Table 10.12 Educational effects of groups on the users

Apprenticeship groups	Educational effects
GROUP1	$(0.9, 0.1, 0.1, -0.9, -0.1, -0.1)$
GROUP2	$(0.5, 0.2, 0.1, -0.5, -0.2, -0.1)$
GROUP3	$(0.9, 0.1, 0.2, -0.9, -0.1, -0.2)$
GROUP4	$(0.8, 0.1, 0.2, -0.8, -0.1, -0.2)$
GROUP5	$(0.5, 0.5, 0.1, -0.5, -0.5, -0.1)$
GROUP6	$(0.2, 0.1, 0.1, -0.2, -0.1, -0.1)$

It can be noted that GROUP1 has 60% didactical behavior, which is maximum among all other groups, 10% indeterminacy and 20% is irrelevant to its objectives. Moreover, it owns −60% of prenicious behavior, which is minimum as compared to all other groups and −20% of extraneous behavior. Thus, GROUP1 can be a most appropriate choice for an anonymous user. Secondly, one should do his research on the powerful impacts of all under consideration groups on their members. Degrees of membership of all group vertices depict their effects on their members as given in Table 10.12.

Note that, GROUP1 has maximum positive effects and minimum negative effects on its members. Thirdly, a person can detect the influence of a group by calculating its in-degree and out-degree as in-degrees interpret the percentage of users joining the

Table 10.13 Educational effects of groups on the users

Apprenticeship groups	In-degrees	Out-degrees
GROUP1	$(0.6, 0.1, 0.2, -0.6, -0.1, -0.2)$	$(0, 0, 0, 0, 0, 0)$
GROUP2	$(0.2, 0.1, 0.1, -0.2, -0.1, -0.1)$	$(0.5, 0.1, 0.2, -0.5, -0.1, -0.2)$
GROUP3	$(1.2, 0.2, 0.6, -1.2, -0.2, -0.6)$	$(0.5, 0.1, 0.2, -0.5, -0.1, -0.2)$
GROUP4	$(0.6, 0.1, 0.3, -0.6, -0.1, -0.3)$	$(0, 0, 0, 0, 0, 0)$
GROUP5	$(0.5, 0.1, 0.4, -0.5, -0.1, -0.4)$	$(0.2, 0.1, 0.1, -0.2, -0.1, -0.1)$
GROUP6	$(0.2, 0.1, 0.3, -0.2, -0.1, -0.3)$	$(0.4, 0.2, 0.2, -0.4, -0.2, -0.2)$

group and out-degrees interpret the percentage of users leaving that group. In-degrees and out-degrees of all Apprenticeship groups are given in Table 10.13.

Thus, an Apprenticeship group having maximum in-degrees and minimum out-degrees will be the most suitable choice. It can be noted that GROUP1 and GROUP4 both have minimum out-degrees. To handle such type of situations, we will then compare the in-degrees of these two groups. GROUP1 and GROUP4 both have same positive truth membership and positive indeterminacy but the falsity membership value is minimum in case of GROUP1 and the conditions are same in case of negative membership values. Hence GROUP1 will be more suitable than GROUP4. Note that, in all above cases that we have discussed, GROUP1 is the most appropriate choice with the given data. So if any user wants to join an Apprenticeship group, by following the above procedure, he should be affiliate with GROUP1, as this group has maximum positive effects on the didactical behavior of its members and is more closely to the educational objectives. The guide will help to think about selecting a group based on the purpose of someone communication and understanding of the users. It will also help to consider what information is best communicated through different groups. The method of searching out the most beneficial group is described in the following Algorithm 10.6.1.

Algorithm 10.6.1 *Algorithm to search out the most beneficial group*

1. Input the degree of membership of all vertices(users) $x_1, x_2, \ldots, x_n$.
2. Find the positive and negative contributions of users toward groups by calculating the degree of membership of all directed hyperedges as

$$T_\rho^+(E_k) \leq \min\{T_{\mu_i}^+(x_1), T_{\mu_i}^+(x_2), \ldots, T_{\mu_i}^+(x_m)\},$$

$$I_\rho^+(E_k) \leq \min\{I_{\mu_i}^+(x_1), I_{\mu_i}^+(x_2), \ldots, I_{\mu_i}^+(x_m)\},$$

$$F_\rho^+(E_k) \leq \max\{F_{\mu_i}^+(x_1), F_{\mu_i}^+(x_2), \ldots, F_{\mu_i}^+(x_m)\},$$

$$t_\rho^-(E_k) \geq \max\{t_{\mu_i}^-(x_1), T_{\mu_i}^-(x_2), \ldots, T_{\mu_i}^-(x_m)\},$$

$$I_\rho^-(E_k) \geq \max\{I_{\mu_i}^-(x_1), I_{\mu_i}^-(x_2), \ldots, I_{\mu_i}^-(x_m)\},$$

$$F_\rho^-(E_k) \geq \min\{F_{\mu_i}^-(x_1), F_{\mu_i}^-(x_2), \ldots, F_{\mu_i}^-(x_m)\}.$$

3. Obtain the most appropriate group as

$$\max T_\rho^+(E_k), \max I_\rho^+(E_k), \min F_\rho^+(E_k), \min T_\rho^-(E_k), \min I_\rho^-(E_k), \max F_\rho^-(E_k)$$

4. Find the group having strong educational impacts on the users as

$$\max T_{\mu_i}^+(x_k), \max I_{\mu_i}^+(x_k), \min F_{\mu_i}^+(x_k), \min T_{\mu_i}^-(x_k), \min I_{\mu_i}^-(x_k), \max F_{\mu_i}^-(x_k),$$

where all x_k are the vertices representing the different groups.
5. Find the positive influence of groups x_k on the users by calculating the in-degrees $D^-(x_k)$ as

$$\left(\sum_{x_k \in H(E_k)} T_\varepsilon^+(E_k), \sum_{x_k \in H(E_k)} I_\varepsilon^+(E_k), \sum_{x_k \in H(E_k)} F_\varepsilon^+(E_k), \sum_{x_k \in H(E_k)} t^- \varepsilon(E_k),\right.$$

$$\left.\sum_{x_k \in H(E_k)} I^- \varepsilon(E_k), \sum_{x_k \in H(E_k)} F^- \varepsilon(E_k)\right)$$

6. Find the negative influence of groups x_k on the users by calculating the out-degrees $D^+(x_k)$ as

$$\left(\sum_{x_k \in T(E_k)} T_\varepsilon^+(E_k), \sum_{x_k \in T(E_k)} I_\varepsilon^+(E_k), \sum_{x_k \in T(E_k)} F_\varepsilon^+(E_k), \sum_{x_k \in T(E_k)} t^- \varepsilon(E_k),\right.$$

$$\left.\sum_{x_k \in T(E_k)} I^- \varepsilon(E_k), \sum_{x_k \in T(E_k)} F^- \varepsilon(E_k)\right)$$

7. Obtain the most effective group as, $(\max D^-(x_k), \min D^+(x_k))$.

10.6.2 Portrayal of Compatible Chemicals

The formal concept of "isomorphism" captures the informal notion that some objects have "the same structure" if one ignores individual distinctions of "atomic" components of objects. A hypergraph can exists in different forms having the same number of vertices, hyperedges, and also the same connectivity. Such hypergraphs are called isomorphic. Appropriate chemical storage plans are designed to control health and physical dynamite associated with laboratory chemical storage. There are

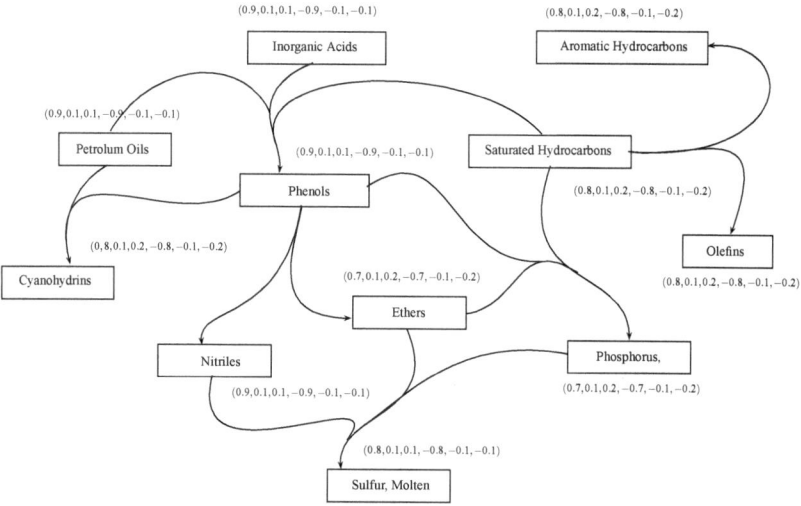

Fig. 10.12 Safe combinations of compatible chemicals

many chemicals which are not compatible with each other and react when they are mixed. The recants can be dangerous in such cases, so care must be taken when attempting to mix or store the chemicals. Here, we describe that if a model representation of compatible groups is given, then by using the isomorphism property we can represent any type of chemicals. Consider the groups of incompatible chemicals, which cannot interact with each other, as the vertices of bipolar neutrosophic directed hypergraph G. Directed hyperedges between the groups represent the safe combinations and absence of hyperedges depicts that the combinations are unsafe. A bipolar neutrosophic directed hypergraph model illustrating the safe combinations is given in Fig. 10.12.

Membership degrees of each chemical group represent that how they react positively or negatively when are mixed. For example, membership degree of Inorganic Acids $(0.9, 0.1, 0.1, -0.9, -0.1, -0.1)$ depict that these chemicals are 90% compatible, 10% have indeterminacy and 10% have chances to explode. Similarly, negative membership degrees describe the incompatibility of this group.

Now, if we have to represent the compatibilities of chemicals {Phosphoric acid, Cyclohexane, Dicylcopentadiene, Gasolines, Carbolic oil, Acetone cyanohydrin, Acetonitrile, Diethyl ether, Phosphorus, Sulfur, Benzene} belonging to different groups as mentioned in the above bipolar neutrosophic directed hypergraph, we will find out a bipolar neutrosophic directed hypergraph isomorphic to above. A bipolar neutrosophic directed hypergraph G' isomorphic to the above is given in Fig. 10.13. Define a bijective mapping $h : G \rightarrow G'$, such that

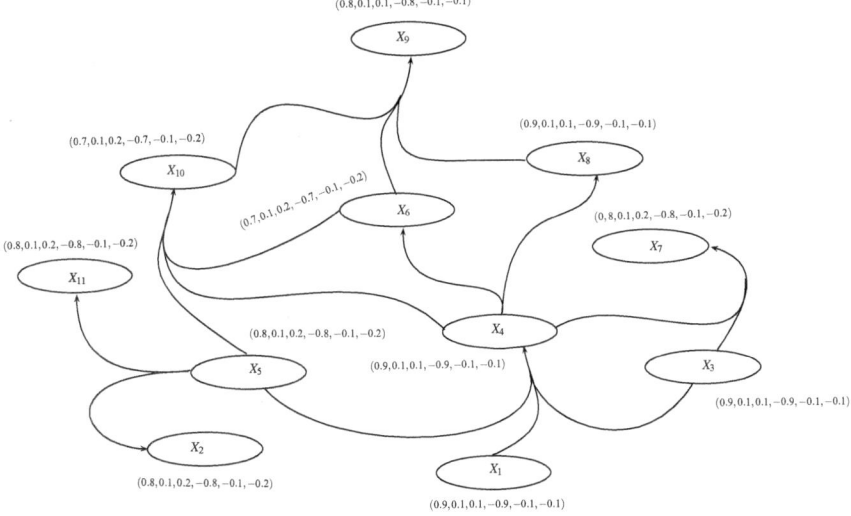

Fig. 10.13 Isomorphic bipolar neutrosophic directed hypergraphs G'

h(Inorganic Acids)	$= X_1$,	h(Aromatic Hydrocarbons)	$= X_2$,
h(Petroleum Oils)	$= X_3$,	h(Phenols)	$= X_4$,
h(Saturated Hydrocarbons)	$= X_5$,	h(Ethers)	$= X_6$,
h(Cyanohydrins)	$= X_7$,	h(Niriles)	$= X_8$,
h(Sulfur, molten)	$= X_9$,	h(Phosphorus)	$= X_{10}$,
h(Olefins)	$= X_{11}$.		

It can be noted that

$$T_G^+(InorganicAcids) = T_{G'}^+(X_1), \quad I_G^+(InorganicAcids) = I_{G'}^+(X_1),$$

$$F_G^+(InorganicAcids) = F_{G'}^+(X_1), \quad T_G^-(InorganicAcids) = T_{G'}^-(X_1),$$

$$I_G^-(InorganicAcids) = I_{G'}^-(X_1), \quad F_G^-(InorganicAcids) = F_{G'}^-(X_1).$$

Similarly, membership degrees of all groups are equal to their images.

Now determine the relative groups of all given elements and put that elements in corresponding images boxes. For instance, Phosphoric acid belongs to the group of Inorganic acids and the image of Inorganic Acids is X_1. Hence, Phosphoric acid will be kept in X_1 box. Similarly, Cyclohexane is an element of Saturated Hydrocarbons, Dicylcopentadiene belongs to Olefins, Gasolines belongs to Petroleum Oils, Carbolic oil is an element of Phenols, Acetone cyanohydrin belongs to Cyanohydrins, Acetonitrile is in Nitriles, Diethyl ether belongs to Ethers, Benzene belongs to

Aromatic Hydrocarbons and Phosphorus, Sulfur are mapped onto themselves. Thus, these elements will be positioned at the places of X_5, X_{11}, X_3, X_4, X_7, X_8, X_6, X_2, X_{10}, and X_9, respectively. Hence, by using the isomorphism property of bipolar neutrosophic directed hypergraphs, we can check the compatibility of chemicals by considering their preimages.

References

1. Akram, M.: Single-valued neutrosophic planar graphs. Int. J. Algebra Stat. **5**(2), 157–167 (2016)
2. Akram, M.: Single-valued neutrosophic graphs. Infosys Science Foundation Series in Mathematical Sciences, vol. 9, pp. 1–373. Springer (2018)
3. Akram, M.: Bipolar fuzzy graphs. Inf. Sci. **181**, 5548–5564 (2011)
4. Akram, M.: Bipolar fuzzy graphs with applications. Knowl. Based Syst. **39**, 1–8 (2013)
5. Akram, M., Dudek, W.A., Sarwar, S.: Properties of bipolar fuzzy hypergraphs. Ital. J. Pure Appl. Math. **31**, 141–160 (2013)
6. Akram, A., Luqman, A.: Certain concepts of bipolar fuzzy directed hypergraphs. Mathematics **5**(1), 17 (2017)
7. Akram, M., Luqman, A.: A new decision-making method based on bipolar neutrosophic directed hypergraphs. J. Appl. Math. Comput. **57**(1–2), 547–575 (2017)
8. Akram, M., Luqman, A.: Intuitionistic single-valued neutrosophic hypergraphs. OPSEARCH **54**(4), 799–815 (2017)
9. Akram, M., Luqman, A.: Certain network models using single-valued neutrosophic directed hypergraphs. J. Intell. Fuzzy Syst. **33**(1), 575–588 (2017)
10. Akram, M., Luqman, A.: Bipolar neutrosophic hypergraphs with applications. J. Intell. Fuzzy Syst. **33**(3), 1699–1713 (2017)
11. Akram, M., Shahzadi, S., Borumand Saeid, A.: Single-valued neutrosophic hypergraphs. TWMS J. Appl. Eng. Math. **8**(1), 122–135 (2018)
12. Atanassov. K.T.: Intuitionistic fuzzy sets. VII ITKR's Session, Sofia (deposed in Central Science-Technical Library of Bulgarian Academy of Science, 1697/84) (1983) (in Bulgarian)
13. Berge, C.: Graphs and Hypergraphs. North-Holland, Amsterdam (1973)
14. Bhowmik, M., Pal, M.: Intuitionistic neutrosophic set. J. Inf. Comput. Sci. **4**(2), 142–152 (2009)
15. Bhowmik, M., Pal, M.: Intuitionistic neutrosophic set relations and some of its properties. J. Inf. Comput. Sci. **5**(3), 183–192 (2010)
16. Deli, I., Ali, M., Smarandache, F.: Bipolar neutrosophic sets and their application based on multi-criteria decision making problems. In: International Conference IEEE on Advanced Mechatronic Systems (ICAMechS), pp. 249–254 (2015)
17. Dubios, D., Kaci, S., Prade, H.: Bipolarity in reasoning and decision, an introduction. In: International Conference on Information Processing and Management of Uncertainty, IPMU, vol. 04, pp. 959–966 (2002)
18. Gallo, G., Longo, G., Pallottino, S.: Directed hypergraphs and applications. Discret Appl. Math. **42**, 177–201 (1993)
19. Karaaslan, F.: Neutrosophic soft sets with applications in decision making. Int. J. Inf. Sci. Intell. Syst. **4**(2), 1–20
20. Kaufmann, A.: Introduction a la Thiorie des Sous-Ensemble Flous, vol. 1. Masson, Paris (1977)
21. Lee, K.M.: Bipolar-valued fuzzy sets and their basic operations. In: Proceedings of the International Conference, Bangkok, Thailand, pp. 307–317 (2000)
22. Lee, K.-M.: Comparison of interval-valued fuzzy sets, intuitionistic fuzzy sets and bipolar-valued fuzzy sets. J. Fuzzy Logic Intell. Syst. **14**(2), 125–129 (2004)

23. Mordeson, J.N., Nair, P.S.: Fuzzy Graphs and Fuzzy Hypergraphs, 2nd edn. Physica Verlag, Heidelberg (2001)
24. Parvathi, R., Karunambigai, M.G.: Intuitionistic fuzzy graphs. In: Reusch, B. (eds.) Computational Intelligence, Theory and Applications. Springer, Berlin, Heidelberg (2006)
25. Parvathi, R., Thilagavathi, S., Karunambigai, M.G.: Intuitionistic fuzzy hypergraphs. Cybern. Inf. Technol. 9(2), 46–53 (2009)
26. Parvathi, R., Thilagavathi, S.: Intuitionistic fuzzy directed hypergraphs. Adv. Fuzzy Sets Syst. 14(1), 39–52 (2013)
27. Rosenfeld, A.: Fuzzy graphs. In: Zadeh, L.A., Fu, K.S., Shimura, M. (eds.) Fuzzy Sets and Their Applications, pp. 77–95. Academic Press, New York (1975)
28. Samanta, S., Pal, M.: Bipolar fuzzy hypergraphs. Int. J. Fuzzy Log. Syst. 2(1), 17–28 (2012)
29. Smarandache, F.: Neutrosophy: Neutrosophic Probability, Set and Logic. American Research Press, Rehoboth, USA (1998)
30. Smarandache, F.: A Unifying Field in Logics. Neutrosophy: Neutrosophic Probability, Set and Logic. American Research Press, Rehoboth (1999)
31. Smarandache, F.: Neutrosophic set-a generalization of the intuitionistic fuzzy set. Int. J. Pure Appl. Math. 24(3), 287 (2005)
32. Wang, H., Smarandache, F., Zhang, Y.Q., Sunderraman, R.: Single valued neutrosophic sets. Multispace Multistructure 4, 410–413 (2010)
33. Yang, H.L., Guo, Z.L., Liao, X.: On single-valued neutrosophic relations. J. Intell. Fuzzy Syst. 30(2), 1045–1056 (2016)
34. Zadeh, L.A.: Fuzzy sets. Inf. Control 8(3), 338–353 (1965)
35. Zhang, W.R.: Bipolar fuzzy sets and relations: a computational framework for cognitive modeling and multiagent decision analysis. Proc. IEEE Conf. 305–309 (1994)
36. Zhang, W.R.: YinYang Bipolar fuzzy sets. In: Fuzzy Systems Proceedings, IEEE World Congress on Computational Intelligence, pp. 835–840 (1998)

Index

© Springer Nature Singapore Pte Ltd. 2020

M. Akram and A. Luqman, *Fuzzy Hypergraphs and Related Extensions*,
Studies in Fuzziness and Soft Computing 390,
https://doi.org/10.1007/978-981-15-2403-5